CHEMISTRY OF THE NATURAL ATMOSPHERE

This is Volume 41 in
INTERNATIONAL GEOPHYSICS SERIES

A Series of Monographs and Textbooks

Edited by RENATA DMOWSKA AND JAMES R. HOLTON

The complete listing of books in this series appears at the back of this volume.

CHEMISTRY OF THE NATURAL ATMOSPHERE

Peter Warneck

Max-Planck-Institut für Chemie
Mainz, Federal Republic of Germany

ACADEMIC PRESS, INC.
Harcourt Brace Jovanovich, Publishers

San Diego   New York   Berkeley   Boston
London   Sydney   Tokyo   Toronto

ACADEMIC PRESS, INC.
1250 Sixth Avenue, San Diego, California 92101

United Kingdom Edition published by
ACADEMIC PRESS INC. (LONDON) LTD.
24–28 Oval Road, London NW1 7DX

Library of Congress Cataloging in Publication Data

Warneck, Peter.
    Chemistry of the natural atmosphere.

    (International geophysics series)
    Bibliography: p.
    Includes index.
    1. Atmospheric chemistry.   I. Title.   II. Series.
QC879.6.W37   1987   551.5   87-1333
ISBN 0–12–735630–4   (alk. paper)

PRINTED IN THE UNITED STATES OF AMERICA

88 89 90 91   10 9 8 7 6 5 4 3 2 1

To

Christian E. Junge,
pioneer in the exploration
of
atmospheric trace substances

CONTENTS

PREFACE

Atmospheric chemistry deals with chemical compounds in the atmosphere, their distribution, origin, chemical transformation into other compounds, and finally, their removal from the atmospheric domain. These substances may occur as gases, liquids, or solids. The composition of the atmosphere is dominated by the gases nitrogen and oxygen in proportions that have been found invariable in time and space at altitudes up to 100 km. All other components are minor ones with many of them occurring only in traces. Atmospheric chemistry thus deals primarily with trace substances.

As an interdisciplinary field of science, atmospheric chemistry has its main roots in meteorology and chemistry, with additional ties to microbiology, plant physiology, oceanography, and geology. The full range of the subject was last treated by C. E. Junge in his 1963 monograph "Air Chemistry and Radioactivity." The extraordinarily rapid development of the field in the past two decades has added much new knowledge and insight into atmospheric processes so that an updated account is now called for. To some extent, the new knowledge has already been incorporated into the recent secondary literature. Most of these accounts, however, address specifically the problems of local air pollution, whereas the natural atmosphere has received only a fragmentary treatment even though it provides the yardstick for any assessment of air pollution levels. The recognition that man has started to perturb the atmosphere on a global scale is now shifting the attention away from local toward global conditions, and this viewpoint deserves a more comprehensive treatment.

Atmospheric chemistry is now being taught in specialty courses at departments of chemistry and meteorology of many universities. The purpose of this book is to provide a reference source to graduate students and other interested persons with some background in the physical sciences. In preparing the text, therefore, I have pursued two aims: one is to assemble and review observational data on which our knowledge of atmospheric process is founded; the second aim is to present concepts for the interpretation of the data in a manner suitable for classroom use. The major difficulty that I encountered in this ambitious approach was the condensation of an immense volume of material into a single book. As a consequence, I have had to compromise on many interesting details. Observational data and the conclusions drawn from them receive much emphasis, but measurement techniques cannot be discussed in detail. Likewise, in dealing with theoretical concepts I have kept the mathematics to a minimum. The reader is urged to work out the calculations and, if necessary, to consult other texts to which reference is made. As in any active field of research, atmospheric chemistry abounds with speculations. Repeatedly I have had to resist the temptation to discuss speculative ideas in favor of simply stating the inadequacy of our knowledge.

The first two chapters present background information on the physical behavior of the atmosphere and on photochemical reactions for the benefit of chemists and meteorologists, respectively. Chapter 3, which deals with observations and chemistry of the stratosphere, follows naturally from the discussion in Chapter 2 of the absorption of solar ultraviolet radiation in that atmospheric region. Chapter 4 develops basic concepts for treating tropospheric chemistry on a global scale. Methane, carbon monoxide, and hydrogen are then discussed. Subsequent chapters consider ozone, hydrocarbons, and halocarbons in the troposphere. Chapters 7 and 8 are devoted to the formation, chemistry, and removal of aerosol particles and to the interaction of trace substances with clouds and wet precipitation. These processes are essential for an understanding of the fate of nitrogen and sulfur compounds in the atmosphere, which are treated in Chapters 9 and 10. The last two chapters show the intimate connection of the atmosphere to other geochemical reservoirs. Chapter 11 introduces the underlying concepts in the case of carbon dioxide; Chapter 12 discusses the geochemical origin of the atmosphere and its major constituents.

A major problem confronting me, as for many other authors, was the proper choice of units. Atmospheric scientists have not yet agreed on a standard system. Values often range over many orders of magnitude, and SI units are not always practical. I have used the SI system as far as possible but have found it necessary to depart from it in several cases. One is the use of molecules per cubic centimeter as a measure of number density since rate coefficients are given in these units. Another example is the use of grams per cubic meter for the liquid water content of clouds. I also have retained moles per liter instead of moles per cubic decimeter, and mbar instead of hPa for simplicity and because of widespread usage, although the purist will disapprove of it.

Literature citations, although extensive, are by no means complete. A comprehensive coverage of the literature was neither possible nor intended. In keeping with the aim of reviewing established knowledge, the references are to document statements made in the text and to provide sources of observational data and other quantitative information. On the whole, I have considered the literature up to 1984, although more recent publications were included in some sections.

Thanks are due to many colleagues of the local science community for advice and information. S. Dötsch and C. Wurzinger have compiled the list of references. I. Bambach, G. Feyerherd, G. Huster, and P. Lehmann have patiently prepared the illustrations. To all of them I am grateful for essential help in bringing this volume to completion.

PETER WARNECK

Chapter

1   Bulk Composition, Physical Structure, and Dynamics of the Atmosphere

1.1 Observational Data and Averages

Our knowledge about atmospheric phenomena derives largely from observations. In dealing with atmospheric data one must make allowance for the unsteadiness of the atmosphere and a tendency for all measureable quantities to undergo sizable fluctuations. The variability of meteorological observables like temperature or wind speed and wind direction is common knowledge from daily experience. In this volume the interest is focused on atmospheric trace constituents. Their concentrations often are found to exhibit similar fluctuations. Part of the variation arises from temporal changes in the production mechanisms of trace materials that we call sources and the removal processes that we call sinks. The irregularities of air motions responsible for the spreading of a trace substance within the atmosphere impose additional random fluctuations on the local concentrations. It is difficult to evaluate short-term, random variations except by statistical methods, so that for most purposes we are forced to work with mean values obtained by averaging over a suitable time interval. This situation is different from that in the laboratory, where it usually is possible to keep parameters influencing experimental results sufficiently well controlled. In addition to random fluctuations, atmospheric data often reveal periodic variations, usually in the form of diurnal or seasonal cycles. The proper choice of

1

averaging period obviously becomes important to bring out suspected periodicities, and this point deserves some attention.

To illustrate the variability of observational data and the applied averaging procedures, consider volume mixing ratios of carbon dioxide at Mauna Loa, Hawaii, as reported by Pales and Keeling (1965). The measurements were performed by infrared analysis with a response time of about 10 min. Figure 1-1 shows from left to right (a) the sequence of hourly means for the day of July 29, 1961; (b) averages of hourly means derived by superimposition of all days of measurement for the month of July 1961; and (c) monthly means for the year 1961. The first frame indicates the variability of CO_2 mixing ratios during a single day. The second frame shows that the afternoon dip observed on July 29 is not an isolated event but occurs fairly regularly, although generally with smaller amplitude. In fact, the afternoon dip is a frequent year-round feature. It is interpreted as an uptake of CO_2 by island vegetation on the lower slopes of Mauna Loa, made manifest by the local circulation pattern (upslope winds during the day, subsiding air at night). The assimilation of CO_2 by plants during the daytime is a well-known process, as is its partial release at night due to respiration. The third frame in Fig. 1-1 demonstrates the existence of a seasonal variation of CO_2 mixing ratios, with a maximum in May and a minimum in September. In contrast to the diurnal variation, which is a local phenomenon, the annual cycle has global significance. The decline of CO_2 in the northern hemisphere during the summer months is believed to result from the incorporation of CO_2 into plants during their growth period, whereas the rise of CO_2 in winter and early spring must then be due to the decay of leaf litter and

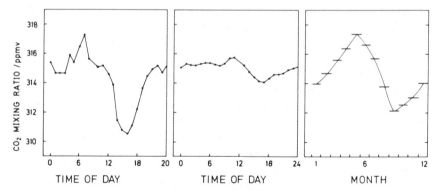

Fig. 1-1. Volume mixing ratio of CO_2 at Mauna Loa, Hawaii, according to Pales and Keeling (1965). Left: Hourly means during July 29, 1961. Center: Averages for each hour of the day for the month of July 1961. Right: Monthly means for 1961.

other dead plant material. Carbon dioxide is one of the few atmospheric trace gases for which measurements have been extended over many years, and Fig. 1-2 shows the record of monthly means since 1959. It is apparent that the seasonal oscillation occurs with similar amplitude each year, but in addition one finds a steady increase of CO_2 averaging about 0.2% per year. This long-term variation originates from the combustion of fossil fuels by humans and the accumulation of CO_2 in the atmosphere–ocean system. A more detailed discussion of CO_2 will be given in Chapter 11.

A second example for the variability inherent in observational data is provided by total ozone. This quantity represents the column density of ozone in the atmosphere. It is determined by optical absorption in the ultraviolet spectral region, with the sun as the background source. Figure 1-3 presents results of measurements in Arosa, Switzerland. Shown from left to right are (a) the variation of total ozone on a single day; (b) daily means for the month of February 1973; (c) February averages for the period 1958–1971; (d) monthly means for the year 1962; and (e) the annual variation averaged over a greater number of years. The day-to-day variability of the individual data is impressive. It is caused by the influence of the meteorological situation, mainly the alternation of high and low pressure systems. And yet, by using monthly means it is possible to show the existence of an annual cycle, with a maximum in early spring and a minimum in the fall. Its origin lies in part in a varying strength of the meridional circulation. A detailed discussion of it and other aspects of total ozone will be given in Chapter 3. It should be noted, however, that the annual oscillation has the same magnitude as the fluctuation of monthly means, exemplified in Fig. 1-3 for the month of February, and that the annual cycle thus undergoes itself considerable variations. Several years of observations are required to obtain a useful average of it.

The foregoing examples of atmospheric data and their behavior suggest a hierarchy of averages, which are summarized in Table 1-1. The lengths of averaging periods are chosen to be hour, day, month, and year, and these units are essentially predetermined by the calendar. The extension of each period by one unit increases the averaging time by about $1\frac{1}{2}$ orders of magnitude. Hourly means require a sufficiently fast instrument response time. It is clear that an analytical technique necessitating the collection of material for 24 h cannot reveal any diurnal variations, but it would still disclose seasonal changes. The fourth column of Table 1-1 indicates averages derived from the superimposition of several periods. This procedure, as we have seen, is useful to enhance periodicities or trends. Diurnal or annual cycles ultimately result from periodic changes associated with the earth's revolution about itself and around the sun. Also known are longer periodicities, such as the 11-yr sunspot cycle.

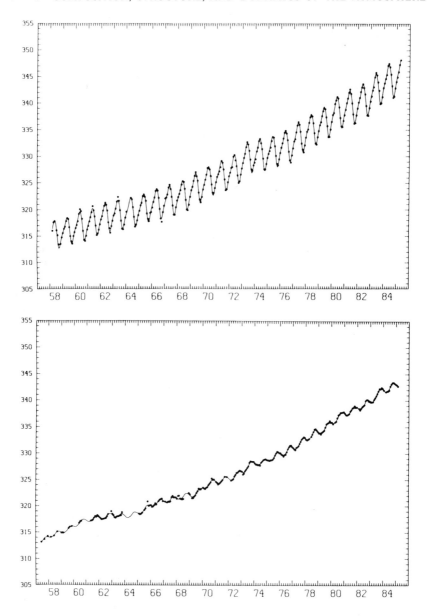

Fig. 1-2. Secular rise of CO_2 mixing ratios in the atmosphere. Dots represent monthly mean values in ppmv as observed with a continuously recording nondispersive infrared analyzer. The smooth curve represents a fit of data to a fourth harmonic annual cycle, which increases linearly with time, and a spline fit of the interannual component of variation. Top: Mauna Loa Observatory, Hawaii; bottom: South Pole. [From C. D. Keeling *et al.* (1982) and unpublished data courtesy C. D. Keeling.]

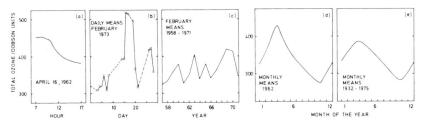

Fig. 1-3. Total ozone at Arosa, Switzerland (in Dobson units; 1 D.U. corresponds to a 10^{-2} mm thick layer of pure ozone at 273 K and 1 bar pressure). (a) Variation in the course of 1 day. (b) Day-to-day fluctuation in February 1973. (c) Variation of February means 1957–1971. (d) Monthly means for 1962. (e) Annual variation averaged over many years. [Adapted from Dütsch (1980).]

Long-term trends observed by recent measurements cover a few decades, at most. For much longer periods, the supposition of a quasi-stationarity of the atmosphere makes it likely that a longer trend will not continue indefinitely, but that eventually a new steady state is reached or the trend reverses. In recent years it has become possible to investigate the abundance of trace substances and their variations in past epochs by the exploration of deep ice cores from the great ice sheets of Greenland and Antarctica. These provide a record of atmospheric conditions dating back at least 70,000 years.

Finally, a few words must be added about spatial averages. They are important in all global considerations of atmospheric trace constituents. A continuous three-dimensional representation of data is neither practical nor

Table 1-1. *Hierarchy of Averages*[a]

Time scale	Time average, $\bar{y}_k$	k	Superimposed epoch average, $\bar{\bar{y}}_k$	Cycle or trend
Minutes	Not considered	—	—	Short-term fluctuations
Hour	Hourly mean	24	Hour by hour for many days ($n \approx 10$)	Diurnal cycle
Day	Daily mean	30	Day by day for several months ($n \approx 3$)	Seasonal variation
Month	Monthly mean	12	Month by month for several years ($n \geqslant 3$)	Annual cycle
Year	Annual mean	—	—	Long-term trend

[a] Time averages are defined as $\bar{y}_k = \int_\tau y(t)\, dt / \tau$ where τ is a fixed length of time and k designates an individual mean value within a sequence of averages over the next longer period. Associated with each period is a superimposed epoch average $\bar{\bar{y}}_k = \Sigma_{i=1}^{i=n} \bar{y}_{ki}/n$ where n is the number of periods available for averaging. The variability in each case can be described by the standard deviation from the mean.

Table 1-2. *Spatial Averages Commonly Applied in Global Considerations of Data*

Type of average	Averaging coordinate	Resulting data representation	Applicability
Zonal means	Longitude	Two-dimensional, meridional cross section	General
Zonal + meridional	Longitude + latitude	One-dimensional, versus height	Average height profile
Zonal + height	Longitude + height	One-dimensional, versus latitude	Selected altitude domains, usually lower atmosphere
Hemispheric mean	All three coordinates	Average value for each hemisphere	Mass balance
Global average	All three coordinates	Both hemispheres combined	Mass balance

are the data available in sufficient detail. Table 1-2 lists common spatial averages based on longitude, latitude, and height as space coordinates. Most important, because readily justifiable due to the comparatively rapid circumpolar atmospheric circulation, are zonal means, which are defined by an average along latitudinal circles. Thereby one obtains a two-dimensional representation of data in the form of a meridional cross section. This treatment is always desirable for data that depend on both latitude and height. Often, however, the data base is not broad enough or variations with either latitude or height are small. In this case, a one-dimensional representation of the data is more appropriate. Table 1-2 includes the two resulting cases. If the averaging process is carried still further, one obtains global means, which are important for budget considerations. Since the northern and the southern hemispheres are to some extent decoupled, it is sometimes necessary to consider averages for each hemisphere separately, and Table 1-2 makes allowance for this possibility.

1.2   Temperature Structure and Atmospheric Regions

Meteorological sounding balloons (radiosondes) are used routinely to measure from networks of stations on all continents the vertical distribution of pressure, temperature, wind velocity, and to some extent also relative humidity. Radiosondes reach altitudes of about 30 km. Atmospheric conditions at greater heights have been explored by means of rocket sondes and, in recent years, also by infrared sounding techniques from satellites.

Figure 1-4 shows zonal mean temperatures thus obtained as a function of latitude and height. In this form, the temperature distribution is intro-

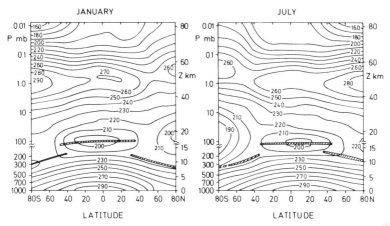

Fig. 1-4. Atmospheric distribution of zonal mean temperatures (in K) for January and July. [Data from Newell *et al.* (1972) for the lower 20 km and from Labitzke and Barnett (1979) for the upper atmosphere are combined.] The tropopause levels are indicated by shaded bars.

duced here mainly for illustration, and the details will not be discussed. Alternatively, one may set up mean annual temperature–altitude profiles at several latitudes and derive, by interpolation, an average profile valid for middle latitudes. The result of this procedure is shown in Fig. 1-5. The solid curve represents a so-called standard profile composed of straightline sections, which allow the delineation of standard pressures and densities. The need for a standard or model atmosphere arose originally with the advent of commercial airflight, but atmospheric chemistry also benefits from having available a model atmosphere that provides average values for the principal physical parameters as a function of height.

The right-hand side of Fig. 1-5 gives the nomenclature adopted by international convention to distinguish different altitude regimes. The division is based on the prevailing sign of the temperature gradient in each region. The troposphere is the region nearest to the earth surface. Here, the temperature decreases with height in a fairly linear fashion up to the tropopause. Occasional temperature inversions are restricted to narrow layers. The negative temperature gradient or lapse rate averages 6.5 K/km. A well-defined tropopause is marked by an abrupt and sustained change in the lapse rate to low values. Examples of radiosonde data illustrating this point are contained in Fig. 3-3. The altitude level of the tropopause depends considerably on latitude, as Fig. 1-4 shows, and varies with the seasons and meteorological conditions. The variations are important for the exchange of air masses between the troposphere and the stratosphere, which is the next higher atmospheric region. Here, the temperature rises again reaching

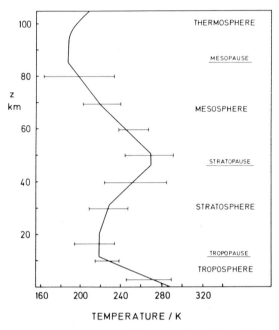

Fig. 1-5. Vertical temperature structure of the atmosphere. The solid curve represents the U.S. Standard Atmosphere, and the horizontal bars indicate the range of monthly means observed between equator and north pole. [Adapted from U.S. Standard Atmosphere (1976).]

a peak at the stratopause. The stratosphere is followed by the mesosphere, where the temperature declines toward an absolute minimum of 185 K, on average. Toward still greater heights lies the thermosphere, which has no well-defined upper boundary. In the thermosphere the temperature increases to values of the order of a thousand degrees at 250 km altitude but varies widely with solar activity. The outermost region of the atmosphere before it merges with the interplanetary medium is called the exosphere. Here, the density is so low and the mean free path of gas molecules between two collisions so great that the molecules describe ballistic orbits in the earth's gravitational field. Altogether, the physical and chemical conditions of the thermosphere differ radically from those existing in the lower atmosphere. The differences are due, on one hand, to the exposure of the outer atmosphere to very energetic radiation and, on the other hand, to the low densities, which promote the separation of lighter and heavier components by molecular diffusion. A treatment of the outer atmosphere is beyond the scope of this book, but one physical phenomenon connected with it, namely, the escape of hydrogen and helium to space, will be considered in Chapter 12.

Temperature is a measure of the internal energy of air, and the temperature attained by an atmospheric layer derives from the quasi-stationary energy balance (input versus loss). The energy spectrum of the sun has its maximum in the visible region of the electromagnetic spectrum. Only a small portion of the incoming solar radiation is absorbed directly in the atmosphere. Most of it reaches the ground surface, is largely absorbed there, and thus heats the atmosphere from below. The global energy budget of the atmosphere is balanced by outgoing infrared (thermal) radiation at wavelengths greater than 3 μm. Two processes are important in the vertical transfer of heat. One is absorption and reemission of thermal radiation, and the other is local convection. The optically active atmospheric components effecting radiative energy transfer are H_2O, CO_2, O_3, and aerosol particles. The main constituents of air, nitrogen and oxygen, cannot emit dipole radiation and thus do not participate in the thermal radiation budget. Local convection arises from the heating of the earth surface and the transfer of heat to adjacent layers of air. As a heated air parcel rises, it expands due to the pressure decrease with height, thereby undergoing adiabatic cooling. A subsiding air parcel contracts and its temperature is correspondingly raised. This mechanism leads to a rapid vertical exchange of air in the troposphere and causes the temperature to decrease with height.

Convection ceases at the tropopause level, and the temperature in the stratosphere and mesosphere is determined strictly by radiation balance. At altitudes above 20 km the absorption of solar ultraviolet radiation becomes increasingly important. The temperature peak at the stratopause has its origin in the absorption of near-ultraviolet radiation by stratospheric ozone. In fact, the existence of the ozone layer is in itself a consequence of the ultraviolet (UV) irradiation of the atmosphere. The enormous temperature increase in the thermosphere is due to the absorption of extremely shortwaved and thus energetic radiation coupled with the tenuity of the atmosphere, which prevents an effective removal of heat by thermal radiation. Instead, the heat must be carried downward by conduction toward denser layers of the atmosphere, where H_2O and CO_2 are sufficiently abundant to permit the excess energy to be radiated into space.

1.3 Pressure, Density, and Mixing Ratios

At the pressures and temperatures prevailing in the atmosphere, air and its gaseous constituents have the properties of an ideal gas to a very good approximation. Even water vapor can be treated in this way. Absolute temperature T, total pressure p, and air density ρ thus are related by the ideal gas law

$$p = \nu R_g T / V = \rho R_g T / M_{air} \qquad (1\text{-}1)$$

where $R_g = 8.31$ m^3 Pa/mol K $= 8.31 \times 10^{-5}$ m^3 bar/mol K is the gas constant, $\nu/V = \Sigma \, \nu_i/V$ denotes the total number of moles per volume V derived from all the gaseous constituents i, and

$$M_{air} = \Sigma \, \nu_i M_i / \nu = \Sigma \, m_i M_i \qquad (1\text{-}2)$$

is the molecular weight of air obtained from the individual molecular weights and the volume mixing ratios

$$m_i = \nu_i / \nu = p_i / p \qquad (1\text{-}3)$$

Since pressure and density are additive properties of ideal gases, one finds that Eq. (1-1) applies also to each gaseous constituent separately:

$$p_i = (\nu_i / V) R_g T = \rho_i R_g T / M_i \qquad (1\text{-}4)$$

The quantities ν_i / V and ρ_i are called concentrations. Although they may be used to specify the abundance of a constituent in a parcel of air, it is preferable to use mixing ratios for this purpose because they are independent of pressure and temperature. For a constituent that mixes well in the atmosphere, the mixing ratio will be constant.

Table 1-3 summarizes units of mixing ratios that are currently in use. We shall have to distinguish between mixing ratios by volume and by mass. The former is given by Eq. (1-3) and will be indicated by an added v. The latter is sometimes used to describe mixing ratios in the liquid or solid phases of particles suspended in the air. If it is to specify the abundance of material in the atmosphere, it defines the ratio of the mass of a constituent in a volume of air to the mass of air in the same volume. It is then convenient to normalize the volume to standard conditions of temperature and pressure (273 K and 1.013 bar). The units are then kg/m^3 stp, whereas normally the mixing ratio is dimensionless. In this form, the mass mixing ratio of a gas is related to its volume mixing ratio by

$$m_{i,mass} = \frac{\nu_i M_i}{V_{stp}} = \rho_{stp} \frac{M_i}{M_{air}} m_{i,volume} \qquad (1\text{-}5)$$

where $\rho_{stp} = 1.293$ kg/m^3 and Eq. (1-1) has been applied. We shall reserve

Table 1-3. *Units of Mixing Ratios*

Percent (%)	1 in 100	1×10^{-2}
Per mille (‰)	1 in 1000	1×10^{-3}
Parts per million (ppm)	1 in 10^6	1×10^{-6}
Parts per hundred million (pphm)	1 in 10^8	1×10^{-8}
Parts per billion (ppb)	1 in 10^9	1×10^{-9}
Parts per trillion (ppt)	1 in 10^{12}	1×10^{-12}

the mass mixing ratio in units of $\mu g/m^3$ stp to describe the abundance of particulate matter and aerosol-forming trace gases in the atmosphere. Concentrations rather than mixing ratios determine the rates of chemical reactions. The traditional unit of concentration in chemistry is the number of moles per volume, ν_i/V. Atmospheric chemists have adopted instead the number density n_i, that is, the number of molecules of type i contained in a volume of air,

$$n_i = \nu_i N_A / V = p_i N_A / R_g T = p_i / k_B T \qquad (1\text{-}6)$$

Here, $N_A = 6.02 \times 10^{23}$ molecules/mol is Avogadro's number and $k_B = R_g / N_A$ is Boltzmann's constant. The customary unit for n_i is molecules/cm^3, and we shall adhere to it although it does not conform to SI units.

In the literature, the expressions mixing ratio and concentration are frequently used indiscriminately. This may give rise to confusion. As long as one talks loosely about the abundance of a gaseous constituent in surface air both terms are justifiable, but when units are given the two terms must be distinguished.

Table 1-4 shows the mixing ratios of the main constituents of dry air: nitrogen, oxygen, argon, and carbon dioxide. The average molecular weight of air as calculated from Eq. (1-2) is $M_{air} = 28.97 \approx 29$ g/mol. This value will be markedly lowered in the presence of water vapor, whose abundance in the atmosphere is highly variable. It is determined essentially by the saturation vapor pressure, which is a strong function of temperature. In the middle and upper atmosphere the H_2O mixing ratios are so low that M_{air} remains unaffected. The highest H_2O mixing ratios occur in the surface air of the tropics. An H_2O mixing ratio of 5% lowers the molecular weight of air to 28.4 g/mol. The sum of nitrogen, oxygen, and argon alone amounts to 99.96% of the total contributions from all constituents. The mixing ratios of these three gases are constant with time and show no variation in space up to about 100 km altitude. Gas analytical procedures in use since the last century and modern instrumental techniques provide comparable

Table 1-4. *Bulk Composition of Dry Air[a]*

Constituent	M (g/mol)	m (volume %)	Remarks
N_2	28.02	78.084	Permanent gases,
O_2	32.01	20.946	variability
Ar	39.96	0.946	not measurable
CO_2	44.02	0.032	Somewhat variable
$\bar{M} = 28.97$		$\Sigma = 99.998$	

[a] From Glueckauf (1951).

accuracies. Measurements performed within the last 100 years have not detected deviations from the mixing ratios given in Table 1-4 outside the error limits (Glueckauf, 1951). A recent reinvestigation of oxygen by Machta and Hughes (1970) showed no change from previous values. Owing to their invariability, nitrogen, oxygen, and argon are called permanent gases in the atmosphere. This should not be taken to imply that the present levels of abundance have existed throughout the entire history of the earth atmosphere. We shall discuss the evolution of these gases in Chapter 12. Their small variability is the consequence of a long atmospheric residence time.

According to the hydrostatic principle, the pressure decreases with height due to the decrease in the column density of air overhead. The differential pressure decrease for an atmosphere in hydrostatic equilibrium is given by

$$dp = g\rho(-dz) = -(gM_{air}/R_gT)p\,dz \qquad (1\text{-}7)$$

where g is the acceleration due to gravity and the expression on the right-hand side is obtained by virtue of Eq. (1-1). Up to 100 km altitude, g decreases by less than 3%. Its variation with latitude is even smaller, so that it may be taken as constant. The standard value is $g = 9.807$ m/s^2. Equation (1-7) is integrated stepwise over height z, starting at the ground surface and using the linear sections of the standard temperature profile in Fig. 1-5. In regions of constant temperature one obtains

$$p/p_0 = \rho/\rho_0 = \exp(-z/H) \qquad (1\text{-}8)$$

where the subscript zero refers to the bottom of the atmospheric layer considered and $H = R_gT/gM_{air} = 29.2\,T$ (in meters) is called the *scale height*. Its value ranges from 5.8 to 8.7 km for temperatures between 200 and 300 K. In regions with linearly increasing or decreasing temperatures the scale height varies with z. Integration of Eq. (1-7) then yields

$$p/p_0 = (T/T_0)^{-\beta} \qquad (1\text{-}9)$$

$$\rho/\rho_0 = (T/T_0)^{-(1+\beta)}$$

with

$$\beta = gM_{air}/(R_g\,dT/dz)$$

It is an interesting fact that with any prescribed vertical temperature distribution the column density of air above a selected altitude level z is given by the product of air density and scale height at that level,

$$W = \rho_z H_z = p_z/g \qquad (1\text{-}10)$$

The substitution $\rho_z H_z = p_z/g$ follows from Eq. (1-7) and the definition of the scale height. Proof of Eq. (1-10) is obtained by integrating Eqs. (1-8) and (1-9) for all layers from the level z up to infinite heights.

The application of Eq. (1-10) makes it convenient to estimate the air masses contained in the atmospheric domains of principal interest here, namely, the troposphere and the stratosphere. The total mass of the atmosphere is obtained as follows. The air mass over the oceans is $(p_0/g)A_{sea}$, where the zero subscript refers to sea level and $A_{sea} = 3.61 \times 10^{14} \, m^2$ is the total ocean surface area. The average elevation of the continents was estimated by Kossina (1933) as 874 m above sea level (a.s.l.). From Eq. (1-9) one estimates with $T_0 = 284$ K that the average column density of air over the land areas is reduced to 0.91 of that over the oceans. The surface area of the continents is $A_{cont} = 1.49 \times 10^{14} \, m^2$. The total mass of the atmosphere thus sums to

$$G = (p_0/g)(A_{sea} + 0.91 A_{cont}) = 5.13 \times 10^{18} \, kg \qquad (1-11)$$

The pressure at sea level varies somewhat with the seasons and with latitude. In addition it is slightly lower, on average, in the southern hemisphere compared with the northern. Trenberth (1981) reviewed available data, including a revision of the average height of the continents, and calculated the mass of the atmosphere as 5.137×10^{18} kg with an annual cycle of amplitude of 1×10^{15} kg.

The air mass of the troposphere cannot be derived with the same confidence as the total air mass, because of the greatly varying tropopause height. The variation with latitude may be taken into account by adopting different average tropopause levels in the tropical and extratropical latitude belts. It is convenient to subdivide the tropospheric air space in each hemisphere along the 30° latitude circle, because equatorial and subpolar regions then cover equal surface areas. Temperatures at the average tropopause can be estimated from Fig. 1-4, and pressures and densities are then obtained from Eq. (1-9). The resulting values are given in Table 1-5. The residual air masses above the tropopause in the equatorial and subpolar latitude regions of both hemispheres combined are 3.12×10^{17} and 5.98×10^{17} kg, respectively. The sum must be deducted from the total mass of the atmosphere to give the air mass of the troposphere, 4.22×10^{18} kg. The air mass of the stratosphere can be calculated similarly from the difference of the air masses above the tropopause and the stratopause. The results are

Table 1-5. *Tropopause Temperatures, Pressures, Densities, and Number Densities*

	T_0 (K)	T_{tr} (K)	p_{tr} (mbar)	ρ_{tr} (kg/m^3)	ρ_{tr}/ρ_0	ρ_{tr}/ρ_{stp}	n (molecules/cm^3)
Tropical	295	195	120	0.21	0.17	0.166	4.4×10^{18}
Extratropical	285	215	230	0.37	0.30	0.280	7.5×10^{18}

Table 1-6. *Air Masses (in kg) Contained in Various Regions of the Lower Atmosphere*

Total atmosphere	5.13×10^{18}
Tropical troposphere	2.25×10^{18}
Extratropical troposphere	1.97×10^{18}
Total troposphere	4.22×10^{18}
Lower stratosphere (<30 km)	8.48×10^{17}
Upper stratosphere (30–50 km)	5.80×10^{16}
Total stratosphere	9.06×10^{17}
Remaining atmosphere	4×10^{15}

summarized in Table 1-6. Only about 1% of the entire mass of the atmosphere resides above 30 km and less than 0.1% in the regions above the stratopause. The troposphere contains roughly 80% of the air mass and the lower stratosphere most of the remainder. The seasonal variation of the tropopause at middle and high latitudes has been estimated by Reiter (1975) and Danielsen (1975) to cause a variation of 10–20% in the stratospheric air mass. The corresponding variation of tropospheric air mass is about 4%, which is almost negligible.

1.4   Global Circulation and Transport

The outstanding feature of atmospheric motions is that they are turbulent. Wind speed and direction fluctuate considerably, giving rise to small- and large-scale eddies. These are responsible for the intermingling of air parcels and the spreading of trace substances by turbulent mixing. From the observed wind patterns it is possible to derive the mean wind field averaged over longitude. The procedures were described, for example, by Newell *et al.* (1972). Thereby one obtains the mean zonal circulation along latitude circles and the mean meridional motion, as a function of both latitude and height. Mean motions and turbulent mixing combined determine the time scale for the spreading of a tracer in the atmosphere.

1.4.1   ATMOSPHERIC MEAN MOTIONS

Figure 1-6 illustrates the mean zonal wind field for the northern hemisphere. The situation in the southern hemisphere mirrors that of the northern, although not quantitatively. In midlatitudes the dominant wind direction is from west to east, with maximum velocities in the vicinity of the subtropical jet stream near 30° latitude. The location of the polar jet stream associated with the polar front is more diffuse, and it does not show up in the averaged wind field. The westerlies encircle the globe in a wave-like

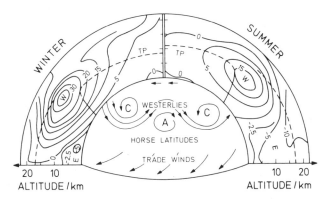

Fig. 1-6. Mean zonal circulation in the northern hemisphere, 0–20 km. Distribution of wind velocities (in units of m/s) was taken from Labitzke (1980). W, Mean winds from the west; E, mean winds from the east; the heavy lines indicate the approximate location of the polar front, the broken lines the tropopause. The maximum wind speed coincides approximately with the subtropical jet stream. The location of the polar jet fluctuates considerably and does not show up in the average. The center is to illustrate wind directions near the earth surface (trade winds and westerlies); cyclones (C) and anticyclones (A) imbedded in the westerlies are only sketched; the frontal systems associated with cyclones cannot be shown in this extremely simplified diagram.

pattern (Rossby waves). Perturbations are due to the buildup and decay of temporary as well as quasi-stationary cyclones and anticyclones. Toward the equator, in the trade wind region, the main wind direction is from the east to the southwest. The trade winds are caused by the uprising air motion in the tropics and the associated Hadley cell circulation. In summer, the region of the easterlies extends far into the stratosphere and overlays the domain of the westerlies at midlatitudes. In winter, the Hadley cell is located farther to the south and the easterlies are confined to the lower atmosphere. In the region of the westerlies the travel time of an air parcel circling around the globe is about 3 weeks. This gives an impression about the time scale involved in the transport of a tracer by the zonal circulation.

Figure 1-7 shows the mean meridional circulation patterns and their annual variations. The meridional component of the mean wind field is comparatively weak, yet it is extremely important in the north–south transport of heat and angular momentum. In the tropics the energy budget is positive; near the poles it is negative. Here, more energy is radiated toward space than the earth receives from the sun. Meridional circulation serves to maintain the heat flux required to balance the global energy budget. The major features of meridional circulation are the two tropical Hadley cells. They are separated by a region of ascending air called the interhemispheric tropical convergence zone (ITCZ). It forms a barrier to the exchange of air

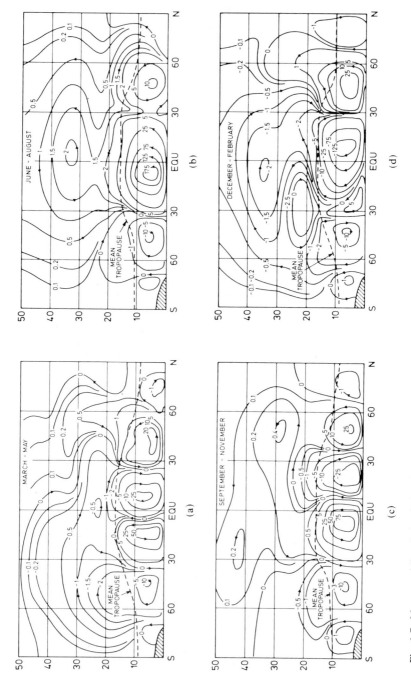

Fig. 1-7. Mean meridional circulation. [Adapted from Louis (1975); below 15 km from Newell *et al.* (1972).] (a) December–February. (b) March–May. (c) June–August. (d) September–November; mass stream lines in units of Tg/s.

between the two circulation systems and thus divides the troposphere into a northern and a southern part. In December–February, and again in July–August, one large thermally driven Hadley cell dominates the tropical circulation, while the second Hadley cell diminishes in strength. This causes the ITCZ to oscillate northward or southward about an average position near the equator. The exchange of air masses between the two hemispheres is largely due to the seasonal relocation of the ITCZ.

Mean motions in the lower stratosphere also are driven by the Hadley circulation, because its ascending branch causes tropospheric air to enter the stratosphere. The direction of motion in the lower stratosphere is northward in December–February, southward in July–August. The maximum value of the integrated mass flux associated with the Hadley cell is 200 Tg/s in the middle troposphere and 6 Tg/s in the stratosphere.

The descending branch of the Hadley cell gives rise to a high-pressure belt at about 30° latitude. Further poleward, the meridional circulation is opposite in direction to that required for thermally driven motion. These features, which are called Ferrel cells, must be indirect circulation systems. Yet another pair of cells is evident in the polar regions. They are very weak systems.

1.4.2 EDDY DIFFUSION

Turbulence in the atmosphere results primarily from the fact that air motions generally are discontinuous, often occurring in a succession of brief bursts, so that the wind speed fluctuates considerably about a mean value. Due to shear stresses, small vortices or eddies develop that carry momentum, heat, and trace materials in the direction perpendicular to the mean air flow. The eddies can be made visible by a suitable tracer such as smoke particles. Any observer of the plume emerging from a smoke stack will be impressed by the apparently random dispersal of material by small eddies in a manner resembling the spreading of particles by Brownian motion. Although individual eddies never move very far before viscous drag dissipates them into the surroundings, new eddies are constantly generated in a statistically homogeneous turbulent flow field so that the dispersal of material continues.

Although turbulent mixing may be treated mathematically in different ways (Sutton, 1955; Pasquill, 1974), the most common approach in air chemistry is by analogy to Fickian diffusion, that is, the flux of a substance contained in the air is assumed to derive from the gradient of mixing ratio established by the injection (or removal) of material into (or from) the atmosphere. Due to this analogy, turbulent mixing generally is referred to as eddy diffusion. Let $\bar{u}$ denote the mean velocity of turbulent air motion and m_s the local mean value of the volume mixing ratio of a gaseous constituent s. The flux of s can then be written as a sum of two terms, the first describing the mean flux in the direction of the air flow, the second the eddy flux caused by the turbulence:

$$F_s = \bar{u}nm_s - \mathbf{K}n \text{ grad } m_s \qquad (\text{molecules}/\text{m}^2\text{ s})$$

$$F_s = \frac{M_s}{M_{\text{air}}}(\bar{u}\rho m_s - \mathbf{K}\rho \text{ grad } m_s) \qquad (\text{kg}/\text{m}^2\text{ s})$$

$$(1\text{-}12)$$

As indicated, the flux may be expressed either in units of molecules/m^2 s or in units of kg/m^2 s. Here, ρ and n are the density and number density of air, respectively, and $\mathbf{K}$ is called the eddy diffusion coefficient. This quantity must be treated as a tensor because atmospheric diffusion is highly anisotropic due to gravitational constraints on the vertical motion and large-scale variations in the turbulence field. Eddy diffusivity is a property of the flowing medium and not specific to the tracer. Contrary to molecular diffusion, the gradient is applied to the mixing ratio and not to number density, and the eddy diffusion coefficient is independent of the type of trace substance considered. In fact, aerosol particles and trace gases are expected to disperse with similar velocities.

Mass conservation requires that the flux vector equation [Eq. (1-12)] be used in conjunction with the continuity equation

$$n\frac{\partial m_s}{\partial t} = -\text{div } \mathbf{F}_s + q_s - s_s$$

$$= \text{div}(\mathbf{K}n \text{ grad } m_s - \bar{u}nm_s) + q_s - s_s \qquad (1\text{-}13)$$

which describes the change of local concentration with time due to the presence of local sources q_s and sinks s_s. For simplicity, the steady-state assumption $dm_s/dt = 0$ is usually made, which requires that appropriate temporal averages are applicable to all quantities on the right-hand side of Eq. (1-13). These equations may be used to describe the dispersal of material from a power-plant plume as well as the global transport of a tracer, provided allowance is made for the difference in scales and a corresponding change in the magnitude of the eddy coefficients. The reason for their dependence on scale is that the atmosphere harbors eddies of many sizes. In the troposphere, turbulent mixing on a global scale is effected by large cyclones, whereas local mixing is due to small-sized eddies. For a more detailed discussion of this problem, see Bauer (1974). The main topic of interest here is the global situation, and we shall therefore consider only large-scale eddy diffusion. For practical purposes it is generally assumed that the zonal mean circulation is sufficiently rapid to make transport by turbulent mixing important only in the meridional plane. Accordingly, the averaging schemes of Table 1-2 are applicable.

Our knowledge of eddy diffusion coefficients derives largely from observations of the spreading of natural or artificial tracers in the atmosphere and an evaluation of these data by eddy diffusion models based on Eqs. (1-12) and (1-13). The two-dimensional representation requires

$$\mathbf{K} = \begin{pmatrix} K_{yy} & K_{yz} \\ K_{zy} & K_{zz} \end{pmatrix}$$

where K_{yy} and K_{zz} are the components in north–south and vertical directions, respectively. In the lower stratosphere the off-diagonal elements of the $\mathbf{K}$ tensor are nonzero because mixing occurs preferentially in a direction roughly parallel to, though somewhat more inclined than, the tropopause. The term $K_{yz} = K_{zy}$ vanishes when the vertical and horizontal motions are statistically uncorrelated. Empirical values for K_{yy}, K_{yz}, and K_{zz} depend on latitude and height, and to some extent also on season. In one-dimensional models one must set u_y or u_z equal to zero; otherwise it is difficult to satisfy the continuity equation. This is particularly necessary in the vertical one-dimensional model due to the change of air density with altitude. In the one-dimensional models the effects of transport by mean motions and eddy diffusion must be combined into the $\mathbf{K}$ value. It is clear that these $\mathbf{K}$ values differ from the corresponding ones in the two-dimensional model.

Order-of-magnitude estimates for K_y and K_z may be obtained from wind variance data on the basis of the mixing-length hypothesis, which is again made in analogy to molecular diffusion. The hypothesis assumes that an air parcel moves a distance l before it mixes suddenly and completely with the surroundings, and that during the displacement the mixing ratio of a tracer (or another property such as heat) is conserved. The momentary flux F_s' resulting from the variance u' of the wind vector is given by

$$F_s' = nu'[m_s(x_1) - m_s(x_2)] = -nu'(x_2 - x_1)\, \text{grad}\, m_s$$

$$= -nu'l'\, \text{grad}\, m_s \tag{1-14}$$

where x_1 and x_2 are the space coordinates connected with the displacement l'. Averaging over all such displacements and combining the average flux due to turbulent mixing with that due to the mean motion gives

$$F_s = nm_s\bar{u} - \overline{nu'l'}\, \text{grad}\, m_s \tag{1-15}$$

Comparison with Eq. (1-12) then shows that $\mathbf{K} = \overline{u'l'}$ with the components $K_{jk} = \overline{u_j'l_k'}$. In contrast to the wind variance u', it is not generally possible to observe the mixing lengths directly in the atmosphere. Accordingly, the expression derived for **K** has mainly a heuristic value. It is possible, however, to observe the time periods associated with the displacement of an air parcel, for example, from the motions of a small balloon. From such observations, Lettau and Schwerdtfeger (1933) obtained one of the first estimates for K_z, for which they found a value of $5\ m^2/s$ in the lower troposphere. Since l' and u' are in the same direction, one may set $l' = u'\bar{\tau}$ so that the eddy coefficient becomes proportional to the mean square of the wind fluctuation

$$\mathbf{K} = \begin{pmatrix} \overline{u'_y u'_y} & \overline{u'_y u'_z} \\ \overline{u'_z u'_y} & \overline{u'_z u'_z} \end{pmatrix} \bar{\tau} \qquad (1\text{-}16)$$

According to F. Möller (1950), $\bar{\tau}$ may be regarded as an average time constant for the rate at which the property transferred by individual air parcels adjusts to the mean value of the surroundings. In statistical turbulence theory, $\bar{\tau}$ is identical with the time integral over a correlation coefficient (Pasquill, 1974). From 13 years of meteorological observations in Europe, F. Möller (1950) showed that $\bar{\tau}$ is quite constant with a value of 12 h $(4.2 \times 10^4\ s)$ for horizontal motions. The magnitude of the time constant confirms that turbulent mixing by large-scale horizontal transport is dominated by cyclone activity and waves. If $\bar{\tau}$ is assumed to be roughly independent of latitude and height, typical wind variances $u'_y = 7\ m/s$ and $u'_z = (0.5\text{-}1.5) \times 10^{-2}\ m/s$ yield the following order-of-magnitude values for the eddy coefficients:

$$K_{yy} = 2 \times 10^6 \qquad |K_{yz}| \leqslant 4 \times 10^3 \qquad K_{zz} = 1\text{-}10$$

in units of m^2/s. We shall now describe the variation of **K** with latitude and height.

Czeplak and Junge (1974), who discussed a one-dimensional horizontal diffusion model for the troposphere, derived the dependence of K_y with latitude from wind-variance data of Flohn (1961) and Newell et al. (1972) using $\bar{\tau} = 4.2 \times 10^4\ s$ as given by Möller (1950). The results are shown in Fig. 1-8. They represent considerable improvements over earlier estimates, which were reviewed by Junge and Czeplak (1968). All data show that K_y is greatest in midlatitudes where cyclone activity is most prominent. This results in a very efficient mixing by turbulence in the region of the westerlies. The much lower K_y values near the equator are due, in part, to the increase in tropopause height, as K_y is an average over the vertical extent of the troposphere. The low equatorial K_y values are mainly responsible for the relatively slow transport of trace substances across the interhemispheric tropical convergence zone.

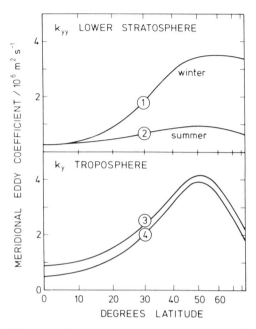

Fig. 1-8. Eddy diffusion coefficients in the northern hemisphere. Upper frame: Lower stratosphere (100 mbar, about 16 km altitude); curve 1 for December–January, curve 2 for June–August, approximated from Luther (1975) and Hidalgo and Crutzen (1977). Lower frame: One-dimensional K_y values for the troposphere after Czeplak and Junge (1974); curve 3 based on wind variance data of Flohn (1961) and Newell *et al.* (1966); curve 4 based on data of Newell *et al.* (1972), annual average.

Figure 1-8 includes for comparison values for K_{yy} in the lower stratosphere at the 100-mbar pressure level. These data also are based on the mixing-length hypothesis and observational data of heat flux, temperature gradients, and wind variances (Luther, 1975). The first estimates for K_{yy} in the stratosphere were made by Reed and German (1965) and Gudiksen *et al.* (1968). Both groups of authors sought to simulate the observed spreading of tungsten-185 debris injected into the tropical stratosphere by atomic weapons tests in the late 1950s. Observations and model calculations agreed remarkably well and showed that eddy diffusion in the north–south direction was more important in spreading the tungsten-185 cloud than was the relatively weak Hadley circulation. Current two-dimensional models (e.g., Gidel *et al.*, 1983) use K_{yy} values that are slightly modified from those of Luther (1975). As in the troposphere, the stratospheric K_{yy} values are higher in midlatitudes compared with those near the equator. Figure 1-8 illustrates further a seasonal dependence such that turbulent mixing in the north–south direction is greatly enhanced during the winter months. A similar situation probably exists in the troposphere.

Values for the off-diagonal component of the **K** tensor usually are scaled to those of K_{yy} by means of heat-flux data. The K_{yz} values determine the slope of the mixing surface. In the lower stratosphere of the northern hemisphere the values are negative—that is, the mixing surface is slanted downward as one goes from equator to pole. Since the K_{yz} terms do not contribute directly to the rate of transport, we shall not discuss them further.

Values for K_z have been derived mainly from tracer studies. To illustrate the application of the one-dimensional diffusion equation in a simple case, consider radon as a tracer. The half-life time of radon-222 is 3.8 days. Its main source is continental soils. The oceans are not a significant source. In the middle of a continent the concentration of radon decreases with altitude, because it decays as it is transported upward by vertical mixing. Figure 1-9 shows a few measurements of vertical profiles of the radon mixing ratio. Individual profiles exhibit a considerable variability, which is in part due to short-term changes in meteorological conditions, such as enhanced convection during the day. Nevertheless, the average decline is quasi-exponential with a scale height of 3.7 km, that is,

$$m(\mathrm{Rn}) = m_0(\mathrm{Rn}) \exp(-z/h)$$

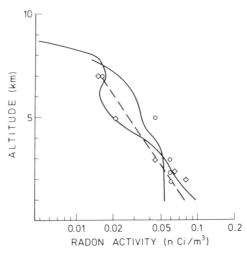

Fig. 1-9. Decay of radon activity with height: two profiles from Wexler *et al.* (1956) (solid lines) as given by Machta (1974), obtained from short-lived radon-daughter activity over Ohio; and averages of three profiles each taken at 6 a.m. (circles) and 2 p.m. (diamonds) by Wilkening (1970) over New Mexico. His average profile is shown by the dashed line. This gives an average scale height of 3.7 km over the continent. In coastal and marine areas the profiles are inverted because the ocean is a sink for terrestrial radon.

The steady-state solution of Eq. (1-13) for a constant K_z and $s_s = nm_s\lambda^*$ in the one-dimensional case, where $\lambda^* = 2.1 \times 10^{-6} \, s^{-1}$ is the radon decay constant, then yields an algebraic equation for h from which one obtains

$$K_z = \lambda^* h / (1/H + 1/h) = 20 \quad m^2/s \qquad (1-17)$$

Here, $H = 8500$ m denotes the atmospheric scale height. Radon obviously would be a good tracer for turbulent mixing in the troposphere. Unfortunately, it is hard to measure. It should be noted that the assumption of a constant K_z in the troposphere is a reasonable approximation only for the region outside the planetary boundary layer ($z \geqslant 2$ km). The air motions close to the earth surface are greatly influenced by friction, which causes the eddy coefficients to decrease from their values in the free troposphere.

Table 1-7 summarizes vertical eddy diffusion coefficients for the troposphere and the lower stratosphere as derived from various observations, mainly of tracers. For the troposphere the K_z data of Davidson et al. (1966) are the least trustworthy, because they are based on tracers originating in the stratosphere. The best values probably are those of Bolin and Bischof (1970), derived from the seasonal variation of CO_2. Above the tropopause

Table 1-7. Values for the Eddy Diffusion Coefficient K_z (Vertical Transport) Derived Mainly from Tracer Observations

Authors		Remarks
	Troposphere	
Davidson et al. (1966)	1–10	From the fall-out of bomb-produced [185]W and ^{90}Sr; source in the stratosphere
Bolin and Bischof (1970)	14–26	From the seasonal oscillation of $m(CO_2)$ imposed by the biosphere
Machta (1974)	40	Coarse estimate from radon decay data (see Fig. 1-9)
Present	20	From the radon decay data of Fig. 1-9
	Stratosphere	
Davidson et al. (1966)	0.1–0.6	From the distribution of [185]W and [90]Sr
	0.1	Best value
Gudiksen et al. (1968)	0.15–3.6	Two-dimensional model including mean motions of [185]W tracer distribution
Wofsy and McElroy (1973)	0.2	From one-dimensional diffusion model and observed altitude profile of methane in the stratosphere
Luther (1975)	0.2–2.0	From heat flux, temperature, and wind-variance data
Schmeltekopf et al. (1977)	0.3–0.4	From one-dimensional diffusion model and measured altitude distribution of N_2O averaged over one hemisphere

the eddy coefficient K_z decreases quite abruptly, from high values in the troposphere to low values in the range 0.2–0.4 m²/s. In the stratosphere, methane and nitrous oxide have served as useful tracers. Both gases are well mixed in the troposphere, but above the tropopause their mixing ratios decrease with height due to losses by chemical reactions. Figure 1-10 shows the dependence of K_z on altitude as determined by a number of studies from the observed vertical distribution of CH_4 and N_2O. For both trace gases the K_z values were low adjacent to the tropopause, but the values increase with height and eventually reach a magnitude similar to or greater than that of K_z in the troposphere.

Finally, it is necessary to assess the importance of transport by eddy diffusion relative to that by mean meridional motions. For this purpose, we define time constants for the transport over a characteristic distance d_c as follows

$$\text{Mean motion:} \qquad \tau_{\text{trans}} = d_c / \bar{u}_d \qquad (1\text{-}18\text{a})$$

$$\text{Eddy diffusion:} \qquad \tau_{\text{trans}} = \overline{d_c^2}/2K_d = \pi \bar{d}_c^2 / 4K_d \qquad (1\text{-}18\text{b})$$

where d_c equals y_c in the meridional and z_c in the vertical direction. The characteristic distance parallel to the Earth surface is taken as 3300 km, which corresponds to one-third of the distance between equator and pole. For z_c the atmospheric scale height H is used. The results are assembled in Table 1-8. The data indicate that a tracer is transported along a meridian about three times faster by eddy diffusion than by mean motions. Stratosphere and troposphere behave rather similarly in this regard. The characteristic time for the spreading of a tracer in north–south direction is about

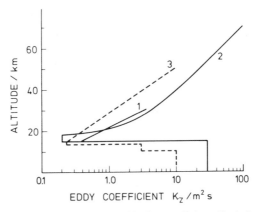

Fig. 1-10. One-dimensional vertical eddy diffusion coefficient K_z derived from trace gas observations in the stratosphere: (1) from nitrous oxide, (2,3) from methane. (1) Schmeltekopf *et al.* (1977), (2) Wofsy and McElroy (1973), (3) Hunten (1975).

Table 1-8. *Time Constants for Horizontal and Vertical Transport by Atmospheric Mean Motions and Eddy Diffusion*[a]

Altitude level	Horizontal transport					Vertical transport				
	y_c (km)	$\bar{u}_y$ (m/s)	τ_{trans} (months)	K_{yy} (10^6 m²/s)	τ_{trans} (months)	z_c (km)	$\bar{u}_z$ (m/s)	τ_{trans} (months)	K_z (m²/s)	τ_{trans} (months)
Troposphere, 800 mbar, 2 km	3300	0.20 w 0.05 s	6 25	1.8 w 0.5 s	1.8 ⎫ 6.6 ⎬	8.5	0.00015	20	15	1.2
Stratosphere, 100 mbar, 16 km	3300	0.20 w 0.12 s	6 10	3.0 w 0.6 s	1.1 ⎫ 5.5 ⎭	6	0.00012	19	0.4	27

[a] Data for $\bar{u}_y$, $\bar{u}_z$, and K_{yy} were taken from Hidalgo and Crutzen (1977) for 40° northern latitude.

3 months, on average, in both regions. Vertical transport in the troposphere is dominated by eddy diffusion, with a characteristic time of about 4 weeks. This applies to average meteorological conditions. High-reaching convection of cyclone activity may achieve a much shorter exchange time. In the lower stratosphere, mean motions and eddy diffusion contribute about equally to vertical transport. The direction of the mean motion thus becomes important. The time constants are such that a substance injected into the stratosphere will spend 1–2 years there before it is drained to the troposphere by downward transport, but during this time it will become meridionally well mixed. The time constants shown in Table 1-8 are important for all considerations of chemical reactivity in the atmosphere. If the rate of chemical consumption is greater than that of transport, the mixing ratio of the chemical compound will decrease with distance from the point of injection. Unreactive substances, in contrast, spread to fill the entire atmosphere within a time period determined solely by the rate of transport.

1.4.3 Molecular Diffusion

The coefficient for molecular diffusion D is determined by the mean free path between collisions of molecules, which is inversely proportional to pressure. In the ground-level atmosphere the value for the diffusion coefficient typically is $D \approx 2 \times 10^{-5}\ m^2/s$, which makes molecular diffusion unsuitable for large-scale transport. Turbulent mixing is faster. The pressure decreases with height, however, causing the diffusion coefficient to increase until it reaches a value of $100\ m^2/s$ at an altitude near 100 km. Above this level molecular diffusion becomes the dominant mode of large scale transport.

In the troposphere, molecular diffusion is important in the size range of 1 mm or less. This may be gleaned from Eq. (1-18b) by substituting $D = 2 \times 10^{-5}\ m^2/s$ for K_d, which gives a time constant for the displacement $d_c \leqslant 10^{-3}$ m of 0.1 s or less. Molecular diffusion is responsible for the growth of cloud drops by the condensation of water vapor, and for the exchange of water-soluble gases between cloud or rain drops and the surrounding air. In the case of a spherical drop there is no dependence on angular coordinates due to the isotropic nature of molecular diffusion. In the absence of volume sources or sinks the diffusion equation reduces to

$$\partial n_s / \partial t = D \text{ div grad } n_s = \frac{D}{r^2} \frac{\partial}{\partial r} r^2 \frac{\partial n_s}{\partial r} \qquad (1\text{-}19)$$

Solutions to this equation for various boundary conditions have been assembled by Crank (1975). For a molecular flux that varies slowly with time, the steady-state approximation may be made, which in the case of an infinite supply of molecules of substance s leads to a total flux into (or out of) a drop with radius r of

$$F_{s,tot} = 4\pi Dr(n_{sr} - n_{s\infty}) \qquad (1\text{-}20)$$

where $n_{s\infty}$ is the number density of s at a distance far from the drop. We shall discuss applications in Sections 7.3.1 and 8.4.2.

1.5   Air Mass Exchange between Principal Atmospheric Domains

A simple yet very useful concept in treating material transport in geochemistry and in atmospheric chemistry as well is to consider a small number of reservoirs that communicate with each other across common boundaries. Ideally, the contents of each reservoir should be well mixed; that is, the rate of internal mixing should exceed the rate of material exchange between adjoining reservoirs. In these so-called box models, the exchange is treated kinetically as a first-order process.

The relatively slow rate of transport across the interhemispheric tropical convergence zone naturally subdivides the troposphere into a northern and a southern hemispheric part. In a similar manner one may consider the stratosphere and the troposphere separate reservoirs with the tropopause serving as the common boundary. Vertical and horizontal mixing in each part of the troposphere is faster than the transport of air across the ITCZ or across the tropopause, so that box-model conditions are met. The stratosphere, in contrast, is not an ideal reservoir, because its upper boundary is not well defined. In addition, in its lower part the rate of vertical mixing is much reduced from that in the troposphere. These shortcomings must be kept in mind when one interprets observational data from the stratosphere in terms of the box model. The rate of exchange between stratosphere and troposphere inferred from the model will reflect to some degree the rate of transport of material from greater heights downward to the tropopause, rather than the true exchange rate across the tropopause. We discuss below two examples for the application of the four-reservoir atmospheric box model and provide numerical values for the exchange times as derived from tracer studies.

The model is illustrated in Fig. 1-11. The exchange coefficients κ are distinguished by subscripts whose sequence designates the direction of mass flow. The flux of material across a boundary is given by the product of the respective exchange coefficient and the mass content of the material in the donor box. If the mass flux is given in units of kg/yr, the exchange coefficients have the units yr^{-1}. The inverse of each κ represents the time constant for the exponential decay of the reservoir's content in the case of no return

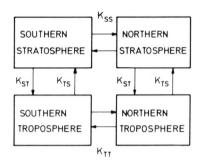

Fig. 1-11. Four-box model of the atmosphere; common boundaries are the tropopause and the interhemispheric tropical convergence zone. The rate of air exchange between individual reservoirs is expressed in terms of the exchange coefficients shown and the air mass content of the donor reservoir.

flow. The coefficients for the exchange between the two tropospheric reservoirs are equal because the air masses are equal. The coefficients for the exchange between stratosphere and troposphere differ, because the air mass in the troposphere is by a factor of 5.2 greater than that in the stratosphere, $\kappa_{ST} = 5.2\kappa_{TS}$. The model of Fig. 1-11 assumes for simplicity that κ_{ST} has the same value in the northern and southern hemispheres. This condition is an approximation and need not be true. In fact, estimates of the influx of stratospheric ozone into the troposphere suggest a smaller flux in the southern when compared with the northern hemisphere.

Tracers are again used to obtain information on the exchange coefficients. To illustrate the approach in the case of interhemispheric air mass exchange, consider the rise of CO_2 in the atmosphere due to the combustion of fossil fuels. During the period 1958–1965 the rise was nearly linear in both hemispheres. This can be seen from Fig. 1-2 if the seasonal variation is subtracted. The linear increase means that similar amounts of CO_2 were added to the atmosphere each year. Since over 90% of the addition occurs in the northern hemisphere, we may, for simplicity, ignore the sources in the southern hemisphere. The small fluxes into the stratosphere and into the ocean will be neglected. Let m_N and m_S denote the mixing ratios of CO_2, and $G_N(CO_2) = m_N G_T/2$ and $G_S(CO_2) = m_S G_T/2$ the masses of CO_2, in the northern and southern hemispheres, respectively. Further, let $P_a G_T$ be the constant annual rate of the CO_2 increase in the atmosphere. The behavior of CO_2 with time should then be described by the two coupled equations

$$
\begin{aligned}
dG_N(CO_2)/dt &= \kappa_{TT}[G_N(CO_2) - G_S(CO_2)] + P_a G_T \\
dG_S(CO_2)/dt &= \kappa_{TT}[G_N(CO_2) - G_S(CO_2)]
\end{aligned}
\tag{1-21}
$$

After subtracting the second equation from the first and dividing through by $G_T/2$, one obtains

$$
d(m_N - m_S)/dt = -2\kappa_{TT}(m_N - m_S) + 2P_a
\tag{1-22}
$$

The last equation has the solution

$$\Delta m = \Delta_0 m \exp(-2\kappa_{TT}t) + (P_a/\kappa_{TT})[1-\exp(-2\kappa_{TT}t)] \qquad (1\text{-}23)$$

which after a time of adjustment of the order of $1/\kappa_{TT}$ relaxes to the steady state,

$$\Delta m = P_a/\kappa_{TT} \qquad \text{so that} \qquad \kappa_{TT} = P_a/\Delta m \qquad (1\text{-}24)$$

From Fig. 1-2 one estimates that $\Delta m = 0.66$ ppmv during the period considered. The value for P_a is given by the slope of the rise curve, or more precisely by the average slope $(\Delta m_N + \Delta m_S)/2\Delta t = 0.64$. The exchange coefficient thus is $\kappa_{TT} = 0.64/0.66 = 0.97$ yr^{-1}. The corresponding exchange time is $\tau_{TT} = 1/\kappa_{TT} \approx 1$ yr.

The inventory of strontium-90 in the stratosphere provides another set of observational data amenable to box model analysis. Strontium-90 is an artificial radioactive isotope with a half-life due to beta decay of 28.5 yr. It was brought into the atmosphere by atomic weapons tests and has been monitored more or less continuously since the late 1950s by means of aircraft and balloon sampling. Figure 1-12 shows the integrated activity of ^{90}Sr for the northern and southern stratospheric reservoirs separately, and for both regions combined. In the south the amount of ^{90}Sr rises initially as material enters from the north, where most of it was originally deposited. Eventually, the activity declines in both hemispheres due to losses to the troposphere. Strontium combines with other elements in the atmosphere to form inorganic salts, and once it reaches the troposphere it is quickly removed by wet precipitation on a time scale shorter than 4 weeks. For a treatment of the observational data, the four-box model must be used, although it suffices

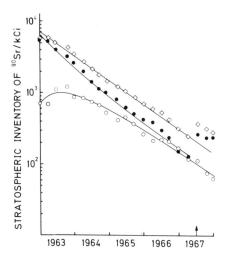

Fig. 1-12. Inventory of strontium-90 in the stratosphere following the injection by nuclear weapons tests. Redrawn from Reiter (1975), originally Krey and Krajewsky (1970) and Krey et al. (1974). The arrow indicates a new injection due to Chinese tests in June 1967. ●, Northern Stratosphere; ○, Southern Stratosphere; ◇, Sum of these.

to consider only the stratospheric reservoirs. Since the mass or atom content of ^{90}Sr in both reservoirs is directly proportional to the integrated activities plotted in Fig. 1-12, we may use activities to set up the budget equations

$$da_N/dt = -\kappa_{ST}a_N - \kappa_{SS}(a_N - a_S)$$
$$da_S/dt = -\kappa_{ST}a_S + \kappa_{SS}(a_N - a_S) \tag{1-25}$$

The solution to these equations is derived most conveniently from two new equations obtained from the above by addition and subtraction, respectively:

$$d(a_N + a_S)/dt = -\kappa_{ST}(a_N + a_S)$$
$$d(a_N - a_S)/dt = -(\kappa_{ST} + 2\kappa_{SS})(a_N - a_S) \tag{1-26}$$

Integration is now straightforward and yields

$$a_N + a_S = (a_N^0 + a_S^0)\exp(-\kappa_{ST}t) \tag{1-27a}$$

$$a_N - a_S = (a_N^0 - a_S^0)\exp[-(\kappa_{ST} + 2\kappa_{SS})t] \tag{1-27b}$$

where a_N^0 and a_S^0 denote the initial activities at the time $t = 0$. In the semilog plot of Fig. 1-12, the data for the sum of a_N and a_S follow a straight line, thereby verifying the exponential decay predicted by Eq. (1-27a). From the slope of the curve one obtains $\kappa_{ST} = 0.77$ yr^{-1}, corresponding to an exchange time of $\tau_{ST} = 1.3$ years. The second exchange coefficient can be obtained by fitting Eq. (1-27b) to the data for the ^{90}Sr activity in the southern stratosphere. Such a fit is shown in Fig. 1-12 for $\kappa_{SS} = 0.28$ yr^{-1}. The corresponding exchange time is $\tau_{SS} = 3.5$ yr. The loss of strontium-90 by radioactive decay was neglected in the above treatment. The decay constant of $\lambda^* = 0.0248$ yr^{-1} is sufficiently small to permit this approximation. If the decay is taken into account, one finds that κ_{ST} is slightly reduced to $\kappa_{ST} = 0.74$ yr^{-1}, whereas κ_{SS} remains unaffected.

Table 1-9 summarizes exchange times derived from observations of a variety of tracers. Most investigators made use of radioactive isotopes brought into the atmosphere by atomic weapons tests in 1958 and again in 1962. Especially the increase of carbon-14 has been studied in detail. The analysis of radiocarbon is complicated by the fact that CO_2 exchanges with the oceans and the biosphere, the latter giving rise to seasonal variations. A four-box model is inadequate to describe the temporal behavior of $^{14}CO_2$ in the atmosphere, and more intricate models must be applied. The values obtained for the exchange times nevertheless are in good agreement with those derived from other tracers. The results of Walton et al. (1970) are an exception. Their measurements were made in 1967/1968, at a time when the differences of activities between different atmospheric reservoirs had already become small. The value deduced by Pannetier (1970) for τ_{TT} from the increase of krypton-85 also appears too high. All the other data indicate time constants of about 1 yr for the intertropospheric air exchange, 1.4 yr for the exchange from stratosphere to troposphere, and 3.5 yr for the exchange between the two stratospheric reservoirs. Two estimates based on meteorological data are in agreement with the results from tracer studies.

The exchange of air between the northern and southern troposphere is caused to some extent by eddy diffusion in the equatorial upper troposphere, and by the seasonal displacement of the interhemispheric tropical convergence zone which lies north of the equator in July and south of the equator in January. The displacement is greatest in the region of the Indian Ocean, where the ITCZ is relocated over the Indian subcontinent in July. By tracing the transport of fission products from Chinese and French weapons tests, Telegadas (1972) has demonstrated the importance of monsoon systems for the interhemispheric air exchange in the Indian Ocean.

E. R. Reiter (1975) has reviewed in detail the various mechanisms contributing to the air exchange across the tropopause. Four processes must be distinguished: (a) the seasonal adjustment of the average tropopause level; (b) organized large-scale mean motions due to meridional circulation; (c) turbulent exchange processes via tropopause gaps or folds associated with jet streams; and (d) small-scale eddy transport across the entire tropopause. The last process does not contribute to the exchange of bulk air, but it is important in the exchange of trace constituents if a vertical gradient of the mixing ratio exists. A few remarks are necessary to expose the underlying concepts.

The tropopause is not fixed but readjusts to meteorological conditions on a time scale of 1-2 days on account of radiative balancing. Double tropopauses (i.e., two breaks in the temperature lapse rate, each qualifying as a tropopause according to the meteorologist's rule book) are frequently evident in radiosonde data. Under normal conditions one of the two may be considered a remnant, which will subsequently disappear. In the vicinity of a jet stream, multiple tropopauses are a regular phenomenon (see Fig. 1-13). Although in midlatitudes the tropopause level fluctuates somewhat with high- and low-pressure systems, these fluctuations are not considered important in the exchange of air between troposphere and stratosphere. The seasonal displacement is more substantial. Poleward of 30° latitude the tropopause is relocated to a lower level in winter, and equatorward it is relocated to a higher level compared with summer conditions. The relocation incorporates tropospheric air into the stratosphere when the tropopause is lowered, and stratospheric air into the troposphere when it is raised. E. R. Reiter (1975) estimated that 4×10^{16} kg of air are displaced in this manner

Table 1-9. *Values for the Exchange Times between Different Atmospheric Reservoirs Reported by Various Authors*[a]

Authors	Tracer	τ_{TT}	τ_{ST}	τ_{SS}	Remarks
Czeplak and Junge (1974)	CO_2	>0.7			From difference in the annual variation in the two hemispheres
Czeplak and Junge (1974)	CO_2	1.0			From difference in secular increase between the hemispheres (treated here)
Münnich (1963)	$^{14}CO_2$	<1			From difference in the increase of bomb-produced ^{14}C in both hemispheres
Lal and Rama (1966)	$^{14}CO_2$	1.2	0.8 ± 0.3		From tropospheric increase of bomb-produced ^{14}C injected from the stratosphere
Feely et al. (1966)	$^{14}CO_2$		2.2		From stratospheric inventory change with time
Young and Fairhall (1968)	$^{14}CO_2$		1.5		From stratospheric inventory change with time
Nydal (1968)	$^{14}CO_2$	1.0 ± 0.2	2.0 ± 0.5	5.0 ± 1.5	Detailed box-model consideration of ^{14}C variations in the atmosphere
Walton et al. (1970)	$^{14}CO_2$	4.4	2.1		Difference of ^{14}C in different reservoirs from 1967/1968 data

Reference	Tracer	τ_{TT}	τ_{ST}	τ_{SS}	Remarks
Czeplak and Junge (1974)	$^{14}CO_2$	1.0			Using the data of Münnich (1963)
Gudiksen et al. 1(968)	^{185}W		1.2 ± 0.5		From stratospheric inventory change with time
Peirson and Cambray (1968)	^{144}Ce, ^{137}Cs, ^{90}Sr		1.37 ± 0.05	3.5 ± 1.0	From ratio of ^{144}Ce to ^{137}Cs fission products in surface air and ^{90}Sr in stratosphere and fallout
Feely et al. (1966)	^{54}Mn, ^{90}Sr		1.2		From stratospheric inventory change with time
Fabian et al. (1968)	^{90}Sr		1.56 ± 0.13	3.3 ± 0.3	From surface fallout data
This book	^{90}Sr		1.35	3.5	Using data of Krey and Krajewski (1970) and Krey et al. (1974)
Pannetier (1970)	^{85}Kr	2			From latitudinal ^{85}Kr profile and ^{85}Kr increase
Czeplak and Junge (1974)	^{85}Kr	1.8			Using data of Pannetier (1970)
Newell et al. (1969)	Meteorological data	0.9			Calculated from mean motion across the equator
Reiter (1975)	Meteorological data		1.4	6.6	Estimated from mean and eddy motions
Averaged values		1.0	1.4	4.0	Results of Walton and Pannetier omitted

[a] τ_{TT}, Troposphere–troposphere; τ_{ST}, stratosphere–troposphere; τ_{SS}, stratosphere–stratosphere (in years).

in the nothern hemisphere, which corresponds to about 10% of the air mass residing in the lower stratosphere.

The contribution of mean meridional motion to the air-mass exchange across the tropopause arises from the Hadley circulation shown in Fig. 1-7. Air enters the stratosphere primarily with the equatorial uprising branch of the Hadley cell, gets directed toward the winter pole, and leaves the stratosphere partly through the tropopause gaps associated with the subtropical jet stream. From the mass streamlines shown in Fig. 1-7, one estimates that 10^{10} kg/s of air migrates across the equatorial tropopause during northern hemispheric winter (December–February). About 80% of this flux is directed northward. The flux entering the northern hemisphere in spring is 4×10^9 kg/s; in autumn it is 6×10^9 kg/s; during summer the flux is negligible. The total annual air mass transfer is obtained by summing and amounts to 1.4×10^{17} kg, or 32% of the stratospheric air mass.

The physical structure of a jet stream can be detailed from radiosonde and aircraft observations. Figure 1-13 shows, somewhat schematically, a cross section through a jet stream. Isotachs (i.e., regions of constant wind velocities) are shown by dashed lines. Isotherms of potential temperature are shown by solid lines. In this representation, the stratosphere is indicated by the crowding of isotherms. The location of the tropopause is shown by

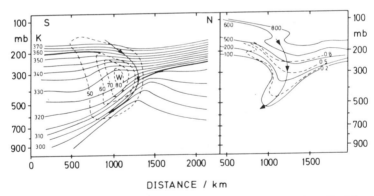

Fig. 1-13. South–north cross section through a jet stream, simplified from Danielsen *et al.* (1970). Shown on the left are isotachs (dashed lines, in units of m/s) and isotherms (solid lines) of potential temperature θ (temperature that an air parcel would assume if it were compressed adiabatically to a pressure of 1 bar). The heavy lines indicate the tropopause. Shown on the right is the extrusion of ozone-rich air from the stratosphere [dashed lines, $m(O_3)$ in ppm]. Shown for comparison are isolines of potential vorticity (solid lines) $(-Q_z \, \partial\theta/\partial z$, where Q_z is the vertical component of absolute vorticity). Potential vorticity correlates well with tracer observations. Both may be used to discriminate stratospheric from tropospheric air. The arrows in both parts of the figure indicate the direction of the air flow. A certain return flow takes place on the opposite side of the jet stream above its core.

the heavy lines. In the center of the jet stream, the tropopause cannot be defined and a gap appears in the tropopause surface. A frontal zone extends from the tropopause section on the right into the lower troposphere. The mean circulation relative to the position of the jet stream causes an extrusion of stratospheric air into the frontal zone, as indicated by the curved arrow. Several tracers have been used to identify such extrusions. Briggs and Roach (1963) used humidity and ozone; Danielsen (1968), Danielsen et al. (1970), and Danielsen and Mohnen (1977) used strontium-90, zirconium-95, and ozone as tracers. The distribution of ozone in the vicinity of the jet stream is shown on the right-hand side of Fig. 1-13 by the dashed lines. A tongue of ozone-rich air is seen to extend from the stratosphere into the frontal zone in direction of the arrow. This demonstrates most directly the outflow of stratospheric air through the tropopause gap.

Jet streams occur fairly regularly in the 30–35° latitude belt (subtropical jet stream), and somewhat less regularly near 50–60° latitude where they are associated with cyclone activity. From a number of case studies by aircraft, Reiter and Mahlmann (1965) estimated the mass transfer of stratospheric air for a cyclone of average intensity over North America as 6×10^{14} kg. About 22 such events are estimated to occur over North America each year. By extrapolation to all longitudes, the polar jet-stream belt would cause a total mass transfer of 4×10^{16} kg annually. If the subtropical jet stream is assumed to contribute equally, the total mass transfer in the northern hemisphere due to jet-stream activity would be 8×10^{16} kg annually or about 19% of the stratospheric air mass. Since all this is inferred from a few case studies, there are large uncertainties associated with the global exchange rate due to jet streams. Moreover, the magnitude of the return flow above the jet's core is not well known, so that a part of the outflow from the stratosphere may be fed by the Hadley circulation.

Table 1-10 summarizes the individual exchange rates as given by E. R. Reiter (1975). The total exchange rate resulting from the three processes discussed above is 0.68 yr^{-1}, which corresponds to an exchange time of 1.45 yr. For comparison with the strontium-90 data one would have to add the flux of strontium atoms due to eddy transport. List and Telegadas (1969) have given the distribution of ^{90}Sr with latitude and height as determined in 1965. The data are somewhat coarse-grained, and gradients near the tropopause were not specifically measured. Nevertheless, the vertical gradient is reasonably well defined as 50 disintegrations per minute per 1000 ft^3 per 2 km in the region north of 30° latitude, which translates to a gradient of the mixing ratio of 6.9×10^{-22} m^{-1}. In the equatorial region the eddy flux is largely counterbalanced by upward mean motions, so that one needs to consider only the extratropical latitude region. According to Eq.

Table 1-10. *Individual Processes and Their Contributions (Fraction of Stratospheric Air Mass) to the Air Exchange across the Tropopause (Stratosphere-Troposphere)*

	Rate (yr^{-1})	
Type of process	Reiter (1975)	Text
Seasonal relocation of the tropopause	0.10	0.15
Large-scale meridional circulation	0.38	0.32
Tropopause folding events	0.20	0.19
Small-scale eddy diffusion (for ^{90}Sr)	0.01	0.16
Total rate	0.69	0.82
Stratospheric residence time for ^{90}Sr (yr)	1.45	1.22

(1-13) the local eddy flux is

$$F(^{90}\text{Sr}) = -n_T K_z \, dm(^{90}\text{Sr})/dz$$

$$= (-7.5 \times 10^{24})(0.3)(6.9 \times 10^{-22})$$

$$= -1.6 \times 10^3 \quad \text{atoms/m}^2 \, \text{s}$$

where n_T, the number density of air molecules at the tropopause, was taken from Table 1-5. The global area north of 30° latitude is $1.25 \times 10^{14} \, \text{m}^2$ and the number of seconds per year is 3.15×10^7. This leads to an annual flux of 8.3×10^{24} ^{90}Sr atoms by eddy diffusion. The total inventory of ^{90}Sr in the northern stratosphere in March 1965 was 8×10^2 kCi, which corresponds to 1.77×10^{18} disintegrations per minute or 3.74×10^{25} atoms. Accordingly, the eddy flux of ^{90}Sr into the troposphere accounted for 16% of the strontium-90 inventory in the stratosphere. Comparison of the individual entries in Table 1-10 shows that all processes are important but that the meridional mean motion contributes the lion's share.

Transport of a trace substance by eddy diffusion across the intact tropopause will be significant only if a vertical gradient exists in the lower stratosphere. An assessment of the flux requires that the gradient be measured. Figure 1-14 illustrates the behavior of several trace gases in the vicinity of the tropopause. The mixing ratio of CO declines with height above the tropopause, indicating that the flux is directed from the troposphere into the stratosphere. The mixing ratio of ozone increases with height, so that in this case the flux is directed from the stratosphere into the troposphere. The mixing ratio for hydrogen exhibits no significant gradient, which means that the flux in either direction must be negligible. To the troposphere, the stratosphere thus represents a source region with regard to ozone, and a sink region with regard to carbon monoxide.

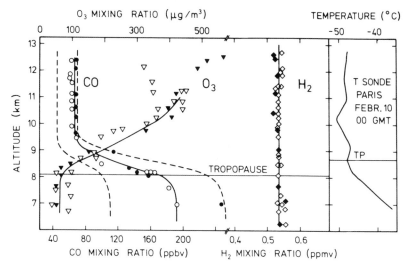

Fig. 1-14. Vertical profiles of mixing ratios for ozone, carbon monoxide, and hydrogen in the region of the tropopause as derived from aircraft observations over western France on February 9–10, 1972; open symbols indicate ascents, filled symbols represent descents. [Composed from data of Warneck *et al.* (1973) and Schmidt (1974).] Dashed curves show the range of CO mixing ratios according to Seiler and Warneck (1972). The right-hand side gives the temperature profile from a radiosonde ascent at Paris. Note that the tropopause level derived from the ozone and the temperature profiles differ somewhat, presumably due to the difference in location.

1.6   The Planetary Boundary Layer

The importance of the Earth surface to atmospheric chemistry lies in the fact that it acts as an emitter and/or as a receiver of atmospheric trace substances. In this section we discuss concepts that have been found useful in describing these processes. The transport of gases across the ocean surface is reasonably well understood. The release of gases from soils, in contrast, cannot be modeled and must be explored by field studies. In the literature, much emphasis has been placed on the absorption of gases at the ground surface, a process referred to as dry deposition.

The physical conditions of the atmosphere in the vicinity of the Earth surface are determined by friction and by heat transfer. Friction reduces wind velocity and turbulence as one approaches the surface, so that the eddy transport rate declines. Heating of the surface by solar radiation imparts energy to the overlying air, causing an enhancement of vertical motions and eddy transport owing to convection. An increase of temperature with height in the atmosphere for a certain distance instead of the normal, adiabatic decrease is called a temperature inversion layer. Temperature inversions often occur a few hundred meters above the ground, due to radiative cooling of the air below, or the advection of warmer air above the inversion level. Such inversion layers impede vertical mixing in the atmosphere because they stabilize buoyant air parcels, and they cause an accumulation of pollutants in the atmospheric boundary layer. As a consequence of these phenomena, the conditions in the boundary layer vary considerably, making it impossible to establish a working average that is universally applicable.

Boundary layer conditions are called labile when convection predominates in the vertical air exchange; stable when the vertical air motions are dampened, such as by a temperature inversion; and indifferent or neutral when the vertical turbulence is induced solely by the horizontal wind field. In the last case, the horizontal wind speed is generally found to increase with height in accordance with a logarithmic function

$$\bar{u} = u_*/k_* \, \ln\left(\frac{z + z_0}{z_0}\right) \qquad (1\text{-}28)$$

where z_0 is an empirical parameter called the surface roughness length, $k_* \approx 0.4$ is von Karman's constant, and u_* is the friction velocity, which is a measure of the drag exerted on the wind at the surface. Because the expression is simple, neutral stability conditions frequently are assumed when considering the vertical flux of a trace gas in the boundary layer.

If a trace gas in the atmosphere undergoes irreversible absorption or chemical reaction at the ground surface, the process will set up a vertical gradient of the mixing ratio leading to a downward flux:

$$F_s = -K_z(z) n \, dm_s/dz = -K_z(z) \, dn_s/dz \qquad (1\text{-}29)$$

Here the number density of air molecules, n, may be taken as constant within the first 100 m above the ground, whereas $K_z(z)$ must be treated as a function of height z. Since under steady-state conditions the flux remains constant, the equation can be integrated over height and one obtains

$$-F_s = [n_s(z) - n_s(0)] \bigg/ \int_0^z \frac{dz}{K_z(z)} = [n_s(z) - n_s(0)]/r_g(z) \qquad (1\text{-}30)$$

where $r_g(z)$ represents the gas-phase resistance to material transfer. It includes the resistance by turbulent transport as well as the resistance due to molecular diffusion in the laminar layer adjacent to the surface. An additional resistance r_s occurs at the surface itself in the uptake of the trace gas. If the surface resistance is defined as $r_s = n_s(0)/F_s$, the flux takes the form

$$-F_s = n_s(z)/[r_g(z) + r_s] = v_d n_s(z) \qquad (1\text{-}31)$$

where $v_d = [r_g(z) + r_s]^{-1}$ is called the deposition velocity. The term v_d is usually determined from concentrations measured at a height of 1 m above the ground, or above forests at a similar height above the canopy. Since $n_s(z)$ varies slowly with height, $v_d(z)$ is not strongly dependent on the choice of reference height.

Wind tunnel measurements combined with turbulence theory for hydrodynamic flows influenced by surface friction have led to semi-empirical expressions for the vertical fluctuation of wind speed $u' = l' \, d\bar{u}/dz$ and average mixing length $|\bar{l}'| = k_* z$ (Sutton, 1953). When these formulations are used in conjunction with Eqs. (1-16), (1-28), and (1-30), one obtains $r_g(z)$ in terms of wind and friction velocities:

$$r_g(z) = \bar{u}(z)/u_*^2 + B/u_* \tag{1-32}$$

Here, the first term represents the resistance due to eddy transport, and the second term that of molecular diffusion. The term $B \approx 0.1$ m/s depends on the type of surface, but it is only mildly dependent on u_* (Chamberlain, 1966; Garland, 1977). In this manner, $r_g(z)$ becomes amenable to calculation, and Table 1-11 presents some numerical values that were given by Garland (1979). The results are maximal velocities, as they do not yet include the effect of the surface resistance r_s. For the uptake of gases by soils and vegetation, r_s must be considered an empirical quantity to be determined by experiments.

Garland (1979) has summarized various techniques for the determination of deposition velocities. We consider here only the disappearance of a trace gas from the volume of air inside a box or tent placed open face down tightly onto the ground surface. The walls of the enclosure must be inert to the gas being studied, and the air volume must be stirred to ensure an adequate mixing. It is clear that the method is unsuitable for the investigation of high-growing vegetation, but it has the advantage of showing most directly

Table 1-11. *Turbulent Vertical Transport under Neutral Stability Conditions*[a]

Type of surface	$\bar{u}$ (m/s)	u_* (m/s)	$r_g(z=1)$ (10^2 s/m)	$v_d(z=1)$ (10^{-2} m/s)
Grass	3	0.13	1.25	0.8
	10	0.45	0.41	2.44
Cereal crop	3	0.18	0.87	1.15
	10	0.59	0.32	3.12
Forest	3	0.27	0.26	3.85
	10	0.89	0.11	9.09

[a] Aerial resistance r_g and corresponding maximum deposition velocity v_d for gases, according to calculations of Garland (1979) for three surfaces and two wind speeds $\bar{u}$ 100 m above ground.

whether the soil under study leads to a consumption of the trace gas or its release. Figure 1-15 illustrates this point for the trace gases CO, H_2, and N_2O. In all three cases the gases undergo consumption when the initial mixing ratios are high, whereas the gases are released when the initial mixing ratios are low. For longer observation times the values tend toward a steady state, indicating that a balance is reached between production and consumption in the soil. For CO, H_2, and N_2O the balance is largely controlled by microbiological processes. The soil thus acts as a sink when the normal mixing ratio of the trace gas in ambient air is higher than the steady-state level (CO, H_2); otherwise the soil provides a source (N_2O). Measurements of the uptake rates under actual field conditions yield values for the surface resistance r_s. For CO, H_2, and a number of other trace gases, $r_s \gg r_g$, so that $v_d \approx 1/r_s$. For several important trace gases in the atmosphere, notably sulfur dioxide and ozone, the resistances r_s and r_g have a comparable magnitude. In these cases, both quantities must be combined in determining the true deposition velocity.

Table 1-12 lists selected values for deposition velocities determined by field experiments. McMahon and Denison (1979) have provided a more comprehensive compilation. A certain variability of the data must be tolerated. Water-soluble gases are absorbed better when the surfaces are wetted than under dry conditions, and acidic gases have a higher affinity for calcareous than for acidic soils. The uptake of gases by vegetation takes

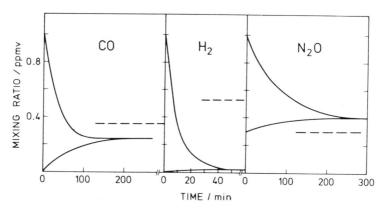

Fig. 1-15. Temporal development of CO, H_2, and N_2O mixing ratios in a fixed volume of air in contact with natural soils. An increase in the mixing ratio indicates that the trace gas is released from the soil, whereas a decline indicates that it is absorbed. In all cases shown, a temperature-dependent steady-state level is reached after a certain time regardless of the initial mixing ratio. The dashed lines indicate typical mixing ratio in ambient continental air. When this is higher than the steady-state level, the soil acts as a sink; otherwise it provides a source for the trace gas. [Data were taken from Seiler (1978) and Seiler and Conrad (1981).]

Table 1-12. *Deposition Velocities for a Number of Atmospheric Trace Gases, as Derived from Field Measurements*[a]

Trace gas	Type of surface	v_d (m/s)	Resistance (s/m)	Method	Authors
H$_2$	Soil, bare and grass-covered	$(2-12) \times 10^{-4}$	$r_s = (0.8-5) \times 10^3$	Flux box	Liebl and Seiler (1976)
		7×10^{-4}		Flux box	Conrad and Seiler (1985a)
CO	Soil, bare and grass-covered	$(2-7) \times 10^{-4}$	$r_s = (1.4-5) \times 10^3$	Flux box	Liebl and Seiler (1976)
		$(0.8-5) \times 10^{-4}$	$(2-12.5) \times 10^3$	Flux box	Conrad and Seiler (1985a)
SO$_2$	Grass-covered soil	$(3-12) \times 10^{-3}$	$r_g + r_s = 83-330$	Gradient	Shepherd (1974), Garland (1977)
		8×10^{-3}			
	Wheat	7.4×10^{-3}	$r_g + r_s = 135$	Gradient	Fowler (1978)
	Pine forest	$(2-20) \times 10^{-3}$	$r_g + r_s = 50-500$	Gradient, tracer	Garland (1977), Garland and Branson (1977)
	Water	4.5×10^{-3}	$r_g + r_s = 222$	Gradient	Garland (1976, 1977)
O$_3$	Soil, bare and grass-covered	$(4-8) \times 10^{-3}$	$r_g + r_s = 125-250$	Flux box	Aldaz (1969), Garland (1977) Galbally and Roy (1980)
		$(6-10) \times 10^{-3}$			
	Plants	$(0.3-3.3) \times 10^{-3}$	$r_g + r_s = 300-3300$	Flux box	Turner et al. (1974), Thorne and Hanson (1972)
HNO$_3$	Water, snow	$(2-4) \times 10^{-4}$	$r_s = (2.5-5) \times 10^3$	Flux box	Aldaz (1969), Galbally and Roy (1980)
	Grass-covered soil	$(2-3) \times 10^{-2}$	$r_g + r_s = 33-50$	Gradient+ heat flux	Huebert and Robert (1985)
NO$_2$	Pine needles	$(3-8) \times 10^{-3}$	$r_s = 125-250$	Flux box	Grennfelt et al. (1983)
	Soils	$(3-6) \times 10^{-3}$	$r_s = 167-360$	Flux box	Böttger et al. (1978)
	Vegetation	$(0.5-6) \times 10^{-3}$	$r_g + r_s = 167-2000$	Eddy-correlation	Weseley et al. (1982)

[a] See McMahon and Denison (1979) for a more comprehensive compilation.

place primarily via the stomata (leaf breathing pores), followed by destruction within the plant tissue. Plants control the rate of water transpiration by opening or closing the stomata, and the resistance to the uptake of gases increases when the stomata are closed. This causes a diurnal variation of the deposition rate, in addition to that resulting from the diurnal variation of aerial turbulence.

Finally, we consider the situation at the ocean surface. The exchange of gases across the gas–liquid interface is often treated in terms of the thin-film model depicted in Fig. 1-16 (Danckwerts, 1970; Liss and Slater, 1974). The resistances due to turbulent transport in both media are here considered small compared with those in the laminar layers, where the transfer must occur by molecular diffusion. Accordingly, assuming steady-state conditions, the flux through the interface is given by

$$F_s = \frac{D_g}{\Delta z_g}(c_{gi} - c_g) = \frac{D_L}{\Delta z_L}(c_L - c_{Li}) \qquad (1\text{-}33)$$

The notation is explained in Fig. 1-16, and D_g and D_L are the molecular diffusion coefficients for air and for seawater, respectively. The flux is maintained by the concentration differences across the molecular diffusion layers. The flux is directed from the ocean into the atmosphere when the concentration in seawater is greater than that at the interface; the flux is directed from the atmosphere into the ocean when the concentration difference is negative. If the exchanging gas obeys Henry's law, the concentrations at the gas–liquid interface are connected by

$$c_{gi} = \mathbf{H} c_{Li} \qquad (1\text{-}34)$$

where $\mathbf{H}$ is the dimensionless Henry coefficient that results when gas and

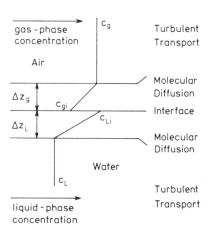

Fig. 1-16. Two-layer thin-film model of the air-sea interface, adapted from Liss and Slater (1974). The concentrations at the interface, c_{gi} (gas-phase) and c_{Li} (liquid-phase), are related by Henry's law. Material transport across the interface occurs by molecular diffusion within the film layers.

liquid concentrations are expressed in the same units of mass per volume. This Henry coefficient is related to that used in Section 8.4 by $H = HM_w/R_g T\rho_w$. Here, M_w is the molecular weight and ρ_w the density of seawater, R_g is the gas constant, and T is the absolute temperature. With the substitutions

$$\frac{D_g}{\Delta z_g} = k_g \quad \text{and} \quad \frac{D_L}{\Delta z_L} = k_L \qquad (1\text{-}35)$$

one obtains by combining the preceding equations

$$F_s = (c_L - c_g/H)\bigg/\left(\frac{1}{k_g H} + \frac{1}{k_L}\right) = \frac{1}{r}(c_L - c_g/H) \qquad (1\text{-}36)$$

with

$$r = r_g + r_L = \frac{1}{k_g H} + \frac{1}{k_L}$$

representing the sum of gas- and liquid-phase resistances to the transfer. For water molecules crossing the air–sea interface, the liquid-phase resistance is nil, so that in this case the resistance to transfer is given by that of the gas phase. Measurements of the transfer rate indicate a dependence on wind velocity. Liss and Slater (1974) have used the field data of Schooley (1969) to derive a mean value for $k_g(H_2O)$ of 8.3×10^{-3} m/s. For other gases the different molecular diffusivity must be incorporated; that is, $k_g(H_2O)$ must be multiplied by the ratio of the square roots of the molecular weights for H_2O and the other gas.

Gas exchange rates across the sea surface have been studied by various techniques, such as by tracers like radon or $^{14}CO_2$. The results indicate a liquid film thickness in the range of 25–60 μm, depending on wind speed and surface roughness (Liss, 1973; Broecker and Peng, 1971, 1974; Peng et al., 1979). For a diffusion coefficient of $D_L \approx 2 \times 10^{-9}$ m^2/s and $\Delta z_L = 36$ μm one finds $k_L \approx 5.5 \times 10^{-5}$ m/s. In the case of gases that form ions upon dissolving in water, the transport across the liquid film is complicated by the presence of several diffusing species. For SO_2, Liss and Slater (1974) estimated that the effect enhances k_L several thousand fold. We need not discuss this problem in detail because SO_2 is fairly rapidly oxidized in the slightly alkaline environment of seawater so that the concentration of SO_2 in the ocean surface waters must be nearly zero.

Table 1-13 presents a few examples for the exchange of trace gases at the sea surface with data partly taken from the compilation of Liss and Slater (1974). SO_2 is readily absorbed by the ocean. In this case the gas-phase resistance exceeds that of the liquid phase, leading to a deposition velocity

Table 1-13. *Air–Sea Exchange of Various Gases*[a]

Trace gas	k_g (m/s)	k_L (m/s)	H	r_g (s/m)	r_L (s/m)	m_s	$(c_L - c_g/\mathbf{H})$ (μm/m³)	F_s (μg/m² s)	Remarks
SO$_2$	4.4×10^{-3}	9.6×10^{-2}	0.02	1.1×10^{4}	10.4	45 pptv	-3.4	-5.7×10^{-4}	$c_L = 0$, $v_d = 4.4\times10^{-3}$
O$_3$	5.1×10^{-3}	4.9×10^{-4}	3	6.5×10^{1}	2.0×10^{3}	25 ppbv	-17.9	-8.5×10^{-3}	$v_d = 5\times10^{-4}$
CCl$_4$	2.9×10^{-3}	3.0×10^{-5}	1.1	3.2×10^{2}	3.3×10^{4}	71 pptv	-0.04	-1.2×10^{-6}	According to measurements of
CCl$_3$F	3.0×10^{-3}	3.1×10^{-5}	5	6.6×10^{1}	3.2×10^{4}	50 pptv	-0.015	-4.7×10^{-7}	Lovelock *et al.* (1973)
N$_2$O	5.3×10^{-3}	5.5×10^{-5}	1.6	1.2×10^{2}	1.8×10^{4}	0.3 ppmv	$+68$	$+3.7\times10^{-3}$	
CO	6.7×10^{-3}	5.5×10^{-5}	50	3.0	1.8×10^{4}	0.15 ppmv	$+113$	$+6.2\times10^{-3}$	For measurements see Chapters 4,
CH$_4$	8.8×10^{-3}	5.5×10^{-5}	42	3.7×10^{-1}	1.8×10^{4}	1.6 ppmv	$+9.53$	$+5.3\times10^{-4}$	9, and 10
(CH$_3$)$_2$S	4.5×10^{-3}	3.1×10^{-5}	0.3	7.4×10^{2}	3.2×10^{4}	100 pptv	$+217$	$+6.5\times10^{-3}$	

[a] Given are the gas- and liquid-phase transfer coefficients, and the corresponding resistances to transfer as defined in the text, according to Liss and Slater (1974), except for ozone (Garland *et al.*, 1980). Approximate mixing ratios of the gases in surface air and liquid phase concentration differences resulting from the consumption (−) or production (+) in seawater determine the fluxes through the air–sea interface.

of $v_d = 4.4 \times 10^{-3}$ m/s, which agrees with the observational data of Table 1-12. Seawater absorbs ozone less well than SO_2, but the reaction of ozone with iodide reduces the liquid-phase resistance by an order of magnitude as Garland *et al.* (1980) have shown. The resulting deposition velocity is in reasonable accord with the field measurements reported in Table 1-12. For the halocarbons CCl_4 and CCl_3F, the flux also is directed from the atmosphere into the ocean owing to the fact that their concentrations are higher in the gas than in the liquid phase. The other gases shown in Table 1-13 are released from the ocean to the atmosphere. In these cases, the gases are produced in seawater by biological processes so that a positive concentration difference is set up.

2 | Photochemical Processes and Elementary Reactions

Chemical reactions that occur spontaneously upon mixing together two or more stable chemical reagents are quite rare. Usually, a reaction proceeds only if stimulated by the addition of energy. In the laboratory, the commonest source of energy is heat; other forms of energy, such as visible or ultraviolet light, electric discharges, or ionizing radiation, are used less frequently. In the atmosphere, in contrast, most chemical processes are initiated by solar radiation, because the heat content of ambient air is insufficient for thermal activation except in a few special cases, and lightning discharges and cosmic ionizing radiation are globally comparatively insignificant in stimulating chemical reactions.

The description of complex photochemical reactions is facilitated by subdividing the total reaction into a series of elemental steps, of which the primary (or initiation) process follows the absorption of a photon, and the subsequent processes are thermal (or dark) reactions. The interaction of solar radiation with the atmosphere and its constituents, and the ensuing primary photochemical processes are examined in this chapter. Dark reactions will be discussed whenever appropriate. The identification of the individual reaction steps and their characterization by rate laws are the subject of chemical kinetics. Some familiarity with reaction kinetics and the behavior of elementary reactions is essential to the discussion of atmospheric chemistry. Hence, a brief summary of the basic concepts will be provided initially in this chapter. Only homogeneous gas-phase reactions are considered, however. Reactions occurring in the aqueous phase of clouds are discussed in Chapter 8.

Chemistry is basically a practical science, and most of our knowledge about atmospheric chemical reactions has been established by laboratory experiments. Despite considerable advances in laboratory techniques it is often difficult to find experimental conditions whereby the reaction of interest can be sufficiently isolated from other concurrent ones so that its rate behavior can be unambiguously determined. Rate data thus are considered reliable only if two independent experimental techniques lead to the same results. As a consequence, the accumulation of dependable information on reactions deemed important to atmospheric chemistry has been a slow process. Nevertheless, a fairly large body of data now exists, at least for homogeneous gas-phase reactions. Tables A-4 and A-5 provide a compilation of reaction rate data as a reference for all subsequent discussions. The numbering of the reactions is used throughout this book.

2.1 Fundamentals of Reaction Kinetics

Chemical reactions manifest themselves by a change in the composition of a mixture of chemical substances. In the course of time, some substances disappear while others are formed. The former are called reactants, and the latter products. On the molecular level, chemical reactions lead to a rearrangement of the atoms among the molecules making up the reacting mixture. Products may arise from the rearrangement of the atoms within an individual molecule, from the disintegration or combination of molecules, from atom interchange among two or more reactant molecules, etc. In chemistry the integrity of the atoms is preserved. The law of mass conservation then leads to an important restriction: the number of atoms of each kind taking part in the chemical reaction remains constant, no matter how extensive a rearrangement of atoms takes place. To give a simple example, the complete oxidation of methane to carbon dioxide and water must be written

$$CH_4 + 2O_2 \rightarrow CO_2 + 2H_2O$$

in order that the same number of carbon, hydrogen, and oxygen atoms appear on the reactant and the product sides of the equation. This requirement is known as the law of stoichiometry.

The oxidation of methane proceeds at measureable rates at temperatures between 400 and 500°C. Although the final products are CO_2 and H_2O, one should not assume that the reaction occurs in one single step as written above. When the reaction is studied in a closed vessel, several intermediate products can be discerned, as Fig. 2-1 shows. Formaldehyde and hydroperoxide appear first, followed by carbon monoxide, before carbon dioxide emerges. Thus, the reaction proceeds in stages. The above way of writing the reaction indicates only the overall chemical change, but it says nothing about the detailed pathways by which the final products are formed, nor does it account for the intermediate products occurring along the way.

Oxidation reactions, like many other chemical reactions, are complex processes involving whole arrays of individual reaction steps. A minimum set of individual reactions required to explain the experimental observations is called a reaction mechanism. The elucidation of reaction mechanisms is one of the principal aims of reaction kinetics. Mechanisms usually evolve from initial postulates, which are modified as a better understanding of the reaction system is obtained by further studies. Data such as those shown in Fig. 2-1 do not contain sufficient information to support a specific reaction mechanism. But over the years kineticists have learned to isolate many of the postulated reactions for individual study. In addition, experiments may be devised to verify or reject postulated mechanisms. In this way, many of the reactions participating in the methane oxidation have been identified. They are listed in Table 2-1. The experimental evidence for their occurrence was reviewed by Hoare (1967).

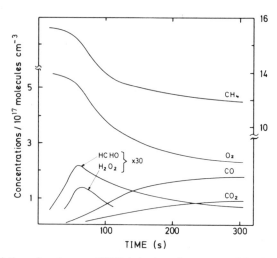

Fig. 2-1. Oxidation of methane at 500°C: behavior of reactant and product concentrations with time. Adapted from measurements of Blundell *et al.* (1965). Initial number densities were $n(O_2) = 6.8 \times 10^7$, $n(CH_4) = 1.56 \times 10^{18}$, $n(N_2) = 2.63 \times 10^{18}$, in units of molecules/cm^3. The number densities for H_2O_2 were scaled from the data of Egerton *et al.* (1956).

Table 2-1. *Reaction Mechanism for the Thermal Oxidation of Methane*[a]

(a)	$CH_4 + O_2 \rightarrow HO_2 + CH_3$	Initiation step
(b)	$CH_3 + O_2 \rightarrow CH_2O + OH$	} First chain
(c)	$OH + CH_4 \rightarrow H_2O + CH_3$	
(d)	$OH + CH_2O \rightarrow H_2O + CHO$	
(e)	$CHO + O_2 \rightarrow HO_2 + CO$	} Second chain
(f)	$HO_2 + CH_2O \rightarrow H_2O_2 + CHO$	
(g)	$H_2O_2 + M \rightarrow 2OH + M$	Chain branching
(h)	$OH + CO \rightarrow CO_2 + H$	
(i)	$H + O_2 + M \rightarrow HO_2 + M$	} Third chain
(j)	$HO_2 + CO \rightarrow CO_2 + OH$	
(k)	$OH + HO_2 \rightarrow H_2O + O_2$	
(l)	$HO_2 + HO_2 \rightarrow H_2O_2 + O_2$	} Chain termination
(m)	Loss of HO_2 and H_2O_2 at the walls of the vessel	

[a] M indicates a chemically inert constituent, like N_2, which acts mainly as an energy-transfer agent.

The reactions in Table 2-1 are elementary in the sense that they cannot or need not be broken up further. It is evident that all the reactions listed involve molecular fragments, either as reactants or as products. Species such as CH_3, CHO, OH, or HO_2 are called radicals. By their nature they are very reactive, so that reactions involving radicals usually are fast. The mechanism of methane oxidation contains three sequences of reactions in which a radical consumed in one reaction is regenerated in a following reaction. Such sequences are called chain reactions. The rate of methane consumption is almost entirely due to the first of the three reaction chains shown in Table 2-1. The direct interaction of methane with oxygen is slow. It serves to initiate one of the reaction chains by generating radicals in the first place. From Fig. 2-1 one may glean that the rate at which methane is consumed accelerates as the reaction progresses. This kind of behavior is known as autocatalysis. It is observed frequently when in the course of a reaction the rate of formation of one or more of the reactive intermediates increases. In the oxidation of methane the rate acceleration is due to reaction (g). The decomposition of H_2O_2 leads to the formation of two OH radicals in place of one HO_2 radical consumed in reaction (f) to form H_2O_2. The propagation of reaction chains is stopped by the termination reactions (k)–(m). These reactions convert the chain carriers OH and HO_2 to less reactive products.

The occurrence of many of the reactions listed in Table 2-1 is not restricted to methane oxidation. OH and HO_2 are important chain carriers in most oxidation processes involving organic compounds. Whenever form-aldehyde appears as an intermediate product, it will enter into the same reactions as those shown in Table 2-1. Methane and carbon monoxide are removed by OH radicals in the atmosphere as well. The reactions contained in Table 2-1 thus have a more universal significance than appears on first sight. In fact, by tacit agreement, universality is considered a requirement that any genuine elementary reaction must meet. This feature makes it possible to acquire knowledge about an elementary reaction from one particular experiment and apply it to another experimental situation, or even to an entirely different environment such as the atmosphere.

In the foregoing, mention has been made of reaction rates. This quantita-tive aspect of reaction kinetics is now discussed. The rate of change for any component of a reaction mixture is defined by the derivative of its concentra-tion with regard to time. The derivative is negative for reactants and positive for products. Explicit mention of the sign is rarely made, however. Instead, it is customary to talk about rates of consumption and rates of formation, respectively. Intermediates in complex chemical reactions are first products, then reactants, so that their rates change from positive to negative during the reaction. Formaldehyde in Fig. 2-1 provides an example for this behavior.

For complex reactions such as the oxidation of methane no mathemati-cally simple rate laws can be formulated. The rates of any of the species involved are complicated functions of the concentrations of all the other participating reactants and intermediates, temperature, pressure, sometimes also wall conditions of the vessel, and other parameters. Isolated elementary reactions, in contrast, obey comparatively simple rate laws. Table 2-2 sum-marizes rate expressions for several basic types of homogeneous elementary reactions. Let us single out for the purpose of illustration the bimolecular reaction $A + B \rightarrow C + D$. On the left-hand side of the rate expression one has equality between the rates of consumption of each reactant and the rates of formation of each product, in accordance with the requirements of stoichiometry. On the right-hand side, the product of reactant concentrations expresses the notion that the rate of the reaction at any instant is proportional to the number of encounters between reactant molecules of type A and B occurring within unit time and volume. The rate coefficient k_{bim} is still a function of temperature, but it is independent of the concentrations. The same is assumed to hold for the rate coefficients of the other types of elementary reactions in Table 2-2. At a constant temperature the rate coefficients are constants and the equations can be integrated to yield the concentrations of reactants and products as a function of time.

In studies of elementary reactions in the laboratory one will strive for conditions by which the reaction is isolated so that simple rate laws apply. In complex reaction systems a chemical species usually is involved in several

Table 2-2. *Rate Laws for Three Common Types of Elementary Chemical Reactions*[a]

Reaction	Type	Rate law	Dimension of k
$A \to B + C$	Unimolecular decomposition	$-\dfrac{dn_A}{dt} = \dfrac{dn_B}{dt} = \dfrac{dn_C}{dt} = k_{uni} n_A$	s^{-1}
$A \to B + B$	Unimolecular decomposition	$-\dfrac{dn_A}{dt} = \dfrac{1}{2}\dfrac{dn_B}{dt} = k_{uni} n_A$	s^{-1}
$A + B \to C + D$	Bimolecular	$-\dfrac{dn_A}{dt} = -\dfrac{dn_B}{dt} = \dfrac{dn_C}{dt} = \dfrac{dn_D}{dt} = k_{bim} n_A n_B$	$cm^3/molecule\ s$
$A + A \to B + C$	Bimolecular	$-\dfrac{1}{2}\dfrac{dn_A}{dt} = \dfrac{dn_B}{dt} = \dfrac{dn_C}{dt} = k_{bim} n_A^2$	$cm^3/molecule\ s$
$A + B + M \to C + M$	Termolecular	$-\dfrac{dn_A}{dt} = -\dfrac{dn_B}{dt} = \dfrac{dn_C}{dt} = k_{ter} n_A n_B n_M$	$cm^6/molecule^2\ s$
$A + A + M \to B + M$	Termolecular	$-\dfrac{1}{2}\dfrac{dn_A}{dt} = \dfrac{dn_B}{dt} = k_{ter} n_A^2 n_M$	$cm^6/molecule^2\ s$

[a] M signifies an inert constituent that acts as a catalyst but whose concentration is not changed by the reaction.

reactions simultaneously. The appropriate rate expression for this species then contains on the right-hand side as many terms as there are reactions in which the species is involved. Formaldehyde in the methane oxidation mechanism may again be cited as an example. According to Table 2-1, it enters into reactions (b), (d), and (f), so that the rate is given by the sum of the three individual rates,

$$dn(HCHO)/dt = k_b n(CH_3) n(O_2) - k_d n(OH) n(HCHO)$$

$$- k_f n(HO_2) n(HCHO) \qquad (2\text{-}1)$$

with the $n(x)$ denoting the number densities of the various reactants. The rate equations for the other species appearing in the mechanism of the methane oxidation can be set up in a similar manner, but attention must be paid to apply the correct stoichiometric factors associated with each term. There are altogether 11 species, so that one has also a total of 11 rate equations. This result may be generalized: for any complex chemical reaction mechanism, the number of rate equations is equal to the number of chemical species participating in the reaction. A complete set of rate equations forms a mathematically well defined nonlinear system of differential equations, which together with the initial conditions can be solved to yield the individual concentrations as a function of time. It is clear that closed solutions will exist only in special cases, but numerical solutions can always be obtained with the aid of modern computers. In this way the main features of any reaction mechanism, even a hypothetical one, can be exposed. Reaction kineticists make liberal use of this technique to verify reaction mechanisms by comparing computed with experimental results. With appropriate modifications, the same procedure of computer simulation is used in atmospheric chemistry to interpret observational data in terms of specific reaction mechanisms. The success of such models depends on a sufficient knowledge of elementary reactions, the associated rate coefficients, and their temperature and pressure dependence. The last aspect will be considered next.

2.2   Properties of Rate Coefficients

Temperature and pressure represent external variables that can be controlled in experimental studies of reaction rates. It has been found that the rate coefficients associated with the three basic types of elementary reactions in Tables 2-2 behave quite differently with regard to both variables, so that it is convenient to treat each reaction type separately. Bimolecular reactions will be discussed first, and decomposition and recombination reactions subsequently.

Bimolecular reactions are basically independent of pressure. If an influence of pressure is observed, it is an indication that the reaction is not truly elementary but involves more than one chemical process. This complication will be ignored here. The rate of a bimolecular reaction increases as the temperature is raised. Laboratory experience shows that the increase of the rate coefficient with temperature over not too wide a range can be expressed almost always by an equation of the form

$$k_{\text{bim}} = A \exp(-E_a / R_g T) \tag{2-2}$$

where T is the absolute temperature, R_g is the gas constant (here expressed in J/mol degree), and A and E_a are constants. The preexponential factor A is related to the collision frequency between reactants. In the gas phase, its upper limit is $\sim 10^{-10}$ cm^3/molecule s, but usually it is smaller. The constant E_a has the dimensions of energy. This quantity is referred to as the activation energy following the proposal first made by S. Arrhenius in 1889 that the reaction proceeds only via a small fraction of all collisional encounters between reactant molecules, namely, those involving molecules that are thermally activated to energies sufficient to overcome an energy barrier E_a. The exponential form of the temperature dependence follows from the Boltzmann distribution of energies among the various degrees of freedom of the molecules.

Figure 2-2 illustrates the modern concept of activation energy for the simplest bimolecular reaction, the atom exchange process $A + BC \rightarrow AB + C$. The diagram on the left shows a map of the potential-energy surface

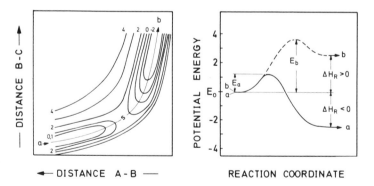

Fig. 2-2. Left: Map of potential-energy surface established in the interaction of three atoms A, B, and C. The reaction $A + BC \rightarrow AB + C$ proceeds from point a via a saddle point s to point b. Energy units are arbitrary and relative to the starting point. Right: Potential energy along the reaction coordinate for the forward reaction (solid line) and the reverse reaction (dashed line) relative to the starting points a and b, respectively. The heat of the reaction is negative in the first case and positive in the second.

established by the interaction of the three atoms involved in the colinear A–B–C collision complex. The preferred reaction path, shown by the solid line, starts out from point (a) in the reactant valley, passes a saddle point, and then proceeds into the product valley towards points (b). The diagram on the right shows the local potential energy along the reaction path relative to the energy at the starting position. The activation energy E_a corresponds to the height of the potential barrier at the saddle point. In the example chosen, the bottom of the product valley ultimately reaches an energy level lower than that at the starting point. Thus, energy is set free during the reaction. The excess energy appears temporarily in the vibrational, rotational, and translational modes of the products before it is dissipated to the surroundings. Reactions of this kind are called exothermic. The total energy made available, that is, the heat of reaction, ΔH_R, is negative, as Fig. 2-2b shows. The opposite case, of a reaction requiring an energy input, is called endothermic. One can visualize this case by reversing the reaction path in Fig. 2-2a to proceed from point (b) to point (a). The reaction then reads $C + AB \rightarrow BC + A$. The behavior of potential energy along the reaction path relative to the energy at the starting point (b) is shown in Fig. 2-2b by the dashed line. In this case ΔH_R is positive. The activation energy is now the sum of the energies E_a and ΔH_R—that is, the activation energy is somewhat greater than the endothermicity of the reaction. Owing to the exponential dependence of the rate coefficient on the activation energy, it is apparent that the endothermic pathway of our model reaction will be much slower at ambient temperatures than the exothermic pathway. At low temperatures, the reaction will proceed from point (a) to point (b), whereas the reverse reaction can be neglected. At elevated temperatures the reverse reaction must be taken into account.

In the atmosphere, reactions with appreciable endothermicities are unimportant because temperatures are not high enough. It turns out that also many exothermic reactions between stable atmospheric species are negligibly slow due to high activation energies. The reaction $CO + NO_2 \rightarrow CO_2 + NO$ may serve as an example. The heat of the reaction is $\Delta H_R = -225.9 \, kJ/mol$, but the activation energy is $E_a = 132.3 \, kJ/mol$ (Johnston et al., 1957), leading to an exponential factor of 4×10^{-27}. Only reactions with activation energies lower than about $50 \, kJ/mol$ need be considered in the atmosphere. Reactions involving radicals owe their importance largely to the fact that the associated activation energies lie predominantly in the low range, 5–$40 \, kJ/mol$.

Thermal decomposition reactions may be considered briefly. The breakage of a chemical bond requires an energy input so that these reactions are endothermic. Again, the rate coefficient displays an exponential temperature dependence. The activation energy is related to the endothermicity of the reaction. In the atmosphere only a few molecules with low bond dissociation energies are capable of thermal decomposition, among them nitrogen pentoxide ($E_a = 88$ kJ/mol) and peroxyacetyl nitrate ($E_a = 104$ kJ/mol). For temperatures and pressures existing near the earth surface, 293 K and 1000 mbar, these substances decompose with time constants $\tau = 1/k$ of 10 and 6500 s, respectively (Cox and Roffey, 1977; Hendry and Kenley, 1977; Connell and Johnston, 1979). In the upper troposphere and in the stratosphere where temperatures and pressures are lower, both substances are considerably more stable. Note that admitted activation energies for unimolecular reactions of importance in the atmosphere may be higher than those for bimolecular reactions because of more favorable preexponential factors.

In the gas phase, thermal decomposition reactions generally follow unimolecular rate behavior only at sufficiently high pressures (high-pressure limit). Then the number of collisions effecting the transfer of energy from the carrier gas to the reactant molecules is high enough to establish an equilibrium of activated reactant molecules with the surrounding heat bath. Activating and deactivating collisions are then in balance, and their rates exceed the decomposition rate. At reduced pressures the equilibrium may be disturbed in that the rate of decomposition of activated molecules becomes faster than that of deactivating collisions. For such conditions, the decomposition rate is collision-limited and a bimolecular rate law applies. In the intermediate region between both limits the first-order rate coefficient depends on both the pressure and the nature of the carrier gas molecules. An example for the behavior of the rate coefficient as a function of pressure and temperature is shown in Fig. 2-3 for the thermal decomposition of nitrogen pentoxide. Johnston (1966) and Connell and Johnston (1979) have discussed formulas by which the functional relationship $k(p, T)$ for the decomposition of N_2O_5 can be approximated.

Recombination reactions are the inverse of unimolecular dissociation processes, and the associated rate coefficients correspondingly exhibit also a pressure dependence. Like thermal decomposition processes, recombination reactions require an energy transfer by collision. The pressure dependence results from the change in efficiency with which the excess energy is removed from the incipient product molecule by the third body M. The situation can be made clearer by writing the reaction as a sequence of two steps

$$A + B \rightleftharpoons AB^* \qquad k_q, k_r$$

$$AB^* + M \rightarrow C + M \qquad k_s$$

where AB^* is an energy-rich intermediate capable of redissociating to the original reactants. Only a collision with a chemically inert molecule M

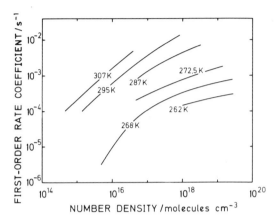

Fig. 2-3. Empirical first-order rate coefficients for the thermal decomposition of N_2O_5 in nitrogen as a function of the effective number density and temperature. Interpolated data from Connell and Johnston (1979).

stabilizes the AB^* product. If one treats the concentration of AB^* molecules as stationary so that the rates of formation and consumption are equal, one has

$$dn(AB^*)/dt = k_q n(A)n(B) - [k_r + k_s n(M)]n(AB^*) = 0 \qquad (2\text{-}3)$$

and the rate for the formation of the final products becomes

$$\frac{dn(C)}{dt} = k_s n(AB^*)n(M) = \frac{k_s k_q n(A)n(B)n(M)}{k_r + k_s n(M)} \qquad (2\text{-}4)$$

It is apparent that two limits exist. At low pressures when $k_s n(M) \ll k_r$, the overall rate law of the reaction will follow a termolecular rate law with an effective rate coefficient $k_0 = k_s k_q / k_r$, whereas at high pressures, when $k_s n(M) \gg k_r$, the $n(M)$ cancel and the formation of the product C proceeds in accordance with a bimolecular rate law. The effective rate coefficient then is $k_\infty = k_q$. In the intermediate pressure region neither rate law applies, and the rate coefficient becomes pressure-dependent. Usually, the rate of the reaction depends somewhat on the nature of the third body M as well.

Recombination reactions differ from bimolecular and thermal decomposition reactions also by their weak and usually negative dependence on temperature. Recombination reactions do not require an activation energy. Rather, the excess energy has to be dissipated. If the temperature dependence of the rate coefficient is expressed nonetheless by an exponential factor, the exponent $-E_R/R_g T$ is positive and E_R constitutes just a parameter with no physical meaning. Frequently, the temperature dependence of the rate coefficient k_R for a recombination reaction is written in the form

$$k_R = k_R^{300}(300/T)^m \qquad (2\text{-}5)$$

with k_R^{300} referring to the value of k_R at 300 K. Values for m range from 0.5 to 5. For recombination coefficients with a strong pressure dependence, Zellner (1978) has given an approximation formula based on the theoretical work of Troe (1974):

$$k_R(M, T) = \left[\frac{k_0(T)n(M)}{1 + k_0(T)n(M)/k_\infty(T)}\right] \times 0.8^a$$

$$a = (1 + \{\log_{10}[k_0(T)n(M)/k_\infty(T)]\}^2)^{-1} \qquad (2\text{-}6)$$

which is well applicable to atmospheric conditions. The terms $k_0(T)$ and $k_\infty(T)$, respectively, are the low- and high-pressure limiting values for the rate coefficient k_R. Their temperature dependencies are given by Eq. (2-5). Figure 2-4 compares recombination rate coefficients for the reaction OH + $NO_2 \rightarrow HNO_3$ measured as a function of temperature and N_2 carrier gas pressure by Anastasi and Smith (1976), with rate coefficients interpolated using Eq. (2-6). Zellner (1978) has compiled experimental data for several recombination reactions of importance to atmospheric chemistry to compute the effective rate coefficients as a function of altitude. The data are shown in Table 2-3. Further data on recombination reactions occurring in the atmosphere are summarized in Table A-5.

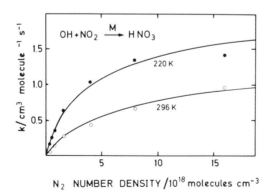

Fig. 2-4. Bimolecular rate coefficient for the reaction OH + $NO_2 \rightarrow HNO_3$ in nitrogen as a function of number density and temperature. Points indicate measurements of Anastasi and Smith (1976); solid curves were calculated with $k_0 = 2.6 \times 10^{-30}(300/T)^{2.9}$ and $k_\infty = 1.5 \times 10^{-11}(300/T)^{1.3}$ using the approximation formula of Zellner (1978).

Table 2-3. *Effective Bimolecular Rate Coefficients (cm^3/ molecule s)[a] for Several Recombination Reactions as a Function of Altitude in the Atmosphere* (Zellner, 1978)

z (km)	p (mb)	T (K)	n_M (molecules/cm^3)	Effective rate coefficient k_{bim}				
				$OH + NO_2$	$OH + NO$	$OH + SO_2$	$ClO + NO_2$	$HO_2 + NO_2$
0	1013	288	2.5 (19)	1.2 (−11)	4.9 (−12)	8.9 (−13)	2.3 (−12)	1.2 (−12)
5	487	268	1.3 (19)	1.0 (−11)	4.1 (−12)	8.2 (−13)	1.8 (−12)	1.0 (−12)
10	237	233	7.4 (18)	9.6 (−12)	3.7 (−12)	7.8 (−13)	1.6 (−12)	9.6 (−13)
15	121	220	4.0 (18)	7.8 (−12)	2.8 (−12)	6.8 (−13)	1.1 (−12)	7.7 (−13)
20	55	217	1.8 (18)	5.4 (−12)	1.7 (−12)	5.0 (−13)	5.8 (−13)	5.1 (−13)
25	26.7	222	8.5 (17)	3.1 (−12)	8.9 (−13)	3.1 (−13)	2.6 (−13)	2.8 (−13)
30	11.9	227	3.8 (17)	1.5 (−12)	4.1 (−13)	1.7 (−13)	1.1 (−13)	1.4 (−13)
35	5.7	235	1.8 (17)	7.2 (−13)	1.9 (−13)	8.2 (−14)	4.9 (−14)	6.3 (−14)
40	2.9	250	8.5 (16)	3.0 (−13)	7.9 (−14)	3.5 (−14)	1.9 (−14)	2.6 (−14)
45	1.47	260	4.0 (16)	1.3 (−13)	3.4 (−14)	1.5 (−14)	8.2 (−15)	1.1 (−14)
50	0.67	270	1.8 (16)	5.6 (−14)	1.4 (−14)	6.5 (−15)	3.4 (−15)	4.6 (−15)

[a] Powers of 10 are indicated in parentheses.

2.3 Photochemical Processes

As stated earlier, chemical reactions in the atmosphere are initiated mainly by photochemical processes rather than by thermal activation. It becomes necessary, therefore, to supplement the preceding survey of thermal reactions with a brief account of the principles of photochemistry.

The basic ideas underlying photochemical reactions are as follows. Light can become photochemically effective only if it is absorbed—that is, radiative energy must be incorporated into a molecule. The mere interaction of light with matter, such as in scattering, is not sufficient to initiate a chemical reaction. Moreover, energy is quantized. A ray of light may be described as a flux of photons, each of which represents an energy quantum $E = h\nu$ proportional to the frequency ν of light and inversely proportional to the wavelength $\lambda = c/\nu$, with c being the velocity of light. In the act of optical absorption, exactly one photon is taken up by the absorbing molecule, whose internal energy, in turn, is raised by exactly the amount of energy supplied by the photon. A quantum requirement is that the energy of the photon coincides with the energy level of one of the many higher states of the molecule relative to its initial state. If the molecular states are widely spaced, the spectrum of absorption as a function of wavelength will consist of discrete lines. A dense spacing will give the appearance of a continuous absorption spectrum. The strength of the absorption is determined by the probability for such transitions between the energy states.

It is customary to distinguish energy levels due to rotational, vibrational, and electronic motions within the molecule. Under normal conditions, the excitation of rotational and vibrational levels does not provide enough energy for photochemical action. Due to spectroscopic selection rules, high-lying vibrational levels are also not directly accessible from the vibrational ground state. Thus, photochemical processes generally require transitions to electronic states. In the excitation process a part of the energy usually is channeled also into vibrational energy modes. Commonly, energies in excess of 100 kJ/Einstein are involved corresponding to light in the very near infrared, visible, and ultraviolet spectral regions.

The fate of the excitation energy depends on the nature of the molecule and on the amount of energy is receives. The excited molecule may give off the energy as radiation (fluorescence), dissipate it by collisions (quenching), utilize the energy for chemical transformations (isomerization, dissociation, ionization, etc.), transfer all or part of the energy to other molecules that then react further (sensitization), or enter into chemical reactions directly. Several of these processes are written in Table 2-4 in the form of chemical reactions. They are considered primary processes in the sense that they all involve the excited molecule formed initially by photon absorption.

Table 2-4. *Types of Primary Photochemical Processes Following the Act of Photon Absorption $AB + h\nu \rightarrow AB^*$*

(1)	$AB^* \rightarrow AB + h\nu'$	Fluorescence
(2)	$AB^* \rightarrow A + B$	Dissociation
(3)	$AB^* + M \rightarrow AB + M$	Quenching
(4)	$AB^* + C \rightarrow AB + C^*$	Energy transfer
(5)	$AB^* + C \rightarrow A + BC$	Chemical reaction

Any subsequent thermal chemical reactions or physical energy dissipation processes are of a secondary nature. Specific examples for both primary and secondary photochemical reactions will be given in later chapters.

The quantitative assessment of photochemical activity is facilitated by introducing the quantum yield. In the practice of photochemistry a variety of quantum yield definitions are in use depending on the type of application. There are quantum yields for fluorescence, primary processes, and final products, among others. In atmospheric reaction models, the primary and secondary reactions usually are written down separately, so that the primary quantum yields become the most important parameters, and only these will be considered here. Referring to Table 2-4, it is evident that an individual quantum yield must be assigned to each of the primary reactions shown. The quantum yield for the formation of the product P_i in the *i*th primary process is the rate at which this process occurs in a given volume element divided by the rate of photon absorption within the same volume element.

$$\phi_i = \frac{\text{product molecules } P_i \text{ formed/cm}^3 \text{ s}}{\text{photons absorbed/cm}^3 \text{ s}} \qquad (2\text{-}7)$$

The quantum yield thus represents the probability for the occurrence of a selected process compared to the total probability for all the primary processes taken together. The sum of the individual primary quantum yields is unity.

A complete set of primary and secondary photochemical processes constitutes the mechanism of the overall photochemical reaction. The resulting changes of concentrations with time are described by the appropriate system of rate equations, as explained in the second section of this chapter. In setting up the rate equations, the primary processes can be treated as elementary reactions similar to thermal elementary reactions. The rate of chemical change associated with an individual primary process can be quantified in the following manner. Let light of wavelengths within a range $\Delta\lambda$ pass through a not necessarily homogeneous layer containing different kinds of absorbing molecules. If we single out molecules of type a, present with number density n_a, they contribute to the local rate of light absorption at each wavelength the term

$$\left[-\frac{dI(\lambda)}{dl} \right]_a = \sigma_a(\lambda) n_a I(\lambda) \qquad (2\text{-}8)$$

where $I(\lambda)$ is the local photon flux at the wavelength being considered, l is the absorption path parameter, and $\sigma_a(\lambda)$ is the absorption cross section. The local rate of product formation at the same wavelength λ, resulting from the ith primary process associated with the absorption of type a molecules, is obtained by multiplying $[-dI(\lambda)/dl]_a$ with the appropriate quantum yield $\phi_i(\lambda)$. The total local rate for the formation of a specific product from light of all wavelengths is then given by the integral over the entire wavelength range,

$$\frac{dn_i}{dt} = \int_{\Delta\lambda} \phi_i(\lambda) \left[-\frac{dI(\lambda)}{d} \right]_a d\lambda$$

$$= n_a \int_{\Delta\lambda} \phi_i(\lambda) \sigma_a(\lambda) I(\lambda) \, d\lambda = j_i n_a \qquad (2\text{-}9)$$

As the right-hand side shows, the primary photochemical process can be treated as a unimolecular reaction with the integral, denoted by j_i, taking the place of a rate coefficient. If the primary photochemical process is photodissociation, j_i is called the photodissociation coefficient. Its determination requires an evaluation of the integral and the parameters that enter into it. The terms $\sigma_a(\lambda)$ and $\phi_i(\lambda)$ are molecular properties, which can be measured in suitable laboratory experiments. Both parameters may depend also on pressure. The photon flux $I(\lambda)$ is a local parameter, which in the laboratory can be confined to a narrow wavelength band. For optically thin conditions it equals the incident light flux. In the atmosphere, one must take into account the attenuation of solar radiation and corresponding changes in the spectral intensity distribution. Solar radiation additionally undergoes scattering that gives rise to a diffuse flux component. Both the direct radiation flux and the diffuse radiation flux contribute to the photochemical effects. In the visible and ultraviolet spectral regions, attenuation by absorption is confined to the upper atmosphere, whereas scattering of radiation becomes important in the lower atmosphere. The next section will give a brief summary of the solar spectrum and its attenuation in the atmosphere.

2.4 Attenuation of Solar Radiation in the Atmosphere

Figure 2-5 shows the spectral distribution of solar radiation outside the earth's atmosphere and the direct flux reaching the ground surface. Table

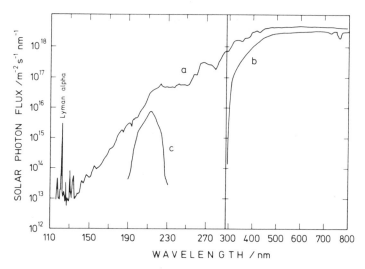

Fig. 2-5. The solar flux spectrum in the 120–800 nm wavelength region (a) outside the earth atmosphere, (b) at sea level, and (c) at about 30 km altitude to show the atmospheric window in the 185–215 nm wavelength region.

A-3 of the Appendix lists the numerical data. In the visible and near-infrared spectral regions the fluxes were mainly derived from ground-based measurements at mountain stations supplemented by aircraft and balloons as platforms. The ultraviolet portion of the solar spectrum has been explored with instruments on board rocket sondes and satellites. These data are of special interest here. Measurements before 1980 have been reviewed by Ackerman (1971), Simon (1978), and Nicolet (1981). More recent data have been reported by Mount and Rottman (1981, 1983, 1985) and Mentall *et al.* (1985).

The solar spectrum consists of a continuum superimposed by line structure. Emission as well as (Fraunhofer-type) absorption lines are evident. The continuum is dominant in the visible and near-ultraviolet regions of the spectrum. At wavelengths above 400 nm the continuum flux distribution corresponds to that of a blackbody at 5900 K. Below 400 nm the solar flux decreases more strongly than a blackbody temperature of 5900 K would predict. At wavelengths between 200 and 250 nm the flux corresponds to a blackbody temperature of 5100 K, and at wavelengths between 130 and 170 nm to 4600 K. With decreasing wavelength below 200 nm there is an increasing contribution of solar emission lines, of which the hydrogen Lyman alpha line at 121.6 nm is the strongest one. Most of the other lines cannot be shown isolated from the continuum because the absolute intensity measurements on which the spectrum in Fig. 2-5 is based were made at moderate spectral resolution, causing the line structure to be smeared out.

An adequate knowledge of the solar ultraviolet flux is essential for an understanding of stratospheric photochemistry. For the calculation of photodissociation coefficients a spectral resolution of 1 nm usually suffices. A higher resolution is required in a few important cases involving molecules subject to predissociation, such as oxygen and nitric oxide. Then, the fine structure of the absorption bands and their overlap with solar lines must be considered in detail. Simon (1978) has emphasized that photochemical calculations require fluxes obtained by integration over the entire solar disk. Measurements made with high spatial resolution at the center of the disk give only upper-limit fluxes due to the effect of limb darkening. Partly for this reason the early measurements of Detwiler *et al.* (1961) gave results that are now considered too high. All the data, however, are uncertain by about ±50% due to problems of calibration. In addition, the solar flux exhibits a certain natural variability. This aspect has been discussed in some detail by Brueckner *et al.* (1976) and by Heath and Thekaekara (1977). The ultraviolet spectrum is more variable than the visible. Solar emission lines originate mainly from the sun's chromosphere and corona and are influenced to a greater extent by sun spots and related phenomena than the continuum at wavelengths above 300 nm, which is emitted from the photosphere. Variations occur on time scales of minutes, due to solar flares; days, caused by the rise and decay of active regions; 27 days, due to the sun's rotation; and 11 years, on account of the different phases of the solar cycle. Satellite measurements have shown that at wavelengths near 120 nm the peak-to-peak variations of the solar flux amount to 35% during one solar rotation (Heath, 1973; Vidal-Madjar, 1975, 1977). Individual flares can change the flux of Lyman alpha by up to 16%. At wavelengths near 190 nm the flux varies by less than 10%, whereas at wavelengths greater than 210 nm the variation is less than 1%. These intensity variations occur in addition to the cyclic modulation of ±3.3% of the entire solar flux due to the eccentricity of the earth's orbit around the sun. All fluctuations generally are neglected when one considers photochemical processes in the atmosphere.

With regard to absorption processes in the atmosphere, the spectrum of the sun may be subdivided into three wavelength regions: <120 nm, 120–300 nm, and 300–1000 nm. Radiation within the extreme ultraviolet portion of the spectrum is absorbed at altitudes above 100 km, which will not be considered here. Note, however, that owing to the high photon energy the absorption leads in part to photoionization and to the formation of the earth's ionosphere (see, e.g., Banks and Kockarts, 1973, for details). Radiation within the second spectral region, 120–300 nm, is absorbed mainly in the mesosphere and stratosphere. The associated attenuation of solar radiation and the resulting photochemical processes are discussed below. Radiation within the third spectral region penetrates into the troposphere but is partly reduced by Rayleigh scattering. Radiation with wavelengths longer than 1000 nm does not supply enough energy for photochemical action and thus need not be considered. As was noted earlier, however, this infrared radiation is of prime importance to the heat balance of the atmosphere.

Nitrogen, although the major constituent of the atmosphere, is not an effective absorber at wavelengths greater than 120 nm because its principal absorption system, the Lyman–Birge–Hopfield bands, covers only the range shortward of 145 nm and the absorption is extremely weak. The principal absorbers of solar radiation in the ultraviolet spectral region are oxygen and ozone. Figure 2-6 shows the absorption cross sections of oxygen as a function of wavelength. Prominent features are the Herzberg continuum, the Schumann–Runge bands, and the Schumann continuum. A discussion of the O_2 absorption spectrum and the associated photochemical processes is facilitated by considering the energy level diagram of oxygen shown in Fig. 2-7. The absorption spectrum results from transitions from the ground state to electronic states at higher energies. Transitions to the low-lying excited states $^1\Delta_g$ and $^1\Sigma_g^+$ are forbidden by spectroscopic selection rules,

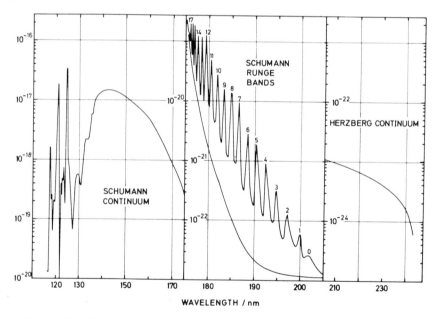

Fig. 2-6. Ultraviolet absorption spectrum of molecular oxygen, with cross sections given in units of cm^2/molecule. Data compiled from Ackerman and Biaumé (1970), Ditchburn and Young (1962), Hudson et al. (1966), Ogawa (1971), Shardanand (1969), Shardanand and Rao (1977), and Watanabe et al. (1953).

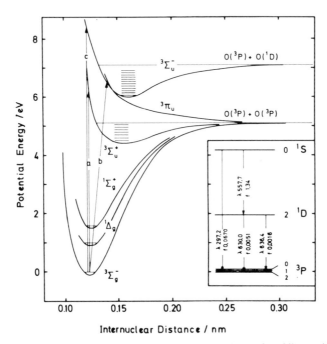

Fig. 2-7. Potential energy diagram for the lowest states of O_2, after Gilmore (1965). The indicated transitions produce the following absorption spectra: (a) Herzberg continuum, (b) Schumann-Runge bands, (c) Schumann continuum. The insert shows the lowest energy levels of the oxygen atom, the wavelengths of the emission lines resulting from the transitions indicated, and the associated oscillator strengths. Energy units: 1 eV = 96.5 kJ/mol.

so that the transition probabilities are low. The corresponding weak absorption bands lie in the near-infrared spectral region, where they are observed in the solar spectrum at long atmospheric pathlengths. The transitions to the discrete levels of the $^3\Sigma_u^+$ state give rise to the weak Herzberg bands in the wavelength region 250–300 nm. These bands are shielded by the much stronger absorption of ozone in the atmosphere, which makes them also unimportant. At wavelengths below 242.4 nm (i.e., at energies exceeding 5.11 eV or 492 kJ/mol), the spectrum becomes continuous, because the transition to the $^3\Sigma_u^+$ state now leads to the dissociation of O_2. A portion of the Herzberg continuum is apparent in Fig. 2-6 at wavelengths above 190 nm.

The transition to the next higher $^3\Sigma_u^-$ state is the origin of the Schumann-Runge bands, which with decreasing wavelengths merge into the Schumann continuum when the energy becomes high enough to reach the second dissociation limit of O_2 at 175 nm. Another transition in the same energy range is to the repulsive $^3\Pi_u$ state. This absorption process probably is responsible for the continuum underlying the Schumann–Runge bands. The intensity of such transitions is governed largely by the Franck–Condon principle, which states that the positions of the atomic nuclei in the molecule should change little if the transition is to be favorable. The rise in the absorption cross section with decreasing wavelength (increasing energy) must be ascribed mainly to the increase in the Franck–Condon probability. The absorption cross section attains a maximum value near 145 nm for a nearly vertical transition. At still shorter wavelengths the O_2 absorption spectrum consists of a great number of diffuse bands associated with transitions to higher, not well characterized dissociative states.

Ozone absorbs radiation throughout the near-infrared, visible, and ultraviolet spectral regions. The low energy requirement for the dissociation process $O_3 \rightarrow O_2 + O$ of 100 kJ/mol, corresponding to the wavelength limit of 1190 nm, admits photodissociation to be effective in all these absorption regions. In the visible and near infrared, the diffuse Chappuis bands represent a broad yet weak dissociation regime. Owing to the weakness of the absorption, the attenuation of the solar flux is comparatively minor (see Fig. 2-5). Absorption cross sections for ozone in the spectral region below 350 nm are shown in Fig. 2-8. At the long-wavelength limit the absorption

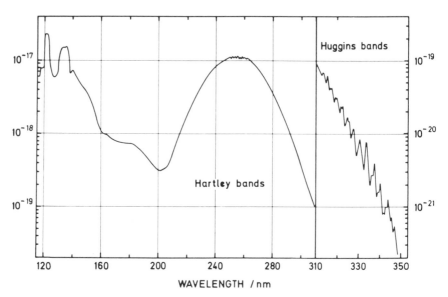

Fig. 2-8. Absorption spectrum of ozone in the wavelength region 115–350 nm, with cross sections given in units of cm^2/molecule. Data used are from Inn and Tanaka (1953), Tanaka *et al.* (1953), and Griggs (1968).

sets in with the Huggins bands, then blends into the Hartley bands, which has the appearance of a continuum topped by weak discrete structure. Here, the cross section reaches a first maximum at 255 nm. Toward shorter wavelengths there follows a region of lower absorption intensity before the cross section rises again toward a maximum at 135 nm and further peaks at still shorter wavelengths. This absorption domain is of lesser interest here, because it is dominated by oxygen in the atmosphere, and ozone contributes little to the attenuation of the solar flux.

In those spectral regions where the attenuation of incoming solar radiation is due primarily to absorption, the intensity of the solar flux having penetrated to the altitude level z_0 follows from the generalized form of Beer's law,

$$I(\lambda, z_0, \chi) = I_0(\lambda) \exp\left\{ -\frac{1}{\cos \chi} \int_{z_0}^{\infty} \left[\sum_k \sigma_k(\lambda) n_k(z) \right] dz \right\} \quad (2\text{-}10)$$

where $I_0(\lambda)$ is the incident solar flux outside the earth atmosphere, χ is the solar zenith angle, and the summation

$$\sum_k \sigma_k(\lambda) n_k(z) = \sigma_{O_2}(\lambda) n_{O_2}(z) + \sigma_{O_3}(\lambda) n_{O_3}(z) \quad (2\text{-}11)$$

involves only the number densities and absorption cross sections of oxygen and ozone. The contributions of all other constituents of the earth atmosphere can be neglected.

The wavelength dependence of the attenuation of solar radiation as it penetrates into the atmosphere is most conveniently illustrated by plotting the altitude at which the solar flux is diminished to one-tenth its initial value at normal incidence. Such a plot is shown in Fig. 2-9. The peak absorption cross section of oxygen in the Schumann region and at shorter wavelengths attenuate the flux already at altitudes above 90 km, but a number of windows exist allowing some wavelengths to penetrate to much lower altitude levels. Most notably, the Lyman alpha line falls exactly into one of the windows. In the wavelength region below 190 nm, where absorption cross sections of oxygen and ozone are comparable in magnitude, oxygen is the principal absorber because the mixing ratio of ozone is always much smaller than that of oxygen. At wavelengths beyond 210 nm, the absorption cross section of oxygen falls to very low values so that ozone becomes the principal absorber. The number density of ozone does not decrease quasi-exponentially with height like that of oxygen. Ozone arises as a photodissociation product of oxygen and attains its maximum number density at altitudes near 25 km. In this region ozone forms a layer that shields the lower atmosphere from solar radiation at wavelengths between 210 and 300 nm. Between the two principal absorption regimes of oxygen

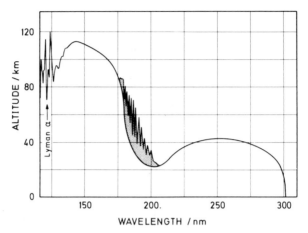

Fig. 2-9. Altitude at which incoming solar radiation from an overhead sun is attenuated to one-tenth the initial intensity.

and ozone, respectively, an optical window exists at wavelengths 190–210 nm where solar radiation reaches deeper into the atmosphere. Radiation in the window region is responsible for the photodissociation of several minor constituents in the atmosphere that absorb in this wavelength region but not at wavelengths above 300 nm. Water vapor, H_2O, nitrous oxide, N_2O, and the halocarbons CF_2Cl_2 and $CFCl_3$ are such constituents. Specifically, N_2O, CF_2Cl_2, and $CFCl_3$ are chemically quite inert, and processes removing these substances from the troposphere are not known. Upward transport into the photodissociation region of the stratosphere appears to provide the only loss process.

Beginning in the lower stratosphere, but becoming most important in the lower troposphere, is the phenomenon of radiation scattering. The principal processes involved are Rayleigh scattering of light by air molecules and Mie scattering by air-borne solid and liquid particles. The attenuation of incoming solar radiation due to scattering is again expressed by Eq. (2-10), where now

$$\sum \sigma_k(\lambda) n_k(z) = \sigma_{N_2}^R(\lambda) n_{N_2}(z) + \sigma_{O_2}^R(\lambda) n_{O_2}(z) + \sigma^M(\lambda) n_p(z) \quad (2\text{-}12)$$

Here, the superscripts R and M signify Rayleigh and Mie scattering, respectively, and n_p is the number density of particles. Rayleigh cross sections vary with λ^{-4} and have values ranging from about 2×10^{-25} to less than 10^{-26} cm^2/molecule in the wavelength region 300–700 nm. The scattering cross section for particles rises with the square of particle radius. In the absence of clouds, only particles in a narrow range of the aerosol size spectrum, about 0.2–2 μm, are effective as scatterers because larger particles are not sufficiently numerous. Number densities of particles maximize near the ground level and decrease rapidly with height (see Fig. 7-25). The Mie scattering coefficient is roughly independent of wavelength in the 300–800 nm spectral region. At sea level, for urban conditions (i.e., with particle concentrations that are elevated compared with normal background conditions), Rayleigh scattering dominates the attenuation of solar radiation in the wavelength region 300–500 nm, whereas Mie scattering takes over at longer wavelengths.

Unlike absorption, scattering does not result in a loss of radiation. While the direct flux of incoming solar radiation is attenuated, the diverted light undergoes multiple scattering and is available as a diffuse component, which must be added to the direct flux. In addition, some radiation is reflected at the Earth surface (or from clouds, although clouds are rarely considered). Thus, the total flux may be written

$$I_{total} = I_{direct} + I_{diffuse} + I_{reflected} \tag{2-13}$$

The reflected light also undergoes scattering. A photochemically active volume of the lower atmosphere evidently receives radiation from all direction, and not just from the preferred direction pointing toward the sun. Accordingly, it is necessary to integrate over fluxes being incident from all directions to determine the photochemically effective or actinic flux. This integrated flux must then be inserted in Eq. (2-9) to derive the photodissociation coefficients.

The diffusive component of the total flux and its contribution to photodissociation rates can be calculated from radiative transfer theory with the aid of modern high-speed computers (Braslau and Dave, 1973; Peterson, 1976; Luther and Gelinas, 1976). To save computing time, simplified models of radiative transfer of multiply scattered light are usually invoked. Leighton (1961), who first recognized the problem in conjunction with photochemical air pollution, devised a simple routine for calculating the actinic flux at the Earth surface. For more complex photochemical models of the lower atmosphere involving many photodissociation processes, Isaksen et al. (1977) argued that because Rayleigh scattering directs light preferentially forward and backward with equal probability, it would simplify the calculations if scattering is assumed to take place primarily in the direction of the sun's beam. While this is an oversimplification, the methods appears to give reasonably accurate photodissociation coefficients. A more realistic model is that of Peterson (1976), who included scattering due to aerosol particles. These scatter light preferentially in the forward direction. Figure 2-10 shows some results of Peterson (1976) to illustrate the relative contribution of upward scattered radiation to the total flux. Generally, the upward component increases with height near the earth's surface, partly because aerosol

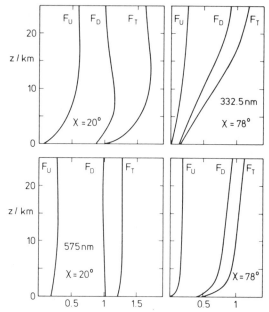

Fig. 2-10. Upward directed flux F_U, downward directed flux F_D, and total actinic flux F_T as a function of altitude in the lower atmosphere. Values are given relative to a solar constant of unity for two wavelengths, 332.5 and 575 nm, and for two zenith angles, 20° and 78°. The calculations of Peterson (1976) took into account Rayleigh scattering, absorption by ozone, and scattering and absorption by aerosol particles.

particles were assumed to absorb some of the radiation and the concentration of the particles is greatest near the surface. The upward-scattered flux is largest at short wavelengths and for small zenith angles. For large zenith angles the upward-scattered flux is comparatively small, because less radiation reaches the earth surface, but downward scattering now makes a fairly large contribution to the total flux.

A further complication of an already difficult subject is caused by the presence of clouds in the troposphere. Small clouds such as fair-weather cumuli produce much forward scattering so that the situation differs little from that for a clear sky. An extended cloud cover reduces photochemical activity considerably, however. The modification of the actinic radiation field by clouds, both inside and outside of the clouds, will be an important research subject of the future.

2.5 Photodissociation of Oxygen and Ozone

In continuing the discussion of ultraviolet photon absorption processes, we consider now the production of oxygen atoms from the photodecomposition of oxygen and ozone. Both ground-state and electronically excited oxygen atoms are involved. The lowest energy levels of atomic oxygen and the spectroscopic term symbols are shown in Fig. 2-7. The ground state is a triplet, because the coupling of electron spin and angular momentum vectors permits three quantum states of total angular momentum. The two lowest excited states, singlet D and singlet S, are highly metastable: that is, the transitions ^1D-^3P, ^1S-^1D, and ^1S-^3P have low transition probabilities due to spectroscopic selection rules. Radiative lifetimes for $O(^1D_2)$ and $O(^1S_0)$ are 148 and 0.71 s, respectively. In the atmosphere the excitation will be removed by collisional quenching rather than by radiation except at high altitudes. Table 2-5 summarizes data regarding production and loss processes for excited oxygen atoms in the atmosphere. From the quenching coefficients one estimates that collisional deactivation of $O(^1D_2)$ becomes important at altitudes below 250 km, and that of $O(^1S_0)$ below 95 km. Emissions from both excited states are observed at the appropriate altitude levels.

As Fig. 2-7 shows, the photoexcitation of O_2 in the wavelength region of the Herzberg and Schumann continua leads to the repulsive portions of the corresponding potential energy curves. Dissociation then takes place within the time span of one vibrational period, that is, within some 10^{-12} s, unaffected by collisions. This leads to a quantum yield of two for oxygen atom production. Excitation of O_2 by radiation in the wavelength region of the Schumann–Runge band system populates the discrete levels of the $^3\Sigma_u^-$ state. Here, one would not expect a dissociation to take place, but there exists a crossing and perturbation by the repulsive $^3\Pi_u$ state that gives rise to predissociation from vibrational levels of the $^3\Sigma_u^-$ state with $v \geq 3$. As a consequence, absorption within the Schumann–Runge bands can also produce oxygen atoms. Predissociation in the region of the Schumann–Runge bands is evidenced by the absence of fluorescence and by a broadening of rotation lines to an extent implying a lifetime of the $^3\Sigma_u^-$ state of the order of 10^{-11} s (Hudson et al., 1969; Ackerman and Biaumé, 1970; Frederick and Hudson, 1979a). This may be compared with a radiative lifetime of 2×10^{-9} s. Since the lifetime for predissociation is still shorter than the collision frequency, a quantum yield of unity is indicated. While for the wavelength region of the Schumann continuum laboratory studies have confirmed quantum yields of unity for the photodissociation of O_2 (Sullivan and Warneck, 1967; Lee et al., 1977), similar studies for the wavelength region of the Schumann–Runge bands are more difficult to do. Washida et al. (1971) explored ozone formation in oxygen at 184.9 and 193.1 nm and found O_3 quantum yields of 2.0 and 0.3, respectively. The first value agrees with expectation according to the preceding discussion; the second does not.

Table 2-5. *Data Regarding Excited Oxygen Atoms: (a) Known Production Processes; (b) 298 K Quenching Coefficients, Radiative Lifetimes τ_R, and Altitudes for Equivalent Loss Rates for Quenching by Collisions versus Radiative Emission; (c) Emission Altitudes*

Excitation level	Type of process		Altitude regime
	(a) Production		
$O(^1D)$	Photodissociation	$O_2 + h\nu \rightarrow O + O(^1D)$	$\lambda < 175$ nm	>85 km
		$O_3 + h\nu \rightarrow O_2(^1\Delta_g) + O(^1D)$	$\lambda < 310$ nm	Entire atmosphere
	Ion–electron recombination	$O_2^+ + e \rightarrow O + O(^1D)$		
	Electron impact	$O + e_{fast} \rightarrow O(^1D) + e_{slow}$		
$O(^1S)$	Photodissociation	$O_2 + h\nu \rightarrow O + O(^1S)$	$\lambda < 133$ nm	
	O-atom recombination	$O + O + O \rightarrow O_2 + O(^1S)$		>80 km

(b) Quenching coefficients (cm³/molecule s)[a]

	N_2	O_2	CO_2	O_3	H_2O	τ_r (s)	$z(\tau_R = \tau_Q)$ (km)
$O(^1D)$	2.4 (−11)	4.0 (−11)	1.0 (−10)	2.4 (−10)	2.0 (−10)	148	~250
$O(^1S)$	<5 (−17)	2.6 (−13)	3.7 (−13)	5.8 (−10)	>1 (−10)	0.71	96

(c) Observed emissions

$O(^1D)$	$\lambda = 630, 636.4$ nm	>110 km (day) > 160 km (night)
$O(^1S)$	$\lambda = 557.7$ nm	85–110 km

[a] Powers of 10 are given in parentheses.

The knowledge of quantum yields, absorption cross sections, and solar intensities, all as a function of wavelength, allows the calculation of photo-dissociation coefficients for oxygen from Eq. (2-9). The calculations present no particular difficulties for those spectral regions featuring absorption continua, but in the Schumann–Runge bands region the calculations become exceedingly difficult, for an exact evaluation of the integral makes it necessary to take into account each rotational line including effects of temperature, line width, and line overlap. The calculations of the past thus have made use of measured absorption or transmission data averaged over separate bands (Hudson and Mahle, 1972; Kockarts, 1976; Blake, 1979). The situation was reviewed by Nicolet and Peetermans (1980), who also supplied extensive numerical data on O_2 photodissociation coefficients. Figure 2-11 shows their results for an overhead sun as a function of altitude. The present interest is directed mainly at the atmospheric region below 50 km. Here, only the tail of the Schumann–Runge band system and the Herzberg continuum remain effective. Shorter wavelengths within the solar spectrum are too much attenuated.

The behavior of the total O_2 photodissociation coefficient in the strato-sphere is shown in Fig. 2-12. The strong decrease of the value toward lower altitudes by more than five orders of magnitude is due to the attenuation of solar radiation by both oxygen and ozone. In fact, $j(O_2)$ depends sensi-tively on the amount and the distribution of ozone assumed in the calcula-tion. Uncertainties arise also from computational problems at high degrees of attenuation in the lower stratosphere. Figure 2-12 thus indicates a range of values.

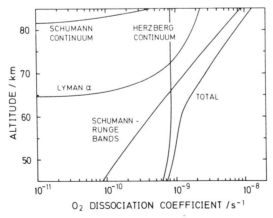

Fig. 2-11. Photodissociation coefficients for oxygen in the mesosphere for an overhead sun. From calculations of Nicolet and Peetermans (1980).

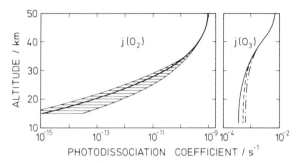

Fig. 2-12. Photodissociation coefficients for oxygen and ozone in the stratosphere. For $j(O_2)$ a range of values is given corresponding to data from various authors (Dütsch, 1968; Nicolet, 1975 and private communication; Luther and Gelinas, 1976; Johnston and Podolske, 1978); the upper limit is obtained for minimum ozone number densities. The dashed lines for $j(O_3)$ are obtained when backscattering is taken into account: - - - scattering only, - · - · - surface albedo 0.25. The $j(O_2)$ is not affected by backscattering. The heavy lines are used to calculate ozone number densities in Fig. 3-5.

The uneven distribution of rotational lines in the Schumann–Runge band system admits radiation that falls in between the lines to leak through to lower altitudes. Cicerone and McCrumb (1980) have pointed out that a portion of this radiation is absorbed by the ^{18}O ^{16}O isotope, which is present in the atmosphere with a mixing ratio of 0.00408 and whose rotational lines are shifted from those of the normal $^{16}O_2$ molecule. Although the ^{18}O ^{16}O molecule features twice as many lines and thus absorbs radiation more effectively than $^{16}O_2$, the mixing ratio is insufficient to increase the total O_2 photodissociation rate significantly.

The photochemistry of ozone has been studied in detail in the laboratory, and quantum yields for the formation of oxygen atoms are known for the most important wavelength regions. Schiff (1972) and Welge (1974) have reviewed the data. For the weak Chappuis bands appearing at wavelengths near 600 nm, the structure is diffuse, suggesting a predissociation process. The dissociation products are $O_3 \rightarrow O_2(^3\Sigma_g^-) + O(^3P)$. The work of Castellano and Schumacher (1962) established a quantum yield of unity for O-atom formation. In pure ozone, the quantum yield for O_3 loss is two, because of the subsequent reaction $O + O_3 \rightarrow 2O_2$. The Huggins bands in the wavelength region 310–350 nm likewise are sufficiently diffuse, even when sharpened by lowering the temperature, to imply the occurrence of predissociation. At 334 nm the quantum yield for O_3 removal in pure O_3 is four. This suggests a quantum yield of unity for oxygen atom production and the formation of $O_2(^1\Delta_g)$ or $O_2(^1\Sigma_g)$, which are both capable of dissociating ozone by energy transfer whereby another oxygen atom is formed (Jones and Wayne, 1970; Castellano and Schumacher, 1972). For the Hartley band at wavelengths between 200 and 300 nm there is now much evidence that the main dissociation process is $O_3 \rightarrow O_2(^1\Delta_g) + O(^1D)$, which becomes energetically possible at wavelengths near 310 nm. Figure 2-13 shows primary quantum yields for $O(^1D)$ as a function of wavelength. The values are nearly unity throughout most of the spectral region. The associated production of $O_2(^1\Delta_g)$ has been observed at 253.7 nm to occur also with a quantum yield of unity (Jones and Wayne, 1971; Gauthier and Snelling, 1971). Fairchild *et al.* (1978), however, using molecular-beam photofragment spectroscopy, have found a production of up to 10% of $O(^3P)$ in the wavelength region 270–300 nm. The $O(^1D)$ quantum yield thus must be somewhat lower than unity. Uncertainties of this magnitude are common in photochemical quantum yield determinations.

Figure 2-12 includes photodissociation coefficients for ozone as a function of altitude calculated with the assumption of quantum yields of unity. In the stratosphere, due to the occurrence of the ozone layer there, wavelengths falling into the spectral region of the Hartley bands are filtered out and $j(O_3)$ decreases accordingly toward the lower atmosphere. Photodissociation in the spectral regions of the Huggins and Chappuis bands is still effective in the troposphere.

The threshold to $O(^1D)$ formation in the photodecomposition of ozone is of considerable interest to tropospheric chemistry. On one hand it is located at wavelength above the cutoff for solar radiation so that $O(^1D)$ is formed also in the troposphere; on the other hand the $O(^1D)$ atoms can react with water vapor (which $O(^3P)$ atoms do not, for energetic reasons), a reaction that leads to the production of OH radicals. They are of immense importance in the initiation of oxidation processes. As Fig. 2-13 shows, the quantum yield for $O(^1D)$ decreases at wavelengths greater than 302 nm, but its tail extends beyond the nominal energy threshold for $O(^1D)$ formation at 310 nm due to the thermal excitation of rotational levels of the O_3 ground state. The quantum yield curve near threshold thus depends on temperature. Moortgat *et al.* (1977) have shown, however, that the temperature dependence is greater than that predicted from the additional energy content due to the thermal population of ground-state rotational levels. Since the threshold for $O(^1D)$ formation lies so close to the cutoff wavelength for solar radiation penetrating into the troposphere, it is clear that a very narrow wavelength range is responsible for $O(^1D)$ formation. In addition, the rate varies considerably with the attenuation of solar radiation by the ozone layer overhead. The natural distribution of stratospheric ozone (see Fig. 3-4) features greater ozone column densities at high latitudes, which favors higher rates of $O(^1D)$ formation near the equator than at the poles.

The production of $O(^1S)$ in the photodissociation of ozone becomes energetically possible at wavelengths less than 237 nm, if a ground-state O_2

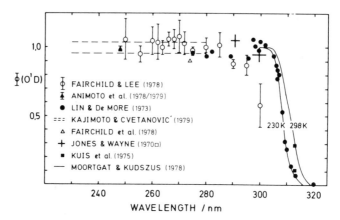

Fig. 2-13. Quantum yields for the formation of $O(^1D)$ atoms in the photodecomposition of ozone as a function of wavelength. The threshold behavior near 310 nm is of considerable importance to tropospheric chemistry. Note the temperature dependence of the quantum yield in the threshold region.

molecule is the second product. This process does not conserve spin, so that the probability for its occurrence is expected to be low. The spin-allowed process $O_3 \rightarrow O_2(^1\Delta_g) + O(^1S)$ has its threshold at 199 nm. Although the onset of a new continuum shortward of the Hartley region coincides with the threshold wavelength, it appears not to be associated with the production of $O(^1S)$ atoms. Lee *et al.* (1980) reported quantum yields less than 0.1% for the 170-240 nm wavelength region. As Table 2-5 indicates, the rate coefficients for $O(^1S)$ quenching by oxygen and nitrogen are small compared with their reactions with H_2O and O_3. This contrasts with the behavior of $O(^1D)$. Thus, if $O(^1S)$ atoms were formed in the upper stratosphere, a good deal of them would react with H_2O, causing the production of OH radicals in addition to those resulting from the reaction of $O(^1D)$ with H_2O. The small $O(^1S)$ production rate from the photodissociation of ozone makes this possibility unlikely.

2.6 Photochemistry of Minor Atmospheric Constituents

The atmosphere contains many trace gases that are photochemically active. Few, however, attain a significance similar to oxygen or ozone in driving atmospheric chemistry. Table 2-6 gives an overview on the photochemical behavior of atmospheric constituents. The last column indicates whether the photodecomposition of the substance is important to atmospheric chemistry or not. The molecules listed may be subdivided into three groups. (1) For a number of gases such as methane or ammonia that enter into other reactions photodecomposition is relatively unimportant because the photochemically active wavelength region is shielded by oxygen and/or ozone. (2) A second group includes gases that are fairly unreactive in the troposphere but undergo photodecomposition in the stratosphere, primarily in the 185–215 nm wavelength region of the atmospheric window. Nitrous oxide and the halocarbons $CFCl_3$, CF_2Cl_2, and CCl_4 belong to this group. (3) The third group comprises all gases that are subject to photodecomposition by radiation reaching the troposphere. Examples are nitrogen dioxide and formaldehyde. Whether for such gases photodecomposition is important as a loss process depends on the extent of other competing loss reactions. Even if photolysis does not cause an appreciable loss of the trace gas itself, the photoproducts may still initiate important chemical reactions.

In this section we discuss absorption spectra, quantum yields, and photodissociation coefficients for several atmospheric trace species. DeMore *et al.* (1985) and Baulch *et al.* (1980, 1982, 1984) have critically reviewed data relevant to atmospheric processes. Table 2-7 summarizes important photochemically active molecules and associated photodissociation coefficients. These were computed with the full, unattenuated solar spectrum so that they represent maximum values. But an attempt has been made to separate critical wavelength regions as a discriminator for the significance of the process in the stratosphere and troposphere, respectively.

Figure 2-14 shows absorption spectra for CO_2, H_2O, N_2O, HNO_3, H_2O_2, and N_2O_5 in the 170–370 nm wavelength region. At the longer wavelengths,

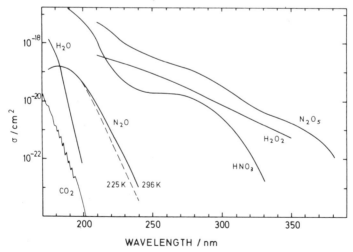

Fig. 2-14. Absorption spectra of CO_2, H_2O, N_2O, HNO_3, H_2O_2, and N_2O_5. From data assembled by De More *et al.* (1982) and Baulch *et al.* (1980, 1982). Data for CO_2 are from Ogawa (1971).

Table 2-6. *Photochemical Behavior of Trace Gases in the Atmosphere*[a]

	Absorption spectrum		Photochemistry		Significance to the atmosphere
Constituent	Type and approximate long-wavelength limit (nm)	Dissociation limit (nm)	Main primary process	Quantum yield	
O_3	Hartley bands	320	310	$O_3 \rightarrow O(^1D) + O_2(^1\Delta_g)$	$\phi \approx 0.9$	Important in stratosphere; see Figs. 2-8 and 2-13
	Huggins bands	360	410	$O_3 \rightarrow O(^3P) + O_2(^3\Sigma_g)$	$\phi \approx 1.0$	Important in the troposphere
	Chappuis bands	850	1180	$O_3 \rightarrow O(^3P) + O_2(X^3\Sigma_g^-)$	$\phi \approx 1.0$	
CO_2	Continuum over-lapped by bands	200	226	(a) $CO_2 \rightarrow CO + O(^3P)$	$\phi \approx 1.0$	(a) Mesosphere and stratosphere, 175-200 nm;
			167	(b) $CO_2 \rightarrow CO + O(^1D)$		(b) shielded by O_2
CO	Resonance bands 4th positive system	155	111	Fluorescence/quenching		Not significant
CH_4	Continuum, some band structure	160	277	Dissociation		Not significant; shielded by O_2
H_2	Lyman bands	111	84.4	Fluorescence/quenching		Not significant
H_2O	Continuum, some band structure	200	242	$H_2O \rightarrow OH + H$	$\phi \approx 1.0$	Mesosphere and stratosphere in the atmospheric absorption window (185-215 nm)
			177	$H_2O \rightarrow H_2 + O(^1D)$	Small	
H_2O_2	Continuum	350	578	$H_2O_2 \rightarrow OH + OH$	$\phi \approx 1.0$	Mainly strato-sphere and meso-sphere, >185 nm

N_2O	Continuum, some diffuse band structure	230	741.5	(a) $N_2O \rightarrow N_2 + O(^3P)$	Spin forbidden	(b) Stratosphere and in the atmospheric absorption window, >185 nm
			340.7	(b) $N_2O \rightarrow N_2 + O(^1D)$ Dissociation	$\phi \approx 1.0$	
NH_3	Bands overlying a continuum	235	281			Not significant
NO	Resonance bands Continuum+bands	230	190.8	Fluorescence/quenching (a) Predissociation, mainly the bands 0-0 and 0-1	$\phi \approx 1.0$	(a) Mesosphere and upper stratosphere in the atmospheric absorption window, (b) by Ly α in the upper mesosphere
		135	133.8	(b) Photoionization		
NO_2	Bands, diffuse structure / Continuum below 250	700	397.8	(a) $NO_2 \rightarrow NO + O(^3P)$	See Fig. 2-16	(a) Very important throughout entire atmosphere (see Fig. 2-16)
			244	(b) $NO_2 \rightarrow NO + O(^1D)$		
NO_3	Diffuse bands	690	580	$NO_3 \rightarrow NO_2 + O$	$\phi \approx 0.77$	Important
			900	$NO_3 \rightarrow NO + O_2$	$\phi \approx 0.07\text{-}0.23$	
HNO_2	Diffuse bands	390	475	$HNO_2 \rightarrow OH + NO$	$\phi \approx 0.9$	Important
			366	$HNO_2 \rightarrow H + NO_2$	$\phi \approx 0.1$	
HNO_3	Continuum	330	599	$HNO_3 \rightarrow OH + NO_2$	$\phi = 1$	Mainly in the stratosphere
N_2O_5	Continuum	380	1340	$N_2O_5 \rightarrow NO_2 + NO_3$	See text	Mainly in the stratosphere
			495	$N_2O_5 \rightarrow N_2O_4 + O$		
HO_2NO_2	Continuum	330	1340	$HO_2NO_2 \rightarrow HO_2 + NO_2$	Expected	Stratosphere
HCl	Continuum	220	280	$HCl \rightarrow H + Cl$	$\phi = 1.0$	Possibly in the mesosphere > 175 nm
$HOCl$	Continuum	390	513	$HOCl \rightarrow OH + Cl$	$\phi = 1.0$	Stratosphere
CH_3Cl	Continuum	220	347	$CH_3Cl \rightarrow CH_3 + Cl$ (expected)	$\phi = 1.0$	Stratosphere
CH_3Br	Continuum	300	417	$CH_3Br \rightarrow CH_3 + Br$	$\phi = 1.0$	Not competitive
CH_3I	Continuum	380	512	$CH_3I \rightarrow CH_3 + I$	$\phi = 1.0$	Important

(continued)

Table 2-6. (*Continued*)

Constituent	Absorption spectrum		Photochemistry		
	Type and approximate long-wavelength limit (nm)	Dissociation limit (nm)	Main primary process	Quantum yield	Significance to the atmosphere
CF_4	Continuum + bands, 103	220	$CF_4 \rightarrow CF_3 + F$ (presumably)		Only known loss process
$CFCl_3$	Continuum, 230	375	$CFCl_3 \rightarrow CFCl_2 + Cl$ $CFCl_3 \rightarrow CFCl + 2Cl$	See Fig. 2-15	Important in the stratosphere in the atmospheric absorption window, 185–215 nm
CF_2Cl_2	Continuum, 220	354	$CF_2Cl_2 \rightarrow CF_2Cl + Cl$ $CF_2Cl_2 \rightarrow CF_2 + 2Cl$	See Fig. 2-15	Important in the stratosphere in the atmospheric absorption window, 185–215 nm
CCl_4	Continuum, 240	407	$CCl_4 \rightarrow CCl_3 + Cl$ $CCl_4 \rightarrow CCl_2 + 2Cl$	See Fig. 2-15	Important in the stratosphere in the atmospheric absorption window, 185–215 nm
$ClONO_2$ $ClONO$	Continuum, 450 Continuum, 400	509 400	$ClONO_2 \rightarrow ClONO + O(^3P)$ $ClONO \rightarrow ClO + NO$ (expected)	$\phi \approx 1.0$	Stratosphere Not established
$CH_3CO_3NO_2$	Continuum, 300	1086 914	$CH_3CO_3NO_2 \rightarrow CH_3CO_3 + NO_2$ $CH_3CO_3NO_2 \rightarrow CH_3CO_2 + NO_3$		Not established
$CH_3O_2NO_2$	Continuum, 315	1493	$CH_3O_2NO_2 \rightarrow CH_3O_2 + NO_2$ $CH_3O_2NO_2 \rightarrow CH_3O + NO_3$		Not established

	Type	λ (nm)	λ (nm)	Photochemical process	Quantum yield / notes	Significance
HF	Continuum	162	211	HF → H + F		Shielded by O₂
C₂–C₅ alkanes	Continuum	170		Photodecomposition		Not significant
C₂–C₆ alkenes	Continuum	205		Photodecomposition		Not significant
C₂H₂	Continuum, weak band structure	237	230	$C_2H_2 \rightarrow C_2H + H$		Not established
HCHO	Bands	360	335	$HCHO \rightarrow H + HCO$ $HCHO \rightarrow H_2 + CO$	See Fig. 2-19	Important in the entire atmosphere
CH₃CHO	Quasi-continuum, some structure	340	350.5	$CH_3CHO \rightarrow CH_3 + CHO$ (300 nm)	ϕ pressure dependent	Somewhat significant in the troposphere
CH₃OOH	Continuum	350	647.5	$CH_3OOH \rightarrow CH_3O + OH$	Expected $\phi \approx 1.0$	Possibly stratosphere
SO₂	Very weak bands	390		Excited state quenching and reaction		Not significant because of quenching
	Stronger bands	340				
	Continuum + bands	220	220	$SO_2 \rightarrow SO + O$		Stratosphere
COS	Continuum	255	397	$COS \rightarrow CO + S_1(^3P)$ $COS \rightarrow CO + S(^1D)$	$\phi = 0.27$ $\phi = 0.67$	Stratosphere
CH₃SH	Continuum	280	311	$CH_3SH \rightarrow CH_3S + H$	$\phi \approx 0.9$	Not significant
CS₂	Two strongly structured bands	220	223	$CS_2 \rightarrow CS + S(^3P)$		Not important
		350	281	$CS_2 \rightarrow CS + S(^1D)$		

[a] Shown are type and wavelength regions of spectra, dissociation limits, photochemical main processes, and an assessment of their significance to atmospheric chemistry; data taken mainly from Okabe (1978).

Table 2-7. *Some Photochemically Active Trace Constituents of the Atmosphere and Associated Photodissociation Coefficients Calculated from the Solar Radiation Flux Outside the Earth Atmosphere (Nicolet, 1978)*[a]

Constituent	Wavelength region (nm)	j (s^{-1})	Constituent	Wavelength region (nm)	j (s^{-1})
O_3	<310	8.75 (−3)	HO_2NO_2	<300	6.3 (−4)
	>310	4.6 (−4)		>300	1.5 (−4)
CO_2	Ly α	2.2 (−8)	HCl	>175	1.2 (−6)
	>175	1.9 (−9)	HOCl	<300	9.2 (−5)
H_2O	Ly α	4.2 (−6)		>300	3.7 (−4)
	>175	8.6 (−7)	CH_3Cl	>175	1.0 (−7)
H_2O_2	<300	1.0 (−4)	$CFCl_3$	>175	1.5 (−5)
	>300	1.3 (−5)			
			CF_2Cl_2	>175	6.3 (−7)
N_2O	>175	7.3 (−7)	CCL_4	>175	3.5 (−5)
NO_2	175–240	5.9 (−5)	ClO	<300	6.5 (−3)
	240–307	2.3 (−4)		>300	5.5 (−4)
	>310	8.0 (−3)			
$NO_3 \rightarrow NO_2 + O$	>400	1 (−1)	$ClONO_2$	<310	2.0 (−5)
$NO_3 \rightarrow NO + O_2$		4 (−2)		>310	9.5 (−4)
N_2O_5	<310	5.8 (−4)	ClONO	<310	2.0 (−5)
	>310	3.6 (−5)		>310	9.5 (−4)
HNO_3	<200	7.1 (−5)	HCHO → H + HCO	<300	4.5 (−5)
	200–307	7.2 (−5)		>300	1.1 (−5)
	>307	2.0 (−6)	HCHO → H_2 + CO	<300	4.0 (−5)
HNO_2	<310	3.4 (−4)		>300	1.1 (−5)
	>310	2.5 (−3)	CH_3O_2H	>300	1.2 (−5)

[a] Powers of 10 are shown in parentheses.

the absorption cross sections are low, but they rise with decreasing wavelength and attain apreciable values in the atmospheric window region near 200 nm. On the basis of the nature of the spectra, which except for CO_2 are continuous, one expects photodissociation to occur in all cases. This has largely been confirmed by laboratory studies. For water vapor, the major dissociation process in the wavelength region considered is $H_2O \rightarrow OH + H$, which becomes energetically feasible at wavelengths below 242 nm. A second spin-allowed process, $H_2O \rightarrow H_2 + O(^1D)$, requires more energy. It may occur at wavelengths below 177 nm, although Chou *et al.* (1974), who studied it by using the HTO isotope, established a quantum yield for this process at 175 nm of less than 0.3%. Okabe (1978) has discussed the photodissociation of water vapor in more detail.

The main photodissociation process for nitrous oxide is $N_2O \rightarrow N_2 +$ $O(^1D)$. It occurs with a quantum yield of nearly unity (Cvetanović, 1965; Greiner, 1967a; Paraskepopoulos and Cvetanović, 1969). The photoproducts $N_2 + O(^3P)$ and $NO + N(^4S)$, although energetically allowed, would violate the spin conservation rule and are expected to be generated with low probability. For the second of these processes Preston and Barr (1971) have established a quantum yield of less than 2%.

The photodecomposition of HNO_3 in the 200–300 nm wavelength region has been studied by Johnston et al. (1974). In the presence of excess CO and O_2 to scavenge OH radicals so as to prevent their reaction with HNO_3, nitrogen dioxide is formed with a quantum yield of unity. This suggests that the principal primary photodissociation process is $HNO_3 \rightarrow OH + NO_2$. It represents a reversal of the recombination reaction between OH and NO_2 discussed earlier.

Hydrogen peroxide generally is assumed to undergo photodecomposition to form two OH radicals. If this is true, the studies of Volman (1963) suggest a quantum yield of unity for the loss of H_2O_2, and a quantum yield of two for the production of OH radicals. Wavelengths below 365 nm provide enough energy for the formation of the products $H_2O + O(^1D)$ in addition to two OH radicals. Greiner (1966) has pointed out that the data of Volman (1963) do not entirely preclude O-atom formation. This possibility thus requires further study.

Nitrogen pentoxide is difficult to study experimentally because of its thermal instability, and the exact products resulting from its photodecomposition are not known. Since the thermal dissociation leads to NO_2 and NO_3 as fragments, one expects photodissociation to give the same products. For the primary quantum yield a value near unity may be assumed, but it would require experimental confirmation. As an alternative, the photodecomposition of N_2O_5 may produce an O atom in addition to two NO_2 molecules. If, however, NO_3 is formed it would also undergo photolysis in the atmosphere at a rate several orders of magnitude greater than that for N_2O_5, as Table 2-7 shows. NO_3 features a series of diffuse absorption bands in the visible portion of the spectrum where the solar flux has its maximum. The dissociation products from NO_3 are mainly oxygen atoms and NO_2. Accordingly, the photolysis of N_2O_5 leads to the same final products, regardless of whether the oxygen atom is ejected directly or formed via NO_3 as an intermediate.

The photochemical behavior of methyl chloride, CH_3Cl, and the halocarbons CF_2Cl_2, $CFCl_3$, and CCl_4 is illustrated in Fig. 2-15. These gases begin to absorb radiation at wavelengths below 250 nm. The absorption spectra are again quasi-continuous and the cross sections rise with decreasing wavelength. Thus, photodecomposition occurs mainly in the 185–215 nm

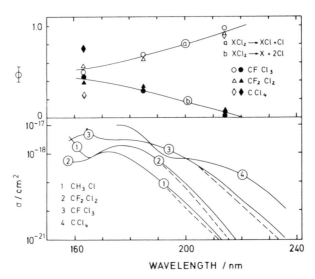

Fig. 2-15. Absorption spectra and quantum yields for the photodecomposition of CH_3Cl, CF_2Cl_2, $CFCl_3$, and CCl_4. Absorption cross sections from Chou *et al.* (1977) and Hubrich *et al.* (1977); quantum yields from Rebbert and Ausloos (1975, 1977).

wavelength region of the atmospheric window. The main photodecomposition process for the chlorine-containing gases is the release of a chlorine atom. The halocarbons admit the release of a second chlorine atom from the remaining fragment, provided additional energy is imparted to the molecule. The upper half of Fig. 2-15 shows quantum yields for both processes. At the long-wavelength limit the quantum yield for the process $XCl_2 \rightarrow X + 2Cl$ is nearly zero, whereas that for the process $XCl_2 \rightarrow XCl + Cl$ is close to unity. The first process increases in significance at the expense of the second as the wavelength is lowered (i.e., the photon energy is raised), until at wavelength near 166 nm the probability for the occurrence of both processes is about the same.

Figure 2-16 shows absorption spectrum and quantum yields for the photodissociation of nitrogen dioxide. The photoactivity of this molecule is of major importance to atmospheric chemistry in addition to that of ozone. NO_2 absorbs radiation throughout the entire visible and near ultraviolet spectral region. The spectrum is extremely complex with little regularity in rotational and vibrational structure. The excitation of NO_2 at wavelengths above 430 nm leads to fluorescence, which is quenched by collisions with air molecules. The first dissociation limit lies at 397.9 nm according to the energy requirement of the process $NO_2 \rightarrow NO + O(^3P)$. At shorter wavelengths the rotational structure becomes diffuse and fluorescence vanishes, indicating that predissociation takes place. Simultaneously the

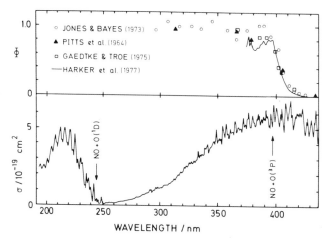

Fig. 2-16. Absorption spectrum and $O(^3P)$ quantum yields for nitrogen dioxide. Absorption cross sections adapted from Bass *et al.* (1976), quantum yields from the authors listed. Threshold wavelengths for the formation of $O(^3P)$ and $O(^1D)$ are indicated.

quantum yield for oxygen atom production rises to values close to unity. At wavelengths above the dissociation threshold the O-atom quantum yield decreases gradually to zero. The falloff curve of the quantum yield is explained by the thermal energy content of the molecule, residing mainly in rotation, which supplements the energy provided by the absorbed photon. Toward shorter wavelengths the absorption cross section for NO_2 first decreases to low values and then rises again below 244 nm. This value coincides with the second dissociation limit, leading to the production of $O(^1D)$ atoms. In the wavelength range 214–242 nm Uselman and Lee (1976) have found an $O(^1D)$ quantum yield of 0.5 ± 0.1 indicating that other primary photochemical processes also are at work. Table 2-7 shows, however, that this wavelength region makes a subordinate contribution to the total photodissociation coefficient of NO_2 in the atmosphere, so that a precise knowledge of the photodissociation products is of less importance to atmospheric chemistry.

Nitrous acid, HNO_2, exhibits a series of diffuse absorption bands in the spectral region 300–400 nm (Cox and Derwent, 1976; Stockwell and Calvert, 1978). The photochemical behavior of HNO_2 is difficult to study because the substance is in thermochemical equilibrium with NO, NO_2, and H_2O, and when one attempts to isolate HNO_2 from the mixture it is unstable. Cox (1974) and Cox and Derwent (1976) have studied the photolysis of such mixtures at 365 nm wavelength. The main photoproducts were NO and NO_2, which suggests that HNO_2 photodecomposes to give NO + OH followed by the reaction $OH + HNO_2 \rightarrow H_2O + NO_2$. By comparing rates of photon absorption by NO_2 and HNO_2 with the measured rate of product formation, Cox and Derwent (1976) established a quantum yield for the dissociation process of 0.92 ± 0.16. Accordingly, a quantum yield of close to unity is assumed to apply throughout the 300–400 nm wavelength region.

Formaldehyde, HCHO, is another important photoactive trace component of the atmosphere. The absorption spectrum of formaldehyde is shown in Fig. 2-17. It consists of a series of fairly sharp bands extending at wavelengths from 355 nm on downwards. The system leads to the excitation of the first electronic singlet state. The photodecomposition of formaldehyde (and other aldehydes) proceeds by internal energy transfer to highly excited vibrational levels. Photodissociation occurs along two routes: $HCHO \rightarrow H_2 + CO$, and $HCHO \rightarrow H + HCO$. The first of these processes occurs throughout the entire absorption region, the second has its energy threshold at 335 nm and becomes effective below that wavelength. The quantum yields for both decomposition channels are shown in Fig. 2-17. At room temperature the onset of H-atom formation is shifted toward 340 nm, presumably because of the thermal energy content of the molecule. At wavelengths below 330 nm the total quantum yield for both decomposition processes combined is about unity. At wavelengths above 330 nm the quantum yield is pressure dependent. The values shown in Fig. 2-17 refer

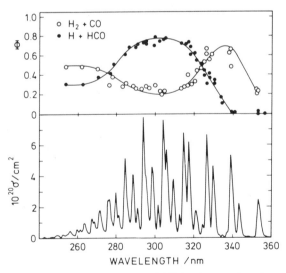

Fig. 2-17. Absorption spectrum and quantum yields for the two photodecomposition channels of formaldehyde in air at 1000 mbar pressure; assembled from data of Clark *et al.* (1978), Horowitz and Calvert (1978), Tang *et al.* (1979), and Moortgat *et al.* (1979, 1983).

to an ambient air pressure of 1000 mbar. Decreasing the pressure increases the quantum yield until at low pressures it tends toward unity. The reason for this behavior is that excited formaldehyde molecules undergo collisional quenching. This process causes a complication in calculating photodissociation coefficients for the two product channels.

Like formaldehyde, acetaldehyde, CH_3CHO, and acetone, CH_3COCH_3, are products resulting from the oxidation of hydrocarbons. Both molecules have absorption spectra with maximum cross sections near 250 nm and tails extending beyond the 300 nm limit for solar radiation reaching the troposphere (Calvert and Pitts, 1966). The long wavelength onset of absorption for acetaldehyde occurs near 340 nm, and that for acetone near 330 nm. The photodecomposition pathways in this wavelength region are $CH_3CHO \rightarrow CH_3 + HCO$ and $CH_3COCH_3 \rightarrow CH_3CO + CH_3$. In the second case it is not clear whether other processes are additionally involved. Both acetaldehyde and acetone are subject to collisional deactivation after photoexcitation, so that quantum yields for photodissociation are low at ground-level ambient pressures (Horowitz and Calvert, 1982; Meyrahn et al., 1982, 1986; Gardner et al., 1984). The consequence is photodissociation coefficients that are at least an order of magnitude smaller than that for formaldehyde.

Figure 2-18 shows the altitude dependence of photodissociation coefficients for several important photoactive atmospheric trace constituents discussed above. These data should be compared with those given in Table 2-7. For the group II gases H_2O, N_2O, and CF_2Cl_2, whose absorption spectra

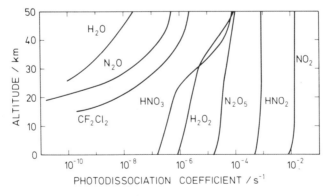

Fig. 2-18. Photodissociation coefficients for several photochemically active atmospheric trace components as a function of altitude. Data for H_2O are from Park (1974) for an overhead sun, for N_2O and HNO_3 from Johnston and Podolske (1978) for global average conditions, for CF_2Cl_2 from Rowland and Molina (1975) for an overhead sun, and for H_2O_2, N_2O_5, HNO_2, and NO_2 from Isaksen et al. (1977) for 60° solar zenith angle.

are located at wavelengths below 300 nm, photochemical activity is restricted to the upper atmosphere. The gases are photolysed mainly in the region of the atmospheric absorption window, and at altitudes below 20 km the photodissociation coefficients fall rapidly to very low values. The absorption continua of HNO_3, H_2O_2, and N_2O_5 extend beyond the 300-nm limit, so that these molecules are photodissociated also in the troposphere. Here, however, their photochemical activity is of lesser importance because of a high affinity toward liquid water. HNO_3, H_2O_2, and N_2O_5 are readily scavenged by clouds. The photodissociation coefficients for NO_2 and HNO_2 show fairly little dependence on altitude. The photodecomposition of these molecules in the atmosphere is dominated by the absorption of radiation in the 300–400 nm wavelength region. Shorter wavelengths contribute in a minor way to the total j value as Table 2-7 shows.

Figure 2-19 is added to show the dependence on altitude of the photodissociation coefficients for the two photodecomposition pathways of formaldehyde:

$$HCHO + h\nu \rightarrow H + HCO \quad J_1$$
$$\rightarrow H_2 + CO \quad J_2$$

The overall behavior is similar to that of N_2O_5 except that the rise of the H_2 production rate with increasing altitude is due to the decrease in pressure and a corresponding growth of the quantum yield for the process. At

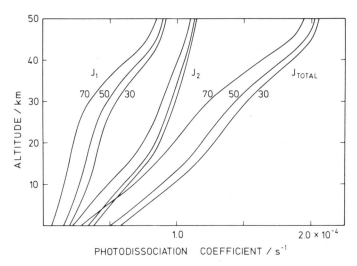

Fig. 2-19. Photodissociation coefficients for the two photodecomposition channels of formaldehyde as a function of altitude and solar zenith angle (30°, 50°, and 70°); calculated by the method of Isaksen *et al.* (1977) by G. K. Moortgat and P. Warneck (unpublished).

altitudes above 30 km the rate of H-atom formation increases by at least a factor of two, since ozone no longer shields the wavelength region below 300 nm and additional bands within the HCHO absorption spectrum begin to absorb radiation.

The calculated photodissociation coefficients shown in Figs. 2-18 and 2-19 are to indicate orders of magnitude only. Even for a cloudless sky the exact values depend critically on the solar zenith angle, on the thickness of the ozone layer overhead, which varies with season and latitude, on the reflective properties of the earth surface, and on the particle load in the lower atmosphere, which determines the scattering properties of the atmosphere. This has led to studies in which important photodissociation coefficients were derived directly from field measurements by chemical actinometry. A transparent vessel containing the substance of interest is exposed to sunlight, and the rate of photodecomposition is measured together with the solar flux. Backreactions must be suppressed in a suitable way. Using chemiluminescence detection of NO, Stedman *et al.* (1975), Harvey *et al.* (1977), Bahe *et al.* (1980), Dickerson *et al.* (1982), and Madronich *et al.* (1983) have studied the photodissociation of NO_2 in the lower atmosphere. Bahe *et al.* (1979) and Dickerson *et al.* (1979) have studied the formation of $O(^1D)$ atoms from the photodissociation of ozone

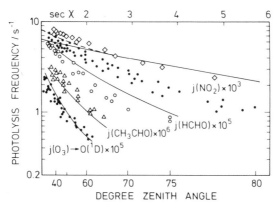

Fig. 2-20. Photodissociation coefficients for NO_2, HCHO, CH_3CHO, and $O(^1D)$ produced from ozone, for ground-level, clear-sky conditions. Nitrogen dioxide: diamonds are measurements of Madronich *et al.* (1983), solid points are data of Marx *et al.* (1984), the solid line represents calculations of Madronich *et al.* (1983) using the method of Isaksen *et al.* (1977), and the dashed line is for isotropic scattering. Formaldehyde: open points are measurements of Marx *et al.* (1984), and the solid line represents calculations of Calvert (1980). Acetaldehyde: triangles are measurements of Marx *et al.* (1984), and the solid line represents calculations of Meyrahn *et al.* (1982). Ozone: solid points are measurements of Dickerson *et al.* (1979) and Bahe *et al.* (1980) normalized to 325 Dobson units of total ozone overhead, and the solid line represents calculations of Dickerson *et al.* (1979).

in the ground-level atmosphere using N_2O to scavenge $O(^1D)$ followed by the detection of the products, either N_2 or NO. Marx *et al.* (1984) further included formaldehyde and acetaldehyde in their studies. By and large, the experimental data are in reasonable agreement with those calculated from the solar radiation field. Figure 2-20 compares ground-based, clear-sky measurements of the photolysis frequencies for NO_2, HCHO, and $O(^1D)$ formed from ozone with calculated values, both as a function of solar zenith angle. Madronich *et al.* (1983) found that for NO_2, calculations on the basis of isotropic Rayleigh scattering provided a better fit to the observational data than the columnar scattering approximation of Isaksen *et al.* (1977) at solar zenith angles less than 60°. The data of Marx *et al.* (1984) for NO_2 fall below this line. They measured the loss of NO_2, whereas Madronich *et al.* (1983) determined the pressure increase resulting from the dissociation of NO_2. Generally, the *j* values decline with solar zenith angle more strongly as the onset of the photodissociations process is shifted toward the 300-nm cutoff wavelengths of solar radiation reaching the troposphere. Thus, the steepest curve is found for the formation of $O(^1D)$ from ozone. This process also depends considerably on the thickness of the ozone layer overhead.

3 | Chemistry of the Stratosphere

Chemical processes in the stratosphere are intimately connected with the phenomenon of the ozone layer in the 15- to 35-km altitude regime. The significance of the ozone layer is twofold: on one hand it absorbs ultraviolet solar radiation at wavelengths below 300 nm, so that this biologically harmful radiation is prevented from reaching the Earth surface; on the other hand it dissipates the absorbed energy as heat, thereby giving rise to the temperature peak at the stratopause. Because of its importance to the terrestrial radiation budget, ozone has become one of the most thoroughly studied trace components of the atmosphere. Recent research, however, has emphasized trace gases other than ozone and their influence on chemical reaction mechanisms in the stratosphere. In contrast to ozone, which is formed within the stratosphere, the other trace gases have their origin in the troposphere. They are cycled through the stratosphere in the sense that they are brought into this region by upward diffusion, undergo chemical modification, and finally leave the stratosphere via the tropopause as a different chemical entity. In this chapter we first describe the behavior of stratospheric ozone, then discuss the chemistry of other trace components that are considered important.

3.1 Historical Survey

Hartley (1881) appears to have first identified ozone as the cause for the abrupt cutoff at 300 nm of the solar spectrum observable in the troposphere. Forty years later, Fabry and Buisson (1921) confirmed Hartley's conclusion and obtained the first reliable estimate for total ozone overhead from optical measurements with the sun as the background source. At about the same time, Strutt (1918) had shown in a milestone experiment involving a 6.5-km optical path of 237-nm mercury radiation that the ground level O_3 mixing ratio was 40 ppbv or less which was insufficient to explain the column density of ozone if its vertical distribution were uniform. The discrepancy gave rise to the idea of an ozone layer occurring at greater heights. In 1929, on an expedition to Spitsbergen, Götz (1931) studied the zenith sky radiation as a function of solar elevation and discovered that the ratio of intensities at two neighboring ultraviolet wavelengths exhibited a reversal of the trend toward large solar zenith angles (Umkehr effect). He attributed the effect to the layered structure of atmospheric ozone, and after evaluating further Umkehr measurements in Arosa in collaboration with Meetham and Dobson (Götz *et al.*, 1934) he located the maximum ozone density at 22 km altitude. His conclusions were confirmed in the same year by the first balloon sounding of ozone by Erich and Victor Regener (1934).

In 1930, Chapman offered an explanation for the origin and nonuniform distribution of atmospheric ozone in terms of a photochemical theory involving the photodissociation of oxygen. Owing to uncertainties in the rate parameters, solar intensities, and physical conditions of the stratosphere, Chapman could draw few quantitative conclusions, yet his results and the subsequent more detailed calculations of Wulf and Deming (1937) revealed the formation of an ozone layer at about the correct height. At the same time, the chemistry of ozone was studied in the laboratory. Schumacher (1930), who had studied the photodecomposition of ozone, succeeded in interpreting the data in terms of three reactions that are inherent also in the Chapman mechanism. Schumacher (1930, 1932) was unaware of Chapman's work and the significance of the laboratory results to the atmosphere, but Eucken and Patat (1936) fully recognized the implications when they redetermined the ratio of the two rate coefficients associated with the dark reactions in the Chapman mechanism. Thirty-five years later, after chemical kineticists had mastered experimental difficulties in measuring rate coefficients of individual reactions, the values obtained by Eucken and Patat (1936) were found incorrect by a factor of six, while their temperature dependence was essentially confirmed. In 1950, Craig comprehensively reviewed all the important observations and laboratory data entering into the calculations. He concluded that calculated and observed verical distributions of atmospheric ozone were in reasonable agreement.

A problem that had come to light early was the latitude distribution of total ozone, which was first explored by Dobson and Harrison (1926). Total ozone was found to increase from the equator toward higher latitudes, whereas the Chapman theory predicted an opposite behavior. Equally at variance with the theory was the observation of a winter maximum at high latitudes. It was recognized that in the lower stratosphere the photochemical activity is so weak that ozone would be influenced more by transport than by photochemical processes, and eventually a poleward flux of ozone due to meridional motions was postulated. The measurements of stratospheric water vapor by Brewer (1949) and his interpretation of the low mixing ratios by a meridional circulation model were taken to provide support for the transport of ozone toward high latitidues. It now appeared that ozone in the stratosphere was more a problem of meteorology than of photochemistry, and this view was to persist for the following two decades.

During this period considerable progress was made in the exploration of the upper atmosphere, which, in turn, stimulated extensive laboratory investigations of reactions considered important in the atmosphere. The installation of a global network of measurement stations brought forth better data on total ozone, the vertical distribution, and seasonal variations. A really global coverage was achieved only more recently, however, by means of satellite observations. Ironically, it was not so much the ozone measurement program that provided the most clear-cut evidence for the existence of meridional transport in the lower stratosphere, but the behavior of radioactive debris from the nuclear weapons tests in the early 1960s. Finally, the space exploration produced for the first time realistic information on solar radiation intensities in the ultraviolet spectral region. The improved data base stimulated new calculations of the ozone density profile in the altitude regime where the Chapman equations were deemed applicable, namely, above 25 km in the tropics. By comparison with observations, it was found that the theory overestimated the ozone density by at least a factor of two. At this time, in view of the improved data base, the discrepancy was considered serious and one began to look into possible causes, specifically chemical reactions other than those few contained in the Chapman mechanism.

Laboratory experience had convinced chemists earlier that the Chapman mechanism needed a supplement of additional reactions. In 1960, McGrath and Norrish discovered the formation of OH radicals in the reaction of water vapor with $O(^1D)$ atoms generated by the photolysis of ozone, and they proposed a chain decomposition of ozone by water radicals. Meinel (1950) had previously demonstrated the existence of OH in the upper atmosphere by observing its emission spectrum, and Nicolet (1954) had discussed the chemical significance of H, OH, and HO_2 in the mesosphere. Hunt (1966a,b) then showed that a modification of the Chapman mechanism by such reactions might bring it into harmony with observations. When, however, laboratory measurements of the rate coefficients were accomplished in the early 1970s, the water radical reactions were found too slow for an effective reduction of the stratospheric ozone concentration. But Hunt's study had set a new trend, and from now on stratospheric trace gases other than ozone came under closer scrutiny.

An important step forward was made by Crutzen (1970, 1971) and Johnston (1971), who drew attention to the catalytic nature of the nitrogen oxides NO and NO_2 with regard to ozone destruction. At that time, aircraft companies advocated the construction for commercial use of high-flying supersonic airplanes with a projected cruising altitude of about 20 km. Atmospheric scientists became concerned about a possible pollution of the stratosphere by nitrogen oxides contained in the engine exhaust gases, a concomitant reduction of stratospheric ozone, and a corresponding increase of ultraviolet radiation at the Earth surface, which would lead to an increased incidence of skin cancer. The stratosphere thus became a political issue, which led the U.S. Congress to block subsidies for the development of commercial supersonic aircraft, and the Department of Transportation to initiate a research program for the study of conceivable perturbations of the stratosphere. By 1974, exploratory measurements had traced the vertical concentration profiles of N_2O, NO, NO_2, and HNO_3, in addition to methane and hydrogen, and chemical reaction models were able to explain most of these data. The issue of high-flying jet liners had hardly been settled when, in a dramatic new development, Molina and Rowland (1974) discovered that a man-made chlorofluoromethanes have an adverse effect on stratospheric ozone. These compounds are widely used as refrigerants and aerosol propellants. In the atmosphere they are degraded primarily by photodecomposition in the upper stratosphere. In the process, chlorine atoms are formed, which can then enter into a similar catalytic ozone destruction cycle as the nitrogen oxides. The worst aspect of the halocarbons is their predicted long-term effect. Continued production causes their accumulation in the atmosphere, but they diffuse into the upper stratosphere only slowly, so that their full impact on stratospheric ozone is delayed. This threat eventually led to a ban of nonessential uses of chlorofluoromethanes in the United States.

Yet another problem that recently has been brought closer to a solution is the origin of the stratospheric aerosol layer discovered by Junge et al. (1961). The layer centers at altitudes near 20 km and consists mainly of particles representing a mixture of sulfuric acid and water. A number of observations based on optical and direct sampling techniques have shown that high-reaching volcanic eruptions contribute significantly to the strength of the aerosol layer due to the injection of ash particles, SO_2, and H_2O into the stratosphere. But even during extended periods of volcanic quiescence, long after the excess material has cleared the stratosphere, a persistent layer of sulfate particles remains. This background layer can be explained only if a gaseous sulfur compound diffuses upward from the trophosphere, undergoing oxidation and conversion to H_2SO_4 not before having penetrated to the 20 km altitude level. Initially, SO_2 was thought to meet the conditions. The rate of SO_2 photooxidation increases markedly when the exciting wavelength is lowered toward 200 nm, and these wavelengths become available somewhat above 20 km. In the mid-1970s, when stratospheric chemistry became better understood, the reaction of SO_2 with OH radicals came into focus. It was then realized [by Moortgat and Junge (1977)] that the rapid rate of this reaction might prevent SO_2 from actually reaching the photochemically active region. At the same time, carbonyl sulfide was detected in the atmosphere, leading Crutzen (1976) to explore its behavior at high altitudes. He was able to show that COS is converted to SO_2 in the 20- to 25-km altitude regime, and that the resulting flux of sulfur into that region is sufficient to maintain the Junge layer. In addition, OH radicals were found to react with COS much more slowly than with SO_2. Finally, Inn *et al.* (1981) obtained direct evidence for the presence of COS in the stratosphere from *in situ* measurements.

The last decade has seen a consolidation of earlier results, both experimental and theoretical, with much attention being paid to details. Stratospheric chemistry has developed into an extremely complex subject, as about 100 individual chemical reactions are now known to be involved. Of these, only the most important ones can be discussed in this chapter.

3.2 Ozone Observations

Measurements of stratospheric ozone fall into two classes: total ozone observed from the ground or by means of satellites, and vertical distributions derived from balloon or rocket sondes. The data are too numerous for individual discussion; only the most obvious features can be treated. The subject has been reviewed by Craig (1950), Paetzold and Regener (1957), Vassy (1965), and by Dütsch (1971, 1980). These reports trace the progress achieved in our understanding of the behavior of stratospheric ozone.

Ground-based measurements of total ozone use the Dobson spectrometer. This instrument measures the column density of ozone in the atmosphere by optical absorption, with the sun or the moon as background source. The data are expressed as an equivalent column height at standard pressure and temperature, in units of 10^{-2} mm (Dobson unit, DU). Owing to the characteristic vertical distribution of ozone, it is mainly stratospheric ozone that is determined. The contributions of tropospheric and mesospheric ozone to the total column content are small. As discussed in Section 1-1, total ozone at middle and high latitudes displays a considerable day-to-day fluctuation requiring extensive smoothing of the data by averaging. Figure 1-3 showed data from Arosa, Switzerland, that reveal that the monthly means undergo an almost sinusoidal annual variation, with a maximum in spring and a minimum in the fall. A similar seasonal variation is observed at a number of other stations at sufficiently high latitudes, whereas in the tropical region the seasonal variation of total ozone is small.

Figure 3-1 shows the distribution of total ozone with latitude and its temporal variation for two consecutive years. These data were derived from *Nimbus* 4 satellite observations by averaging over narrow latitude belts. The results agree quite well with those deduced from the network of Dobson stations around the world, but the satellite data are preferred here because they provide a more complete global coverage. As Fig. 3-1 indicates the lowest values of total ozone are found in the equatorial region, whereas the highest values occur in spring near the North Pole. Here, the annual peak-to-peak variation amounts to as much as 40% of the yearly average. In the southern hemisphere, the spring maximum occurs first at about 60° latitude, then migrates toward the Antarctic where it arrives 1–2 months later. The maximum in the southern hemisphere is lower than that of the northern. It is evident that the cyclic behavior of total ozone is similar in the two hemispheres but not identical. The differences have been known

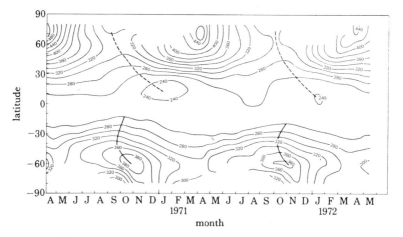

Fig. 3-1. Total ozone versus time and latitude as derived from 10° zonal means of *Nimbus* 4 satellite data. [From Hilsenrath *et al.* (1979), with permission.]

for some time from ground-based measurements, but until the advent of the satellite data their reality had been in doubt because of the sparsity of Dobson stations in the southern hemisphere. As noted previously, the spring maximum is thought to arise from meridional transport of ozone from low to high latitudes, mainly in the winter hemisphere. The dissimilarity of behavior in the two hemispheres thus points to differences in the circulation patterns.

Satellite and ground-based observations reveal interhemispheric differences also with regard to the distribution of total ozone with longitude, which is shown in Fig. 3-2. In the northern hemisphere, total ozone has three maxima. They coincide approximately with the quasi-stationary low-pressure region near 60° latitude (wave number 3). The maxima are located over Canada, northern Europe, and eastern Russia. In the southern hemisphere, in contrast, only one broad maximum is seen at 60° latitude (wave number 1), located southeast of Australia. The occurrence of these maxima may be explained at least in part in the same way as the day-to-day variation of total ozone due to the ever-changing meteorological situation. A correlation is known to exist between total ozone and the pressure at the tropopause level. The height of the tropopause serves only as an indicator of cyclonic activity, however. Low-pressure systems are associated with a lower than average tropopause level, whereas total ozone is above average. Upper-tropospheric troughs are generally displaced westward of the surface center of cyclones, and it is there that the highest amounts of total ozone are found. The reason for the increase in total ozone must be sought in the local dynamics of the lower stratosphere. Low-pressure systems induce a

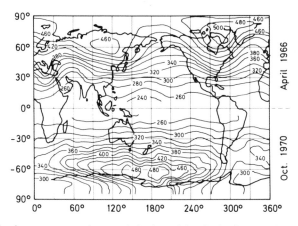

Fig. 3-2. Total ozone averaged versus latitude and longitude. [From Dütsch (1980), with permission.]

convergence and subsidence of air in the midstratosphere, forcing ozone-rich air from the 15- to 20-km altitude regime downward into the lower stratosphere. The ozone column density thus increases. A reversal of the situation occurs for high-pressure systems. The velocity of vertical air motions can reach 1 km/day, which is sufficient to change total ozone by 25% within a few days. As described by Dütsch (1980), the process has been followed by successive balloon soundings.

Figure 3-3 shows two examples for the vertical distribution of ozone. The data were selected to present typical concentration profiles for low and high latitudes. Both examples clearly show the existence of the ozone layer but reveal also latitudinal differences. In the tropics, due to the high-lying tropopause, the layer is fairly narrow with a maximum at 26 km altitude. In the middle and high latitudes the distribution is broader, much more ragged, and the maximum occurs at heights between 20 and 23 km. At greater heights, above 35 km, the rocket-sonde data of Krueger (1973) indicate that the ozone density decreases with a scale height of about 4.6 km regardless of latitude. The steep rise of ozone above the tropopause at high latitudes is evidence for the lower rate of vertical eddy diffusion in the stratosphere compared with that in the troposphere. The resulting vertical gradient of mixing ratios gives rise to a downward flux of ozone toward the tropopause and thence into the troposphere. The gradient may be enhanced by large-scale mean motions, which at the higher latitudes usually are directed downward. The periodic relocation of the tropopause due to heat-exchange processes causes a portion of ozone-rich air to enter the troposphere when the tropopause level is raised. Such an event appears to have taken place just before the data on the right-hand side of Fig. 3-3 were

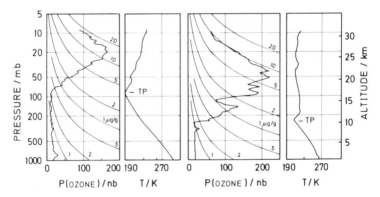

Fig. 3-3. Balloon soundings of ozone (in units of partial pressure). Left: Low latitude, Right: High latitude. Vertical temperature profiles are included to indicate the tropopause levels. [Adapted from Hering and Borden (1967).]

taken. This is made evident by the spike of ozone at 8 km altitude, and by the abrupt cutoff of the stratospheric ozone profile at the tropopause.

In the tropics the situation is somewhat different. Here, the mean motion is directed upward as it follows the ascending branch of the Hadley cell, thus counteracting the downward flux of ozone by eddy diffusion. The ozone concentration rises more gradually, and the gradient at the tropopause is essentially negligible. Since in addition the tropopause level is fairly stable, there is practically no influx of ozone into the troposphere. Instead, ozone undergoes lateral, poleward transport and can then enter the troposphere at midlatitudes, in part via the subtropical tropopause breaks

Balloon soundings at higher latitudes often reveal a remarkable pastry-like structure of the vertical ozone distribution. Layers rich in ozone are interspersed with layers containing lesser amounts. The right-hand side of Fig. 3-3 may serve to demonstrate the extent of the stratification. Simultaneous measurements with two instruments (Attmannspacher and Dütsch, 1970) have proven beyond doubt that the phenomenon is real and that it is not caused by instrumental errors. Dobson (1973) has performed a statistical analysis of the frequency with which a laminated structure occurs, using data available for the northern hemisphere. He found the effect to be most frequent in spring and in the region poleward of 30° latitude, whereas in the equatorial stratosphere the ozone profiles generally had a smooth appearance. At midlatitudes a particularly well-developed minimum of stratospheric ozone was often encountered at or near the 15-km height level. Figure 3-3 also displays this feature. The minimum appears to derive from the injection of tropospheric air into the stratosphere at the location of the subtropical tropopause break. On its journey poleward the air is then gradually enriched with ozone from adjacent layers. According to Table 1-8, horizontal transport by eddy diffusion is much faster than vertical transport. The phenomenon thus shows most directly, although qualitatively, the relatively slow rate of vertical mixing versus a fairly rapid transport in quasi-horizontal direction.

Figure 3-4 is added to show the meridional distribution of ozone derived by averaging over many data. The left-hand side gives the distribution in units of partial pressure. It illustrates again the gradual broadening of the layer and its displacement toward lower heights as one goes from the equator to the poles. The highest densities occur at high lattides and in the winter hemisphere. The right-hand side of Fig. 3-4 shows the ozone distribution in units of mixing ratios. In this representation one finds the ozone maximum in the 30- to 40-km altitude regime of the equatorial region, that is, in the domain of greatest photochemical activity. The mixing ratios decline toward lower heights and toward the poles. This distribution must give rise to an eddy flux of ozone in poleward direction. Transport toward higher latitudes

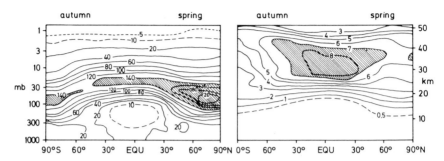

Fig. 3-4. Meridonal cross section of ozone in the atmosphere. Left frame: Expressed as partial pressure in units of nbar. Right frame: Expressed as mixing ratio in units of ppmv. [From Dütsch (1980), with permission.]

is thus seen to arise at least in part of eddy diffusion and not solely from meridional mean motions.

3.3 The Chapman Model

The formation of the stratospheric ozone layer can be understood most simply on the basis of a reaction model composed of a minimum set of four elementary processes: (a) the dissociation of oxygen molecules by solar radiation in the wavelength region 180–240 nm; (b) the attachment of oxygen atoms to molecular oxygen, leading to the formation of ozone; (c) the photodissociation of ozone in the Hartley band between 200 and 300 nm; and (d) the destruction of ozone by its reaction with oxygen atoms. The reactions may be written

$$O_2 + h\nu \rightarrow O + O \qquad j_a \qquad \text{(a)}$$

$$O + O_2 + M \rightarrow O_3 + M \qquad k_b \qquad \text{(b)}$$

$$O_3 + h\nu \rightarrow O_2 + O \qquad j_c \qquad \text{(c)}$$

$$O + O_3 \rightarrow O_2 + O_2 \qquad k_d \qquad \text{(d)}$$

where the rate coefficients are indicated on the right. The reactions are here identified by letters rather than by the numbers of Table A-4 so as to keep the following discussion simple. In reaction (b), M denotes as usual an inert third body, which in the atmosphere is supplied by either nitrogen or oxygen. The efficiencies of N_2 and O_2 as third bodies in reaction (b) are almost equal. Therefore it is not necessary to distinguish between them. The temperature dependence of reaction (b) differs somewhat for N_2 and O_2 as third bodies, but this effect may be neglected to a first approximation. The excited $O(^1D)$ atoms formed in reaction (c) are fairly rapidly deactivated by collisions with nitrogen or oxygen, so that the oxygen atoms may be taken to react in the ^3P ground state. Let n_1, n_2, and n_3 represent the number densities of the three species O, O_2, and O_3, respectively, at a selected altitude, and let n_M be the number density of N_2 and O_2 combined. Further, assume that the influence of transport processes can be neglected. The local variations with time of the number densities of interest are then given by the rate equations

$$dn_1/dt = 2j_a n_2 - k_b n_1 n_2 n_M + j_c n_3 - k_d n_1 n_3 \qquad (3\text{-}1)$$

$$dn_2/dt = -j_a n_2 - k_b n_1 n_2 n_M + j_c n_3 + 2k_d n_1 n_3 \qquad (3\text{-}2)$$

$$dn_3/dt = +k_b n_1 n_2 n_M + j_c n_3 - k_d n_1 n_3 \qquad (3\text{-}3)$$

As a check for the correct application of the stoichiometry factors, we multiply the second equation by two, and third by three, and obtain after summing

$$dn_1/dt + 2\, dn_2/dt + 3\, dn_3/dt = 0 \qquad (3\text{-}4)$$

Integration then yields the mass balance equation

$$n_1 + 2n_2 + 3n_3 = \text{constant} = \Sigma\, n_2^0 \qquad (3\text{-}5)$$

which demonstrates that mass conservation is obeyed and that the stoichiometry factors were applied correctly. The observations of ozone further establish that $n_2 \gg n_1 + n_3$ so that $n_2 = n_2^0$ may be taken as constant. The first and the third of the rate equations then suffice to determine n_1 and n_3. These equations form a system of coupled differential equations whose exact solutions can be obtained only by numerical integration.

Fortunately, a number of simplifications are possible. First, it may be noted that the O-atom concentration approaches the steady state much faster than the ozone concentration. At 50 km altitude, the time constant for the adjustment to steady state of n_1 is $\tau_0 = 1/k_b n_2 n_M \approx 20$ s, and the value decreases as one goes toward lower altitudes. The time constant for the approach to steady state of ozone, in contrast, is much longer, and it increases with decreasing altitude (see below). It is thus reasonable to assume that oxygen atoms are always in steady state, that is, $dn_1/dt = 0$. In addition, it turns out that the second and third terms in Eq. (3-1) are dominant compared with the other two. Accordingly, one has approximately

$$n_1 = j_c n_3 / k_b n_2 n_M \qquad (3\text{-}6)$$

The first and last terms of Eq. (3-1) cannot be neglected, of course, because they are ultimately responsible for the production and the removal of the sum of oxygen atoms and ozone. This fact may be taken into account by summing Eqs. (3-1) and (3-3), and by using the sum equation in conjunction with Eq. (3-6). With these simplifications one obtains

$$dn_1/dt + dn_3/dt \approx dn_3/dt = 2j_a n_2 - 2k_b n_1 n_3$$

$$= 2(j_a n_2 - k_d j_c n_3^2 / k_b n_2 n_M) \qquad (3\text{-}7)$$

This equation can be integrated without difficulty. We assume for simplicity that the ozone density is intitially zero and obtain

$$n_3 = \left(\frac{B}{A}\right)^{1/2} \frac{1 - \exp[-2(AB)^{1/2}t]}{1 + \exp[-2(AB)^{1/2}t]} \qquad (3\text{-}8)$$

where $A = 2k_d j_c / k_b n_2 n_M$ and $B = 2j_a n_2$.

The number densities n_2 and n_M are connected by the mixing ratio of oxygen $m_2 = n_2/n_M = 0.2095$. After a sufficiently long period of time, the number density of zone tends toward a steady state with the value

$$n_3 = \left(\frac{B}{A}\right)^{1/2} = n_2 \left(\frac{k_b n_M j_a}{k_d j_c}\right)^{1/2} = m_2 n_M \left(\frac{k_b r_2}{k_d j_c m_2}\right)^{1/2} \qquad (3\text{-}9)$$

with $r_2 = j_a n_2$ denoting the O_2 photodissociation rate. The simplicity of Eq. (3-9) is only apparent. The photodissociation coefficients are

$$j_a = \int \sigma_2 I_0 \exp(-\sec) \left[\int_z^\infty (\sigma_2 n_2 + \sigma_3 n_3) \, dz \right] d\lambda$$

$$j_c = \int \sigma_3 I_0 \exp(-\sec) \left[\int_z^\infty (\sigma_2 n_2 + \sigma_3 n_3) \, dz \right] d\lambda \qquad (3\text{-}10)$$

where, as above, the subscripts 2 and 3 refer to oxygen and ozone, respectively, and quantum yields of unity are assumed. Since the expressions for j_a and j_c involve the number density of ozone, it is not possible to calculate n_3 directly from Eq. (3-9). A stepwise approximation procedure is required, starting from an assumed ozone density profile and improving it by iteration, working from high altitudes on downward. For the purpose of illustration, the photodissociation coefficients shown in Fig. 2-12 may be used to calculate ozone number densities as a function of altitude. The ratio of rate coefficients $k_b/k_d = k_1/k_2$ may be gleaned from Table A-4. In the altitude region 15–50 km the temperature increases from 210 to 280 K, and k_b/k_d decreases in magnitude from 3.4×10^{-18} to 1.6×10^{-19} cm^3/molecule. The number density of oxygen decreases from 8.1×10^{17} to 4.8×10^{15} molecules/cm^3. Figure 3-5 shows the number densities of ozone calculated with these data. Included for comparison is the measured density profile at low latitudes. The calculations give approximately the correct altitude distribution, but the absolute concentrations are greater than those observed by at least a factor of two. In part, this is due to the assumption of an overhead sun. More realistically,

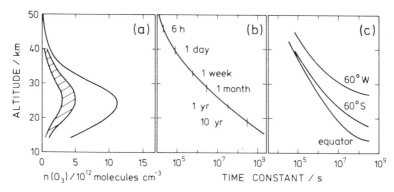

Fig. 3-5. (a) Vertical profile of ozone number densities calculated from Eq. (3-9). The hatched area shows for comparison the range of observations from data of Krueger (1969), Randhawa (1971), and Mauersberger *et al.* (1981), at low latitudes. (b) Time constant for the approach to photostationary state of ozone calculated from Eq. (3-11). (c) Ozone replacement times calculated from Eq. (3-12) by Johnston and Whitten (1973), here for 60° northern latitude, summer and winter, and at the equator.

a diurnal average of the photodissociation coefficients should be taken. Johnston (1972) has made such calculations and found a maximum ozone number density of 9.7×10^{12} molecules/cm^3. Another part of the discrepancy arises from uncertainties in the O_2 absorption cross sections. Recent data discussed by Herman and Mentall (1983) suggest lower cross sections in the 200- to 240-nm region than were used previously.

According to Eq. (3-8), the photochemical steady state is approached with a time constant

$$\tau_{ss} = \tfrac{1}{2}(AB)^{-1/2} = \tfrac{1}{4}\left(\frac{k_b n_M}{k_d j_a j_c}\right)^{1/2} \tag{3-11}$$

Again, the appropriate diurnal averages of the photodissociation coefficients must be inserted. For the purpose of illustration, half values of those for an overhead sun are used here. The resulting time constants are plotted in Fig. 3-5 as a function of height. At altitudes above the ozone density maximum, the time constants are sufficiently small to justify the assumption of steady state in the ozone concentrations. Below the ozone maximum the time constants become longer than a year. For such conditions it is no longer realistic to neglect transport by eddy diffusion and large-scale mean motions, so that a photostationary state is not established.

Instead of a time constant for the approach to steady state one may also consider the time required to replenish the average amount of ozone observed at a given altitude by the photodissociation of oxygen. This replacement time is

$$\tau_{\text{replacement}} = \frac{n_{3,\text{obs}}}{2j_a n_2} \qquad (3\text{-}12)$$

Johnston and Whitten (1973, 1975) have calculated ozone replacement times from zonally averaged O_2 photodissociation rates as a function of latitude and height. The right-hand side of Fig. 3-5 shows results for 60° latitude and for the equatorial region to provide a comparison with the time constants for the approach to photostationary state. Both time constants exhibit a similar behavior with regard to altitude, so that both are suitable indicators for the stability of ozone in the stratosphere. At high latitudes, the replacement time is larger in winter than in summer, in agreement with expectation if one considers the greater solar inclination during the winter months. The same statement applies to the time constant for the approach to steady state. An ozone replacement time of 4 months or less may be taken to characterize the main region of ozone formation. This region coincides approximately with the region of high O_3 mixing ratios shown in Fig. 3-4. In the other regions where the replacement time exceeds 1 year, ozone must be considered completely detached from the region of O_2 photodissociation.

Transport of ozone over large distances into those regions of the lower stratosphere where ozone is no longer replaced requires a sufficiently long life-time against photolytic destruction. This time constant differs from the time constants discussed above. Within the framework of the Chapman theory, the photochemical life time of ozone is obtained from Eq. (3-7) by setting $j_a = 0$. Integration then gives the decay of the ozone density with time,

$$n_3 = n_3^0/(1 + An_3^0 t) \qquad (3\text{-}13)$$

which leads to a half-life time of

$$\tau_{1/2} = 1/An_3^0 = \frac{k_b n_2 n_M}{2k_d j_c n_3^0} \qquad (3\text{-}14)$$

For an altitude of 15 km or less one finds half-life times greater than 100 y. This value must be considered an upper limit, however, because it ignores the catalytic influence on ozone destruction by other trace gases. Up to 80% of ozone destruction may take place by such catalytic cycles. Accordingly, the lifetime of ozone in the lower stratosphere will be shortened to about 20 years. The discussion in Chapter 1 and the ozone observations show that each year the poleward transport of ozone takes place mainly during the winter months, and that a large portion of the stratospheric ozone content at high latitudes is discharged into the troposphere in spring and early summer. Accordingly, the residence time of ozone in the lower

Table 3-1. *Globally Integrated Rates of Ozone Formation and Ozone Destruction (Johnston, 1975)*

Process	January 15 (10^6 mol/s)	March 22 (10^6 mol/s)
Gross rate of O_3 formation by O_2 photodissociation	83.3	81.0
Chemical loss from Chapman mechanism	14.3	14.8
Transport to the troposphere	1.0	1.0
Imbalance between production and losses	68.0	65.2
Additional chemical losses due to water radical reactions	9.3	9.0
Unbalanced difference	58.7	56.2

stratosphere cannot be longer than a few years, at most. This is much shorter than the photochemical life time derived above. Consequently, ozone can be considered chemically stable in those regions of the stratosphere that lie outside the principal domain of ozone formation and where the influence of transport is preponderant.

The importance of transport has led to the suggestion that the excess of ozone derived from the calculations (see Fig. 3-5) is an artifact of the steady-state assumption resulting from the neglect of losses due to transport. Although this possibility cannot be entirely discounted, it must be recognized that the main effect of transport is to displace ozone from one region of the stratosphere to another. The global budget of ozone should not be affected. Johnston (1975) has calculated globally integrated rates of (a) ozone formation due to the dissociation of oxygen, (b) ozone losses due to reaction with oxygen atoms, and (c) transport to the troposphere. The results are shown in Table 3-1. It is apparent that the global loss of ozone to the troposphere is but a small fraction of the total budget of stratospheric ozone. Table 3-1 shows also that, on a global scale, the total loss rate does not balance the production rate, so that there must be losses due to other processes not included in the Chapman mechanism. This comparison provides the most compelling argument for the necessity of expanding the Chapman reaction scheme by additional reactions involving trace constituents other than ozone. These are discussed below.

3.4 Trace Gases Other than Ozone

Processes leading to the loss of ozone in addition to its reaction with atomic oxygen involve substances which contain nitrogen (NO_x, odd nitrogen), hydrogen (HO_x, odd hydrogen), and chlorine (ClO_x) or other halogens. The various species derive from source gases entering into the stratosphere mainly from the troposphere. The most prominent source gases are nitrous oxide (N_2O), water vapor and methane (H_2O, CH_4), and the chlorofluoromethanes (CH_3Cl, CCl_4, CCl_2F_2, $CFCl_3$). These gases are essentially unreactive toward ozone and must first be converted to the reactive species. The following discussion will be based primarily on the reports of Hudson and Reed (1979) and Hudson (1982). It will be convenient to treat each class of compound separately. It should be noted, however, that some interaction exists between the three groups, so that they are not wholly independent. The origin of the source gases and their behavior in the troposphere will be discussed in later chapters.

3.4.1 NITROGEN OXIDES

Figure 3-6 shows that the mixing ratio of N_2O in the stratosphere declines with altitude from 300 pptv at the tropopause to about 20 pptv at 40 km. The decline is somewhat more pronounced in midlatitudes due to the lower-lying tropopause there compared with the tropics. Evidently, N_2O is destroyed at heights above 25 km. This establishes a vertical gradient of mixing ratios that maintains an eddy flux of N_2O from the tropopause on up into the destruction region. The principal loss process is photodecomposition,

$$N_2O + h\nu \rightarrow N_2 + O(^1D)$$

which becomes effective at wavelengths below 220 nm, but occurs mainly in the atmospheric absorption window near 200 nm (see Fig. 2-9). A second

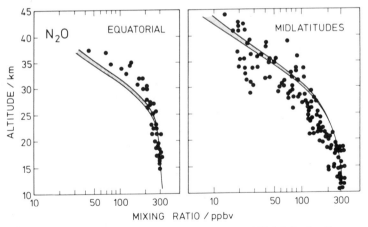

Fig. 3-6. Vertical profiles of the N_2O mixing ratio at low and high latitudes. From measurements of Heidt *et al.* (1980), Tyson *et al.* (1978), Vedder *et al.* (1978, 1981), Fabian *et al.* (1979b, 1981), Goldan *et al.* (1980, 1981). The solid lines were calculated by Gidel *et al.* (1983) with a two-dimensional model.

process contributing to the N_2O loss is the reaction

$$O(^1D) + N_2O \rightarrow N_2 + O_2$$
$$\rightarrow 2\,NO$$

where the $O(^1D)$ oxygen atoms arise primarily from the photodissociation of ozone in the 200–300 nm wavelength region. The branching ratio of the reaction is about unity. Only the second process yields reactive nitrogen oxides, however. Table 3-2 lists a number of recent estimates for the global loss rate of N_2O and the production of NO. The latter makes up about 20% of the former. The principal region of this photochemical activity lies at altitudes of 30–35 km in the tropics. Other mechanisms of NO production have been considered. They include ionization of air by cosmic radiation (Warneck, 1972; Nicolet, 1975) and the injection into the atmosphere of solar protons resulting from sunspot activity (Crutzen et al., 1975; Frederick, 1976). An intercomparison of various sources by Jackman et al. (1980) has shown that nitrous oxide is by far the largest source of NO on a global scale.

Nitric oxide is quite rapidly oxidized to NO_2 by reaction with ozone. Nitrogen dioxide, in turn, is subject to photolysis whereby NO is regenerated. This leads to the following reaction sequence

$$NO_2 + h\nu \rightarrow NO + O$$
$$O + O_2 + M \rightarrow O_3 + M$$
$$NO + O_3 \rightarrow NO_2 + O_2$$

which leaves the sum of $NO + NO_2 = NO_x$ unaffected. At altitudes below 45 km, the time constant for the approach to photostationary state between NO and NO_2 is less than 100 s, and the steady state is rapidly established. After sunset, NO decays to low values, and it comes up again following sunrise. The diurnal variation of NO has been documented by Ridley et al.

Table 3-2. *Recent Estimates for the Global Loss of N_2O Due to Its Photodissociation in the Stratosphere, and the Production of NO Due to the Reaction $O(^1D) + N_2O \rightarrow 2NO$*

Authors	Loss of N_2O		Production of NO	
	(kmol/s)	(TgN/yr)	(kmol/s)	(TgN/yr)
Schmeltekopf et al. (1977)	16.7	15	3.6	1.6
Johnston et al. (1979)	10	9	2.3	1.0
Levy et al. (1980)	—	—	1.2	0.5
Jackman et al. (1980)	—	—	2.8	1.0
Crutzen and Schmailzl (1983)	7.5–11.7	6.8–10.3	0.9–1.5	0.4–0.7

(1977) and Ridley and Schiff (1981) from *in situ* measurements on board a balloon floating at 26 km altitude.

Figure 3-7 shows mixing ratios of NO_2 and NO as a function of altitude. NO_2 was determined primarily by infrared spectroscopy and NO by a chemiluminescence technique, with instruments carried aloft by balloons or rockets. The vertical gradient of the mixing ratios of these species is opposite to that for N_2O, indicating that they are transported downward from the stratosphere into the troposphere. The degree of NO_2 dissociation increases with height due to the fact that the number density of ozone declines at altitudes above 30 km, whereby the $O_3 + NO$ reaction becomes less and less capable of reverting NO to NO_2. In the mesosphere, NO_x exists mainly as NO. Into this region NO is supplied from the ionosphere (where it is produced by ion reactions) rather than the stratosphere, and it is destroyed by predissociation at wavelength shorter than 200 nm due to the reactions

$$NO + h\nu \rightarrow N + O$$

$$N + O_2 \rightarrow NO + O$$

$$N + NO \rightarrow N_2 + O$$

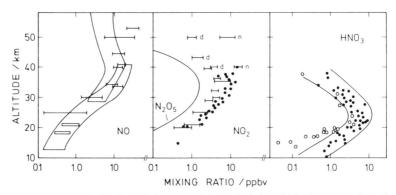

Fig. 3-7. Vertical distribution of nitrogen oxides and nitric acid in the stratosphere. Left: Nitric oxide in the sunlit atmosphere; the fields enclose data obtained with the chemiluminescence technique (Horvath and Mason, 1978; Roy *et al.*, 1980; Ridley and Schiff, 1981; Ridley and Hastie, 1981); horizontal lines represent measurements by infrared optical techniques (Drummond and Jarnot, 1978; Roscoe *et al.*, 1981; Loewenstein *et al.*, 1978a,b). Center: Nitrogen dioxide as observed by optical measurement techniques, day (d) and night (n); points indicate data from Murcray *et al.* (1974), Goldman *et al.* (1978), Blatherwick *et al.* (1980); horizontal bars are from Drummond and Jarnot (1978) and Roscoe *et al.* (1981). The N_2O_5 profile was obtained by Toon *et al.* (1986) at sunrise. Right: Nitric acid observed by *in situ* filter sampling (open points) (Lazrus and Gandrud, 1974) and by infrared spectroscopy and mass spectroscopy (solid points) (Fontanella *et al.*, 1975; Harries *et al.*, 1976; Evans *et al.*, 1978; Arnold *et al.*, 1980; Murcray *et al.* as quoted by Hudson, 1982; Fischer *et al.*, 1985). The envelope gives the error range.

The situation has been discussed by Strobel (1971), Brasseur and Nicolet (1973), Frederick and Hudson (1979b), and Jackman *et al.* (1980). It appears that the flux of NO at the stratopause, either into or out of the mesosphere, is essentially negligible compared with the stratospheric NO production rate.

The significance of NO_x to stratospheric ozone is due to the competition between NO_2 and O_3 for oxygen atoms:

$$O + NO_2 \rightarrow NO + O_2$$
$$O + O_3 \rightarrow 2O_2$$

and the fact that the rate coefficient for the first of these reactions is much greater than that for the second. The first reaction, therefore, is favored even though the mixing ratios of NO_2 are about two orders of magnitude smaller than those of ozone. This leads to the reaction sequence

$$O_3 + h\nu \rightarrow O + O_2$$
$$O + NO_2 \rightarrow O_2 + NO$$
$$\underline{NO + O_3 \rightarrow NO_2 + O_2}$$
$$\text{net} \quad O_3 + O_3 \rightarrow 3O_2$$

In these reactions NO_x is not consumed while it destroys ozone. Rather, NO_x acts as a catalyst to ozone destruction in a pure oxygen atmosphere. Because it is faster, the catalytic cycle proceeds several times during the same time interval in which the O_3 loss reaction of the Chapman mechanism occurs once.

In addition to reactions leading to the formation of NO, nitrogen dioxide undergoes oxidation to N_2O_5 and to nitric acid. The formation of N_2O_5 follows from the relatively slow reaction of NO_2 with ozone,

$$NO_2 + O_3 \rightarrow NO_3 + O_2$$
$$NO_3 + NO_2 \rightleftharpoons N_2O_5$$

where NO_3 serves as an intermediate. The second equation shows that NO_3 and N_2O_5 are in thermal equilibrium. At the low temperatures in the stratosphere, the equilibrium tends toward the right (see Table 9-9). While these reactions occur at all times, they are supplemented during the day by the photodecomposition reactions

$$N_2O_5 + h\nu \rightarrow NO_2 + NO_3$$
$$NO_3 + h\nu \rightarrow NO_2 + O$$

The photodecomposition of NO_3 has a large rate with $j(NO_3) \approx 0.1 \text{ s}^{-1}$, since it occurs mainly by radiation in the visible region of the spectrum. This prevents the production of N_2O_5 during the day. The absorption of light by N_2O_5 becomes appreciable at wavelengths shorter than 350 nm, although it is most effective in the wavelength band of the atmospheric window near 200 nm. N_2O_5 thus builds up during the night and undergoes photodecomposition during the day (with a time constant of about 7 h; see Fig. 2-18). The consequence is a diurnal variation of NO_x, which attains low values in the morning and higher ones (by a factor of two) in the late afternoon. Such a variation was first observed by Brewer and McElroy (1973) from twilight measurements of NO_2, and it was subsequently confirmed by other measurements made from balloons and from the ground (Kerr and McElroy, 1976; Evans et al., 1978; Noxon, 1979). Clear-cut evidence for the presence of N_2O_5 in the stratosphere has been established only recently via the infrared spectrum of N_2O_5 (Toon et al., 1986). The altitude profile deduced is included in Fig. 3-7. It appears that N_2O_5 represents a relatively small fraction of total NO_x.

Figure 3-7 further shows mixing ratios for nitric acid in the stratosphere. Nitric acid arises from the interaction of NO_2 with water radicals. The relevant reactions are

$$NO_2 + OH + M \rightarrow HNO_3 + M$$

$$HNO_3 + h\nu \rightarrow NO_2 + OH$$

$$HNO_3 + OH \rightarrow H_2O + NO_3$$

The absorption spectrum of nitric acid also has a long-wavelength tail reaching beyond 300 nm, so that HNO_3 is somewhat susceptible to photo-decomposition in the lower stratosphere and in the troposphere. In these regions, $j(HNO_3) \approx 5 \times 10^{-7} s^{-1}$ (see Fig. 2-18)—that is, the time constant for photodestruction is of the order of 25 days. At altitudes below 25 km, therefore, HNO_3 is relatively stable and provides the major reservoir of NO_x. Nitrogen dioxide may then be viewed as the photodissociation product of HNO_3. Owing to the long time constant for photolysis, it is doubtful whether a photostationary state between HNO_3 and NO_2 is actually achieved. Above 30 km altitude, $j(HNO_3)$ increases by more than an order of magnitude and $n(OH)$ increases as well, so that the time constant for the approach to steady state reduces to less than 1 day. In the upper stratosphere HNO_3 and NO_2 thus are in photostationary state, but in this region HNO_3 becomes less important as an NO_x reservoir. Finally, it should be noted that pernitric acid, HO_2NO_2, has been proposed as yet another NO_x reservoir. Pernitric acid arises from the $HO_2 + NO_2 = HO_2NO_2$ re-combination–dissociation equilibrium, but no detection of its presence in the stratosphere has been reported.

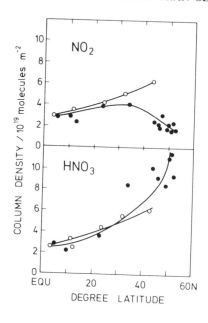

Fig. 3-8. Seasonal variation of the total column densities of NO_2 and HNO_3 in the stratosphere as a function of latitude ($\bigcirc$, summer; $\bullet$, winter). [From measurements of Coffey *et al.* (1981).]

Similar to ozone, the nitrogen oxides and nitric acid undergo poleward transport and partly accumulate in the lower stratosphere at high latitudes before they are discharged into the troposphere. Figure 3-8 shows column densities of NO_2 and HNO_3 as a function of latitude as reported by Coffey *et al.* (1981) from infrared optical observations by aircraft. The general tendency is an increase of NO_x as one goes from the equator to the pole, but NO_2 shows a ridge near 35°N in winter and then declines further northward. The decline is partly compensated by an increase in HNO_3, which suggests that the photostationary state between NO_2 and HNO_3, as far as it exists, is shifted in favor of HNO_3 due to the low elevation of the sun. The appearance of the ridge was first observed by Noxon (1975, 1979). According to his analysis, a sharp ridge appears when there exists a strong polar vortex in northern hemispheric winter that blocks meridional transport of NO_x into the Arctic. The ridge disappears as the vortex breaks down in early spring. While the origin of the phenomenon is as yet uncertain, it is the cause of a pronounced seasonal variation of the NO_2 column density at high latitudes, with low values in winter and high ones in summer.

3.4.2 WATER VAPOR, METHANE, AND HYDROGEN

In the troposphere, the mixing ratio of water vapor decreases with height due to the lowering of the saturation vapor pressure with decreasing temperature and condensation of the excess. The coldest temperatures between troposphere and stratosphere occur at the tropical tropopause. Since most of the air entering the stratosphere passes through this region, one expects the mixing ratio of water vapor in the stratosphere to adjust accordingly. Table 1-5 gave temperature and pressure at the tropical tropopause as 195 K and 120 mbar, respectively. The equilibrium vapor pressure of H_2O at 195 K is roughly 5×10^{-7} bar, corresponding to an H_2O mixing ratio of 5 ppmv. Harries (1976) and Elsaesser et al. (1980) have reviewed a great number of statospheric water vapor measurements. The bulk of the data, as far as they are considered reliable, falls into the range 2-7 ppmv, the average rising slightly with altitude up to 40 km. This is approximately in agreement with expectation. The data exhibit a considerable scatter, however. In the lower stratosphere, variations may be caused by tropopause relocations and high-reaching convective clouds. The origin of the fluctuations in the upper stratosphere is unknown. If water vapor enters the stratosphere via the Hadley circulation as originally proposed by Brewer (1949), it must also return to the troposphere by meridional motions in order to maintain a global balance. Moreover, additional water vapor is brought into the stratosphere by the oxidation of methane. This process presumably is the origin of the weak rise of H_2O mixing ratio with altitude up to the stratopause. The flux of methane into the stratosphere exceeds by two orders of magnitude the planetary escape flux of hydrogen, so that the excess water produced must also return to the troposphere. Stanford (1973) and Elsaesser (1974) have pointed out that the winter temperatures of the polar stratospheres, especially that of the Antarctic, fall below that of the tropical tropopause. The winter pole might thus serve as a sink region for stratospheric water

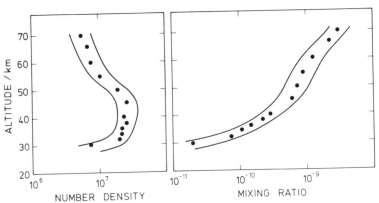

Fig. 3-9. Vertical distribution of OH radicals in the stratosphere. Left: Number density in molecules/cm^3. Right: Mixing ratio. Points are averages from balloon and rocket measurements of Anderson (1971, 1976a, 1980) and Heaps and McGee (1982); the envelope indicates the range of data.

vapor via the formation and precipitation of ice crystals. This would keep the system in a steady state.

The reaction of water vapor with $O(^1D)$ atoms is the major source of hydroxyl radicals in the stratosphere

$$O(^1D) + H_2O \rightarrow 2OH$$

Ground-state oxygen atoms do not react with water vapor, and the photodissociation of H_2O does not become effective until one reaches the mesosphere because at lower altitudes the wavelength region below 190 nm, where H_2O starts to absorb is shielded by oxygen (Nicolet, 1971). Figure 3-9 shows the altitude profile of day time mixing ratios of OH radicals as derived from fluorescence measurements with balloon and rocket-borne instrumentation. The OH mixing ratio is controlled by a greater number of chemical reactions. Among them are the chain reactions

$$O + OH \rightarrow O_2 + H$$

$$H + O_3 \rightarrow O_2 + H$$

$$H + O_2 + M \rightarrow HO_2 + M$$

$$O + HO_2 \rightarrow OH + O_2$$

which are of major importance in the upper stratosphere and mesosphere. These reactions shuffle odd hydrogen back and forth between OH and HO_2 radicals, and in the process they lead to the destruction of odd oxygen representing the sum of atomic oxygen and ozone. The odd hydrogen cycle is terminated by the recombination reaction

$$OH + HO_2 \rightarrow H_2O + O_2$$

A similar recombination of HO_2 radicals to form H_2O_2 is not very effective in removing odd hydrogen because H_2O_2 is subject to photodissociation whereby OH radicals are reconstituted

$$HO_2 + HO_2 \rightarrow H_2O_2$$

$$H_2O_2 + h\nu \rightarrow 2OH$$

The process is more important in the troposphere since in that region H_2O_2 can dissolve in cloud water and be lost by rainout.

In the lower stratosphere the interaction of OH and HO_2 with oxygen atoms becomes relatively unimportant, whereas the direct reactions with ozone gain in significance. These are

$$OH + O_3 \rightarrow HO_2 + O_2$$

$$HO_2 + O_3 \rightarrow OH + 2O_2$$

$$\text{net} \quad O_3 + O_3 \rightarrow 3O_2$$

The second of these reactions is fairly slow, and it must compete with the reaction of HO_2 with nitric oxide,

$$HO_2 + NO \rightarrow NO_2 + OH$$
$$NO_2 + h\nu \rightarrow NO + O$$
$$O + O_2 + M \rightarrow O_3 + M$$

which regenerates the ozone lost by reaction with OH and therefore causes no net loss. In the lower stratosphere, odd hydrogen is removed by the reaction of OH with HNO_3 discussed previously, and possibly by the reaction of OH with pernitric acid. The reaction $OH + HO_2$ becomes comparatively unimportant.

Hydroxyl radicals also are responsible for the oxidation of methane in the stratosphere by virtue of the reaction

$$OH + CH_4 \rightarrow H_2O + CH_3$$

The mechanism for the oxidation of methane will be discussed in more detail in Section 4.2. Here, it suffices to note the first nonradical product following the attack of OH radicals upon methane is formaldehyde. This compound is subject to photodecomposition in the near-ultraviolet spectral region. There are two decomposition channels,

$$HCHO + h\nu \rightarrow H + HCO$$
$$\rightarrow H_2 + CO$$

one of which yields hydrogen and carbon monoxide as stable products. Both enter into reactions with OH radicals:

$$OH + CO \rightarrow CO_2 + H$$
$$OH + H_2 \rightarrow H_2O + H$$

The reaction with CO is by two orders of magnitude faster than that with H_2 so that CO is kept at a fairly low level of mixing ratios, about 10 ppbv according to measurements of Farmer et al. (1980) and Fabian et al. (1981b), at altitudes between 20 and 30 km. In the mesosphere, CO is produced from the photodissociation of CO_2 and the CO mixing ratio rises again. In the lower stratosphere, where the main fate of H atoms is attachment to O_2, the generation of hydrogen atoms by the reactions $OH + CO$ and $OH + H_2$ represents a source of ozone, because the HO_2 radicals formed react with NO to produce NO_2, which subsequently photodissociates to form O atoms. This process is similar to that discussed in Section 5.2 for the formation of ozone by photochemical smog.

The vertical distributions of methane and hydrogen in the stratosphere are shown in Fig. 3-10. The loss of methane is made evident from the decrease of the CH_4 mixing ratio with increasing altitude. In the lower stratosphere, reaction with OH is the sole loss process. In the upper stratosphere it is supplemented by reactions of methane with $O(^1D)$ and Cl atoms. Crutzen and Schmailzl (1983) estimated from two-dimensional model calculations that OH and $O(^1D)+Cl$, respectively, contribute about 50% each to the total loss of methane. The global rate of stratospheric CH_4 removal according to their estimate is 42 Tg/yr. In Section 4.3 a different method will be employed to derive a value of 55 Tg CH_4/yr.

The chemical reactions discussed above suggest a yield of two H_2O molecules for each molecule of methane that is oxidized. This leads to a global stratospheric production rate for H_2O of $2(M_{H_2O}/M_{CH_4})F(CH_4) =$ 95 Tg/yr. In Section 1.5 the air flow into the stratosphere due to the Hadley motion was given as 1.4×10^{17} kg/yr. If one adopts an average mixing ratio for H_2O at the tropical tropopause of 4 ppmv = 2.5 ppmw, the flux of water vapor cycled through the stratosphere is 350 TgH_2O/yr. The oxidation of methane thus adds about 25% to the flux of water vapor entering the stratosphere directly through the tropopause.

In contrast to methane, the vertical distribution of hydrogen shows little change with altitude, although reaction rates for H_2 and CH_4 interacting with OH and $O(^1D)$ are approximately equivalent. An explanation of the H_2 altitude profile thus requires a source of H_2. At altitudes below 40 km,

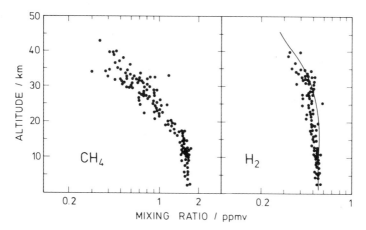

Fig. 3-10. Vertical distribution of methane and hydrogen in the stratosphere at midlatitudes (40-60° N). [From observations of Bush *et al.* (1978), Ehhalt and Heidt (1973), Ehhalt *et al.* (1974, 1975), Pollock *et al.* (1980), Heidt and Ehhalt (1980), Fabian *et al.* (1979, 1981b), and Volz *et al.* (1981b).]

oxidation of methane is the only known source of molecular hydrogen. Above 40 km one would have to add the reaction

$$H + HO_2 \rightarrow H_2 + O_2$$

in order to maintain the hydrogen mixing ratio, since at high altitudes methane declines to low values and can no longer provide an adequate supply. Figure 3-10 includes a vertical H_2 profile derived from one-dimensional eddy diffusion calculations. The calculated distribution shows a slight bulge near 20 km, indicating that in this region the production of hydrogen exceeds the losses. The measurements do not generally exhibit this feature, even though some individual observations seem to agree with it (Schmidt *et al.*, 1980). Measurements and calculations nevertheless are reasonably consistent.

3.4.3 HALOCARBONS

Important source gases for stratospheric chlorine are methyl chloride, CH_3Cl, the chlorofluoromethanes F11, $CFCl_3$, and F12, CF_2Cl_2, carbon tetrachloride, CCl_4, and methylchloroform, CH_3CCl_3. Of these, only methyl chloride has a natural origin, whereas the others are released into the atmosphere from anthropogenic sources. Figure 3-11 gives an overview on the vertical distributions of halocarbons in the stratosphere. Their behavior is similar to that of nitrous oxide or methane—that is, the halocarbons diffuse upward from the tropopause, suffering destruction somewhere in the middle and upper stratosphere. The hydrogen-bearing species are lost mainly by reacting with OH and $O(^1D)$, and the others undergo photo-decomposition. F11, F12, and CCl_4 begin to absorb solar radiation at wavelengths at or below 230 nm. Absorption and photodissociation become most effective in the wavelength region 170–190 nm, which is partially shielded by the Schumann–Runge bands of oxygen. The photolysis of the halocarbons leads to the generation of atomic chlorine and ClO, for example,

$$CFCl_3 + h\nu \rightarrow Cl + CFCl_2$$
$$CFCl_2 + O_2 \rightarrow COFCl + ClO$$

The carbonyl halides thus formed also are subject to photodecomposition whereby further chlorine (or fluorine) atoms are released. Their interaction with ozone results in the catalytic cycle

$$Cl + O_3 \rightarrow ClO + O_2$$
$$ClO + O \rightarrow Cl + O_2$$

net $O + O_3 \rightarrow 2O_2$

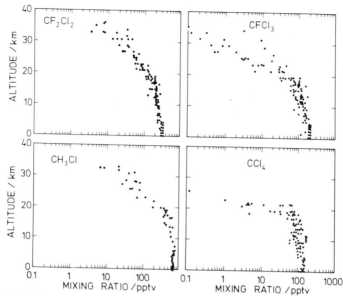

Fig. 3-11. Vertical distribution of CF_2Cl_2, $CFCl_3$, CH_3Cl, and CCl_4 in the stratosphere. [Data compiled by Fabian (1986) from Lovelock (1974a), Heidt *et al.* (1975), Schmeltekopf *et al.* (1975), Krey *et al.* (1977), Robinson *et al.* (1977), Seiler *et al.* (1978b), Tyson *et al.* (1978), Vedder *et al.* (1978, 1981), Fabian *et al.* (1979), Goldan *et al.* (1980), Penkett *et al.* (1980a), Rasmussen *et al.* (1980), Fabian *et al.* (1981a), Leifer *et al.* (1981), Schmidt *et al.* (1981), Rasmussen *et al.* (1982a), and Borchers *et al.* (1983).]

which represents an important process in the consumption of ozone. The ClO_x and NO_x systems are coupled by the reactions

$$ClO + NO \rightarrow NO_2 + Cl$$

$$Cl + O_3 \rightarrow ClO + O_2$$

$$NO_2 + h\nu \rightarrow NO + O$$

$$O + O_2 + M \rightarrow O_3 + M$$

In this reaction sequence ozone is reconstituted so that there is no net loss. Yet another link between the ClO_x and NO_x families is the formation of chlorine nitrate and its photodecomposition:

$$ClO + NO_2 \rightarrow ClONO_2$$

$$ClONO_2 + h\nu \rightarrow Cl + NO_3$$

$$Cl + O_3 \rightarrow ClO + O_2$$

$$NO_3 + h\nu \rightarrow NO_2 + O$$

$$O + O_2 + M \rightarrow O_3 + M$$

This reaction sequence also is seen to produce no net loss of ozone, but it temporarily ties up some chlorine in an inactive reservoir. A tentative upper-limit mixing ratio for $ClONO_2$ of 1 ppbv near 30 km altitude has been set by Murcray *et al.* (1979) from infrared measurements. The calculated daytime average mixing ratio is 0.2 ppbv.

An interaction occurs also between the chlorine and HO_x families in that ClO reacts with HO_2 to form hypochlorous acid. This leads to the reaction sequence

$$ClO + HO_2 \rightarrow HOCl + O_2$$
$$HOCl + h\nu \rightarrow OH + Cl$$
$$Cl + O_3 \rightarrow ClO + O_2$$
$$OH + O_3 \rightarrow HO_2 + O_2$$
$$\text{net} \quad O_3 + O_3 \rightarrow 3O_2$$

which again represents an ozone loss process. Maximum $HOCl$ mixing ratios of about 0.1 ppbv are predicted to occur at 35 km altitude in the summer hemisphere. Stratospheric measurements have not been reported.

The principal sink for active chlorine species is conversion to hydrochloric acid. The reactions involved are

$$Cl + CH_4 \rightarrow HCl + CH_3$$
$$Cl + HO_2 \rightarrow HCl + O_2$$
$$Cl + H_2 \rightarrow HCl + H$$

with methane representing the main pathway for HCl formation. From HCl the chlorine atoms are regenerated, at a much slower rate, by reaction with OH radicals

$$OH + HCl \rightarrow H_2O + Cl$$

and this process provides the principal source of ClO_x in the upper stratosphere. Fluorine atoms are similarly converted to hydrogen fluoride. Its reaction with OH is endothermic, however. The photolysis of HF further is shielded by oxygen so that fluorine atoms do not play the same role in stratospheric chemistry as chlorine atoms.

Figure 3-12 shows the vertical distribution of HCl and HF in the stratosphere in addition to that of (daytime) ClO. The mixing ratio of ClO, although appreciable, generally falls below that of HCl, so that HCl is the major chlorine reservoir in the upper stratosphere. The gradients of mixing ratios are positive, which indicates that HCl and HF are transported down-

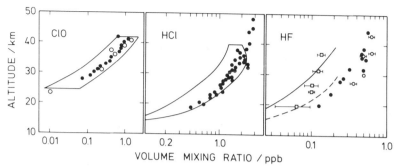

Fig. 3-12. Vertical distribution of ClO, HCl, and HF in the stratosphere. Left: Filled circles give the averages of eight altitude profiles for ClO measured in 1976–1979 by *in situ* resonance fluorescence; the envelope indicates the range of values (Weinstock *et al.*, 1981); two additional high-mixing-ratio profiles are not included. The open circles are from balloon-borne infrared remote measurements by Waters *et al.* (1981) and Menzies (1983). Center: The envelope encompasses observational data for HCl obtained by balloon-borne infrared measurement techniques (Farmer *et al.*, 1980; Buijs, 1980; Raper *et al.*, 1977; Eyre and Roscoe, 1977; Williams *et al.*, 1976; Zander, 1981); filled circles represent more recent preliminary data cited in WMO (1985). Right: Balloon-borne infrared remote measurements of HF [——— Farmer *et al.* (1980), - - - Buijs (1980), ○ Zander (1975, 1981), ● preliminary data cited in WMO (1985); and observations of HF by *in situ* filter collections (Mroz *et al.*, 1977).

ward into the troposphere. Girard *et al.* (1978, 1983) and Mankin and Coffey (1983) have measured total column densities of HCl and HF by infrared absorption and reported an increase from the equator toward high latitudes by about a factor of four. This indicates that HCl and HF are transported poleward and accumulate in the lower stratosphere to some extent before they enter into the troposphere. The behavior is qualitatively similar to that of ozone and nitric acid and shows that HCl and HF represent indeed the terminal sinks for chlorine and fluorine.

At the stratopause, the mixing ratio of HCl extrapolates to 2–3 ppbv, whereas that of HF extrapolates to a value exceeding 0.5 ppbv. Table 3-3 gives mixing ratios for the main Cl- and F-containing source gases at the tropopause level. By summing over the individual contributions, the mixing ratio for total chlorine is found to add to 2.7 ppbv, and that for total fluorine to 0.7 ppbv. The abundances of other chlorine- and fluorine-containing compounds not listed in Table 3-3 are relatively minor and may be neglected to a first approximation. Comparison then shows that the mixing ratios for total chlorine and fluorine at the stratopause are the same as those at the tropopause—that is, the mixing ratios are conserved as required by material balance. Berg *et al.* (1980) have used ultrapure charcoal traps to collect all chlorine compounds at altitudes of 15, 19, and 21.5 km. Neutron activation analysis gave total chlorine mixing ratios in the range 2.7–3.2 ppbv, which confirms the preceding conclusion.

Since the concentration of ClO is directly proportional to that of HCl, total chlorine determines the rate of ozone destruction in the upper stratosphere by the reaction of ClO with O atoms. Table 3-3 thus shows that chlorine from human-made halocarbons now overrides natural chlorine from CH_3Cl by a factor of four. The fluxes of chlorine, however, are not apportioned in the same way. The fluxes are determined by the global loss rates of the source gases in the stratosphere. The estimates in Table 3-3 suggest that CH_3Cl contributes about 30% to the total chlorine flux. CF_2Cl_2, $CFCl_3$, and CCl_4 make up most of the remainder. It should be emphasized that for the last three components photolytic destruction in the stratosphere is the only known atmospheric loss process, whereas CH_3Cl and CH_3CCl_3 react with OH radicals also in the troposphere.

Bromine is potentially able to interact with stratospheric ozone in the same manner as chlorine (Wofsy et al., 1975). The catalytic cycle for bromine is expected to be quite efficient, because its reaction with methane is slower than that of Cl atoms; in addition, the reaction of OH with HBr is faster than that of OH with HCl. The major bromine compound in the troposphere is methyl bromide, which has a natural origin and occurs with a mixing ratio of about 10 pptv (see Table 6-14). This seems small enough to neglect bromine to a first approximation.

3.4.4 Carbonyl Sulfide and Sulfur Dioxide

Most sulfur compounds in the troposphere are short-lived and do not reach the stratosphere. A notable exception is COS, which is present with a mixing ratio of about 500 pptv and represents the major reservoir of sulfur in the troposphere. Sulfur dioxide also appears to be relatively stable in the upper troposphere, where it occurs with mixing ratios in the range 40–90 pptv. Both components can be traced into the stratosphere. This is illustrated in Fig. 3-13. The mixing ratio of COS decreases with increasing altitude, whereas that of SO_2 remains approximately constant up to 25 km. Such a behavior suggests that SO_2 is a product resulting from the oxidation of COS. In the stratosphere, COS is subject to photodissociation and reaction with oxygen atoms. The relevant processes are

$$COS + h\nu \rightarrow CO + S(^3P, {}^1D)$$
$$S + O_2 \rightarrow SO + O$$
$$O + COS \rightarrow SO + CO$$
$$SO + O_2 \rightarrow SO_2 + O$$
$$SO + O_3 \rightarrow SO_2 + O_2$$

Table 3-3. *Halocarbons: Mixing Ratios of Total Chlorine and Fluorine in the Troposphere (in 1980; Hudson, 1982) and Global Fluxes[a] of Chlorine and Fluorine into the Stratosphere (Courtesy U. Schmailzl, Private Communication)*

Halocarbon	M (g/mol)	m(HC) (pptv)	m(Cl) (pptv)	m(F) (pptv)	Flux into the stratosphere		
					Tg/yr	mol Cl/yr	mol F/yr
CH_3Cl	50.5	650	650	—	0.096	19.1×10^8	—
CF_2Cl_2	121	285	570	570	0.034	5.6×10^8	5.6×10^8
$CFCl_3$	137.5	180	540	180	0.057	12.4×10^8	4.1×10^8
CCl_4	154	130	520	—	0.063	16.5×10^8	—
CH_3CCl_3	133.5	140	420	—	0.075	17.0×10^8	—
Total Cl, F	—	—	2700	750	—	70.4×10^8	9.7×10^8

[a] The fluxes carry an uncertainty of up to 50%.

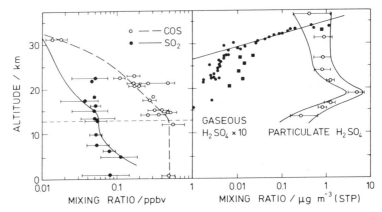

Fig. 3-13. Left: Vertical distribution of carbonyl sulfide and sulfur dioxide in the stratosphere. [From data of Maroulis *et al.* (1977), Sandalls and Penkett (1977), Torres *et al.* (1980),
Mankin *et al.* (1979), Inn *et al.* (1979, 1981) for COS, and from Jaeschke *et al.* (1976),
Maroulis *et al.* (1980), Georgii and Meixner (1980), Inn *et al.* (1981) for SO$_2$.] Curves represent
calculations of Turco *et al.* (1980, 1981a) for an assumed cutoff of COS photodissociation of
312 nm. Right: Vertical distribution of gaseous and particulate sulfuric acid. Solid squares and
circles are from mass spectrometric measurements of Arijs *et al.* (1982) and Viggiano and
Arnold (1983), respectively. Open circles with error bars (one standard deviation) are from
filter collections of Lazrus and Gandrud (1977). The range given by the thin lines indicates
the seasonal variability of particulate sulfate. The solid line indicates the vapor pressure of
H$_2$SO$_4$ over a 75% H$_2$SO$_4$/25% H$_2$O mixture.

These reactions convert COS to sulfur dioxide. In addition, it can be seen
that odd oxygen is both produced and consumed. At 30 km altitude and
below, the loss rate for COS reacting with oxygen atoms is an order of
magnitude smaller than that of photodissociation. At the same altitude, the
reaction of SO with O$_2$ contributes about 70% to the total rate of SO to
SO$_2$ conversion. Generally it appears, therefore, that more odd oxygen is
formed by the above reaction sequence than it removes.

Sulfur dioxide originating from the oxidation of COS enters into the
reactions

$$SO_2 + h\nu \rightarrow SO + O$$
$$SO + O_2 \rightarrow SO_2 + O$$

which are seen to generate odd oxygen, and

$$SO_2 + OH + M \rightarrow HOSO_2 + M$$
$$HOSO_2 + O_2 \rightarrow HO_2 + SO_3$$
$$SO_3 + H_2O \rightarrow H_2SO_4$$

which convert SO$_2$ permanently to sulfuric acid. No reactions are known by which sulfur is reactivated from H_2SO_4. The photodissociation of SO_2 becomes energetically possible at wavelengths below 228 nm. Thus, it is effective in the atmospheric radiation window near 200 nm. The SO_2 absorption spectrum in this wavelength region consists of a continuum overlapped by absorption bands. Only the continuum leads to dissociation, whereas excitation due to discrete absorption features is subject to collisional quenching (Warneck et al., 1964; Driscoll and Warneck, 1968). At 30 km altitude, taking into account only the continuum, the photodissociation coefficient is $j(SO_2) \approx 10^{-5} \, s^{-1}$. The rate of the competing reaction of SO_2 with OH radicals at the same altitude is one order of magnitude smaller, so that SO_2 undergoes about 10 cycles generating oxygen atoms before it is permanently stored in the form of H_2SO_4. For the purpose of illustrating the effects of COS and SO_2 on stratospheric odd oxygen production, we shall assume that the conditions at 30 km altitude are representative of the entire COS destruction region.

The column loss rate for COS may be estimated as $(0.5-1.0) \times 10^{11}$ molecules/m^2 s (Inn et al., 1981; Crutzen and Schmailzl, 1983). The corresponding global rate of SO_2 production in the stratosphere is 83 mol/s. If SO_2 undergoes 10 photolytic cycles, the global production rate of odd oxygen would be 1.7×10^3 mol/s. The gross rate of odd oxygen formation from the photodissociation of O_2 according to Table 3-1 is 81×10^6 mol/s, which is 45,000 times greater than that caused by the injection of COS into the stratosphere. The comparison thus shows that the effect of carbonyl sulfide on the production (and destruction) of stratospheric ozone is negligible.

Sulfuric acid in the stratosphere exists as a vapor as well as a condensation aerosol. Figure 3-13 includes data for gaseous H_2SO_4 and for particulate sulfate. The mass mixing ratio of the sulfate layer has its peak at altitudes between 20 and 25 km (compare this with Fig. 7-25, which shows the altitude profile of aerosol particles in terms of number densities). The maximum occurs markedly below the altitude region where SO_2 is most effectively converted to H_2SO_4. Junge et al. (1961) noted that particles with radii near 0.3 μm have a tendency to accumulate at the 20 km altitude level in the approach to a steady-state between sedimentation and eddy diffusion. This prediction is roughly in agreement with observations. The time constant, however, for the system to come to steady state is appreciable for particles smaller than 1 μm. The size distributions that have been observed usually exhibit maxima at radii near 0.1 μm, according to the review by Inn et al. (1982), so that a steady state is unlikely to be established. At altitudes above 30 km the concentration of the H_2SO_4 vapor rises, whereas that of sulfate particles declines. As Fig. 3-13 shows, the data for gas-phase H_2SO_4 in this region coincide approximately with the saturation vapor pressure curve, which suggests the existence of an equilibrium between vapor and condensate. Owing to the temperature rise with altitude above 30 km, the saturation vapor pressure rises as well, so that with increasing height more and more particles evaporate.

Chemical analysis of particles collected *in situ* indicate that the stratospheric aerosol consists basically of a solution of 75% sulfuric acid in water with an admixture of dissolved nitrosyl sulfates and solid granules containing silicates (Cadle *et al.*, 1973; Rosen, 1971; Farlow *et al.*, 1977, 1978). The nonsulfate constituents derive partly from the influx of micrometeorites (Ganapathy and Brownlee, 1979), and to a larger extent from the sporadic injection of solid ash particles from high-reaching volcanic eruptions. The solid particles presumably assist in the formation of sulfate aerosol in that they serve as condensation nuclei. Model calculations by Hamill *et al.* (1982) suggests that the homogeneous, vapor-phase formation of sulfate particles is quite inefficient in the stratosphere. Mechanisms of homogeneous and heterogeneous nucleation are discussed in Chapter 7.

In addition to ash particles, high-reaching volcanic eruptions carry also sulfur dioxide into the stratosphere. Cadle *et al.* (1977) estimated that the 1963 eruption of Mt. Agung in Bali, which was a major event, introduced into the stratosphere 12 Tg of sulfur dioxide. This was 100 times the amount that the stratosphere receives annually by the oxidation of COS. The eruption of the volcano Fuego in Guatemala in 1974 was estimated to have injected about 2 Tg SO_2. These events have caused considerable perturbations of the stratospheric sulfur budget. Castleman *et al.* (1973, 1974), from a study covering the period 1963–1973, have found variations of sulfate concentrations from 0.1 to 40 $\mu g/m^3$ at altitudes near the peak of the sulfate layer. The highest values occurred in the southern hemisphere about 1 yr after the Mt. Agung eruption. Subsequently, the values decayed to a level of quiescence that was reached in 1967, with only mild pertubations due to other, less violent volcanic eruptions. The time lag observed for sulfate to reach its maximum concentration after the Mt. Agung eruption demonstrates that sulfate is not injected directly into the stratosphere, but requires a precursor such as SO_2, which is converted toward H_2SO_4 with a certain time constant. Castleman *et al.* (1974) further studied the $^{34}S/^{32}S$ isotope ratio and its variation. Invariably the isotope ratio decreased after a major volcanic event and then increased again. This aspect shows perhaps most clearly that the stratospheric sulfate layer arises from two sources, one that is present all the time and has its origin in the diffusion of source gases like COS into the stratosphere, and a second input due to high-reaching, violent volcanic eruptions that carry additional sulfur gases into the stratosphere.

3.5 The Budget of Ozone in the Stratosphere

Having discussed the chemistry associated with various trace gases in the stratosphere we can now return to the problem of balancing the ozone budget. The impact of this additional chemistry upon stratospheric ozone is most conveniently assessed by considering the results of model calculations. Both one- and two-dimensional eddy diffusion models incorporating the reactions introduced in the preceding sections are currently able to reproduce the vertical distributions of trace components that have been observed in the stratosphere. Accordingly, these models presumably achieve a reasonably quantitative simulation of stratospheric chemistry, although the built-in uncertainties are such that one should not expect models to describe the complex behavior of the stratosphere in every detail. As discussed previously, all diffusion models are based on the continuity equation [Eq. (1-13)], which must be supplemented with appropriate boundary conditions such as fixed mixing ratios at the tropopause and vanishing fluxes at high altitudes. Flow and temperature fields and rate coefficients of elementary reactions are specified parameters. Solar fluxes are calculated as needed from the extraterrestrial radiation field.

The interest in this section is directed at loss rates for odd oxygen due to the catalytic cycles, which may be summarized schematically by

$$X + O_3 \rightarrow XO + O_2$$

$$XO + O_{2n+1} \rightarrow X + (n+1)O_2$$

$$\text{net} \quad O_3 + O_{2n+1} \rightarrow (n+2)O_2$$

where $X = H$, OH, NO, or Cl, and $n = 0$ or 1. Altogether there are 10 such reactions. Figure 3-14 shows individual loss rates as a function of altitude according to the one-dimensional model of Logan et al. (1978; Wofsy, 1978) after the incorporation of more recent reaction rate data. The figure makes apparent that not all the reactions are equally effective and that their relative significance depends on the altitude regime. Above 45 km the Chapman reaction $O + O_3$ remains the most important loss process for odd oxygen, although it is somewhat assisted by the reactions $O + OH$ and $O + HO_2$. In the 35–45 km altitude region the Chapman process and the loss cycles involving NO_x and ClO are about equally important. In the lower stratosphere odd oxygen is predominantly lost by catalytic reactions with NO_x. The reaction $OH + O_3$ comes to prominence at altitudes below 18 km, and the remaining reactions $HO_2 + O_3$, $HO_2 + ClO$, and $H + O_3$ are relatively unimportant. In this model, the total $O + O_3$ destruction rate conicides, of course, with the rate of odd oxygen production.

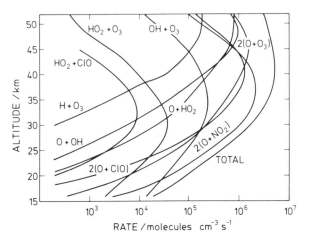

Fig. 3-14. Rates for the loss of odd oxygen in the stratosphere due to various reactions, averaged over a 24-hr period. [Adapted from Wofsy and Logan (1982).]

For a comparison with the data in Table 3-1, one must integrate the individual loss rates over height, from 15 to 50 km, and extrapolate the column loss rates to a global scale. The results are given in Table 3-4. The calculated loss rate associated with the Chapman reaction is 21.1×10^6 mol/s. This is 42% higher than the value derived from the observed distribution of ozone. The difference may be caused by uncertainties in the model or in the measurements, but it is not a serious discrepancy. The total rate of 81.4×10^6 mol/s compares well with the global rate of odd oxygen production due to photodissociation of O_2. The model indicates that odd oxygen is consumed to 25% by the Chapman reaction, 33.8% by NO_x, 21.7% by HO_x, and 18.7% by chlorine radicals. The losses thus are apportioned fairly evenly, although NO_x has the greatest effect. Losses due to NO_x are net losses, since as discussed earlier the reaction $HO_2 + NO$ leads to a production of odd oxygen, which must be subtracted from the loss term.

Table 3-4 includes loss rates for odd oxygen derived from the two-dimensional model of Gidel *et al.* (1983; Crutzen and Schmailzl, 1983). These results are based mainly on observational data for ozone and other trace gases. In this case the loss rate due to the Chapman reaction is close to that of Johnston (1975), who used a similar procedure. With regard to the other data in Table 3-4, the results of both groups of investigators are in good agreement even though the two-dimensional model shows a more pronounced effect of the HO_x cycle at high altitudes.

There can be no doubt that the inclusion of loss reactions other than the Chapman reaction has greatly improved the ozone balance in the stratosphere, but the consistency of the data in Table 3-4 should not be taken to

Table 3-4. *Globally Integrated Stratospheric Loss Rates Due to the Chapman Reaction and Catalytic Destruction Cycles Involving ClO, NO_x, and HO_x* [a]

Processes	Column loss rate (10^{14} molecules/m² s)		Global loss rate (10^6 mol/s)		Percent of total loss	
	L&W	C&S	L&W	C&S	L&W	C&S
$O + O_3 \rightarrow 2O_2$	253	191	21.1	15.9	25.9	20.4
$O + ClO \rightarrow O_2 + Cl$						
$Cl + O_3 \rightarrow ClO + O_2$	182	187	15.2	15.6	18.7	20.0
$O + NO_2 \rightarrow NO + O_2$						
$NO + O_3 \rightarrow NO_2 + O_2$	330	289	27.5	24.1	33.8	30.9
$OH + O_3 \rightarrow HO_2 + O_2$						
$HO_2 + O_3 \rightarrow OH + 2O_2$	51	[b]	4.2	[b]	5.2	[b]
$O + OH \rightarrow H + O_2$						
$H + O_3 \rightarrow OH + O_2$						
$O + HO_2 \rightarrow OH + O_2$	161	269	13.4	22.4	16.5	28.7
Total loss rate	977	936	81.4	78.0	100	100

[a] According to the one-dimensional model of Logan *et al.* (1978; Wofsy and Logan, 1982) (L&W) as shown in Fig. 3-14, and the two-dimensional model of Crutzen and Schmailzl (1983) (C&S).

[b] Included in the data of the next line.

imply that the balance is perfect. Crutzen and Schmailzl (1983) have addressed uncertainties in several parameters, primarily the O_2 absorption cross sections in the Herzberg continuum, to demonstrate the occurrence of ozone production rates up to 60% in excess of odd oxygen losses in the altitude region below 35 km. The imbalance is much greater than can be accounted for by a transport of ozone toward the troposphere. The discrepancy is difficult to remove except by the assumption of additional unknown and thus speculative catalytic loss reactions, which would have to be active primarily in the lower stratosphere. In addition, it should be noted that the great variability of measured ozone has so far defied an adequate reproduction of the observed vertical profiles by modeling. The best simulation appears to have been achieved by the two-dimensional model of Miller *et al.* (1981), of which Fig. 3-15 shows an example.

The various influences on stratospheric ozone caused by trace gases having their origin in the troposphere make the ozone layer susceptible to considerable perturbations, both natural and human. Currently of greatest interest are the effects of the chlorofluorocarbons, but there are others as well, such as variations in the mixing ratios of N_2O and methane in the troposphere, or the rise of CO_2 in the atmosphere. The last constituent does not enter into chemical reactions, but it cools the stratosphere, thereby lowering the rates of important reactions such as that of OH with HCl. The various

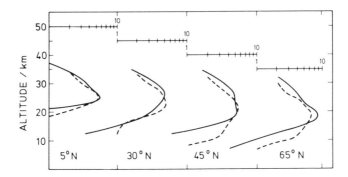

Fig. 3-15. Comparison of calculated with observed vertical ozone distributions at four latitudes (in units of 10^{12} molecules/cm^3). Solid lines: Calculated by Miller *et al.* (1981) with a two-dimensional model. Broken lines: Observed by Wilcox *et al.* (1977). Note that the agreement of ozone profiles by itself does not provide validation of the model; agreement with other trace components must also be accomplished. Miller *et al.* (1981) reported this to have been achieved.

influences may be either positive or negative, making it difficult to estimate potential ozone reductions (or increases) except by complicated models. Moreover, the current situation depends on the past history of trace gases in the trophosphere, which is not fully known in every case.

With regard to the industrial production and release of halocarbons, a number of model predictions have been made for two cases: (a) the reduction of total ozone by 1980 based on the previous release of F11 and F12 as far as known, and (b) the reduction of total ozone for the hypothetical steady state arising 100 yr from now, if the mid-1970 release rates continue at the same level. The data are shown in Table 3-5. They are based on the most recent knowledge of reaction-rate coefficients. Prior to 1980 the percentage changes of total ozone underwent continual revisions as the rate data were improved. Table 3-5 suggests that the emissions of the chlorofluorocarbons into the atmosphere have caused a depletion of total ozone of somewhat less than 1%, but that in the future the losses will increase and eventually reach about 6%. Gidel *et al.* (1983) included carbon tetrachloride and methyl chloroform in their computations and found a depletion of 1.3% to have occurred already before 1970, mainly due to CCl$_4$. The greatest local depletion occurs in the 35–45 km altitude region (see Fig. 3-14), where it is now estimated to amount to 6%. An immediate consequence is that solar ultraviolet (UV) radiation can penetrate deeper into the atmosphere to cause an increase of the local ozone concentration in the 15–25 km altitude region. In this domain, the self-healing effect is amplified by the interaction of chlorine species with HO$_x$ and NO$_x$, leading to a reduction in the efficiency of odd oxygen recombination cycles. Thus, the current

Table 3-5. *Percentage Changes of Total Ozone in* 1980 *and at Steady State Resulting from the Continued Release of F*11 *and F*12 *at Mid-*1970s *to* 1980 *Rates, as Compared to the Situation Prior to* 1970[a]

Investigators	Type of model	Percent change by 1980	Percent change at steady state	
			Without temperature feedback	With temperature feedback
Liu (1982)[b]	1-D	−0.5	−6.5	—
Sze (1983)	1-D	−0.6	−5.3	—
Wuebbles (1983)	1-D	−0.6	−7.0	−7.2
Prather et al. (1984)	1-D	—	−5.3	—
Owens et al. (1985)	1-D	−0.6	−4.9	−6.1
Brasseur (1985)	1-D	−1.1	−4.8	−7.9
Miller et al. (1981) Steed et al. (1982)	2-D	−0.9	−6.2	—
Gidel et al. (1983)	2-D	−3.0[c]	−6.0[d]	—
Crutzen (1985)[e]	2-D	−1.9	−9.1	—
Ko et al. (1984, 1985)	2-D	—	−8.5	—

[a] From Hudson (1982), with supplements.
[b] As quoted by Hudson (1982).
[c] Includes CCl_4 and CH_3CCl_3.
[d] By 1995; not yet in steady state.
[e] Private communciation.

estimate of 1% depletion of total ozone represents a sum of a much larger depletion at high altitudes and an increase at lower altitudes. The two-dimensional models (e.g., Pyle and Derwent, 1980; Steed et al., 1982; Gidel et al., 1983) also predict a dependence on latitude. At polar latitudes the depletion of total ozone is twice as large as in the tropics. The increase in UV radiation penetrating to the ground surface in the wavelength region causing DNA damage, according to Steed et al. (1982), is about 2.5 times the decrease in total ozone, essentially independent of latitude.

So far, the predicted changes in total ozone have remained largely hypothetical (for observations in the Antarctic, see further below). The predictions generally neglect other changes by natural and human activities, such as the increase in methane, carbon dioxide, or tropospheric ozone, which may mask the effect due to the chlorofluorocarbons. Long-term trends of total ozone as derived from observations within the network of Dobson stations have been subjected to statistical analyses by Komhyr et al. (1971), Angell and Korshover (1976, 1978), London and Oltmans (1979), St. John et al. (1981), and by Reinsel (1981). These studies indicated an increase of total ozone during the 1960s and fairly little change during the 1970s. The recognition that the greatest depletion of ozone would occur at altitudes above 30 km has led to a search for such changes in the data record available from ozone soundings by ground-based optical (Umkehr) and by balloon-borne *in situ* measurements. Far fewer data of this type exist in comparison with total ozone. Their analysis by Bloomfield *et al.* (1982) and Reinsel *et al.* (1983), among others, again did not reveal any definite trend since 1970. More recently, Reinsel *et al.* (1984) have applied corrections for the effects of the stratospheric aerosol layer on the Umkehr measurements, wherein they used a series of atmospheric transmission measurements at Mauna Loa, Hawaii, to infer the influence of aerosols on the vertical ozone profiles obtained with the optical inversion technique. The analysis indicated a statistically significant trend in the upper Umkehr layers (38–43 and 43–48 km altitude), which amounted to −0.2 to −0.3% per year during the 1970s. This appears to have the predicted order of magnitude.

Satellite ultraviolet backscatter observations of ozone also have been examined with regard to an ozone reduction in the 35–45 km altitude region. Data from *Nimbus* 4, which was operational in 1970–1972, and *Nimbus* 6 (1975–1976) and *Nimbus* 7 (since 1978) have been compared. In principle, satellites provide a better global coverage than the network of Dobson stations. Unfortunately, as discussed by Hudson (1982), the satellite measurements have suffered from instrumental drift and other deficiencies, which make the early data less suitable for a trend analysis than would be desirable. Generally, the ozone concentration at 35 km altitude is subject to various other perturbations, such as the variation in the ultraviolet solar flux, and of NO_x produced by cosmic radiation, by the 11-yr solar cycle; or the penetration of solar protons, which cause a temporary decrease of ozone due to the additional production of odd nitrogen. These effects require more study before the reported depletion of ozone in the 35–45 km altitude region can be regarded as well established.

In the Antarctic, total ozone shows a relative minimum in spring when the dark period ceases. Recent observations by Farman *et al.* (1985) have established a decrease of springtime total ozone by approximately 40% since the mid 1970s. Such large ozone variations are not seen over the Arctic region. They must be connected with the very stable polar vortex existing in the Antarctic during the winter months. A plausible mechanism to explain the change in total ozone has not yet been offered, however. Current models are unable to simulate the effect. Thus, it is not known whether it is an early indicator of future changes of global ozone, or perhaps a cyclic phenomenon confined to the Antarctic region because of the special conditions there.

Chapter

4 | Chemistry of the Troposphere:
The Methane Oxidation Cycle

A fairly general treatment of trace gases in the troposphere is based on the concept of the tropospheric reservoir introduced in Section 1.6. The abundance of most trace gases in the troposphere is determined by a balance between the supply of material to the atmosphere (sources) and its removal via chemical and biochemical transformation processes (sinks). The concept of a tropospheric reservoir with well-delineated boundaries then defines the mass content of any specific substance in, its mass flux through, and its residence time in the reservoir. For quantitative considerations it is necessary to identify the most important production and removal processes, to determine the associated yields, and to set up a detailed account of sources versus sinks. In the present chapter, these concepts are applied to the trace gases methane, carbon monoxide, and hydrogen. Initially, it will be useful to discuss a steady-state reservoir model and the importance of tropospheric OH radicals in the oxidation of methane and many other trace gases.

4.1 The Tropospheric Reservoir

The lower bound to the troposphere is the earth surface, and its upper bound is the tropopause. Exchange of material occurs via these boundaries with the ocean, the terrestrial biosphere, and the stratosphere. The justification for treating the tropopause as a boundary layer derives from the fact that the rate of vertical transport by turbulent mixing changes abruptly over a short distance, from high values in the troposphere to low values in the stratosphere. This limits the exchange of trace species with the stratosphere while at the same time keeping the troposphere vertically reasonably well mixed. The behavior of trace gases in the tropospheric reservoir is to some extent determined by the competition between losses due to chemical reactions and the rate of vertical and horizontal mixing. The assumption of a well-mixed reservoir is frequently made, but it is not a necessary precondition for the application of reservoir theory, as the following treatment will show.

For trace gases whose mixing ratio varies within the troposphere the local number density is, as always, determined by the continuity equation

$$\frac{dn_s}{dt} = \operatorname{div} f + q - s$$

where q denotes the local production rate and s the local consumption rate due to chemical reactions, and the flux f is given in the usual way by the gradient of the mixing ratio augmented by an advection term. For the present purpose we shall not seek possible solutions to this equation, but integrate it over the entire tropospheric air space. After multiplying with the molecular weight M_s of the trace species considered and dividing by Avogadro's number ($N_A = 6.02 \times 10^{23}$ molecules/mol), we obtain the time derivative of the total mass of the trace substance in the troposphere,

$$\frac{M_s}{N_A} \int \frac{\partial n_s}{\partial t} \, dV = \frac{dG_s}{dt} = \frac{M_s}{N_A} \left(\int (\operatorname{div} f) \, dV + \int q \, dV + \int s \, dV \right) \quad (4\text{-}1)$$

which is equivalent to the budget equation

$$\frac{dG_s}{dt} = F_{in} - F_{out} + Q - S \quad (4\text{-}2)$$

Capital letters are now used to denote the integrated terms. Integration of the divergence term yields the difference of the mass fluxes entering and leaving the tropospheric reservoir through its boundaries. The terms Q and S denote the sum of the internal sources and sinks, respectively. The global source and sink strengths are given by the sums

$$Q_T = F_{in} + Q \qquad S_T = F_{out} + S \quad (4\text{-}3)$$

The mass content G_s of the trace constituent in the tropospheric reservoir may be derived from measurements of the volume mixing ratio at various locations. Let $\bar{m}_s$ denote the spatial average of the mixing ratio; G_s is then obtained from

$$G_s = \frac{M_s}{M_{air}} \bar{m}_s G_T \qquad (4\text{-}4)$$

where $M_{air} = 29$ g/mol is the molecular weight of air and $G_T = 4.25 \times 10^{18}$ kg is the total tropospheric air mass. Of the four quantities on the right-hand side of Eq. (4-2), the outgoing flux and the volume sink are functions of G_s, whereas the input flux and volume source are independent of G_s. The flux leaving a well-mixed chemical reactor is proportional to G_s. This relation is expected to hold also when the condition of complete mixing is not achieved. Accordingly, we set

$$F_{out} = k_F G_s \qquad (4\text{-}5)$$

where k_F must be treated as an empirical parameter. The functional dependence of S on G_s is less transparent. The local consumption rate of a trace gas depends on the type of reaction into which it enters. For photodecomposition the loss rate is given by $j_s n_s$. For bimolecular reactions the loss rate is proportional to the number density of the second reactant n_r and the associated rate coefficient k_r. The local volume sink including all such processes thus is given by

$$s = n_s \left(\sum_r k_r n_r + j_s \right) \qquad (4\text{-}6)$$

which is proportional to n_s. A rate proportional to n_s^2 would be obtained if the trace gas reacted mainly with itself, but no such case is known. In the process of integrating s over the tropospheric volume one may express the summation over individual reaction rates by a suitable tropospheric average rate:

$$S = \frac{M_s}{N_A} \int n_s \left(\sum_r k_r n_r + j_s \right) dV = \left(\overline{\sum_r (k_r n_r + j_s)} \right) \frac{M_s}{N_A} \int n_s \, dV$$

$$= k_V \frac{M_s}{M_{air}} \bar{m}_s \frac{M_{air}}{N_A} \int n_T \, dV = k_V G_s \qquad (4\text{-}7)$$

Thus, if s is proportional to n_s, the integrated sink strength is also proportional to G_s. It may happen that the number densities n_r depend, in part, on n_s due to chemical feedback processes, and when this occurs k_V will depend to some extent on G_s. We cannot pursue this problem here and

must assume that the influence of G_s on k_V is negligible to a first approximation. If this is so, one obtains the rather general budget equation

$$\frac{dG_s}{dt} = F_{in} + Q - (k_F + k_V)G_s \qquad (4\text{-}8)$$

in which all quantities are functions of time. It is useful to distinguish two
time scales: a short-term component such as diurnal or seasonal variations,
and a long-term component on a time scale greater than a year. The
short-term variations may be smoothed by averaging over a suitable time
period. If the long-term variation is small during this period, one has
approximately $\overline{dG_s/dt} \approx 0$; that is, G_s will be nearly constant. This leads to
the steady-state condition

$$\bar{F}_{in} + \bar{Q} - (\bar{k}_F + \bar{k}_V)G_s = \bar{Q}_T - \bar{S}_T \approx 0 \qquad (4\text{-}9)$$

For such conditions the budget is balanced, global source and sink strengths
have the same magnitude, and G_s is constant:

$$G_s = \frac{\bar{F}_{in} + \bar{Q}}{\bar{k}_F + \bar{k}_V} \qquad (4\text{-}10)$$

The mean residence time of the trace substance in the tropospheric reservoir
is

$$\tau = \frac{G_s}{\bar{F}_{in} + \bar{Q}} = \frac{G_s}{\bar{F}_{out} + \bar{S}} = \frac{1}{\bar{k}_F + \bar{k}_V} \qquad (4\text{-}11)$$

The two expressions on the left are independent of any assumptions concerning the functional relationship between $\bar{S}$ and G_s, but if there exists a
first-order dependence, τ is the inverse of $\bar{k}_F + \bar{k}_V$. In the literature τ is
frequently designated the "lifetime" of the substance in the troposphere.
This term implies that losses are due exclusively to chemical reactions within
the tropospheric reservoir; that is, it does not account for the transport of
material across the boundaries. The term residence time encompasses both
types of losses; hence it is preferrable. However, we shall retain the
expression lifetime when gas-phase reactions are the only loss processes.

If the residence time is longer than the short-term averaging period and
long-term variations are negligible, the system will always tend toward a
steady state. After a perturbation i.e., when any of the fluxes, sources, or
sinks take on a new value sufficiently rapidly, G_s gradually adjusts to the
new situation. The change of G_s with time after the event is obtained by
integration of Eq. (4-8), taking the new values of F_{in}, F_{out}, Q and S as
constant. The result is

$$G_s = G_s^0 \exp[-(\bar{k}_f + \bar{k}_V)t] + \frac{F_{in} + Q}{\bar{k}_F + \bar{k}_V}\{1 - \exp[-(\bar{k}_F + \bar{k}_V)t]\} \qquad (4\text{-}12)$$

The first term on the right describes the decay of the initial mass of G_s with time due to the consumption processes. The second term describes the rise of G_s with time due to the input of new material. A new steady state is reached after sufficient time has passed so that the exponential has decayed to small values. The time constant for the approach to steady state is identical to the residence time of the substance in the tropospheric reservoir. This equivalence holds only by virtue of the assumed proportionality of S_T with G_s and not in other cases.

Gradients of mixing ratios arise whenever the residence time of a trace substance in the troposphere is shorter than the time scale for transport by turbulent mixing. The longest mixing time in the troposphere is that associated with the exchange of air between the two hemispheres, which requires about 1 yr. A uniform abundance of a trace gas in both hemispheres thus implies a residence time much greater than 1 yr, provided, however, that both sources and sinks are not extremly evenly distributed. Alternatively, the existence of a latitudinal gradient of the mixing ratio within one hemisphere indicates a residence time less than 1 yr. In that case, the northern and southern hemispheres should be treated as two separate, coupled reservoirs, with the interhemispheric tropical convergence zone as a common boundary.

Similar arguments lead one to expect that, in general, residence time and spatial variation of mixing ratio are inversely related for any tropospheric constituent. Junge (1974) has explored the quantitative aspects of this relationship. Let $\bar{m}$ be the mixing ratio of a constituent averaged over space and time taking the time period long enough to eliminate daily and seasonal variations. The local values of m are then

$$m = \bar{m} + m'(x_i, t) \qquad i = 1, 2, 3 \qquad (4\text{-}13)$$

where $m'(x_i, t)$ is the momentary deviation from $\bar{m}$ as a function of the space coordinates x_i and time t. A suitable measure for the variability of $m'(x_i, t)$ is the relative standard deviation

$$\sigma_s = \sigma_s^*(m')/\bar{m} \qquad (4\text{-}14)$$

where $\sigma_s^*(m')$ is the absolute standard deviation. Junge (1974) has used available information on a number of atmospheric trace gases to estimate σ_s and τ. His results are shown in Fig. 4.1. The sizes of the boxes shown in the figure indicate the estimated uncertainties for τ and σ_s. The solid line indicates the approximate relationship $\sigma_s^* \tau = 0.14$. The fact that helium does not fall on the solid line is almost certainly due to the limitations of the measurement techniques required to establish the natural variability for helium. Oxygen probably is a borderline case. The scatter of the other data must be accepted as real, because the residence time is only one of the

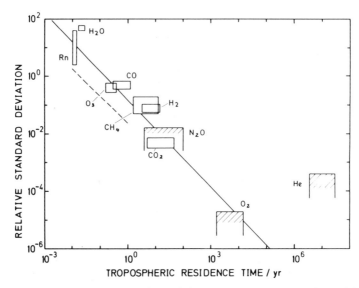

Fig. 4-1. Double-logarithmic plot of the relative standard deviation σ_s of the mixing ratio versus the residence time in the troposphere for several atmospheric constituents. [Adapted from Junge (1974).] The solid line shows the relation $\sigma \cdot \tau = 0.14$; the dashed line indicates results from model calculations.

factors having an influence on τ. Junge (1974) has rationalized the results of Fig. 4-1 by means of a steady-state model providing the distribution of a tracer in the air passing alternately over continents and oceans of equal widths (6000 km), the continents serving as source areas. The results are shown in Fig. 4-1 by the broken line. Its slope is almost identical to that derived from the empirical data, but the absolute values are smaller.

4.2 Hydroxyl Radicals in the Troposphere

The recognition that OH radicals might be important in the troposphere has come comparatively late and then by indirect reasoning. One problem was faced in early work directed at understanding the origin of photo-chemical smog by simulations in large photochemical reactors (smog chambers). The rapidity of hydrocarbon oxidation in the presence of nitric oxide and its concomitant rapid conversion to NO_2 had baffled researchers for almost two decades. In 1969, Weinstock (1971; see also Niki *et al.*, 1972) and Heicklen *et al.* (1971) independently suggested that the unexplained high rate of hydrocarbon consumption is due to reaction with OH radicals and that these are regenerated by a chain mechanism of the type

$$OH + CO \rightarrow CO_2 + H$$
$$H + O_2 + M \rightarrow HO_2 + M$$
$$HO_2 + NO \rightarrow NO_2 + OH$$

where M is a suitable third-body molecule. The last reaction, which continues the chain, is now known to be fast and the reaction scheme is generally accepted.

Another concurrent development has concerned the fate of carbon monoxide in the atmosphere. A survey of measurements (Pressman and Warneck, 1970) had shown that the CO mixing ratio in air had stayed fairly constant at about 0.1 ppmv during the period 1953–1969, despite a sizable increase in the rate of anthropogenic emissions, mainly from automobile exhaust. Weinstock (1969) considered the formation of ^{14}CO by cosmic radiation and utilized measurements of the $^{14}CO/^{12}CO$ ratio in air reported by McKay et al. (1963) to derive an order of magnitude estimate of 0.1 yr for the residence time of CO in the troposphere. Losses of CO to the stratosphere turned out insufficient as a sink for CO. The possible role of OH radicals as a sink in the troposphere gained ground as other conceivable oxidation reactions, such as that of CO with ozone (Arin and Warneck, 1972), were found too slow. The breakthrough came when Levy (1971) identified a route to the formation of OH radicals in the troposphere, namely, photolysis of ozone by ultraviolet radiation in the vicinity of 300 nm to produce $O(^1D)$ atoms, which subsequently react with water vapor

$$O_3 + h\nu \rightarrow O(^1D) + O_2(^1\Delta_g)$$
$$O(^1D) + N_2 \,(\text{or } O_2) \rightarrow O(^3P) + N_2 \,(\text{or } O_2)$$
$$O(^1D) + H_2O \rightarrow 2OH$$

A good deal of the $O(^1D)$ atoms are deactivated by collisions with N_2 or O_2 molecules, but the remainder generates enough OH radicals to provide a sink for CO of the required magnitude. In addition, they oxidize methane to formaldehyde and by photolysis of the aldehyde provide a source of CO. The actual presence of OH in ground-level, continental air during daylight was confirmed fairly unequivocally by its characteristic ultraviolet absorption feature using a long-path laser beam as the background light source (Perner et al., 1976; Hübler et al., 1982, 1984). Attempts to determine absolute concentrations by laser-induced fluorescence (Wang and Davis, 1974; Wang et al., 1981; Davis et al., 1976, 1979) have met with difficulties because the laser pulse generates additional OH via the photodissociation of ozone (Ortgies *et al.*, 1980; Davis *et al.*, 1981). Our knowledge of average OH concentrations in the troposphere thus relies mainly on model calculations and indirect determinations rather than direct observations.

The local primary OH production rate according to the above mechanism is

$$P_0(OH) = \frac{2k_{35}m(H_2O)j(^1D)n(O_3)}{k_{32}m(N_2) + k_{33}m(O_2) + k_{35}m(H_2O)} \tag{4-15}$$

Diurnally averaged values $\bar{P}_0(OH)$ are given in Table 4-1 as a function of latitude and height of the northern hemisphere. Production rates at noon are by about a factor of five higher (Warneck, 1975). The rates are greatest in the lower troposphere because of the highest H_2O mixing ratios there, and in the equatorial region where the solar UV flux is least attenuated by stratospheric ozone. These production rates are augmented by chain reactions in which OH is regenerated from HO_2 (see below).

Hydroxyl is unreactive toward the main atmospheric constituents N_2, O_2, H_2O, and CO_2, but reacts readily with many trace gases. Table 4-2 lists trace gases, their approximate mixing ratios, and rate coefficients for reaction with OH for three tropospheric conditions. The reactions with methane and

Table 4-1. *Diurnally Averaged Primary OH Production Rate (Not Including Amplification by Secondary Chain Reactions) as a Function of Latitude and Height in the Troposphere (Northern Hemisphere)[a]*

	Degrees latitude									
z (km)	0	10	20	30	40	50	60	70	80	90
					July					
0	716	677	1080	865	530	386	186	75.2	34.9	10.9
2	515	543	847	589	379	316	154	66.7	30.9	8.21
4	311	360	541	348	209	187	89.1	41.9	19.6	5.45
6	137	190	291	156	87.5	77.6	35.9	17.1	8.39	2.24
8	53.2	68.8	113	60.7	34.3	39.2	15.5	9.41	6.81	1.84
					January					
0	504	444	318	110	18.6	1.34	0.105	—	—	—
2	357	259	171	52.4	10.7	0.72	0.054	—	—	—
4	218	115	73.3	23.3	5.20	0.32	0.020	—	—	—
6	105	43.7	27.4	7.83	1.99	0.11	0.006	—	—	—
8	41.2	15.9	8.92	3.26	0.83	0.085	0.005	—	—	—

[a] The values were calculated according to Eq. (4-15) using observational data for $n(O_3)$, $m(H_2O)$, and solar radiation flux, as well as laboratory data for (O_3), $q(O^1D)$, and $O(^1D)$ rate coefficients as outlined by Warneck (1975). The numerical values are given in units of 10^3 molecules/cm³ s.

Table 4-2. *Removal of OH Radicals by Reactions with Trace Gases: Mixing Ratios, Rate Coefficients, and Relative Efficiencies for Various Tropospheric Constituents under Three Conditions: (A) Continental Boundary Layer ($n_M = 2.5 \times 10^{19}$ molecules/cm³, T = 288 K), (B) Marine Boundary Layer, or (C) Upper Troposphere, $z \approx 7$ km ($n_M = 1.2 \times 10^{19}$ molecules/cm³, T = 240 K)*[a]

Trace gas	m (ppbv)			k (cm³/molecules s)		Rate factor $k n_M m$			Percent of total		
	A	B	C	A, B	C	A	B	C	A	B	C
CH₄	1.5 (3)	1.5 (3)	1.5 (3)	7.7 (−15)	1.9 (−15)	2.8 (−1)	2.8 (−1)	3.4 (−2)	9.1	18	8.6
CO	2.5 (2)	1 (2)	1 (2)	2.7 (−13)	1.9 (−13)	1.7	9.9 (−1)	3.2 (−1)	55	65	81
H₂	6 (2)	5 (2)	5 (2)	7.5 (−15)	1.2 (−15)	1.1 (−1)	1.1 (−1)	7.9 (−3)	3.6	7.1	2.0
O₃	25	25	30	6.8 (−14)	3.2 (−14)	4.4 (−2)	4.4 (−2)	1.9 (−2)	1.4	2.9	4.8
NO₂	2	0.03	0.03	1.1 (−11)	5.5 (−12)	5.7 (−1)	8.1 (−3)	2.0 (−3)	18	0.5	0.5
SO₂	2	0.05	0.05	9.0 (−13)	8.0 (−13)	4.4 (−2)	1.1 (−3)	4.8 (−4)	1.4	0.1	0.1
NH₃	4	0.1	0.1	1.5 (−13)	8.2 (−14)	1.5 (−2)	3.6 (−4)	1.0 (−4)	0.5	—	0.1
HCHO	2	0.3	0.1	1.2 (−11)	1.2 (−11)	3.1 (−1)	1.0 (−1)	1.3 (−2)	10	6.5	3.3

[a] Numbers in parentheses indicate powers of ten.

carbon monoxide are dominant in the entire troposphere; nitrogen dioxide and formaldehyde are important in the continental boundary layer. The other trace gases are less significant with regard to losses of OH. The inverse of the sum of the individual reaction rates gives the lifetime of OH, which falls into the range 0.3–2.5 s. Before the concept of OH reactions gained ground, researchers thought that radicals would be quickly scavenged by aerosol particles so that gas-phase reactions would be less important. Warneck (1974) investigated this possibility and found that OH radicals are not thus affected. Collision frequencies of molecules with particles of all sizes fall into the range 9×10^{-4} to $6 \times 10^{-2}\,\mathrm{s}^{-1}$. The corresponding time constants (see Table 7-5) are 16 and 1150 s, respectively. While the aerosol obviously is not a good scavenger for OH in comparison with gas-phase reactions, other radicals with lesser gas-phase reactivities may suffer such losses. Thus it is necessary to treat this problem separately for each type of radical.

The reactions following the attack of OH on methane were first discussed by Levy (1971) and McConnel *et al.* (1971). They include

$$OH + CH_4 \rightarrow H_2O + CH_3$$

$$CH_3 + O_2 + M \rightarrow CH_3O_2 + M$$

$$CH_3O_2 + NO \rightarrow CH_3O + NO_2$$

$$CH_3O + O_2 \rightarrow HCHO + HO_2$$

Formaldehyde is the immediate stable product from this reaction sequence. In addition, one HO_2 radical is formed for each OH radical consumed. This is similar to the reaction of OH with CO discussed earlier. The reaction of OH with nitrogen dioxide

$$OH + NO_2 \xrightarrow{\ M\ } HNO_3$$

by contrast, does not produce HO_2 because it is an addition reaction leading to a stable product. The photodecomposition rate of HNO_3 in the troposphere is small, and according to present knowledge HNO_3 is mainly precipitated.

The photodissociation of formaldehyde or its reaction with OH constitutes another source of HO_2 radicals:

$$HCHO + h\nu \xrightarrow{a} H_2 + CO$$

$$\xrightarrow{b} HCO + H$$

$$HCHO + OH \rightarrow HCO + H_2O$$

$$HCO + O_2 \rightarrow HO_2 + CO$$

$$H + O_2 + M \rightarrow HO_2 + M$$

The relative quantum yield of HO_2 from formaldehyde, $2j_b(HCHO)/j(HCHO)$, is about 0.8 at ground level. Since photodecomposition is the dominant loss process for formaldehyde in the atmosphere, the oxidation of methane ultimately produces more than one HO_2 radical for each OH radical entering into reaction with methane, providing enough NO is present to convert all the CH_3O_2 radicals to formaldehyde. Losses of formaldehyde from the atmosphere due to in-cloud scavenging and wet precipitation amount to less than 15% of those caused by photolysis and reaction with OH radicals (Warneck *et al.*, 1978; Thompson, 1980).

The main reactions of HO_2 radicals that must be considered are

$$HO_2 + NO \rightarrow NO_2 + OH$$

$$HO_2 + O_3 \rightarrow 2O_2 + OH$$

$$HO_2 + HO_2 \rightarrow H_2O_2 + O_2$$

$$HO_2 + CH_3O_2 \rightarrow CH_3OOH + O_2$$

$$HO_2 + OH \rightarrow H_2O + O_2$$

Only the first two of these reactions convert HO_2 back to OH. The other reactions terminate the reaction chain. The fate of the hydroperoxides H_2O_2 and CH_3OOH in the atomosphere is not precisely known. Both substances may react with OH radicals or undergo photodecomposition (thereby creating new OH radicals). Compared with formaldehyde, however, the rates for these processes are smaller, whereas the solubilities of H_2O_2 and CH_3OOH in water are greater (cf. Table 8-2). This suggests that hydroperoxides are preferably lost by precipitation or reactions in aqueous solution.

The source of nitric oxide in the unpolluted troposphere is the photodissociation of nitrogen dioxide, which leads to the reactions

$$NO_2 + h\nu \rightarrow NO + O$$

$$O + O_2 + M \rightarrow O_3 + M$$

$$O_3 + NO \rightarrow NO_2 + O_2$$

$$HO_2 + NO \rightarrow NO_2 + OH$$

$$RO_2 + NO \rightarrow NO_2 + RO$$

The reactions with ozone, HO_2, CH_3O_2, and other RO_2 radicals cause NO to be rapidly reverted back to NO_2 so that steady-state conditions are set up for NO. The lifetimes of HO_2 and CH_3O_2 are quite sensitive to this stationary NO concentration. The tropospheric background level of NO_2 is of the order of 30 pptv in marine air, and higher over the continents. Measurements of NO in the free troposphere indicate daytime mixing ratios of about 10 pptv or less (see Table 9-12). In rural continental air, values of up to a few ppbv are encountered. The assumption of an NO mixing ratio of 10 pptv leads to lifetime estimates for HO_2 and CH_3O_2 of 500 s; increasing the NO mixing ratio to 500 pptv reduces the lifetime to about 10 s. The values are high enough to make losses of radicals by collisions with aerosol particles important, but this process is usually ignored.

The reactions presented above form the basis of a tropospheric chemistry model in which radicals adjust to the photostationary state within seconds, formaldehyde within hours, and hydroperoxides within days. Input parameters are the solar radiation flux and the concentrations of ozone, water vapor, methane, carbon monoxide, and nitrogen oxides. A two-dimensional analysis is called for, since most of the input data vary with latitude and height. The model yields the distribution of diurnal averages of OH number densities in the troposphere, its seasonal variation, and annual and spatial averages. Figure 4-2 shows as an example the seasonally averaged OH distribution according to the calculations of Crutzen (1982). It makes it again evident that the maximum OH concentrations occur near the equator. Table 4-3 summarizes a number of estimates for the globally and seasonally averaged OH concentration. The value for this quantity, $\bar{n}(OH) = 5 \times 10^5$ molecules/cm^3, is now known to within a factor of two. The OH concentration in the southern hemisphere is somewhat higher than that in the northern hemisphere due to the uneven latitudinal distribution of CO mixing ratios. The early estimates by Levy (1972, 1974), McConnel et al. (1971), and Wofsy et al. (1972), which were based on a one-dimensional analysis at midlatitudes of the northern hemisphere, gave much higher values mainly because the latitudinal dependence of the radiation field was

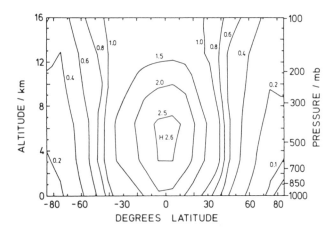

Fig. 4-2. Meridional cross section of annual average OH number densities in the troposphere, calculated with a two-dimensional model. [Adapted from Crutzen (1982).]

Table 4-3. *Annual Average of OH Number Densities (in Units of 10^5 molecules/cm^3) in the Troposphere as Derived from Two-Dimensional and Box Models*

Authors	Northern hemisphere	Southern hemisphere	Global	Remarks
Warneck (1975)	7	—	—	Based on observational data, primary OH production rate emphasized; major uncertainty, NO_x distribution
Crutzen and Fishman (1977)	2.6	7	4.8	Mainly based on observational data
Singh (1977b)	2.6	6.5	5.2	From production and distribution of CH_3CCl_3 and $CHCl_3$
Neely and Ploncka (1978)	4.8	18	11	From production and distribution of CH_3CCl_3
Chameides and Tan (1981)	2.9–7.6	2.1–8.1	3–8	Continental and marine regions considered separately
Volz et al. (1981a)	6.5	6.5	6.5	Based on ^{14}C and ^{12}C budgets
Derwent (1982)	6.3	5.7	6	Constrained by ^{14}CO, ^{12}CO, and halocarbon data
Crutzen and Gidel (1983)	4.9	6.3	5.5	NO_x distribution derived from budget considerations
Singh et al. (1983a)	—	—	5.2 ± 1.7	From production and distribution data for CH_2Cl_2, CH_3CCl_3, and C_2Cl_4
Prinn et al. (1983b)	—	—	5 ± 2	From production and trend data for CH_3CCl_3

neglected, but also because the adopted NO_2 concentrations were too high. Even today, the NO_2/NO concentration field is poorly known. Crutzen and Gidel (1983) thus calculated it from the source strengths of nitrogen oxides due to industrial inputs, production by lightning, etc., and appropriate loss processes.

In addition to model computations, the average OH concentration can be derived independently from tracer measurements, provided the source strength of the tracer gas is known and reaction with OH is the dominant sink. The associated rate coefficient must be known and the abundance of the tracer should be small enough to preclude any interference with the major routes of tropospheric OH chemistry. Singh (1977a) has pointed out that the halocarbons CH_3CCl_3 and $CHCl_3$ are useful as tracers because they are entirely human-made and their source strengths can be estimated from industrial production data. Singh (1977a) represented the source data for the period 1955–1975 by an exponential growth function $Q_s(t) = a \exp(bt)$. It is then simple to integrate the budget equation [Eq. (4-8)] with the result

$$G_s(t) = \frac{a\tau}{b\tau + 1} \exp(bt)\{1 - \exp[-(b + 1/\tau)t]\} \qquad (4\text{-}16)$$

where τ is the residence time of the tracer in the troposphere. It is convenient to introduce the ratio R_s of the total mass of the tracer in the troposphere to the cumulative mass emitted by the source:

$$\Delta G_s(t) = \int_0^t Q_s(t) \, dt = (a/b)[\exp(bt) - 1] \qquad (4\text{-}17)$$

For sufficiently large times t the ratio approximates to

$$R_s = G_s(t)/\Delta G_s(t) = b\tau/(b\tau + 1) \qquad (4\text{-}18)$$

and one obtains for the residence time

$$\tau = [3.15 \times 10^7 \bar{k}\bar{n}(OH)]^{-1} = R_s/b(1 - R_s) \qquad (4\text{-}19)$$

where the factor 3.15×10^7 is the number of seconds per year. From measurements of the mixing ratios of CH_3CCl_3 and $CHCl_3$ performed in 1976 in both hemispheres, and from emission estimates that had become available at the same time, Singh (1977a) derived the data assembled in Table 4-4. The calculation of $\bar{n}(OH)$ from the residence time requires a suitable choice of temperature to determine the effective rate coefficient $\bar{k}$. OH number density, trace gas concentration, and temperature all decline with altitude, so that the reaction rate is heavily weighted in favor of the lower troposphere. A temperature of 280 K is probably most appropriate. The corresponding values for the rate coefficients are indicated in Table 4-4. Singh (1977a)

Table 4-4. *Emission Growth Factor b, Cumulative Emissions to the Atmosphere in 1976 ΔG_s, Total Mass of Trace Constituent Still Present in the Troposphere in 1976 G_s, Ratio $G_s/\Delta G_s$, Residence Times, and Average OH Number Density Deduced from Budgets of CH_3CCl_3 and $CHCl_3$ According to Singh (1977a, b)*

Trace gas	b (yr^{-1})	ΔG_s (Tg)	G_s (Tg)	R_s	τ (yr)	$\bar{n}$(OH) (10^5 molecules/cm^3)	
CH_3CCl_3	0.166	3.0	1.76	0.58	8.3	4.7^b	(5.1)^c
$CHCl_3$	0.121	1.1^a	0.2	0.15	1.7	2.3^b	(3.0)^c

^{a} Less certain than the emission for CH_3CCl_3.
^{b} $k(CH_3CCl_3) = 8.1 \times 10^{-15}$, $k(CHCl_3) = 8.2 \times 10^{-14}$ cm^3/molecules s for $T = 280$ K.
^{c} As given by Singh (1977a).

used an average tropospheric temperature of 265 K, which is too low, but with the preliminary data for the rate coefficients available to him he obtained k values similar to those given in Table 4-4. The average OH concentrations derived in this manner compare favorably with the results of model calculations shown in Table 4-3.

The mixing ratio of CH_3CCl_3 is a factor of 1.43 greater in the northern hemisphere compared with the southern. Singh (1977b) has used this difference to estimate the average OH number densities separately for each hemisphere. For this purpose it is necessary to consider two coupled budget equations, one for each hemisphere:

$$dG_{sn}/dt = a \exp(bt) - (1/\tau_n)G_{sn} - (1/\tau_e)(G_{sn} - G_{ss})$$
$$dG_{ss}/dt = -(1/\tau_s)G_{ss} + (1/\tau_e)(G_{sn} - G_{ss}) \tag{4-20}$$

where it is assumed that the anthropogenic emissions occur primarily in the northern hemisphere. The terms τ_n and τ_s are the residence times of CH_3CCl_3 resulting from reaction with OH in the northern and southern hemispheres, respectively, and τ_e is the time constant for the exchange of air masses between the two hemispheres. As is usual for such equations, the solutions are series of exponential functions of time, but similar to the simpler case of a unified tropospheric reservoir treated above the solutions for sufficiently long times t, starting from $G_{sn} = G_{ss} = 0$ at $t = 0$, yield expressions for the residence times that are

$$\tau_s = \tau_e/[(G_{sn}/G_{ss}) - (1 + b\tau_e)]$$
$$\tau_n = R_s(G_{sn}/G_{ss})/[b(1 - R_s)(1 + G_{sn}/G_{ss}) - R_s/\tau_s] \tag{4-21}$$

Here R_s is again the total mass of methyl chloroform present in the troposphere divided by the cumulative emission up to time t. The residence time in the southern hemisphere depends on the interhemispheric exchange time,

Table 4-5. *Hemispheric and Average Tropospheric Residence Times and Corresponding Average OH Number Densities Obtained from the Detailed Budget of CH_3CCl_3 (Singh, 1977b)*

R_s	τ_n (yr)	τ_s (yr)	τ (yr)	$\bar{n}_n(OH)$	$\bar{n}_s(OH)$	$\bar{n}_s(OH)/\bar{n}_n(OH)$
0.522	7.5	5.6	6.6	4.4	5.9	1.3
0.58	12.6	5.6	8.3	2.6	5.9	2.2
0.638	28.2	5.6	10.6	1.2	5.9	5.0

which is fairly well known (see Table 1-9). Singh (1977b) used the value $\tau_e = 1.25 \pm 0.08$ yr. The values that one obtains for τ_n are rather sensitive to the ratio R_s, so that this quantity must be evaluated as precisely as possible if the results are to be meaningful. The initial data reported by Singh (1977a) were not adequate for such a purpose. The later data given by Singh (1977b) for $G_s = G_{sn} + G_{ss}$ and ΔG_s are shown in Table 4-5. They yield $R_s = 0.58$, but there should be no doubt that the errors are appreciable. Rather than varying τ_e as Singh has done, one should consider a range of values for R_s keeping τ_e constant at $\tau_e = 1.25$. Table 4-5 shows individual residence times for $R_s = 0.58$ plus or minus 10%, as well as the average tropospheric residence time defined by

$$\tau = (1 + G_{sn}/G_{ss})(1/\tau_s) + (1/\tau_n)(G_{sn}/G_{ss}) \tag{4-22}$$

Table 4-5 shows also the corresponding average OH number densities and their ratios for the southern and northern hemispheres. It is evident that a moderate change of R_s leads to an appreciable change in the values of τ_n and $\bar{n}(OH)$. The average tropospheric residence time is much less affected by such changes. Neely and Ploncka (1978) have subsequently extended the analysis of the two-box model for methyl chloroform using an updated emissions inventory and incorporating losses to the stratosphere and the ocean. Their results are included in Table 4-3, in addition to two later evaluations by Singh *et al.* (1983a) and Prinn *et al.* (1983b) based on more recent measurements. Neely and Ploncka's average OH concentrations are much higher than all the others. In the following, a globally averaged OH concentration of 5×10^5 molecules/cm^3 will be used for the troposphere whenever lifetimes of gases due to reaction with OH are discussed.

4.3 The Budget of Methane, CH$_4$

The discovery of methane in the atmosphere is due to Migeotte (1948). He observed the characteristic absorption features of CH_4 in the infrared spectrum of the sun and estimated a mixing ratio of 1.5 ppmv. The majority of all recent measurements since the late 1960s have employed gas chromatographic techniques. Wilkness *et al.* (1973), who first explored the latitudinal distribution of methane in the marine atmosphere of the Pacific Ocean, found an almost uniform distribution of 1.4 ppmv, with a small excess of about 6% in the northern hemisphere. This difference has been confirmed by others, and Table 4-6 summarizes the results.

When Ehhalt (1974) reviewed the early measurements, he concluded that methane was not only evenly distributed in the troposphere, but its mixing ratio also had stayed constant with time. This viewpoint now requires revision. Improvements in the precision of measuring techniques and a more systematic monitoring of atmospheric methane since 1978 have shown that its mixing ratio is rising at a rate of about 20 ppbv per year (Rasmussen and Khalil, 1981a; Fraser *et al.*, 1981, 1984; Blake *et al.*, 1982) and now approaches a value of 1.7 ppmv in the northern hemisphere. The long-term trend of atmospheric methane has been explored by the analysis of air incorporated into the great ice sheets of Greenland and Antarctica. Figure 4-3 shows the results of three separate studies. The individual data indicate a CH$_4$ mixing ratio of about 0.7 ppmv during the period before 1700 A.D. and an approximately exponential increase during the last 300 yr. Khalil and Rasmussen (1985) have pointed out that the increase parallels the growth of human population, which suggests that anthropogenic activities are responsible for the trend.

A clue about the origin of atmospheric methane can be obtained from measurements of its radiocarbon content. Carbon-14 is produced continuously in the atmosphere by the interaction of cosmic ray–produced neutrons with nitrogen (Lal and Peters, 1967). The halflife of ^{14}C due to beta decay is about 5700 yr. Methane originating from biological processes must have a ^{14}C content close to that of recent wood, whereas methane derived from fossil fuels, volcanic activity, and other ancient sources contains practically no radiocarbon. Enhalt (1974) and Ehhalt and Schmidt

Table 4-6. *Recent Measurements of the Average Tropospheric Mixing Ratio $\bar{m}$ of Methane and the Ratio $\bar{m}_n / \bar{m}_s$ for the Mixing Ratios in the Northern and Southern Hemispheres*

Authors	$\bar{m}$	$\bar{m}_n / \bar{m}_s$
Wilkness *et al.* (1973)	1.40	1.06
Ehhalt and Schmidt (1978)	1.35	1.06
Mayer *et al.* (1982)	1.54	1.05
Singh *et al.* (1979)	1.41	1.03
Heidt *et al.* (1980)	1.65	1.04
Rasmussen and Khalil (1982)	1.54	1.04
Khalisl and Rasmussen (1983)	1.57	1.07

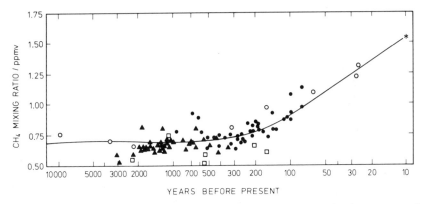

Fig. 4-3. Long-term trend of methane in air bubbles trapped in the great ice sheets, corrected for firn closure time. Squares are results of Robbins *et al.* (1973) from Greenland and the Antarctic; open circles are results of Craig and Chou (1982) from the Antarctic; filled circles and triangles are data of Rasmussen and Khalil (1984) from Greenland and the South Pole, respectively. The asterisk indicates an average global value for the late 1970s.

(1978) have assembled data for the ^{14}C content of methane collected from air liquefaction plants. Most of the samples were taken before the atmosphere became contaminated with ^{14}C from large-scale nuclear weapons tests (see Fig. 11-5). The average of these data indicate that before 1960 the ^{14}C content of atmospheric methane was $83 \pm 13\%$ of that of recent wood. This figure does not yet take into account the isotope separation in the biological release of methane in favor of the lighter isotope ^{12}C. The fractionation can be estimated from the depletion of ^{13}C in methane derived from bacterial fermentation (Rosenfeld and Silverman, 1959; Dana and Deevy, 1960). For ^{14}C it amounts to about 7%, so that $(0.07 \times 83) + 83 = 89\%$ of CH_4 in the atmosphere may be taken to arise biogenically (Ehhalt, 1979). In reality this is still a lower limit, because air liquefaction plants usually are located in heavily industrialized regions, which are contaminated by air pollution.

The production of methane in nature occurs by the bacterial degradation of organic matter, primarily plant residues in the anaerobic sludges or sediments of lakes and swamps, but also in fermentation tanks of sewage disposal and in the rumen of cattle. Figure 4-4 illustrates the situation for methane production at the bottom of a lake. Methanogenic bacteria belong to a small group of strict anaerobes living in symbiosis with other bacteria, which derive their energy needs from the fermentation of cellulose and other organic material. By disproportionation the material is broken down into carbon dioxide, alcohols, fatty acids, and hydrogen. A small group of methanogens produces methane from CO_2 and H_2. Other bacteria utilize

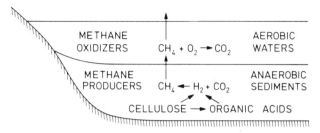

Fig. 4-4. Methane production and consumption in lake sediments and waters. [Adapted from Schlegel (1974).]

methanol, acetic acid, or butyric acid. The resulting methane escapes the anaerobic sediments and enters into the overlying lake waters. As soon as the environment turns aerobic, methane becomes subject to oxidation by over a hundred different bacteria, most of which are obligately bound to methane as a substrate. Methane obviously can rise to the lake surface and enter the atmosphere only if the lake is sufficiently shallow. Otherwise there is a complete consumption by methane oxidizers. In freshwater lakes with depths greater than 10 m, the release of methane becomes negligible. Most important as sources of atmospheric methane are the shallow waters of swamps, marshes, and rice-paddy fields.

The early estimates of CH$_4$ production from such sources were based on a laboratory study of Koyama (1963, 1964), who measured the release rates from nine Japanese paddy soils mixed with water and incubated at different temperatures. Figure 4-5 shows that the production rate increases approximately with the square of the temperature. More recently, Baker-Blocker *et al.* (1977) studied the escape of methane from swampy wetlands. A portion of their data is included in Fig. 4-5 to show that they agree with those of Koyama despite the difference in the experimental techniques. From statistical data about the area of rice-paddy fields in Japan (2.9×10^4 km^2), the average depth of paddy soils (15 cm), the annual average of the temperature (16°C), and the fact that Japanese rice paddies are water-logged for 4 months of the year, Koyama (1963) estimated an annual production rate for methane of 80 g/m^2 yr. In a similar manner, he derived a rate of 0.44 g/m^2 yr for upland fields and grasslands, and of 8×10^{-3} g/m^2 yr for forest soils. These data led to world production estimates of 1900 Tg/yr from rice paddies 0.9×10^6 km^2), 10 Tg/yr from other cultivated soils (30×10^6 km^2), and 0.4 Tg/yr from forest areas (44×10^6 km^2).

In reviewing these estimates, Ehhalt (1974) noted in an increase in the global area of rice-paddy fields to 1.35×10^6 km^2, according to United Nations statistics in 1970. In addition, he adopted a global area for swamps and marshes of 2.6×10^6 km^2 from Twenhofel (1951) to estimate a CH$_4$

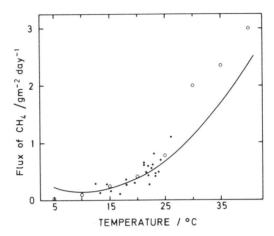

Fig. 4-5. Methane emission rates measured as a function of temperature: open circles, from incubated paddy soils (Koyama, 1963); filled circles, from wetlands (Baker-Blocker *et al.*, 1977). The solid curve $F = 0.46 - 0.00056\,T + 0.0026\,T^2$ g/m^2 day is the best fit to the observational data according to Baker-Blocker *et al.* (1977).

production from this source of 130–260 Tg/yr. He further pointed out that water-logged tundra may be a source of methane, and he provided an estimate of 1.3–13 Tg/yr from this source. Freshwater lakes and forests are negligible contributors by comparison.

Recently, Cicerone *et al.* (1983), Cicerone and Shetter (1981), and Seiler *et al.* (1984a) have studied the production of methane from rice paddies under actual field conditions using the flux box method. This technique employs a closed-off collector box that is placed over part of the field. The rise of the CH$_4$ mixing ratio in the air inside the box is then monitored. Rice fields in California, Italy, and Spain were investigated. The studies revealed that the release of methane proceeds primarily by diffusive transport via the stems of the rice plants, not so much through the open water space. The direct emission by rising gas bubbles occurs only in the absence of plants. The release rates showed a strong seaonal variation with maximum values toward the end of the flowering stage. Diurnal variations also were observed, presumably due to the diurnal temperature cycle. The average release rates during the growing period until harvest were 0.1–0.25 g CH$_4$/m^2 day. On the basis of these data, a growth period of 120–150 days per year, and a world harvest area of 1.45×10^6 km^2, Seiler *et al.* (1984a) estimated a CH$_4$ production rate from rice paddy fields of 35–59 Tg/yr globally. This value is much lower than the earlier estimates.

The surface waters of the oceans are slightly supersaturated with dissolved methane compared with the concentrations expected from the CH$_4$ mixing ratio in the atmosphere (Swinnerton and Linnenboom, 1967; Swinnerton *et al.*, 1969; Lamontagne *et al.*, 1974). Accordingly, there exists a concentration gradient that gives rise to a flux of methane from the ocean into the atmosphere. The flux can be estimated from the stagnant film model discussed in Section 1.6. This model assumes that the transport rate is determined by molecular diffusion across the ocean–air liquid interface (Broecker and Peng, 1971, 1974; Peng *et al.*, 1979). The local flux is given by

$$F = D_L(c_L - c_0)/\Delta z = D_L c_0[(c_L/c_0)-1]/\Delta z \qquad \text{in kg/m}^2 \text{ s}$$

where $D_L = 2.5 \times 10^{-9}$ m^2/s is the diffusion coefficient, $\Delta z = 35$ μm is the film thickness, c_L and c_0 are the actual and the equilibrium aqueous concentrations of methane, respectively, and $c_L/c_0 = 1.35$ is the measured saturation ratio. The equilibrium concentration is given by Henry's law (see Section 8.4.1). The value is $c_0 = 0.02 \, m(CH_4) = 3 \times 10^{-8}$ kg/m^3 at 20°C. With these data, Ehhalt (1974) derived a flux of methane from the open ocean (3.61×10^{14} m^2) of 4–7 Tg/yr, whereas Seiler and Schmidt (1974b) obtained 16 Tg/yr. The difference arises from uncertainties regarding the film thickness, which varies somewhat with agitation of the sea's surface and thus depends on the wind speed. In view of the fact that methanogenic bacteria are strict anaerobes, it is somewhat surprising to find an excess of methane in the aerobic surface waters of the ocean. Pockets of anaerobic regions must exist in or around organic particles undergoing bacterial degradation in order that methane can be produced in this environment.

Finally, we discuss the release of methane to the atmosphere from domestic herbivores. Enteric fermentation as a source of methane has been recognized relatively early, and Hutchinson (1948, 1954) provided estimates for the associated daily production rates. His values for cattle, horses, sheep, and goats were 200, 106, 15, and 15, respectively, in units of grams per head per day. From these data Hutchinson derived a global CH$_4$ production of 45 Tg/yr. Ehhalt (1974) used 1970 United Nations statistics that indicated an increase in the number of heads to, in millions, 1198 cattle, 125 horses, 1026 sheep, and 384 goats. The corresponding global production rate is 100 Tg CH$_4$/yr. This figure has been adopted in most subsequent budget estimates.

Methane is also produced in the digestive tract of herbivorous insects (e.g., termites) by symbiotic microorganisms. Termites as a source of methane have been studied by Zimmerman *et al.* (1982) and by Rasmussen and Khalil (1983) in the laboratory, and by Seiler *et al.* (1984b) in the fields of South Africa. Relative production rates of CH$_4$ versus CO$_2$ ranged from 7×10^{-3}% up to 1%, depending on the termite species. The lowest ratios were observed for fungus-growing termites, and the highest for grass-feeding termites. The average value is about 0.4%. Although the observations are consistent, the investigators came to different conclusions with regard to the global CH_4 release rate. Zimmerman et al. (1982) suggested 150 Tg/yr, Rasmussen and Khalil (1983) 10-100 Tg/yr, and Seiler et al. (1984b) 2-5 Tg/yr. The discrepancy arises from differing assumptions about the consumption and utilization of biomass by the total world population of termites. Zimmerman et al. (1982) estimated that termites consume 33 Pg of biomass per year (1 Pg = 10^{15} g), that it is almost entirely converted to CO_2, and that the CH_4/CO_2 ratio is 0.77% on a carbon-to-carbon basis. Biomass consumption that high would amount to 50% of total annual biomass production (Section 11.2.4). This appears questionable. Collins and Wood (1984) calculated a consumption of 7.3 Pg/yr instead, which corresponds to about 10% of biomass production. Seiler et al (1984b) pointed out that the CO_2 released from termite nests represents only a fraction of total carbon ingested, about 30% for grass-feeding termites. This lowers the average ratio of methane produced per biomass consumed to $(4.8-12.4) \times 10^{-4}$. A carbon content of biomass of 60% was adopted. The resulting global CH_4 release rate of 2-5 Tg/yr makes a relatively small contribution to the total methane budget.

Anthropogenic sources of methane comprise emissions from coal and lignite mining fields, chemical and petroleum industries, and automobile exhaust, and losses of methane from natural-gas wells. Estimates for the corresponding source strengths are shown in Table 4-7. A good statistical estimate for the leakage of methane from natural-gas consumption is not available. The current global consumption rate is about 1 Pg/yr. A leakage rate of 2% would yield 20 Tg/yr. The total anthropogenic contribution should not exceed the limit set by the ^{14}C content of atmospheric methane, about 10% of natural emissions, that is, 50 Tg/yr.

Table 4-7. *Anthropogenic and Other Nonbiogenic Sources of Methane and Global Strength Estimates*

Source	Global production rate Tg/yr
Coal mining	20^a, $6.3-22^b$
Lignite mining	$1.6-5.7^b$
Automobile exhaust	0.5^b, 35^c, 25^d
Industrial losses	$7-21^b$, 14^c
Volcanoes	0.2^b

[a] Koyama (1963).
[b] Hitchcock and Wechsler (1972).
[c] Robinson and Robbins (1968b).
[d] Junge and Warneck (1979).

An anthropogenic source unrelated to the others is the generation of methane by incomplete combustion in forest fires, the burning of agricultural wastes, and the clearance of forest and grass lands by fire for purposes of agriculture. The last operation is particularly important in the slash, burn, and shift agricultural practices of the tropics. The associated emission rates for various trace gases relative to that of carbon dioxide have been discussed by Crutzen et al. (1979). For methane the volume ratio CH$_4$/CO$_2$ ranges from 1 to 2.3%. The annual consumption of biomass by burning is of the order of 2 Pg C (cf. Table 11-10). These data led Crutzen et al. to estimate a CH$_4$ production from this source of 30–110 Tg/yr.

Table 4-8 summarizes methane emission rates as estimated by different authors. Most source estimates sum to a total of about 500 Tg/yr. An exception is the value derived by Sheppard et al. (1982) from a classification of natural ecosystems similar to that described in Section 11.2.4 for the turnover of biogenic carbon. Even this estimate, however, agrees with the others within a factor of 2–3. We think that is too high, mainly because it cannot be compensated by the known methane sinks, which are discussed next.

Two sinks for methane in the troposphere are fairly well known. The major one is reaction with OH radicals as discussed in Section 4.2, and the other sink is a loss of methane to the stratosphere. According to Eq. (4-7),

Table 4-8. *Summary of CH$_4$ Emission Rates (Tg/yr) from Individual Sources*

Type of source	Ehhalt (1974)	Sheppard et al. (1982)	Crutzen (1983)	Khalil and Rasmussen (1983)	Seiler (1984)
Domestic animals	101-220	90	60	120	72-99
Rice paddy fields	280	39	30-60	95	30-75
Swamps/marshes	130-260	39	30-220	150	13-57
Ocean/lakes	5.9-45	65	—	23	1-7
Other biogenic	—	817[a]	150[b]	100[c]	6-15
Biomass burning	—	60	30-110	25	53-97
Natural gas leakage		50	20	—	18-29
Coal mining	15.6-49.4	-	—	40	30
Other nonbiogenic	—	50[d]	—	—	1-2[e]
Total source strength	533-854	1210	170-620	553	225-395

[a] From considerations of biomass turnover in natural ecosystems.

[b] Production by termites.

[c] Includes 88 Tg/yr from termites, 12 Tg/yr from tundra.

[d] Fossil-fuel sources other than gas leakage.

[e] Fossil-fuel combustion.

the global sink strength for destruction by OH is

$$S_{OH} = 3.15 \times 10^7 \bar{k}_{72} \bar{n}(OH) G_{CH_4}$$

where the numerical factor converts seconds to years and

$$G_{CH_4} = \bar{m}(CH_4)(M_{CH_4}/M_{air}) G_T = 3700 \quad Tg$$

Since the spatial distribution of OH and temperature favors a destruction of methane in the lower troposphere, an average rate coefficient of $5 \times 10^{-15} \, cm^3/molecules$ may be applied, corresponding to an effective temperature of 280 K. With $\bar{n}(OH) = 5 \times 10^5$ molecules/cm^3, one then obtains $S_{OH} = 300$ Tg/yr. Crutzen and Fishman (1977) used a more detailed integration procedure to obtain $S_{OH} = 334$ Tg/yr.

The flux of methane into the stratosphere is determined by eddy diffusion in the lower stratosphere. Here, the vertical profile of CH_4 mixing ratios may be approximated by an exponential decrease with altitude

$$m(z) = m_{tr} \exp(-z/h)$$

where m_{tr} denotes the mixing ratio at the tropopause and h is an empirical scale height. At middle and high latitudes $h \approx 25$ km, whereas at tropical latitudes $h \approx 50$ km. The local upward-directed flux is given by

$$F_{CH_4} = -K_z n_T (dm/dz)|_{tropopause}$$

$$= -K_z \rho_{tr}(M_{CH_4}/M_{air}) m_{tr}/h$$

Here, n_T is the number density of air molecules at the tropopause and ρ_{tr} the corresponding air density. These data may be taken from Table 1-5. Values for the eddy diffusion coefficients in the lower stratosphere are 1 m^2/s in the equatorial updraft region and 0.3 m^2/s at higher latitudes. The second of the above equations gives the flux in units of kg/m^2 s. The global flux is obtained after converting from seconds to years and integrating over the Earth's surface area. This yields loss rates for the tropical and extratropical latitudes of 29 and 31 Tg/yr, respectively.

The total flux of methane into the stratosphere, 60 Tg/yr, combines with the rate of CH_4 oxidation by OH to give a total sink strength of 400 Tg/yr. This agrees approximately with the global emission estimates of Table 4-7. Although methane has been observed to increase in the troposphere, the budget must be about balanced since the increase is slow. If we take a value of 400 Tg/yr as representative for $Q_{CH_4} \approx S_{CH_4}$, the tropospheric residence time for methane is

$$\tau_{CH_4} = G_{CH_4}/S_{CH_4} = 3700/400 = 9 \quad yr$$

Between 50 and 90% of the emissions listed in Table 4-7 are human-made, and the largest of these are associated with agriculture, namely, the growing of rice, the raising of cattle, and the burning of biomass for purposes of land clearance. Statistics indicate that these activities have increased from 1940 to 1980 by factors of 1.8 for the growing of rice, 1.6 for raising cattle, and 1.6 for the burning of biomass. During the same period the area of swamps has decreased by 40%, whereas the production of natural gas has increased 16-fold (Seiler, 1984). Thus, it appears that the current upward trend of atmospheric methane is due to human activities.

4.4 Formaldehyde, HCHO

This compound is the first relatively stable product resulting from the oxidation of methane. In the marine atmosphere methane is expected to be the major precursor of formaldehyde, whereas over the continents multiple other sources exist. Among them are direct emissions from industries, automotive exhaust, stationary combustion, biomass burning, and secondary formation of HCHO by the photooxidation of nonmethane hydrocarbons, both natural and human-made. The great number of sources makes a critical assessment difficult. In the atmosphere, formaldehyde undergoes photo-decomposition to form CO so that both budgets are coupled. Uncertainties in the continental budget of formaldehyde thus carry over into the budget of CO. The following brief discussion of formaldehyde deals primarily with the marine atmosphere.

Wet-chemical analysis and optical methods have been used to determine the abundance of formaldehyde in ambient air. Table 4-9 shows mixing ratios observed at various locations. The highest values are associated with urban air, especially under conditions of photochemical smog. Maximum values may then reach 70–100 ppbv. Mixing ratios in rural areas are of the order of a few ppbv, and still lower values are found in marine air masses.

Lowe and Schmidt (1983) have measured formaldehyde over the Atlantic Ocean as a function of latitude. Their results are shown in Fig. 4-6. In the region 33°S–40°N the data scatter around a mean mixing ratio of $nm(HCHO) = 0.22$ ppbv. Further northward they decline. The lifetime of formaldehyde due to photodecomposition by sunlight is about 5 h at midlatitudes, and 15 h when averaged over a full day (Warneck et al., 1978). In addition, formaldehyde reacts with OH. Where the oxidation of methane is the sole source, formaldehyde mixing ratios are expected to be in steady state with methane $[dm(HCHO)/dt = 0]$:

$$m(HCHO) = \frac{k_{72}\bar{n}(OH)m(CH_4)}{j_{HCHO} + k_{80}\bar{n}(OH)}$$

$$= \frac{(7 \times 10^{-15})(5 \times 10^5)(1.5 \times 10^{-6})}{(1.8 \times 10^{-5}) + (1.5 \times 10^{-11})(5 \times 10^5)} = 0.2 \quad \text{ppbv}$$

Table 4-9. *Mixing Ratios of Formaldehyde Observed at Urban, Rural Continental, and Maritime Locations*

Authors	Location	m (ppbv)
	Urban air	
Altshuller and McPherson (1963)	Los Angeles, California	150 (Highest maximum)
Cleveland *et al.* (1977)	New York City	2 Average, 10 maximum
	Bayonne, New Jersey	3 Average, 20 maximum
Fushimi and Miyake (1980)	Tokyo, Japan	1–24
Tuazon *et al.* (1981)	Claremont, California	16–49 (Daily average), 23–71 (daily maximum)
Kuwata *et al.* (1983)	Osaka, Japan	1.6–8.5 (34 Highest maximum)
Grosjean *et al.* (1983)	Los Angeles, California	2–40
	Claremont, California	3–48
Lipari *et al.* (1984)	Warren, Michigan	1.3–6.5
	Rural air	
Platt *et al.* (1979), Platt and Perner (1980)	Jülich, Federal Republic of Germany	0.1–6.5
Fushimi and Miyake (1980)	Mt. Norikura, Japan	1–4
Lowe *et al.* (1980)	Eifel region, Federal Republic of Germany	0.03–5
Neitzert and Seiler (1981)	Mainz-Finthen, Federal Republic of Germany	0.7–5.1
Schubert *et al.* (1984)	Deuselbach, Federal Republic of Germany	2.3–3.9
	Jülich, Federal Republic of Germany	0.9–5
	Marine air	
Platt and Perner (1980)	Dagebüll, Federal Republic of Germany	0.2–0.5
	Loop Head, Ireland	0.3
Fushimi and Miyake (1980)	Pacific Ocean	0.2–0.8
	Indian Ocean	0.8–11
Zafiriou *et al.* (1980)	Enewetal Atoll, Pacific	0.3–0.6
Lowe *et al.* (1980)	Irish west coast	0.1–0.42
Neitzert and Seiler (1981)	Cape Point, South Africa	0.2–1.0
Lowe and Schmidt (1983)	Central Atlantic Ocean	0.12–0.33

Comparison with Fig. 4-6 shows that the steady-state prediction is in very good agreement with the observational data. This justifies the assumption that methane is the dominant source of formaldehyde in the marine atmosphere. Figure 4-6 includes the results from a two-dimensional model calculation (Derwent, 1982). The values are somewhat higher than the measured ones, but they confirm the constancy with latitude in the equatorial region. The decline of $m(\text{HCHO})$ toward high latitudes, which is evident

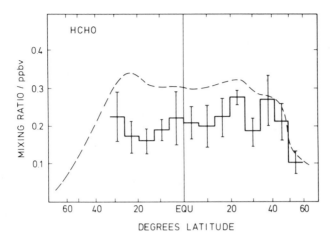

Fig. 4-6. Distribution with latitude of formaldehyde over the Atlantic Ocean; vertical bars indicate two-sigma variances. [Adapted from Lowe and Schmidt (1983).] The dashed curve shows the results of two-dimensional model calculations (Derwent, 1982).

in both model and measurements, arises from the fact that the average OH concentration decreases faster toward the poles than the HCHO photodissociation coefficient. The reason is the increasing optical density of the ozone layer.

The photochemical lifetime of formaldehyde is short enough to allow some diurnal variation of its mixing ratio. Lowe and Schmidt (1983) have found that such diurnal variations do occur indeed, although not to the extent predicted by model calculations. The model on which the prediction was based included losses of HCHO due to dry decomposition onto the ocean surface. This process was not considered in the above equation because it is less important than daytime photochemistry, but it would continue during the night, thereby causing the HCHO mixing ratios to decrease below the steady-state value. In seawater, formaldehyde is biologically consumed so that the ocean is not a source of it (Thompson, 1980).

In the free troposphere the steady-state HCHO mixing ratios are expected to decline with increasing height. On one hand, the OH production rate decreases with height due to the falloff in the H_2O mixing ratio; on the other hand, the rate coefficient for the reaction of OH with methane decreases as the temperature declines. In addition, the HCHO photodissociation coefficient increases because of the pressure dependence of the quantum yield for dissociation (see Fig. 2-19). Exploratory measurements by Lowe et al. (1980) over western Europe have confirmed the expected trend. The HCHO mixing ratio was found to decline from high values near 1 ppbv at the ground level to values less than 0.1 ppbv at 5 km altitude. Values in the lower pptv range are estimated for formaldehyde mixing ratios in the upper troposphere.

4.5 Carbon Monoxide, CO

Like many other gases in the atmosphere, carbon monoxide was discovered as a terrestrial absorption feature in the infrared solar spectrum (Migeotte, 1949). A variety of sporadic measurements until 1968, reviewed by Pressman and Warneck (1970), established a CO mixing ratio of the order of 0.1 ppmv. Robinson and Robbins (1968a) and Seiler and Junge (1970) then developed a continuous registration technique for CO in air based on the reduction of hot mercury oxide to Hg and its detection by atomic absorption spectrometry. In addition, gas chromatographic techniques have been utilized.

In the cities, the CO mixing ratios are in the range of 1–10 ppmv due to nearby sources, mainly emissions from automobiles. In clean-air regions, the mixing ratios decrease to values of about 200 ppbv in the northern hemisphere and 50 ppbv in the southern. The existence of a latitudinal gradient was initially documented by Robinson and Robbins (1968b, 1970a) for the Pacific Ocean, and by Seiler and Junge (1970) for the Atlantic Ocean. Other studies have subsequently confirmed it from measurements onboard of ships and by aircraft (Swinnerton and Lamontagne, 1974a; Seiler, 1974; Heidt *et al.*, 1980; Seiler and Fishman, 1981; Robinson *et al.*, 1984). Figure 4-7 depicts the distribution of CO with latitude over the Atlantic Ocean,

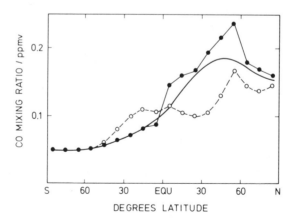

Fig. 4-7. Latitudinal distribution of carbon monoxide over the Atlantic Ocean. Open circles: in the upper troposphere, from aircraft measurements. Filled circles: at sea level, from measurements onboard of ships. Solid curve: air mass–weighted average CO mixing ratio. [Adapted from Seiler and Schmidt (1974a).]

both at the sea surface and at about 10 km altitude. From his measurements, Seiler constructed a somewhat idealized two-dimensional distribution of CO with latitude and height, which is reproduced in Fig. 4-8. Although only data from marine air were considered, the CO mixing ratio shows a surface maximum at midlatitudes of the northern hemisphere, indicating a source of considerable magnitude in this region. By following the gradient of CO mixing ratios one may trace the direction of transport from the surface on up into the upper troposphere and then on into the southern hemisphere. This is shown by the arrow. The figure thus indicates that the principal pathway of air exchange between the two hemispheres is via the upper troposphere. This agrees with conclusions of Newell *et al.* (1974) from air motions driven by the Hadley cells. The fact that CO gradients exist points toward a CO residence time in the troposphere of less than 1 yr, according to the discussion in Section 4.1. The prediction is confirmed by the CO budget, to be considered later.

In the southern hemisphere, south of 50° latitude, the CO mixing ratio assumes a fairly constant value of about 50 ppbv. McConnel *et al.* (1971) have pointed out that a photostationary state will be set up for CO, if the oxidation of CH_4 is the sole source of CO and its reaction with OH radicals is the sole sink. In this case one has

$$dm(CO)/dt = k_{72}n(OH)m(CH_4) - k_{132}n(OH)m(CO) = 0$$

and

$$m(CO) = (k_{72}/k_{132})m(CH_4)$$

$$m(CO) = [(7 \times 10^{-15})/(2.7 \times 10^{-13})]1.5 \times 10^{-6} = 35 \quad ppbv$$

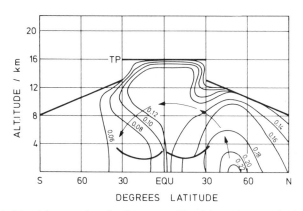

Fig. 4-8. Meridonal cross section of carbon monoxide mixing ratios over the Atlantic Ocean. Contours give CO mixing ratios in units of ppmv. [Adapted from Seiler (1974) and Seiler and Schmidt (1974a).]

Note that the OH concentration cancels. Comparing calculated with observed mixing ratios shows that methane is indeed a major source of CO but not the only one. The addition of a second source q_{CO}/n_T yields

$$m(CO) = \frac{k_{72}m(CH_4)n(OH) + q_{CO}/n_T}{k_{132}n(OH)}$$

Seiler *et al.* (1984c) have observed a seasonal variation of CO mixing ratios at Cape Point, South Africa, 34°S. The data were obtained during a 3-yr period (1978–1981) and are summarized in Fig. 4-9. Maximum values near 80 ppbv occurred in September/October, minimum values of 50 ppbv in January (southern hemispheric summer). The trend of OH number densities is opposite to that of CO. In principle, the variation of CO might be explained by the variation of $n(OH)$. Using the above equation one then finds that q_{CO}/n_T contributes 48% to the total CO source strength but that it should be seasonally independent. This condition is hard to understand because most CO sources that one can think of would be subject to seasonal variations. It may be that these partly compensate each other.

The rate of CO production from the oxidation of methane in the troposphere is $(M_{CO}/M_{CH_4})\,[S_{OH}(CH_4)] = (28/16)340 = 600$ Tg CO/yr. At least 50% of it, or 300 Tg/yr, is produced in the southern hemisphere. From the distribution of CO with latitude as shown in Fig. 4-7 and the air masses

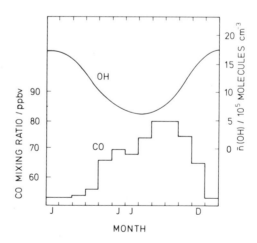

Fig. 4-9. Seasonal variation of monthly mean CO mixing ratios observed at Cape Point, South Africa, averaged over the period 1978–1981; from Seiler *et al.* (1984c). Added is the seasonal variation of OH number density taken from calculations of Logan *et al.* (1981) and Crutzen (1982).

given in Table 1-6 for tropical and extratropical latitudes, one estimates values of 310 and 150 Tg for the mass contents of CO in the northern and southern hemispheres, respectively. The difference causes an influx of CO from the northern into the southern hemisphere of $(1/\tau_{exch})(310-150) = 160$ Tg CO/yr. This is sizable but less than needed to balance the additional source in the southern hemisphere. Biomass burning and photooxidation of nonmethane hydrocarbons have been suggested to make up for the difference.

Anthropogenic emissions appear to be mainly responsible for the bulge of CO mixing ratios in the northern hemisphere. A large share of it is associated with the combustion of fossil fuels and industrial activities, which are concentrated in the northern hemisphere because 90% of all people live there. The usual procedure for estimating the global anthropogenic source strength is to combine statistical data for the consumption or production of a commodity with observed emission factors, which are defined as the amount of CO produced per unit of fuel consumed or material produced. Typical emission factors are summarized in Table 4-10, and fuel-use patterns are shown in Table 4-11. The data are combined in Table 4-12 to obtain the global rate of CO production due to energy- and industry-related activities. Automobiles account for 80% of CO from the combustion of

Table 4-10. *Emission Factors for CO from Combustion and Industrial Processes*[a]

Combustion of hard coal (g/kg)		Natural gas (g/m^3)	
Utility boilers	0.5–1.0	Utility boilers	0.04–0.27
Coke ovens	0.6	Industrial boilers	0.08–0.27
Industrial boilers	0.5–1.0	Residential units	0.32–2.73
Residential units	5–290	Industrial processes (g/kg of product)	
Combustion of lignite (g/kg)		Pig iron production	80
Utility boilers	1.0	Steel production	70
Coke ovens	1.0	Cracking of crude oil	39
Industrial boilers	1.0	Carbon black production	2000
Residential units	45–137	Ammonia production	100
Combustion of oil (g/l)		Pig iron foundry	72
Utility boilers	0.1–63	Biomass burning (volume % CO/CO$_2$)	
Industrial boilers	0.4–63	Forest wild fires	11–25
Industrial gasoline engines	472	Grass, stubble, straw	2.9–14.9
Industrial diesel engines	12	Agricultural wastes	3–16
Residential boilers	0.1–63	Burning of wood for fuel	6–18
Transportation (g/l)			
Gasoline engines	180–360		
Diesel engines	15–30		

[a] Compiled from Logan *et al.* (1981) and Crutzen *et al.* (1979).

Table 4-11. *Fossil Fuel Consumption Patterns, 1976, in Units of Tg Carbon[a]*

	United States	Europe	Rest of the world
Energy conversion			
Hard coal	275.3	152.6	360.4
Lignite	—	390.5	43.5
Oil	93.8	125.8	141.3
Natural and manufactured gas	108.2	85.3	30.5
Industrial consumption			
Hard coal	115.7	156.0	237.1
Lignite	—	65.1	7.2
Oil	53.6	188.7	135.8
Natural and manufactured gas	127.9	100.4	35.9
Residential uses			
Hard coal	8.0	30.5	34.8
Lignite	—	9.3	1.0
Oil	134.0	207.2	214.8
Natural and manufactured gas	91.8	65.3	24.6
Transportation (oil-derived)			
Gasoline and diesel oil	388.6	218.3	108.0
Total consumption rates	1397.0	1795.0	1375.0

[a] Data were compiled by Logan *et al.* (1981) from United Nations statistics (1974–1978). This takes into account the percentages of carbon in different grades of coal and lignite. The carbon content of solid fuel is 70%, that of liquid fuel 84%, and that of natural gas 0.536 g/l, according to Keeling (1973a).

fossil fuels, and for 50% of all emissions from these sources. The other large source, which contributes about 30% to the total, is due to industrial processes, mainly the production of steel and the catalytic cracking of crude oil. The individual emission estimates of Logan *et al.* (1981), which are shown in Table 4-12, combine to a total of 445 Tg CO/yr. A similar value was previously derived by Jaffe (1973), whereas Seiler (1974) gave a slightly higher estimate of 640 Tg/yr. He considered CO emissions from home heating to be underestimated, because Baum (1972) had shown for West Germany that residential heating contributed 30% of the annual production of CO by motor vehicles. Indeed, the emission factors shown in Table 4-10 for this source admit a large range of uncertainty.

The annual rate of biomass destruction by fire is estimated from the area cleared or burned per year, the quantity of fuel consumed, and the production of CO per unit of fuel. Rather detailed estimates for the first two factors have been furnished by Seiler and Crutzen (1980). Their data are shown in Table 4-13. The total mass of carbon released annually from biomass burning

Table 4-12. *Anthropogenic CO Emissions from Fossil Fuel Combustion and Industrial Activities in Various Parts of the World*[a]

Source type	North America	Europe	Rest of the world	Total
Combustion				
Coal	1	24	23	48
Lignite	—	3	—	3
Gas	0.2	0.3	0.1	0.6
Oil	2.2	4.6	3.7	11
Total combustion	3.4	32	27	62
Transportation	94	71	66	233
Industries				
Pig iron production	7.8	16.2	18.0	42
Steel production	9.5	19.8	17.7	47
Cracking of crude oil	4.5	5.0	4.1	13.6
Miscellaneous	10.5	9.6	7.3	27.4
Total from industries	32.3	50.6	47.1	130
Waste disposal	3	6	11	20
Sum total	137	155	152	445

[a] Compiled by Logan *et al.* (1981) from United Nations statistics (1974–1978), in units of Tg CO/yr.

[b] The distribution was taken from the yearbook of industrial statistics (United Nations, 1978), since Logan *et al.* (1981) did not give it. The total from industries thus differs slightly from that of Logan *et al.* (1981).

as $CO + CO_2$ is 3 Pg. According to Crutzen *et al.* (1979), and as shown in Table 4-10, the CO/CO_2 volume ratio in the plume of fires is 0.14, on average, so that the fraction of carbon appearing as CO is 12%. This leads to a CO emission rate of $(M_{CO}/M_C)(3 \times 10^3)(0.12) = 840$ Tg CO/yr with a range of 300–1600 Tg/yr. Roughly 50% of the emissions are associated with the agricultural practices of the tropics, namely, slash, burn, and shift practices and the burning of grass lands at the end of the dry period. Forest fires in temperate climate zones make a relatively small contribution, but the disposal of agricultural wastes is important. Biomass burning furnishes a source of CO similar in magnitude to fossil fuel combustion, but it has a greater range of uncertainty due to the indirect means of assessment.

Yet another anthropogenic CO source results from the oxidation of hydrocarbons that are released into the atmosphere, mostly from automobiles and the use of industrial solvents. Logan *et al.* (1981) combined emission data for the United States (Environmental Protection Agency, 1976, 1978) with fuel-use patterns to estimate a global strength for this CO source of 90 Tg/yr. It was assumed that 50% of carbon in volatile hydrocar-

Table 4-13. *Area Cleared or Burned, and Amount of Biomass Burned Annually, in Units of 10^6 Hectares and 100 Tg Dry Matter, Respectively[a]*

Process	Burned or cleared area	Biomass coverage (kg/m²)	Total biomass cleared	Biomass exposed to fire	Biomass burned	Above-ground biomass remaining unburned	Carbon converted to $CO+CO_2$
Burning due to shifting agriculture	41 (21–62)	15	62 (31–92)	48 (24–72)	17 (9–25)	44 (16–72)	7.6 (4–11.2)
Deforestation due to population increase and colonization	12 (8.8–15.1)	22	26.5 (20–33)	20.5 (16–25)	7.2 (5.5–8.8)	13.3 (10.5–16)	3.2 (2.5–4.0)
Burning of savanna and bushland	600	0.3	18 (12–23.8)	18 (12–23.8)	11.9 (4.8–19)	3.6 (2.4–4.8)	5.4 (2.2–8.5)
Wildfires in temperate forests	4.1 (3.0–5.0)	35	14 (10.5–17.5)	10.3 (7.7–13)	2.1 (1.5–2.6)	8.2 (6.2–10.2)	1.0 (0.7–1.2)
Prescribed fires in temperate forests	2.5 (2.0–3.0)	60	1.5 (1.2–1.8)	0.4 (0.3–0.5)	0.2 (0.1–0.2)	0.3 (0.2–0.3)	0.1 (0.04–0.09)
Wild fires in boreal forests	1.3 (1.0–1.5)	25	3.2 (2.5–3.8)	2.3 (1.8–2.7)	0.5 (0.4–0.6)	1.8 (1.4–2.1)	0.23 (0.18–0.27)
Burning of industrial wood and fuel wood	—	—	31.5 (31–32)	11.5 (11–12)	10.5 (10–11)	1[b]	4.7 (4.5–4.9)
Burning of agricultral wastes	—	—	—	21 (19–23)	19 (17–21)	2.1 (1.9–2.3)	8.5 (7.6–9.4)
Totals	660 (630–690)		180 (130–250)	132 (92–172)	68 (48–88)	74 (40–109)	31 (22–40)

[a] The last column gives the amounts of carbon converted to CO_2 (conversion factor = 0.45). Estimate ranges are shown in parentheses. From Seiler and Crutzen (1980).

[b] Excluding wood used in long-lasting structures.

bons is converted to CO. Even if the conversion efficiency were higher, it is clear that hydrocarbons make a relatively minor contribution to all anthropogenic CO sources. Hydrocarbons, however, are also released from plants and soils. This contribution to the CO budget is more substantial.

Hydrocarbon emissions from vegetation have been studied by Zimmerman *et al.* (1978). The principal hydrocarbons appear to be isoprene and terpenes. The former is the dominant compound from deciduous plants, notably oak, but it is emitted only during daylight. The latter emissions are associated primarily with conifers. Terpene emissions are independent of light conditions but vary with temperature, and thus with season and latitude. Experimentally observed emission rates in units of milligrams carbon per kilogram leaf biomass per hour ranged from 3.5 to 8.9 for terpenes from conifers and from 2.4 to 24.7 for isoprene from deciduous trees (see Table 6-5). Zimmerman *et al.* (1978) constructed a natural hydrocarbon inventory for the United States in four different temperature and biotope regions and compared it with the annual net primary productivity. This quantity is the amount of carbon from atmospheric CO_2 fixed as plant matter and not quickly respired (see Section 11.2.4). The comparison suggested a ratio of hydrocarbon emission to net primary productivity of 0.7%. The ratio was then adopted to represent globally averaged conditions. The total net primary productivity of the world is 55 Pg C/yr, or about 120 Pg/yr of plant matter, which produces a global hydrocarbon emission rate of 830 Tg/yr. Roughly 60% of it or 500 Tg/yr was estimated to occur in the tropics. Total isoprene emissions were estimated as 350 Tg/yr, and terpenes as 480 Tg/yr.

In the atmosphere, hydrocarbons are subject to attack by OH radicals and ozone which initiate an oxidation mechanism whereby the materials are first converted to oxygenated compounds and then partly to CO. Hydrocarbon oxidation mechanisms are discussed in Section 6.3. Here we note that not every carbon atom is converted to CO. Accordingly, a yield estimate is required if one wishes to utilize the above data in estimating the production of CO from the oxidation of hydrocarbons. For isoprene the oxidation mechanism has been staked out and one expects a conversion yield of 80% CO, 20% CO_2. A laboratory study of Hanst *et al.* (1980) has essentially confirmed these yields. Terpenes, by contrast, pose much large uncertainties, because a substantial portion of the material may be converted to low-volatility products, which condense onto aerosol particles (see Section 7.4.3). The experiments of Hanst *et al.* (1980) on α-pinene indicated a total yield of $CO + CO_2$ of 30% and a CO/CO_2 ratio of 0.7. Thus, about 20% of carbon in α-pinene was converted to CO. If the conversion efficiencies for other terpenes were similar, one would obtain the following CO production rates from the oxdiation of natural hydrocarbons:

From isoprene: $0.8[(M_{CO}/(M_{C_5H_8}/5)]350 = 576$ Tg/yr

From terpenes: $0.2[(M_{CO}/(M_{C_{10}H_{16}}/10)]480 = 198$ Tg/yr

The combined rate of 774 Tg CO/yr is similar in magnitude to that of CO production from methane, but it is confined to the continents. The world ocean is a source of light (C_2-C_8) and heavier (C_9-C_{30}) hydrocarbons, but the emission rate is not competitive on a global scale with that from continental vegetation. Emission estimates for marine hydrocarbons are of the order of 30–50 Tg/yr (Rudolph and Ehhalt, 1981; Eichmann *et al.*, 1980). Hydrocarbons also are emitted from open fires. The observations of Darley *et al.* (1966) and Gerstle and Kemnitz (1967) regarding the burning of agricultural wastes and municipal or landscape refuse suggests an emission factor of 14 g/kg material burned, which translates to a molar carbon to carbon ratio of 2.5%. This is about one-fifth of that for the direct emission of CO from open fires, so that hydrocarbon emissions as a source of CO may be neglected to a first approximation in comparison with direct emissions.

In addition to the secondary formation of CO from the oxidation of hydrocarbons, CO is emitted directly also from the oceans and from land plants. The first type of emission is due to biological activity; which causes a supersaturation of CO in seawater with respect to CO in the atmosphere. It was Swinnerton *et al.* (1970) who discovered the supersaturation and recognized the ocean to be a source of atmospheric CO. Subsequent measurements of Swinnerton and Lamontagne (1974a), Seiler (1974), Seiler and Schmidt (1974a,b), and Conrad *et al.* (1982) have established that excess CO concentrations occur in the surface waters of the Pacific and the Atlantic Oceans fairly independent of latitude. The CO concentrations maximize near the sea's surface. Strong diurnal variations have been observed with maxima in the early afternoon. This points toward a photo-active source mechanism. Swinnerton *et al.* (1974a) assumed that the production is caused by algae, but Conrad *et al.* (1982; Conrad and Seiler, 1980d) have applied filtration to remove algae and bacteria from the water samples. The production of CO continued in sunlit vessels but not in the dark. To explain this finding, Conrad *et al.* (1982) adopted an earlier idea of Wilson *et al.* (1970), who had suggested that the production of CO arises from the photooxidation of organic compounds that algae release into the seawater environment. The rapid diurnal changes of dissolved CO can be understood only if CO is consumed as well as produced. Losses by molecular diffusion are slow and can be ignored. Conrad *et al.* (1982; Conrad and Seiler, 1980a) have carried out laboratory and field experiments to show that the consumption is due to bacteria. The combined action of photoproduction and microbial consumption leads to a steady-state CO concentration that depends on light intensity and dissolved organic matter content of seawater. The observational data indicate an average supersaturation of CO by a factor of 30 ± 20. The associated flux rate of CO into the atmosphere may be derived by the method outlined in Section 4.3 for methane. In this manner, Linnenbom et al. (1973), Seiler (1974), Seiler and Schmidt (1974b) and Conrad et al. (1982) have obtained values of 220, 100, 70, and 75 Tg CO/yr, respectively. The possible range of values given by the last authors was 10–180 Tg/yr.

Green plants may both absorb carbon monoxide and release it. Wilks (1959) found alfalfa, sage plants, and cedar, among other species, to produce CO, whereas Krall and Tolbert (1957), Bidwell and Fraser (1972), and Peiser et al. (1982) reported an uptake of CO by barley, bean, and lettuce leaves, respectively. Part of the results were obtained by the ^{14}C method, which is very sensitive. Seiler et al. (1978a) and Bauer et al. (1979) have studied the behavior of CO in enclosures surrounding either parts of or whole plants. The species studied were Fagus silvatica, Pinus silvestris, Vicia faba, and Platanus acerifolia. A net production of CO was found in all cases. The production rate was proportional to leaf area, and it rose almost linearly with radiation intensity. There were no signs of saturation even at $800 \, W/m^2$, an intensity at which net photosynthesis was already inhibited. Apart from the dependence on light intensity, the CO emission was essentially independent of season, even in fall when the leaves showed discoloration. The net production rate for a radiation flux of $50 \, W/m^2$ was $3 \times 10^{-9} \, g \, CO/m^2 \, s$. Bauer et al. (1979) used these results, in combination with average values for global leaf surface area and light intensity, to estimate a global source strength for CO from plants of 75 ± 25 Tg/yr. The mechanism for the production of CO in plant leaves has not been clearly identified. CO may be produced as a metabolite during the degradation of porphyrins such as chlorophyll (Troxler and Dokos, 1973); it may be formed by the degradation of glycolate, a product of photorespiration (Fischer and Lüttge, 1978; Lüttge and Fisher, 1980); or it may arise from the photooxidation of plant cellular material. Bauer et al. (1980) have shown that the last pathway occurs in phototrophic microorganisms.

The preceding survey of CO sources is summarized in Table 4-14, which presents emission estimates from three independent research groups. The total source strength is of the order of 3000 Tg CO/yr. Anthropogenic activities consisting of fossil-fuel combustion, industries, and biomass burning contribute roughly 45% to the total. The oxidation of hydrocarbons, both methane and others, makes up about 50%, whereas emissions from ocean and plants are relatively minor. Table 4-14 includes estimates for CO sinks. The dominant sink, leading to a consumption rate of about 2000 Tg/yr,

Table 4-14. *The Global Budget of Carbon Monoxide in the Troposphere (Estimates in Tg CO/yr)*

Type of source or sink	Seiler (1974)	Logan et al. (1981)			Volz et al. (1981a)	Seiler and Conrad (1987)
		Global	Northern hemisphere	Southern hemisphere		
Sources						
Fossil fuel combustion and industrial activites	640	450	425	25	640	640 ± 200
Biomass burning	60	655	415	240	300–2200	1000 ± 600
Oxidation of human-made hydrocarbons	—	90	85	5	—	—
Oxidation of natural hydrocarbons	60	560	380	180	200–1800	900 ± 500
Ocean emissions	100	40	13	27	100	100 ± 90
Emissions from vegetation	—	130	90	40	50	75 ± 25
Oxidation of methane	1500–4000	810	405	405	600–1300	600 ± 300
Total source strength	2360–4860	2735	1813	922	2800 ± 900	3315 ± 1700
Sinks						
Reaction with OH radicals	1940–5000	3170	1890	1280	1650–3550	2000 ± 600
Consumption by soils	450	250	210	40	320	390 ± 140
Flux into the stratosphere	110	—	—	—	—	110 ± 30
Total sink strength	2500–5560	3420	2100	1320	2800 ± 800	2500 ± 770

is the oxidation of CO by OH radicals. This process does not quite suffice to balance the sources, however. Two other sinks that have to be considered are the stratosphere and microbiological processes in soils.

The stratosphere as a sink was first identified by Seiler and Junge (1969). The flux of CO into the stratosphere is caused by a decline of CO mixing ratios above the tropopause toward a steady-state level lower than that normally found in the upper troposphere (see Fig. 1-14). In the lower stratosphere, CO is produced from methane and other long-lived hydrocarbons, and it is consumed by reaction with OH as in the troposphere, but the rate of vertical mixing is much slower (Seiler and Warneck, 1972; Warneck et al., 1973). The flux of CO from the troposphere into the stratosphere can be derived from the observed gradient of the CO mixing ratio above the tropopause in a manner described in Section 4.3 for methane. The loss rate obtained, 110 Tg/yr, is small compared with that for the reaction of CO with OH radicals.

The importance of soil surfaces as a sink for atmospheric CO was first pointed out by Inman et al. (1971; see also Ingersoll et al., 1974). Their experiments were carried out with rather high CO mixing ratios, however, which led them to overestimate the sink strength due to soils. Liebl and Seiler (1976) then showed that soils may represent either a source of CO or a sink, because the mixing ratio of CO in the air above soils confined by an enclosure adjusts to a steady-state value determined by conditions within the soils (see Fig. 1-15). The steady-state value is strongly dependent on temperature, moisture content, population of bacteria, and other soil parameters. A soil will act as a source of CO when the steady-state CO mixing ratio is higher than that of ambient air, and it will act as a sink when these conditions are reversed. Usually, atmospheric CO is lost at the soil surface, and only arid soils serve as a source of CO (Conrad and Seiler, 1982a, 1985a).

Field and laboratory studies leave no doubt that the consumption of CO in soils is a biological process (Ingersoll et al., 1974; Bartholomew and Alexander, 1979, 1982; Conrad and Seiler, 1980b, 1982b), probably involving microorganisms. For example, Conrad and Seiler (1982b) have shown that the activity for CO consumption follows Michaelis–Menten kinetics, that it is removed by filtration through a 0.2-μm filter but not a 0.3-μm filter, and that it is inhibited by antibiotics. Uncertainties still exist about the types of microorganisms responsible for the CO losses. It is not known whether the microbes are capable of growing on atmospheric CO or just oxidize it in a nonutilitarian manner. The production of CO in soils, in contrast to its consumption, is not related to microorganisms, even though a number of bacteria and fungi are known to produce CO from certain substrates. It appears that CO is produced from the oxidation of organic compounds present in soil humus. Evidence for the abiotic nature of the process has been obtained by Conrad and Seiler (1980b, 1985b). They inhibited the microbial activity by antibiotics or heat treatment of the soil, but this did not eliminate the evolution of CO. The rate of CO production was stimulated when soil pH was raised, or when the air was replaced by pure oxygen, whereas a replacement by nitrogen decreased the rate. Sterile mixtures of humic acids in quartz sand produced CO, while pure quartz sand did not. Field measurements of the steady-state CO mixing ratio and its variation with temperature above arid soils enabled Conrad and Seiler (1985a) to separate the effects of CO consumption and production. The consumption may be expressed in terms of a deposition velocity, which was found to vary little with temperature. The CO production rate, in contrast, was strongly dependent on temperature, following an Arrhenius law with an apparent heat of activation in the range 57–110 kJ/mol. The rate was also proportional to the organic carbon content of the soil. From these and other observations, Conrad and Seiler (1985a) deduced a globally averaged deposition velocity of $(2-4) \times 10^{-4}$ m/s, which is equivalent to a global CO loss rate of 190–580 Tg CO/yr with an average of 390 Tg/yr. The global emission rate from soils is more difficult to estimate because the local rate follows the diurnal variation of soil temperature, and it depends on the soil organic matter content. Conrad and Seiler (1985a) derived an estimate of 17 ± 15 Tg CO/yr. These emissions occur mainly from tropical and desert soils. The global emission rate is much smaller, however, than the consumption rate, so that it may be ignored to a first approximation.

The various estimates in Table 4-14 for the total CO production and loss rates show that the CO budget is only approximately balanced. Individual source and sink estimates carry uncertainties that presently are too large to allow an improvement of the budget. The average residence time for CO in the troposphere is

$$\tau = G(\text{CO})/S(\text{CO}) = 460/2800 = 0.18 \quad \text{yr}$$

or about 2 months. This is short enough to cause the spatial variation of CO within the troposphere shown in Fig. 4-8 and the difference of CO mixing ratios between the northern and sourthern hemispheres.

4.6 Hydrogen, H_2

The presence of hydrogen in the atmosphere appears to have been discovered as a by-product of the liquefaction of air. An early report by Paneth (1937) indicated an H_2 mixing ratio of about 0.5 ppmv as derived from liquefied air in 1923. Schmidt (1974) has summarized a variety of measurements made during the period 1950–1970 that led to similar mixing ratios. During this time the interest was concentrated on the tritium content of atmospheric hydrogen and its rise due to atomic weapons tests and emissions from nuclear industries (Ehhalt *et al.*, 1963; Begemann and Friedman, 1968). According to Suess (1966), it was generally assumed that hydrogen in the air originated from the photodissociation of water vapor in the upper stratosphere and mesosphere. This idea had to be abandoned, however, when Ehhalt and Heidt (1973b) and Schmidt (1974) showed that the H_2 mixing ratio above the tropopause lacked a vertical gradient. Accordingly, the flux of H_2 between stratosphere and troposphere must be negligible. In the early 1970s when the methane oxidation cycle was explored, it was recognized that hydrogen is a product of the photodissociation of formaldehyde (Calvert *et al.*, 1972). A search for other H_2 sources indicated emissions from the oceans and from anthropogenic activities. In addition, the reaction of hydrogen with OH radicals and its uptake by soils have been identified as sinks. These studies have led to a budget of hydrogen in the troposphere that is largely controlled by the biosphere. The budget of hydrogen in the stratosphere is detached from that in the troposphere.

The spatial distribution of hydrogen in the troposphere has been investigated by Schmidt (1974, 1978) and Ehhalt *et al.* (1977). The distribution is rather uniform except for a 2% excess in the northern hemisphere, as shown in Table 4-15. Aircraft measurements over the continents of the northern hemisphere further showed a slight decline of the mixing ratio with height, indicating that the continents are a net source of H_2. In the southern hemisphere the vertical gradient is insignificant. Much higher than average H_2 mixing ratios occur in the cities. Hydrogen is an important by-product of combustion processes, and it seems reasonable to assume that much of the excess H_2 in the northern hemisphere is of anthropogenic origin.

Table 4-15. *Mixing Ratio (ppbv) of Hydrogen in Unpolluted Air of the Northern and Southern Hemispheres and in Mainz (F.R.G.) (Standard Deviation in Parentheses) (Schmidt 1974, 1978) and the Mass Content (Tg) of Hydrogen in the Two Hemispheres*

	Northern hemisphere	Southern hemisphere
Surface	584 (27)	552 (10)
Upper troposphere	559 (17)	551 (13)
Average $m(H_2)$	563	552
Mainz, Germany	800 (168)	—
$G(H_2)$	82.5	80.9

Scranton *et al.* (1980) have studied the diurnal variation of hydrogen in Washington, D.C. In winter, the mixing ratio attained maximum values of up to 1.2 ppmv twice daily. The maxima concided with the rush-hour traffic, indicating that motor vehicle exhaust is a major source of H_2. Indeed, hydrogen correlated well with CO and nitrogen oxides. From aircraft measurements in the plume of the city of Munich, Germany, Seiler and Zankl (1975) have shown that the H_2/CO volume ratio of the emissions is approximately unity. Combining this result with the emission rate for CO from transportation (Table 4-12) yields a source strength for anthropogenic hydrogen of 17 Tg/yr. Liebl and Seiler (1976) derived a somewhat higher value of 20 Tg/yr, whereas Schmidt (1974), who did not yet have this information, came up with 13 Tg/yr. Since there may be other anthropogenic sources besides motor-vehicle exhaust, a source strength of 20 Tg/yr appears reasonable. To it must be added the emission of hydrogen associated with the burning of biomass, which Crutzen *et al.* (1979) estimated in the same manner as discussed previously for CO. The value obtained was 15 Tg/yr with a possible range of 9–21 Tg/yr.

In order to estimate the rate of hydrogen production from the oxidation of methane by OH radicals it is convenient to assume that photodissociation is the dominant loss process for formaldehyde. From the photodissociation coefficients (see Fig. 2-19), one finds that the channel leading to $H_2 + CO$ as dissociation products contributes roughly 68% to the overall photo-decomposition process. The rate of H_2 formation thus is

$$\alpha S_{OH}(CH_4)(M_{H_2}/M_{CH_4}) = 28.9 \quad Tg/yr$$

with $\alpha = 0.68$ and $S_{OH}(CH_4) = 340$ Tg/yr. More difficult to estimate is the rate of hydrogen production from the oxidation of nonmethane hydrocarbons. It should be recognized that formaldehyde is the only aldehyde whose decomposition yields H_2, although all of them generate CO. Of the hydrocarbons emitted naturally, the terpenes are not expected to yield much formaldehyde when oxidized, and we shall disregard these compounds. Isoprene is more effective, because its oxidation is known to produce about three molecules of formaldehyde per molecule of isoprene that becomes oxidized. With a 68% yield of hydrogen from formaldehyde, one thus finds a source strength of

$$(3/5)(0.68)Q(C_5H_8)M_{H_2}/(M_{C_5H_8}/5) = 21 \quad Tg/yr$$

where $Q(C_5H_8) = 350$ Tg C/yr, as discussed in Section 4.5.

The oceans represent a source of atmospheric hydrogen due to the fact that the surface waters of the oceans are supersaturated with H_2—that is the concentrations are higher than those expected from simple solution equilibrium of atmospheric hydrogen. Saturation factors observed by

Schmidt (1974) in the northern and southern Atlantic varied from 0.8 to 5.4 with an average of 2.5. From these data, extrapolated to other oceans, Schmidt (1974) and Seiler and Schmidt (1974b) estimated a global emission rate of 4 Tg/yr. In contrast to CO, the aqueous concentrations of H$_2$ do not undergo diurnal variations. This does not preclude a photochemical production of H$_2$, as was shown for carbon monoxide, but other production mechanisms for hydrogen, such as by bacteria or in the digestive tracts of zooplankton, may be more effective.

Finally, we must discuss the behavior of hydrogen in soils. In water-logged, anaerobic soils, hydrogen is generated in large amounts due to the bacterial fermentation of organic matter. Hydrogen emissions from such soils nevertheless are marginal because nearly the entire production is consumed by other bacteria, mainly denitrifiers, sulfate reducers, and methanogens, which, as oxygen becomes scarce, successively utilize nitrate, sulfate and CO$_2$ as an oxygen supply. After the onset of methanogenesis, hydrogen occurs only in traces in such soils although the turnover rate is high. Thus methane is emitted rather than hydrogen.

Aerobic environs further the consumption of H$_2$ by oxidizing bacteria. Aerobic soils thus generally act as a sink for atmospheric hydrogen, except in the presence of legumes. Members of this plant family live in symbiosis with *Rhizobium* bacteria. Contained in root nodules, these bacteria are able to fix atmospheric N$_2$ (see Section 9.1) by means of the enzyme nitrogenase, which catalyzes the reduction of N$_2$ to NH$_3$. In addition, nitrogenase is able to reduce protons to hydrogen, and it appears that this process occurs always in parallel to N$_2$ reduction. About 30% of the electron flow needed for N$_2$ fixation is diverted to reduce protons to H$_2$ (Schubert and Evans, 1976). Most of the N$_2$ reducers have developed enzymes that recycle the H$_2$ so formed internally (Robson and Postgate, 1980) so that hydrogen does not leave the cell. *Rhizobium* species, in contrast, lack these enzymes causing hydrogen to be released to the environment. Conrad and Seiler (1980c) have studied the release of hydrogen from soils planted with clover, in comparison with soils overgrown exclusively with grasses. Similar to CO, hydrogen was observed to approach a steady-state mixing ratio within 60 min after the plot was covered with an airtight box. Hydrogen as well as CO thus is subject to simultaneous production and loss. From the rate of change of the H$_2$ mixing ratio inside the box, Conrad and Seiler (1980c) were able to separate the contributions of both processes. The results indicated consumption rates that were approximately equal for all soil plots and similar to those derived earlier by Liebl and Seiler (1976). The corresponding H$_2$ deposition velocities fell into the range $(2-10) \times 10^{-4}$ m/s, with a minimum in winter and a maximum in early autumn. Hydrogen emissions occurred only from plots planted with clover, indicating that at least part of the hydrogen resulting from nitrogen fixation is released to the atmosphere. There was a strong seasonal variation of the release rate, with a maximum during the main growth period of the plants approximately coinciding with the maximum of nitrogen fixation activity.

Burns and Hardy (1975) have estimated the rate of N_2 fixation by agricultural legumes as 35 Tg N/yr, or about 25% of the global nitrogen fixation rate of 139 Tg N/yr. The remainder is mainly due to grasslands (45 Tg N/yr) and forest and woodlands (40 Tg N/yr). If the volume ratio of hydrogen produced to nitrogen consumed is 30%, one finds that the total rate of hydrogen production is

$$0.3 \times 139(M_{H_2}/M_N) = 6 \quad Tg/yr$$

Not all of it is released, however. Conrad and Seiler (1980c) have considered the various contributions more carefully and, using their observational data, derived a global H_2 release rate of 3 Tg/yr with a possible range of 2.4–4.9 Tg/yr.

The rate of hydrogen consumption by soils increases with temperature toward an optimum at 30–40°C (Liebl and Seiler, 1976). Soil moisture is another parameter affecting the deposition rate. Conrad and Seiler (1980c, 1985a), however, found comparable H_2 deposition velocities for soils in subtropical and temperate latitude regions, with an average of 7×10^{-4} m/s. Accordingly, the value may be used as a global average. In estimating the effective land area, one must take into account the winter season at high latitudes and subtract urban areas, deserts, and regions covered with perpetual ice. Conrad and Seiler (1980c, 1985a) derived an effective area of 90×10^6 km^2, which leads to a global rate of hydrogen uptake by soils of 90 Tg/yr with a possible range of 70–110 Tg/yr. This is much larger than the potential rate of hydrogen emissions due to nitrogen fixation.

Table 4-16 summarizes the known sources and sinks for atmospheric hydrogen. The total budget amounts to about 90 Tg H_2/yr. Roughly 40% of it is due to anthropogenic activities, mainly automotive exhaust and biomass burning. The major natural sources are the oxidation of methane and other hydrocarbons. Emissions from oceans and legumes are relatively minor. The most important sink is the uptake of hydrogen by soils. A much smaller fraction undergoes chemical oxidation by the reaction $OH + H_2 \rightarrow H_2O + H$. The global distribution of sources and sinks favors the continents. The uniformity of the mixing ratio requires that hydrogen has a residence time in the troposphere sufficient for transport to even out any gradients that might develop. From the budget one obtains the residence time

$$\tau = G(H_2)/Q(H_2) = 163/90 = 1.8 \quad yr$$

The value is sufficiently large to ensure a fairly uniform distribution within

Table 4-16. *Budget of Hydrogen ($Tg\ H_2/yr$) in the Troposphere According to Different Estimates*

Type of source or sink	Schmidt (1974)	Conrad and Seiler (1980c)	Present
Sources			
Anthropogenic emissions	13–25.5	20 ± 10	17^a
Biomass burning	—	20 ± 10	15^b
Oceans	4	4 ± 2	4^c
Methane oxidation	4.6–9.2	15 ± 5	29^a
Oxidation of nonmethane hydrocarbons	—	25 ± 10	21^a
Biological N_2 fixation	—	3 ± 2	3^d
Volcanoes	0.1	—	0.2^e
Total sources	21.6–38.7	87 ± 38	89
Sinks			
Oxidation by reaction with OH radicals	3.7–7.3	8 ± 3	11
Uptake by soils	12–31	90 ± 20	78^f
Total sinks	15.7–38.3	98 ± 23	89

[a] See text.

[b] Crutzen *et al.* (1979).

[c] Seiler and Schmidt (1974a,b).

[d] Identical to average given by Conrad and Seiler (1980).

[e] From Section 12.4.1.

[f] Prorated to balance sources; this is within the uncertainty range given by Conrad and Seiler (1980a).

each hemisphere. It is not large enough, however, compared with the interhemispheric exchange time to guarantee *a priori* a uniform distribution between the northern and southern hemispheres. The fact that the average H_2 mixing ratios in both hemispheres are nevertheless almost the same suggests that the sources and sinks in each of the two hemispheres are closely balanced.

Chapter

5 | Ozone in the Troposphere

5.1 Introduction

Mixing ratios of ozone in the troposphere fall into the range of, roughly, 10-100 ppbv. The presence of ozone in air was already known to Schönbein (1840, 1854), the discoverer of ozone. He exposed to the atmosphere paper strips saturated with a starch–iodide solution and found that a characteristic blue color of the starch–iodine complex developed as iodide was oxidized to iodine. Paper ozonometry became very popular in central Europe in the middle of the last century, due to the widespread belief that ozone-rich air is beneficial to health because ozone is a disinfectant. The development of a reliable quantitative analytical technique based on potassium iodide was accomplished comparatively recently (Cauer, 1935). The correct order of magnitude of the ozone content of surface air was first established by Strutt (1918) by long-path optical absorption at 254 nm. He found a mixing ratio of 0.27 mm STP or $m(O_3) = 40$ ppbv. A few years earlier, Pring (1914) had employed potassium iodide to measure ozone in the Swiss Alps and in England and claimed to have found mixing ratios of the order of 1 ppmv, a value that is excessive even for extremely polluted air.

Ehmert (1949, 1951) introduced a technique utilizing the current derived from the ozone–iodide reaction in an electrochemical cell. The signal can be amplified and a continuous record obtained. The principle was adopted by Brewer and Milford (1960) to construct a simple balloon ozone sonde, which is still in use today for stratospheric ozone measurements. The potassium iodide reaction is somewhat subject to interferences by other oxidants in polluted air, but the effect is thought negligible in clean-air regions and in the free troposphere. More recently, commercial instruments were developed that determine ozone either by ultraviolet absorption or by gaseous chemiluminescence. These instruments have a greater dynamic range and are relatively free of interference effects caused by other trace gases in the air.

Ultraviolet radiation capable of photodissociating oxygen does not penetrate into the troposphere, so that ozone in that atmospheric domain must have a different origin. The classical concept of tropospheric ozone considers the stratospheric reservoir the main source (Junge, 1963). From here ozone enters the troposphere via tropopause exchange processes (see Section 1.5) and is then carried downward by turbulent mixing until it reaches the ground surface, where it suffers destruction.

While this concept was being developed, a phenomenon termed photochemical smog made its appearance, first in Los Angeles, and later in many other populous cities. In photochemical smog, ozone is formed as a by-product of the photooxidation of hydrocarbons. For many years, photochemical smog was considered an air-pollution problem with only local significance, perhaps affecting clean-air sites somewhat by transport of polluted air. Crutzen (1973, 1974a,b) first made the much further-reaching suggestion that smog-like reactions accompanying the oxidation of methane and other natural hydrocarbons in the atmosphere might induce a photochemical production of ozone also in the unpolluted troposphere. Chameides and Walker (1973, 1976) have gone even further in presenting a model in which ozone in the lower troposphere is controlled entirely by photochemistry. Their proposal has met with considerable skepticism among proponents of an ozone budget dominated by downward transport (Fabian, 1974; Chatfield and Harrison, 1976). Other aspects of the model also have been criticized by Dimitriades et al. (1976) and by Jeffries and Saeger (1976). But observations such as the seasonal variation with a maximum in summer can often be interpreted to support either transport from above or local photochemical production of ozone. The controversy has not been settled and, by necessity, will permeate the following discussions. The current great interest in tropospheric ozone has two roots. One is the global importance of ozone in generating OH radicals by the photoproduction of $O(^1D)$ and its reaction with water vapor. A quantitative assessment of OH production rates requires a reasonably sound knowledge of the distribution of ozone in the troposphere. The other aspect deals with the classification of ozone as a pollutant that should be brought under effective control. The specific issue here is the great variability of the natural background level of ozone in the boundary layer, which makes it difficult to define a threshold concentration above which ozone becomes a pollutant.

5.2 Photochemical Air Pollution

Los Angeles was the first city that, starting in the mid-1940s, became afflicted with the phenomenon of photochemical smog. The frequency and severity of its occurrence soon made it unbearable and called for study and abatement. Although currently many other cities in the world are burdened with photochemical smog, Los Angeles is still the prime example for a region thus affected, and despite considerable efforts to bring the problem under control it has been lingering on almost undiminished.

The term smog combines the words fog and smoke. It was originally applied to the episodes of London-type fogs containing abnormally high concentrations of smoke particles and sulfur dioxide. This type of smog occurs in winter, caused by an accumulation of products resulting from the combustion of fossil fuels in home heating, industry, and power stations under adverse meteorological conditions, and is severely taxing to the bronchial and pulmonary system. The Los Angeles type of smog is more a haze than a fog, characteristic of a low-humidity aerosol. Its origins are photochemical processes involving nitrogen oxides and hydrocarbons emitted from automobiles, petroleum industries, dry cleaning, etc. The major feature of photochemical smog is the high level of oxidant concentration, mainly ozone and peroxidic compounds, produced by the photochemical reactions. The principal effects are eye and bronchial irritation as well as plant and material damage.

The favored season for the occurrence of photochemical smog is summer, when solar intensities are high, but meteorological conditions are important, too. Specifically, the formation of a temperature inversion layer appears to be a prerequisite for the accumulation of smog precursors. Temperature inversions obstruct the vertical exchange of air by convective mixing and so prevent the dispersal of polluted air into the free troposphere. In Los Angeles, temperature inversions are frequent, occurring on about 300 days of the year. They are caused by the inflow of cool marine air during the daytime, the air moving inland close to the ground underneath warmer continental air. The stratification is not easily broken up because mountains surrounding the Los Angeles basin impede an horizontal advection of continental air. As a result, the polluted air is pushed inland by the sea breeze until in the late afternoon solar heating of the continental surface subsides and the local circulation reverses. At this time the pollution has been carried into many neighboring communities and fills almost the entire basin.

Photochemical air pollution arises largely from gas-phase chemistry. The reaction mechanisms are extremely complex, involving many different chemical species, and have not been fully elucidated. Mechanisms of hydrocarbon oxidation will be treated in Section 6.2. Here, we shall concentrate on the chemistry of ozone production. The various aspects of the subject have been reviewed by Leighton (1961), Altshuller and Bufalini (1965, 1971), Finlayson and Pitts (1976), and Demerjian et al. (1974), among others. Two important facts emerged early: nitrogen oxides are important as catalysts of hydrocarbon oxidation in the atmosphere; and automobiles are a major source of both hydrocarbons and nitrogen oxides. It is appropriate, therefore, to begin the discussion of ozone formation in photochemical smog with a brief description of the chemical composition of emissions derived from automobiles.

Gasoline-powered vehicles account for most of the emissions from mobile sources, and for the purpose of the present discussion we can neglect the contributions from diesel and jet engines. The bulk composition of exhaust gases from piston engines run on gasoline consists of, roughly, 78% nitrogen, 12% carbon dioxide, 5% water vapor, 1% unused oxygen, and about 2% each of carbon monoxide and hydrogen. The remainder is about 0.08% hydrocarbons, 0.06% nitric oxide, and several hundred ppmv partially oxidized hydrocarbons, mostly aldehydes with formaldehyde making up the largest fraction. The precise composition clearly depends on the type of engine and the way the engine is run.

Nitric oxide is formed in combustion engines by the interaction of oxygen and nitrogen in air at the high temperatures reached during the combustion cycle. The percentages of NO found in the exhaust gas is close to that calculated from the chemical equilibrium $N_2 + O_2 = 2NO$ at peak temperatures near 2500 K. Once NO is formed, its abundance seems to be effectively frozen in. The mechanism of NO formation is not precisely known. Most authors have adopted the reaction chain first proposed by Zeldovich et al. (1947):

$$O + N_2 \rightarrow NO + N$$

$$N + O_2 \rightarrow NO + O$$

but the first of these reactions is quite endothermic (by 316 kJ/mol), and it is questionable whether it can proceed sufficiently rapidly, even at 2500 K, to bring the NO level up to equilibrium values within the millisecond time scale available during the combustion cycle. The problem has been discussed by Palmer and Seery (1973) and by Campbell (1977). In the combustion of coal and heavy fuel oils, organic nitrogen compounds already present in the fuel contribute appreciably to nitrix oxide formation. From gasoline the contribution is negligible.

The approximate composition of hydrocarbons in the exhaust of gasoline-powered cars is given in Table 5-1. Also listed for comparison are typical compositions of gasoline, of fuel evaporates, and of Los Angeles city air sampled in the morning. In the table, the hydrocarbons are grouped as alkanes, alkenes, alkynes, and aromatics, but a differentiation between the various isomers is not made. The data were obtained primarily in the 1960s, but should still be applicable today. Gasoline contains mostly saturated compounds in the C_4 to C_9 range of carbon numbers plus varying amounts of aromatics and unsaturated hydrocarbons. The exact composition depends greatly on the supplier and is rarely known. In fact, gas chromatographic separation of all isomeric components in gasoline appears to have been achieved quite late (Maynard and Sanders, 1969). The "typical" composition shown in Table 5-1 consists of 60% alkanes, 30% aromatics, and 10% alkenes. Formerly, the percentage of the alkenes was greater. The reduction was made on account of their high reactivity in the atmosphere.

The lower members of the hydrocarbon chains have a tendency to evaporate, and column 2 of Table 5-1 gives the composition of vapor in the head space of a fuel tank. Starting in 1971, U.S. cars were equipped with evaporation control systems whereby the direct emission of fuel into the atmosphere from tanks and carburetors was greatly reduced. Fuel evaporation remains important in the handling of fuel at filling stations.

The composition of hydrocarbons in the engine exhaust gases is partially correlated with fuel composition inasmuch as a small portion of the fuel leaves the combustion chamber unburned. The lighter hydrocarbons, such as methane, ethane, ethylene, and acetylene, are formed by the fragmentation of the fuel components during the combustion process. Certain other species—for example, benzene—also are present with a higher relative abundance than in the fuel, which indicates that they must be formed by combustion reactions.

Essentially all hydrocarbons—that is, also the isomers found in the exhaust gases—show up again in the urban air. Relative concentrations are different, however. A striking example is methane. After correcting for the approximately 1.5 ppmv contribution by natural methane, one finds that CH_4 accounts for 70% of all hydrocarbons in urban air, although its relative abundance in automobile exhaust is only 15%. In fact, practically all alkanes appear to be enriched compared with the alkenes and aromatic compounds. One possible reason is that because of their lower reactivity, and hence longer lifetime in the atmosphere, the alkanes carry over from previous days. Note also that propane is present at 46 ppbv in Los Angeles air, although automobile exhaust contains negligible amounts of this substance. It is obvious that propane must originate from a different source. Possible origins were first discussed by Neligan (1962) and Stephens and Burleson

Table 5-1. *Hydrocarbons Derived from Automobiles*[a]

Hydrocarbon	Fuel (%)	Evaporate (%)	Exhaust ppmv	Exhaust %[b]	Los Angeles air ppbv	Los Angeles air %[b]
Alkanes						
C_1	—	—	86	15.0	3220	70[c]
C_2	—	—	14	2.4	100	4.1
C_3	—	—	1	0.1	47	1.9
C_4	7.0	33	23	4.0	74	3.0
C_5	19.0	35	23	4.0	76	3.1
C_6	11.0	7.8	14	2.4	42	1.7
C_7	12.0	—	13	2.3	35	1.4
C_8	8.0	—	48	8.4	26	1.1
$\geq C_9$	3.5	—	3	0.5	?	—
Alkenes						
C_2	—	—	97	17.0	67	2.7
C_3	—	—	40	7.0	19	0.8
C_4	0.5	3.2	17	3.0	15	0.6
C_5	4.5	10.8	16	2.8	12	0.5
$\geq C_6$	5.0	—	9	1.6	2.5	0.1
Alkynes						
C_2	—	—	55	9.7	55	2.2
C_3	—	—	7	1.2	10	0.4
Aromatics						
C_6	1.0	0.4	14	2.4	23	0.9
C_7	16.5	0.9	43	7.5	39	1.6
C_8	8.5	—	18	3.1	57	2.3
$\geq C_9$	3.5	—	29	5.1	28	1.1

[a] Comparison of hydrocarbon composition of gasoline, fuel evaporate, exhaust, and ambient air sampled in Los Angeles. Adapted from data presented by Glasson and Tuesday (1970), Heaton and Wentworth (1959), Kopczynski *et al.* (1972), Leach *et al.* (1964), Lonneman *et al.* (1974), Maynard and Sanders (1969), McEwen (1966), Stephens *et al.* (1958), Stephens and Burleson (1967), and Tuesday (1976).

[b] Percent of total hydrocarbons.

[c] After deducting the contribution due to natural methane.

(1967), who considered natural gas, which contains about 1.3% of propane besides 6% of ethane, as a like source of propane. Propane is an extremely minor constituent of gasoline, of the order of 0.05 ppmw, although its vapor pressure is high. Mayrsohn and Crabtree (1976) have given a reasonable assignment to various sources of hydrocarbons in Los Angeles air (see Table 6-5).

Consider now the interaction of nitrogen oxides with hydrocarbons in sunlit urban air and the generation of ozone from this mixture. Figure 5-1 shows the variation with time of the nitrogen oxides, hydrocarbons, and

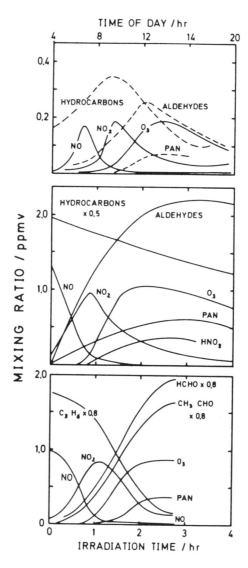

Fig. 5-1. Variation with time of hydrocarbons, aldehydes, nitrogen oxides, ozone, and peroxyacetyl nitrate (PAN) for three conditions. Top: Downtown Los Angeles during the course of a day with eye irritation; composed of data presented in Leighton (1961) and Air Quality Criteria for Photochemical Oxidants (1970). Center: Irradiation of automobile exhaust diluted with air in a smog chamber of a plastic bag exposed to sunlight; composed of data from Leighton (1961, originally Schuck *et al.*, 1958), Kopzcynski *et al.* (1972), Wilson *et al.* (1973), Miller and Spicer (1975), Jeffries *et al.* (1976), and Wayne and Romanofsky (1961). Bottom: Smog-chamber irradiation of a mixture of propene, nitric oxide, and air. [Adapted from data presented by Altshuller *et al.* (1967) and Pitts *et al.* (1975).] Note differences in the time scales.

ozone in Los Angeles during the cycle of 1 day. In the city, a great deal of NO and hydrocarbons is deposited in the early morning during the rush-hour traffic. In the course of time, NO is converted to NO_2, hydrocarbons to aldehydes, and ozone rises as a product peaking in midday before decaying again due to reactions of its own as well as dispersal processes. Figure 5-1, center, shows results of experiments with automobile exhaust, diluted with clean air, filled into transparent bags or vessels, and irradiated with either natural or artificial sunlight. The data resemble rather closely the behavior of nitrogen oxides, hydrocarbons and aldehydes, and ozone in the urban atmosphere. The striking similarity provides an important piece of evidence for the belief that automobiles are a prime source for photochemical smog precursors. Figure 5-1, bottom, is added to show that equivalent results are obtained when mixtures of individual hydrocarbons with air containing also nitric oxide are irradiated in smog chambers. The example given, with propene as the hydrocarbon, has been selected from many such results in the literature because it clearly demonstrates the autocatalytic nature of hydrocarbon oxidation in photochemical smog. The rate of hydrocarbon consumption accelerates with time, indicating the occurrence of a chain reaction accompanied by a multiplication of chain carriers (compare with Fig. 2-1). The common features of all the data assembled in Fig. 5-1 are the formation of aldehydes as intermediate products of hydrocarbon oxidation and, more importantly for the present discussion, the rapid conversion of NO to NO_2 preceding the formation of ozone. These results lead to the conclusion that the reaction mechanisms at work follow a common scheme, even though the individual hydrocarbon species involved may differ.

Figure 5-1 suggests that the formation of ozone is the result of two consecutive reaction steps. The first is the oxidation of NO to NO_2, and the second a reaction that converts NO_2 to ozone. The latter process has been recognized quite early to result from the photochemical activity of NO_2 (Leighton, 1961). As discussed in Chapter 2, nitrogen dioxide is an effective absorber of visible and near ultraviolet radiation. At wavelengths below about 400 nm, NO_2 undergoes photodissociation giving rise to the following reaction sequence:

$$NO_2 + h\nu \rightarrow NO + O \qquad J_{NO_2} \approx 5 \times 10^{-3} \, s^{-1}$$

$$O + O_2 + M \rightarrow O_3 + M \qquad k_1 n(O_2) n(M) \approx 10^5 \, s^{-1}$$

$$O_3 + NO \rightarrow NO_2 + O_2 \qquad k_{60} n(O_3) \approx 10^{-2} \, s^{-1}$$

The photodissociation of NO_2 is the only definitely established process for the formation of ozone in the troposphere and must be held responsible also for the generation of ozone in photochemical smog. The reaction of alkylperoxy radicals with oxygen, $ROO + O_2 \rightarrow RO + O_3$ has occasionally been invoked—for example, by Cadle and Allen (1970)—but according to current thermochemical knowledge the RO–O bond is stronger than the O_2–O bond in ozone, so that the reaction would be endothermic and thus slow.

Without knowing the individual rate coefficients associated with the above reactions, one might interpret the rise of ozone at the expense of NO_2 apparent in all data in Fig. 5-1 to result from a slow approach to photochemical steady state. This conclusion is unjustified, however, because all three reactions are rapid. Approximate rates are indicated on the right-hand side of each process for $m(O_3) = 25$ ppbv and $p = 1000$ mbar (i.e., normal ground-level conditions). The numbers show that the $O_3 + NO$ reaction is rate-determining and that the steady state is reached within about 200 s. In the absence of other reactions interfering with the above scheme, the photostationary state should prevail during the entire day after NO_2 has attained its maximum. If one sets up the kinetic equations to calculate steady-state mixing ratios of NO, NO_2, and ozone for comparison with observations, one often finds agreement within a factor of two or so. This result unfortunately gives a false impression about the applicability of the reaction mechanism. It is necessary to take into account the stoichiometric relations

$$m(O_3) = m(NO_x) - m(NO_2) = m(NO)$$

where $m(NO_x)$ denotes the sum of the mixing ratios of NO and NO_2. Even a cursory inspection of Fig. 5-1 shows that the expected stoichiometric balance is not obeyed. The amount of O_3 rises while that of NO declines, and if one ascribes the decay of NO_2 to a loss such as oxidation to nitric acid, the sum of NO and NO_2 should decline as well. Ozone then should follow suit, whereas, in fact, it continues to rise. It is clear, therefore, that the three reactions coupling NO, NO_2, and O_3 cannot by themselves describe the formation of ozone in photochemical smog. Additional reactions must be involved; specifically, a reaction must be invoked that converts NO to NO_2 without affecting ozone. The same reaction might then explain the rapid oxidation of NO to NO_2 that initiates the formation of ozone.

It is significant to note that the reaction $NO + O_3$ cannot be used to explain the conversion of NO to NO_2 during the first stage of smog formation, because initially the mixing ratio of ozone is too small to have much effect. Chemists generally are familiar with the Bodenstein (Bodenstein and Milford, 1918) reaction

$$2NO + O_2 \rightarrow 2NO_2$$

which oxidizes nitric oxide directly. In the atmosphere, however, this process is extremely slow at the usual concentrations of NO, since the rate of the reaction involves the NO concentration squared:

$$-\frac{dn(NO)}{dt} = 2k_{67}n(O_2)n^2(NO)$$

where $n(O_2)$ and $n(NO)$ are the number densities of oxygen and nitric oxide, respectively. For this second-order process, the time $t_{1/2}$ after which the NO number density has been reduced to one-half its initial value $n_0(NO)$ is given by

$$t_{1/2} = 1/[2k_{67}n(O_2)n_0(NO)]$$

$$= 1/[2k_{67}n_T^2 m(O_2)m_0(NO)]$$

Mixing ratios are introduced for convenience in the second expression, with n_T denoting the total number density of all molecules in the NO–NO_2–air mixture. The rate coefficient, gleaned from Table A-5 in the Appendix, is $k_{67} = 2 \times 10^{-38}$ cm^6/molecule^2 s for $T = 300$ K, and taking $m(O_2) = 0.21$ and $n_T = 2.5 \times 10^{19}$ molecules/cm^3 one calculates the following NO half-life times for various initial NO mixing ratios:

$m_0(NO)$ (ppmv)	0.1	1.0	10	100
$t_{1/2}$ (h)	530	53	5.3	0.53

Mixing ratios in excess of 10 ppmv are required in order to achieve a reasonably rapid conversion of NO to NO_2 by the 2NO + O_2 reaction. Such high values of NO mixing ratios may occur in the immediate vicinity of emission sources, but are unusual in urban air, where they ordinarily are in the parts per billion range. This fact makes the Bodenstein reaction an almost negligible process in the atmosphere.

The high rate at which NO is converted to NO_2 in photochemical smog has baffled atmospheric scientists for many years, but it is now believed to result from reactions of peroxidic radicals of the general type ROO· formed by the association of alkyl or other hydrocarbon radicals with molecular oxygen. The reaction sequence can be written

$$R· + O_2 + M \rightarrow ROO· + M$$

$$ROO· + NO \rightarrow NO_2 + RO·$$

At atmospheric pressure, the rate coefficients for the R + O_2 association reactions are already close to their second-order limit. The high concentration of oxygen makes these reactions occur almost instantaneously, so that other conceivable reactions of R radicals are negligible by comparison. Reactions of ROO· with NO are known to be fast for HO_2 and CH_3O_2 (see Table A-4), and other alkylperoxy radicals are assumed to react with NO at similar rates. A justification for this assumption is derived from model calculations designed to interpret smog-chamber data. The calculated NO–NO_2 conversion rates generally are found to agree well with observations. Examples can be found, among others, in the early work of Niki et al. (1972), the detailed review of smog-chamber reactions mechanisms by Demerjian et al. (1974), and the more recent comprehensive study by Carter et al. (1979) of the oxidation of butane and propene in smog chambers.

The hydrocarbon radicals required to form alkylperoxy radicals must derive from the breakup of hydrocarbons during their oxidation, and we turn now to the processes initiating hydrocarbon decomposition. Historically there was a period when the consumption of hydrocarbons in photochemical smog was treated in terms of reactions with oxygen atoms and ozone, the oxygen atoms arising mainly from the photolysis of NO_2, and ozone coming in during the later stages of smog formation. The steady-state number density of O atoms resulting from NO_2 photodissociation can be fairly reliably estimated as about 2×10^5 molecule/cm^3 in smog chambers, and the numbers density of ozone is obtained directly from measurement as about 2.5×10^{12} molecules/cm^3. When rate coefficients for their reactions became available and rates for the consumption of hydrocarbons were calculated, it was found that they fall far short of explaining the loss rates of hydrocarbons observed. Niki et al. (1972) were among the first to point out the inadequacy of O-atom and ozone reactions, and they suggested that hydrocarbons are preferentially attacked by OH radicals. The realization that hydroxyl radicals are important intermediates and that they are more effective than other species in reacting with hydrocarbons came relatively late, presumably because rate coefficients for OH reactions were initially not well known. The development of the flash photolysis technique for the laboratory generation of OH radicals from H_2O greatly improved the situation with respect to determinations of OH rate coefficients. Greiner (1967b, 1970) furnished the first reliable rate data for reactions with CO and several alkanes. Stuhl and Niki (1972) then perfected the detection of OH by means of resonance fluorescence, and the combination of these methods provided a wealth of data during the subsequent years. Atkinson et al. (1979) have reviewed these data.

Table 5-2 compares rate coefficients for reactions of oxygen atoms, ozone, and hydroxyl radicals with selected hydrocarbons, and the corresponding rates for the loss of the same hydrocarbons calculated with appropriate steady-state values for the number densities of O, O_3, and OH. The rates of OH radicals exceed those of the other reactants for almost every hydrocarbon listed. The OH number density adopted, 2×10^6 molecules/cm^3, corresponds to that expected at noon on a clear summer day in the natural atmosphere at 30° northern latitude.

Table 5-2. *Rate Coefficients for Reactions of Oxygen Atoms, Ozone, and OH Radicals with Selected Hydrocarbons*[a]

Hydrocarbon	k_O	k_{O_3}	k_{OH}	R_O	R_{O_3}	R_{OH}
Methane, CH_4	1.7 (−17)	<7 (−24)	7.7 (−15)	2.6 (−12)	<1.7 (−11)	1.5 (−8)
Ethane, C_2H_6	9.1 (−16)	<1 (−20)	2.5 (−13)	1.4 (−10)	<2.5 (−8)	5.0 (−7)
Propane, C_3H_8	1.1 (−14)	<1 (−20)	1.1 (−12)	1.8 (−9)	<2.5 (−8)	2.2 (−6)
n-Butane, C_4H_{10}	2.2 (−14)	<1 (−20)	2.7 (−12)	3.5 (−9)	<2.5 (−8)	5.4 (−6)
Ethylene, C_2H_4	6.6 (−13)	1.9 (−18)	2.5 (−12)	1.0 (−7)	4.2 (−6)	5.0 (−6)
Propylene, C_3H_6	3.3 (−12)	1.1 (−17)	2.5 (−11)	5.2 (−7)	2.7 (−5)	5.0 (−5)
cis/trans-Butene, C_4H_8	2.0 (−11)	2.0 (−16)	6.0 (−11)	3.2 (−6)	5.0 (−4)	1.2 (−4)
Acetylene, C_2H_2	1.4 (−13)	3.8 (−20)	8.5 (−13)	2.2 (−8)	9.5 (−8)	1.7 (−6)
Benzene, C_6H_6	2.4 (−14)	7 (−23)	1.2 (−12)	3.8 (−9)	1.7 (−10)	2.4 (−6)
Toluene, C_7H_8	7.5 (−14)	1.5 (−22)	6.4 (−12)	1.2 (−8)	3.7 (−10)	1.2 (−5)

[a] The corresponding hydrocarbon consumption rates $R = -dn(HC)/n(HC)\,dt$ are also given, assuming steady-state number densities of 1.6×10^5, 2.5×10^{12}, and 2×10^6 molecules/cm^3 for oxygen atoms, ozone, and OH radicals, respectively. Orders of magnitude are shown in parentheses.

By itself, the high reactivity of OH radicals does not constitute sufficient evidence for the hypothesis that they are the principal species responsible for the degradation of hydrocarbons in photochemical smog. A more tangible support is provided by the finding that hydrocarbon reactivities in smog chambers correlate closely with the rate coefficients of OH reactions. This aspect seems to have been recognized first by Morris and Niki (1971) for a number of alkenes and aldehydes. The effectiveness of individual hydrocarbons in producing photochemical smog may be defined in several different ways. Examples are the rates of hydrocarbon consumption, NO to NO$_2$ conversion, or oxidant formation, all measured under controlled experimental conditions. Biological effects may provide another, quite different, standard, with which we shall not be concerned here, however. Of the first category, only hydrocarbon consumption will be considered as a measure of reactivity. The existence of a coarse relationship between hydrocarbon consumption and oxidant formation rates has been established by Farley (1978) and by Winer et al. (1979). The fact that hydrocarbon reactivities differ has been known since the earliest days of smog investigations and can be understood on the basis of differences in the reaction rates. Table 5-2 may serve to indicate the range. Low-weight alkenes react faster than benzene and toluene, and these react faster again than the light saturated hydrocarbons. The differences in reactivities had given rise to hopes that one might curb oxidant formation through a stringent emission control of the more reactive hydrocarbons, and for this purpose much effort has gone into setting up relative reactivity scales by means of smog-chamber experiments. Altshuller and Bufalini (1971) have reviewed the results of such studies carried out during the 1960s. Unfortunately, the data scattered widely due to different and sometimes poorly controlled experimental conditions, in addition to influences of wall and memory effects. More recently, Pitts et al. (1978) have been able to reduce the uncertainties to a tolerable level by studying hydrocarbon disappearance rates relative to a compound whose OH rate constant was well known and that thus served as an internal standard. By scaling the results it was found that the hydrocarbon loss rates agreed favorably with OH rate coefficients obtained in independent investigations using the flash photolysis–resonance fluorescence method. Figure 5-2 compares results presented by Pitts et al. (1978) with similar data reported by Wu et al. (1976), who used cis-2-butene as an internal standard. Pitts et al. (1978) used n-butane. The correlation is so good that the smog chamber has been proposed as a device for the determination of OH rate coefficients for those hydrocarbons that cannot be investigated by other experimental techniques (Darnall et al., 1976a). It must be emphasized that hydrocarbon loss rates correspond to OH reaction rates only during the initial stage of the reaction in smog chambers,

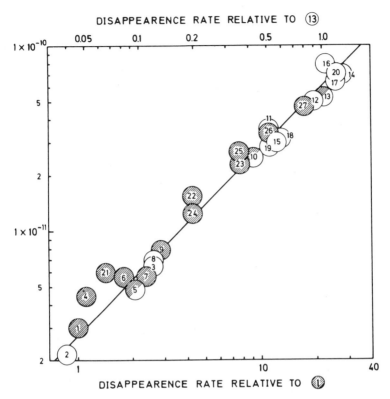

Fig. 5-2. Plot of relative disappearance rates for hydrocarbons observed in smog chambers versus OH rate coefficients (in units of cm^3/molecules s) from independent measurements. Key: (1) n-butane, (2) isobutane, (3) n-pentane, (4) isopentane, (5) n-hexane, (6) 2-methylpentane, (7) 3-methylpentane, (8) cyclohexane, (9) ethane, (10) propene, (11) 1-butene, (12) isobutene, (13) cis-2-butene, (14) trans-2-butene, (15) 1-pentene, (16) methyl-1-butene, (17) cis-2-pentene, (18) 1-hexene, (19) 3,3-dimethylbutene, (20) cyclohexene, (21) toluene, (22) o-xylene, (23) m-xylene, (24) p-xylene, (25) 1,2,3-trimethylbenzene, (26) 1,2,4-trimethylbenzene, (27) 1,3,5-trimethylbenzene. Open circles from Wu et al. (1976) relative to cis-2-butene, upper scale; hatched circles from Lloyd et al. (1976) and Pitts et al. (1978) relative to n-butane, lower scale.

that is, during the NO to NO_2 conversion stage. Once the concentration of ozone rises to appreciable levels, one must take into account the reaction of the hydrocarbon with ozone. This qualification applies specifically to the alkenes that are most reactive toward ozone. Table 5-2 may again serve to demonstrate this point. Although rate coefficients for reactions between alkenes and ozone are not particularly large on an absolute scale, the reactions do become competitive with OH reactions when the concentration of ozone builds up.

The preceding considerations provide the basis for a general mechanism for ozone production in photochemical smog by the oxidation of hydrocarbons. Early in the morning the rush-hour traffic leads to high NO concentrations, which prevent appreciable amounts of ozone being present because of the reaction $O_3 + NO \rightarrow NO_2 + O_2$. Before ozone can appear, NO must first be converted to NO_2 by the degradation of hydrocarbons. As an example for such a reaction scheme, consider the oxidation of a straight-chain alkane $R_1CH_2R_2$ where R_1 and R_2 stand for suitable alkyl groups. The reaction is started by the abstraction of a hydrogen atom from the carbon skeleton, preferentially at sites of secondary hydrogen atoms. The chain is propagated by reactions of alkyl peroxy radicals as noted earlier.

$$R_1CH_2R_2 + OH \rightarrow R_1\dot{C}HR_2 + H_2O$$

$$R_1R_2\dot{C}H + O_2 \rightarrow R_1R_2CHOO\cdot$$

$$R_1R_2CHOO\cdot + NO \rightarrow R_1R_2CHO\cdot + NO_2$$

$$R_1R_2CHO\cdot + O_2 \rightarrow R_1COR_2 + HO_2$$

$$R_1R_2CHO\cdot \rightarrow R_1\cdot + R_2CHO$$

$$R_1\cdot + O_2 \rightarrow R_1OO\cdot$$

$$R_1OO\cdot + NO \rightarrow R_1O\cdot + NO_2$$

$$R_1O\cdot + O_2 \rightarrow R_1'CHO + HO_2$$

$$HO_2 + NO \rightarrow NO_2 + OH$$

net: $R_1CH_2R_2 + 3NO + O_2 \rightarrow 3NO_2 + R_1COR_2 + R_1'CHO + H_2O$

$$NO_2 + h\nu + O_2 \rightarrow NO + O_3$$

Here, R_1' represents an alkyl group containing one carbon atom less than R_1 so that $R_1O\cdot = R_1'CH_2O\cdot$. The net reaction converts one molecule of *n*-alkane into one molecule of ketone and aldehyde each, and it oxidizes three molecules of NO to NO_2. The subsequent photodissociation of NO_2 is the source of ozone in photochemical smog. Note that the OH radical initiating the reaction sequence is regenerated so that it can continue the chain reaction.

Aldehydes are considerably more reactive toward OH radicals than alkanes, so that the aldehydes produced are subject to further oxidation. Thereby additional NO is converted to NO_2. To give an example, let R_1' stand for CH_3 so that $R_1'CHO$ represents acetaldehyde. The mechanism for acetaldehyde oxidation may be written as follows:

$$CH_3CHO + OH \rightarrow CH_3CO + H_2O$$

$$CH_3CO + O_2 \rightarrow CH_3(CO)O_2$$

$$CH_3(CO)O_2 + NO \rightarrow CH_3 + CO_2 + NO_2$$

$$CH_3 + O_2 \rightarrow CH_3O_2$$

$$CH_3O_2 + NO \rightarrow CH_3O + NO_2$$

$$CH_3O + O_2 \rightarrow HCHO + HO_2$$

$$HO_2 + NO \rightarrow NO_2 + OH$$

net: $CH_3CHO + 3NO + 3O_2 \rightarrow HCHO + 3NO_2 + CO_2 + H_2O$

$$NO_2 + h\nu + O_2 \rightarrow NO + O_3$$

Here again, the OH radical initiating the reaction is regenerated. In the mechanism shown, three molecules of NO are oxidized to NO_2 and an equivalent number of ozone molecules is subsequently formed. This raises the total number of product ozone molecules to six. Additional NO_2 and ozone arises from the photoxidation of formaldehyde, HCHO, produced from acetaldehyde, and from the photooxidation of the ketones occurring as products of alkane oxidation.

The full potential of ozone production by hydrocarbon oxidation will be utilized only if the oxidation is carried to completion. Locally, this is rarely the case, however. The buildup of aldehydes during a day in the Los Angeles atmosphere and in smog chambers, displayed by the data in Fig. 5-1, makes clear that the time of exposure to solar radiation is too short for all the aldehydes to react. It is useful, therefore, to distinguish two time scales, a short one covering only the initial oxidation stage leading to the formation of carbonyl compounds, and a longer time scale during which the oxidation of carbonyls also is completed. The first time scale indicates a minimum, and the second a maximum potential for ozone formation. Table 5-3 shows for several alkanes and alkenes the number of NO molecules converted to NO_2 in the course of the oxidation of one molecule of hydrocarbon. The yields of NO_2 (and ozone) are based on reaction mechanisms similar to those shown above. A more detailed discussion will be given in Section 6.2. The intermediate aldehydes and ketones are shown in the second column of Table 5-3. The third column indicates the number of NO_2 molecules formed during the initial oxidation state, and the fourth column the number of NO_2 molecules formed in the subsequent oxidation of carbonyls. In the first stage the number is two for alkenes and three for alkanes, except for ethane. This agrees with measurements of Niki *et al.* (1978) for ethene, propene, and *trans*-2-butene. The extent of NO conversion during the second stage grows roughly with the carbon number of the hydrocarbon being oxidized, with the alkanes having a slightly greater potential than the alkenes. It is doubtful whether in the atmosphere the maximum potential for ozone formation can be realized, since aldehydes and ketones are removed not only by photooxidation but also by physical scavenging processes such as wet and dry deposition. In addition, the oxidation of higher hydrocarbons

Table 5-3. *Potential for Ozone Formation by Several Hydrocarbons Following Their Reaction with OH Radicals*[a]

Compound	Intermediate aldehydes and ketones	Number of NO molecules converted		
		Initial	From carbonyl compounds	Total
Ethene	$2CH_2O$	2	2	4
Propene	CH_3CHO, CH_2O	2	5	7
1-Butene	CH_3CH_2CHO, CH_2O	2	8	10
cis/trans-2-Butene	$2CH_3CHO$	2	8	10
Isobutene	CH_3COCH_3, CH_2O	2	5	7
Ethane	CH_3CHO	2	4	6
Propane	CH_3CHO, CH_2O, CH_3COCH_3	3	5	8
n-Butane	$2CH_3CHO$	3	8	11
Isobutane	CH_3COCH_3, CH_2O	3	5	8
n-Pentane	$CH_3(CH_2)_2CHO$, CH_2O CH_3CH_2CHO, CH_3CHO	3	11	14

[a] The number of NO molecules converted to NO_2 per hydrocarbon molecule is given for two stages of the oxidation: from the initial stage leading only to the formation of aldehydes and ketones, and for the further oxidation of the carbonyl compounds.

and aromatics may lead to the formation of condensable products that attach to aerosol particles, thus escaping further oxidation. Finally, it should be noted from Fig. 5-1 that an appreciable portion of product NO_2 gives rise to peroxyacetyl nitrate and nitric acid whereby the yield of ozone is further diminished.

The preceding discussion has eluded the question about the origin of OH radicals that are necessary to trigger hydrocarbon oxidation. In the natural atmosphere hydroxyl radicals are believed to originate from $O(^1D)$ atoms formed in the near ultraviolet photolysis of ozone and their subsequent reaction with water vapor (cf. Section 4.2). While the same process will be important in the urban atmosphere after ozone has built up to sufficient levels, it cannot serve to initiate photochemical smog in the early morning because at that time the high NO concentrations keep ozone at low levels. Hydroxyl radicals may of course be formed by hydrogen abstraction from hydrocarbons due to reactions of oxygen atoms resulting from the photo-decomposition of NO_2, but again it is first necessary to convert NO to NO_2 before this source attains significance. Other conceivable sources of OH are photolysis of H_2O_2 or HNO_3. Their absorption cross sections at wavelengths above 300 nm are very low, however, as Fig. 2-14 showed, so that these processes are rather ineffective.

The most promising process for the generation of OH radicals in the morning appears to be the photolysis of nitrous acid,

$$HNO_2 + h\nu \rightarrow NO + OH \qquad \lambda = 395 \quad nm$$

Nitrous acid has been observed in the atmosphere by long-path optical absorption (Platt and Perner, 1980; Platt *et al.*, 1980b; Pitts *et al.*, 1984c). Substantial evidence for the importance of HNO_2 as an OH source was obtained at Riverside, California, a community located about 65 miles east of Los Angeles. Riverside is almost daily subjected to a plume of smog-laden air carried eastward from Los Angeles. The observational data for HNO_2 at this site are shown in Fig. 5-3. The mixing ratio of nitrous acid increases during the night until it reaches a level of about 2 ppbv. After daybreak, HNO_2 disappears at a rate in agreement with that estimated from the HNO_2 photodissociation coefficient. During the day the mixing ratio of HNO_2 is kept at a low value due to its photolysis.

In the morning, the efficiency of HNO_2 as an OH source is greater than that of formaldehyde, which is emitted from automobiles simultaneously with NO and hydrocarbons and provides yet another source of OH, albeit an indirect one. As discussed in Section 2.5, one of the photodecomposition channels of formaldehyde leads to the formation of radicals. This causes the following reaction sequence:

$$HCHO + h\nu \rightarrow HCO + H$$

$$HCO + O_2 \rightarrow HO_2 + CO$$

$$H + O_2 + M \rightarrow HO_2 + M$$

$$NO + HO_2 \rightarrow NO_2 + OH$$

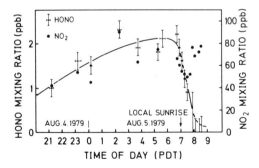

Fig. 5-3. Behavior of nitrous acid and nitrogen dioxide during the night of August 4–5, 1979, at Riverside, California. PDT = Pacific Daylight Time. [From Platt *et al.* (1980b) with permission.]

Comparison of the data in Table 2-7 (and of Figs. 2-18 and 2-19) shows that the photodissociation coefficient for radical formation from formaldehyde is by almost two orders of magnitude smaller than that for nitrous acid. Formaldehyde is definitely more abundant in urban air than nitrous acid, but in the morning the HCHO mixing ratio is not much greater than 50 ppbv if one accepts the data of Fig. 5-1. At sunrise, formaldehyde will be an inferior source of OH compared with HNO_2. Later in the day the mixing ratio of formaldehyde increases whereas that of nitrous acid declines to low values. Formaldehyde then becomes more important as a radical source. At the same time, the mixing ratio of ozone rises as well so that it contributes to OH formation via $O(^1D)$ production by photolysis at wavelengths below 310 nm. These processes sustain the smog reactions during the day.

A continuous source of radicals is required, in addition to that resulting from the regeneration of OH in hydrocarbon oxidation chains, in order to replace radicals that are lost by chain-terminating reactions. No particular attention has been paid to chain-terminating reactions in the foregoing description of smog chemistry, but that they should not be neglected has been made clear in the general discussion of chain reactions in Section 2.1. Two specific reactions leading to identifiable products are $2HO_2 \rightarrow H_2O_2 + O_2$ and $NO_2 + OH \rightarrow HNO_3$. Both hydrogen peroxide and nitric acid have been observed in the urban atmosphere as well as in smog chambers (Bufalini et al., 1972; Lonneman et al., 1976; Miller and Spicer, 1975). Other products from termination reactions are thermally unstable and may redissociate when temperatures are high. Peroxyacetyl nitrate resulting from the recombination of $CH_3(CO)OO$ radicals with NO_2 is such a compound. It is stable enough to be transported over some distance so that it may serve as an indicator of photochemical air pollution outside the source regions.

5.3   Distribution and Behavior of Tropospheric Ozone

Measurements of ozone in the troposphere fall into three categories: balloon soundings, observations on board instrumented aircraft, and surface measurements. The great variability of ozone requires longer measurement series in order to determine average mixing ratios, seasonal trends, and other features. Only two measurement programs have given information on the meridional distribution of ozone in a systematic fashion: the North American Ozonesonde Network, which was operative during the 1960s (Hering and Borden, 1967), and the network of stations for the measurement of surface ozone established by Fabian and Pruchniewicz (1976) during the 1970s in western Europe and on the African continent. This program was supplemented by aircraft measurements at high altitudes. Data from both programs cover only part of the troposphere, and supplementary data from individual stations at other locations should be added to complete the overall picture.

5.3.1 BALLOON SOUNDINGS

Balloon sondes for the measurement of stratospheric ozone generally employ electrochemical instruments based on the ozone–iodide reaction (Brewer and Milford, 1960; Komhyr, 1969). The North American Ozonesonde Network used for a number of years a different measurement technique, based on the dry chemiluminescence from an organic dye (Regener, 1964). The sensor requires calibration in the field, which is achieved most conveniently by Dobson spectrophotometry. The electrochemical sonde is potentially an absolute instrument. In practice, it requires a number of corrections, and comparison with independent measurements of total ozone is always made. A field intercomparison of both sondes (Hering and Dütsch, 1965) gave good agreement in the strato- sphere. In the troposphere the values obtained with the electrochemical sonde were often higher than those indicated by chemiluminescence. The origin of the differences has not been revealed. Both detectors may be influenced by trace constituents other than ozone, particularly in the lowest layers of the atmosphere where concentrations are high. During a certain period in 1963/1964 many chemiluminescent sondes were found to experience a loss in sensitivity during ascent. The problem was traced to a modified procedure for the fabrication of the dye-covered chemiluminescing disk and was subsequently corrected, but it undermined confidence in the validity of the data. One would nevertheless think that a lowering of the sensitivity during ascent affects stratospheric ozone values more than tropo- spheric ones, so that after normalization to total ozone higher rather than lower concentrations of ozone should be determined in the troposphere compared with the electrochemical sonde. Chatfield and Harrison (1977a, b) have performed a critical analysis of both data sets. They confirmed that in the lower troposphere the Regener sonde measured 35% less ozone, on average, than the Brewer sonde. They also noted that the former showed larger fluctuations than the latter, leading to greater standard deviations among the observed values. For this reason they considered the chemiluminescence device less reproducible and hence less reliable. This ignores its very rapid response, however. The Regener sonde indicates natural fluctuations of ozone almost instantaneously, whereas the Brewer sonde has a response time of 20 s, which tends to smear out local variations in the ozone mixing ratio. One should further appreciate that the sensitivities

of both sensors are adjusted to be optimal in the stratosphere and that
readings in the troposphere are low and correspondingly more uncertain.

Figure 5-4 compares averaged latitude profiles to tropospheric ozone for
the summer season. The differences between both sensors are clearly displayed. While in the lower troposphere the electrochemical instrument gives
the higher values, there is a tendency for a reversal of the trend in the upper
troposphere. The variances also are considerable. The horizontal bars superimposed on the profile at Wallops Island in Fig. 5-4 are to indicate one
standard deviation. Variances at other stations are similar except in the
tropics, where they are smaller. Note the increase of variability as one
approaches the tropopause height. The effect is caused mainly by the partial
sampling of stratospheric air and less by the variability due to the influx
of stratospheric onzone into the troposphere. Gradients of ozone mixing
ratios have a similar magnitude at extratropical latitudes. The increase in
the upper troposphere must again be assigned to tropopause variability. In
the tropics the gradients are smaller. To the extent that the upward motion
of the Hadley circulation impedes a steady injection of ozone across the
tropical tropopause, the data are consistent with the viewpoint that tropospheric ozone has its origin in the stratosphere. The gradient then is a
consequence of the continuous eddy flux of ozone from the tropopause

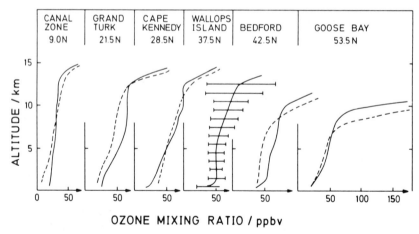

OZONE MIXING RATIO / ppbv

Fig. 5-4. Averaged altitude profiles of ozone in the troposphere above six stations of the
North American Ozonesonde Network. Solid curves represent data for July obtained with the
electrochemical sonde in the years 1966-1969 (Chatfield and Harrison, 1977b). The dashed
curves represent data for the summer season obtained with the chemiluminescence sonde in
the years 1963-1965 (Hering and Borden, 1967). Since they do not include data for Cape
Kennedy, the data for Tallahassee, Florida, at 30.4°N were used instead. Horizontal bars for
the data of Wallops Island indicate one standard deviation.

toward the ground surface. The existence of a gradient as such does not provide evidence for the dominance of transport, however, since the local generation and loss of ozone by photochemical processes may be fast enough to overrule transport processes.

Chatfield and Harrison (1977b) have analyzed the seasonal dependence of the averaged ozone mixing ratios above the six stations from where regular balloon soundings have been made with electrochemical sondes. Time histories are given in Fig. 5-5 as a function of altitude, the contours tracing ozone mass mixing ratios $(1\ \mu g/g = 1.29 \times 10^3\ \mu g/m^3\ STP = 604\ ppbv)$. The most striking feature is the annual wave in middle and high latitudes. Again, the data in the uppermost altitude range refer in part to

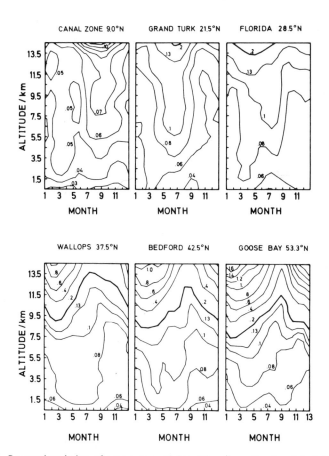

Fig. 5-5. Seasonal variation of ozone mass mixing ratios (in units of $\mu g/g$) at six stations of the North American Ozonesonde Network. [Adapted from Chatfield and Harrison (1977b).]

stratospheric air. Chatfield and Harrison (1977) did not delineate the boundary region, but one may obtain an indication for it from the variances
since they have a tendency to maximize near the tropopause due to the
natural fluctuation of the tropopause level. The appropriate curves are
shown in Fig. 5-5 by the heavy lines. They may be taken to represent monthly
mean tropopause heights. The seasonal behavior is consistent with this
interpretation, low levels occurring in winter and high levels late in summer.
Ozone mixing ratios in adjacent regions are seen to covary with $m(O_3)$ at
the tropopause, in accordance with expectation if the stratosphere serves
as an ozone source. Ozone-rich air penetrates into the troposphere deeper
in spring than at any other time of the year at all locations north of Florida.
The spring maximum extends all the way down to the ground surface,
although it experiences a time lag with increasing distance from the
tropopause. In the surface layer the maximum thus occurs in early summer
rather than in spring. This feature provides strong support for a downward
transport of ozone into the lower troposphere. It is not possible to explain
the phase shift alternatively in terms of a photochemical mechanism.

It is considerably more difficult to interpret the data in the equatorial
and subtropical regions. Near the equator, a semiannual wave is indicated,
but the amplitude is low, the variability is high, and the data may be taken
equally well to show no seasonaly dependence at all. At Grand Turk a
vertical tongue of ozone-rich air reaches downward from the upper troposphere in July, apparently unrelated to tropopause height. A similar feature,
much weaker in magnitude, appears later in the year at the Canal Zone
station. Chatfield and Harrison have noted that the tongue occurs roughly
synchronously with the maximum of annual rainfall activity associated with
the upward branch of the Hadley cell, which approaches its northernmost
position in July and thereafter retreats southward again. It is possible that
high-reaching cumulus towers coupled with cyclonic activity disturb the
tropopause sufficiently to admit the injection of ozone from the stratosphere.
Stallard *et al.* (1975) have found evidence for a locally significant influx of
stratospheric ozone via mesoscale squall systems. Alternatively, one must
take into account a transport of ozone from higher toward lower latitudes.
The most likely injection regime near the station at Grand Turk is the
subtropical jet stream, which in the summer is located at 40° northern
latitude. This location is at a considerable distance to the north of the
station, but during the summer months the meridional circulation system
would favor a southward dispersion of ozone injected, as opposed to a
pronounced northward transport during the winter months.

Fishman and Crutzen (1978a,b) have combined data from the North
American Ozonesonde Network with observations at other stations in the
northern and southern hemisphere to derive an approximate meridional

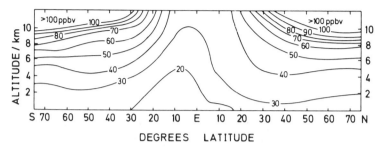

Fig. 5-6. Seasonally averaged meridional distribution of ozone according to the analysis of Fishman and Crutzen (1978a,b).

distribution for both hemispheres. Data for the southern hemisphere were taken primarily from Canton Island (3°S), La Paz (16°S), Aspendale (38°S), Christchurch (43°S), and Syowa (69°S). The more than 700 soundings at Aspendale were analyzed previously by Pittock (1974, 1977). Data from the other stations are less numerous, so that results for the southern hemisphere overall must be considered tenuous. The meridional distribution of tropospheric ozone obtained by Fishman and Crutzen is shown in Fig. 5-6 as an annual average. Values exceeding 100 ppbv may be assumed to represent stratospheric air. The distribution indicates an excess of ozone in the northern hemisphere at altitudes below about 5–6 km. From an analysis of average vertical ozone profiles for summer and winter conditions at similar latitudes in both hemispheres, Fishman and Crutzen (1978b) and Fishman *et al.* (1979) deduced that in addition to greater absolute amounts, more ozone is present in the northern hemisphere in summer compared with winter. This is in contrast to the southern hemisphere. Noting that carbon monoxide mixing ratios are higher in the northern hemisphere compared with the southern, and using reactions discussed in Section 4.2, Fishman *et al.* (1979) argued that the excess of ozone might result as a by-product from the oxidation of carbon monoxide via the OH reaction chain. A sufficiently large concentration of NO is required for photochemical O_3 formation to be effective, however, and since the distribution of nitrogen oxides in the troposphere is still not well established, the authors were forced to conclude that the relative importance of photochemistry in the O_3 budget cannot be quantified. Other uncertainties carry equal weight, however, in particular the inadequacy of the existing data base in the southern hemisphere.

5.3.2 AIRCRAFT OBSERVATIONS

Early aircraft measurements of vertical ozone profiles were reviewed by Junge (1963). Owing to the great variability of ozone, isolated aircraft observations add little to our knowledge of the large-scale distribution of tropospheric ozone, so that more comprehensive measurement programs are required. In recent years, commercial airliners have been utilized for this purpose. Most observations were made at cruising altitudes, that is, 10–12 km. Sampling of outside air can be done conveniently via the fresh-air system. A number of studies by Tiefenau *et al.* (1973) have shown that ozone remains unaffected by compression of air within the air duct system of the aircraft. Nastrom (1977) compared mean ozone values obtained on board of commercial airliners with those provided by the North American Ozonesonde Network and found them compatible. This established additional confidence in the aircraft data. Fabian and Pruchniewicz (1977) made a series of measurements within a flight corridor between 20° western and 35° eastern longitude extending all the way from Norway to South Africa. A continuous record of ozone mixing ratios was obtained on each flight, and data pertaining to the stratosphere were subsequently removed if necessary. A typical example for an individual flight record for the upper troposphere is shown in the uppermost frame of Fig. 5-7. Fabian and Pruchniewicz (1977) have averaged data from a total of about 50 such flights to derive seasonal means as a function of latitude. The results are reproduced in the central portion of Fig. 5-7. To minimize systematic errors, the authors have chosen a relative scale of ozone mixing ratios that detracts somewhat from the usefulness of the data for the purpose of comparison with others.

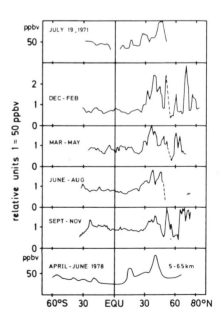

Fig. 5-7. Ozone mixing ratios in the upper troposphere as a function of latitude, from aircraft observations. Top: Data from one flight on July 19, 1971 (Fabian and Pruchniewicz, 1977), flight altitude 11–12 km. Center: Seasonally averaged data from about 40 individual flights between Norway and South Africa (Fabian and Pruchniewicz, 1977) at altitudes of 11–12 km. Bottom: Average ozone mixing ratios at altitudes of 5–6.5 km over the North American continent and the Pacific Ocean (Routhier *et al.*, 1980).

Routhier *et al.* (1980) have reported midtropospheric mixing ratios of ozone at latitudes between 58°S and 70°N over the Pacific ocean and the North American continent. An interpolation of this set of data is included in Fig. 5-7.

Two aspects of the data in Fig. 5-7 invite comment. The absence of a significant meridional gradient between 25°S and 25°N indicates an efficient horizontal equilibration by eddy diffusion in the equatorial upper troposphere. This is in agreement with the notion that in this altitude regime an interhemispheric exchange of air masses is most effective. Pruchniewicz *et al.* (1974) noted a slight seasonal dependence of ozone mixing ratios, with maximum values in May and minimum values in November–December, the variation being essentially in phase on both sides of the equator. The behavior is not compatible with the balloon sounding data in Fig. 5-5, however, and may be an overinterpretation of a limited data set. The second aspect of the data in Fig. 5-7 is the peaks of ozone-rich air at those latitudes where ozone intrusions via the tropopause gaps are expected to occur with high frequency. If the peak heights are taken to indicate the extent of ozone injection, the ozone flux is seen to be strongest in the first quarter of the year and weakest during the third quarter, in coincidence with the buildup of the stratospheric ozone reservoir during the winter months and its depletion during the summer season, and also with the strength of subtropical jet-stream activity. A similarly intensive activity of ozone injections is not apparent during the winter season of the southern hemisphere. The data of Fabian and Pruchniewicz (1977) unfortunately do not extend beyond the 35° southern latitude circle. Yet on the basis that stratospheric injection rates through the tropopause gaps are the same in both hemispheres, a steeper rise of ozone mixing ratios is expected at 30°S than is actually observed. The midtropospheric data of Routhier *et al.* (1980) confirm the existence of ozone intrusions in the northern hemisphere and the comparative weakness of those in the southern hemisphere, although it should be noted that their southernmost data were obtained during the quiet season in that region. Nevertheless, both data sets indicate a smaller injection rate of stratospheric ozone into the southern hemisphere compared with the northern during the same season.

5.3.2 SURFACE MEASUREMENTS

Ozone mixing ratios in surface air are highly variable, due to losses of the trace gas upon contact with the ground surface and its coverage. Such losses set up gradients of ozone mixing ratios, which for a constant deposition flux decrease with height avove the ground due to the increase in eddy diffusivity (see Section 1.6). The local behavior of ozone, however, depends greatly on the topography and micrometeorology of the sampling site. Flatland stations show a characteristic diurnal variation of the surface ozone mixing ratio. Maximum values generally occur in the early afternoon, when convective turbulence due to solar heating is strongest and tropospheric ozone is rapidly carried downward. Minimum values are observed during the early morning hours when stable conditions prevail in the continental boundary layer. Diurnal variations at mountain stations, by contrast, often show small amplitudes, with maxima occurring preferentially late in the evening. The former observation indicates that the influence of ground losses is weak, and the latter observation is explained by the typical mountain circulation pattern with air directed upslope during the day and downslope during the night. The upslope circulation brings with it air depleted in ozone; the downslope circulation imports air from the free troposphere which is richer in ozone. A similar situation exists at coastal sites, where convective cells established during the day give rise to surface winds directed inland, whereas at night when the land surface cools subsidence sets in that reverses the air flow. Again, maximum ozone mixing ratios generally are found late in the evening. Figure 5-8 shows frequency distributions for the occurrence of the daily ozone maximum at three stations, each representing a typical measurement site: Hermanus (South Africa) is a continental flatland station, Zugspitze (Germany) is a mountain station, and Westerland (Germany) is a coastal station. Although the frequency distributions are fairly broad in all three cases, it is obvious that in Hermanus the ozone maximum occurs preferentially in midday, whereas Zugspitze and Westerland show it shortly before midnight.

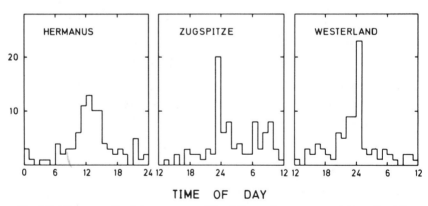

TIME OF DAY

Fig. 5-8. Histograms for the occurrence of the daily maximum ozone mixing ratio at three typical measurement sites. Left: Flatland station (Hermanus, South Africa), 90 days observation time. Center: Mountain station (Zugspitze, Germany), 90 days observation time. Right: Coastal station (Westerland, Germany), 88 days observation time. [From Fabian and Pruchniewicz (1977).]

Figure 5-9 is added to show, somewhat schematically, ozone mixing ratios as a function of height above the ground at a flatland site. In the middle of the day mixing ratios are high and the vertical gradient is small, because turbulent mixing maintaining the downward ozone flux is appreciable. Later in the afternoon the intensity of turbulence weakens, the eddy diffusivity decreases, and the gradient of ozone mixing ratios becomes more pronounced. The surface mixing ratio then is markedly lowered due to ground losses, while the effect at 12 m height is still minimal. In the evening the situation is qualitatively still the same, but the supply of ozone from above now ceases and an accelerated decay of ozone takes place at all height levels. At midnight, the ozone profile has almost stabilized and the further decrease of the O_3 mixing ratio is slow. At daybreak the activity of turbulence in the surface boundary layer resumes and eventually the ozone mixing ratios rise again.

The preceding discussion suggests that in unpolluted air the daily maximum values of surface ozone are due to the downflow of ozone from greater heights, in which case the influence of surface destruction of ozone is weakest. Junge (1962) therefore argued that the daily maxima are most likely to represent ozone mixing ratios existing in the free troposphere. Fairly convincing evidence for this concept has been obtained by Pruchniewicz et al. (1974). They measured ozone as a functions of height with the help of the 120-m tower at Tsumeb (South Africa), together with meteorological parameters that furnished information on the stability of the atmospheric boundary layer. The dimensionless Richardson number was used as a characteristic turbulence indicator to differentiate between labile conditions and stable stratification. In the first case, the intensity of turbulence is high and the ozone mixing ratio near the surface correlates well with that observed at 100 m above the ground. A one-to-one relationship

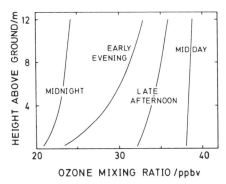

Fig. 5-9. Vertical profiles of ozone mixing ratio at a flatland station during the course of 1 day. Schematic representation of data obtained by Regener (1957).

is closely approached. In the second case of stable stratification the correlation is poor and proportionality does not obtain. Pruchniewicz *et al.* (1974) further pointed out than within the entire 2-month period while these measurements were made, the daily maxima were always found to occur during labile conditions, in accordance with Junge's hypothesis. It must be recognized that a midday maximum of ozone may be interpreted equally well to arise from photochemical ozone production as from turbulent downward mixing, since both maximize in midday. It is possible, therefore, that at flatland stations ozone being transported toward the measurement site originates from photochemical processes in the boundary layer and does not represent the ozone mixing ratio of the free troposphere. Maximum ozone values at mountain stations, however, cannot be interpreted in this way.

Occasionally, rather high mixing ratios of surface ozone are encountered with values exceeding 100 ppbv. The first explicit report of this kind was made by Attmannspacher and Hartmannsgruber (1973), who had found values up to 390 ppbv at a hilltop station in southern Germany during an episode associated with a passing cold front. Lamb (1977) observed ozone mixing ratios above 100 ppbv for several hours during a rainy night in Santa Rosa, California, with a maximum hourly value of 230 ppbv. Reiter *et al.* (1977) described an event with peak O_3 mixing ratios exceeding 145 ppbv at the Zugspitze mountain. These reports caused excitement because to provide an air pollution standard the U.S. Environmental Protection Agency had set a limit of 80 ppbv (now relaxed to 120 ppbv) not to be exceeded for longer than an hour for more than 1 day per year. None of the events can be related to photochemical air pollution, however. They are attributed to local intrusions of stratospheric ozone via the tropopause folds connected with frontal jet streams (see Fig. 1-9). Under favorable conditions the ingested ozone-rich air parcels are brought straight down to the surface layer with the frontal air flow. The meteorology of intrusions associated with jet streams has been discussed by Reiter *et al.* (1969), Danielsen *et al.* (1970), Danielsen and Mohnen (1977), and others. In the case of the Zugspitze event the stratospheric origin of ozone was demonstrated by the simultaneous measurement of high levels of beryllium-7, a spallation product of cosmic radiation interacting with the stratosphere. The question of the frequency at which such events occur has not been settled. The analysis by Reiter *et al.* (1975b, 1977) of the Zugspitze data indicates that 0.2% of the hourly averages exceed 80 ppbv. Dutkiewicz and Husain (1979) examined ozone mixing ratios at Whiteface mountain in New York State in conjunction with beryllium-7 data and found for the months of June and July 1977 an occurrence frequency of stratospheric ozone on 11 days out of 61. On these 11 days fresh stratospheric ozone contributed between 26 and 94% to the total 24-h ozone mixing ratio.

If the daily maxima are assumed to represent unperturbed tropospheric ozone mixing ratios, one may use them to derive monthly means and thereby determine the seasonal variation. A number of results from measurement series extending over at least 3 years are shown in Fig. 5-10. The first of these was obtained at Arosa, a Swiss resort town at an elevation of 1860 m. The data were discussed by Junge (1962), who first noted the cyclic annual behavior. Junge interpreted the seasonal cycle to arise from the injection of stratospheric ozone into the troposphere, the flux being proportional to the amount of total ozone, and a phase shift resulting from the finite residence time of ozone in the troposphere until it reaches the surface boundary layer. The balloon sounding data in Fig. 5-5 demonstrate that Junge's interpretation is basically correct, at least for middle and high latitudes. One expects, therefore, that surface ozone at other measurement sites shows a similar seasonal variation, and we may inspect the other data in Fig. 5-10 from this viewpoint. The observations on the Brocken, a middle-range mountain in Germany with an elevation of 1140 m, are due to Warmbt (1964). The data represent average daily ozone mixing ratios rather than diurnal maxima, but the seasonal cycle is nevertheless clearly displayed. The remaining data are from the meridional network of Fabian

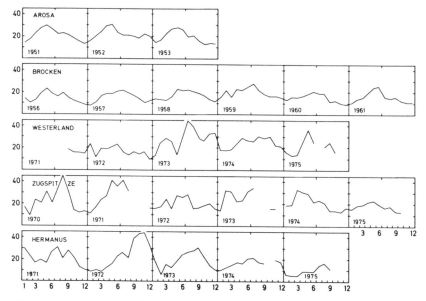

Fig. 5-10. Monthly means of daily 1-h ozone maxima (daily average for Brocken); observations over at least 3 years at several stations in Europe and South Africa. Data for Arosa, Switzerland, from Junge (1962); for Brocken, Germany, from Warmbt (1964); others from Fabian and Pruchniewicz (1977).

and Pruchniewicz (1977). The stations selected here are the same as characterized earlier as typical coastline, mountain, and flatland sites. These data
are much more ragged than the earlier results from Arosa and the Brocken.
The considerable variability at the Zugspitze mountain is particularly surprising because the distance to Arosa is only 130 km, and at an elevation
of 2960 m the Zugspitze mountain should be even more favorable for
observations of tropospheric ozone than other sites in Europe. Unfortunately, neither Zugspitze nor Westerland can be classified as remote sites,
since both are tourist attractions and may suffer from local contaminations.
Westerland is in addition subject to long-distance pollution carried over
the North Sea from industrial centers on the British Isles. It is possible,
therefore, that anthropogenic effects spoil the data at both locations. The
results obtained at Hermanus, a fairly clean air site, are more regular
exhibiting the seasonal cycle quite well. Note that Hermanus lies near 34°
southern latitude. Here, the annual maximum of surface ozone occurs in
September or October, again about 2 months later than the maximum of
total ozone.

The seasonal dependence of surface ozone can be made more distinct by
averaging monthly means over a number of years (superimposed epoch
averages; see Table 1-2). Results for the data discussed above are shown
in Fig. 5-11. In each case, the seasonal behavior is found to fit simple
sinusoidal wave patterns, which are shown by the solid curves. This approximation holds also for the results obtained at Zugspitze mountain and at
Westerland, despite the large scatter of the individual data points. At Arosa
a long-term record for total ozone is available, and the appropriate long-term
average is included in Fig. 5-11 for comparison with the surface data and
to demonstrate the phase shift between both data sets. Other results included
in Fig. 5-11 are observations at Mauna Loa, Hawaii, and at two remote
sites in the Rocky Mountains. These data were analyzed by Singh et al.
(1978). The seasonal variation at Mauna Loa and at Rio Blanco is consistent
with a simple annual cycle. The pattern is in harmony with that of the other
observations. The results for White River are different in that they indicate
a bimodal behavior. White River is located 160 km to the west of Rio Blanco
and there is no obvious reason for the difference. One possibility is that
the record of data at both sites covers somewhat less than 2 years and the
period may be too short to yield representative monthly averages. Singh et
al. (1978) have interpreted the summer maximum as being caused by a
photochemical generation of ozone. The mixing ratio of NO_2 was measured
simultaneously with that of ozone and was found to rise during the summer
months. In clean-air regions the mixing ratio of NO_2 generally is smaller
than that of O_3, but at White River they reached comparable levels. No
record of NO_2 is available at Rio Blanco, so that Singh's hypothesis cannot

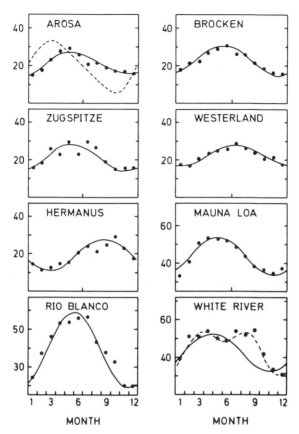

Fig. 5-11. Monthly means of daily 1-h ozone maxima averaged over a number of years (at least two). Data for Arosa, Brocken, Zugspitze, Westerland, and Hermanus as given in Fig. 5-10. Data for Mauna Loa, Hawaii (1974–1975), Rio Blanco, Colorado (1974–1976), and White River, Utah (1975–1976) as presented by Singh *et al.* (1978). The solid curves represent sinusoidal fits to the data points. The White River data cannot be fitted in this way. A better fit is obtained with a harmonic function composed of an annual and a semiannual wave (Falconer *et al.*, 1978), as indicated by the dashed curve. Total ozone at Arosa is shown by the dotted curve to indicate the phase shift between total and surface ozone.

be tested. If NO_2 mixing ratios were higher than expected, however, the levels of hydrocarbons acting as ozone precursors presumably were raised as well.

The mesoscale transport of urban plumes carrying ozone precursor substances into rural and clean air regions has become a well-studied issue (Coffey and Stassink, 1975; Cleveland *et al.*, 1976; Dimitriades and Altshuller, 1976; Robinson, 1977; Wolff *et al.*, 1977). The evidence indicates elevated ozone concentrations as far as 300 km downwind of urban centers. Even remote sites such as Whiteface Mountain in upper New York State are known to be thus affected (Coffey *et al.*, 1977; Singh *et al.*, 1978). These observations unfortunately suggest that in the industrialized countries it is already difficult to find measurement sites free from anthropogenic influences. Nevertheless, if the pollution is discontinuous, that is limited to sporadic episodes, one may hope to eliminate its influence by long-term averaging. The results obtained from the meridional network of stations organized by Fabian and Pruchniewicz provides some support for the validity of this procedure.

Finally, Fig. 5-12 shows the distribution of surface ozone as a function of latitude. Individual points with vertical bars are annual averages reported by Fabian and Pruchniewicz (1977) for the 16 stations of their network, and for a number of additional stations operated by other investigators. The lowest O_3 values of about 15 ppbv are again found in the equatorial region. At extratropical latitudes the mixing ratios average to about 25 ppbv, but the data scatter considerably and individual variances are high. These long-term averages are in reasonable accord with the distribution shown in

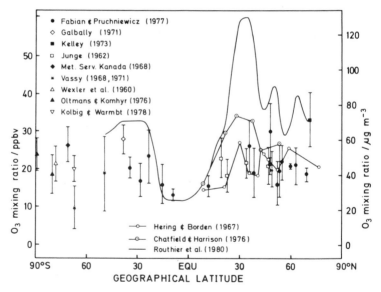

Fig. 5-12. Annual average ozone mixing ratios at various stations of the world plotted versus latitude. Vertical bars indicate one standard deviation. Included for comparison are data for the 0–2 km height level from the North American Ozonesonde Network according to Hering and Borden (1967) and Chatfield and Harrison (1977b). Further included for comparison are results from the aircraft observations of Routhier *et al.* (1980) at about 2 km altitude.

Fig. 5-6, although they do not indicate a sizable difference between the two hemispheres. An important question concerns the assumption that the data represent ozone values of the free troposphere. Hence, results from the North American Ozonesonde Network are included in Fig. 5-12 for comparison with the surface measurements. Here, averaged data for the 0–2 km height region are shown. The early results obtained with the chemiluminescence detector lie well within the spread of the surface data. The later results obtained with the electrochemical sonde are somewhat higher, especially at latitudes near 30°N, but the differences are not great enough for a clear distinction between free tropospheric and surface ozone. By and large, it appears that the surface measurements when carried out in the way described do represent realistic mixing ratios of tropospheric ozone.

5.4 Budget Considerations

If one accepts the classical viewpoint and assumes photochemical ozone production and loss reactions negligible on a global scale, the budget of ozone in the troposphere will be dominated by the injection of ozone from the stratosphere and its destruction at the ground surface. Clearly, this is a minimum budget. Injection and destruction rates are examined below. For steady-state conditions both rates must balance, and if they do not, one would have an indication for the importance of additional sources or sinks of tropospheric ozone. The following discussion will show, however, that within a rather wide margin of error, the two rates are indeed compatible.

5.4.1 INFLUX OF OZONE FROM THE STRATOSPHERE

As discussed in Section 1.5, the principal air-exchange mechanisms between stratosphere and troposphere are seasonal adjustment of the tropopause level, large-scale organized mean motion in the lower stratosphere, turbulent transport associated with jet streams, and small-scale eddy diffusion. The first three of these processes transfer air in bulk at a rate of about 3×10^{17} kg/yr in the northern hemisphere. The contribution of the fourth process is determined by the vertical gradient of O_3 mixing ratios immediately above the tropopause. From Fig. 5-5, an annual average gradient of 90 ppbv/km is estimated for the region poleward of 30° northern latitude. In the tropics, downward eddy diffusion is counteracted by the upward branch of the Hadley motion.

The ozone flux due to bulk air exchange is obtained by multiplying the exchange rate with an appropriate ozone mass mixing ratio in the lower stratosphere. Figures 3-4 and 5-5 suggest an average value of 1 μg/g. Both quantities are subject to seasonal variation, however. From observations of radioactive tracers such as strontium-90 it has been shown that the maximum injection rate in spring is about three times the minimum rate in the fall (Danielsen, 1964, 1968; Reiter, 1975). Similarly, ozone in the lower stratosphere reaches a maximum in spring with concentrations about 30% higher than the annual average, and a minimum in fall with concentrations 30% lower. Accordingly, the annual flux of ozone into the troposphere of the northern hemisphere is estimated as

$$F(O_3) = m(O_3)F_{air}$$
$$= [(1.3 \times 10^{-6})(2 \times 10^{17})] + [(0.7 \times 10^{-6})(10^{17})]$$
$$= 3.3 \times 10^{11} \text{ kg } O_3/\text{yr}$$

The contribution of small-scale eddy diffusion that must be added is

$$F(O_3) = K_z \rho_{tr}(M_{O_3}/M_{air})[dm(O_3)/dz](A/4)\gamma$$
$$= 0.64 \times 10^{11} \quad \text{kg } O_3/\text{yr}$$

where $K_z = 0.3 \text{ m}^2/\text{s}$ is the eddy diffusion coefficient, M_{O_3} and M_{air} are the molecular weights of ozone and air, respectively, and $A = 5 \times 10^{14} \text{ m}^2$ is the surface area of the earth. A factor $\gamma = 3.15 \times 10^7$ must be applied to convert seconds to years. Adding both fluxes gives a total of 394 Tg/yr. Similar estimates were obtained by Mohnen (1977) on the basis of ozone intrusions into the troposphere that occur in association with tropopause foldings in the vicinity of jet streams. Values in the range 470–560 Tg/yr were obtained for an ozone mixing ratio of 1.3 μg/g in the lower stratosphere.

    Another method of estimating the influx of ozone into the troposphere makes use of the observed gradient of ozone mixing ratios in the middle and upper troposphere in conjunction with the eddy diffusion equation. The data of Hering and Borden (1967) as analyzed by Chatfield and Harrison (1977b) may be exploited for this purpose. As Fig. 5-4 shows, the average gradients increase with latitude from a typical value of $1 \times 10^{-12} \text{ m}^{-1}$ in the tropics to $7 \times 10^{-12} \text{ m}^{-1}$ at high latitudes. It is assumed that the gradients are zonally representative. Zonal fluxes are calculated by multiplying the gradients with the vertical eddy diffusion coefficient ($K_z = 17 \text{ m}^2/\text{s}$ in the troposphere), the appropriate air density at the altitude considered, and the area of the latitude belt. The results are shown on Table 5-4. The flux above each station should be independent of altitude, but this condition is met only within a fairly wide range of scatter. Values for the total hemispheric flux, however, derived by summation over all latitude belts, are in good agreement for both altitude layers adopted for the calculations. The total flux also agrees with the values given above.

Table 5-4. *Downward Flux of Ozone in the Middle Troposphere of the Northern Hemisphere*[a]

Latitude zone	Area (10^{13} m^2)	Station	Gradient 5.5–6.5 km (10^{-12} m^{-1})	Flux (Tg/yr)	Gradient 6.5–7.5 km (10^{-12} m^{-1})	Flux (Tg/yr)
50–90°	5.85	Goosebay (53.3°)	7.0	235	10.0	299
40–50°	3.08	Bedford (42.5°)	2.9	51	7.2	113
30–40°	3.57	Wallops (37.5°)	5.4	110	2.7	49
25–30°	1.93	Grand Turk (28.5°)	1.9	21	2.3	22
0–25°	10.56	Canal Zone (9°)	1.3	79	0.9	48
Total flux	(Tg O$_3$/yr)			497		532

[a] Estimated from average gradients established by the North American Ozonesonde Network (Hering and Borden, 1967; Chatfield and Harrison, 1977b), assuming $K_z = 17$ m^2/s.

Table 5-5 summarizes several estimates for the influx of stratospheric ozone into the troposphere. Aside from the two methods indicated above, two additional procedures are of interest. One is based on the flux of potential vorticity. In Section 1.5 it was pointed out that there exists a good correlation between ozone mixing ratios and potential vorticity. Mahlman and Moxim (1978) further showed that any quasi-conservative tracer in the lower stratosphere eventually assumes a positive correlation with potential vorticity, even if the tracer originated from an instantaneous point source. Gidel and Shapiro (1980) made use of this relationship to estimate the flux of ozone in conjunction with a general circulation model similar to that of Mahlman and Moxim. The average rates for the injection of ozone into the troposphere computed by Gidel and Shapiro are 310 and 157 Tg O_3/yr for the northern and southern hemispheres, respectively. The latter value is only about half as large as the former. This result is in harmony with the aircraft observations discussed earlier (see Fig. 5-7) of ozone in the upper troposphere, which had indicated a reduced rate of ozone injection in the region of the southern subtropical jet stream compared to the northern.

Yet another method for estimating the flux of ozone into the troposphere is based on a model for the seasonal behavior of ozone first proposed by Junge (1962). The model assumes that the flux of ozone across the tropopause is proportional to total ozone and that the spring maximum of

Table 5-5. *Estimates for the Influx of Stratospheric Ozone into the Troposphere (Annual Average in Units of* Tg O_3/yr)

Authors	Northern hemisphere	Southern hemisphere	Remarks
Mohnen (1977), based on earlier work of Danielsen (1964, 1968) and Reiter (1975)	320 −470	—	Stratospheric–tropospheric air mass exchange; tropopause folding events plus large-scale Hadley circulation; $m(O_3) = 1.3$ μg/g assumed.
Present (see text)	390	—	Based on the data of Table 1-10
Nastrom (1977)	250	—	Aircraft measurements of O_3 near the tropopause combined with average vertical air velocity, only for 30°N
Gidel and Shapiro (1980)	310	157	From the flux of potential vorticity in the stratosphere
Present (see text)	490 −530	—	From gradient in the upper troposphere, Table 5-4
Fabian and Pruchniewicz (1976, 1977)	425	275	From the phase shift of maxima between total and surface ozone observed at 27 stations

the flux is responsible for the annual maximum of tropospheric ozone near the earth surface. The second maximum occurs somewhat later in the year compared with the first, due to the finite time required for the downward transport of ozone. From the observed amplitudes and the phase shift between both (sinusoidal) functions, one can calculate ozone injection rates above the station where the observations are made. Fabian and Pruchniewicz (1976, 1977) have treated data from various stations of their network in this manner. The results are included in Table 5-5. Although the method would fail if photochemical sources and sinks of tropospheric ozone were important, the data of Fabian and Pruchniewicz are in accord with the other estimates of Table 5-5. Most interestingly, their evaluation also leads to a smaller O_3 injection rate in the southern hemisphere compared with the northern. Average rates of ozone transport from the stratosphere into the troposphere, as suggested by the data of Table 5-5, are 400 and 210 Tg/yr for the northern and southern hemispheres, respectively.

5.4.2 DESTRUCTION OF OZONE AT THE EARTH'S SURFACE

The removal of a trace gas from the atmosphere due to absorption at the ground surface was treated in Section 1.6. The flux toward the surface may be expressed in terms of a deposition velocity that is an empirical parameter determined by a series of transfer resistances associated with aerial transport and uptake of material at the surface. Data specific to ozone have been reviewed by Galbally and Roy (1980) and are summarized in Table 1-12.

Resistances to the uptake of ozone by soil, grass, water, and snow were first measured by Aldaz (1969). He observed the rate of disappearance of ozone inside a large box placed open-face down onto the test surface. The deposition velocities derived were 0.006 m/s for land, 0.0004 m/s for water, and 0.0002 m/s for ice and snow. These values must be regarded as upper limits, however, because they neglect the additional resistance associated with aerial transport. Aldaz (1969) nevertheless used his results to estimate a global rate of ozone destruction of 1300–2100 Tg/yr, a value four to five times greater than previous, albeit less direct, estimates. Fabian and Junge (1970) then developed a procedure based on the neutral stability model discussed in Section 1.6 by which they corrected the deposition velocities given by Aldaz for the additional resistance due to turbulent transport in the boundary layer and for surface roughness. The correction is most significant for continental conditions because uptake rates for ozone by bare and plant-covered soils are fairly high. The solubility of ozone in water is low so that the uptake resistance for seawater is higher than the aerial resistance. In this case the correction is a minor one. Altogether, the corrections reduced the global loss rate of Aldaz by a factor of three.

Galbally and Roy (1980) discussed a variety of subsequent measurements of ozone dry deposition rates performed with the flux box technique, by means of wind-tunnel experiments, and by observations of the vertical gradient of ozone mixing ratios in the field. By and large, the results coincided with the earlier data. Their own observations, however, led Galbally and Roy (1980) to the conclusion that the surface resistance for the uptake of ozone by grass-covered soils showed a nocturnal increase, from a daytime average of 100 s/m to a value of 300 s/m at night. The reason for this behavior presumably is the closure of leaf stomata more than any variation of other parameters. Armed with this new information, Galbally and Roy (1980) recalculated the global rate of ozone destruction at the earth's surface. The results are shown in Table 5-6. Unlike Fabian and Junge (1970), who allowed for the different boundary conditions but took the ozone mixing ratio to be that of the free troposphere at 2 km height, Galbally and Roy (1980) incorporated the resistance associated with turbulent transport in the atmosphere into the ozone mixing ratio 1 m above the surface. Accordingly, their $m(O_3)$ is lower over the continents than over the ocean, where the aerial resistance is essentially negligible compared with that of the sea's surface. Galbally and Roy also adopted a larger uptake rate of ozone by seawater, although several measurements had confirmed the Aldaz (1969) original value, because agitation of the water surface was found to decrease the uptake resistance by a factor of about two. This agreed with values derived by Tiefenau and Fabian (1972, as analyzed by Regener, 1974) from profiles of O_3 mixing ratios over the North Sea.

Table 5-7 presents global ozone destruction rates as derived by a number of authors. The results are fairly consistent if one excludes those of Aldaz (1969), which are too high for the reasons discussed earlier. The results of Fishman and Crutzen (1978b) are based on a less detailed evaluation compared with estimates of either Galbally and Roy (1980) or Fabian and Pruchniewicz (1977). The last authors applied the procedure of Fabian and Junge (1970) but used updated deposition rates similar to the data employed by Galbally and Roy (1980). The difference in the results of both investigators thus indicates the uncertainty inherent in the applied methods. The accuracy claimed in these studies is about 50%. All authors obtained higher deposition rates for the northern hemisphere compared with the southern. The ratio is roughly 3:2. Fishman and Crutzen (1978b) obtained a higher ratio of about 3:1. The reason for the difference is not entirely obvious. Partly it derives from their using a lower destruction rate for the oceans, but they appear to have overestimated also the loss rate in the northern hemisphere.

The ozone deposition rates in each hemisphere compare well with the corresponding ozone injection rates shown in Table 5-5, even though the

Table 5-6. *Ozone Destruction Rates at the Earth's Surface in Different Latitude Bands, Subdivided According to the Type of Surface and the Corresponding Average Deposition Velocity v_d, and the Ozone Mixing Ratio $m(O_3)$*[a]

Latitude	Surface type	Surface area %	Surface area 10^{12} m²	$m(O_3)$ ($\mu g/m^3$)	v_d (m/s)	Flux (Tg/yr)
60–90°S	Snow/ocean	7	35.7	50	0.001	56.2
30–60°S	Ocean	17	86.7	50	0.001	136.6
	Grassland	0.9	4.6	35.6	0.0048	24.7
	Forests	0.1	0.6	35.6	0.0063	3.5
15–30°S	Ocean	9	45.9	37.6	0.001	54.4
	Grassland	2.5	12.8	26.4	0.0043	45.8
	Forests	0.5	2.6	26.4	0.0063	13.6
0–15°S	Ocean	10	51.0	29.6	0.001	47.5
	Grassland	1.4	7.1	20.8	0.0043	20.0
	Forests	1.6	8.2	20.8	0.0063	34.0
Total flux in the southern hemisphere						436.3
0–15°N	Ocean	10	51.0	28.0	0.001	44.9
	Grassland	1.8	9.2	19.6	0.0043	25.0
	Forests	1.2	6.1	19.6	0.0063	22.7
15–30°N	Ocean	8	40.8	38.4	0.001	49.3
	Grassland	3.8	19.4	26.4	0.0043	69.4
	Forests	0.2	1.0	26.4	0.0063	5.2
30–60°N	Ocean	9	45.9	44.8	0.001	64.8
	Grassland	5.8	29.6	31.6	0.0047	138.5
	Forests	3.2	16.3	31.6	0.0063	102.2
60–90°N	Ocean	3.8	19.4	47.2	0.001	28.8
	Grassland	2.2	11.2	33.2	0.0047	55.0
	Forests	1.0	5.1	33.2	0.0063	33.6
Total flux in the northern hemisphere						639.4

[a] Data from Galbally and Roy (1980).

Table 5-7. *Estimates for the Global Rate of Ozone Dry Deposition in Each Hemisphere ($Tg\ O_3/yr$)*

Authors	Northern hemisphere	Southern hemisphere
Aldaz (1969)	(905–1308)	(255–658)
Fabian and Junge (1970)	(285–491)	(159–242)
Fabian and Pruchniewicz (1977)	(302–554)	(176–378)
	425 average	275 average
Fishman and Crutzen (1978b)	784	270
Galbally and Roy (1980)	648	440

latter are somewhat smaller than the former. In comparing both data sets, however, one must take into account that rates for the transport of ozone across the tropopause are more difficult to estimate and hence carry larger uncertainties. Within the ranges of uncertainty both rates are equivalent so that the budget of ozone in the troposphere is fairly well balanced by these two fluxes alone. Note that the asymmetry between the two hemispheres occurs in both data sets—that is, deposition as well as influx rates are reduced in the southern hemisphere. The data of Fabian and Pruchniewicz (1977) are somewhat biased in this regard, because in their model the two rates are artificially balanced. The injection rate for the southern hemisphere thus is still not as well known as one would wish. But the results of Gidel and Shapiro (1980) leave no doubt that the injection rate in the southern hemisphere is only about 60% of that in the northern hemisphere.

Finally, we consider the residence time of ozone in the troposphere that is in accordance with the minimum budget discussed here. Owing to the uneven distribution of ozone, its mass content in the troposphere can be estimated only approximately. For the present purpose we make use of the data in Fig. 5-6 to estimate average ozone mixing ratios in tropical latitudes of 35 ppbv in the northern and 30 ppbv in the southern hemisphere. Combining this with the tropospheric air mass of 1.13×10^{18} kg in the 0–30° latitude belt leads to 65 Tg ozone for this region in the northern and 55 Tg in the southern part. For the region poleward of 30° latitude we take an average mixing ratio of 45 ppbv in both hemispheres, which adds 75 Tg ozone each. This procedure gives a total of 140 Tg ozone in the northern troposphere, and 120 Tg in the southern. Further, using the deposition fluxes of Fabian and Pruchniewicz (1977) and Galbally and Roy (1980), one obtains residence times in the range 79–120 days for the northern troposphere, and of 99–159 days for the southern. These values correspond to residence time of the order of 3–5 months. The data suggest a slightly shorter residence time in the northern hemisphere compared with the southern, but this result would be in harmony with the land–sea distribution, which favors higher deposition rates in the northern hemisphere.

5.5 Photochemical Production and Loss of Ozone in the Unperturbed Troposphere

The tacit assumption made in the preceding discussion that ozone is photochemically stable in the troposphere is basically incorrect, and we must finally consider this aspect. It is true that the photodissociation of ozone as far as it leads to $O(^3P)$ atoms causes no losses, because their subsequent attachment to molecular oxygen regenerates ozone. In Section 4.2 it was shown, however, that a part of the $O(^1D)$ atoms produced in the long-wavelength tail of the Hartley band ($\lambda < 320$ nm) reacts with water vapor to form OH radicals, and this process represents a loss of tropospheric ozone. Losses are most severe in the surface boundary layer, where the concentrations of water vapor are high. In the upper troposphere the losses are essentially negligible. The magnitude of the losses may be estimated from the OH production rates given in the literature. For example, the data of Warneck (1975) indicate an average primary OH column production rate of 2.4×10^{14} molecules/m^2 s in the northern hemisphere, whereas Fishman *et al.* (1979) calculated from the ozone distribution shown in Fig. 5-6 column production rates of 10.5×10^{14} and 5.0×10^{14} OH molecules/m^2 s for the northern and southern hemispheres, respectively. The associated ozone loss rates are 148, 668, and 321 Tg O_3/yr, respectively, indicating a very severe perturbation of the minimum budget of ozone discussed previously. Because the production of OH radicals maximizes in the tropics, the loss of ozone would also maximize there. Additional losses of ozone occur due to the reactions

$$OH + O_3 \rightarrow O_2 + HO_2$$

$$HO_2 + O_3 \rightarrow 2O_2 + OH$$

The second of these is more important than the first one, since the OH radicals are converted to HO_2 mainly by reacting with methane and carbon monoxide rather than with ozone. All these losses are at least partially retrieved, however, by the formation of ozone from the photodissociation of NO_2 when NO is converted to NO_2 in smog-like reactions involving the oxidation of methane and carbon monoxide, as was first pointed out by Crutzen (1973). In fact, more ozone may be formed in this manner than is lost by the conversion of $O(^1D)$ to OH radicals, if conditions are favorable. The oxidation of methane and carbon monoxide was discussed in Section 4.2. Here, we will reconsider it from the viewpoint of its ozone generating potential.

If it is assumed that peroxidic radicals react only with nitrix oxide, the reaction of OH with methane induces the following reaction sequence:

$$CH_4 + OH \rightarrow CH_3 + H_2O$$
$$CH_3 + O_2 + M \rightarrow CH_3O_2 + M$$
$$CH_3O_2 + NO \rightarrow CH_3O + NO_2$$
$$CH_3O + O_2 \rightarrow HCHO + HO_2$$
$$HO_2 + NO \rightarrow NO_2 + OH$$
$$\left. \begin{array}{c} NO_2 + h\nu \rightarrow NO + O \\ O + O_2 + M \rightarrow O_3 + M \end{array} \right\} \times 2$$

net: $\quad CH_4 + 4O_2 \rightarrow HCHO + H_2O + 2O_3$

In this scheme two molecules of ozone are generated for each molecule of methane consumed. The subsequent photodissociation of formaldehyde produces equivalent amounts of carbon monoxide, which also undergoes oxidation by OH radicals:

$$OH + CO \rightarrow CO_2 + H$$
$$H + O_2 + M \rightarrow HO_2 + M$$
$$HO_2 + NO \rightarrow NO_2 + OH$$
$$NO_2 + h\nu \rightarrow NO + O$$
$$O + O_2 + M \rightarrow O_3 + M$$

net: $\quad CO + 2O_2 \rightarrow CO_2 + O_3$

Here, one ozone molecule is formed for each CO molecule that is oxidized. Altogether, the methane oxidation mechanism is capable of generating three ozone molecules for each methane molecule undergoing oxidation. Still not considered in this account is the formation of HO_2 radicals by the photolysis of formaldehyde. These additional HO_2 radicals convert further NO to NO_2, thereby increasing the total yield of ozone.

There should be no doubt that it is quite unrealistic to assume that CH_3O_2 and HO_2 radicals react solely with NO. Radical termination reactions must compete with chain propagation reactions lest the chain continues indefinitely. Termination reactions divert CH_3O_2 and HO_2 into other reaction channels, so that the yield of ozone does not attain maximum values. The principal termination reactions are

$$HO_2 + HO_2 \rightarrow H_2O_2 + O_2$$
$$HO_2 + OH \rightarrow H_2O + O_2$$
$$HO_2 + CH_3O_2 \rightarrow CH_3OOH + O_2$$

They are in competition with the ozone production and loss reactions

$$HO_2 + NO \rightarrow NO_2 + OH$$
$$HO_2 + O_3 \rightarrow 2O_2 + OH$$

A production of ozone occurs only by that fraction of HO_2 radicals entering into reaction with NO, whereas the fraction entering into reaction with O_3 causes a loss of ozone. The relative probability of each of the above reaction steps is determined by the steady-state concentrations of the corresponding HO_2 reactions partners. Apart from the photochemical production rate, the concentrations of the radicals depend critically on the extent of the recycling reactions of HO_2 with NO and O_3, that is, on the concentrations of ozone and nitric oxide. Since the former may be considered fixed, it is the latter that becomes a crucial parameter. In order to provide a more quantitative assessment of the situation, we make use of the numerical data of Fishman *et al.* (1979), who calculated concentrations and reaction probabilities for radicals for representative boundary layer conditions.

Figure 5-13 shows the fraction of HO_2 radicals undergoing reaction with NO and O_3 as a function of NO number density. The first reaction predominates when NO number densities are high, and the second when they are low. The crossover point is reached when both reactions proceed at equal rates according to the condition

$$n(NO) = (k_{53}/k_{54})n(O_3) = (1.8 \times 10^{-15}/8.0 \times 10^{-12})8 \times 10^{11}$$

$$= 1.8 \times 10^8 \quad molecules/cm^3$$

The corresponding mixing ratio for ground-level conditions is about 10 pptv. The probability of CH_3O_2 radicals entering into reaction with nitric oxide can be estimated similarly from the three competing reactions of CH_3O_2 with NO, HO_2, and with itself. The dependence on NO number density shown in Fig. 5-13 was obtained with the assumption that the rate coefficients are the same as those for the HO_2 reactions. In this manner one obtains, as a function of $n(NO)$, the relative probability f_1 for the reaction of CH_3O_2 with NO, and the relative probabilities f_2 and f_3 for the reactions of HO_2 with NO and O_3, respectively.

In order to determine the yield of ozone from reactions of OH with methane and carbon monoxide, it is instructive to trace the pertinent reaction

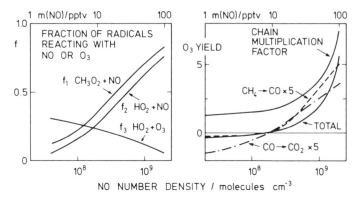

Fig. 5-13. Left: Fraction of CH_3O_2 and HO_2 radicals reacting with NO, respectively, and of HO_2 reacting with O_3, as a function of NO number density. The corresponding NO mixing ratio at ground level is shown by the upper scale. Right: Net production of ozone in numbers of O_3 molecules formed per molecules of CO oxidized to CO_2 and per CH_4 molecules oxidized to CO, both expanded by a factor of five; and the total number of O_3 molecules formed per OH radical that has reacted, taking chain multiplication into account. Negative values show ozone loss. The chain multiplication factor is shown as well.

pathways by means of the flow diagram shown in Fig. 5-14. A yield of 0.8 is assumed for HO_2 radicals resulting from the photolysis of formaldehyde. The mixing ratios of CH_4 and CO in the northern hemisphere are taken to be 1.5 and 0.1 ppmv, respectively. For these conditions, about 23% of the OH radicals react with methane, 72% with carbon monoxide, and the remainder with HO_2 radicals. The appropriate branching factors are indicated in Fig. 5-14. The number of ozone molecules resulting from the oxidation of one molecule of CO is $f_2 - f_3$, and the number of ozone molecules resulting from the oxidation of one molecule of methane to CO is $1.8 f_1 (f_2 - f_3)$. These individual ozone yields are shown as a function of NO number density on the right-hand side of Fig. 5-13. The average yield of ozone—that is, the yield obtained for one OH radical starting the reaction chain—is given by $(0.72 + 0.23 \times 1.8 f_1) (f_2 - f_3)p$, where p is a multiplier determined by the effective chain length. The fraction of OH radicals regenerated in each cycle is $f_4 = (0.72 + 0.41 f_1) \times (f_2 + f_3)$. The total number of OH radicals that have reacted after the cycle is repeated an infinite number of times is given by the geometric progression $p = \Sigma_s^\infty f_4^s = 1/(1 - f_4)$, so that the total yield of ozone obtained from one primary OH radical becomes

$$Y(O_3) = (0.72 + 0.41 f_1)(f_2 - f_3)/(1 - f_4)$$

The dependences of the multiplier p and the total ozone yield on the NO number density also are shown on the right-hand side of Fig. 5-13. Both increase strongly with the NO concentration.

Since two OH radicals are originally formed for each $O(^1D)$ atom having reacted with H_2O, the loss of ozone due to this reaction is compensated when the ozone yield from the reaction of OH with methane is 0.5. A yield of 0.5 is obtained for an NO number density of 4×10^8 molecules/cm^3, which corresponds to an NO mixing ratio of about 20 pptv. At lower NO

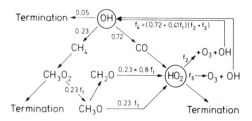

Fig. 5-14. Flow chart for the formation of ozone from the reactions of OH with CH_4 and CO present with mixing ratios of 1.5 and 0.1 ppmv, respectively. The term f_1 is the relative probability for the reaction of CH_3O_2 with NO, f_2 that of HO_2 reacting with NO, and f_3 that of HO_2 reacting with ozone; f_4 is the fraction of OH radicals regenerated in each cycle.

concentrations the formation of OH from $O(^1D)$ represents an ozone loss, whereas at higher NO concentrations it causes a net production of ozone. In the unpolluted troposphere NO is believed to derive entirely from the photodissociation of NO_2. In addition to reacting with CH_3O_2 and HO_2, NO is oxidized to NO_2 also by ozone. For steady-state conditions, as explained previously, the NO/NO_2 ratio should be given by

$$\frac{n(NO)}{n(NO_2)} = j(NO_2)/[k_{60}n(O_3) + k_{74}n(CH_3O_2) + k_{54}n(HO_2)]$$

Using numerical values of $j(NO_2) = 0.005$, $n(O_3) = 8 \times 10^{11}$, $n(CH_3O_2) = 3 \times 10^8$, and $n(NO_2) = 6 \times 10^8$ molecules/cm^3, which are mainly gleaned from Fishman *et al.* (1979), one obtains $n(NO)/n(NO_2) = 0.29$ or $n(NO)/[n(NO) + n(NO_2)] = 0.225$. Accordingly, approximately one-quarter of the total amount of nitrogen oxides is present as NO under daylight conditions. The total concentration of nitrogen oxides needed to reach the break-even point between photochemical production and destruction of ozone then becomes $4 \times 10^8/0.225 = 1.8 \times 10^9$ molecules/cm^3 or 90 pptv. The value corresponds roughly to the mixing ratio currently believed to exist in tropospheric background air of the northern hemisphere (see Section 9.4.2). It appears that photochemical production and loss rates for ozone are just balanced. For the northern hemisphere, Fishman *et al.* (1979) calculated an ozone production (loss) rate of 1320 Tg/yr in the case of balance. Since the O_3 dry deposition rate according to Table 5-7 is about 650 Tg/yr, twice as much ozone appears to be formed and destroyed by photochemical pathways. The residence time of tropospheric ozone must be lowered accordingly.

Figure 5-13 is misleading in the sense that it suggests a continuous rise of the ozone production rate with increasing nitrogen oxide concentration. The mechanism considered above does not yet include reactions of radicals with NO_2, however. The neglect of these reactions is permitted for NO_2 mixing ratios less than 200 pptv. The reaction of foremost importance is the termination process

$$OH + NO_2 \rightarrow HNO_3$$

but other reactions such as the addition of HO_2 to NO_2 also impede chain propagation in that they trap radicals. Thereby the rate of ozone formation becomes limited. This aspect has been explored by Stewart *et al.* (1977) and Hameed *et al.* (1979), who calculated steady-state concentrations of ozone, OH, and HO_2 resulting from the photochemical oxidation of CH_4 and CO for NO_x mixing ratios in the range 0.1–10 ppbv. The results are shown in Fig. 5-15. With increasing mixing ratio of NO_x, ozone first rises

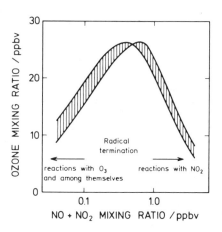

Fig. 5-15. Photochemical steady-state ozone mixing ratios resulting from the oxidation of CH_4 and CO in the boundary layer, calculated as a function of the $NO_x =$ $NO + NO_2$ mixing ratio by Stewart *et al.* (1977). The range of O_3 mixing ratios indicated by the hatched area corresponds to relative humidities between 20 and 80%. Rate coefficients and mechanism employed differ somewhat from those used by Fishman *et al.* (1979).

as discussed above, then reaches a maximum of 26 ppbv when $m(NO_x) =$ 0.5 ppbv, and finally declines again for still higher NO_x mixing ratios. For low NO_x mixing ratios steady-state ozone is controlled by loss reactions of radicals among themselves, whereas for high NO_x mixing ratios the control is exerted by radical termination with NO_2. The reaction mechanisms used contain some speculative elements, and the rate coefficients differed also somewhat from those employed by Fishman *et al.* (1979). Both groups of workers thus derived numerically different radical concentrations but found a qualitatively similar behavior. According to Fig. 5-15, the maximum steady-state ozone mixing ratio is 26 ppbv. This is comparable to mixing ratios usually observed in the boundary layer in middle latitudes of the northern hemisphere. Over the continents, where the abundance of NO_x is sufficient, it evidently is not possible to distinguish from mixing ratios alone whether the ozone that one observes has its origin in the stratosphere or has resulted from *in situ* photochemical production. The impact of hydrocarbons other than methane on ozone production or loss remains to be explored.

Chapter

6 | Volatile Hydrocarbons and Halocarbons

Hydrocarbons, as the name suggests, are compounds consisting of the elements hydrogen and carbon. Methane, the simplest and most abundant hydrocarbon in the atmosphere, was discussed in Section 4.3. Here, we consider other members of the family as far as they are volatile enough to reside primarily in the gas phase. This emphasizes low-molecular-weight compounds. Halocarbons are derivatives of hydrocarbons in which the hydrogen atoms are replaced by fluorine, chlorine, bromine, or iodine. Under the same heading we include in Section 6.3 a number of compounds in which hydrogen is only partially substituted by halogens.

6.1 Hydrocarbons

The number of hydrocarbons in the atmosphere is potentially very large, since vapor pressures are favorable and the heavier species admit many isomers. In urban areas several hundred different hydrocarbons have been identified by gas chromatography (Appel *et al.*, 1979; Louw *et al.*, 1977). They include saturated compounds (alkanes) unsaturated species with one carbon–carbon double bond (alkenes) or two double bonds (alkadienes), acetylene type compounds (alkynes), and benzene derivatives or aromatic compounds (arenes). To separate that many different compounds requires

quite sophisticated analytical techniques. Even more difficult are the sampling and analysis of hydrocarbons in the remote atmosphere, because it is
necessary to accumulate the material and then transfer it to the laboratory
for analysis. This involves elaborate precautions against contamination in
the transfer process. The diversity of hydrocarbon species has further
hampered a systematic exploration so that our knowledge and understanding
of this class of compounds in the atmosphere is as yet unsatisfactory.

6.1.1 HYDROCARBON LIFETIMES

All hydrocarbons except methane react rapidly with OH radicals. Unsaturated hydrocarbons (i.e., compounds containing at least one carbon–carbon
double bond) react also with ozone at a rate competing with the OH reaction.
Table 6-1 lists a number of hydrocarbon species of interest and the rate
coefficients for their reactions with OH and O_3. The corresponding atmospheric reaction rates may be estimated on the assumption that average
number densities for OH and O_3 are 5×10^5 and 6.5×10^{11} molecules/cm^3,
respectively. Atmospheric lifetimes are then obtained by summing the
individual rates and taking the inverse value. These data are included in
Table 6-1. The exercise shows that hydrocarbon lifetimes in the troposphere
are of the order of days—that is, they are considerably shorter than the
tropospheric residence time of the trace gases treated so far. Owing to their
reactivity hydrocarbons are expected to develop large gradients of mixing
ratios between the source regions and the remoter troposphere. With the
sources being concentrated at the earth surface, the gradients will be directed
mainly in the vertical but also from the continents toward the ocean for
those hydrocarbons whose sources are located on the continents. Figure
6-1 shows vertical profiles for ethane, ethene, and propane over Europe to
indicate that the expected behavior is actually observed.

Another interesting aspect is the filter effect associated with long-range
transport. Since the more reactive hydrocarbons are removed at a faster
rate than the less reactive ones, the abundance spectrum changes in favor
of the less reactive species. The longest-lived compounds are ethane and
acetylene. They have lifetimes exceeding 1 month and thus are expected to
spread around the globe even if their sources were unevenly distributed
(which they are). At the other end of the reactivity scale one finds the
higher alkenes, alkadienes, and terpenes. The lifetimes of these very reactive
compounds are of the order of hours, so that they cannot spread very far
beyond the boundaries of the source regions.

Table 6-1 includes rate coefficients for the reactions of OH radicals with
a few oxygenated organic compounds that are potential products resulting
from the oxidation of hydrocarbons. The corresponding lifetimes are also

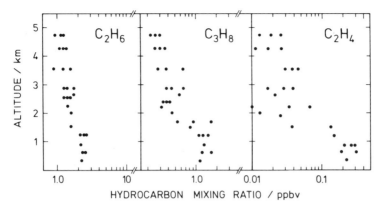

Fig. 6-1. Vertical distribution of ethane, propane, and ethene over Europe according to measurements by Ehhalt *et al.* (1986).

shown. Our knowledge about oxygenated species in the atmosphere is too fragmentary to discuss them in detail. They are included in Table 6-1 to show that their lifetimes with regard to interaction with OH have similar magnitudes as those of many hydrocarbons, so that the same conclusions are applicable. Note, however, that at least formaldehyde, which was treated in Section 4.4, and acetone undergo photodecomposition in addition to reaction with OH, so that the lifetimes shown are upper limits.

While rate coefficients for reactions with OH radicals with the lower members of the alkane and alkene series are well known from laboratory measurements, there exist only a few sporadic data for the higher homologues. It is possible, however, to estimate rate coefficients by an extrapolation of existing data where needed, and this allows us to derive lifetimes also for compounds for which OH rate coefficients have not been determined. Specifically, for the alkanes Greiner (1970a,b) has proposed a formula based on his studies of hydrogen abstraction by OH radicals from saturated hydrocarbons. The expression is

$$k_{OH} = 10^{-12}[1.02 N_1 \exp(-823/T) + 2.34 N_2 \exp(-428/T)$$
$$+ 2.09 N_3 \exp(95.6/T)] \tag{6-1}$$

which for $T = 298$ K leads to

$$k_{OH} = 10^{-13}(0.65 N_1 + 5.57 N_2 + 28.8 N_3)$$

Here, N_1, N_2, and N_3, respectively, are the number of primary, secondary, and tertiary hydrogen atoms associated with the alkane, and k_{OH} is given in units of cm^3/molecule s. In the range C_2-C_6 the accuracy of the formula

Table 6-1. *Hydrocarbon Reactivities: Rate Coefficients (at 298 K) for Reactions with OH Radicals and Ozone; The Corresponding Rates for $\bar{n}(OH) = 5 \times 10^5$, $n(O_3) = 6.5 \times 10^{11}$ molecules/cm³; And the Associated Lifetimes of the Hydrocarbons in the Troposphere*

Compound	$10^{12} k_{OH}{}^a$ (cm³/molecule s)	$10^{17} k_{O_3}{}^b$ (cm³/molecule s)	$10^6 R_{OH}$ (s⁻¹)	$10^6 R_{O_3}$ (s⁻¹)	τ (days)
Alkanes					
Ethane, C_2H_6	0.28	c	0.14	—	83
Propane, C_3H_8	1.6	c	0.80	—	15
n-butane, C_4H_{10}	2.7	c	1.35	—	8.6
Isobutane, $(CH_3)_2CHCH_3$	2.5	c	1.25	—	9.3
n-Pentane, C_5H_{12}	3.7	c	1.85	—	6.2
Isopentane, $(CH_3)_2CHC_2H_5$	3.1	c	1.55	—	7.5
n-Octane, C_8H_{18}	8.4	c	4.20	—	2.7
Alkenes and alkadienes					
Ethene, C_2H_4	7.9	0.19	3.95	1.23	2.2
Propene, C_3H_6	21	1.3	10.5	8.45	0.6
1-Pentene, C_5H_{10}	31	1.1	15.5	7.15	0.5
2-Pentene, C_5H_{10}	20	2.1	10.0	13.6	0.5
Butadiene, C_4H_6	68	0.8	34.2	5.2	0.3
Isoprene, C_5H_8	78	1.7	39.0	10.4	0.23
Acetylene, C_2H_2	0.7	0.004	0.35	0.02	31
Aromatic compounds					
Benzene, C_6H_6	1.2	c	0.6	—	19.3
Toluene, $C_6H_5CH_3$	6.4	c	3.2	—	3.6
ortho-Xylene, $C_6H_5(CH_3)_2$	14	c	7.0	—	1.6
para-Xylene, $C_6H_5(CH_3)_2$	15	c	7.6	—	1.5
meta-Xylene, $C_6H_5(CH_3)_2$	24	c	12.0	—	0.96
Ethylbenzene, $C_6H_5C_2H_5$	7.9	c	3.9	—	2.9

Terpenes,[e] $C_{10}H_{16}$				
α-Pinene	58	14	29	91	0.1
β-Pinene	67	0.4	33.5	26	0.19
d-Limonene	150	64	75	420	0.02
Myrcene	230	120	115	780	0.01
3-Carene	86	10	43	65	0.11
Aldehydes and ketones					
Formaldehyde, HCHO	14	c	7.0	—	1.6[d]
Acetaldehyde, CH_3CHO	15	c	7.5	—	1.5[d]
Benzaldehyde, C_6H_5CHO	13	c	6.5	—	1.8[d]
Acetone, CH_3COCH_3	0.5	c	0.25	—	46[d]
Ethyl methyl ketone, $C_2H_5COCH_3$	3.4	c	1.7	—	6.8[d]
Alcohols					
Methanol, CH_3OH	1.1	c	0.55	—	21
Ethanol, C_2H_5OH	3.3	c	1.65	—	7.0
n-Butanol, C_4H_9OH	7.6	c	3.8	—	3.0

[a] Rate coefficients are from the compilation of Atkinson et al. (1979), and for acetone from Cox et al. (1980).

[b] Rate coefficients for reactions of ozone with alkenes are from Japar et al. (1974), and for ozone with terpenes from Grimsrud et al. (1975).

[c] Rate coefficients for these reactions are less than 10^{-22} cm^3/molecule s.

[d] Upper-limit lifetimes, inasmuch as aldehydes and ketones also are subject to photodecomposition; this applies specifically to formaldehyde and acetone.

[e] The structures of individual terpenes are shown in Fig. 6-2.

227

is about 20%, and this is expected to hold also for the higher homologues. Primary hydrogen atoms are those associated with CH_3 groups, secondary ones refer to $-CH_2-$ links in the hydrocarbon chain, and tertiary H atoms are those occurring at chain branching points. In agreement with the experimental results, the expression shows that OH radicals abstract tertiary H atoms faster than secondary ones, and these are more readily removed than primary H atoms. The reason for this behavior lies in the difference of carbon–hydrogen bond strengths, which increase by about 25 kJ/mol when going from tertiary toward primary H atoms. The activation energy for the abstraction reactions increases accordingly, whereupon the rate coefficient declines. The above formula assumes that hydrogen atoms within each of the three groups react with equal probability. For the longer chains, therefore, the rate for the abstraction of secondary hydrogen atoms predominates over that for primary H atoms.

The reaction of OH radicals with alkenes and aromatic compounds proceeds by addition to the double bond and to the benzene ring. In these cases the above formula is not applicable. This does not preclude the occurrence of abstraction of H atoms with a certain probability from long-chained alkenes and aromatic compounds with longer side chains.

6.1.2 SOURCES OF ATMOSPHERIC HYDROCARBONS

Hydrocarbons are known to be released from anthropogenic sources, from vegetation and soils, and from the oceans. Duce et al. (1983) have reviewed available emission rates. These refer almost entirely to bulk emissions and not to individual compounds, because information on the distribution of hydrocarbons being emitted is in many cases incomplete. Ehhalt et al. (1986) have prepared a more detailed global source estimate for the C_2-C_5 hydrocarbons. Source distributions for the light hydrocarbons are now reasonably well known. The following presentation includes information on source distributions as far as possible.

Anthropogenic sources dominate in the cities. Table 6-2 shows mixing ratios of the major hydrocarbons observed in several North American cities and in Sidney, Australia. The data for Los Angeles in Table 5-1 may be added to the list. A notable feature is the similarity of hydrocarbon compositions in different urban regions. This is not really surprising, however, since source distributions in large cities are fairly much the same. Major sources are vehicular exhaust, gasoline evaporation and spillage, leaks of commercial natural gas, emissions from petrochemical manufacturing plants and refineries, and chemical solvents. Source apportionments have been reported by Mayrsohn and Crabtree (1976) for Los Angeles, and by Nelson et al. (1983) for Sidney. Mayrsohn and Crabtree (1976) estimated the following

Table 6-2. *Mixing Ratios (ppbv) of Major Hydrocarbons in Several U.S. Cities (Arnts and Meeks,* 1981; *Sexton and Westberg,* 1984) *and in Sidney, Australia (Nelson et al.,* 1983)

Compound	Houston	Philadelphia	Boston	Tulsa	Milwaukee	Sidney
Ethane	12.5	6.5	4.0	4.5	4.5	7.5
Propane	17.0	9.7	3.0	3.1	3.3	5.9
n-Butane	16.0	11.5	7.2	12.5	7.0	7.4
Isobutane	8.2	5.2	3.0	3.1	1.7	4.7
n-Pentane	7.6	5.4	3.2	8.2	2.4	5.0
Isopentane	13.4	8.4	7.0	13.2	4.6	9.0
2-Methylpentane	3.3	2.7	2.0	3.6	1.2	2.6
3-Methylpentane	2.5	1.8	1.3	2.2	0.8	1.6
n-Hexane	3.3	2.2	1.5	2.4	1.1	2.1
Ethene	—	—	—	3.4	—	12.5
Propene	5.7	3.3	1.3	1.0	1.0	7.4
1-Butene	1.0	0.25	0.2	—	0.25	1.0
Isobutene	1.5	0.25	0.2	1.1	0.5	1.4
trans-2-Butene	1.2	0.25	0.25	1.1	0.25	1.1
cis-2-Butene	—	—	—	0	—	1.0
Acetylene	7.5	3.0	4.5	4.5	2.5	10.1
Benzene	3.0	2.2	1.3	—	0.7	2.6
Toluene	6.8	4.1	4.0	2.1	2.3	8.9
Ethylbenzene	1.9	0.75	0.5	0.35	0.4	1.3

relative contributions for the city of Los Angeles (in weight percent): car exhaust, 48.8; gasoline, 16.3; gasoline evaporation, 13.3; commercial natural gas, 5.3; and natural gas from oil-field or production operations, 15.3. Industrial and solvent-related emissions were not considered. For the city of Sidney, Australia, Nelson *et al.* (1983) derived the following source apportionment (in weight percent): car exhaust, 35.8; gasoline, 16.3; gasoline evaporation, 16.0; solvents, 22.9; commercial natural gas, 3.7; and industrial process emissions, 5.4. Thus, approximately 73% of all emissions arises from the operation and supply of automobiles.

Table 6-3 gives a breakdown of sources for individual hydrocarbons according to the estimates of Mayrsohn and Crabtree (1976) and Nelson *et al.* (1983). The results show that the major source of ethane is natural gas, either from the commercial network or from the direct geological exploitation of resources. Propane appears to derive from natural gas as well as from petrochemical industries. The principal sources of butane and pentane are automotive exhaust and gasoline, although contributions from natural gas and industrial processes are not entirely negligible. Alkane solvent emissions become increasingly important for the higher members of this group, so that nonane and decane originate almost wholly from

Table 6-3. *Percentage Contribution of Various Sources to Hydrocarbons in (a) Los Angeles, California (Mayrsohn and Crabtree, 1976) and (b) Sidney, Australia (Nelson et al., 1983)*

Compound	Location	Car exhaust	Gasoline spillage	Gasoline evaporation	Natural gas	Industrial processes	Solvents
Ethane	a	7.9	—	—	90.9[a]	—	—
	b	18.2	—	—	82.2	—	—
Propane	a	—	—	3.6	96.4[a]	—	—
	b	1.2	—	7.9	26.6	64.4	—
n-Butane	a	24.0	7.0	40.5	28.2[a]	—	—
	b	14.6	9.3	59.3	4.0	12.8	—
Isobutane	a	16.3	3.5	33.3	46.9[a]	—	—
	b	11.4	6.1	56.2	4.3	22.0	—
n-Pentane	a	47.5	13.3	23.3	15.9[a]	—	—
	b	26.6	24.2	43.7	1.7	—	2.9
Isopentane	a	37.5	14.0	37.3	10.7[a]	—	—
	b	22.6	22.3	53.6	0.9	—	0.9
2-Methylpentane	b	31.1	29.8	21.8	0.8	—	15.7
3-Methylpentane	b	33.1	30.4	20.4	0.8	—	17.3
n-Hexane	b	32.9	28.3	15.0	1.3	—	22.6
n-Nonane	b	7.6	6.5	—	—	—	85.1
n-Decane	b	18.1	8.7	—	—	—	73.1
Ethene	b	98.9	—	—	1.4	—	—
Propene	b	49.9	—	—	0.3	49.8	—
1-Butene	b	67.3	3.3	29.4	—	—	—
Isobutene	b	77.4	2.6	17.7	—	—	—
trans-2-Butene	b	23.3	10.6	65.8	—	—	—
cis-2-Butene	b	22.7	11.3	63.9	—	—	—
Acetylene	a	100	—	—	—	—	—
	b	100	—	—	—	—	—
Benzene	b	77.0	17.8	6.0	—	—	—
Toluene	b	38.7	16.3	1.7	—	—	43.0
Ethylbenzene	b	45.4	17.5	1.1	—	—	33.7

[a] Includes both natural gas emanating from the ground before processing and commercial natural gas.

solvents. The alkenes, especially ethene, derive primarily from car exhaust. Nelson *et al.* (1983), however, estimated 50% of propene to be released from industrial processes. The sole source of acetylene is combustion. For this reason acetylene has been suggested to serve as a useful trace for automotive emissions. Small amounts of acetylene, however, are produced also in the burning of agricultural wastes and other biogenic materials. For the aromatic compounds, Table 6-3 suggests that three-quarters of benzene arise from automobile exhaust and the rest from automotive fuels. Toluene and ethylbenzene originate partly from car exhaust and gasoline and partly from solvent emissions.

As discussed earlier for carbon monoxide (see Section 4.5), anthropogenic emission rates can be estimated by combining statistical data for the production and usage of fuels with appropriate emission factors. For hydrocarbons, anthropogenic emissions were discussed at length in "Vapor Phase Organic Pollutants" (National Research Council, 1976) with specific reference to the situation in the United States. In this region roughly 25 Tg of hydrocarbons are emitted annually. Table 6-4 shows a breakdown for various sources. If the source distribution is assumed to be the same worldwide, one may extrapolate the emission rate linearly on the basis of fuel consumption rates. Logan *et al.* (1981) assembled data for the world fuel consumption from United Nations statistics, which show that the consumption of petroleum-derived fuels in the United States is about 33% of the entire world production. This suggests a global emission rate for anthropogenic hydrocarbons of 81 Tg/yr. A similar value of 88 Tg/yr was first given by Robinson and Robbins (1986b, 1972). Both values include emissions of methane. Following Peterson and Junge (1971), Duce (1978) assumed that 75% of the total emissions consisted of nonmethane hydrocarbons. He thus obtained a global

Table 6-4. *Hydrocarbon Emissions from Anthropogenic Sources: (a) In the United States, Including Methane (Environmental Protection Agency, 1976; Mann, 1981); (b) On a Global Scale, Nonmethane Hydrocarbons Only (Ehhalt et al., 1986); (c) Source Apportionment in Sidney, Australia (Nelson et al., 1983)*

Type of source	United States		Worldwide		Sidney (percent)
	Tg/yr	percent	Tg/yr	percent	
Transportation	11.3	41.7	29	51.8	68.1
Stationary combustion	1.6	5.9	4	7.1	—
Chemical industry and petroleum refineries	2.4	8.8	3	5.3	5.4
Oil and gas production	3.7	13.6	5	9.0	3.7
Organic solvents	8.1	30.0	15	26.8	22.9
Total	27.1		56		

production rate for this category of 65 Tg/yr. Ehhalt *et al.* (1986) derived a completely independent estimate for the emission rate of nonmethane hydrocarbons by the technique of detailed source accounting. Their value of 56 Tg/yr is in good agreement with that derived by Duce (1978). Table 6-4 includes a source breakdown for the global emission rate according to Ehhalt *et al.* (1986) to show that the assumption made above—namely, that the source distribution for the world is similar to that for the United States—is quite reasonable. Table 6-4 further includes the source distribution for Sidney, Australia, according to Nelson *et al.* (1983). In the cities, a larger than average share of hydrocarbons arises from transportation, whereas the contribution from oil and natural gas fields is smaller, of course. All three data sets indicate a sizable contribution of hydrocarbon emissions from solvents.

Stephens and Burleson (1969) were among the first to investigate the composition of hydrocarbons generated by brush fires. Their data indicated the production of alkanes, mainly ethane and propane, acetylene, and alkenes, with ethene and propene being the major unsaturated hydrocarbons emitted. More recently, Greenberg *et al.* (1984) studied hydrocarbon emissions from fires in Brazilian grassland and tropical forest regions. The composition of hydrocarbons emitted from biomass burning in both regions were similar. Ethane made up about 15 weight percent of total nonmethane hydrocarbons, propane 4–15%, ethene 25%, propene 12%, alkynes 1–10%, benzene, 8%, toluene 3.5%, furan 4.4%, and 2-methylfuran 1–4%. Altogether, these compounds comprised 81–86% of total nonmethane hydrocarbons produced by biomass burning. Relative to CO_2, 0.6–15 weight percent of hydrocarbons, with an average of 1.8%, was estimated to be released from fires. This compares well with earlier data of Darley *et al.* (1966), Bouble *et al.* (1969), and Sandberg *et al.* (1975) for the burning of grass stubbles and straw, agricultural wastes, and brushwood.

In the industrialized countries, the contribution of biomass burning to the total anthropogenic emissions is comparatively minor. In the tropics, however, the practices of slash, burn, and shift agriculture, in addition to deforestation, make biomass burning a major source of hydrocarbons, at least during the preferred burning season of the year. On the basis of an estimate of 3100 Tg C/yr (Seiler and Crutzen, 1980) for the global production of CO_2 from biomass burning, Greenberg *et al.* (1984) derived a global emissions rate for hydrocarbons from this source of 34 Tg C/yr, which corresponds to about 40 Tg/yr in terms of hydrocarbon mass. Methane, which is not included in this estimate, would add another 37 Tg/yr. The amounts of nonmethane hydrocarbons released from biomass burning thus are comparable to those of other anthropogenic emissions. Duce *et al.* (1983) have estimated the emission rate of hydrocarbons from biomass burning to range from 3.5 to 87 Tg/yr; whereas Ehhalt *et al.* (1986) derived a value of 14 Tg/yr.

Went (1955, 1960) pioneered the idea that plants may release substantial amounts of hydrocarbons to the atmosphere, partly by the volatilization of essential oils. Using data on the amount of leaf oils in sagebrush, Went (1960a) estimated a global emission rate from this source of 175 Tg/yr. The estimate was later revised to 200–400 Tg/yr by Rasmussen and Went (1965) after they had demonstrated the ubiquitous presence of terpenes in the air surrounding natural vegetation and dying leaves. Subsequently, Sandadze and Kalandadze (1966) and Rasmussen (1970) directed attention to isoprene as a forest emittant. It appears that isoprene is the predominant hydrocarbon emitted from forest species such as oak, poplar, sycamore, willow, cotton-wood, and eucalyptus (Rasmussen, 1972; Westberg, 1981; Isidorov *et al.*, 1985), whereas many other plants, especially conifers, emit primarily monoterpenes.

Figure 6-2 shows chemical structures for nine of the most frequently observed monoterpenes. These are alpha-pinene, beta-pinene, myrcene, delta-3-carene, *d*-limonene, terpinolene, alpha- and beta-phellandrene, and camphene. Flyckt (1979) and Westberg (1981) found the first six plus beta-phellandrene to account for most of the mass of volatile organic compounds emitted from Ponderosa pine. Isidorov *et al.* (1985) reported the relative terpene content in hydrocarbon emissions from Scots pine as alpha-pinene, 42%; beta-pinene, 21.5%; limonene, 20%; camphene, 6.7%; alpha-phellandrene, 5%; and myrcene, 4.8% by volume. Relative and absolute emissions vary considerably among plant species and also with external conditions. Following up on the work of Rasmussen and Jones (1973), Tingey (1981; Tingey *et al.*, 1979, 1980) measured emission rates as

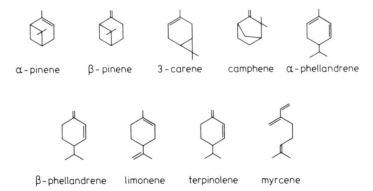

α-pinene β-pinene 3-carene camphene α-phellandrene

β-phellandrene limonene terpinolene myrcene

Fig. 6-2. Skeletal structures of several monoterpenes that have been identified in the atmosphere.

a function of light intensity and temperature. The emission rate of isoprene from oak saplings increased with increasing light intensity and with temperature. In the dark, the emission rates were low regardless of temperature. Emissions of monoterpenes, in contrast, were independent of light intensity, but increased log-linearly with rising temperature. This result is consistent with the behavior of monoterpene vapor pressures.

Volatile emissions from vegetation include hydrocarbons other than isoprene and terpenes. Altshuller (1983) has compiled emission data available to him [mainly from Zimmerman (1979a,b)]. The emissions contained C_2-C_6 alkanes, various alkenes, and C_6-C_{12} volatile organic compounds. Practically every deciduous plant and all the grasses studied emitted alkanes with ethane and propane dominating the mixture. Twenty to 50% of total hydrocarbon emissions, on average, consisted of alkanes.

Zimmerman (1979a,b; Zimmerman et al., 1978) has developed biogenic emission inventories for the United States and for global conditions based primarily on the release of isoprene and the monoterpenes. Emission rates were measured in the field by the bag enclosure technique. It involves placing a Teflon bag around one or several branches of the plant, replacing the air inside the bag with clean air, and determining the rate of increase of C_2-C_{12} hydrocarbons in the enclosed volume. Table 6-5 shows the emission rates derived, standardized to a temperature of 30°C. The region of the United States was divided into four geographic latitude zones. Area, total leaf biomass, the composition of leaf biomass, and representative monthly temperature ranges were estimated for each zone. As Table 6-5 indicates, isoprene emissions were assumed to be zero in winter except in the southernmost latitude zone because of the shedding of leaves from deciduous trees. This led to the following emission rates for isoprene (in units of Tg/yr): zone I (45–50°N), 0.18; zone II (40–45°N), 0.58; zone III (35–40°N), 4.3; zone IV (25–35°N), 15. The individual values sum to a total

Table 6-5. *Emission Rates for Hydrocarbons from Vegetation, Standardized to 30°C (Zimmerman et al., 1978)*

Classification	Summer		Winter
	Day	Night	
Conifers[a]	8.9	8.9	3.5
Oaks[a]	24.7	4.7	0
Nonconifer non-isoprene[a]	4.3	4.3	0
Nonoak isoprene[a]	10.3	2.4	0
Leaf litter and pasture[b]	162	162	0

[a] In μg C/(g leaf biomass) h.
[b] In μg C/(m² surface) h.

of 20 Tg/yr of isoprene emissions. For the monoterpenes the same procedure gave annual emission rates of 4.6, 10, 16, and 19 Tg/yr for zones I-IV, respectively, which sum to a total of 50 Tg/yr. Note that due to the dependence on light intensity and temperature the emissions occur primarily in the summer season.

In order to derive an estimate for the global emission rate, Zimmerman *et al.* (1978) first extrapolated their results for the United States to other continental regions in temperate latitude zones. This gave rates of 96 and 230 Tg/yr for isoprene and terpenes, respectively. For the tropics, Zimmerman *et al.* (1978) argued that the emission should resemble most closely those in the southern United States, where isoprene was found to contribute roughly 50% to the total emission rate. The ratio of hydrocarbon production to net annual CO_2 fixation was estimated as 0.7% for the United States. Zimmerman *et al.* (1978) used this value to extrapolate hydrocarbon emissions derived for the United States to global conditions. The procedure, which was already described in Section 4.5, gave 830 Tg/yr, with 500 Tg/yr of it occurring in the tropics. These values are similar in magnitude to the earlier estimate derived by Rasmussen and Went (1965). It appears that hydrocarbon emissions from vegetation dwarf those from anthropogenic sources and biomass burning. It should be reemphasized, however, that the composition of hydrocarbons being emitted from vegetation is considerably different from that of other sources. On the other hand, it must be cautioned against accepting isoprene and monoterpenes to be the dominant hydrocarbons emitted from vegetation everywhere in the world. As discussed earlier, Altshuller (1983) presented field data from several regions in the United States where alkanes comprised about 20% of emissions, and nonisoprene alkenes another 20%, from deciduous foliage regardless whether isoprene was emitted or not. These observations admit deciduous plants to serve as major sources of alkanes and alkenes in the atmosphere.

Table 6-5 includes average fluxes of hydrocarbons from grasses. According to Table 11-5, roughly 60×10^{12} m^2 of the continental surface may be characterized as grassland or land covered jointly by grasses and other vegetation. If one assumes that the emissions are negligible during the winter (or dry) season, the global rate of hydrocarbon emissions from this source would be 47 Tg/yr. The rate is small compared with that for hydrocarbons released from foliage, but it is competitive with anthropogenic emissions. Since 40-50% of hydrocarbons from grasses is alkanes, about 20 Tg/yr of alkanes would be produced from this source. The rate is greater than that for alkanes released from natural gas and oil fields. Thus, vegetation may provide the dominant source of ethane, propane, and higher alkanes in the atmosphere.

It is possible that a part of the emissions from gasses are in effect due to emanations from rotting leaves or from soil. Van Cleemput *et al.* (1981) have looked into this possibility. They found the production of hydrocarbons in various incubated soil samples small compared with that of methane. Relative production rates for ethane and propane were 0.07% and 0.04%, respectively, that of ethene was 0.5% by volume. Adopting from Table 4-8 a rate of 300 Tg/yr for the global production of methane from microbial sources, one obtains production rates for ethane of 0.4, propane 0.33, and ethene 2.6 Tg/yr.

Finally, one must consider the ocean as a source of nonmethane hydrocarbons. Lamontagne *et al.* (1974) and Swinnerton and Lamontagne (1974b) have reported that ethane, propane, ethene, and propene occur dissolved in marine surface waters with concentrations exceeding those expected if the gases were in equilibrium with atmospheric concentrations. Supersaturations accordingly exist causing the hydrocarbons to escape from the ocean into the atmosphere. For certain ocean areas the fluxes can be estimated. Rudolph and Ehhalt (1981) have used the aqueous concentration data of Swinnerton and Lamontagne (1974b) in conjunction with their own measurements of atmospheric mixing ratios and the solubility data of McAuliffe (1966) to derive emission rates in units of $\mu g/m^2 h$ of 0.3 for ethane, 0.4 for propane, 0.6 for ethene, and 0.5 for propene. These values refer to a region of the North Atlantic Ocean at latitudes near 70°N. A linear extrapolation to the entire ocean surface area suggests global emission rates of 1.0, 1.3, 1.9, and 1.6 Tg/yr for ethane, propane, ethene, and propene, respectively. There are indications, however, that the tropical ocean contains somewhat higher supersaturations of ethene and propene (Swinnerton and Lamontagne, 1974b). Seawater concentrations a factor of three higher in the tropics would raise the global emission of ethene and propene from marine sources to 3.8 and 3.2 Tg/yr, respectively.

The atmosphere above the ocean contains C_9-C_{28} *n*-alkanes in addition to low-weight hydrocarbons. Eichmann *et al.* (1980) have estimated that the ocean might emit 26 Tg of higher alkanes each year. The value is based on the measured total atmospheric concentration for C_9-C_{28} *n*-alkanes of 300–400 ng/m^3 over the Indian and the North Atlantic Oceans, coupled with an assumed residence time of the alkanes of 5 days resulting from their reactions with OH radicals. From measurements at Enewetak Atoll (Marshall Islands), however, Duce *et al.* (1983) reported atmospheric concentrations of $C_{12}-C_{28}$ alkanes that were an order of magnitude lower than those found by Eichmann *et al.* (1980). It is questionable, therefore, whether their data can be extrapolated to the whole ocean area, so that the emission rate derived must be regarded an upper limit.

Table 6-6 summarizes the hydrocarbon emission data discussed above. Although numerous sources have been identified, and several of them contribute substantially to atmospheric hydrocarbons at least on a local scale, no individual source can match global emissions from vegetation.

Table 6-6. *Summary of Global Emission Rates of Nonmethane Hydrocarbons from Various Sources*

Type of source	Emission rate (Tg/yr)	Remarks
	Anthropogenic sources	
Combustion and chemical industry	36	Mainly automobiles, alkanes, alkenes, and aromatic compounds
Natural gas	5	Mainly light alkanes
Organic solvents	15	Higher alkanes and aromatic compounds
Biomass burning	40	Mainly light alkanes and alkenes
	Biogenic sources	
Foliar emissions	830	Mainly isoprene and monoterpenes, some alkenes and alkanes
Grasslands	47	Light alkanes and higher hydrocarbons
Soils	<3	Mainly ethene
Ocean waters	6–10	Light alkanes and alkenes
	<26	C_9–C_{28} Alkanes

6.1.3 HYDROCARBON MIXING RATIOS OVER THE CONTINENTS

Despite a growing interest in hydrocarbons, stimulated by the recognition of their importance to atmospheric chemistry, there still are very few data available for hydrocarbon abundances outside urban areas. Table 6-7 summarizes mixing ratios of C_2–C_5 alkanes determined at rural measurement sites, mainly in the United States. Table 6-8 adds data for several alkenes, acetylene, and a few aromatic compounds. Altshuller (1983) has reviewed a number of other sporadic measurements made on the North American continent. Mixing ratios range from less than 0.1 ppbv for the higher hydrocarbons to about 2 ppbv for ethane and acetylene. The alkanes, and to some extent also the alkenes, show a fairly regular distribution in that their mixing ratios decrease with increasing carbon number. Two factors may be responsible for this pattern: the source distribution, and the greater activity of compounds with higher carbon numbers. The second factor applies mainly to hydrocarbons from distant sources such as urban centers, and less to local sources. This leads to the question whether alkanes produced in the cities can explain the distribution of alkanes observed at rural sites. Table 6-2 makes it evident that in the cities the abundances of C_2–C_5 alkanes are very similar. For the present discussion one may assume equal mixing ratios for the six species involved. During transport the mixing ratios are reduced by dilution and by reaction with OH radicals. Adopting the lifetimes of Table 6-1 and a transport time of two days, one obtains a distribution relative to ethane of 0.9, 0.81, 0.83, 0.74, 0.78 for propane, *n*- and isobutane, and *n*- and isopentane, respectively. If the OH number

Table 6-7. *Average Mixing Ratios of Alkanes (ppbv) in the Continental Boundary Layer at Rural Measurement Sites*

Authors	Location	C_2H_6	C_3H_8	$n\text{-}C_4H_{10}$	$i\text{-}C_4H_{10}$	$n\text{-}C_5H_{12}$	$i\text{-}C_5H_{12}$	Others (ppbC)[e]	Total alkanes (ppbC)[e]
Rasmussen et al. (1977)	Elkton, Missouri, August 1975	2.23	1.37	0.61	0.35	0.43	n.g.[f]	n.g.	14.6
Lonneman et al. (1978)	Florida, May 1976, citrus groves	1.50	1.03	1.87	0.65	0.73	1.80	16.0	48.6
Arnts and Meeks (1981)	Everglades	1.15	0.13	0.15	0.07	0.08	0.18	11.1	16.0
	Great Smoky Mtn. National Park, Tennessee, September 1978	5.70	2.72	1.20	0.64	0.59	0.97	8.9	48.2
	Rio Blanco, Colorado, July 1978	2.60	1.63	0.30	0.25	0.40	0.38	30.4	48.1
Sexton and Westberg (1984)	Belfast, Maine, summer 1975	1.75	0.67	0.50	0.12	0.20	0.20	n.g.	10.0
	Miami, Florida, summer 1976	2.00	0.67	0.12	0.10	0.20	0.30	n.g.	9.5
Greenberg and Zimmerman (1984)	Central Brazil, 1979/1980	2.09[a]	0.45	0.24	n.g.	n.d.	n.g.	1.6	8.0
		2.97[b]	0.42	0.34	n.g.	n.d.	n.g.	1.6	12.3
Colbeck and Harrison (1985)	Lancaster, England early summer 1983	3.20[c]	2.20	0.33	0.20	0.74	n.g.	2.6	21.4
		14.7[d]	4.97	0.85	3.52	0.98	n.g.	17.0	83.8
Greenberg et al. (1985)	Kenya, July/August 1983 Lake Baringo	0.68	0.12	0.09	0.05	0.04	0.03	0.7	2.7
	Marigat	1.85	0.32	0.07	0.03	0.03	n.d.	1.6	6.9

[a] All alkanes measured below canopy.
[b] Above canopy as determined by aircraft sampling.
[c] On days when ozone levels were near background values.
[d] On days when ozone levels exceeded background values.
[e] In ppbv of carbon.

238

Table 6-8. *Average Mixing Ratios (ppbv) of Several Alkenes, Acetylene, and Some Aromatic Compounds in the Continental Boundary Layer at Rural Measurement Sites*

Authors	Location	Ethene	Propene	Butenes	Alkadienes	Acetylene	Benzene	Toluene	Ethyl-benzene
Rasmussen et al. (1977)	Elkton, Missouri, August 1975	0.79	0.21	n.g.[g]	1.21[e]	0.34	n.g.	0.27	0.06
Lonneman et al. (1978)	Florida, May 1976								
	Citrus groves	1.85	0.32	0.25	0.10[f]	2.22	n.g.	1.11	0.16
	Everglades	0.20	0.20	n.d.	n.d.	0.60	n.g.	0.36	0.25
Arnts and Meeks (1981)	Great Smoky Mtn. National Park, Tennessee, September 1978	2.14	0.74	0.23	0.60[e]	2.39	0.89	0.96	0.12
	Rio Blanco, Colorado, July 1978	0.63	0.30	n.d.	0.75[e]	1.00	n.g.	0.84	0.12
Sexton and Westberg (1984)	Belfast, Maine, summer 1975	1.0	0.17	0.12	n.g.	0.25	n.g.	n.g.	n.g.
	Miami, Florida, summer 1976	0.25	0.17	0.12	n.g.	0.25	n.g.	n.g.	n.g.
Greenberg and Zimmerman (1984)	Central Brazil, 1979—1980	1.89[a]	0.31	0.47	2.44[e]	0.88	0.50	0.12	0.08
		2.70[b]	0.68	1.08	2.50[e]	1.08	1.15	0.14	0.15
Colbeck and Harrison (1985)	Lancaster, England early summer 1983	0.75[c]	0.93	0.40	n.g.	0.25	n.g.	n.g.	n.g.
		6.00[d]	0.70	0.57	n.g.	2.70	n.g.	n.g.	n.g.
Greenberg et al. (1985)	Kenya, July/August 1983								
	Lake Baringo	3.18	0.84	0.40	0.11[e]	1.14	0.27	0.15	0.04
	Marigat	0.87	0.18	0.32	0.03[e]	0.58	0.17	0.14	0.06

[a] Below canopy.
[b] Above canopy as determined by aircraft sampling.
[c] On days when ozone levels were near background values.
[d] On days when ozone levels exceeded background values.
[e] Mainly isoprene.
[f] Mainly butadiene.
[g] n.d., Not detected; n.g., not given.

239

density were twice as high as that inherent in the data of Table 6-1, one would obtain the distribution 0.80, 0.66, 0.68, 0.55, 0.61. Most of the data in Table 6-7 show alkane distributions that are different. The decline of mixing ratios from ethane toward propane is steeper and that from propane toward the butanes and pentanes is flatter. Thus it appears that urban emissions are not the sole source of alkanes observed at rural sites. At least ethane, but probably propane as well, requires additional sources if the butanes and pentanes are assumed to arise from urban emissions and perhaps local automobile traffic. Altshuller (1983) suggested for the United States that emissions from natural gas wells are responsible for the excess alkanes. As discussed previously, however, copious amounts of ethane and propane are emitted from vegetation, and these emissions represent a more likely source for excess alkanes even over the North American continent.

Acetylene has been suggested to provide a tracer of automotive emissions, and according to Table 6-3 the butanes and pentanes also result largely from the operation of automobiles. The average ratios relative to acetylene of n-butane, isobutane, n-pentane, and isopentane in the cities are 1.0, 0.38, 0.49, and 0.85, respectively, according to Table 6-3. These ratios would not be greatly altered by individual hydrocarbon reactivities during two days of transport. The corresponding average ratios for the first seven entries of Table 6-7 and 6-8 are 1.0, 0.46, 0.64, and 0.75. They do not exactly match the ratios expected from automotive emissions, but the variability is considerable, so that an assignment of the butanes, the pentanes, and acetylene to automotive emissions appears justified. In the southern hemisphere, automotive emissions are expected to contribute less to total hydrocarbons. The data obtained by Greenberg and Zimmerman (1984) over the Brazilian forests are unfortunately too imcomplete to provide much of a test, although it is interesting to note that the mixing ratio of n-pentane is much smaller than that of n-butane. The data for the butanes and pentanes obtained by Greenberg et al. (1985) in Kenya, when expressed as ratios of mixing ratios versus that for acetylene, are similar to those in the United States. In both regions, ethane is the major nonmethane alkane.

Alkenes and aromatic compounds have atmospheric lifetimes shorter than those of the low-weight alkanes. After 2 days of transport, propene and the butenes are reduced to a few percent of their original abundances relative to acetylene. If propene and the butenes are found at rural measurement sites, they must have local sources. Ethene, toluene, and ethylbenzene are less reactive. Their abundances relative to acetylene are reduced by a factor of about two after 2 days of transport. Toluene and ethylbenzene are primarily anthropogenic compounds. Both may serve as tracers for urban hydrocarbons in the same way as acetylene. In the cities the abundance ratio of ethylbenzene to toluene is about 1.5. The ratio should not change much during transport because both compounds have comparable lifetimes. A portion of the data in Table 6-8 shows ethylbenzene to toluene ratios similar to those found in the cities. For the remaining data the ratio is greater. The reasons for this behavior are not known.

The average mixing ratios of ethene, propene, and the butenes relative to toluene for the first five entries in Table 6-8 are 1.65, 0.55, and 0.23, respectively. The relative abundance of ethene is found to fluctuate more than those of the other compounds. In the cities, the ratios are 1.0, 0.61, and 0.46. After 2 days of transport they would be 0.72, 0.04, and 0.01. Thus, it is clear that local sources must exist at rural sites for all the alkenes, although the excess observed for propene and the butenes is greater than that for ethene. The similarity of ratios with those in the cities suggests vehicular exhaust emissions as an important component. Indeed, for the region of the United States, automobile emissions appear to provide an acceptable explanation for the presence of alkenes in rural areas. The data from Brazil and Kenya, however, still show ethene, propene, and butene to be present in roughly the same proportions as in the United States despite an about tenfold lower abundance of toluene. These observations indicate additional sources of alkenes in regions with little or no automobile traffic. In the tropics, biomass burning may contribute significantly to ethene and propene in the atmosphere. The generation of butenes from this source has not been reported. Thus, it is currently more appealing to assign the observation of low-weight alkenes to emissions from vegetation.

Isoprene is the major alkadiene at rural measurement sites. As discussed earlier, foliar emissions provide the major source of isoprene and monoterpenes. In the United States alone, 15 Tg of isoprene and 50 Tg of monoterpenes are released annually from vegetation according to an estimate of Zimmerman *et al.* (1978), versus 27 Tg of anthropogenic hydrocarbons. Table 6-9 shows atmospheric mixing ratios of isoprene and common terpenes determined in or near forests. The abundances of the compounds are low compared with the large emission rates that have been projected to arise from vegetation, especially in view of the fact that most of the measurements were made in summer when the emissions maximize. One possible reason for the low mixing ratios may be the high reactivity of isoprene and terpenes in the atmosphere. Their reactions with ozone and OH radicals are extremely rapid, and Table 6-1 shows that the corresponding lifetimes are in the order of hours. Accordingly, one expects a fairly rapid decline of mixing ratios with distance from the source. Several observers have indeed reported measurable concentrations of isoprene and terpenes underneath the canopy, but much lower values, frequently below the detection limit, have been found outside and downwind of a forest (Whitby and Coffey, 1977;

Table 6-9. *Average Mixing Ratios (ppbv) of Isoprene and Terpenes in the Continental Boundary Layer at Rural Measurement Sites, Mainly Underneath Forest Canopies*

Authors	Location	Isoprene	α-Pinene	β-Pinene	Δ-3-Carene	Camphene	d-Limonene	Others
Rasmussen et al. (1977)	Elkton, Missouri, August 1975, hardwoods	1.21	0.033	n.g.[f]	n.g.	n.g.	n.g.	n.g.
Holdren et al. (1979)	Moscow Mtn., Idaho, 1976/1977, pine and fir	Trace	0.113	0.086	0.064	n.g.	0.01	n.g.
Cronn and Harsch (1980)	Great Smoky Mtn. National Park, Tennessee, September 1978, deciduous	1.20	0.40	0.20	0.50	n.g.	n.g.	n.g.
Arnts and Meeks (1981)	Great Smoky Mtn. National Park, Tennessee, September 1978, deciduous	0.60	0.10	n.g.	n.g.	n.g.	n.g.	n.g.
Roberts et al. (1983)	Niwot Ridge, Colorado, 1981/1982, fir and spruce	n.g.	0.054	0.097	0.051	0.038	0.03	n.g.
Shaw et al. (1983)	Mt. Kanobili, Abastumani forest, Georgia, Soviet Union, July 1979, pine and spruce	1.4	0.8	0.43	0.90	0.09	0.08	c
Ciccioli et al. (1984)	Outskirts of Rome, Italy, pine forest	n.g.	1.50	0.18	0.06	n.g.	0.04	d
Greenberg and Zimmerman (1984) 1979/1980	Central Brazil,	2.40[a]	n.g.	0.27	0.24	n.g.	n.g.	e
		2.27[b]	n.g.	0.15	0.82	n.g.	n.g.	e

[a] Below canopy.

[b] Above canopy as determined by aircraft sampling.

[c] Myrcene, 0.68.

[d] p-Cymene, 0.046.

[e] Below canopy: Δ-4-carene, 0.39; myrcene, 0.19; α-phellandrene, 0.18; α-terpinene, 0.49; plus others; above canopy: Δ-4-carene, 2.08; myrcene, 1.09; α-phellandrene, 1.17; α-terpinene, 0.12; plus others.

[f] n.g. Not given.

Yokouchi *et al.*, 1983). Micrometeorological conditions also are important in controlling the concentrations of forest emissions. For example, Isidorov *et al.* (1985) noted that a stable atmospheric stratification in the evening caused a considerable enhancement of terpene concentrations.

Altshuller (1983) has approached the problem from a different viewpoint. Arguing that propene has a lifetime in the atmosphere similar to that of isoprene and alpha-pinene, and that in the United States propene should result primarily from vehicular emissions, he computed the ratio of isoprene to propene implied by the relative emission rate estimates given above and found a ratio of 25:1, on average, with a maximum of 74:1 in July. Similar values were derived for alpha-pinene. Comparison with the data in Table 6-8 and 6-9 clearly shows that ratios that high are observed nowhere in the United States. The highest ratios were observed by Greenberg and Zimmerman (1984) in and above forests of central Brazil, where propene is not expected to arise from automotive emissions. The inconsistency of measured and projected ratios of isoprene and alpha-pinene versus propene led Altshuller (1983) to conclude that either biogenic emission rates were grossly overestimated or anthropogenic emission rates were underestimated, or both. It has already been noted, however, that rural mixing ratios of propene are too high to be assigned exclusively to automotive exhaust and that vegetation must provide an additional source of propene. In this case the method of accounting employed by Altshuller (1983) would fail.

Arnts *et al.* (1982) have measured the dispersion of alpha-pinene from a loblolly pine forest with the help of SF_6, which was released inside the forest and served as a tracer for air motions. Both SF_6 and alpha-pinene were measured downwind at a number of receptor sites. An atmospheric plume dispersion model was used to describe the observations and to infer emission rates. Losses of alpha-pinene due to reactions with ozone and OH radicals were incorporated into the model. Emission rates derived ranged from 11 to 19 $\mu g/m^2$ min. No direct comparison was made with the rate of alpha-pinene emissions from three branches by the bag enclosure technique. Instead, available data from Rasmussen (1972) and Zimmerman (1979b) were used to estimate the rate expected for this particular forest site at the temperatures encountered during the measurements. The values obtained were 14–35 $\mu g/m^2$ min and 47 $\mu g/m^2$ min, respectively. Although the alpha-pinene fluxes inferred from the tracer release approach are lower, the differences are not great enough in view of the large variability of actual emission rates to raise doubts about the applicability of the bag enclosure technique.

While the controversy about large emissions of isoprene and terpenes versus their low atmospheric mixing ratios has not been fully settled, there exists no stringent support for Altshuller's (1983) contention that the ratio of natural to anthropogenic hydrocarbon emissions has been greatly over-estimated. At the same time, it is clear that our understanding of natural hydrocarbons in the atmosphere is still fragmentary. In addition, there is very little information on the oxidation products. Obviously, the subject requires much more study.

6.1.4 HYDROCARBONS IN THE MARINE ATMOSPHERE

Measurements of hydrocarbons over the oceans were reviewed by Duce *et al.* (1983). The data for the light hydrocarbons include alkanes, alkenes, and acetylene. Measurements of the heavier hydrocarbons refer mainly to *n*-alkanes. Compounds with carbon numbers up to C_{32} have been observed. Table 6-10 gives an overview on total concentrations. Figure 6-3 is added to show the distribution of alkane concentrations with carbon number. Individual concentrations range over several orders of magnitude, from micrograms per cubic meter or ppbv for ethane to nanograms per cubic meter or less than 0.1 pptv for C_{30} alkanes. The steepest decrease in relative abundance occurs in the C_2–C_{12} range of carbon numbers. A slight secondary maximum appears to exist for C_{15}–C_{21} hydrocarbons. Much lower concentrations of C_{12}–C_{32} alkanes were observed at Enewetak Atoll in the tropical northern Pacific than either in the Atlantic Ocean or at Gape Grim,

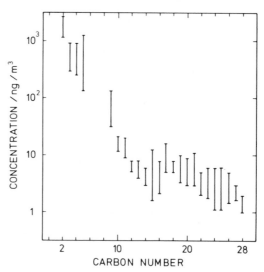

Fig. 6-3. Alkanes in marine air near the ocean surface. Shown is the distribution of concentration with carbon number according to data of Rudolph and Ehhalt (1981) for C_2–C_5 alkanes over the Atlantic Ocean; and of Eichmann *et al.* (1979, 1980) for C_9–C_{28} alkanes in the air of the North Atlantic and South Indian Oceans.

Table 6-10. Concentrations of Hydrocarbons in the Marine Atmosphere near the Ocean Surface

Authors	Location	Carbon numbers	Mean	Range	Remarks
Robinson et al. (1973)	Point Reyes, California	$C_4–C_7$	8.8	—	2.2 $\mu g/m^3$ of acetone is also reported
	Brethway, Washington	$C_4–C_7$	4.3	—	
Rasmussen (1974, as reported by Duce et al., 1983)	Hawaii and North Atlantic	$C_2–C_{12}$	8^a	$4–16^a$	
Wade and Quinn (1974)	Bermuda	$C_{14}–C_{32}$	0.18^a	$0.05–0.5^a$	
Eichmann et al. (1979)	Loop Head, Ireland	$C_9–C_{17}$	0.06	0.04–0.16	Only n-alkanes
		$C_{18}–C_{28}$	0.2	—	
Eichmann et al. (1980)	Cape Grim, Tasmania	$C_9–C_{28}$	0.34	—	Only n-alkanes in pure marine air masses except C_{17} where all data are included
Rudolph et al. (1980)	Loop Head, Ireland	$C_2–C_5$	9.1	—	Includes alkanes, alkenes, and acetylene
	North Atlantic	$C_2–C_5$	4.0		
Rudolph and Ehhalt (1981)	North Atlantic	$C_2–C_5$	7.11	2.3–28.4	
	Equatorial Atlantic	$C_2–C_5$	2.5	1.8–3.3	
Singh and Salas (1982)	Eastern Pacific Ocean	$C_2–C_5$	7.58^b	$3.9–9.6^b$	
			3.39^c	$2.1–4.2^c$	
Duce and Gagosian (1982, from E. Atlas and C. Giam, unpubl.)	Enewetak Atoll, Pacific	$C_{13}–C_{27}$	2.5×10^{-3}	—	n-Alkanes
Duce et al. (1983, from O. C. Zafiriou et al., unpubl.)	Enewetak Atoll, Pacific	$C_{21}–C_{32}$	3.1×10^{-4}	—	n-Alkanes

Range of mixing ratio ($\mu g/m^3$ STP)

[a] In μg C/m^3 STP.
[b] Northern hemisphere.
[c] Southern hemisphere.

Tasmania. A similar difference was found for *n*-alkanes on aerosol particles (see Section 7.5.2). The reasons for the differences are not known. They may be characteristic of the measurement site or may arise from artifacts associated with the collection and analysis of air samples.

The light hydrocarbons have the highest abundances and thus contribute most to the total hydrocarbon mass budget, even though they span a small range of carbon numbers. Figures 6-4 to 6-6 show meridional distributions of ethane, propane, and acetylene. All three compounds show a similar behavior in that low mixing ratios are found in the southern hemisphere and high ones in the northern, maximizing near 40°N latitude. The behavior is qualitatively similar to that of carbon monoxide. Since reaction with OH radicals is the major sink for hydrocarbons, the uneven distribution indicates an excess of sources in the northern hemisphere. For ethane and acetylene the data show a considerable variability. In addition, the aircraft measurements indicate mixing ratios somewhat lower than those at the surface. This suggests the existence of a vertical gradient similar to that occurring over the continents (see Fig. 6-1). The surface data for the Pacific and the Atlantic Oceans are in reasonable agreement, however, which shows that ethane and acetylene are transported around the globe with the general circulation.

In the southern hemisphere the oceans cover about two-thirds of the global surface area. In Section 6.1.2 estimates for emission rates from the ocean were given as 0.3 $\mu g/m^2$ h for ethane and 0.4 $\mu g/m^2$ h for propane. This leads to global source strengths of $Q(C_2H_6) = 0.63$ Tg/yr and $Q(C_3H_8) = 0.84$ Tg/yr for the southern hemisphere. If the ocean provided

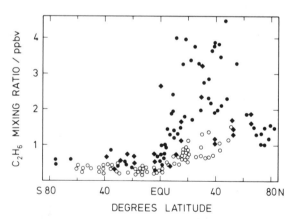

Fig. 6-4. Latitudinal distribution of ethane; open circles indicate aircraft data over the Pacific from Rasmussen (Aiken *et al.*, 1982), diamonds are data for surface air over the Pacific (Singh *et al.*, 1979b), and filled circles are data for surface air over the Atlantic Ocean from Ehhalt *et al.* (1986).

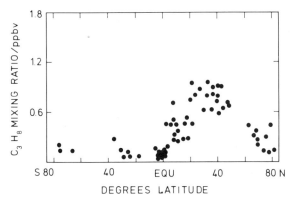

Fig. 6-5. Latitudinal distribution of propane in the surface air over the Atlantic Ocean (Ehhalt *et al.*, 1986).

the only source and reaction with OH radicals the sole sink, one would estimate from the lifetimes shown in Table 6-1 average mixing ratios of 0.065 and 0.016 ppbv for ethane and propane, respectively. Although the mixing ratios observed for both compounds appear to be fairly uniform, they have values of $m(C_2H_6) \approx 0.4$ ppbv and $m(C_3H_8) \approx 0.2$ ppbv, which are much higher than anticipated on the basis of the above argument. Evidently, additional sources must exist on the continents for both compounds. The observed mixing ratios suggest instead source strengths of at least 4 Tg/yr for ethane and 12 Tg/yr for propane in the southern hemisphere alone. In

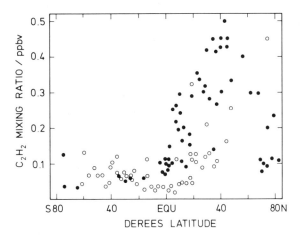

Fig. 6-6. Latitudinal distribution of acetylene; open circles indicate aircraft data over the Pacific from Rasmussen (Robinson, 1978), and filled circles are data for surface air over the Atlantic from Ehhalt *et al.* (1986).

the northern hemisphere such estimates are more difficult to make because of the nonuniform distribution of ethane and propane and the larger scatter of the data. Nevertheless, source strengths of the order of 20 Tg/yr for ethane and 25–30 Tg/yr for propane are indicated. According to the discussion in Section 6.1.2, most of the material appears to have a natural origin. Anthropogenic sources emit relatively little ethane and propane.

Acetylene may be assumed to originate entirely from combustion processes. The mixing ratios for acetylene over the oceans are lower than those for either ethane or propane. In the southern hemisphere the mixing ratio of acetylene averages about 0.065 ppbv. Assuming this value to be representative for the entire southern hemisphere and combining it with an atmospheric lifetime of 31 days as shown in Table 6-1 gives a source strength of 1.5 Tg/yr. In the northern hemisphere the acetylene mixing ratio is a factor of four higher than in the southern, which suggests a global source strength of 7.5 Tg/yr. According to Table 6-6, anthropogenic emissions of hydrocarbons from automobiles, industries, and biomass burning sum to 76 Tg/yr. Total hydrocarbons produced by combustion processes contain about 10% acetylene, so that the production for this compound should be about 7.6 Tg/yr. The two figures thus match. In fact, considering the usual uncertainties, the agreement looks almost too good.

Far fewer data exist for the butanes and pentanes in the marine atmosphere. Average mixing ratios for the Atlantic and Pacific Oceans are shown in Table 6-11. The mixing ratios observed by Singh and Salas (1982) over the Eastern Pacific are much higher than those found by Rudolph and Ehhalt (1981) in various parts of the Atlantic Ocean. It is not known whether the butanes and pentanes are released from the ocean, but presumably they are remnants from continental sources. Ethene and propene also occur in the marine atmosphere. Table 6-11 includes average mixing ratios observed for these two compounds. In contrast to the alkanes, ethene and propene are fairly evenly distributed between both hemispheres. In Section 6.1.2 it was pointed out that both alkenes are released from the ocean at rates indicating global source strengths of 1.9 Tg/yr for ethene and 1.6 Tg/yr for propene. The lifetimes of C_2H_4 and C_3H_6 in the atmosphere are considerably shorter than those of alkanes, making it unlikely that the alkenes survive long-range transport in significant amounts. Accordingly, it appears that ethene and propene observed in marine air have an oceanic origin. Owing to the short lifetime, an appreciable vertical gradient is expected to exist from the ocean surface on up to greater altitudes, similar to that evident from Fig. 6-1 for ethene over the continent. The average mixing ratios for C_2H_4 and C_3H_6 in the troposphere thus must be smaller than those occurring at the ocean surface. If these factors are taken into account, the inferred fluxes may suffice to explain the abundances of ethene and propene. Note,

Table 6-11. *Mixing Ratios (pptv) of C_4 and C_5 Alkanes, Ethene, and Propene in the Marine Atmosphere Near the Ocean Surface*

Authors		n-C_4H_{10}	i-C_4H_{10}	n-C_5H_{12}	i-C_5H_{12}	C_2H_4	C_3H_8
Rudolph and Ehhalt (1981)	Equatorial Atlantic	75	23	29	16	250	120
	North Atlantic and Barents Sea[a]	98	54	57	40	180	110
Singh and Salas (1982)	Eastern Pacific, northern hemisphere	197	530	232	370	95	545
	Eastern Pacific, southern hemisphere	100	153	143	273	73	207

[a] Two data sets with much higher mixing ratios are omitted.

however, that the data of Singh and Salas (1982) for the Eastern Pacific indicate higher mixing ratios for propene than for ethene, in spite of a shorter lifetime of propene in the atmosphere but equivalent release rates. This is an inconsistency that must be resolved in the future.

Finally, we consider the behavior of the higher hydrocarbons in the marine atmosphere. Figure 6-7 shows data for the C_{10}-C_{28} n-alkanes obtained at Loop Head, Ireland, and at Cape Grim, Tasmania. At both locations, trajectory analyses were used to select pure marine air masses that had no immediate land contact before reaching the measurement sites. In addition, ethylbenzene and the xylenes were used as convenient indicators of contaminated air. As Figure 6-7 shows, the concentrations are similar in the middle latitudes of the northern and southern hemispheres. Somewhat lower concentrations have been observed in the central Atlantic. Concentrations lower by up to two orders of magnitude have been determined at Enewetak Atoll in the tropical North Pacific (see Duce *et al.*, 1983). Figure

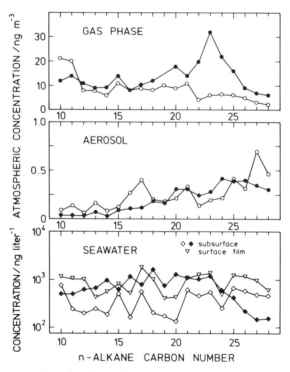

Fig. 6-7. Concentrations of n-alkanes in pure marine air masses and in seawater observed at Loop Head, Ireland (filled symbols) and at Cape Grim, Tasmania (open symbols) according to measurements of Eichmann *et al.* (1979, 1980).

6-7 further shows that the hydrocarbons observed in the air are present also in seawater at appreciable concentrations. A comparison of water samples taken 15 cm below the sea surface with surface film material collected with a screen clearly indicates an enrichment of hydrocarbons at the air–sea interface. Wade and Quinn (1975), who measured total hydrocarbon concentrations of 13–239 μg/l (subsurface) in the Sargasso Sea, reported on a similar enrichment. Roughly 11% of the material could be resolved into the n-C_{13}–n-C_{32} fraction. Barbier et $al.$ (1973) found that C_{14}–C_{32} soluble hydrocarbons occurred at depths down to 4500 m off the coast of Africa with concentrations of 10–140 μg/l. The higher hydrocarbons thus appear to be ubiquitous in the world's oceans. Owing to the enrichment in the surface microlayer and the incomplete understanding of its physicochemical significance to the transfer of hydrocarbons between ocean and atmosphere, it is presently not possible to calculate transfer rates with confidence.

Hydrocarbons originating from terrestrial vegetation, such as the higher n-alkane fraction of plant waxes, show a pronounced predominance of odd over even carbon numbers. In certain cases a preference of odd to even carbon numbers by a factor of 10 has been observed. The gas-phase n-alkanes over the oceans do not show this behavior (even though Fig. 6-7 indicates a possible enhancement of the C_{15} component). This observation combined with the ubiquitous presence of n-alkanes in seawater points toward a marine rather than a terrestrial origin. Additional support for this view is based on the relatively short lifetime of the higher n-alkanes with regard to their reaction with OH radicals. Eichmann et $al.$ (1980) estimated lifetimes of the order of 1–2 days. The seawater concentrations show little variation with the size of the n-alkane, and no pronounced odd-to-even carbon number preference as required if the ocean is to be the main source of the higher n-alkanes in the marine atmosphere.

Finally, it is of interest to compare the observed gas phase concentrations of n-alkanes with their saturation vapor pressures. If these were approached, an appreciable fraction of the n-alkanes would have to be associated with the atmospheric aerosol. Vapor pressures of n-alkanes are available only for the temperature range 200–400°C, so that the data have to be extrapolated toward ambient temperature. Eichman et $al.$ (1979) estimated vapor pressures at 20°C for C_{10}, C_{20}, and C_{28} n-alkanes of 3.6, 2.7×10^{-4}, and 6.7×10^{-8} mbar, respectively. The corresponding saturation concentrations are 2.1×10^{10}, 3.3×10^{6}, and 1.2×10^{3} ng/m^3. The observed concentrations, even for n-C_{28}, are much lower, so that condensation onto aerosol particles is not expected. Nevertheless, as Fig. 6-7 shows, a small fraction of n-alkanes does indeed reside with the aerosol. The fraction increases with carbon number and approaches about 10% at the high end of the scale.

6.2 Hydrocarbon Oxidation Mechanisms

A general scheme for the oxidation of hydrocarbons in the atmosphere was discussed in Section 5.2 in conjunction with the formation of photochemical smog. Here, we shall treat several individual hydrocarbon compounds in more detail. Their oxidation is initiated by reactions with OH radicals or with ozone. The oxidation then proceeds via alkylperoxy ($ROO\cdot$) and alkoxy ($RO\cdot$) radicals toward aldehydes and ketones as products. The presence of nitric oxide is required to convert alkylperoxy toward alkoxy radicals. In urban and regionally polluted air there generally is sufficient NO present to sustain the propagation of the oxidation chain. In the unpolluted, remote troposphere, the concentrations of nitric oxide are low and its effectiveness as a chain catalyst may be greatly reduced by competing reactions. Accordingly, before discussing individual reaction mechanisms, it will be useful to consider first the chemical behavior of alkylperoxy and alkoxy radicals.

6.2.1 ALKYLPEROXY RADICALS

The abstraction of a hydrogen atom from an alkane first produces an alkyl radical. In the atmosphere, however, alkyl radicals have but little choice other than to combine with oxygen to yield an alkylperoxy radical. As mentioned previously, tertiary hydrogen atoms are abstracted more easily than secondary H atoms, and their abstraction, in turn, is more facile than that of primary H atoms. In the higher hydrocarbons the number of secondary H atoms usually exceeds that of primary or tertiary ones, so that secondary alkyl and alkylperoxy radicals are most frequently formed:

$$
\begin{array}{ccc}
R_1 & R_1 & R_1 \\
\diagdown & \diagdown & \diagdown \\
CH_2 \xrightarrow{OH} H_2O + & CH \xrightarrow{O_2} & CHOO\cdot \\
\diagup & \diagup & \diagup \\
R_2 & R_2 & R_2
\end{array}
$$

From rate coefficients for the attachment of CH_3 and C_2H_5 to molecular oxygen (see Table A-4), it can be estimated that their lifetimes in air at ground-level pressure is about 10^{-7} s. Other alkyl radicals undoubtedly convert to alkylperoxy radicals at similarly rapid rates. Occasionally, an abstraction reaction of the type

$$H_3C\dot{C}H_2 + O_2 \rightarrow H_2C{=}CH_2 + HO_2$$

has been invoked for ethyl and other small alkyl radicals, because such reactions appear to be prominent in low-temperature flames (Pollard, 1977). Since abstraction reactions require an activation energy whereas addition reactions do not, the ratio of the two rate coefficients decreases rapidly as the temperature is lowered toward ambient. For C_2H_5, for example, the laboratory data of Dingledy and Calvert (1963) suggest a ratio of less than 10^{-3} at room temperature. For isobutyl radicals, Slater and Calvert (1968) reported a ratio of 0.1. Accordingly, it appears that abstraction of an H atom by oxygen is negligible under atmospheric conditions.

Alkylperoxy radicals in the atmosphere may undergo reactions with other atmospheric constituents or internal rearrangement. These two possibilities are now discussed. Rate coefficients for reactions with several atmospheric constituents are available only for CH_3O_2 radicals, but other alkylperoxy radicals may be assumed to react at similar rates. The reactions of CH_3O_2 may then serve as a guide toward selecting the most prominent routes. Table 6-12 shows pertinent reactions for CH_3O_2, and the associated rate coefficients and relative rates calculated for low and medium level of NO. Boundary-layer and daylight conditions are adopted. The HO_2 radical derives mainly from reactions of OH with CO and to a lesser extent from the reaction of OH with CH_4. This reaction, of course, is the source of CH_3O_2.

Table 6-12 shows that the reactions of CH_3O_2 with NO and NO_2 predominate. When NO_x concentrations are low, the reaction with HO_2 also becomes important, whereas the reaction of CH_3O_2 with itself is of lesser significance. Alkylperoxy radicals may also be scavenged by aerosol particles. The

Table 6-12. *Comparison of Atmospheric Reaction Rates for Methylperoxy Radicals*[a]

Reaction of CH_3O_2 with reactant X	$10^{13}k$	(A)		(B)	
		$n(X)$	$kn(X)$	$n(X)$	$kn(X)$
$CH_3O_2 + HO_2 \rightarrow CH_3OOH + O_2$	65	5 (8)	3.2 (−3)	2 (7)	3.2 (−5)
$CH_3O_2 + CH_3O_2 \rightarrow 2CH_3O$ (33%)	4.6	1 (8)	9.2 (−5)[b]	3 (6)	2.8 (−6)[b]
$\rightarrow HCHO + CH_3OH$					
$CH_3O_2 + NO \rightarrow CH_3O + NO_2$	75	2 (8)	1.5 (−3)	1 (10)	7.5 (−2)
$CH_3O_2 + NO_2 \rightarrow CH_3O_2NO_2$	16	1 (9)	1.6 (−3)	4 (10)	6.4 (−2)
$CH_3O_2 + SO_2 \rightarrow CH_3O_2SO_2$	<0.01	5 (9)	2.5 (−5)	5 (10)	2.5 (−4)
$CH_3O_2 +$ aerosol particles $\rightarrow$ Scavenging products	—	Marine	1.3 (−3)	Rural	1.3 (−2)

[a] Shown are the reactions, the rate coefficients k (at 298 K, in units of cm^3/molecule s), and the relative rates (in units of s^{-1}) for two atmospheric conditions: (A) low and (B) medium number densities of NO, NO_2, and SO_2. The last line gives collision frequencies for CH_3O_2 radicals interacting with marine and rural aerosol particles (see Table 7-5), assuming a scavanging efficient of unity (maximum rate). Numbers in parentheses indicate powers of 10.
[b] $= 2kn(X)$.

scavenging rate is greatly influenced by the efficiency for the accommodation of radicals are the particle surface, which is difficult to estimate. The collision rates entered in Table 6-12 represent upper-limit values for the scavenging rate. They show, however, that this is potentially an important loss process.

Higher alkylperoxy radicals are expected to react analogously to CH_3O_2 radicals, but a few supplemental remarks are in order regarding the products.

1. Alkylnitrate yields observed in smog chambers suggest that alkylperoxy radicals when reacting with NO produce $RONO_2$ as an attachment product in addition to RO· radicals. The formation of alkylnitrates is negligible for methyl and ethylperoxy radicals but appears to become increasingly favorable for sufficiently large alkyl groups:

$$RO_2 + NO \xrightarrow{a} NO_2 + RO$$

$$\xrightarrow{b} RONO_2$$

Darnall et al. (1976b) and Carter et al. (1979) have estimated ratios of k_b/k_a of 0.04 for propylperoxy, 0.09 for butylperoxy, and 0.16 for pentylperoxy radicals. Atkinson et al. (1982a) found similar values rising up to 0.32 for n-octylperoxy radicals. In addition to these authors, Demerjian et al. (1974) and Niki et al. (1972) have shown by computer simulations of smog-chamber data that higher alkylperoxy radicals react with NO at least as rapidly as CH_3O_2.

2. The addition of NO_2 to peroxy radicals leads to alkylperoxy nitrates, $ROONO_2$, which are unstable with decomposition lifetimes of less than 1 min at temperatures of 273 K and higher (Atkinson and Lloyd, 1984). If in analogy to $CH_3O_2NO_2$ and acetylperoxy nitrate (see Section 9.4.1), the decomposition yields mainly ROO and NO_2, there will be no net reaction and the peroxynitrates serve solely as a storage for peroxy radicals. It is possible, however, that large peroxynitrates isomerize before undergoing decomposition and then the product may well be different.

3. The reaction of alkylperoxy radicals with HO_2 leads to the formation of stable alkylhydroperoxides. This is a radical chain termination process.

4. Laboratory studies of the autooxidation of hydrocarbons (Ingold, 1969; Howard, 1971) have revealed that the self-reactions of alkylperoxy radicals among each other are slow with rate coefficients generally smaller than that for the $CH_3O_2 + CH_3O_2$ reaction. These reactions need not be considered in the atmosphere. A possible exception may be the reaction of ROO radicals with CH_3O_2, which is the most abundant peroxy radical in the atmosphere.

In considering the isomerization of alkylperoxy radicals, it must be pointed out that internal hydrogen abstraction exemplified by

$$
\begin{array}{cc}
\text{H} \qquad \text{H} & \text{H} \qquad \text{H}\\
\text{H}_3\text{C}-\text{C}-\text{CH}_2-\text{C}-\text{CH}_3 \;\rightarrow\; & \text{H}_3\text{C}-\text{C}-\text{CH}_2-\text{C}-\text{CH}_3\\
\;\;\text{O}_{\text{O}\cdot} \quad \text{H} & \;\;\text{O}_{\text{OH}}
\end{array}
$$

is the most frequent mode of isomerization, leading to the formation of hydroperoxyalkyl radicals. These would then add another oxygen molecule to form new peroxy radicals, which by reacting with HO_2 would produce dihydroperoxides. Internal hydrogen abstraction was first postulated by Rust (1957) to explain the production of such dihydroperoxides in the low-temperature oxidation of certain hydrocarbons. Internal hydrogen abstraction proceeds via a cyclic transition state where the six-membered transitory ring structure is favored over other configurations. The abstraction of hydrogen atoms by the peroxy group is slightly endothermic, and the associated activation energies are considerably higher than those for OH abstraction reactions. In fact, the isomerization of peroxy radicals in the atmosphere would not be important were it not for the large preexponential factors of rate coefficients for unimolecular reactions. Unfortunately, there is still very little solid information on the associated rate coefficients. Here we make use of the experimental data of Allara et al. (1968) and Mill et al. (1972) on the oxidation of normal and isobutane to estimate activation energies for the abstraction of primary, secondary, and tertiary H atoms by peroxy radicals. The corresponding values are 77.3, 66.0, and 57.6 kJ/mol. Ring strain energies must be added to obtain activation energies associated with intramolecular hydrogen abstraction. Laboratory studies of the oxidation of 2,4-dimethylpentane by Mill and Montorsi (1973) and of n-pentane and n-nonane by VanSickle et al. (1973) have given rate coefficients for the internal abstraction of H atoms that suggest a ring strain energy of about 12 kJ/mol, if one adopts a value of $10^{11}\,\text{s}^{-1}$ for the preexponential factor. The combination of these data lead to the rate coefficients

$$k_{\text{prim}} = 10^{11}\exp(-10{,}770/T) \quad \text{s}^{-1}$$

$$k_{\text{sec}} = 10^{11}\exp(-9418/T) \quad \text{s}^{-1}$$

$$k_{\text{tert}} = 10^{11}\exp(-8408/T) \quad \text{s}^{-1}$$

for the three unimolecular hydrogen abstraction processes. At ambient temperatures in the ground-level atmosphere the rate coefficients assume the following values: $k_{\text{prim}} = 1.4 \times 10^{-5}$, $k_{\text{sec}} = 1.4 \times 10^{-3}$, and $k_{\text{tert}} = 4.2 \times 10^{-2}$ in units of reciprocal seconds. The values would be raised by a factor of two, had we assumed a preexponential factor of $2 \times 10^{10}\,\text{s}^{-1}$. A comparison of these data with the rates shown in Table 6-12 for the other processes shows that an internal abstraction of primary hydrogen atoms is unimportant in the atmosphere, but that the abstraction of secondary and tertiary H atoms can compete effectively with the other reactions of alkylperoxy radicals when NO_x levels are low.

6.2.2 ALKOXY RADICALS

Alkoxy radicals that arise from the reaction of NO with alkylperoxy radicals also may enter into several competing processes. Four such reactions must be considered: thermal decomposition, isomerization, reaction with oxygen, and addition to either NO or NO_2. Falls and Seinfeld (1978) have presented a brief review of the various possibilities. Table 6-13 summarizes current information on rate coefficients and projected rates in the atmosphere for several small alkoxy radicals.

1. Measurements of alkoxy decomposition rates were analyzed by Baldwin et al. (1977) and extrapolated toward atmospheric conditions. The results shown in Table 6-13 are believed to be accurate within a factor of two.
2. The isomerization of alkoxy radicals would occur in a manner similar to that of alkylperoxy radicals, that is, by hydrogen abstraction. In alkoxy radicals, however, the process is more favorable because the reaction is exothermic and activation energies tabulated by Gray et al. (1967) for H-atom abstraction by CH_3O radicals suggest values of 29.7, 19.7, and 15.5 kJ/mol for primary, secondary, and tertiary H atoms, respectively. The ring strain energy must be added. Carter et al. (1976) and Baldwin et al. (1977) assumed ring strain energies of about 2.5 kJ/mol for a six-membered ring and 27.2 kJ/mol for a five-membered ring structure. The first value probably is too low in view of the results for the isomerization of alkylperoxy radicals discussed earlier. Accordingly, the activation energy of 31.8 kJ/mol given by these authors for the isomerization of butoxy radicals must be considered a lower limit, and the resulting rate coefficient at 295 K of $6 \times 10^5 \, s^{-1}$ an upper limit. The application of a ring strain energy of 12.3 kJ/mol that was appropriate in the case of alkylperoxy radicals isomerization leads to an activation energy for internal H abstraction of 42 kJ/mol. Together with a preexponential factor of $10^{11} \, s^{-1}$, which is similar to that estimated by Baldwin et al. (1977), one then obtains a rate coefficient of 4×10^3 for the internal abstraction of primary H atoms in alkoxy radicals. Carter et al. (1979) found from a computer simulation of the oxidation of n-butane in smog chambers that the evolution of butyraldehyde is best fitted by using a rate coefficient of $2 \times 10^4 \, s^{-1}$ for the isomerization of 1-butoxy radicals. This result indi-

Table 6-13. *Left: Rate Coefficient Estimates for Reactions of Alkoxy Radicals in the Atmosphere, Taken Partly from the Review of Atkinson and Lloyd (1984)[a]; Right: Rates Calculated from the Rate Coefficients Assuming Number Densities $n(O_2) = 5 \times 10^{18}$, $n(NO_2) = 2.5 \times 10^{10}$ molecules/cm^3[b]*

Radical	k_{dec}[c]	k_{isom}	k_{O_2}[d]	k_{NO_2}[e]	R_{dec}	R_{isom}	R_{O_2}	R_{NO_2}
CH_3O	—	—	1.3 (−15)	1.3 (−11)	—	—	6.5 (3)	3.5 (−1)
CH_3CH_2O	1 (−1)	—	7 (−15)	2 (−11)	1 (−1)	—	3.5 (4)	5.0 (−1)
$C_2H_5CH_2O$	3 (−1)	8.3[f]	7 (−15)	3 (−11)	3 (−1)	8.3	3.5 (4)	7.5 (−1)
$CH_3CH(O)CH_3$	7 (1)	—	3 (−14)	3 (−11)	7 (1)	—	1.5 (5)	7.5 (−1)
$CH_3(CH_2)_2CH_2O$	5 (−1)	4 (3)–6 (5)[g]	7 (−15)	3 (−11)	5 (−1)	4 (3)	3.5 (4)	7.5 (−1)
$CH_3CH(O)C_2H_5$	3 (3)	8.3–4.5 (2)[g]	5 (−14)	3 (−11)	3 (3)	4.5 (2)	2.5 (5)	7.5 (−1)
$(CH_3)_3CO$	2 (3)	—	1 (−14)	3 (−11)	2 (3)	—	50 (4)	7.5 (−1)

[a] The reactions considered are decomposition and isomerization (k in s^{-1}), and reactions with oxygen and nitrogen dioxide (k in cm^3/molecule s).

[b] Numbers in parentheses indicate powers of 10.

[c] According to the analysis of Baldwin et al. (1977) and Batt (1979, 1980).

[d] Estimates of Gutman et al. (1982).

[e] Estimates based on data of Baker and Shaw (1965), Wiebe et al. (1973), Batt et al. (1975), Batt and Milne (1976, 1977), Barker et al. (1977). Baldwin et al. (1977) gave 1.4×10^2.

[f] Lower limit if the activation energy is 42 kJ/mol, upper limit from Baldwin et al. (1977).

[g]

cates again that the estimate of Baldwin *et al.* (1977) is too high. The internal abstraction of a secondary H atom in 1-butoxy requires a five-membered transitory ring structure. The activation energy for this process is at least $19.7 + 27.2 = 46.9$ kJ/mol. The rate coefficient resulting for this process becomes smaller than that for primary H atom abstraction. Higher alkoxy radicals, starting with 1-pentoxy, admit the abstraction of secondary H atoms via formation of a six-membered ring. The rate coefficient for this process is calculated to be at least $2 \times 10^5 \text{ s}^{-1}$, in agreement with the value derived by Carter *et al.* (1976). Baldwin *et al.* (1977) envisioned that internal hydrogen abstraction can repeat itself due to the successive addition of oxygen to the carbon moiety made available by the abstraction process, and they postulated the formation of compounds having the general structure $(OH)_x CH_{3-x}(CH_2)_2 CH_{3-y}(OH)_y$. Compounds with more than one OH group attached to the same carbon atom are inherently unstable, however, as they split off water to establish a carbonyl. In fact, the abstraction of a hydrogen from a $-CH_2OH$ group to form $-CHOH$ already strengthens the carbon–oxygen bond making the further addition of oxygen unlikely. Instead, the simultaneous weakening of the O–H bond enables the hydrogen atom to be abstracted by oxygen. For the simplest radical of this type, $\dot{C}H_2OH$, Radford (1980) found the abstraction process

$$CH_2OH + O_2 \rightarrow HO_2 + HCHO$$

very rapid. In the same vein, Carter *et al.* (1979), who had studied the photooxidation of ethanol and 2-butanol, concluded from the product distribution that H-atom abstraction from the OH group supercedes oxygen addition also for the higher homologues of CH_2OH. The successive oxidation of 1-butoxy radicals thus terminates with the step

$$HO\dot{C}H(CH_2)_2CH_2OH + O_2 \rightarrow OCH(CH_2)_2CH_2OH + HO_2$$

3. The reaction of oxygen with alkoxy radicals is fairly slow, and it is important in the atmosphere only due to the high concentration of oxygen. In all cases oxygen is assumed to abstract a hydrogen atom from the alkoxy radical to give an aldehyde or a ketone. Measured rate coefficients exist only for methoxy, ethoxy, and propoxy radicals. Gutman *et al.* (1982) have used similarity arguments to estimate rate coefficients for several other alkoxy radicals as shown in Table 6-13.

4. Rate coefficients for the addition of NO to alkoxy radicals have been inferred from the reverse reaction and estimates of the reaction entropies involved. These reactions are fast. Batt *et al.* (1975) reported values of 2×10^{-11} cm^3/molecules s for methoxy and about 5×10^{-11} for

butoxy and propoxy radicals. Reactions of alkoxy radicals with NO_2 occur at similar rates. Baker and Shaw (1965) and Wiebe *et al.* (1973) obtained ratios for the rate coefficients k_{NO}/k_{NO_2} of 2.7 and 1.2, respectively, for methoxy radicals, whereas Baker and Shaw (1965) gave a ratio of 1.7 for butoxy radicals. These data provide the basis for the estimates shown in Table 6-13. In the unpolluted atmosphere the concentration of NO_2 exceeds that of NO. Since the rate coefficients have the same magnitude, only the reaction with NO_2 needs to be considered. These reactions allow the possibility of hydrogen abstraction apart from addition:

$$RCHO \cdot + NO_2 \rightarrow RCHONO_2$$
$$\rightarrow RCHO + HNO_2$$

For methoxy radicals the abstraction pathway has been estimated to contribute between 8 and 23% to the total reaction (Wiebe *et al.*, 1973; Barker *et al.*, 1977). Both pathways are termination reactions.

The right-hand side of Table 6-13 shows relative rates for alkoxy radical reactions in the atmosphere for boundary layer conditions. Comparison of the rates makes it immediately clear that reactions with NO_2 (or NO) are of little importance. For the smaller alkoxy radicals the reaction with oxygen is preponderant, whereas for alkoxy radicals larger than butoxy, decomposition and isomerization reactions become competitive. Tertiary butoxy radicals have no abstractable hydrogen atom and thus cannot react with oxygen. In this case, decomposition is dominant.

6.2.3 OXIDATION OF ALKANES

Using the preceding discussion of alkylperoxy and alkoxy radical reactions as a guide, we present now specific mechanisms for several individual hydrocarbons, starting with alkanes.

6.2.3.1 *Ethane*

The oxidation of ethane is similar to that of methane and consists of the following steps (Aikin *et al.*, 1982):

$$C_2H_6 + OH \rightarrow H_2O + C_2H_5$$
$$C_2H_5 + O_2 \rightarrow C_2H_5OO$$
$$C_2H_5OO + NO \rightarrow C_2H_5O + NO_2$$
$$C_2H_5OO + HO_2 \rightarrow C_2H_5OOH + O_2$$
$$C_2H_5O + O_2 \rightarrow HO_2 + CH_3CHO$$

The further oxidation of acetaldehyde then follows the reaction scheme outlined in Section 5.2 (p. 190).

6.2.3.2   n-Butane

The oxidation mechanism for n-butane is shown in Fig. 6-8. n-Butane contains six primary and four secondary hydrogen atoms. From Eq. (6-1) one would estimate that 15% of all reactive encounters with OH radicals lead to the abstraction of primary and 85% to the abstraction of secondary H atoms. This opens up two pathways for the formation of butyraldehyde and methyl ethyl ketone, respectively. Decomposition of the intermediate alkoxy radicals is important. Secondary butoxy radicals undergo decomposition and reaction with oxygen with about equal probability. The decomposition of sec-butoxy radicals leads to the production of two molecules of acetaldehyde. Indeed, methyl ethyl ketone and acetaldehyde are the major products observed in smog chamber studies. The decomposition of prim-butoxy radicals is a minor pathway according to the data in Table 6-13. Nevertheless, it is the only route to the formation of propionaldehyde, which is an observed product in the smog-chamber oxidation of n-butane (Carter et al., 1979). The isomerization of prim-butoxy radicals would lead to the formation of 1,4-hydroxybutanal. This reaction, which has been suggested by Carter et al. (1979), must still be considered speculative, although the data in Table 6-13 are in favor of it. Another, perhaps even

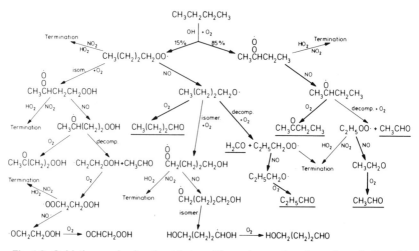

Fig. 6-8. Oxidation mechanism for n-butane. The major reaction channels are indicated by bold arrows. The internal hydrogen abstraction reactions for alkylperoxy and alkoxy radicals are still somewhat speculative.

more speculative, pathway for the isomerization of *n*-butylperoxy radicals is included on the left-hand side of Fig. 6-8 to indicate the potential for the formation of products with mixed aldehydic and hydroperoxidic character. The uncertainties in the oxidation pathways for *prim*-butylperoxy radicals should not be overrated, however, in view of the fact that the formation of *sec*-butylperoxy radicals is dominant.

6.2.3.3 *Isobutane*

The breakdown of isobutane is traced in Fig. 6-9. The initial attack of OH radicals and the subsequent addition of oxygen lead to *prim*-isobutylperoxy radicals in 26% of all reactive encounters, and to *tert*-butylperoxy radicals in 74%. Both radicals are then converted to the corresponding butoxy radicals. *prim*-Isobutoxy has a choice either of reacting with oxygen to form isobutanal, or of decomposing to form formaldehyde and propyl radicals. These add oxygen, whereby they are eventually converted to *sec*-propoxy radicals, which react mainly with oxygen to produce acetone. The decomposition of propoxy radicals is a minor pathway, although the only one yielding acetaldehyde. The principal oxidation pathway for isobutane proceeds via tertiary butoxy radicals. They are unstable and decompose, forming acetone and methyl radicals. The second species enters into the methyl oxidation chain, which leads to formaldehyde and methylhydroperoxide as the final products. The major products resulting from the oxidation of isobutane thus are acetone and formaldehyde.

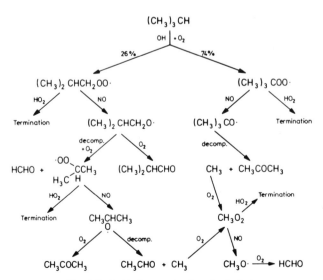

Fig. 6-9. Oxidation mechanism for isobutane.

6.2.4 OXIDATION OF ALKENES

The reaction of OH radicals with alkenes differs from that with alkanes in that OH has a choice of attaching to the C=C double bond versus abstracting a hydrogen atom. Measurements of the temperature dependence of rate coefficients in addition to the observed product distributions have provided ample evidence for the preponderance of OH addition, at least for the smaller alkenes and alkadienes. The preference for addition is due to the fact that hydrogen abstraction reactions require an activation energy in contrast to the process of OH addition. Atkinson *et al.* (1979) have reviewed reactions of OH radicals with a number of alkenes. The large rate coefficients associated with addition reactions leave little doubt that longer side chains are necessary before H-atom abstraction can become competitive. An exception may be the more weakly bonded allylic H atom adjacent to the C=C double bond, as Atkinson *et al.* (1977) have pointed out. The present discussion will concentrate on ethene and propene as suitable examples for the oxidation of alkenes.

6.2.4.1 *Ethene*

The addition of OH to the C=C double bond of ethene creates a free moiety to which an oxygen molecule then attaches. The resulting peroxy radical undergoes the usual reactions. The following mechanism is based on the suggestions of Niki *et al.* (1978, 1981). The initial reactions

$$H_2C=CH_2 \xrightarrow{OH} \overset{H}{\underset{H}{HOC}}-\dot{C}H_2 \xrightarrow{O_2} \overset{H\ H}{\underset{H\ H}{HOC}}-COO$$

are followed by termination as well as O-atom abstraction,

$$HOCH_2CH_2OO + HO_2 \rightarrow HOCH_2CH_2OOH + O_2$$

$$HOCH_2CH_2OO + NO \rightarrow HOCH_2CH_2O + NO_2$$

The hydroxylated alkoxy radical can then either react with oxygen or undergo cleavage

$$HOCH_2CH_2O + O_2 \rightarrow HOCH_2CHO + HO_2$$

$$HOCH_2CH_2O \rightarrow H_2COH + HCHO$$

$$H_2COH + O_2 \rightarrow HCHO + HO_2$$

In the first case glycolaldehyde is formed as the product. In the second case the products are formaldehyde and the CH_2OH radical, which by virtue of its rapid reaction with oxygen then leads to the formation of a second formaldehyde molecule. In laboratory experiments at 100 mbar of air and at room temperature, approximately 80% of the $HOCH_2CH_2O$ radicals undergo decomposition. Glycoaldehyde reacts further with OH, whereby additional formaldehyde is formed.

Ethene, like other alkenes, reacts also with ozone in the atmosphere. The older work on ozone reactions has been reviewed by Leighton (1961) and by Bufalini and Altshuller (1965). More recent work has done much to clarify the principal reaction mechanisms involved. Criegee (1957, 1962, 1975), who had studied the ozonolysis of alkenes in solution, suggested that ozone adds to the $C=C$ double bond, forming an unstable intermediate, which then decomposes toward a carbonyl compound and a zwitterion fragment, for example:

$$H_2C=CH_2+O_3 \rightarrow H_2C\underset{O}{\overset{O}{\underset{|}{\diagup}}}\overset{O}{\underset{|}{\diagdown}}CH_2 \rightarrow HCHO + H_2\overset{+}{C}OO^-$$

In the gas phase the zwitterion would be replaced by the corresponding diradical, that is, a radical with two free valences. Such radicals often are highly energetic. They undergo considerable rearrangement, usually resulting in a further decomposition. In the case of ethene, evidence for the gas-phase formation of the Criegee intermediate has been presented by Lovas and Suenram (1977), who used microwave spectroscopy to identify dioxirane as an intermediate product, and by Martinez et al. (1977), who detected the same product by mass spectrometry. Herron and Huie (1977) found in addition that one molecule of formaldehyde is formed for each molecule of ethene that is consumed. Theoretical studies by Wadt and Goddard (1975) led them to suggest that the diradical formed in the ozonolysis of ethene rearranges to formic acid, which still is highly energetic, however, and decomposes unless it is stabilized by collisions. By mass spectrometry, H_2O, CO, H_2, and CO_2 were identified as the decomposition products. From these studies the mechanism for the reaction of ozone with ethene may be specified as follows:

$$H_2C=CH_2 \xrightarrow{O_3} H_2C\underset{O}{\overset{O}{\underset{|}{\diagup}}}\overset{O}{\underset{|}{\diagdown}}CH_2 \rightarrow H_2\overset{O}{\underset{|}{C}}-\overset{O}{\underset{|}{C}}H_2 \rightarrow HCHO + H_2\dot{C}OO\cdot$$

Primary ozonide   ring opening

$$H_2\dot{C}OO\cdot \rightarrow H_2C\diagup\hspace{-0.3em}\underset{O}{\overset{O}{\diagdown}} \rightarrow \left(HC\diagup\hspace{-0.3em}\underset{O}{\overset{OH}{\diagdown}}\right)^*$$

dioxirane

$\rightarrow CO + H_2O$ (67%)

$\rightarrow CO_2 + H_2$ (18%)

$\rightarrow CO_2 + 2H$ (9%)

$\rightarrow HCOOH$ (6%)

The percentages for the ultimate products are taken from the product distribution reported by Herron and Huie (1977). They refer to comparatively low pressures, but other authors, specifically Scott *et al.* (1957), have found a similarly low yield of formic acid also at higher pressures. Thus, it appears that in the atmosphere decomposition is the main fate of the Criegee radical.

6.2.4.2 Propene

The oxidation mechanism for propene initiated by OH radicals is shown in Fig. 6-10. The addition of OH occurs preferentially at the terminal carbon atom. Sixty-five percent of all reactions go this way, whereas addition to the central carbon atom occurs 35% of the time. Hydrogen abstraction is negligible. The peroxy radicals generated by the sequential addition of OH and O_2 molecules to propene may again either react with HO_2 to form termination compounds or undergo reaction with NO to yield hydroxylated alkoxy intermediates. Their reaction with oxygen produces hydroxypropanone and α-hydroxypropanal. Decomposition of the alkoxy radicals results in equal amounts of acetaldehyde and formaldehyde. The relative significance of decomposition versus H-atom abstraction by oxygen has not yet been clarified. Demerjian *et al.* (1974) considered decomposition dominant. Carter *et al.* (1979) made use of the decomposition rate estimates of Baldwin *et al.* (1977), and by adjusting rates to fit the results of smog-chamber experiments concluded that $HOCH_2CHOCH_3$ predominantly decomposes, whereas $CH_3CH_2OHCH_2O\cdot$ reacts mostly with oxygen.

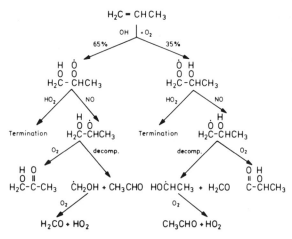

Fig. 6-10. Oxidation mechanism for propene according to Carter *et al.* (1979).

The reaction of propene with ozone leads to the formation of the Criegee radicals $H_2\dot{C}OO\cdot$ and $CH_3\dot{C}HOO\cdot$. In this case the applicability of the Criegee mechanism is supported by the formation of secondary ozonides in a reaction of Criegee radicals with aldehydes, such as

$$\underset{H_3C}{\overset{H}{\diagdown}}COO + CH_3CHO \longrightarrow \underset{H_3C}{\overset{H}{\diagdown}}\underset{O}{\overset{O-O}{\diagdown}}\underset{CH_3}{\overset{H}{\diagup}}$$

Niki *et al.* (1977) have identified this product by long-path infrared spectroscopy in a study of the gas phase ozonolysis of *cis*-2-butene. The occurrence of secondary ozonides in the gas-phase corresponds entirely to their production in the liquid phase at low temperatures (Criegee, 1957). These results led Dodge and Arnts (1979) to derive from a computer simulation of the experimental data of Niki *et al.* (1977) the following percentages

$$\underset{H}{\overset{H_3C}{\diagdown}}\underset{O}{\overset{O}{C}}$$

→ $CH_4 + CO_2$ (15%)

→ $CH_3 + CO + OH$ (34%)

→ $CH_3 + CO_2 + H$ (34%)

→ $HCO + CH_3O$ (8.5%)

→ $H + CO_2 + CH_3$ (8.5%)

for the decomposition of the $CH_3\dot{C}HOO\cdot$ Criegee radical. The formation of H atoms in the ozonolysis of ethene, propene, and butene explains in a natural way the emission of the OH Meinel bands, which were first reported by Finlayson *et al.* (1972, 1974). The Meinel bands arise from the reaction $H + O_3 \rightarrow O_2 + OH^*$, which produces excited OH radicals. The excess energy is released in the form of near-infrared radiation. Finally, it should be noted that the decomposition of both the $CH_3CH(\dot{O})CH_2OO\cdot$ and the $CH_3\dot{C}HOO\cdot$ Criegee intermediates is expected to be faster than their reactions with NO, which would abstract the peroxidic O atoms. Accordingly, the oxidation of alkenes by ozone does not require the presence of nitric oxide in contrast to their oxidation by OH radicals.

6.2.4.3 *Isoprene*

The oxidation of isoprene, $CH_2{=}CHC(CH_3){=}CH_2$, in the atmosphere has not yet been fully delineated and thus remains speculative. Hanst *et al.* (1980) have discussed some of several possibilities. The successive addition of OH and O_2 to either of the two double bonds of isoprene may be modeled in analogy of that to propene and is expected to give formaldehyde and methacroleine or formaldehyde and methyl vinyl ketone as the products. The reaction of isoprene with ozone is expected to produce dioxirane and methylacroleine or methyl vinyl ketone, depending on which of the two double bonds gets involved. In addition to dioxirane two other, more complex, Criegee radicals may be formed, namely, $CH_2=CH\dot{C}(CH_3)OO\cdot$ and $CH_2=C(CH_3)\dot{C}HOO\cdot$. Their reactive behavior has not yet been elucidated.

6.2.5   AROMATIC COMPOUNDS

Our current understanding of the oxidation of aromatic compounds in the atmosphere is quite limited. The reaction with OH radicals may proceed by the addition of OH to the benzene ring or by the abstraction of a hydrogen atom from side chains. The first type of interaction is unique, leading either to a permanent attachment of the OH group to the benzene ring or to ring opening. The second type of reaction may be modeled in analogy to that of OH radicals with alkanes. These possibilities will be illustrated using toluene as an example.

The two principal oxidation pathways for toluene are shown in Fig. 6-11. The relative significance of OH addition to the benzene ring and H-atom abstraction from the methyl group has been determined by Perry *et al.* (1977a) from a study of the temperature dependence of the reaction rate coefficient, and by Kenley *et al.* (1978) from a detailed analysis of the product distribution. Both groups of authors conclude that at ambient temperatures the probability for H-atom abstraction is about 15% of that for the overall reaction. Figure 6-11 shows that the subsequent reactions lead to the formation of benzaldehyde unless the toluenylperoxy radical enters into a termination reaction with HO_2 radicals. In that case toluenylhydroperoxide would be formed.

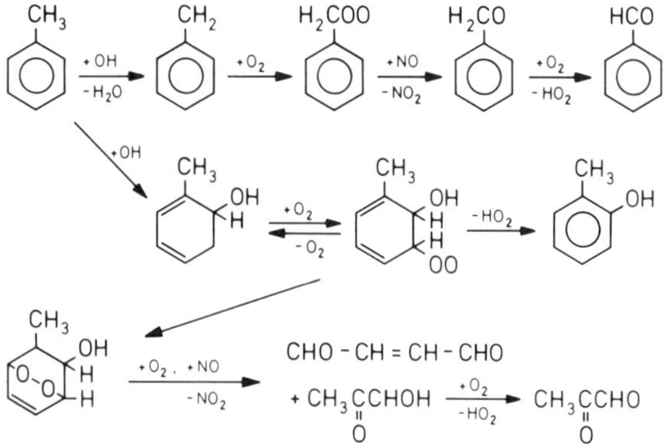

Fig. 6-11. Oxidation mechanism for toluene. The pathway leading to ring opening is not yet fully understood.

The observations of Kenley *et al.* (1978) suggest that the addition of OH to the benzene ring is mainly followed by an abstraction of the excess H atom by O_2, which leads to the formation of cresols. Three possible isomers may be formed, namely, *ortho*-cresol, *meta*-cresol, and *para*-cresol, depending at which position relative to the CH_3 group the OH radical attaches to the benzene ring. The product distribution observed by Kenley *et al.* (1978) gave 77.8% *ortho*-cresol, 8.2% *meta*-cresol, and 13.9% *para*-cresol. Evidently, the formation of *ortho*-cresol is favored. Only this pathway is shown in Fig. 6-11. In the presence of larger amounts of NO_2, Kenley *et al.* (1978) also found nitrotoluene as a product. It appears that NO_2 can add to the benzene ring once it has been attacked by OH and can then displace the OH group by splitting off water.

Cresols represent only a fraction of all the oxidation products from toluene, so that additional reactions must be operative. Not much is known about the other products, but unsaturated 1,4-dicarbonyls and methylglyoxal have been observed (Besemer, 1982; Nojima *et al.*, 1974). The formation of these compounds requires reactions that open the benzene ring. As Fig. 6-11 shows, the formation of the cresols is believed to proceed via a peroxidic intermediate. The addition of O_2 to the OH–toluene adduct is reversible and the reverse reaction has been estimated by Atkinson and Lloyd (1984) to be faster than the interaction of the peroxidic radical with NO under smog-chamber conditions. It is doubtful, therefore, whether O-atom abstraction by NO is effective, and we do not show this pathway in Fig. 6-11, although it would lead to the formation of a carbonyl and ring opening. Atkinson *et al.* (1980) and Atkinson and Lloyd (1984) proposed that instead the peroxidic radical undergoes cyclization to create a new moiety at the carbon atom next to the CH_3 group, which allows the addition of another O_2 molecule. If now the new peroxy group is reduced by interaction with NO, the resulting alkoxy radical becomes highly unstable and rearranges to for 1,4-*d*-butenal and a hydroxylated acetonyl radical, which by reacting further with oxygen stabilizes toward methylglyoxal. This pathway is indicated in Fig. 6-11. Although the course of the reaction is appealing, it must still be considered speculative, and additional work is required for confirmation.

6.3 Halocarbons

The detection and routine observation of halocarbons at unprecedented low levels of abundances was made possible by the invention of the electron-capture detector by Lovelock (1961) and its unique sensitivity for halogen compounds in conjunction with gas chromatography. Lovelock (1971) himself discovered the presence of CCl_3F and SF_6 in the atmosphere, with mixing ratios in the order of 10^{-11} and 10^{-14}, respectively. Subsequently, Lovelock *et al.* (1973) and Wilkness *et al.* (1973) determined the interhemispheric distribution of CCl_3F and reported the first measurements of CCl_4, while Su and Goldberg (1973) demonstrated the presence of CCl_2F_2 in the air. The fully halogenated compounds CCl_3F, CCl_2F_2, and CCl_4 are essentially inert in the troposphere, but they undergo photodecomposition in the stratosphere. This prompted Molina and Rowland (1974, 1975) to look into the effects of human-made chlorine compounds on the chemistry of the stratospheric ozone layer (see Section 3.4.3). Once their importance to stratospheric chemistry had been shown, a search for other chlorine compounds in the atmosphere was made to establish the total burden of chlorine in the stratosphere. Table 6-14 gives an overview on halocarbons that have been observed in the troposphere, their approximate abundances in 1981, and the dominant sources and sinks as far as known. Methyl chloride, CH_3Cl, the chlorofluoromethanes CCl_3F and CCl_2F_2, carbon tetrachloride, CCl_4, and methylchloroform, CH_3CCl_3, have the highest abundances, and these compounds will be discussed.

Methyl chloride, the most abundant halocarbon in the atmosphere, was first detected by Grimsrud and Rasmussen (1975). Despite its favorable abundance, the substance is difficult to analyze because the electron-capture detector is less sensitive to CH_3Cl than to other halocarbons. As a consequence, relatively few measurements exist. Singh *et al.* (1979b) found mixing ratios of 615 ± 100 pptv over the Pacific Ocean essentially independent of latitude in both hemispheres. Aircraft measurements by Rasmussen *et al.* (1980), also over the Pacific Ocean, have confirmed the uniform distribution with latitude, but they suggest that the mixing ratios in the boundary layer are slightly higher than in the free troposphere, namely, 780 versus 620 pptv. Mixing ratios observed over the North American continent, however, were independent of altitude. This suggests that the oceans may act as a source of CH_3Cl and, indeed, Singh *et al.* (1979b) found that the surface waters of the Pacific Ocean are supersaturated with methyl chloride. Further data reported by Rasmussen *et al.* (1980) for several remote locations, such as Alaska, Samoa, or the South Pole, were found consistent with the latitudinal surveys. In addition, no evidence for any temporal trends in CH_3Cl mixing ratios currently exists.

Methyl chloride reacts with OH radicals by hydrogen abstraction in the same way as methane, except that the rate coefficient is about five times greater. From the known rate coefficient and the usual assumption of an average OH number density of 5×10^5 molecules/cm^3 in the troposphere, one infers an atmospheric lifetime for CH_3Cl of about 1.8 yr. The uniform distribution of methyl chloride in the troposphere would be incompatible with such a relatively short lifetime, if the substance were mainly human-made and its sources were located in the northern ,hemisphere. Thus, it must be concluded that CH_3Cl arises primarily from natural sources. Moreover, in view of the short lifetime, the uniform distribution of CH_3Cl requires comparable source strengths in both hemispheres.

The contribution of the oceans to atmospheric CH_3Cl can be estimated from the measured concentrations in seawater. Lovelock (1975) found values of 23 ± 17 ng/l near the English coast, whereas Singh et al. (1979) found averages of 27 ± 31 ng/l at various locations in the open Pacific Ocean. The variability is considerable, even though the average values are nearly identical. Singh et al. (1979b) used the stagnant-film model of Liss and Slater (1974) discussed in Section 1.6 to derive by extrapolation to the total ocean surface area a global CH_3Cl flux of 3 Tg/yr. As usual, the estimate depends on the assumed values for the film thickness and the diffusion coefficient. Watson et al. (1980) used values differing somewhat from those employed by Singh et al. (1979b) and obtained a flux of 8 Tg/yr. A flux of 3 Tg/yr would suffice to compensate losses due to the reaction of CH_3Cl with OH radicals (2.6 Tg/yr). The stratosphere consumes only 0.1 Tg/yr (see Table 3-3). Much larger source strengths would require additional sinks which currently are not known.

Although it is probable that methyl chloride in seawater arises from biological processes, the route to its production is still obscure. Zafiriou (1975) proposed that CH_3Cl is a secondary product resulting from the reaction $CH_3I + Cl^- \rightarrow CH_3Cl + I^-$. The rate coefficient for this reaction was determined by Zafiriou (1975) at two temperatures for seawater conditions. In the Arrhenius form $k = 4.75 \times 10^7 \exp(-9506/T) \text{ s}^{-1}$, at 15°C the value is $2.2 \times 10^{-7} \text{ s}^{-1}$ or 6.9 yr^{-1}. The lifetime of CH_3Cl against conversion to CH_3Cl then is about 2 months, which corresponds approximately to the residence time of CH_3I in the upper-mixed layer of the ocean before it escapes toward the atmosphere. The concentration of methyl iodide in the open ocean is about 0.8 ng/l, according to measurements of Lovelock (1975). Taking the value to represent the concentration of CH_3I in the world ocean and assuming a uniform depth of the mixed layer of 80 m, the total content of CH_3I in the upper portion of the ocean amounts to 0.023 Tg. The corresponding annual production rate of CH_3Cl is (M_{CH_3Cl}/M_{CH_3I}) (6.9) $(0.023) = 0.06$ Tg/yr. The rate supplies only a small fraction of the flux of CH_3Cl escaping from the ocean to the atmosphere.

On the continents, CH_3Cl may be produced microbiologically, anthropogenically, and by biomass burning. Cowan et al. (1973) reported that six species of wood mold of the genus Fomes produce substantial quantities of methyl chloride. The fungi are quite common and may generate appreciable amounts of CH_3Cl, according to an estimate of Watson et al. (1980). A quantitative source assessment is not available, however.

Table 6-14. *Halocarbons in the Troposphere: Approximate Abundances in 1981; Mass Content G_T, Residence Time τ, Major Sources and Sinks*

Compound	Mixing ratio (pptv)		G_T (Tg)	τ (yr)	Sources	Sinks	Remarks
	Northern hemisphere	Southern hemisphere					
CH$_3$Cl	620	620[a]	4.6	1.8	Ocean, biomass burning	Reaction with OH	Uniform distribution
CCl$_2$F$_2$ (F12)	305	285[b]	5.2	~150	Anthropogenic	Photolysis in the stratosphere	Rising at a rate of 26 pptv/yr
CCl$_3$F (F11)	186	172[b]	3.6	~65	Anthropogenic	Photolysis in the stratosphere	Rising at a rate of 15 pptv/yr
CCl$_4$	135	125[b]	2.9	~40	Anthropogenic	Photolysis in the stratosphere, possibly loss in ocean	Rising at a rate of 6 pptv/yr
CH$_3$CCl$_3$	156	116[b]	2.7	8.0	Anthropogenic	Reaction with OH	Rising at a rate of 13 pptv/yr
CF$_4$ (F14)	70	70[c]	0.8	~10^4	Anthropogenic	Photolysis in the upper atmosphere	Very long-lived
CHClF$_2$ (F22)	65	55[d]	0.8	19	Anthropogenic	Reaction with OH	Growth rate 11.7% per year
CH$_2$Cl$_2$	38	21[b]	0.4	0.6	Anthropogenic	Reaction with OH	
CHCl$_3$	21	11[b]	0.28	0.8	Anthropogenic	Reaction with OH	Sources ill-defined
CCl$_2$CCl$_2$	29	5[b]	0.8	0.5	Anthropogenic	Reaction with OH	
C$_2$Cl$_3$F$_3$ (F113)	23	21[b]	0.6	~100	Anthropogenic	Photolysis in the stratosphere	Growth rate estimated as about 15% annually

Compound				Source	Removal process	Comments
C_2H_5Cl	19	5[b]	0.1	0.2	Anthropogenic	Reaction with OH	
CH_2ClCH_2Cl	37	14[b]	0.36	0.3	Anthropogenic	Reaction with OH	
CH_3Br	26	19[e]	~0.2	1.8	Ocean, marine biota	Reaction with OH	Source strength 0.08–0.15 Tg/yr
$C_2Cl_2F_4$ (F114)	14	13[b]	0.34	200	Anthropogenic	Photolysis in the stratosphere	Growth rate estimated as about 6% annually
$CHClCl_2$	12	3[b]	0.11	0.03	Anthropogenic	Reaction with OH	
C_2ClF_5 (F115)	4	—[c]	0.09	400	Anthropogenic	Photolysis in the stratosphere	
C_2F_6 (F116)	4	4[c]	0.08	~1000	Anthropogenic	Photolysis in the upper atmosphere	Very long-lived
$CClF_3$ (F13)	3.1	3.5[c]	0.05	~300	Anthropogenic	Photolysis in the stratosphere	
CH_3I	~2	~2[f]	0.05	0.014	Ocean, marine biota	Photolysis in the troposphere	
$CHCl_2F$ (F21)	1.0	—[g]	0.01	2.5	Anthropogenic	Reaction with OH	
CF_3Br	0.8	0.7[b]	0.02	~100	Anthropogenic	Photolysis in the stratosphere	Used as a fire extinguisher
$CClF_2Br$	1.5	—[h]	0.04	~100	Anthropogenic	Photolysis in the stratosphere	Used as a fire extinguisher

[a] Singh et al. (1979b), Rasmussen and Khalil (1980).
[b] Singh et al. (1983a).
[c] Penkett et al. (1981).
[d] Khalil and Rasmussen (1981).
[e] Singh et al. (1983b), marine atmosphere; Penkett et al. (1981), 10 pptv in the upper troposphere.
[f] Rasmussen et al. (1982c).
[g] Penkett et al. (1980b).
[h] Lal et al. (1985).

Anthropogenic sources may be inferred from the observation of higher than average CH_3Cl background mixing ratios in urban areas (Singh *et al.*, 1977a,b, 1979b). By comparing the enhancement of CH_3Cl mixing ratios with that for CCl_3F, which is similar, Watson *et al.* (1980) estimated an urban source of methyl chloride of 0.3 Tg/yr. The nature of the source has not been identified, however. Crutzen *et al.* (1979) have shown that chlorine contained in wood and other elements of vegetation is partly released as CH_3Cl when the material is burned in forest fires or agricultural slash burnings. They found about 75 mg CH_3Cl/kg material burned and estimated a global production rate of 0.4 Tg/yr, with a considerable margin of uncertainty, from biomass burnings, primarily in the tropics. These sources are much smaller than emissions of methyl chloride from the oceans, although not entirely negligible. The limited data suggest that biomass burning and other anthropogenic sources account for perhaps 15% each of the total source strength. Thus, 70% of CH_3Cl in the atmosphere is of marine origin.

The chlorofluoromethanes CCl_2F_2 and CCl_3F derive entirely from anthropogenic sources. These substances are used as refrigerants, as aerosol spray propellants, and as inflating agents in the manufacture of foam materials. In these applications, their main advantages are chemical stability, nonflammability, and absence of toxicity. During the period 1950–1970, the rate of production of the chlorofluoromethanes rose almost exponentially. In the mid-1970s the production rate leveled off and it is now on the order of 4.5×10^8 and 3.0×10^8 kg annually for CCl_2F_2 and CCl_3F, respectively. Release rates to the atmosphere depend on the end uses of the compounds considered, which makes estimating release rate somewhat difficult. For example, chlorofluorocarbons in domestic refrigerators or in closed-cell foam structures are released slowly in comparison with aerosol propellants. From application patterns McCarthy *et al.* (1977) estimated that 85% of the annual production is emitted into the atmosphere in the same year. The Chemical Manufacturers Association provides updated production and release estimates from time to time. The more recent data indicate that 91% of CCl_2F_2 and 85% of CCl_3F produced are released during the same year.

Photolytic destruction in the stratosphere is the only known removal process for CCl_2F_2 and CCl_3F from the atmosphere. This requires transport by eddy diffusion from the tropical tropopause into the 30–35 km altitude region and leads to tropospheric residence times in the order of 70 yr for CCl_3F and 150 yr for CCl_2F_2. Owing to the long residence times, one expects the chlorofluoromethanes to be uniformly distributed in the atmosphere. Their continuous release, however, has caused the mixing ratios to rise, and since about 90% of the release occurs in the northern hemisphere, an imbalance exists between the northern and southern hemispheres. Figure

6-12 shows the rise of CCl_3F in the last decade. During the period of rapid growth, at the end of 1971, the CCl_3F background mixing ratios in the southern hemisphere were appreciably less than in the northern. In 1980, the north-to-south excess had decreased to about 10%. Fewer data are available for CCl_2F_2, but they show a similar rise as the CCl_3F mixing ratios.

In principle it is possible, as Cunnold et al. (1978) have shown, to use a trend analysis in conjunction with known release rates to derive the tropospheric residence times of chlorofluorocarbons in the atmosphere. The data shown in Fig. 6-12, which are due to many individual measurements, are too scattered for this purpose. Much greater precision was reached in the "Atmospheric Life Time Experiment" described by Prinn et al. (1983a) and Cunnold et al. (1983a,b) for CCl_3F and CCl_2F_2. Four background stations, two in each hemisphere, were selected to measure chlorofluorocarbons four times daily for a period of at least three years (1978-1981). The tropospheric residence time derived for CCl_3F was 83($+73/-27$) yr, which is consistent with the notion that the stratosphere provides the main sink. For CCl_2F_2 it was possible to deduce only a lower limit to the true residence time. The value obtained was 81 yr, which also agrees with other data.

The presence of carbon tetrachloride, CCl_4, in the atmosphere is difficult to explain unless it is of anthropogenic origin. Lovelock et al. (1973, 1974a) thought it necessary to postulate a natural source for CCl_4 because of its

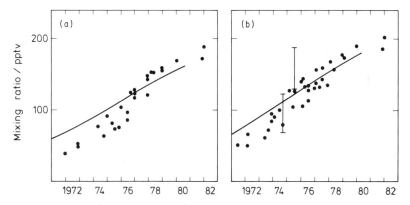

Fig. 6-12. Temporal rise of CCl_3F mixing ratios in (a) the southern and (b) the northern hemispheres according to measurements of Lovelock (1971, 1972, 1974a), Lovelock et al. (1973), Pack et al. (1977), Wilkness et al. (1973, 1975, 1978). Zafonte et al. (1975), Hester et al. (1975), Heidt et al. (1975), Grimsrud and Rasmussen (1975), Singh et al. (1977a, 1979b), Krey et al. (1977), Cronn et al. (1977), Fraser and Pearman (1978), Tyson et al. (1978), Goldan et al. (1980), Leifer et al. (1981), Rasmussen et al. (1981, 1982c), and Pierotti et al. (1978). The solid line shows results of model calculations of Logan et al. (1981) based on the release data of Bauer (1979). Ninety percent of the release was assumed to occur in the northern hemisphere.

rather uniform distribution within and between both hemispheres, and because current anthropogenic emission rates cannot account for the observed atmospheric mixing ratio of about 125 pptv. A closer examination of the history of industrial CCl_4 production and uses by Galbally (1976), Singh et al. (1976), and Parry (1977) revealed, however, that past emissions must have been appreciable and that cumulative emissions are compatible with current levels of CCl_4 in the atmosphere. Before 1950, the main market of CCl_4 was in the United States, where it was used as an industrial solvent, in dry cleaning, as a fire extinguisher, and to a limited extent as a grain fumigant. All these applications must have led to its release to the atmosphere. From 1950 on, carbon tetrachloride was increasingly used as a feedstock for the production of CCl_2F_2 and CCl_3F, whereas dispersive applications declined, partly because of its toxicity. In recent years the release of CCl_4 must have occurred mainly inadvertently during its handling in the production of chlorofluorocarbons.

The early study by Lovelock et al. (1973) of CCl_4 in seawater showed a decline of concentration with depth that was steeper than that for CCl_3F, which suggested that the ocean acts as a sink for atmospheric CCl_4. From measurements of CCl_4 in the surface waters of the Pacific Ocean, Singh et al. (1979b) confirmed that the CCl_4 concentrations are lower than one would expect if they were in equilibrium with CCl_4 in the atmosphere. Singh et al. (1979b) calculated a global flux of CCl_4 into the oceans of 32×10^6 kg/yr. At the time of the measurements, in 1977, the mass content of CCl_4 in the troposphere was 2.7 Tg. The tropospheric residence time due to the marine sink thus would be 85 yr. Since the rate of hydrolysis of CCl_4 in aqueous solution is low, Singh et al. (1979b) suggested that CCl_4 is mainly lost by absorption in the fatty tissue of marine biota.

The stratosphere undoubtedly provides a major sink for tropospheric carbon tetrachloride. Since the absorption spectrum and cross sections for CCl_4 closely resemble those of CCl_3F, their photolytic destruction rates are expected to be nearly the same. Unlike CCl_3F, however, the eddy diffusion profile for CCl_4 should now have reached the steady state, because emissions have been essentially constant for an extended period of time. Comparison with CCl_3F suggests for CCl_4 a tropospheric residence time due to transport into the stratosphere of about 65 yr. The model calculations of Table 3-3 give instead a value of 46 yr. Combining both sinks, the ocean and the stratosphere, gives a residence time in the range of 35–40 yr. Simmonds et al. (1983), who have made time-series measurements of carbon tetrachloride in conjunction with the Atmospheric Life Time Experiment discussed above, derived a residence time of $50(+12/-8)$ years from a trend analysis of their data. The emission estimates on which this result is based carry much uncertainty, however. At the present time, the budget of CCl_4 in the atmosphere must still be considered somewhat tentative.

The presence of methylchloroform in the atmosphere was first reported by Lovelock (1974a). CH_3CCl_3 is used widely as an industrial solvent, and it is rapidly replacing other solvents such as C_2HCl_3 and C_2Cl_4 owing to its lower toxicity. As a consequence, methylchloroform production and release has increased substantially dyring the past two decades at a rate of about 16% per year. The present release rate is on the order of 5×10^8 kg/yr, and thus has reached the magnitude of the release rate for the chloro-fluoromethanes. Atmospheric mixing ratios of CH_3CCl_3 have risen in parallel with the emissions. The measurements of Singh et al. (1979b) during a fairly short period from 1975 to 1977 indicated an annual growth rate of 17%. A longer series of measurements was carried out by Rasmussen et al. (1981). They inferred a growth rate for tropospheric CH_3CCl_3 of 12% per year. Prinn et al. (1983b) found similar growth rates during 1978 and 1979 from data obtained in the Atmospheric Life Time Experiment, whereas the growth rate observed during 1980/1981 was markedly reduced.

Unlike the chlorofluoromethanes, methylchloroform contains abstract-able hydrogen atoms and reacts readily with OH radicals. Thus, CH_3CCl_3 is largely destroyed in the troposphere. The corresponding tropospheric residence time of CH_3CCl_3 is about 8 yr. Between 80 and 90% of CH_3CCl_3 reacts with OH radicals; the rest enters the stratosphere and is destroyed there. Prinn et al. (1983b) used data from the Atmospheric Life Time Experiment to derive a residence time of $10(+5.2/-2.6)$ yr; whereas Singh et al. (1983a), on the basis of a two-box model, found 9 ± 2 years. As discussed in Section 4.2, methylchloroform has served as a key substance in estimating a global average for the OH concentration in the troposphere.

Carbon tetrafluoride, CF_4, and carbon hexafluoride, C_2F_6, are of interest because these substances are chemically so stable in the atmosphere that their photodissociation requires radiation of wavelengths below 100 nm. Thus, CF_4 and C_2F_6 must diffuse upward to altitudes beyond 100 km to undergo photodissociation. The stratospheric altitude profiles for CF_4 and C_2F_6 have been measured by Fabian et al. (1981a). The mixing ratios reveal little change with height up to 33 km, indicating that the lifetimes of these substances in the atmosphere are indeed very long. Cicerone (1979) calcu-lated an atmospheric life time for CF_4 of 10,000 yr; Fabian (1986) estimated a lifetime for C_2F_6 of about 1000 yr. In view of the long lifetime of these compounds, even a very inefficient source may have built up the atmospheric CF_4 and C_2F_6 inventory. From elevated levels of CF_4 and C_2F_6 in the plumes of aluminium plants located along the Columbia River, Penkett et al. (1981) concluded that the aluminum industry is likely to be a major source. The same authors studied samples of volcanic emissions from Mt. Erebus, Antarctica, and found them negligible as a source of CF_4 and C_2F_6. If it is true that anthropogenic emissions are the dominant sources, a gradual increase of the abundances of CF_4 and C_2F_6 must be envisaged.

All halocarbons are active absorbers of infrared radiation. Of special interest are absorption features in the infrared atmospheric window (7-9 μm wavelength) region, because here even relatively small increments in atmospheric abundances will have a pronounced impact on the atmospheric radiation balance, leading to a global temperature increase in the same way as that projected to result from the rise in the mixing ratio of CO_2 (see Section 11.1). This effect is expected to accompany the reduction of stratospheric ozone and calls for more effective emission controls of anthropogenic halocarbons.

Finally, we discuss methyl bromide, CH_3Br, and methyl iodide, CH_3I. These compounds are natural constituents of the atmosphere. Singh et al. (1977a, 1979b) measured CH_3Br background levels of 5 pptv or less, with values as high as 20 pptv in the marine atmosphere. This indicates that the ocean probably is the major source of CH_3Br. Rasmussen and Khalil (1980) reported methyl bromide mixing ratios varying between 5 and 25 pptv, whereas Penkett et al. (1981) found 10 pptv in the upper troposphere. Above the tropopause the mixing ratio of CH_3Br declines rapidly to very low values (Fabian et al., 1981a). In the troposphere, methyl bromide is removed mainly by reaction with OH radicals, which leads to a lifetime of 1.8 yr.

The presence of CH_3I in the atmosphere was discovered by Lovelock et al. (1973) in marine air samples. In addition to aerial concentrations, the authors measured also CH_3I concentrations in seawater, found a large difference, and concluded that the ocean represents a major source. The origin of CH_3I appears to be the production by marine algae and phytoplankton. In the atmosphere, methyl iodide suffers photodecomposition. Lovelock et al. (1973) assumed an atmospheric lifetime of 2 days and estimated a global CH_3I emission rate of 40 Tg/yr. This value is exaggerated, however, for three reasons. One is that the seawater measurements were conducted in regions of high biological productivity, so that an extrapolation of the results to other ocean regions should be viewed with caution. In addition, it appears that the lifetime of CH_3I due to photodecomposition is longer than 2 days—namely, closer to 5 days, if one includes night-time dark periods (Zafiriou, 1974). The photolysis rate depends further on solar elevation and total ozone overhead, and the lifetime increases as one goes toward higher latitudes. A third reason for rejecting an annual CH_3I production rate of 40 Tg is that it would be about 100 times greater than the iodine flux in wet precipitation, according to Miyake and Tsunogai (1963).

Rasmussen et al. (1982b) measured CH_3I in seawater from coastal areas of California and Oregon and from the open Pacific in regions of low and moderate biological productivity. Assuming that 60% 0of the ocean surface area corresponds to moderate, 30% to low, and 10% to high productivity Rasmussen et al. (1982b) then used the thin-film model of Liss and Slater

(1974) to calculate from their own data and those of Lovelock (1975) a global CH_3I flux of 1.3 Tg/yr. This agrees with an independent estimate of 1-2 Tg/yr derived by Zafiriou (1974).

The global measurements by Rasmussen *et al.* (1982b) of methyl iodide in the atmosphere show typical mixing ratios in background air between 1 and 3 pptv in the boundary layer, and a drop-off toward greater heights, which is expected because of the short lifetime. In near-oceanic regions of high biological productivity the mixing ratio of CH_3I may be elevated to 10-20 pptv. The average mixing ratio of CH_3I in the troposphere has been estimated at 0.8 pptv. This leads to a tropospheric mass content of 0.017 Tg for CH_3I, which is in agreement with a source strength of 1.3 Tg/yr and a tropospheric residence time of 5 days.

The photodecomposition of CH_3I generates iodine atoms. Chameides and Davies (1980) have discussed possible subsequent reactions. The most probable pathway is reaction with ozone, $I + O_3 \rightarrow O_2 + IO$, but the fate of the IO radical is uncertain and the consequences cannot yet be assessed.

Chapter

7 | The Atmospheric Aerosol

7.1 Introduction

The concept of air as a colloid and the term aerosol for air containing an assembly of suspended particles were originally introduced by Schmauss and Wigand (1929). Colloids are inherently stable because fine particles are subject to Brownian motion and resist settling by sedimentation. The individual aerosol particles may be solid, liquid, or of a mixed variety, and all types are found in the atmosphere. Solid particles in the air are called dust. They are primarily formed by the erosion of minerals at the earth surface and enter the atmosphere by wind force. Sea spray from the ocean surface provides a prolific source of liquid droplets, which upon evaporation produce sea-salt crystals or a concentrated aqueous solution thereof. Solid and liquid particles also arise from the condensation of vapors when the vapor pressure exceeds the saturation point. For example, smoke from the open and often incomplete combustion of wood or agricultural refuse arises at least in part from the condensation of organic vapors.

For studies of the atmospheric aerosol, the particles are collected on filters or impactor plates. Although the original definition refers to particles and carrier medium taken together, it is now customary to apply the term aerosol more broadly and to include also deposits of particulate matter. We thus speak of aerosol samples, aerosol mass, etc., a usage

278

deriving its justification from the fact that the deposited material was once airborne.

Aerosol particles in the atmosphere usually carry with them some moisture. The amount of water associated with the aerosol depends on the relative humidity. Increasing the relative humidity condenses more water onto the particles, until finally, when the vapor pressure of water exceeds the saturation point, a certain number of particles grows into fog or cloud droplets. Meterologists call these particles condensation nuclei, or simply nuclei. Fogs and clouds are treated as separate systems and are not included in the normal definition of the atmospheric aerosol, even though they represent an assembly of particles suspended in air and thus constitute an atmospheric colloid. The smoothness of the transition from an assembly of aerosol particles to one of cloud elements makes it difficult to define a boundary line between both colloids. Due to the overlap of size ranges of the particles in both systems, any division will be rather arbitrary.

Particulate matter is a truly ubiquitous trace component of the lower atmosphere and is not confined to the region adjacent to the Earth's surface where most of the sources are located. Over the continents, convection currents carry particles into the upper troposphere, where they spread horizontally to fill the entire tropospheric air space. On a global scale, Junge (1963) distinguished three types of aerosol: the continental, the maritime, and the tropospheric background aerosol. By chemical composition, the first two types contain mainly materials from the nearby surface sources, somewhat modified by the coagulation of particles of different origin and by condensation products resulting from gas-phase reactions. The third type represents an aged and much diluted continental aerosol. This fraction is present also at the surface of the oceans. Continental and marine aerosols are blended to some extent in coastal regions to the effect that, depending on wind direction, a continental dust plume may travel far over the ocean, or sea salt may be carried inland for hundreds of kilometers.

As a polydisperse system, the aerosol cannot be fully described without taking into account the particle size spectrum. We shall discuss this aspect in the next section. Two important bulk parameters associated with the size distribution are the total particle number density and the total mass concentration. Table 7-1 presents some typical values for both quantities of ground-level aerosols in different tropospheric regions (3–10 m above ground). Concentrations increase as one goes from remote toward urban regions, but mass and particle concentrations are not linearly correlated. The reason is a shift of the median particle size toward smaller values, as the last column of Table 7-1 shows. It is further instructive to compare observed aerosol mass concentrations with those of atmospheric trace gases. A number of examples are given in Table 7-2. They indicate that the mass

Table 7-1. *Typical Mass and Particle Concentrations for Different Tropospheric Aerosols near Earth's Surface and the Corresponding Mean Particle Radius, Assuming a Mean Density of* 1.8 kg/dm^3 *and Spherical Particles*

Location	Mass concentration (μg/m^3)	Particle concentration (particles/cm^3)	Mean radius (μm)
Urban	~100	10^5–10^6	0.03
Rural continental	30–50	15,000	0.07
Maritime background	~10^a	300–600	0.16
Arctic (summer)	~1	25	0.17

^{a} Includes 8 μg/m^3 sea salt.

Table 7-2. *Comparison of Mass Concentrations of Several Trace Gases with That of Natural Aerosols*

Trace constituent	Concentration (μg/m^3)
Hydrogen (0.5 ppmv)	40
Ozone (30 ppbv)	64
NO$_2$ (0.03–10 ppbv)	0.06–20
CH$_3$Cl (0.5 ppbv)	1
Ethane (0.5–2 ppbv)	0.8–3.2
Aerosol	1–100

concentration of particulate matter is equivalent to that of trace gases having mixing ratios in the ppbv range.

The following sections describe, in turn, particulate size distributions, coagulation and condensation processes, production mechanisms, chemical composition, removal from the atmosphere, and the tropospheric budget of the aerosol.

7.2 Particulate Size Distributions

The size classification of aerosol particles is greatly facilitated if the particles are assumed to have a spherical shape. The size is then defined by their radius or diameter. The assumption clearly is an idealization, as crystalline particles come in various geometrical, but definitely nonspherical, shapes and amorphous particles are rarely perfectly spherical. At sufficiently high relative humidities, however, water-soluble particles turn into concentrated-solution droplets, which are essentially spherical, and therein lies the most convincing justification for the assumption of spherical particles.

A size distribution may refer to number density, volume, mass, or any other property of the aerosol that varies with particle size. The left-hand side of Fig. 7-1 shows somewhat idealized size spectra for the number densities associated with the marine and the rural continental aerosols. The distribution of sea-salt particles that contribute to the marine aerosol is added for comparison. Particle radii and number densities range over many orders of magnitude. In view of the need for a logarithmic representation of the data, it has been found convenient to use the decadic logarithm of the radius as a variable and to define the distribution function by $f(\log r) = dN/d(\log r)$. A conversion of $f(\log r)$ to the more conventional form $f(r)$, if needed, can be done by means of the rule $d(\log r) = dr/(r \ln 10)$ with $\ln 10 = 2.302$, which yields

$$dN/dr = f(r) = f(\log r)/2.302r$$

Traditionally, particles falling into the size range below 0.1 μm are called "Aitken" nuclei. The adjacent region between 0.1 and 1.0 μm contains the "large" particles, whereas components with radii greater than 1 μm are called "giant" particles. Alternatively, one finds in the literature a distinction between "coarse" particles with radii greater than 1 μm and "fine" particles smaller than 1 μm. Sedimentation by gravity becomes important for particles greater than 20 μm in radius. Such giant particles are still observed in the atmospheric aerosol (Delany et al., 1967; Jaenicke and Junge, 1967), although they are rare, occurring with concentrations of less than one particle per cubic meter. Maximum number densities are found at the other end of the size scale, that is, in the Aitken range. Only recently have measurement techniques been developed that enable a more detailed examination of the size spectrum in the Aitken region. The results show fairly convincingly that the size distribution extends all the way down toward the size range of large molecules. The observation of a secondary maximum in the Aitken region is frequently made.

No single instrument is capable of spanning the entire size range from Aitken to giant particles, so that a determination of the full size spectrum is a laborious task. The right-hand side of Fig. 7-1 shows how a size spectrum may be obtained by combining several measurement techniques. Particles greater than 0.1 μm can be separated by means of inertial impactors (Marple and Willeke, 1976). These devices deposit particles larger than a predetermined cutoff radius on a plate positioned behind a nozzle, whereas smaller particles are carried along with the air stream. Two such unit in series, each with its own characteristic cutoff radius, will collect on the second stage only particles within a well-defined size range. The particles can then be counted with a microscope. Direct examination and sizing by microscopy are possible for particles larger than 1 μm, and for all particles if electron

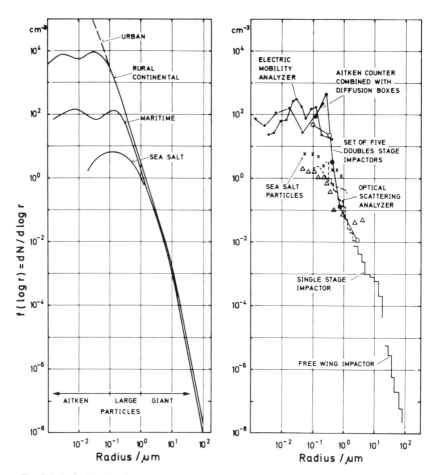

Fig. 7-1. Left: Idealized particle size distributions for the rural continental and the maritime aerosols. The distribution of sea-salt particles that contribute to the maritime aerosol is shown separately. The transition from the rural to the urban aerosol is indicated. Right: Determination of remote tropospheric aerosol size distribution by a combination of instrumental techniques. ⌐ Single-stage and free-wing impactors; O—O set of five double-stage impactors; ⌐ single-particle optical scattering analyzer; these data were obtained at the observatory Izana, Tenerife, 2370 m a.s.l. (Abel *et al.*, 1969). +−+ Aitken counter in conjunction with diffusion battery; - • - • Aitken counter in combination with electric mobility analyzer; ●—● set of double-stage impactors; these data were obtained from measurements on board ships (Jaenicke, 1978a; Haaf and Jaenicke, 1980). △ Sea-salt particles observed by sodium resonance fluorescence (Radke, 1981) and - - - by electron microscopy (Mészarós and Vissy, 1974).

microscopy is employed. Optical (Mie) scattering techniques are limited to the 0.3–3 μm size range, because smaller particles have an inadequate cross section, whereas larger particles are not sufficiently numerous. The Aitken (1923) counter, an instrument resembling a Wilson cloud chamber, is the only detector suitable for the size range below 0.1 μm. The fact that larger particles are counted as well is not critical, since they add little to the total number density. Humidity and volume expansion ratio are adjusted so as to achieve a saturation factor of 1.25, which activates essentially all aerosol particles but excludes small ions and molecules with radii less than 1.2×10^{-3} μm (Haaf and Jaenicke, 1977; Liu and Kim, 1977). The activated particles grow to a size observable by means of optical techniques. To accomplish a size classification of Aitken particles requires the application of mobility analyzers. One method of separation is by diffusion losses of particles in narrow channels of various lengths. The resolution unfortunately is poor, and iterative numerical procedures are required to evaluate the observational data (Maigné *et al.*, 1974; Jaenicke, 1978a). A second method separates charged particles in an electric field (Whitby, 1976; Haaf, 1980). Electric mobility analyzers are endowed with an adequate resolution, but the fraction of particles that can be charged decreases rapidly with decreasing size below 0.01 μm radius. The limiting radius is 0.003 μm.

Size distributions for particle surface, volume, or mass can be derived from that of number density by applying the appropriate weights:

surface: $\quad dA/d(\log r) = 4\pi r^2\, dN/d(\log r)$

volume: $\quad dV/d(\log r) = (4\pi/3)r^3\, dN/d(\log r)$

mass: $\quad dm/d(\log r) = (4\pi/3)r^3\rho(r)\, dN/d(\log r)$

Integration of these equations yields the total particle surface, volume, or mass per unit volume of air. The last quantity is independently accessible by weighing aerosol deposits on filter mats or impactor plates. The variation of density ρ with particle size is difficult to measure and is not generally known. For most purposes the mass distribution is taken to be similar to that of volume and an averaged density is applied. Table 7-3 gives aerosol densities measured by Hänel and Thudium (1977) at several locations. Bulk densities for important salts and quartz are shown for comparison. In the dry state continental aerosols have average densities of about 2×10^3 kg/m³. For relative humidities less than 75%, the water content of the continental aerosol is sufficiently small to be negligible. At higher relative humidities the salts undergo deliquescence, the water content increases, and the density eventually is lowered toward values near 1×10^3 kg/m³. This situation applies specifically to the marine atmosphere.

Table 7-3. *Bulk Densities of Continental Aerosol Samples*[a]

Sampling location	$\bar{\rho} \times 10^{-3}$ kg/m^3	Size range (μm)	Remarks
Mizpeh Ramon, Israel	2.59–2.72	0.15–5.0	Desert aerosol
Jungfraujoch, Switzerland	2.87	0.15–5.0	3600 m Elevation
Mace Head, Ireland	1.93	0.15–5.0	Marine air masses
Mainz, Federal Republic of	1.99	0.15–5.0	Marine air masses advected
Germany (population 200,000)	2.25	>0.1	by strong winds from the Atlantic Ocean
	1.87	0.15–5.0	Urban pollution, inversion layer
Deuselbach, Federal Republic	2.57	0.07–1.0	Winds from rural
of Germany (rural community)	3.32	1.0–10	regions, predominance of soil erosion particles
	1.82	0.07–1.0	High-pressure situation, weak
	1.92	1.0–10	winds, some showers, some fog
Sodium chloride	2.165		
Sea salt	2.25		
Ammonium sulfate	1.77		
Quartz (SiO$_2$)	2.66		

[a] Selected from measurements of Hänel and Thudium (1977). Size ranges were identified by means of impactors. The bulk densities of some inorganic compounds that are believed to be present in aerosols are included for comparison.

Figure 7-2 shows surface and volume distributions for the maritime background (plus sea salt) aerosol, derived from the corresponding number density distributions of Fig. 7-1. In contrast to number density, which is dominated by Aitken particles, aerosol volume and mass are determined primarily by the large and giant particles. Less than 5% of total volume lies in the Aitken range. The volume distribution is inherently bimodal. The shaded area in Fig. 7-2, which indicates the sea-salt fraction, shows that the coarse particle mode represents mainly sea salt. The second mode has it maximum near 0.3 μm, fairly close to that displayed by the distribution of surface area. We shall see later that whenever the aerosol accumulates material by condensation from the gas phase or by coagulation of Aitken particles, the material is deposited in the size range of greatest surface area. Willeke *et al.* (1974) thus designated the peak near 0.3 μm the "accumulation" mode, a term now generally accepted. It should be recognized, however, that coagulation and condensation are not the only processes contributing to the volume in the 0.1–1.0 μm particle size range. To complete this description, note that a third peak exists in the distribution of number density, centered near 10^{-2} μm in the Aitken range. Later discussions will

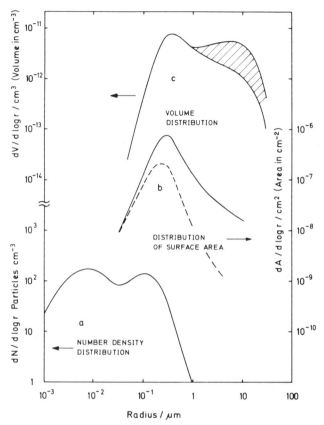

Fig. 7-2. Model size distributions of the marine background aerosol: (a) particle number density, (b) surface area, (c) volume. The contribution of sea salt to the volume distribution is indicated by the shaded area, and arrows indicate the appropriate scale. By integration one obtains a total number density $N = 290$ particles/cm^3, a total surface area $A = 5.8 \times 10^{-7}$ cm^2/cm^3, and a total volume $V = 1.1 \times 10^{-11}$ cm^3/cm^3. For an average density of 10^3 kg/m^3, the mass concentration is 11 μg/m^3 (5 μg/m^3 of sea salt). The dashed curve gives the distribution of the surface area that is effective in collisions with gas molecules. For larger particles the collision rate is lowered by the rate of diffusion.

show that this mode originates as a transient from nucleation processes: that is, the formation of new particles by gas-phase reactions resulting in condensable products. As Fig. 7-2 illustrates, the nucleation mode of the background aerosol has practically no influence on the distributions of surface area and volume.

There is now considerable evidence to show that due to the individual source processes, the atmospheric aerosol as a whole has a size distribution that is basically trimodal, even though it is rare for all three modes to show up concurrently in any particular size representation (Whitby, 1978; Whitby and Sverdrup, 1980). Figure 7-3 gives a number of examples for averaged volume size distributions of continental aerosols. The linear volume scale chosen suppresses the nucleation mode, but accentuates the two peaks associated with fine and coarse particles. On the continents, the coarse particle mode is due to dust from the wind-driven erosion of soils and released plant material. The origin of coarse particles is similar to that of sea salt over the ocean in that it is surface-derived. The fine particles arise mainly from condensation processes, but a mineral component is still present in the submicrometer size range. Curves 1–3 in Fig. 7-3 represent data obtained at Goldstone, California, in the Mojave desert. Whitby and Sverdrup (1980) refer to curve 1 as clean continental background aerosol. It was obtained under somewhat unusual conditions, namely, after a period of rainfalls, and is characterized by low Aitken nuclei counts ($<1000 \text{ cm}^{-3}$). The average distribution at Goldstone is shown by curve 2. In this case, the Aitken nuclei count is perhaps 6000 cm^{-3}, and with the use of a mean density of $2 \times 10^3 \text{ kg/m}^3$ the mass concentration is estimated from the integral

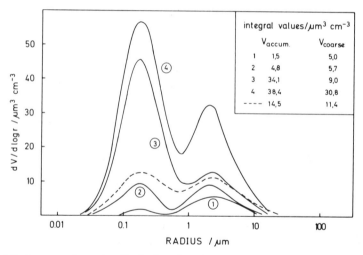

Fig. 7-3. Average volume size distributions for continental aerosols. [Adapted from Whitby and Sverdrup (1980).] The measurement data were smoothed and idealized by fitting to them additive log-normal distributions. (1) Background aerosol, very clean; (2) normal background aerosol; (3) background aerosol disturbed by an urban plume (these data from measurements at Goldstone, California). (4) Average urban aerosol (from data taken at Minneapolis, Minnesota, Denver, Colorado, and various locations in California). The dashed curve gives the volume distribution resulting from the number density distribution for the rural continental aerosol shown in Fig. 7-1. The integrated volumina, given by the area underneath each curve, are shown in the insert.

volume as about 20 $\mu g/m^3$. The dashed curve in Fig. 7-3 gives the volume distribution for the rural continental aerosol whose distribution of number density is shown in Fig. 7-1. The associated population of Aitken particles is 1.5×10^4 cm^{-3} and the mass concentration is about 50 $\mu g/m^3$. Curve 3 depicts the situation arising from the advection of an aged and somewhat diluted urban plume. The term *aging* implies that since their formation the Aitken particles have had sufficient time to coagulate with themselves and with larger particles so that their number density is reduced to normal background values. The accumulation mode is considerably enhanced, however. A volume size distribution typical of an urban aerosol is shown as curve 4. Urban aerosols are characterized by high Aitken counts exceeding 10^5 cm^{-3}. Again, the nucleation mode does not really show up in the volume size distribution except in the vicinity of sources, for example, adjacent to roads with a dense automobile traffic. Cars also stir up dust, whereby the concentration of coarse particles is elevated. A mass concentration of 140 $\mu g/m^3$ is estimated from the integral volume of the urban aerosol.

7.3 The Physicochemical Behavior of Particles

7.3.1 COAGULATION AND CONDENSATION

Aerosol particles tend to coalesce when they collide with each other. Since at normal humidities most particles are sheathed with moisture, the sticking probability is close to unity. Collisions between two particles thus lead to the formation of a new particle of larger size. This process, called coagulation, causes the size distribution to change in favor of larger particles. Coagulation must be distinguished from condensation, which describes the deposition of vapor-phase material onto particulate matter. In the absence of preexisting particles, condensation leads to the formation of new (Aitken) particles, provided the vapor pressure of the condensing substance is sufficiently high. In accordance with current usage, the last process will be termed homogeneous nucleation or gas-to-particle conversion. A more detailed discussion of it is given in Section 7.4.3.

There exists an extensive literature on the theory of coagulation (Fuchs, 1964; Zebel, 1966; Hidy and Brock, 1970; Twomey, 1977), and we can treat here only the most salient features. In the absence of external forces, the aerosol particles undergo collisions with each other due to their thermal (Brownian) motion. The mathematical description of thermal coagulation goes back to the classical work of Smoluchowski (1918) on hydrosols. Application to aerosols seems to have been made first by Whitlaw-Gray and Patterson (1932). Let $dN_1 = f(r_1)\, dr_1$ and $dN_2 = f(r_2)\, dr_2$ describe the number densities of particles in the size intervals $r_1 + dr_1$ and $r_2 + dr_2$, respectively. The collision frequency between particles is then given by the product

$$\mathbf{K}(r_1, r_2)f(r_1)f(r_2)\ dr_1\ dr_2 \tag{7-1}$$

where $\mathbf{K}(r_1, r_2)$ is called the coagulation function. For particles much smaller than the mean free path of air molecules ($\lambda = 6.5 \times 10^{-2}$ μm for atmospheric conditions), the coagulation function is given by the laws of gas kinetics,

$$\mathbf{K}(r_1, r_2) = \pi(r_1 + r_2)^2(\bar{v}_1^2 + \bar{v}_2^2)^{1/2} \tag{7-2}$$

where $\bar{v}_1$ and $\bar{v}_2$ are the mean thermal velocities of the particles. They are given by $\bar{v}_i = [8k_\mathrm{B}T/\pi m(r_i)]^{1/2}$ with $k_\mathrm{B} = 1\cdot38 \times 10^{-23}$ kg m^2/s denoting the Boltzmann constant, T the absolute temperature, and $m(r_i) = 4\pi\rho r_i^3/3$ the mass of a particle. For particles with radii $r \gg \lambda$, whose motion is hampered by friction, the probability of binary encounters is controlled by diffusion. The collision frequency is then calculated from the theory of Smoluchowski by considering the flow of particles of radius r_2 to a fixed particle of radius r_1, with the result that

$$\mathbf{K}(r_1, r_2) = 4\pi(r_1 + r_2)(D_1 + D_2) \tag{7-3}$$

where D_1 and D_2 are the diffusion coefficients associated with the two types of particles. For the transition regime toward small particles where $r \approx \lambda$, Fuchs (1964) has argued that the fixed particle is centered in a concentric sphere $r_a \geqslant r_1 + r_2$ such that outside this sphere the flow of particles is diffusion-controlled, whereas inside the sphere the particles travel freely according to their mean thermal velocities without undergoing collisions with gas molecules. On this basis, Fuchs derived the correction factor

$$\xi = \left[\frac{r_1 + r_2}{r_a} + \frac{4r_a(D_1 + D_2)}{(r_1 + r_2)^2(\bar{v}_1^2 + \bar{v}_2^2)^{1/2}} \right]^{-1} \tag{7-4}$$

to the Smoluchowski expression for the coagulation function. The problem of selecting the most appropriate value for r_a has produced an animated discussion in the literature. A summary of the arguments was given by Walter (1973). He could show that for radii larger than 10^{-3} μm it is permissible to set $r_a = r_1 + r_2$ with reasonable accuracy. This approach has the virtue that in the limit of molecular dimensions the second term in the denominator of ξ becomes large compared with the first term, whereby the gas-kinetic collision rate is recovered. The coagulation function then becomes

$$\mathbf{K}(r_1, r_2) = \frac{4\pi(r_1 + r_2)(D_1 + D_2)}{1 + 4(D_1 + D_2)/(r_1 + r_2)(\bar{v}_1^2 + \bar{v}_2^2)^{1/2}} \tag{7-5}$$

The diffusion coefficients are obtained from the mobilities of the particles, b_i, use being made of the Einstein relation $D_i = k_B T b_i$. For spherical particles the dependence of the mobility on r and λ has been derived empirically from careful measurements by Knudsen and Weber (1911) and by Millikan (1923) in the form

$$b(r) = \frac{1}{6\pi\eta r} \left\{ 1 + \frac{\lambda}{r} [A + B \exp(-Cr/\lambda)] \right\}$$ (7-6)

where $A = 1.246$, $B = 0.42$, and $C = 0.87$ are constants and $\eta = 1.83 \times 10^{-5}$ kg/m s is the viscosity of air. For particles with $r \gg \lambda$ the mobility reduces to the well-known law of Stokes. Table 7-4 lists values of mobilities, diffusion coefficients, thermal velocities, and mean free paths for particles with radii in the range between 10^{-3} and $10\,\mu$m. The last column gives Fuchs's correction factor for collisions of like particles.

The collision of particles with masses m_1 and m_2 leads to the formation of a new particle with mass $m_3 = m_1 + m_2$. If the new particle is assumed to be spherical, its radius is $r_3 = (r_1^3 + r_2^3)^{1/3}$. To ensure mass conservation, the mathematical description of the coagulation process requires an appropriate weighting of the coagulation function, and the delta function is suitable for this purpose. We define

$$W(r_1, r_2/r_3) = \mathbf{K}(r_1, r_2)\, \delta[(r_1^3 + r_2^3)^{1/3} - r_3]$$ (7-7)

as the probability for two particles with radii r_1 and r_2 to form a third particle with radius r_3. The time variation of the distribution function can

Table 7-4. *Mobilities $b(r)$, Diffusion Coefficients $D(r)$, Average Thermal Velocities $\bar{v}(r)$, Mean Free Paths $\lambda(r)$, and Fuchs's Correction Factor $\xi(r)$ for Like Particles, as a Function of Radius for Spherical Particles in Air*[a]

r (μm)	b (s/kg)	D (m²/s)	$\bar{v}$ (m/s)	λ (m)	ξ
1 (−3)	3.172 (14)	1.291 (−6)	4.975 (1)	6.608 (−8)	0.0134
1 (−2)	3.344 (12)	1.361 (−8)	1.573	2.203 (−8)	0.290
1 (−1)	5.467 (10)	2.225 (−10)	4.975 (−2)	1.139 (−8)	0.887
1.0	3.135 (9)	1.276 (−11)	1.573 (−3)	2.065 (−8)	0.977
10	2.922 (8)	1.190 (−12)	4.975 (−5)	6.091 (−8)	0.993

[a] The mean free path of air molecules is $\lambda_{air} = 6.53 \times 10^{-8}$ m at normal pressure and temperature. Powers of 10 are given in parentheses.

then be expressed by a combination of integrals over all possible collisions:

$$\frac{\partial f(r,t)}{\partial t} = \frac{1}{2} \int_0^\infty \int [W(r_1, r_2/r)f(r_1,t)f(r_2,t)]\, dr_1\, dr_2$$

$$- \int_0^\infty \int [W(r, r_2/r_3)f(r,t)f(r_2,t)]\, dr_2\, dr_3 \qquad (7\text{-}8)$$

The first term on the right-hand side describes the gain of particles in the size range $r + dr$ due to coagulation of smaller particles. The factor $\frac{1}{2}$ is required because the integration counts each collision twice. The second term describes the losses of particles in the size range $r + dr$ due to collisions with particles in all other size ranges. After inserting the coagulation probability W, the double integrals can be integrated once, the first over r_2 (by setting $r_1^3 + r_2^3 = x$) and the second over r_3. Thereby one obtains

$$\frac{\partial f(r,t)}{\partial t} = \frac{1}{2} \int_0^r \mathbf{K}(r_1, y)f(r_1,t)f(y,t)\frac{r\, dr_1}{y^2}$$

$$- \int_0^\infty \mathbf{K}(r, r_1)f(r,t)f(r_1,t)\, dr_1 \qquad (7\text{-}9)$$

where $y = (r^3 - r_1^3)^{1/3}$. In the second term on the right, the radii are now renumbered. In the first term, the range of integration can be reduced to $0 - r$, since for $r_1 > r$ the integral is zero. Moreover, because of the symmetry of $\mathbf{K}(r_1, r_2)$, the factor $\frac{1}{2}$ can be dropped if the integration limit is replaced by $r/\sqrt[3]{2}$. Twomey (1977) has given a lucid discussion of the properties of these integrals.

There are no simple solutions to this integrodifferential equation, and numerical procedures must be applied in solving it. Note that Eq. (7-9) describes the change with time of the distribution function $f(r,t)$ when sources of particles are absent. If there is a production of new particles, a source term must be added to the right-hand side.

Model calculations based on the coagulation equation have yielded several important results, which are now considered. The first concerns the time scale for changes of the number density of particles in the Aitken range. Figure 7-4 shows the variation of an assumed size distribution having its maximum particle concentration initially at 0.03 μm radius. In the course of time, the maximum shifts toward larger radii while the concentration of small particles decreases. The simultaneous growth of large particles is hardly noticeable, however, as comparatively little volume is added to this size range. The lifetime of particles with radii below 10^{-2} μm is fairly short.

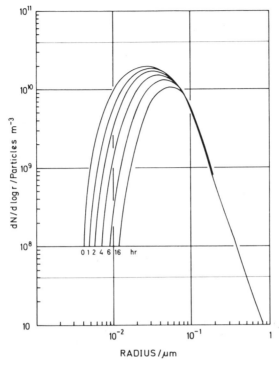

Fig. 7-4. Modification of an aerosol size distribution due to coagulation of small particles as a function of time. The distribution of large particles is not significantly altered by the process. [Adapted from Junge and Abel (1965).]

If such particles are found in the atmosphere, they must be young! This conclusion was reached quite early by Junge (1955) and formed an essential piece of support for his viewpoint that the tropospheric background aerosol as far as it derives from the continental aerosol must have undergone aging by coagulation. The data in Fig. 7-4 suggest lifetimes of the order of 1 h for particles with radii around 10^{-2} μm. The calculation is based on a total number density of 2×10^4 cm^{-3}, which is representative of the rural continental aerosol. Lifetimes at least an order of magnitude greater are obtained if the initial number density is reduced to that characteristic of the tropospheric background aerosol.

Walter (1973) has studied the evolution of the particle size spectrum by coagulation for the case where new particles with a radius of 1.2×10^{-3} μm are continuously generated by nucleation processes. Some of his results are shown in Fig. 7-5. The left-hand side shows the evolution of the size spectrum when other particles are initially absent. A secondary maximum develops

to the right of the production peak. In the course of time, the secondary
peak moves toward larger particles sizes until it eventually reaches the size
range 0.1 μm after about 1 day has passed. At this time the further changes
are minor, indicating that a quasi-steady state is attained. For such conditions, the particles with radii greater than 0.1 μm are recipients of newly
created particle mass, so that the peak near 0.2 μm can be identified as the
accumulation mode.

The right-hand side of Fig. 7-5 shows the effect upon such a steady-state
distribution when the population of large particles is varied. As their number

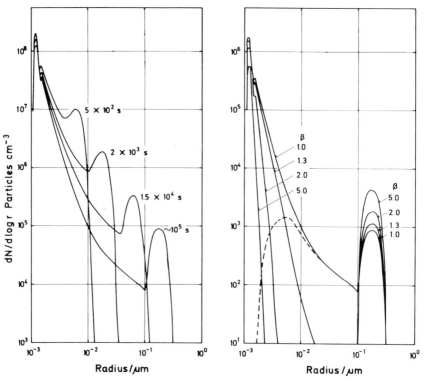

Fig. 7-5. Coagulation behavior of particles produced by nucleation. [Adapted from Walter
(1973).] A continuous generation of embryos with 1.2×10^{-3} μm radius is assumed. Left:
Variation of the distribution function with time in the absence of preexisting particles;
production rate $q = 10^6$ cm^{-3} s^{-1}. Right: Steady-state distributions for different concentrations
of preexisting large particles ($r > 0.1$ μm); production rate $q = 10^2$ cm^{-3} s^{-1}; number density
of preexisting large particles $N_0 = 270\beta$. The dashed curve (for $\beta = 1$) is obtained from the
steady-state distribution 12 h after terminating the source of embryos. The apparent lifetime
of Aitken particles for coagulation is here greater than that indicated in Fig. 7-4 because of
the smaller number density of particles.

density increases from 270 cm^{-3} to five times that number, a deep gap develops in the population of Aitken particles whereby the spectrum is split into two parts. The new particles then coagulate preferentially with the large particles rather than undergoing coagulation with themselves to form particles in the Aitken range. The gap can be filled up if one tolerates an appropriate increase in the particle production rate.

The results are not greatly altered, if one assumes a smaller size of the particles formed or an additional condensation of water vapor onto the coagulating particles up to equilibrium with the environment (relative humidity $\leqslant 90\%$), as Hamill (1975) has done. The existence of a gap such as that appearing in Fig. 7-5 contrasts with our knowledge of natural aerosols that display a minimum in the number density for particles near 3×10^{-2} μm radius but no cleft (cf. Fig. 7-1). Jaenicke (1978b) has tried to overcome the dilemma by suggesting a simultaneous production of particles with different sizes. Maintaining the assumption of a steady state, he calculated the production rates required in various size ranges to simulate an aerosol size distribution observed by him over the Atlantic Ocean. He obtained nonzero production rates throughout the entire Aitken range, although the rates decreased by four orders of magnitude from 10^{-3} to 10^{-1} μm These results are in conflict with current concepts of nucleation by condensation from the gas phase (see Section 7.4.4), even though one cannot off-hand refute the possibility of a rapid growth by heteromolecular condensation of a fraction of the new particles to sizes near 10^{-2} μm. The assumption of steady-state conditions made by Jaenicke (1978b) also is unrealistic. New condensable material such as H_2SO_4 arising from the oxidation of SO_2 is produced mainly photochemically, so that the rate of new-particle production undergoes a diurnal variation. A corresponding variation in the Aitken size spectrum is inevitable in view of the time constants associated with coagulation.

The approximate change in the size distribution that will have taken place after a 12-h intermission of new particle production is shown in Fig. 7-5 by the dashed curve. It results in a bimodal size distribution, which in addition to the accumulation peak now contains a transient peak caused by the incomplete coagulation of Aitken particles. We have previously designated this transient the nucleation mode. The ensuing size distribution gives a better representation of the natural aerosol, even though its resemblance to the size spectra in Fig. 7-1 is still marginal.

We turn now to consider the process of condensation, that is, the deposition of vapor-phase material directly onto preexisting particles. The rate of the process is determined by the number of condensable molecules striking the aerosol particles per unit time. If we assume as previously a sticking probability of unity, the condensation rate is given by an expression similar to Eq. (7-1). For later comparison we use here the aerosol size distribution in the form $dn_2 = f(\log r_2) d(\log r_2)$. Let n_1 denote the number density of condensable molecules and n_s the equilibrium number density for the situation when condensation and reevaporation occur at equal rates (saturation equilibrium). The rate of condensation is then proportional to the difference $n_1 - n_s$ and is given by

$$K(r_1, r_2)(n_1 - n_s)f(\log r_2)\, d(\log r_2) \qquad (7\text{-}10)$$

Referring to Eq. (7-5) for $K(r_1, r_2)$ we note that for molecules compared with particles, $\bar{v}_1 \gg \bar{v}_2$ and $D_1 \gg D_2$, since $r_1 \ll r_2$. These simplifications reduce $K(r_1, r_2)$ to the form

$$K^*(r) = \pi r^2 \bar{v}_1 / (1 + r\bar{v}_1/4D_1) \qquad (7\text{-}11)$$

An asterisk is applied in order to distinguish this collision term from the original coagulation function. In addition, r_2 is replaced by r. The diffusion coefficient can also be replaced by the gas kinetic equivalent expression

$$D_1 = \frac{\pi}{8}\,\bar{v}_1\lambda_1$$

where λ_1 is the mean free path of the condensable molecules in air. Instead of the collision cross section πr^2 of a particle, one may also introduce its surface area $A(r) = 4\pi r^2$. With these substitutions one then obtains

$$K^*(r) = \tfrac{1}{4}A(r)\bar{v}_1 / (1 + 2r/\pi\lambda_1) \qquad (7\text{-}12)$$

The numerator on the right-hand side is the gas-kinetic collision rate. The denominator reduces the surface area available for collisions to the effective surface area,

$$A_{\text{eff}}(r) = \frac{A(r)}{1 + 2r/\pi\lambda_1} \qquad (7\text{-}13)$$

As long as the particle radius is small so that the term $2r/\pi\lambda_1$ is much smaller than unity, the molecules collide with particles at gas-kinetic rates. For larger particles, when $2r/\pi\lambda_1 > 1$, the collision rate becomes diffusion-controlled and increases with r rather than with r^2. Setting approximately $\lambda_1 = \lambda_{\text{air}}$, one finds that the boundary between the two regimes lies near $0.1\,\mu\text{m}$ radius.

It is instructive to compare the distribution of the effective surface area

$$dA_{\text{eff}}(r)/d(\log r) = \frac{4\pi r^2}{1 + 2r/\pi\lambda_1}\,f(\log r) \qquad (7\text{-}14)$$

with the real surface area distribution of the natural aerosol. For this purpose, the function $dA_{\text{eff}}(r)/d(\log r)$ is included in Fig. 7-2 for the maritime background aerosol and is shown by the dashed curve. It is apparent that the effective surface area is greatly reduced when the radius becomes larger than 0.3 μm. The maximum of the distribution curve is shifted only slightly, however, toward smaller radii. When condensation takes place, collisions of condensable molecules with aerosol particles deposit material predominantly in the 0-1-0.5 μm range of radii, that is, in the accumulation range. This is so because the number distribution of particles favors that range. Similar results are also obtained for the continental and the urban aerosols.

The total rate of condensation is obtained by integrating over the entire range of radii and is given by

$$\frac{dn_1}{dt} = -\frac{\bar{v}_1}{4}(n_1 - n_s) \int_0^\infty A_{\text{eff}}(r)f(\log r)\, d(\log r)$$

$$= -\frac{\bar{v}_1}{4}(n_1 - n_s)A_{\text{eff,tot}} \tag{7-15}$$

As in the case of coagulation, a source term must be added to the right-hand side if condensable molecules are continuously produced. In the absence of sources, Eq. (7-15) can be integrated without difficulty and the resulting variation of number density is

$$n_1 - n_s = (n_{10} - n_s)\exp(-t/\tau_{\text{cond}})$$

with $$\tau_{\text{cond}} = 4/\bar{v}_1 A_{\text{eff,tot}}$$

The associated mass increase of particulate matter is

$$\Delta m(t) = (M_1/N_A)(n_{10} - n_s)[1 - \exp(-t/\tau_{\text{cond}})] \tag{7-16}$$

where M_1 is the molecular weight of the condensate and N_A is Avagadro's number. The distribution of the mass increment with particle size

$$d[\Delta m(t)]/d(\log r) = [\Delta m(t)/A_{\text{eff,tot}}]A_{\text{eff}}f(\log r) \tag{7-17}$$

is proportional to the rate of condensation in each size range and corresponds to the distribution of $A_{\text{eff}}(r)$. We are especially interested in the time constant for condensation. Table 7-5 shows values for the integrated surface area, the total effective surface area, and the associated time constants for three important types of atmospheric aerosols. The time constants should be considered minimum values in view of the possibility that the sticking coefficients may be much less than unity.

If $n_s \ll n_1$, the difference $n_1 - n_s$ is large and most of the condensable material is deposited onto the aerosol to stay there. Sulfuric acid is a case in point. In the atmosphere, H_2SO_4 is partially neutralized by ammonia,

Table 7-5. *Total Surface Area, Total Effective Surface Area Available for Collisions with Molecules for Three Important Aerosols, and the Associated Time Constant for the Condensation of Molecules with Molecular Weight $M_1 = 0.1$ kg/mol ($\bar{v}_1 = 2.5 \times 10^2$ m/s)*

Aerosol type	A_{tot} (m^2)	A_{efftot} (m^2/m^3)	τ_{cond} (s)
Marine background aerosol	5.8×10^{-5}	1.4×10^{-5}	1.15×10^3
Rural continental aerosol	3.1×10^{-4}	1.4×10^{-4}	1.15×10^2
Urban aerosol	1.6×10^{-3}	1.0×10^{-3}	1.6×10^1

and the vapor density of the ensuing salts is essentially negligible. If on the other hand $n_1 \approx n_s$, only a fraction of condensable material is transferred onto the aerosol. The amount is controlled by the saturation vapor density n_s, which is a function of temperature. In this case, however, the situation is complicated by the interaction of several condensates, an effect that the above treatment did not include. Water vapor is the most abundant condensable constituent in the atmosphere. The fraction of water present on the aerosol is determined to a large extent by its interaction with water-soluble particulate matter. This effect will be discussed next.

7.3.2 INTERACTION OF AEROSOL PARTICLES WITH WATER VAPOR

The great importance of aerosol particles as condensation nuclei in the formation of fogs and clouds requires that we examine the interaction of the aerosol with water vapor in greater detail. For individual particles the interaction may range from the partial wetting of an insoluble dust particle to the complete dissolution of a salt crystal, such as sodium chloride derived from sea salt. In the course of time, however, coagulation, condensation, and in-cloud modification processes cause even an insoluble siliceous particle to acquire a certain share of water-soluble material. Although there will always be particles that have retained their source characteristics, one may assume that in general the aerosol comprises a mixture of both water-soluble and insoluble matter. Junge (1963) has done much to promote the concept of mixed particles by assembling evidence from a variety of studies, including electron microscopy, to show that mixed particles dominate the continental aerosol. Winkler (1973) has given a more recent discussion of this aspect.

The key to understanding the behavior of aerosol particles in a humid environment was first given by Köhler (1936) on the basis of thermodynamic principles. Dufour and Defay (1963) and Pruppacher and Klett (1980) have provided detailed treatments of all aspects requiring consideration. Here, we discuss only the main result, which arises from two opposing effects: the increase of the water vapor pressure by the curvature of the droplet's surface, and the lowering of the equilibrium vapor pressure with increasing concentration of the solute. Let p_s denote the saturation vapor pressure above a plane surface of pure water at temperature T (in K). The actual vapor pressure above the surface of a droplet with radius r is then given by

$$\ln p/p_s = \frac{2\sigma V_m}{R_g Tr} + \ln a_w \tag{7-18}$$

where σ is the surface tension, V_m the partial mole volume of water, R_g the gas constant, and a_w the somewhat temperature-dependent activity of water in the solution. The first term on the right describes the influence of surface curvature according to Kelvin, and the second term takes account of the influence of solutes in the form of a generalized Raoult's law. For a single solute, the activity of water

$$a_w = \gamma_w x_w = \gamma_w \frac{\nu_w}{\nu_w + \nu_s} = \gamma_w \left(1 + \frac{m_s M_w}{m_w M_s}\right)^{-1} \tag{7-19}$$

is proportional to the mole fraction x_w, which is determined by the mole numbers ν_w and ν_s of water and solute, respectively. They can be replaced by the corresponding ratios of mass to molecular weight. The activity coefficient γ_w is an empirical parameter that is introduced to account for the nonideality of solutions, either because they are not sufficiently dilute to follow Raoult's law or because they contain electrolytes that dissociate into ions. In general, $\gamma_w < 1$, although it tends toward unity for dilute solutions of nondissociable compounds.

Equation (7-19) can also be applied to mixtures of solutes if one replaces ν_s, m_s, and M_s by

$$\nu_s = \sum_i \nu_{si} \qquad m_s = \sum_i m_{si} \qquad \bar{M}_s = \nu_s/m_s$$

No general theory exists that relates the activity coefficients for mixtures to those of the individual components, so that γ_w must be redetermined for each specific mixture that one might wish to consider. We shall assume initially that the aerosol particle contains no insoluble material. Its radius is then related to its mass by

$$m = \frac{4\pi\bar{\rho}r^3}{3} = \nu_s \bar{M}_s + \nu_w M_w = m_s + m_w \tag{7-20}$$

where $\bar{\rho}$, the average density, is determined with the usual and quite realistic assumption (Hänel, 1976) of molar volume additivity,

$$\bar{\rho} = (m_s + m_w) \Big/ \left[\frac{m_s}{\rho_s} + \frac{m_w}{\rho_w}\right] \tag{7-21}$$

By combining the above equations, one obtains for a particle in equilibrium with its environment the ratio of the actual to the saturation vapor pressure of water:

$$\frac{p}{p_s} = \gamma_w \left(1 + \frac{m_s M_w}{m_w \bar{M}_s}\right)^{-1} \exp\left(\frac{2\sigma V_m}{R_g T r}\right)$$

$$= \gamma_w \left[1 + \frac{M_w}{\bar{M}_s} \frac{\rho_0 r_0^3}{(\bar{\rho} r^3 - \rho_0 r_0^3)}\right]^{-1} \exp\left(\frac{2\sigma V_m}{R_g T r}\right) \tag{7-22}$$

Here, r_0 and ρ_0 denote the particle's dry radius and density, respectively. Note that p/p_s represents the relative humidity. Equation (7-22) thus provides an implicit equation for the droplet's radius r as a function of relative humidity. For pure water, $\sigma_w = 7.42 \times 10^{-2}$ N/m and $V_m = 1.8 \times 10^{-5}$ m^3/mol. We take r to be given in micrometers and set $T = 283$ K. The exponential then assumes the value $1.13 \times 10^{-3}/r$, which shows that for radii $r > 0.1$ µm the Kelvin effect is small. Surface-active compounds can reduce the surface tension considerably, making the effect even smaller. Strong electrolytes as solutes may either increase or decrease the surface tension, but their influence amounts to less than 20% within the range of concentrations permitted before the solutions saturate.

The validity of Eq. (7-22) has been tested and confirmed experimentally by Orr et al. (1958a,b) and by Tang et al. (1977) for a number of different salt nuclei with sizes in the large particle range. Figure 7-6 demonstrates the behavior for sodium chloride particles. The curves were calculated and the points were taken from the work of Tang et al. (1977). At very low humidities the salt particle is dry and crystalline. Increasing the relative humidity initially causes some water to be adsorbed on the particle's surface, but the amount is insufficient to dissolve the material and the particle remains largely crystalline. Accordingly, the growth in particle size is small. This portion of the growth curve does not obey Eq. (7-22), of course. At a certain critical relative humidity the crystal deliquesces—that is, it takes up enough water to form a saturated solution. The size of the particle then changes abruptly. The deliquescence of sodium chloride occurs at 75% relative humidity (r.h.). At higher relative humidities the increase in size follows the growth curve described by Eq. (7-22). Upon decreasing the relative humidity again to values below the deliquescence point, the solution becomes supersaturated. Recrystallization does not occur as spontaneously as deliquescence, so that the particle size moves along the extended Köhler

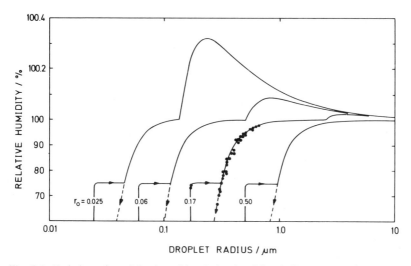

Fig. 7-6. Variation of particle size with relative humidity (Köhler diagram) for sodium chloride particles having different dry radii. Deliquescence occurs at 75% r.h. Note the hysteresis effect when the humidity is raised or lowered beyond the critical value. Curves for relative humidities greater than 75% were calculated. The scale above 100% is expanded. The experimental points refer to observations of Tang *et al.* (1977) on submicrometer-sized monodisperse sodium chloride particles.

curve for a while before it suddenly shrinks to a value near the dry radius. This hysteresis branch of the curve has been observed for airborne particles as well as for bulk material (Winkler and Junge, 1972), but generally it is not very reproducible.

An important feature displayed by each particle growth curve is the maximum in the domain of supersaturation above 100% r.h. Fairly low degrees of supersaturation suffice to reach the peak, so that in Fig. 7-6 an expanded scale is used to bring this feature out more clearly. The rising portion of the curve to the left of the peak results from the Raoult term in Eq. (7-22), whereas the descending branch to the right is due to the Kelvin term. Once a particle has grown to a size just beyond the maximum, it enters into a region of instability and must grow further, provided the water supply in the ambient air suffices. A particle that has passed over the peak is said to have become "activated." This is the process of cloud formation! Note that the height of the maximum decreases with increasing droplet size, so that the largest particles are preferentially activated. At the same time, these particles grow proportionately more than the smaller particles and consume most of the water available. Thereby, the supersaturation is kept at low levels, making it more difficult for small particles to reach the peak and grow into cloud drops. In the continental aerosol the number of large (and giant) particles generally suffices as cloud condensation nuclei. Aitken particles contribute little. In marine air the number density of the background aerosol is less favorable, so that a larger fraction of Aitken nuclei participates.

Because of their comparatively greater mobility, the nonactivated Aitken particles are subsequently incorporated into cloud dops by thermal coagulation. When the cloud eventually evaporates again, the cloud drops have scavenged a large portion of Aitken particles. The aerosol size distribution emerging in the range of large and giant particles is changed little by this process, since Aitken particles do not contribute enough mass. There are other processes, however, that affect the distribution of large and giant particles, namely, the coalescence of cloud drops, the chemical conversion of dissolved gases to nonvolatile products (for example, SO_2 to sulfate), and rainout. A detailed theoretical study of Blifford et al. (1974) has shown that such processes reduce mainly the abundance of giant particles, while the distribution of large particles is least affected. The relative importance of all these activities of in-cloud modification of the aerosol size spectrum depends greatly on the vertical extent and the lifetime of an individual cloud, making it difficult to simulate in-cloud modification in a realistic way.

Growth curves of natural aerosol particles were first measured by Junge (1952a,b) at relative humidities below 100%. He found the growth much less compared with that of pure salts, and attributed the reduction to the presence of insoluble matter. More recently, Winkler and Junge (1972) and Hänel (1976) studied the mass increase with rising humidity of bulk samples of aerosol using a gravimetric method. The procedure is justified because at relative humidities below 95% the Kelvin term can be neglected and the relative mass increase m_w/m_0 of single particles will be identical to that of bulk material of the same composition. Figure 7-7 shows results for the urban, the rural continental, and the maritime aerosol. For the last two, the authors have carried out a coarse size separation by means of impactors.

The growth curves for the continental aerosols are fairly similar. They are much smoother than those for pure salts, lacking the abrupt mass increase associated with deliquescence. Hysteresis loops are still observed for individual samples, although not to the same extent as for pure salts. A large fraction of water-soluble matter consists of sulfates, chlorides, and nitrates, the major cations being NH_4^+, Na^+, K^+, Ca^{2+}, and Mg^{2+}. Table 7-6 lists deliquescence humidities for a number of salts containing these ions. The wide range of deliquescence points, the formation of hydrates, and the interaction of the various constituents in a concentrated solution all contribute to explain the smoothness of the growth curves. Details of composition are important, of course. To show that the observed growth behavior is due to the properties of the aerosol as a mixture of substances,

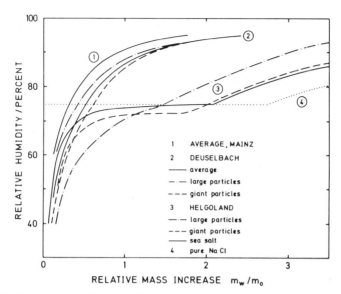

Fig. 7-7. Growth curves for aerosol bulk samples observed by Winkler and Junge (1972). (1) Urban aerosol, average of 28 samples collected at Mainz, Germany; (2) rural continental aerosol, average of 14 samples collected at Deuselbach, Germany; (3) maritime aerosol, average of 10 samples collected at Helgoland Island, located about 50 km off the German coast in the North Sea. In the last two cases, the size ranges $0.1 < r < 1.0$ and $r > 1.0$ μm were separated with a double-stage impactor; the average results for each size range are shown. Growth curves for sea salt and (4) sodium chloride are included for comparison.

Winkler and Junge (1972) prepared an artificial salt mixture having a composition similar to that often found in continental aerosols. Its growth behavior resembled that of natural aerosols, provided allowance was made for the influence of insoluble material.

Figure 7-7 shows further that the marine aerosol features a more pronounced growth with relative humidity than continental aerosols. The reason is the greater content of water-soluble material. For giant particles, the growth

Table 7-6. *Deliquescence Humidities* (%) *for Selected Inorganic Salts*

$MgSO_4$	88	NH_4NO_3	62
$Na_2SO_4 \cdot 10H_2O$	87	$Mg(NO_3)_2 \cdot 6H_2O$	51
KCl	84	$Mg(NO_3)_2 \cdot 4H_2O$	36
$(NH_4)_2SO_4$	80	NH_4HSO_4	39
NH_4Cl	77	$CaNO_3 \cdot 4H_2O$	50
NaCl	75	$CaNO_3 \cdot 2H_2O$	28
$NaNO_3$	75	$MgCl_2$	33
$(NH_4)_3H(SO_4)_2$	69		

curve bears a close resemblance to the curve for sea salt. This observation is easily understood, as the mass of sea salt is concentrated in the size range $r_0 > 1$ μm. The growth behavior of sea salt, in turn, is determined primarily by sodium chloride, which is the dominant component. In the size range of large particles, $0.1 < r_0 < 1$ μm, the growth curve of the maritime aerosol becomes more similar to that of continental aerosols. This indicates the increasing influence of the tropospheric background aerosol.

Consider now the effect of water-insoluble material on the particle growth curve. Let m_u be the mass of the insoluble fraction of the particle's dry mass m_0, and m_s as before the mass of the water-soluble fraction. The total mass of the particle is then given by

$$m = \tfrac{4}{3}\pi\bar\rho r^3 = m_s + m_u + m_w = m_0 + m_w \qquad (7\text{-}23)$$

The ratio m_s / m_w appearing in Eq. (7-19) can be written

$$m_s / m_w = (m_s / m_0)/(m_0 / m_w) = \varepsilon(m_0 / m_w) \qquad (7\text{-}24)$$

where $0 < \varepsilon < 1$. With this relation, Eq. (7-22) changes to

$$p/p_s = \gamma_w \left[1 + \varepsilon \frac{M_w}{\bar M_s} \frac{\rho_0 r_0^3}{(\bar\rho r^3 - \rho_0 r_0^3)} \right]^{-1} \exp\left(\frac{2\sigma V_m}{R_g T r} \right) \qquad (7\text{-}25)$$

The form of this equation is the same as that of Eq. (7-22), but the growth of the particle is reduced because for any fixed relative humidity the mass of water present in solution is determined by m_s and hence by ε and not by $m_0 = m_s + m_u$.

Junge and McLaren (1971) have studied the effect that the presence of insoluble material has on the capacity of aerosol particles to serve as cloud condensation nuclei. Using Eq. (7-25) they calculated the supersaturation needed for an aerosol particle to grow to the critical radius at the peak of the Köhler curve, from where spontaneous formation of cloud drops becomes feasible. The results are shown in Fig. 7-8. They indicate that the difference is less than a factor of two in radius for particles whose soluble fraction is greater than $\varepsilon = 0.1$. The majority of particles can be assumed to meet this condition (see Fig. 7-19). By assuming particle size distributions similar to those of Fig. 7-1 for continental and maritime background aerosols, Junge and McLaren also calculated cloud nuclei spectra as a function of critical supersaturation and compared them with observational data. These results are shown in Fig. 7-8b. We shall not discuss the data in detail. The results make clear, however, that the presence of insoluble matter in aerosol particles does not seriously reduce their capacity to act as cloud condensation nuclei.

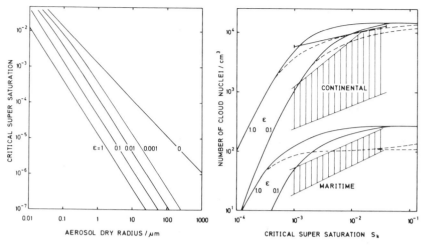

Fig. 7-8. Influence on cloud nuclei formation of the mass fraction ε (water-soluble material/particle dry mass). Left: Critical supersaturation of aerosol particles as a function of particle dry radius. Right: Cloud nuclei spectra calculated for $\varepsilon = 0.1$ and 1 on the basis of two size distributions each for continental and maritime aerosols (solid and dashed curves, respectively). [Adapted from Junge and McLaren (1971).] The curves for the maritime cloud nuclei spectra are displaced downward from the original data to normalize the total number density to 300 cm^{-3} instead of 600 cm^{-3} used originally. The curves for $\varepsilon = 1$ give qualitatively the cumulative aerosol size distributions starting from larger toward smaller particles ($s_k = 10^{-4}$ corresponds to $r_0 \geqslant 0.26$ μm, $s_k = 3 \times 10^{-3}$ to $r \geqslant 0.025$ μm). Similar results were subsequently obtained by Fitzgerald (1973, 1974). The hatched areas indicate the ranges of cloud nuclei concentrations observed in cloud diffusion chambers with material sampled mainly by aircraft [see the summary of data by Junge and McLaren (1971)]; the bar represents the maximum number density of cloud nuclei observed by Twomey (1963) in Australia.

7.4 Aerosol Sources and Global Production Rates

Source mechanisms are the main subject of this section. We describe the production of mineral dust, sea salt, and condensation nuclei, then present estimates of global production rates.

7.4.1 MINERAL SOURCES

Dust emissions from soils caused by wind erosion have been extensively investigated by Gillette (1974, 1978) and collaborators (Gillette and Goodwin, 1974; Gillette and Walker, 1977; Gillette et al., 1980). The following discussion is based primarily on two review articles (Gillette, 1979, 1980).

Soils are formed by the weathering of crustal material of the earth. Rocks, stones, and pebbles slowly disintegrate through the action of water, chemically by the leaching of soluble components, and mechanically by the freeze-thaw cycle of water entering into pores and cracks. In this way, igneous rocks are transformed into clay minerals, carbonates, and quartz grains (sand). Once the material has been broken down to grain sizes less than 1 mm, it can be moved by wind force. With increasing wind speed, mobile particles will first creep or roll before they are temporarily lifted off the ground. The ensuing leaping motion of coarse-grained particles has the effect of sandblasting. Fine particles encrusted onto the coarser grains are loosened, break off, and a fraction of them gets thrown into the air. Most of these particles return quickly to the ground by gravitational settling. Only particles with radii smaller than about 100 μm can remain airborne for a longer period of time, provided they escape the surface friction layer due to turbulent air motions.

To initiate the motion of soil particles requires wind velocities in excess of a threshold value. Experimental studies of threshold velocities for soils consisting of beds of loose, monodisperse particles were described by Bagnold (1941), Chepil (1951), and Greeley et al. (1974) among others. The results indicate that threshold velocities are at a minimum for particles in the size range 25-100 μm radius. To move larger particles requires greater wind forces because the particles are heavier, whereas smaller particles adhere better to other soil constituents so that larger pressure fluctuations are needed to set them into motion.

Natural soils are polydisperse systems of particles, which are rarely present in the form of loose beds. Nevertheless, if loose particles are available, they will be preferentially mobilized. Other factors that are important are the coverage of the soil surface with roughness elements like pebbles, stubbles, bushes, etc., which partially absorb momentum; coherence forces between soil particles due to clay aggregation, organic material, or moisture content; and soil texture, that is, the composition of the soil in terms of particle size classes (see Table 7-7).

Gillette (1980) and Gillette et al. (1980) have presented measured threshold friction velocities for a number of dry soils of different types. The lowest velocities, 0.2-0.4 m/s, were found for disturbed soils having less than 50% clay content and less than 20% pebble cover, or for tilled bare soils. At the upper end of the range ($\geqslant$1.5 m/s) were soils with more than 50% clay content and surface crusts or a cover of coarse (>5 cm) pebbles. The corresponding wind speeds when measured about 2 m above ground were 4-8 and 33 m/s, respectively.

Once the threshold velocity is surpassed, the flux of saltating particles increases rapidly with wind speed. Measurements (Gillette, 1974, 1978) show that the horizontal flux of particles through a plane perpendicular to both ground surface and wind direction increases with $u_*^2(u_* - u_{*0})$ where u_* is the friction velocity and u_{*0} is the threshold value. For large velocities

Table 7-7. *International Standard Classification System for Size Ranges of Soil Particles*[a]

Diameter (μm)	Nomenclature
<2	Clay[b]
2–20	Silt
20–200	Fine sand
200–2000	Coarse sand
>2000	Gravel

[a] Taken from the "Handbook of Chemistry and Physics" (Weast, 1978). Other classifications use slightly different ranges.

[b] The term clay is used here to indicate a size class, not a type of mineral.

the expression tends to u_*^3, a result that was first derived by Bagnold (1941) by dimensional arguments along with the assumption that all the momentum is transferred to the ground surface by saltating sand grains. The corresponding flux of kinetic energy delivered to the soil surface is ρu_*^3. The population of particles at some distance from the ground is fed from the friction layer by eddy diffusion with a flux that for neutral stability conditions, when a logarithmic wind profile exists, is aproximately proportional to u_*^2. Accordingly, the vertical flux of aerosol particles is expected to grow with the fifth power of the friction velocity, at least for sufficiently loose, sandy soils. Figure 7-9 shows vertical fluxes of particles in the 1–10 μm size range measured over a variety of soils. The data indicate that, indeed, a fifth-power law is observed in several cases. For at least one soil, however, the dependence on friction velocity was more pronounced.

Idealized size distributions developing from this mechanism of aerosol generation are summarized in Fig. 7-10. A linear scale is used and normalized mass distributions are shown. The distribution obtained within the dust layer closest to the surface—that is, within the first 2 cm (frame b)—is still fairly similar to that of the parent soil, although much narrower. This result is mainly due to the fact that threshold velocities minimize for particles with radii of 50–100 μm. One meter above the ground (frame c), the mass distribution is shifted toward finer particles. At this height the distribution depends much on soil texture and the prevailing wind force. Two examples are shown. One refers to loamy fine sand with a distribution centered at 10 μm radius. The other was obtained during a sandstorm in the Saharan desert and obviously is weighted toward 50 μm particles. Mixed, bimodal distributions are not uncommon indicating that the assumption of a smooth

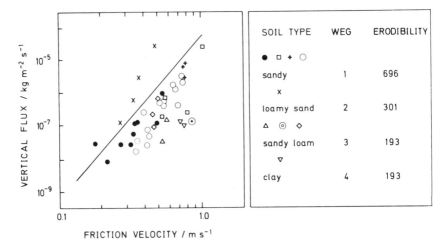

Fig. 7-9. Vertical mass flux of soil-derived aerosol versus the wind friction velocity for different soil types. [Adapted from Gillette (1980).] The solid line indicates a fifth-power dependence. The key on the right correlates soil type with wind erodibility groups (WEG) [after Hayes (1972)] and erodibility in units of 10^3 kg/ha yr [after Lyles (1977)].

size distribution of parent soil is not always warranted. Frame d of Fig. 7-10 shows the average distribution of loess taken from data assembled by Junge (1979). Loess is an aeolian deposit primarily composed of particles in the range 5–50 μm radius. The particles are just small enough to be carried to distances of several hundred kilometers before they fall out by sedimentation. Many loess deposits were formed during the final stages of the last glaciation period up to 300 km from the fringes of the great ice sheets. Further sedimentation of giant particles causes the size distribution to shift further toward smaller particles. The last frame of Fig. 7-10 shows, as an example, the mass distribution of a Saharan aerosol observed at the Cape Verde Island after four days of travel time, carried along by the trade winds of the northern hemisphere. Schütz (1980) has assembled data to show that this aerosol is transported across the Atlantic Ocean, predominantly above the trade inversion layer.

Figure 7-11 compares number density size distributions for soil and aerosol particles at several locations in North Africa to show the dominance of submicrometer particles in all populations studied. The maximum number density was found near 0.1 μm radius in all samples, and particles as small as 0.02 μm were detected.

In the aerosol, dustwind conditions caused an enhancement of concentrations in the size range greater than 3 μm, as expected from the preceding discussion. There is, however, an enhancement of concentrations also in

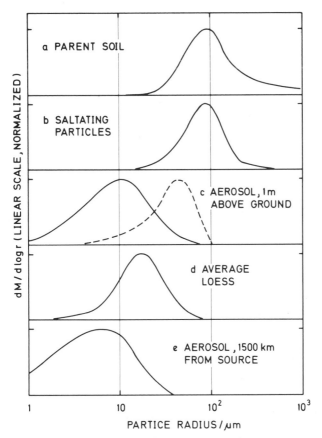

Fig. 7-10. Idealized representation of aerosol formation from mineral sources, based on measurements of Gillette and Walker (1977), Schütz and Jaenicke (1974), Jaenicke and Schütz (1978), and Junge (1979).

the 0.01–1 μm size range relative to that of particles near 1 μm radius. This feature is unexpected and points toward an increase, with increasing wind velocity, in the rate at which small particles are liberated by the sandblasting process.

Another important point is the similarity of size distributions among the different soil samples. It made little difference whether they were taken from sand dunes in Dar Albeida, Mali; alluvial soils in Matan, Senegal; or gravel-covered soils from the Nubian desert in the Sudan or Tamarasset, Algeria. A secondary maximum near 20 μm radius, which is apparent from Fig. 7-11, also was frequently seen. D'Almeida and Jaenicke (1981) hypothesized that the gap between this coarse size fraction and submicron particles,

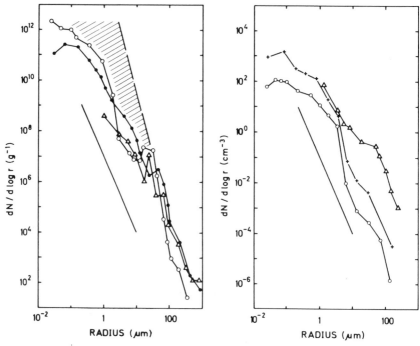

Fig. 7-11. Number density size distributions for selected soil samples (left) and mineral aerosols 2–3 m above the soil (right). Left: ○, Matam, Senegal; △, Camp Derj, Libya; ●, Tamarasset, Algeria. Right: Matam, Segal: ○, usual conditions; +, dustwind; △, Camp Derj, Libya: sandstorm. For comparison, a Junge (r^{-3}) power-law distribution is shown by the solid line. The data are from D'Almeida and Jaenicke (1981) and Schütz and Jaenicke (1974). Characterization of soil samples: Matam, Senegal, alluvial flood plains; Tamarasset, Algeria, gravelly soil; Camp Derj, Libya, desert soil, rock-covered. The hatched area indicates the loss of fine particles to the aerosol according to the interpretation of D'Almeida and Jaenicke (1981).

indicated in Fig. 7-11 by the shaded area, is due to the winnowing process of a continuing loss of fine particles to the atmosphere. It is an unresolved question whether the loss is a permanent one leading to the gradual depletion of fine particles from the soil, or whether these are continuously replenished by the mechanical disintegration of larger particles.

7.4.2 Sea Salt

The production of sea-salt aerosol is due to the agitation of the sea surface by wind force, and in this regard its formation is similar to that of dust aerosol. The mechanism is unique, however, in that sea-salt particles arise from the bursting of air bubbles when they reach the ocean surface. The mechanism is indicated in Fig. 7-12. The surface free energy of the collapsing

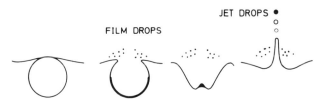

Fig. 7-12. Sea-spray formation from a bursting bubble; schematic representation according to observations of Day (1964) and MacIntyre (1968, 1972). The enrichment of the uppermost jet drop with material from the interior surface of the bubble is indicated. The size of jet drops is about 15% of the bubble diameter (Wu, 1979); film drops are smaller.

bubble is converted to kinetic energy in the form of a jet of water, which, depending on the size of the bubble, ejects between 1 and 10 drops up to 15 cm above the sea surface. Additional drops are produced from the bursting water film covering the bubble when it reaches the surface. A portion of the film drops moves in the direction perpendicular to the jet axis and will be scavenged by the sea. Others are torn off the toroidal rim of the bubble by the escaping air and move upward. The number of film drops increases rapidly with bubble size (Day, 1964) so that bubbles a few millimeters in diameter can produce several hundred film drops. The distinction between jet and film drops is of some importance, because surface-active materials such as certain organic compounds, bacteria, etc. concentrate at the air–sea interface of the bubble to become scavenged by the film drops. A certain enrichment of such materials takes place also in jet drops, however. MacIntyre (1968, 1972) has shown that the jet receives its water from the surface layers surrounding the bubble. The uppermost jet drop derives from water of the innermost shell, the next jet drop from the adjacent outer shell, and so on. Accordingly, one expects the first jet drop to become endowed with most of the surface-active material. Figure 7-12 illustrates this mechanism.

Seawater contains sea salt to about 3.5% by weight, of which 85% is sodium chloride. It can be safely assumed that the sea-salt content of jet and film drops is similar. As the drops enter the atmosphere they experience lower relative humidities and dry up until their water content is in equilibrium with the environment. The particle radius then is about one-quarter of the parent drop's radius. Although parts of sodium chloride may crystallize, one should not expect the particles to dry up completely, because the deliquescence point of magnesium chloride, which is also present in sea salt, lies at 31% r.h. (see Table 7-6). Such low relative humidities are not reached in marine air.

Bubbles are most numerous in the whitecaps associated with breaking waves, where they are formed by the entrainment of air into the surface waters of the ocean during the breaking wave motion. Whitecaps begin to appear at wind speeds of about 3 m/s. At wind speeds of 8 m/s, approximately 1% of the sea is covered with whitecaps (Monahan, 1971; Toba and Chaen, 1973). Bubble sizes in breaking waves range from perhaps a few micrometers to more than several millimeters in diameter. The exact limits are not known. The concentration of giant bubbles is too low to be observable, and the smallest bubbles not only are difficult to detect, but also go rapidly into solution due to surface-curvature effects. Field measurements by Blanchard and Woodcock (1957), Johnson and Cooke (1979), and others revealed bubble size spectra with concentrations maximizing near 100 μm diameter, fairly independent of depth, and decreasing in concentration about inversely to the fifth power of the bubble diameter. Total concentrations decrease quasi-exponentially with depth below the sea surface (scale height ~1 m).

The flux of particles can be estimated from the rate at which bubbles rise to the sea surface. The rise velocity is a function of bubble size, since it is determined by the buoyancy of the bubble and the drag forces acting upon it. The rise velocity increases roughly with the square of bubble diameter. Blanchard and Woodcock (1980) assumed that each bubble produces five jet drops and estimated a production rate of about $2 \times 10^6 \, \mathrm{m^{-2} \, s^{-1}}$ in the whitecap regions of the oceans. The production rate is heavily weighted toward small drop sizes. Several studies reviewed by Blanchard (1963) and Wu (1979) showed that the mean diameter of jet drops is about 15% of that of the parent bubble. Using this information together with a bubble size distribution $(r/r_0)^{-5}$, with $r_0 = 50 \, \mu\mathrm{m}$, and the rise velocities given by Blanchard and Woodcock (1957), one can calculate the production rate of jet drops as a function of drop size. The results are shown in Fig. 7-13 by the solid line. Monahan (1968) has estimated production rates by combining measured concentration–size distributions of jet drops above the water surface with calculated ejection velocities. Figure 7-13 includes results from his own measurements and from data originally presented by Woodcock *et al.* (1963) from the coastal surf region of the Hawaiian beach. The flux spectra of both estimates agree reasonably well. An agreement between the absolute production rates is not expected because they are too dependent on wind speed.

Figure 7-13 gives also size distributions for the concentration of sea-salt particles about 6 m above the ocean surface. They were derived by Blanchard and Woodcock (1980) from mass distributions measured by Chaen (1973) on board ships. Durbin and White (1961) and Woodcock (1972) had earlier obtained similar results by aircraft sampling at elevations between 600 and 800 m. Figure 7-13 shows that the size spectrum of sea-salt particles corresponds to that of sea-spray production, indicating that in the size range

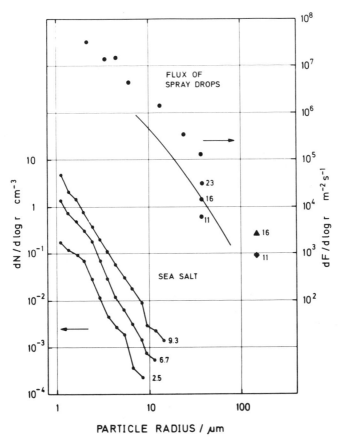

Fig. 7-13. Vertical flux of sea-spray droplets according to Monahan (1968), compared with the number density size distribution of sea-salt particles 6 m above the ocean surface (Chaen, 1973, as reported by Blanchard and Woodcock, 1980). Both follow a Junge (r^{-3}) power law. Numbers next to points indicate the wind speed (m/s). The solid line for the droplet flux was calculated from ejection velocities given by Wu (1979).

below 10 μm the particle size distribution is determined primarily by the production mechanism. The removal rate of sea-salt particles thus must be size-independent. This assumption breaks down for particles larger than 10 μm, since these are increasingly subject to gravitational settling. According to Wu (1979), drops greater than 100 μm in size spend less than 0.5 s in the air before they return to the sea under their own weight.

While the data in Fig. 7-13 cover only particles with radii larger than 1 μm, there can be no doubt that the size spectrum extends toward smaller values. Seasalt particles as small as 0.05 μm have been detected by electron microscopy (Mészarós and Vissy, 1974) and flame scintillation photometry

(Radke, 1981; Radke *et al.*, 1976). Although these small particles contribute little to the total aerosol number density (see Fig. 7-1), their origin is still undetermined. The problem is that bubbles with radii much smaller than 50 μm dissolve too rapidly in seawater. Blanchard and Woodcock (1980) suggest that the absorption of organic matter onto the bubble surface stabilizes some of the bubbles in the 10 μm size range and prevents their dissolution. The possibility that these bubbles are responsible for the production of sea-salt particles in the submicrometer size range must be considered.

Figure 7-14 summarizes our current knowledge about the distribution of sea-salt particles with altitude above the ocean surface. Between 10 and about 500 m elevation the concentration changes comparatively little, but once the cloud level is reached, there is a continuous decay toward negligible concentrations in the 2–3 km height region. Toba (1965) has pointed out that the concentration declines nearly exponentially with a scale height of 0.5 km for particles not exceeding 1 ng in mass (<7 μm radius). For larger particles the scale height is smaller. The decline is attributed to an incorporation of sea salt into clouds, followed eventually by rainout. Figure 7-14 also indicates the occurrence of a sea-salt inversion layer below the level of cloud base. This phenomenon presumably is due to evaporating cloud

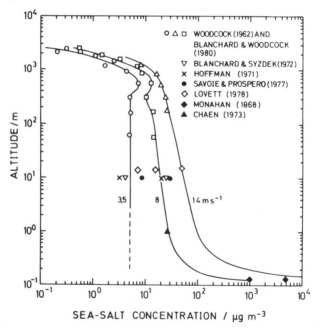

Fig. 7-14. Distribution of sea salt over the ocean as a function of altitude for different wind velocities. [Adapted from Blanchard and Woodcock (1980).]

drops, which by coalescence had grown sufficiently to sediment out. Figure 7-14 shows in addition that the mass concentration of sea salt in the subcloud boundary layer increases with wind speed. This behavior is caused by the increase of whitecap coverage of the sea surface with rising wind force. The observations of Woodcock (1953), Blanchard and Syzdet (1972), and Lovett (1978) indicate an exponential increase of the sea-salt concentration with wind speed in the range 5–35 m/s. The vertical distribution of sea salt appears to adjust quickly to any given wind speed and to the then existing flux of sea salt from the ocean surface into the atmosphere. One should not assume, however, that the distribution represents a steady state. Junge (1957) has numerically solved the time-dependent one-dimensional eddy diffusion equation and has shown that the time constant for the adjustment to steady state is about 12 h, that is, of the same magnitude as that of meteorological changes.

7.4.3   GAS-TO-PARTICLE CONVERSION

Atmospheric gas-phase reactions may lead to the formation of condensable products, which subsequently associate with the atmospheric aerosol. The best-known reaction of this kind is the oxidation of SO_2 to H_2SO_4 and its neutralization by ammonia to form sulfate salts.

Condensation may either cause the formation of new particles in the Aitken range (homogeneous nucleation) or deposit material onto preexisting particles (heterogeneous condensation). In laboratory studies of gas-to-particle conversion one usually starts with air freed from particles by filtration. The development of the particle size spectrum then goes through three successive stages, dominated by nucleation, coagulation, and heterogeneous condensation, in that order (Husar and Whitby, 1973; Friedlander, 1977). In the atmosphere, all three processes take place concurrently. The generation of new particles then requires conditions that allow the growth of molecular clusters by condensation in the face of competition from heterogeneous condensation. These conditions will have to be defined.

Molecular clusters are formed due to weakly attractive forces between molecules, the Van der Waals forces. Except under conditions of low temperature, it is difficult to observe and study in the laboratory clusters containing more than a few molecules, so that details about their properties are sparse. Our understanding of the nucleation process consequently relies mainly on theoretical concepts based largely on the principles of statistical mechanics. The theory of homogeneous nucleation developed by Volmer and Weber (1926), Flood (1934), Becker and Doering (1935), and Reiss (1950) assumes that certain thermodynamic properties, such as the molar volume or the surface tension, that can be determined for bulk material remain valid in the molecular regime. The assumption probably holds for clusters containing more than about 20 molecules (Sinanoglu, 1981), but fails below that number. The error incurred must be tolerated. The theory shows that a steady-state distribution of clusters of various sizes is set up conforming to a quasi-equilibrium between clusters below a certain critical size. In this size range, the abundance of clusters is proportional to

$$\exp[-\Delta G(i)/k_B T]$$

where $\Delta G(i)$, the free energy of formation of the cluster, is a function of the number content i of molecules. At constant temperature, $\Delta G(i)$ is composed of two terms. The first is the difference in the thermodynamic potentials μ_e and μ_g, for the liquid and gas phase, respectively, which is related to the ratio p/p_s of actual to saturation vapor pressure of the compound under consideration. The second term is the surface free energy (Kelvin term):

$$\Delta G(i) = i(\mu_e - \mu_g) + 4\pi\sigma r^2$$
$$= -k_B T i \ln(p/p_s) + \sigma(36\pi)^{1/3}(V_m i/N_A)^{2/3} \qquad (7\text{-}26)$$

Here, k_B is the Boltzmann constant, σ the surface tension, and $V_m = 4\pi r^3 N_A/3i$ the molar volume of the condensed phase. When $p/p_s < 1$, the first term is positive and $\Delta G(i)$ grows monotonically with i. When supersaturation occurs, $p/p_s > 1$, the first term becomes negative, and the free energy has a maximum for clusters with radius $r^* = 2\sigma V_m/k_B T N_A \ln(p/p_s)$. The maximum forms a barrier to nucleation, but once a cluster passes over the hill, it enters into a region of instability and grows further. The situation is similar to the formation of cloud drops depicted in Fig. 7-6. Clusters that have reached the critical size are called embryos. Although the equilibrium distribution must be truncated at that point, it is assumed that the concentration of embryos is still determined by the equilibrium distribution. The nucleation rate can then be calculated from the equilibrium number density of embryos combined with the net rate at which vapor molecules are impinging on them. For the mathematical details of the calculation, the reader is referred to the books of Frenkel (1955), Zettlemoyer (1969), Friedlander (1977), and Pruppacher and Klett (1980).

Generally, the homogeneous nucleation of a single compound is effective only at fairly high degrees of supersaturation. Water vapor, for example, requires values in excess of $p/p_s = 5$ (Pruppacher and Klett, 1980). Such high values are hardly ever reached in the atmosphere because of the presence of aerosol particles giving rise to heterogeneous nucleation at much lower saturation levels ($p/p_s = 1.05$, see Fig. 7-6). In 1961, in a study of the system H_2SO_4–H_2O, Doyle called attention to the conucleation of two vapors, which he found to be more efficient than that of a single component, leading to nucleation even when both vapors are undersaturated. In this case, the equation for the free energy of a cluster contains two pressure terms, one for each component, and $G(i, j)$ is a function of the numbers of both types of molecules, i and j, present in the cluster:

$$\Delta G(i, j) = -k_B T[i \ln(p_1/p_{1s}) + j \ln(p_2/p_{2s})] + 4\pi\sigma r^2 \qquad (7\text{-}27)$$

The free energy can be represented by a surface in three-dimensional space, and the barrier to nucleation now has the shape of a saddle point that the embryos must pass to each the region of instability and further growth. Doyle's (1961) numerical results must be considered preliminary. The work was continued by Kiang et al. (1973), who introduced a simplified kinetic prefactor; by Mirabel and Katz (1974), who corrected sign errors and misprints in Doyle's paper and provided a more complete set of nucleation rates under various conditions; and by Shugard et al. (1974), who refined the model by taking into account the effect of H_2SO_4 hydrates. Figure 7-15 shows nucleation rates due to Mirabel and Katz and compares their results with those of other authors. The results depend critically on the choice of the saturation vapor pressure of sulfuric acid. Mirabel and Katz (1974) used the estimates of Gmitro and Vermeulen (1964), which suggested

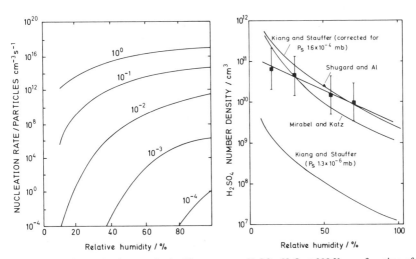

Fig. 7-15. Left: Nucleation rate in the binary system H_2SO_4-H_2O at 298 K as a function of relative humidity and with the H_2SO_4 activity p/p_s as a parameter [from calculations of Mirabel and Katz (1974)]. Right: Steady-state H_2SO_4 number density required for a nucleation rate of 1 particle/cm^3 s as a function of relative humidity. The solid lines and the triangular point are from calculations of the authors indicated; the squares with error bars were obtained experimentally by Boulaud et al. (1977).

$p_s(H_2SO_4) = 4.66 \times 10^{-4}$ mbar at 300 K. A recent experimental determination gave instead 3.33×10^{-5} mbar at 296 K (Roedel, 1979). Kiang *et al.* (1973) had used instead the lower value 1.3×10^{-6} mbar previously applied by Doyle, but their data can be corrected and are then found in accord with the other data, as Fig. 7-15 shows. Boulaud *et al.* (1977) have determined the nucleation rate in the H_2SO_4–H_2O system by laboratory experiments. Their results are included in Fig. 7-15 and provide order-of-magnitude agreement with theory.

The nucleation rate increases strongly with the partial pressure of H_2SO_4. At 60% relative humidity, the ratio $p/p_s = 10^{-3}$ corresponds to an H_2SO_4 number density of approximately 10^{10} molecules/cm^3 at 300 K. It leads to a nucleation rate of 10^2/cm^3 s. Raising the number density by a factor of 10 increases the nucleation rate to almost 10^9/cm^3 s! To maintain such rates in the atmosphere requires a steady production of H_2SO_4 in the face of competition by heterogeneous condensation of H_2SO_4 onto the background aerosol. The rate of the second process can be estimated from the rate at which H_2SO_4 molecules strike the effective aerosol surface, as discussed in Section 7.3.1. Table 7-8 compiles production rates required to maintain the indicated steady-state H_2SO_4 number densities, and the loss rates due to condensation and nucleation, respectively. The first process is dominant when $n(H_2SO_4) = 10^{10}$ molecules/cm^3, and the second when $n(H_2SO_4) = 10^{11}$ molecules/cm^3. The corresponding volume mixing ratios near ground level in the atmosphere are 4×10^{-10} and 4×10^{-9}, respectively. Middleton and Kiang (1978) have presented a model that incorporates nucleation,

Table 7-8. *Steady-State H_2SO_4 Number Density, Associated Production, and Simultaneous Loss Rates for Nucleation and Heterogeneous Condensation onto Various Preexisting Aerosols*[a]

Aerosol type	r.h. (%)	A_{effTot} (m^2/m^3)	$n(H_2SO_4)$ (molecules/cm^3)	$q(H_2SO_4)$[b]	$R(\text{nucl})$[b]	$R(\text{cond})$[b]
Maritime	60	1.4 (−5)	1 (10)	9 (6)	1 (3)[c]	9 (6)[d]
background			1 (11)	1 (10)	1 (10)[c]	9 (7)[d]
Rural	60	1.4 (−4)	1 (10)	9 (7)	1 (3)[c]	9 (7)[d]
continental			1 (11)	1 (10)	1 (10)[c]	9 (8)[d]
Urban[e]	50	1.0 (−3)	*f*	1.2 (7)	7.5 (5)	1.2 (7)[g]
			f	1.2 (8)	1.1 (8)	2.4 (8)[g]

[a] Values in parentheses indicate powers of 10.
[b] In molecules/cm^3 s.
[c] Assuming embryos to contain 10 H_2SO_4 molecules.
[d] From collision rate, sticking probability unity.
[e] According to Middleton and Kiang (1978).
[f] Not given.
[g] Reevaporation included.

coagulation, and heterogeneous condensation on both new and preexisting particles. Their results are included in Table 7-8 and are more favorable than the other data. This is not necessarily a discrepancy, since the model of Middleton and Kiang includes the process of reevaporation of H_2SO_4 whereby the rate of condensation onto the preexisting aerosol is reduced.

Despite the effort that has gone into the calculations, it cannot be claimed that the binary system H_2SO_4–H_2O gives a realistic description of the situation existing in the atmosphere. Aerosol sulfate is found to occur mainly as NH_4HSO_4 and $(NH_4)_2SO_4$, and not so much in the form of sulfuric acid. The ammonium sulfates arise from the partial neutralization of H_2SO_4 by ammonia. There is no reason to assume that the interaction between NH_3 and H_2SO_4 is delayed until sulfuric acid aerosol is actually formed. Rather, the interaction should take place already during the stage of embryo formation. In fact, one would expect NH_3 to react with sulfuric acid hydrates directly at the molecular level, for example, via

$$H_2SO_4(H_2O)_n + NH_3 \rightarrow NH_4HSO_4(H_2O)_{n-1} + H_2O$$

and so forth. In other words, the nucleation of sulfuric acid in the presence of water and ammonia really represents a ternary rather than a binary system and should be treated as such. The vapor pressures of NH_3 and H_2SO_4 in equilibrium with ammonium sulfates are much lower than that of sulfuric acid in the presence of water (Scott and Cattell, 1979), so that the critical cluster size is lowered and the rate of nuclei formation is enhanced.

Laboratory and field experiments indicate that the formation of new Aitken particles in outdoor air is very slow in the dark, whereas irradiation by natural or artificial sunlight increases the rate to observable levels (Bricard et al., 1968; Husar and Whitby, 1973; Haaf and Jaenicke, 1980). Ionizing radiation has a similar effect (Vohra et al., 1970). Figure 7-16 shows some data illustrating this behavior. While these observations demonstrate a photochemical origin of homogeneous nucleation, they do not give any information on the chemistry involved. There are other data, however, that show that the oxidation of SO_2 to sulfate is also mediated by sunlight. An example is the diurnal variation of the SO_2 oxidation rate in power-plant plumes discussed in Section 10.3.3. At wavelengths above 300 nm, the photooxidation of SO_2 in air freed from other trace gases is a slow process, because the excited SO_2 molecules formed initially by absorption of solar radiation lose their energy mainly by collisional deactivation instead of undergoing reaction with O_2. This has been shown by laboratory studies (Cox, 1972, 1973; Friend et al., 1973; Sidebottom et al., 1972; Smith and Urone, 1974), as well as by mechanistic considerations (Calvert et al., 1978). The photooxidation rate increases significantly at wavelengths below the SO_2 dissociation limit near 240 nm (Driscoll and Warneck, 1968; Friend et

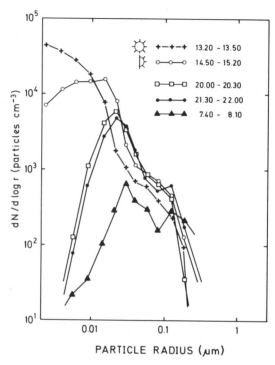

Fig. 7-16. Particulate size distributions at the summit of Mt. Schauinsland (1250 m a.s.l.), May 16-17, 1978. The variation in the Aitken size range is indicative of gas-to-particle conversion during periods of sunlight, and absence of gas-to-particle conversion at night. [Adapted from Haaf and Jaenicke (1980).]

al., 1973), but such short wavelengths do not reach the troposphere. As a consequence one must turn to other reactions, namely, those of SO_2 with transient species that are photochemically generated. Calvert and Stockwell (1984) have provided an updated review of the various possibilities.

Potential SO_2 oxidants are the radicals OH, HO_2, RO_2, and Criegee intermediates. SO_2 oxidation by OH radicals is now well understood to proceed via an addition product:

$$OH + SO_2 \rightarrow HOSO_2$$

$$HOSO_2 + O_2 \rightarrow SO_3 + HO_2$$

$$SO_3 + H_2O \rightarrow H_2SO_4$$

The occurrence of hydrogen abstraction from $HOSO_2$ by oxygen has recently been confirmed by Gleason *et al.* (1987) and Schmidt *et al.* (1985). The reaction of peroxy radicals with SO_2 also appear to proceed via addition products.

$$HO_2 + SO_2 \rightleftharpoons HO_2SO_2 \rightarrow SO_3 + OH$$

$$RO_2 + SO_2 \rightleftharpoons RO_2SO_2 \rightarrow SO_3 + RO$$

In contrast to the reaction with OH, however, the intermediate appears to revert mainly to the original reactants, whereas the forward reaction to produce SO_3 is slow (Calvert and Stockwell, 1984). The direct, thermal reaction of SO_2 with ozone also is negligibly slow. Cox and Penkett (1972) demonstrated that SO_2 is nevertheless converted to H_2SO_4 aerosol at an appreciable rate when SO_2 is added to an ozone–alkene mixture. This suggests that an intermediate of the ozone–alkene reaction is responsible for the oxidation of SO_2 under these conditions. The intermediate might be either the initial adduct between ozone and the alkene or the intermediate formed from it by the split-off of an aldehyde, so that the reactions oxidizing SO_2 are

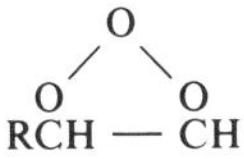

$$\overset{\displaystyle O}{\overset{\displaystyle \diagup \ \diagdown}{\underset{\displaystyle RCH \ - \ CHR}{O \qquad O}}} + SO_2 \rightarrow 2RCHO + SO_3$$

$$R\dot{C}HOO\cdot + SO_2 \rightarrow RCHO + SO_3$$

Niki *et al.* (1977) showed that propylene ozonide formed by the reaction of ozone with *cis*-butene in the presence of formaldehyde (see page 265) was quenched by the addition of SO_2. In addition, the rearrangement of the Criegee intermediate $CH_3\dot{C}HOO\cdot$ to acetic acid was prevented, indicating that one or both of the above reactions with SO_2 did indeed occur. Cox and Penkett (1972) also observed an inhibition of H_2SO_4 formation with increasing concentration of water vapor, which Calvert *et al.* (1978) speculated might be due to a catalytic stabilization of $R\dot{C}HOO\cdot$ to the corresponding acid. In fact, Hatakeyama *et al.* (1981) found that $H\dot{C}HOO\cdot$ produced from the reaction of ozone with ethylene interacted with $H_2^{18}O$ to form labeled formic acid. The atmospheric significance of SO_2 oxidation by products of the alkene–ozone reaction is difficult to assess as long as the precise mechanism and the rate coefficients of the reactions involved are not known. Calvert *et al.* (1978) estimated a relative conversion rate of 0.23% h at 50% relative humidity for a polluted urban atmosphere. This rate seems small compared with that for the reaction of SO_2 with OH, but it will continue during the night time hours so that it might be of importance.

Finally, it should be pointed out that the oxidation of certain hydrocarbons can produce aerosols even in the absence of SO_2. Grosjean (1977) has reviewed organic particulate formation. Smog-chamber experiments have shown that straight-chain alkanes, alkynes, and carbonyl compounds

Table 7-9. *Gas-Phase Reactivity[a] and Aerosol-Formation Efficiency for Selected Hydrocarbons in Smog Chambers, in the Presence and Absence of SO_2, Relative to Cyclohexene (=100)*

Hydrocarbon	Gas-phase reactivity[a] (a)	(b)	Aerosol-forming efficiency Without SO_2	With SO_2	Initial conditions (ppm) HC	NO	NO_2	SO_2	Reference
Alkenes									
Ethene	11	49	2.8	63	10	0	5	2	Prager et al. (1960)
1-Butene	48	83	1.4	81	10	0	5	2	
2-Butene	73	202 (cis)	1.4	86	10	0	5	2	
1-Pentene	40	60	2.8	87	10	0	5	2	
2-Pentene	122	187	0	86	10	0	5	2	
1-Hexene	42	49	1.4	96	10	0	5	2	
3-Heptene	—	134 (trans)	12.6	96	10	0	5	2	
Dienes									
1,3-Butadiene	93	123	33	91	10	0	5	2	
1,5-Hexadiene	—	—	104	116	10	0	5	2	
Cyclic alkenes									
Cyclopentene	—	657	75	—	10	0	5	2	
Cyclohexene	100	100	100	100	10	0	5	2	Groblicki and Nebel (1971)
α-Pinene	75	—	140	137	4	2	0	4	
Aromatics									
Toluene	8.7	37	8.5	8.6	3	0.37	0.37	0.1	Wilson et al. (1973)
Mesitylene	85	146	9	9	3	0.37	0.37	0.1	

[a] (a) k_{OH} from Atkinson et al. (1979); (b) NO–NO_2 conversion efficiency from Glasson and Tuesday (1970).

do not generate aerosols. The trend for alkenes, dienes, and aromatics is indicated in Table 7-9. In spite of their reactivity, most alkenes are not efficient aerosol producers when SO_2 is absent. The ability of alkenes to produce aerosols increases when the carbon number exceeds six. More effective are dienes, cycloalkenes, and terpenes, the latter being represented in Table 7-9 by α-pinene. For these compounds it makes little difference whether SO_2 is present or not. For aromatic compounds there are conflicting reports.

The exceptional ability of cyclic alkenes to form aerosols led Grosjean and Friedlander (1980) to investigate the composition of the particulate products. The results obtained for cyclopentene and cyclohexene are shown in Table 7-10. All the products identified are linear, difunctional compounds

Table 7-10. *Aerosol Products Obtained from the Photooxidation of Cyclopentene and Cyclohexene[a]*

Particulate product	Formula	b.p. (K)[b]	p_s (mbar)
From cyclopentene			
Glutaraldehyde (major)	$CHO(CH_2)_3CHO$	461 (1000)	8.5×10^{-1} [b]
5-Oxopentanoic acid	$CHO(CH_2)_3COOH$	345 (13.3)	
Glutaric acid	$COOH(CH_2)_3COOH$	576 (1000)	1.5×10^{-5} [b]
5-Nitratopentanoic acid	$COOH(CH_2)_3)_3CH_2ONO_2$	473 (26.7)	
4-Hydroxybutanoic acid	$COOH(CH_2)_2CH_2OH$		
4-Oxobutanoic acid	$COOH(CH_2)_2CHO$		
1,4-Butanedial	$CHO(CH_2)_2CHO$	442 (1000)	1.5 [b]
4-Nitratobutanal	$CHO(CH_2)_2CH_2ONO_2$		
From cyclohexene		538 (133)	3×10^{-5} [b]
Adipic acid (major)	$COOH(CH_2)_4COOH$	478 (13.3)	1×10^{-7} [c]
6-Nitratohexanoic acid (major)	$COOH(CH_2)_4CH_2ONO_2$		2×10^{-5} [d]
6-Oxohexanoic acid	$COOH(CH_2)_4CHO$		
6-Hydroxyhexanoic acid	$COOH(CH_2)_4CH_2OH$		
Glutaric acid	$COOH(CH_2)_3COOH$	576 (1000)	1.5×10^{-5} [b]
		473 (26.7)	
5-Nitratopentanoic acid	$COOH(CH_2)_3CH_2ONO_2$		
5-Oxopentanoic acid	$COOH(CH_2)_3CHO$		
5-Hydroxypentanoic acid	$COOH(CH_2)_3CH_2OH$	461 (1000)	8.5×10^{-1} [b]
Glutaraldehyde	$CHO(CH_2)_3CHO$	345 (13.3)	

[a] From Grosjean and Friedlander (1980).

[b] Boiling points, where available, are given in kelvins (at pressures in millibars). They are used to obtain by extrapolation order-of-magnitude estimates for vapor pressures at ambient temperature.

[c] Extrapolation of sublimation pressure; Davies and Thomas (1960).

[d] Estimate by Heisler and Friedlander (1977).

bearing carboxylic acid, carbonyl, hydroxyl, and nitrate ester functional groups. Equilibrium vapor pressures for these compounds are essentially unknown. One can nevertheless estimate from boiling-point temperatures, as far as they are available, that difunctional compounds have vapor pressures far lower than those of the corresponding monofunctional compounds. All the compounds listed in Table 7-10 are highly water-soluble, so that binary homogeneous nucleation with water vapor will take preference. The formation of these compounds can be understood to result from ring opening. For example, the oxidation of cyclohexene by ozone is expected to produce 6-oxohexanoic acid:

Schuetzle *et al.* (1973, 1975) have identified a number of compounds similar to those shown in Table 7-10 in Californian smog aerosols.

Went (1960b, 1964) hypothesized that the blue haze often observed over forested regions is due to aerosols resulting from the photooxidation of terpenes and other organic compounds emitted from vegetation. Schwartz (1974) and Hull (1981) have studied particulate products from the oxidation of α-pinene. Schuetzle and Rasmussen (1978) have similarly studied limonene and terpinolene. Ring opening was observed in all cases. The products from α-pinene are shown in Fig. 7-17. The major products from

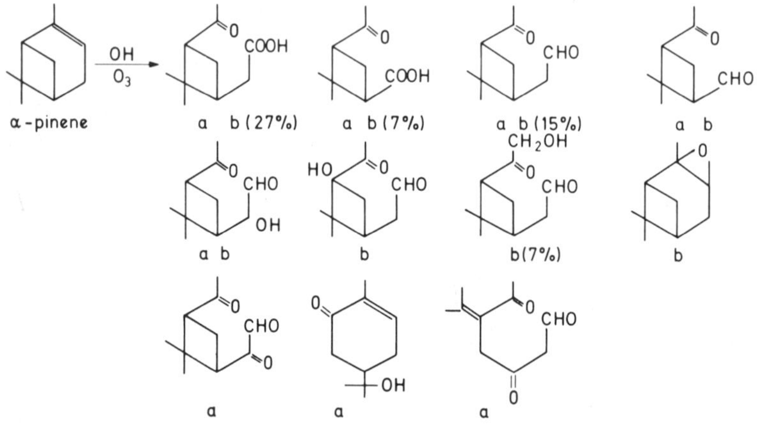

Fig. 7-17. Compounds identified in the aerosol derived from α-pinene (a) by photooxidation in the presence of nitrogen oxides and (b) by ozonolysis; yields in percent are shown in parentheses. [From Schwartz (1974) as reported by Grosjean (1977) and Hull (1981), respectively.]

ozonolysis are *cis*-pinonic acid, which is a solid at 298 K, and pinoneal-dehyde. None of these compounds have as yet been identified in natural aerosols, so that the true nature of the blue haze remains to be determined.

7.4.4 MISCELLANEOUS SOURCES

Biogenic particles are released from plants in the form of seeds, pollen, spores, leaf waxes, resins, etc., ranging in size from perhaps 1 to 250 μm. Delany *et al.* (1967) have used nylon meshes to collect airborne particles greater than 1 μm at Barbados and concluded that most of the material originated from the European and African continents. Besides mineral components, they identified various biogenic objects ranging from microbes to fragments of vascular plants, among them fungus hyphae, marine organisms, and freshwater diatoms, which are also found in Atlantic sediments. Parkin *et al.* (1972) similarly found fragments of humus, dark plant, and fungus debris.

The aerial transport of pollen and microorganisms has received some attention (Gregory, 1973, 1978; R. Campbell, 1977). Bacteria (size < 1 μm) are difficult to discern directly, and their study requires cultural growth techniques. In contrast to fungi spores, they usually occur attached to other aerosol particles because they are mobilized together with dust. Concentrations number several hundred per cubic meter in rural areas and several thousand in the cities. Air is not their natural habitat, so that multiplication does not take place. On the contrary, the atmospheric aerosol appears to have definite germicidal qualities (Rüden and Thofern, 1976; Rüden *et al.*, 1978).

The concentrations and populations of pollen and spores can change rapidly with locality, season, time of the day, and meteorological conditions. On the European continent, concentrations average 12,500 m^3 during the summer months (Gregory and Hirst, 1957; Stix, 1969). The population is dominated by the spores of common moulds, whereas pollen account for less than 1% of the total. Like other aerosol particles, pollen and spores get carried to heights of at least 5 km (Gregory, 1978), and horizontally to distances of many hundreds of kilometers. The spreading of virulent spores such as those of maize and wheat rusts is important to agriculture.

Volcanism constitutes another source of particulate matter in the atmosphere. An eruption of recent times that reached the stratosphere was that of Mt. Agung on Bali in 1963. From radiosonde data, Newell (1970) demonstrated that the dust injection produced, at the 60–80 mbar pressure level, a sudden 6 K temperature increase, which decayed slowly to pre-Agung values during the following years. The effect was observed in a latitude belt 15° north and south of the source.

Cadle and Mroz (1978), Hobbs *et al.* (1977, 1978), and Stith *et al.* (1978) have used aircraft to investigate the effluents from the Alaskan volcano St. Augustine during a period of intensified activity in 1976. The results indicate that in times between eruptions the particles consist of finely divided lava coated with sulfuric acid containing dissolved salts. Sulfur is emitted from volcanoes largely in the form of SO_2, which is subsequently oxidized to H_2SO_4. Paroxysmal events resulting from magmatic movements increase the mass of volcanic ash ejected so much that the acid coating on the particles is greatly reduced. Most of the mass of particles resides in the submicrometer size range. Number densities maximize at radii below 0.1 μm, with the usual fall-off toward larger particle sizes. Only paroxysmal events produce an enhancement in the concentration of giant particles ranging up to 100 μm.

In the recent decade, evidence has been mounting that volcanic emissions lead to an enrichment of so-called volatile elements in the particulate matter ejected, compared with the relative abundances found in bulk material of the earth's crust (Cadle *et al.*, 1973; Mroz and Zoller, 1975; Lepel *et al.*, 1978; Buat-Menard and Arnold, 1978). Figure 7-18 summarizes the results of Buat-Menard and Arnold (1978) for samples from the plume of Mt. Etna, Sicily, and compares them with similar data for aerosols collected

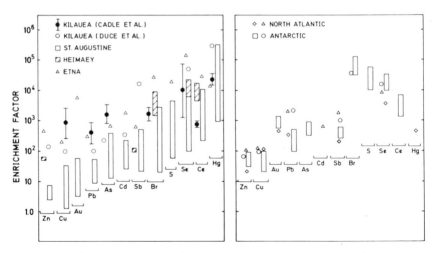

Fig. 7-18. Left: Enrichment factors for volatile elements in aerosols from volcanoes. Data were compiled from Cadle *et al.* (1973) and Duce *et al.* (1976) for the volcano Kilauea; Mroz and Zoller (1975) for Heimaey; Lepel *et al.* (1978) for St. Augustine; Buat-Menard and Arnold (1978) for Etna. Right: Enrichment factors for volatile elements in background aerosols. Data are from Duce *et al.* (1976), triangles, Buat-Menard and Chesselet (1979), diamonds, Zoller *et al.* (1974), circles, and Maenheut *et al.* (1979), vertical bars.

at remote locations. The figure shows enrichment factors (EF) defined by the concentration ratios $EF = [X/Al]_{aerosol}/[X/Al]_{crust}$, aluminum serving as the reference element. Nonvolatile elements have enrichment factors in the vicinity of unity. The values for the elements considered in Fig. 7-18 are all much greater than unity. Sulfur, chlorine, and bromine are emitted mainly in gaseous form as SO_2, HCl, and HBr and are subsequently attached to aerosol particles, so that their enrichment is qualitatively understandable. Mercury is a special case in that its vapor pressure is already appreciable. The enrichment of the other elements cannot be so explained, however, and seems to result from the high-temperature environment during the dispersal. The right-hand side of Fig. 7-18 shows that remote aerosols feature a similar enrichment. This similarity led Lepel et al. (1978), Mroz and Zoller (1975), and Maenhaut et al. (1979) to suggest that emissions from volcanoes might be responsible for the enrichment of trace metals in the natural background aerosol. Other processes that have been considered to explain the observed enrichments were biological volatilization processes, for example, the methylation of trace metals by microorganisms (Wood, 1974; Ridley et al., 1977), and anthropogenic high-temperature combustion processes (Linton et al., 1976; Bolton et al., 1975). Boultron (1980) has been able to show, however, from samples taken from the Antarctic ice shield, that in the past 100 yr concentrations and enrichment factors for Pb, Cd, Cu, Zn, and Ag have remained essentially the same. This confirms that global pollution by these elements is actually negligible in the remote southern hemisphere. In the northern hemisphere, by contrast, data from the Greenland ice shield obtained by Murozumi et al. (1969) indicate quite clearly an increase in the concentration of lead during the years 1750–1950. The dominant sources were shown to be lead smelters. Gasoline lead additives contributed comparatively little. Today, lead concentrations in Greenland snows are well over 500 times higher than natural levels.

Forest fires, like volcanoes, tend to emit particles into a buoyant plume reaching far into the troposphere (not, however, into the stratosphere). In the United States alone, according to Brown and Davis (1973), some 10^6 hectares of forest area, or about 0.4% of the total, are destroyed annually by wildfires. These are still fairly isolated events compared with the more regular or recurrent agricultural burning practices in the tropics. The combustion of wood and other biomaterial proceeds in three stages. The first involves mainly the pyrolysis of the fuel and its partial volatilization. In the second stage the organic vapors produced undergo combustion in the flame zone. Finally, the char produced in the first stage is subjected to a slower oxidation, provided the temperature stays high enough. This last stage is referred to as smoldering combustion. Particulate formation occurs during all three stages by the mechanical release of charcoal particles and by the condensation of organic vapors. The char residue left after the fire dies out is subsequently exposed to wind erosion, which brings additional carbonaceous material into the atmosphere.

The analysis of particulate matter from open fires by Gerstle and Kemnitz (1967) and McMahon and Ryan (1976) indicates that about 50% by weight is benzene-soluble organic compounds, 40% is elemental carbon, and 10% is minerals. The mass of smoke particles from open burnings is concentrated in the submicrometer size range. Radke *et al.* (1978) and Stith *et al.* (1981) have provided size distributions of aerosol particles arising from forest fires. Number density and volume were found to peak near 0.1 and 0.3 μm, respectively, indicating a dominance of the accumulation mode. These results contrast with earlier speculations of Hidy and Brock (1971), who had suggested that combustion would produce preferentially Aitken particles. Eagan *et al.* (1974) have noted that forest fires are prolific sources of cloud condensation nuclei. Combustion aerosols are not expected to be particularly water-soluble, but as Radke *et al.* (1978) have pointed out, they enhance the number of large particles normally present in the atmosphere. Since large particles are quite effective as cloud condensation nuclei, it appears that it is the favorable size distribution of combustion aerosols that causes them to be efficient as cloud condensation nuclei.

7.4.5 Global Aerosol Production Rates

Table 7-11 summarizes global emission and production rates for particulate matter in the troposphere. The table is based on a review by Bach (1976) augmented by a number of additional data. The emission rates refer to all particles that are not immediately returned to the earth surface by gravitational settling. We shall briefly indicate the methods used in deriving the individual estimates.

The estimate for sea salt goes back to a detailed study of Eriksson (1959) of the geochemical cycles of chloride and sulfur. He calculated the rate of dry fallout of sea-salt particles from a vertical eddy diffusion model and then existing measurements of sea-salt concentrations over the ocean. This led to a global rate for dry deposition of 540 Tg/yr. Eriksson then argued that wet precipitation would remove a similar amount annually. It is now known, however, that wet precipitation is more effective than dry deposition in removing aerosol particles from the atmosphere, so that Eriksson's value must be an underestimate. The discussion in Section 10.3.5 suggests a flux rate for sea salt of about 5,000 Tg/yr.

Soil-derived particle fluxes are too much dependent on local conditions and the prevailing wind force to provide reliable estimates. Peterson and Junge (1971) adopted an estimate by Wadleigh (1968) for the emission of wind-blown dust in the United States that included both natural and agricultural sources. Other estimates made use of the measured dust accumulation in glaciers (Windom, 1969) and deep-sea sediments (Goldberg, 1971). The former gives an underestimate, because most glaciers are located 2,000–3,000 m above sea level, where the concentration of aerosol particles is reduced compared with that in the lower atmosphere. The latter must be considered an overestimate, because sediment accumulation is assumed to result exclusively from dustfall. Schütz (1980) has made a detailed study of the dust plume carried from the Saharan desert westward across the Atlantic Ocean. His estimate for the annual mass transport of particles in the plume is 260 Tg/yr. Of this, 80% is deposited into the Atlantic Ocean, and the remaining 50 Tg/yr reaches the Caribbean sea. An additional transport exists from the Sahara toward the Gulf of Guinea, so that the above estimate is again a lower limit. A similar transport of dust occurs from the Desert Gobi via the Pacific Ocean into the Arctic Basin and further toward Spitsbergen (Rahn et al., 1977; Griffin et al., 1968). A global source strength of 500 Tg/yr thus appears reasonable.

The value derived by Peterson and Junge (1971) for the rate of particulate emissions from volcanoes is based on the long-term burden of particulate matter in the stratosphere combined with an assumed stratospheric residence time of 14 months. This gives a lower limit of 3.3 Tg/yr. If 10% of volcanic particulates, on average, reaches the stratosphere, the total emission rate would be 33 Tg/yr. Goldberg (1971) took instead the rate of accumulation of montmorillonite in deep-sea sediments as an indicator for average volcanic activity. His estimate of 150 Tg/yr must be an upper limit. The estimate of 10 Tg/yr adopted by Peterson and Junge (1971) for meteorite debris imparted to the stratosphere is due to Rosen (1969).

All estimates for particle production due to forest fires and biomass burnings are based on experimental emission factors combined with statistical data for the global consumption rates for the materials involved. The most detailed study has been made by Seiler and Crutzen (1980). They deduced a particle production rate from forest fires alone of 100 Tg/yr. Including agricultural biomass burnings raises the rate to about 200 Tg/yr, a value approaching that for mineral dust emissions. This is much higher than all earlier estimates.

In estimating the rate of gas-to-particle conversion involving SO_2 from anthropogenic sources, Peterson and Junge (1971) assumed that 66% is converted to sulfate and the rest is removed by dry deposition. In addition, it was assumed that sulfate is completely neutralized to ammonium sulfate. An emission rate of 160 Tg SO_2/yr from the combustion of fossil fuels then gives a production rate for particulate sulfate of 220 Tg/yr. In 1971 the rate of sulfur compound emissions from natural sources was less well known

Table 7-11. Estimates for Global Production Rates (Tg/yr) of Particulate Matter from Natural and Anthropogenic Sources

Source type	Peterson and Junge (1971)		Hidy and Brock (1971)	SMIC (1971)	Others
	All sizes	r < 2.5 μm			
Direct emissions		Natural sources			
Sea salt	1000	500	1095	300	1300[i], 5200[k]
Mineral sources	500	250	7–365	100–500	500[a], 260[b], 8000[j]
Volcanoes	25	25	4	25–150	
Forest fires	35	5	146	3–150[c]	72–117[d]
Meteorite debris	10	—	0.02–0.2	—	
Biological material	—	—	—	—	80[e], 26.4[f]
Subtotal	1540	780	1610	428–1100	
Gas-to-particle conversion					
Sulfate	244	220	36.8–365	130–200	
Nitrate	75	60	600–620	140–200	
Hydrocarbons	75	75	182–1095	75–200	3000[j]
Subtotal	394	355	2080	345–1100	
Total natural	1964	1135	3690	773–2200	

Anthropogenic sources

Direct emissions						
Transportation	2.2	1.8				
Stationary sources	43.4	9.6				
Industrial processes	56.4	12.4				
Solid-waste disposal	2.4	0.4				
Miscellaneous	28.8	5.4				
Agricultural burnings	—	—				
Subtotal	133.2	29.6	36.8–110	10–90	22[g], 62[h], 130–289[d,i]	98[f]
Gas-to-particle conversion						
Sulfate	220	200	109.5	130–200		
Nitrate	40	35	23	30–35		
Hydrocarbons	15	15	27	15–90		
Subtotal	275	250	159.5	175–325		
Total anthropogenic	408	280	269	185–415		
Sum total	2372	1415	3959	958–2615		

[a] Goldberg (1971).
[b] Schütz (1980), Sahara only.
[c] Includes slash burning debris.
[d] Seiler and Crutzen (1980).
[e] Jaenicke (1978d).
[f] Duce (1978).
[g] SCEP (1970).
[h] Bach (1976).
[i] Includes slash burning but not forest fires.
[j] Petrenchuk (1980), includes man-made emissions.
[k] From Section 10.3.5.

329

than it is today. Peterson and Junge adopted a tentative estimate of 98 Tg S/yr from Robinson and Robbins (1970) to derive a production rate for sulfate particles from natural sources of 244 Tg/yr. Both estimates must be revised. The discussion in Section 10.4 will show that the fraction of SO_2 converted to SO_4^{2-} over the continents is about 47%, and that of dimethyl sulfide in the marine atmosphere 75%. The rate of anthropogenic SO_2 emission has risen to 200 Tg SO_2/yr, whereas the natural emissions amount to about 50 Tg S/yr, of which 36 Tg S/yr is marine dimethyl sulfide. The corresponding sulfate production rates are 217–253 Tg/yr from anthropogenic sources and 28–38 Tg/yr from natural sources.

The fraction of NO_2 that is converted to aerosol nitrate is still extremely uncertain, because a large portion of HNO_3 remains in the gas phase and undergoes dry deposition. Peterson and Junge (1971) thus made use of analytical data collected by Ludwig et al. (1970) at 217 urban and nonurban stations in the United States. The ratio of sulfate to nitrate was 5.5, on average. This value was then combined with the rate of sulfate production to obtain a coarse estimate for the formation of nitrate. Even more uncertain is the rate of gas-to-particle conversion involving organic compounds. Went (1960a, 1966) estimated an emission rate of natural hydrocarbons from vegetation of 154 Tg/yr, mainly terpenes and hemiterpenes. Peterson and Junge (1971) assumed a gas-to-particle conversion efficiency of 50% to derive an aerosol production rate of 75 Tg/yr. More recently, Zimmerman et al. (1978) derived a rate for the emission of natural hydrocarbons of 940 Tg/yr, indicating that the estimate of Peterson and Junge may be a lower limit. Aerosol production from anthropogenic hydrocarbons carries similar uncertainties.

The total rate of aerosol production is of the order of 2000 Tg/yr. Some comparisons are of interest. Anthropogenic sources, both direct and indirect, contribute about 15% to the global aerosol production rate. The percentage of direct emissions from anthropogenic sources is only about 20% of the total contribution, so that most of it derives from gas-to-particle conversion. With regard to natural aerosol formation, the percentage contribution of direct emissions is 60%, so that direct and secondary aerosol production are approximately equivalent.

7.5   The Chemical Constitution of the Aerosol

The chemical composition of particulate matter in the atmosphere is complex and not fully known. For many purposes, therefore, it is useful to distinguish three broad categories: (1) water-soluble inorganic salts (elec-trolytes); (2) water-insoluble minerals of crustal origin; and (3) organic compounds, both water-soluble and insoluble. In order to determine fractions of material present within each group, Winkler (1974) separated bulk samples of European aerosols into soluble and insoluble fractions, using several organic solvents in addition to water. His results are summarized in Table 7-12, which lists averages for urban and rural background aerosols. The large water-soluble fraction comprises about equal parts of organic and inorganic material. The contribution from water-insoluble organics is small, and the fraction of insoluble minerals is about 30%. The composition of individual aerosol samples varies widely. For example, the frequency distribution of the water-soluble fraction ranges from 30 to 80% of total mass. Winkler (1974) showed further that heating the aerosol samples to 150°C caused a mass decrease, which he attributed to the loss of volatile organic compounds in addition to some inorganic salts, such as NH_4HSO_4. Similar effects were observed earlier by Twomey (1971) for continental cloud nuclei and by Dinger et al. (1970) for the maritime aerosol.

Figure 7-19 shows for the rural aerosol how the principal chemical component groups are distributed with particle size. Although only a coarse size classification was accomplished, it indicates that water-soluble matter is present in all size ranges. Giant particles contain roughly 50% each of water-soluble compounds and insoluble minerals. The fraction of water-soluble material increases with decreasing particle size at the expense of the mineral component. If the trend is extrapolated into the Aitken size range, one finds that more than 90% of all particles with radii less than 0.1 μm are water-soluble.

Table 7-12. *Group Classification of European Continental Aerosols: (A) Mainz and (B) Deuselbach, Germany* [a]

Component	Mass fraction (%)	
	(A)	(B)
Insoluble mineral component	35	25
Water-soluble component	58	68
Material soluble in organic solvents	40	42
Water-soluble organic fraction [b]	28	25
Water-insoluble organic fraction	5	6
Water-soluble inorganic salts [b]	30	43

[a] Mass fractions (in percent) of the mineral component, the water-soluble components, and the total organic material soluble in acetone, cyclohexane, ether and/or methanol. From Winkler (1974).

[b] Taking into account that the methanol-soluble fraction dissolves also a number of inorganic salts.

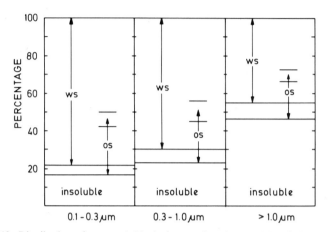

Fig. 7-19. Distribution of water-soluble (ws), organic solvent-soluble (os), and insoluble mass fractions associated with the rural continental aerosol at Deuselbach, West Germany, according to measurements of Winkler (1974). The organic fraction comprises material soluble in cyclohexane, ether, acetone, and part of the methanol-soluble fraction. The uncertainty range assumes that 0–40% of the methanol-soluble fraction contains organic compounds, and the remainder is due to inorganic salts. The water-soluble fraction of organics averages about 0.66 for all size ranges combined.

7.5.1 THE INORGANIC FRACTION, SOLUBLE AND INSOLUBLE

Wet-chemical analyses of aqueous extracts of aerosol samples have established the presence of anions such as sulfate, nitrate, and the halides, and of cations such as ammonium and the ions of the alkali and alkaline earth elements. Table 7-13 shows selected data to illustrate the abundances of important inorganic components in the urban, continental, arctic, and marine aerosols. Included for comparison are the concentrations of silicon, aluminum, and iron, which are the major elements of crustal origin. They occur in oxidized form, such as in aluminosilicates, which are practically insoluble. Taken together, the elements listed in Table 7-13 account for 90% of all inorganic constituents of the atmospheric aerosol.

A reasonably complete analysis of the inorganic chemical composition of the aerosol requires much effort and involves, in addition to wet chemical methods, instrumental techniques such as neutron activation analysis, atomic absorption spectroscopy, or proton-induced X-ray emission (PIXE). These latter techniques yield the elemental composition. They furnish no direct information on the chemical compounds involved, although auxiliary data from mineralogy, chemical equilibria, etc. usually leave little doubt about the chemical form in which the elements occur. Thus, sulfur is present predominantly as sulfate, and chlorine and bromine as Cl^- and Br^-, respectively, whereas sodium potassium, magnesium, and calcium show up as

Na^+, K^+, Mg^{2+}, and Ca^{2+}. Reiter et al. (1976) reported a fraction of calcium as insoluble, and Table 7-13 makes allowance for it. Stelson and Seinfeld (1981) concluded from considerations of equilibria and leaching experiments that CaO is more easily converted to its hydroxide than is MgO, so that it should occur predominantly as Ca^{2+}. Heintzenberg et al. (1981) preferred the PIXE technique in determining calcium. In both cases it was assumed that calcium is completely water-soluble.

The study of Reiter et al. (1976) at the Wank Mountain Observatory distinguished days with strong and weak vertical exchange by means of Aitken nuclei counts and concentrations of radon decay products. The main difference between both meteorological situations is a shift in the total aerosol concentration. Variations in chemical composition are minor. This indicates that at the height of the Wank Mountain [1780 m above sea level (a.s.l.)] the continental aerosol is already well homogenized. Gillette and Blifford (1971) and Delany et al. (1973) have reached similar conclusions from aircraft samples over the North American continent.

In the following discussion we shall use Table 7-13 as a reference in delineating the origins of the various inorganic components of continental and marine aerosols. In this discussion one must keep in mind that the aerosol represents a mixture of substances from several sources, and that any specific component may have more than one origin. The major sources in the natural atmosphere are sea salt, continental soils, and gas-to-particle conversion. To these must be added the various anthropogenic contributions. Comparison of aerosol and source compositions provides important clues with regard to the individual contributions. Below, we discuss first the anions and ammonium, then soluble and insoluble elements.

Sulfate is the most conspicuous constituent of all aerosols considered in Table 7-13, except of the marine aerosol, which is dominated by sodium chloride. SO_4^{2-} mass fractions range from 22–45% for continental aerosols to 75% in the Arctic region. Similarly high fractions of sulfate occur in the Antarctic (Maenhaut and Zoller, 1977). The origin of sulfate over the continents is primarily gas-to-particle conversion of SO_2. The sulfur content of the earth's crust is too low for soils to provide a significant source of sulfate, except perhaps locally in the vicinity of gypsum beds. Figure 7-20 shows mass size distributions for sulfate and several other water-soluble constituents of the rural continental aerosol. Sulfate is concentrated in the submicrometer size range. This important fact was first recognized by Junge (1953, 1954), who had separated SO_4^{2-} into a coarse and a fine fraction. The distribution peaks at radii near 0.3 μm, that is, in that size range where the accumulation of material by condensation and coagulation processes is most efficient (cf. Section 7.3.1). The observed size distribution thus provides further evidence for the idea that gas-to-particle conversion of SO_2

Table 7-13. *Mass Concentrations* ($\mu g/m^3$) *of Water-Soluble Components in Aerosols at Several Locations*[a]

Element or group	Urban aerosol		Continental aerosol, Wank, central Europe			Arctic aerosol, Ny-Ålesund		Marine aerosol, central Atlantic
	Tees-side	West Covina	Average	High	Low	With sea influence	Without sea influence	
SO_4^{2-}	13.80	16.47	3.15	4.90	0.546	1.95	2.32	2.577
NO_3^-	3.00	9.70	0.92	1.335	0.412	0.022	0.055	0.050
Cl^-	3.18	0.73	0.112	0.137	0.076	0.174	0.013	4.625
Br^-	0.07	0.53	—	—	—	—	—	0.015
NH_4^+	4.84	6.93	1.295	1.960	0.351	0.152	0.226	0.162
Na^+	1.18	3.10	0.053	0.084	0.024	0.209	0.042	2.910
K^+	0.44	0.90	0.062	0.121	0.025	0.050	0.023	0.108
Ca^{2+}	1.56	1.93	0.155	0.303	0.042	0.073 (av.)		0.168
Mg^{2+}	0.60	1.37	—	—	—	0.071	0.032	0.402
Al_2O_3	3.63	6.43	0.223	0.389	0.077	—	—	—
SiO_2	5.91	21.1	0.663	1.250	0.167	0.235 (av.)		—
Fe_2O_3	5.32	3.83	0.145	0.365	0.035	0.091 (av.)		0.065
CaO	b	b	0.104	0.182	0.057	b	b	
Sum	43.53	75.70	6.90	11.03	1.81	3.07 (av.)		11.16

Selected mass fractions and ratios

SO_4^{2-} (%)	29.5	21.75	45.65	44.42	30.16	63.5	75.5	22.56
NO_3^- (%)	6.3	12.80	13.3	12.1	22.7	0.72	1.8	0.44
NH_4^+/SO_4^{2-} [c]	1.87	2.24	2.19	2.13	3.43	0.43[d]	0.53	0.47[d]
NH_{4obs}^+/NH_{4equ}^+	0.8	0.77	0.89	0.88	1.0	0.21[d]	0.25	0.25[d]
Si/Al	1.44	2.90	2.63	2.84	1.94	—	—	—
Fe/Al	1.94	0.79	0.86	1.24	0.60	—	—	—
Na/Al	0.61	0.91	0.45	0.41	0.59	—	—	—
K/Al	0.23	0.26	0.52	0.58	0.61	—	—	—
Ca/Al	0.81	0.57	1.94	2.10	2.03	—	—	—

[a] Locations: (1) Tees-side, a conglomerate of towns in the industrial area of Middlesbrough in the northeast coastal area of England. Data taken during June–October 1967, averaged from five sampling stations (Eggleton, 1969). (2) West Covina, greater Los Angeles area, California. Data from three individual days, July and August 1973, averages from three sites (Stelson and Seinfeld, 1981). (3) Wank, a mountain station near Garmisch, Federal Republic of Germany, elevation 1780 m, observation period October 1969–November 1973. Shown are total averages and those for days of strong and minimum vertical exchange (high and low values), respectively (Reiter et al., 1975a,b, 1976, 1978). (4) Ny-Alesund, Spitsbergen, data taken during April and May 1979 (Heintzenberg et al., 1981). Averages for periods with and without advection from the sea are shown. (5) Composite of data obtained in the central Atlantic Ocean on board ships. Na, K, Ca, Mg, SO_4^{2-}, Cl from Buat-Menard et al. (1974), Na, K, Ca, Mg, Fe_2O_3 from Hoffman et al. (1974), Na, SO_4^{2-}, NH_4^+, NO_3^- from Gravenhorst (1975a,b), wind speed range 5–10 m/s. Br is included and taken from airplane data of Duce et al. (1965) made in the area of Hawaii and extrapolated to the ocean surface.

[b] All calcium assumed to be soluble.

[c] Molar ratio.

[d] Corrected for sea salt.

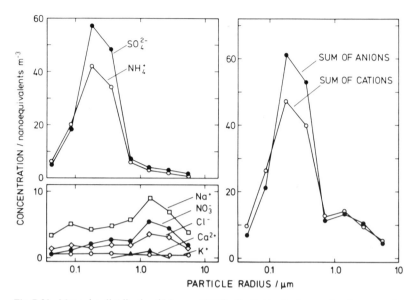

Fig. 7-20. Mass-size distribution ($\Delta \log r = 0.303$) of individual anions and cations associated with the rural continental aerosol at Deuselbach, West Germany (nequiv/m³, = nmol/m³ times the ionic charge number). Left: Individual contributions. Right: Balance between cations and anions. A cation deficit must be balanced by protons leading to a corresponding acidity. [Data from Mehlmann (1986).]

is the principal source of sulfate over the continents. Similar size distributions have been found by Roessler *et al.* (1965), Lundgren (1970), Patterson and Wagman (1976), and Kadowaki (1976) for sulfate in the urban aerosol.

The data in Table 7-13 and Fig. 7-20 show further that ammonium is the principal cation associated with sulfate in the continental aerosol. The observed size distribution leaves no doubt that NH_4^+ arises from a gas-to-particle conversion process, namely, the neutralization of sulfuric acid by gaseous ammonia. The degree of neutralization reached depends on the supply of NH_3 relative to the rate of formation of H_2SO_4. Molar NH_4^+/SO_4^{2-} ratios are expected to range from one to two, corresponding to a composition intermediate between NH_4HSO_4 and $(NH_4)_2SO_4$. Table 7-14 compiles a number of measurements that confirm this expectation. The data of Table 7-13 lead to ratios that exceed a value of two in some cases. This excess may be assigned to the presence of nitric acid, which also binds some ammonia. If nitrate is included in the ion balance, NH_4^+ is found to compensate about 80% of the acids in the continental aerosol. The right-hand side of Fig. 7-20 shows the charge balance between anions and cations. A cation deficit is evident in the size region where SO_4^{2-} and NH_4^+ have their maxima. The major cation missing in the analysis is H^+. Brosset (1978),

Table 7-14. *Molar* NH_4^+/SO_4^{2-} *Ratios Observed by a Number of Investigators for the Continental Aerosol*

Authors	Location	NH_4^+/SO_4^{2-}	Comment
Kadowaki (1976)	Nagoya, Japan	1.3-1.5	Winter
		1	Summer
Moyers *et al.* (1977)	Tucson, Arizona	1.5	Average
Penkett *et al.* (1977)	Gezira, Sudan	1.2	Average
Brosset (1978)	Gothenburg, Sweden	0.9-2.3	(1.5 Average)
Tanner *et al.* (1979)	New York City	1.9	February
		1.2-1.5	August
Pierson *et al.* (1980)	Allegheny Mountains	0.5-2.3	Summer
		0.87	Average

Tanner *et al.* (1979), and Pierson *et al.* (1980) have used the Gran (1952) titration procedure on aqueous extracts of continental aerosols to determine concentrations of protons in addition to those of other cations. They found that including H^+ ions greatly improves the total cation balance. This demonstrates that it is mainly uncompensated acid that accounts for the ammonium deficit, and not so much other cations.

Sulfate in the arctic aerosol also is assumed to derive from the oxidation of SO_2. How much of it is due to local sources, advection from middle latitudes, or subsidence from the stratosphere is not yet clear. Rahn and McCaffrey (1980) have studied the annual variation of sulfate at Barrow, Alaska, and found a pronounced winter maximum. This contrasts with the behavior of the continental aerosol, which shows a maximum in summer (cf. Section 10.3.4). From the covariation of sulfate with vanadium, which is partly a product of pollution, from the concentrations of radon and its decay product ^{210}Pb, and from considerations of air mass trajectories, Rahn and McCaffrey (1980) concluded that the winter Arctic aerosol is associated with aged air masses originating in the middle latitudes, mainly in Europe. The analysis suggests that sulfate is formed from SO_2 during transport into the Arctic. Note that the NH_4^+/SO_4^{2-} ratio according to Table 7-13 is much lower in the Arctic than over the continental mainlands. Evidently, there is even less ammonia available in these remoter regions to balance sulfuric acid. For the aerosol at Spitsbergen, Heintzenberg *et al.* (1981) reported $(H^+ + NH_4^+)/SO_4^{2-}$ ratios near two when marine influences were absent, which indicates that also in remote regions ammonia is the main agent neutralizing sulfuric acid. The Antarctic is not anthropogenically polluted, in contrast to the Arctic. Indeed, Maenhaut and Zoller (1977) have shown that vanadium is not enriched with regard to its crustal abundance. And yet, the concentration of sulfate in the Antarctic is as high as in the Arctic

338

summer aerosol. In the Antarctic, therefore, sulfate must have a natural origin.

The ionic composition of the marine aerosol is quantitatively different from that of the continental aerosol. Figure 7-21 shows the mass size distribution of various components of the marine aerosol. The data were taken in the same manner as those of Fig. 7-20. Sulfate in the marine environment is seen to exhibit a bimodal size distribution. The coarse-particle fraction is associated with sea spray, whereas the submicrometer-size fraction may again be assigned to result from gas-to-particle conversion. Table 7-15 summarizes the major and some minor constituents of sea water. The main components are sodium, chloride, magnesium, sulfate, potassium, and calcium. With the exception of sulfate, these components are all concentrated in the coarse-particle mode of the marine aerosol, as expected. The fraction of sulfate associated with sea salt may be estimated by reference to sodium. In sea water the mass ratio of sulfate to sodium is 0.252. The molar equivalence ratio is $2[SO_4^{2-}]/[Na^+] = 0.12$. The concentration of sulfate in the coarse-particle mode of the data in Fig. 7-21 is somewhat higher, but it covaries with magnesium and potassium as expected if sea salt is the major source. In the submicrometer size range, sulfate is seen to covary with ammonium. The charge balance between cations and anions is

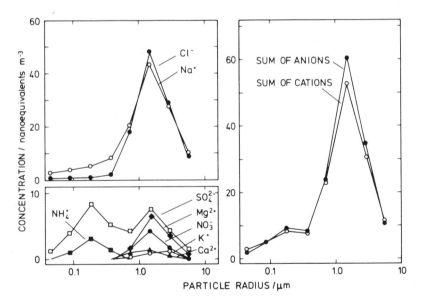

Fig. 7-21. Mass-size distribution ($\Delta \log r = 0.303$) of individual anions and cations associated with the marine aerosol collected over the Atlantic Ocean (nequiv/m³). Left: Individual contributions. Right: Balance of cations versus anions. [Data from Mehlmann (1986).]

Table 7-15. *Major and Some Minor Constituents of Sea Water and Marine Aerosols[a]*

| Element or ion | Sea water | | Marine aerosol | |
	Abundance (mg/kg)	(X)/(Na)	(X)/(Na)	EF
Cl^-	19,344	1.795	1.590	0.88
SO_4^{2-}	2,712	0.252	0.885	3.51
Br^-	67	0.0062	0.0051	0.82
HCO_3^-	142	0.0132	—	—
Na^+	10,773	1	1	1
Mg^{2+}	1,294	0.119	0.138	1.16
Ca^{2+}	412	0.0384	0.058	1.50
K^+	399	0.0370	0.037	1.00
Sr^{2+}	7.9	0.00073	0.0014^b	1.93^b
Si	3	2.8×10^{-4}	—	—
Al	10^{-2}	9.3×10^{-7}	—	—
Fe	10^{-2}	9.3×10^{-7}	1.5×10^{-2}	1.7×10^4
F^-	1.3	1.2×10^{-4}	—	—
I^-	5.4×10^{-2}	5.0×10^{-6}	—	—

[a] Seawater abundances (from Wilson, 1975), (X)/(Na) ratios, and enrichment factors (EF) are shown $[EF = (X)/(Na)_{aerosol}/(X)/(Na)_{seawater}]$. Aerosol data are from Table 7-13.

[b] From Hoffman *et al.* (1974), not included in Table 7-13.

good, even though NH_4^+ contributes less than one-half to the cation balance in the submicrometer size range. The remainder is made up by sodium. The contribution of protons evidently is less than for the continental aerosol.

The major gaseous precursor of sulfate in the unperturbed marine environment is dimethyl sulfide, a biogenic compound emanating from the sea's surface. Oxidation pathways for dimethyl sulfide are discussed in Section 10.2.2. A large fraction of it is converted to sulfuric acid, which partly settles onto the sea-salt aerosol. Figure 7-21 shows the occurrence of a deficit of chloride in the submicrometer size range relative to sodium. The preceding discussion has further shown that there is no significant excess of protons, despite a shortage of NH_4^+. These observations suggest that sulfuric acid liberates chloride and protons in the form of gaseous HCl. The deficit of chloride in submicrometer-sized particles has been observed previously by Martens *et al.* (1973) in Puerto Rico. In their case it was necessary to correct for the land influence, that is, the contribution of sodium from crustal sources. The average chlorine deficit was 12%. Other workers who collected the marine aerosol on filters have found total chlorine deficits in the range 3–12% (Wilkness and Bressan, 1972; Chesselet *et al.*, 1972a; Buat-Menard *et al.*, 1974; Sadasivan, 1978; Kritz and Rancher, 1980). Most of these data refer to the central Atlantic Ocean. The determination of gaseous HCl in air is difficult, and satisfactory data do not exist. By means of filters impregnated with LiOH, which absorbs HCl but lets organic chlorine compounds pass, Rahn *et al.* (1976) demonstrated the presence of inorganic chlorine in both marine and continental air. Kritz and Rancher (1980) used the same technique to confirm the presence of inorganic gaseous chlorine in the Gulf of Guinea. The observed concentration was 1 $\mu g/m^3$, on average, or about one-fifth of that present in particulate form. If it consisted entirely of HCl, the corresponding volume mixing ratio would have been 0.6 ppbv. Kritz and Rancher (1980) assumed all of it to be HCl derived from sea salt and, adopting the steady-state hypothesis, estimated flux rates and residence times (3 days for sea salt, $2\frac{1}{2}$ days for HCl). Although hydrochloric acid has not been identified directly, it is probably the most abundant inorganic chlorine species in the gas phase. Other compounds that have been considered include Cl_2 (Cauer, 1951) and NOCl (Junge, 1963). These are photochemically unstable at wavelengths above 300 nm. Their photodecomposition liberates Cl atoms, which sooner or later react with methane to form HCl as the stable end product.

The possibility that the depletion of chloride in the marine aerosol is due to fractionation during the formation of sea-salt particles by bursting bubbles can be discounted. Laboratory studies of Chesselet *et al.* (1972b) and Wilkness and Bressan (1972) showed no deviation of the Cl^-/Na^+ mass ratio from seawater in the bubble-produced sea-salt particles. It may be mentioned in passing that bromide in marine aerosols shows a deficit similar to chloride, whereas iodide is present in excess. The latter observation is attributed to both chemical enrichment at the sea's surface and scavenging of iodine from the gas phase. A portion of iodine is released from the ocean as methyl iodide, which in the atmosphere is subject to photodecomposition and thereby provides a source of scavengable iodine. The process has been reviewed by Duce and Hoffman (1976). In continental aerosols, chloride and bromide are partly remnants of sea salt, but there exists also a contribution from the gas phase.

The last anion to be discussed is nitrate. In the marine aerosol, nitrate is associated mainly with coarse particles. This fact is apparent in Fig. 7-21, but it has also been observed by Savoie and Prospero (1982). Sea water contains insignificant amounts of nitrate, so that the particulate nitrate must derive from the gas phase, that is, from gaseous nitric acid. As a gas-to-particle conversion process, one would expect the condensation of nitric acid to take place in the accumulation mode. The volatility of HNO_3 is much greater than that of H_2SO_4, however. It appears that the condensation of sulfuric acid prevents the simultaneous condensation of nitric acid in the same size range. In this connection one should remember that with relative humidities in excess of 75%, sea-salt particles are present as aqueous solution droplets. The association of nitrate with the bulk of sea salt thus may result from the dissolution of nitric acid in the sea salt–water droplets. This possibility remains to be explored further, however.

Nitrate in the continental aerosol is distributed over the entire 0.1–10 μm size range, although the coarse-particle mode is again quite prominent, as Fig. 7-20 shows. Kadowaki (1976, 1977) studied the seasonal variation of aerosol nitrate in the urban area of Nagoya, Japan. A bimodal distribution with a clear preference of giant particles was evident during the summer months. An enhancement of nitrate in the accumulation range occurred during winter and spring. An unambiguous explanation for this behavior cannot be offered. In coarse particles nitrate is mainly balanced by sodium, and in submicrometer particles by ammonium. The small nitrate content in the accumulation range during the summer may result from an enhanced photochemical conversion of SO_2 to H_2SO_4 combined with an inadequate supply of NH_3 to neutralize both H_2SO_4 and HNO_3. The simultaneous observation of particulate nitrate and gaseous nitric acid (see Table 9-13) shows that HNO_3 is not fully converted to particulate matter. Instead, an equilibrium between both components appears to be established in the atmosphere. This relationship requires more study.

We turn next to consider the nonvolatile alkali and alkaline earth elements and the insoluble components of mineral origin. Their major natural sources are the Earth's crust and the ocean, respectively. We expect the chemical composition of the aerosol to reflect the relative contributions of elements from both reservoirs, provided other contributions from anthropogenic or volcanic sources are negligible. In Section 7.4.4 it has been noted, however, that his premise does not hold for all constituents of the aerosol. Some trace components are considerably enriched compared with their crustal abundances. It is appropriate, therefore, to inquire whether the observations confirm our expectations at least for the major elements listed in Table 7-13, or whether deviations occur also in these cases. As Rahn (1975a,b) has shown, the problem may be approached in two ways, either by calculating enrichment factors defined by

$$EF(X) = (X)/(Ref)_{aerosol}/(X)/(Ref)_{source} \qquad (7-28)$$

where X is the element under consideration and Ref an appropriate reference element, or by constructing so-called scatter diagrams in which the observational data for the element X are plotted versus those for the reference element. We shall provide examples for both procedures.

An appropriate choice of reference elements is required, as well as a tabulation of elemental compositions of source materials. Elements that are useful as reference elements for crustal material include silicon, aluminum, iron, and titanium. All of them are minor constituents of sea water, whereas they are abundant in rocks. Silicon presents some analytical difficulties, so that it is not regularly determined. For this reason, aluminum is most frequently used as crustal reference element. The best reference element for seawater is sodium. Chlorine has been used occasionally, but in the aerosol it is subject to modification as we have seen, so that it should be avoided. In the following we shall first consider the marine environment. Continental aerosols will be discussed later.

Table 7-15, which was introduced earlier, compares relative element abundances in the marine aerosol with that of sea water and shows the resulting enrichment factors. The deficits for chlorine and bromine and the enrichment of sulfate have been discussed previously. The relative abundances for magnesium, calcium, potassium, and strontium exhibit some enrichment. Early observations of this kind indicated even higher enrichment values and reports of $EF(X) \approx 10$ were not uncommon. This led to the unfortunate and very confusing suggestion that alkali and alkaline earth elements undergo fractionation during the process of seasalt formation. The early studies, however, were conducted at coastal sites without proper precautions against contamination by nonmarine sources. Hoffman and Duce (1972) were the first to show from measurements on Oahu, Hawaii, that the enrichment is caused by the admixture of crustal material. By grouping the data in accordance with wind direction, they found much higher $EF(X)$ values when the winds were variable and came partly from the island as compared to strict on-shore wind conditions. In the latter case the mean element abundance ratios matched those of sea water and calcium was the only element that showed a statistically significant positive deviation. Figure 7-22 shows scatter diagrams for magnesium and calcium according to Hoffman and Duce (1972) for on-shore wind conditions. A linear regression analysis leads to the equations

$$c(\text{Mg}) = 0.002 + 0.117c(\text{Na})$$

$$c(\text{Ca}) = 0.021 + 0.037c(\text{Na})$$

The slope of each line corresponds closely to the respective seawater ratios. For calcium, a positive intercept of the regression line with the ordinate provides evidence for the presence of a nonmarine component. It should not be assumed that its contribution is constant with time, but it appears that the variation is buried in the statistical scatter of the data. It must further be emphasized that the points in a scatter diagram do not generally fall on a straight line. A linear correlation is obtained only if the contribution from one source greatly exceeds that of all others. Only then is a linear regression analysis really warranted.

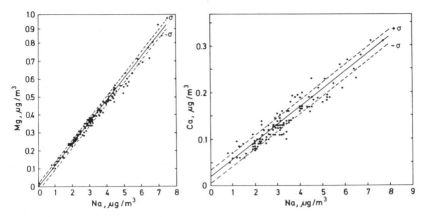

Fig. 7-22. Scatter diagrams for magnesium and calcium versus sodium in marine aerosols. [Adapted from Hoffman and Duce (1972).] Note that the slope of the regression lines gives the mass ratios in sea water ($Mg/Na = 0.119$, $Ca/Na = 0.038$), but that the regression line for calcium does not meet the origin of the plot.

Hoffman *et al.* (1974) found the same procedure applicable to data obtained from measurements on board of ships in the central Atlantic Ocean. Table 7-15 includes mean $(X)/(Na)$ ratios from their work. Shown in parentheses are the values derived from the slopes of regression lines. They are distinctly lower than the averaged data. Hoffman *et al.* (1974) measured also the abundance of iron in the aerosols. Since the samples were taken in a region partly affected by fallout from the Saharan dust plume, iron serves as a convenient indicator for the contribution of material from continental sources. Not surprisingly, the enrichment of the elements Mg, Ca, K, and Sr was well correlated with the iron content. The $(X)/(Na)$ ratios approached those of sea water only when the Fe concentrations were very low. These results demonstrate that materials from both marine and crustal sources are present over the open ocean. In addition, they provide some verification for the existence of a tropospheric background aerosol having the continents as a source, and they confirm the absence of a significant fractionation of alkali and alkaline earth elements in the production of sea salt.

A determination of the contribution of crustal components to the continental aerosol is made difficult by our imprecise knowledge of the composition of material resulting from the wind erosion of soils. Usually a surrogate composition is used. Possibilities include the average composition of crustal rocks, bulk soil, or the aerosol-size fraction of the soil. Although the last choice has advantages, its study has begun only recently and averaged data are not yet available. Most investigators have used either globally averaged rock or soil. Neither is entirely satisfactory as a reference source. Rock is inappropriate because it is unweathered, and bulk soil is inapplicable because the mineral species present in the size range of aerosol particles differ from those making up the bulk of the material. Rahn (1975b) has also considered shales as a reference material, with the argument that clays and silts might better approximate soil substances susceptible to aerosol formation. Table 7-16 compiles abundances of the most frequent elements in crustal rock, bulk soil, and average shales, in addition to the relative composition of fly ash from the combustion of fossil fuels, which is believed to represent an important source of particulate material in densely populated regions. Elemental abundances in soils and shales are somewhat depleted with regard to leachable elements when compared with crustal rock; otherwise, they are fairly similar. Fly ash is depleted in silicon, but not in other elements.

The third section of Table 7-16 shows the relative composition of two continental aerosols to provide a direct comparison with the data in the second part of the table. The first aerosol given represents an average over samples taken at several remote sites in Europe, North America, and Africa (Rahn, 1975b). The second aerosol is the 1974 average obtained by Moyers *et al.* (1977) at two sites in rural Arizona. The $(X)/(Al)$ ratios for the elements Si, Fe, Na, K, and Ca in the aerosols treated in Table 7-13 are shown at the bottom of that table. To further illustrate the behavior of important elements in the atmospheric aerosol, we show in Fig. 7-23 a number of scatter diagrams. The combination of these data will be discussed below.

As Table 7-16 shows, the relative abundances of the major elements in the aerosol do not differ greatly from those in bulk soil, crustal rock, or average shale—that is, the elements are neither greatly enriched nor seriously depleted. A good match with any of the three reference materials is not obtained, however. The differences must be significant, since they are greater than conceivable analytical errors. Consider silicon as an example. Tables 7-13 and 7-16 indicate an average Si/Al ratio of 2.7, which is lower than that for either bulk soil or crustal rock and is more similar to that in shales. Fly ash exhibits a particularly low Si/Al ratio. It is possible that the low aerosol value in heavily industrialized Tees-side (Table 7-13) is due to a mixture of natural and combustion aerosols, but this explanation cannot be extended to the remote continental aerosol. A more likely explanation for the silicon deficiency is the size distribution of the Si/Al ratio in soil particles. The very coarse quartz particles, which are rich in silicon, are not readily mobilized. Since only the fine fraction of soil particles contributes to aerosol formation, the Si/Al ratio in the aerosol will be determined by that of silts and clays (see Table 7-7 for definitions). Common clay

Table 7-16. *Average Absolute and Relative Abundances of Major Elements in Crustal Rock, Soil, and Shale; Relative Abundances of Elements in Fly Ash from Coal and Fuel-Oil Combustion; and Relative Abundances of Major Elements in the Remote Continental Aerosol, with Enrichment Factors (Aerosol)* $EF = (X)/(Al)_{aerosol}/(X)/(Al)_{crustal\ rock}$ [a]

| Element | Elemental abundances (ppmw) | | | | Relative composition (X)/(Al) | | | | | Remote continental aerosol | | | |
	Soil (a)	Crustal rock (b)	(c)	Shale (d)	Soil (a)	Rock (c)	Shale (d)	Fly ash (e) Coal	Oil	(X)/(Al) (f)	(g)	EF (f)	(g)
Si	330,000	277,200	311,000	281,900	4.628	4.018	2.591	1.43	0.48	2.7	3.25	0.67	0.81
Al	71,300	81,300	77,400	108,800	1	1	1	1	1	1	1	—	—
Fe	38,000	50,000	34,300	48,400	0.533	0.443	0.445	0.50	0.50	1.3	0.55	2.9	1.24
Ca	13,700	36,300	25,700	22,500	0.192	0.332	0.207	0.29	0.08	0.80	0.66	2.4	1.99
Mg	6,300	20,900	33,000	11,700	0.088	0.426	0.107	0.057	0.06	0.25	0.15	0.58	0.35
Na	6,300	28,300	31,900	9,720	0.088	0.426	0.089	0.029	0.30	0.20	0.23	0.48	0.56
K	13,600	25,900	29,500	32,000	0.191	0.381	0.294	—	0.02	0.65	0.44	1.70	1.15
Ti	4,600	4,400	4,400	4,440	0.064	0.057	0.041	0.064	6×10^{-3}	0.064	0.083	1.12	1.45
Mn	850	950	670	420	0.012	8.6×10^{-3}	3.8×10^{-3}	1.7×10^{-3}	6×10^{-4}	0.025	0.017	2.9	1.98
Cr	200	100	48	80	2.8×10^{-3}	6.2×10^{-4}	7.3×10^{-4}	2.1×10^{-3}	0.024	8×10^{-3}	2.6×10^{-3}	12.9	9.2
V	100	135	98	106	1.4×10^{-3}	1.3×10^{-3}	9.7×10^{-4}	5.7×10^{-3}	0.50	3.5×10^{-3}	—	2.7	—
Co	8	25	12	13	1.1×10^{-4}	1.5×10^{-4}	1.2×10^{-4}	6.4×10^{-4}	0.03	6.5×10^{-4}	5.8×10^{-4}	4.3	3.9

[a] Sources: (a) Vinogradov (1959); (b) Mason (1966); (c) Turekian (1973); (d) Flanigan (1973), model A; (e) Winchester and Nifong (1971); (f) Rahn (1975b); (g) Moyers et al. (1977).

minerals   include   kaolinite, $Al_4Si_4O_{10}(OH)_8$;   montmorillonite,
$Al_4(Si_4O_{10})_2(OH)_4 \cdot H_2O$; muscovite, $K_2Al_4(S_6Al_2)O_{20}(OH)_4$; and chlorite,
$Mg_{10}Al_2(Si_6Al_2)O_{20}(OH)_6$. The corresponding Si/Al mass ratios are 1.04,
2.07, 1.04, and 1.56, respectively. Frequently, however, aluminum is partially
replaced by other elements, whereby the Si/Al ratio is raised again. Schütz
and Rahn (1982) have essentially confirmed this concept. They investigated
the size distribution of the Si/Al ratio in various soils and found silicon
enrichment factors of 10 in particles with radii larger than 100 μm. Toward
smaller sizes the enrichment factor decreased steadily, until a nearly constant
value of 0.3 was reached in the submicrometer size range.

Iron, in contrast to silicon, is always slighly enriched in the atmospheric
aerosol, with an enrichment factor of $EF(Fe) \approx 1.8$. The Fe/Al ratio at
Tees-side (see Table 7-13) is higher and must be attributed to the vicinity
of steel works in that industrial region. A scatter diagram for iron is shown
in Fig. 7-23. The data include measurements at continental, coastal, and
strictly marine sampling locations. With few exceptions, the points are seen
to parallel the lines representing Fe/Al ratios in soil and crustal rock.
Occasionally a site is found where Fe and Al occur in crustal proportions,
but the average ratio is higher. Since the oceans are negligible as a source
of iron, the scatter of the data must arise from the variable iron content of
different soils. Note that the average Fe/Al ratios in soil, crustal rock, shale,
and fly ashes are all nearly equal. The situation is similar for titanium. In
this case the scatter diagram in Fig. 7-23 shows a much narrower distribution
and the points closely follow the line representing the crustal Ti/Al ratio.
Titanium obviously would serve as a good crustal reference element.
Unfortunately, it presents analytical difficulties, and the data carry larger
uncertainties than the data for the other elements.

Table 7-16 shows that soils and shales are depleted in the elements Na,
K, Ca, and Mg in comparison with crustal rock. The depletion is higher
for sodium and magnesium than for potassium and calcium. The composition of the aerosol for these elements matches none of the reference
materials. The relative abundances of K and Ca are higher than and the
abundances for Na and Mg are intermediate between, those in soil and
crustal rock. It is interesting to note that the ratios Ca/Mg and K/Na
existing in soils are approximately preserved. This result suggests a common
soil–aerosol enrichment process. The true nature of this process has not
been elucidated, however. Schütz and Rahn (1982), who have investigated
elemental abundances as a function of soil particle size for a number of
desert soils, found no evidence for an increase of Na/Al, K/Al, Mg/Al, or
Ca/Al ratios in the size range below 20 μm as compared with larger particles.
Clay mineral evidently cannot be held responsible for the observed enrichment. The composition of shales also argues against this possibility. Rahn

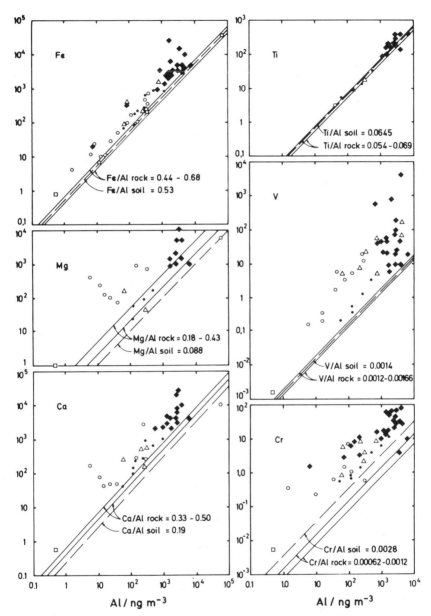

Fig. 7-23. Scatter diagrams for iron, titanium, magnesium, calcium, vanadium, and chromium in the atmospheric aerosol versus aluminum as reference element. Concentrations are given in units of ng/m³. Symbols indicate the type of sampling site: ○ marine, △ marine-influenced, ● remote continental, □ Antarctic. The range of X/Al ratios in crustal rock is shown by the solid lines and X/Al ratios in soil by the dashed lines. [Adapted from Rahn (1975b).]

(1975b) has pointed out that the presence of humus in soil, which usually is neglected in such considerations, might influence the elemental composition of the aerosol. Humus is a ubiquitous, albeit variable, constituent of soils. The active surface of humus exceeds that of clay minerals, and both have a great capacity for binding cations. In its formation from decaying plant material (e.g., dead leaves), humus receives the elements in proportions at which plants incorporate them from the soil. Since the root systems of plants reach down into subsurface layers of the soil, whereas the aerosol arises from the top layer, a certain fractionation and enrichment might be expected. The mechanism is speculative but deserves investigation. The desert soils studied by Schütz and Rahn (1982) presumably contained fairly little humus, so that one must be cautious about generalizing their results.

Figure 7-23 includes scatter diagrams for magnesium and calcium as examples for two leachable elements. These data are selected to provide comparison with Fig. 7-22 and Table 7-13. Airborne magnesium enables a fairly clean distinction between marine and continental sites. The former fall to the left of the line representing Mg/Al ratios for continental soils and crustal rock; the latter scatter around these lines. Sodium, which is not included in Fig. 7-23, shows an almost identical behavior. The scatter diagram for calcium differs from that for magnesium in that marine and continental data points intermingle, making it difficult to identify sites with a dominant marine source character. As discussed earlier, calcium is found enriched with regard to its abundance in seasalt, even in midocean regions. The strong correlation of calcium with aluminum demonstrates the significance of the contribution of continental sources to the atmospheric background aerosol. The average Ca/Al ratio in Fig. 7-23 is 0.8. Similar values occur at Tees-side and West Covina (see Table 7-13), whereas the Ca/Al ratio at the Wank Mountain site is much higher. This station must be influenced locally by the preponderance of limestone in that region.

The remaining elements listed in Table 7-16 have crustal abundances, relative to aluminum, of 1% or less, so that they must be considered trace elements. All other elements not listed in the table, with the exception of oxygen, fall into the same category. Vanadium and chromium may be discussed briefly. Vanadium is of interest because fly ash from fuel oils contains it in disproportionately high amounts. The V/Al scatter diagram indicates two bands of points, each paralleling the lines representing V/Al ratios of crustal rock and soil. In the upper band, vanadium is by an order of magnitude more enriched than in the lower band. The data classified as remote continental lie close to the crustal ratios, although they are somewhat higher (average V/Al ≈ 0.0035). The enrichment factor for these data is 2.7, which is similar to that for other nonvolatile elements such as Fe or Ca and thus may be considered normal. The high enrichment of vanadium in the upper band of points must be ascribed to pollution. The greatest effects clearly occur in the cities, but the observation of high enrichment factors at marine sites (of the northern hemisphere) leaves little doubt that vanadium from anthropogenic sources affects also the tropospheric background aerosol. The Arctic aerosol is similarly influenced by pollution, as Rahn and McCaffrey (1980) have shown. In the Antarctic, however, the effect is nil (Zoller et al., 1974), indicating that neither anthropogenic nor volcanic sources cause an enrichment of vanadium there.

The case of chromium is of interest because the Cr content of soils is by a factor of at least two higher than that of crustal rock. The Cr/Al ratios observed in the aerosol are again consistently higher than those in either soil or crustal rock. The data scatter broadly, without distinction between marine or continental sampling sites. Some of the high values are found in regions where iron and steel industries are located and in the cities. For example, Moyers et al. (1977) have found markedly higher values for chromium, cobalt, and nickel in urban Tucson, Arizona, compared with the surroundings. Data from remote continental sites indicate an average enrichment factor of 12.9, which is the highest for any of the elements listed in Table 7-16, although not as high as in urban regions. If, however, soil rather than crustal rock is used as reference material, the enrichment factor reduces to 2.8, a value comparable to that for other elements discussed above. Chromium evidently does not behave exceptionally, and the comments made earlier in discussing leachable elements apply to chromium as well. The behavior of manganese and cobalt is very similar to that of chromium and needs no further discussion.

Various other trace elements in the atmospheric aerosol are considerably more enriched than the elements treated in the preceding discussion. Enrichment factors observed in the Antarctic, which is not subject to anthropogenic pollution, were presented in Fig. 7-18. The arid soils studied by Schütz and Rahn (1982) revealed enrichment factors betwen 3 and 10 for elements such as Cu, Zn, As, and Sb fairly independent of the size of soil particles. In the aerosol these elements are considerably more abundant. Ag and Au were the only elements for which Schütz and Rahn (1982) found enrichment factors increasing with decreasing particle size, with values approaching 500 and 50, respectively, in the submicrometer size range. The corresponding enrichment factors in the aerosol are greater. Silver in the aerosol is the only element whose enrichment factors come closest to being explicable in terms of submicrometer soil particle enrichment. Soil differentiation, therefore, does not appear to provide a mechanism for the abnormally high enrichment of these elements in the aerosol.

Rahn (1975) has summarized data for many highly enriched elements. From a limited number of elemental mass–size distributions, Rahn (1975a) and, more recently, Milford and Davidson (1985) have calculated mass median diameters for each element. This quantity defines the size for which one-half of the mass of an element is associated with smaller, and the other half with larger, particles. Elements occurring in nearly crustal proportions were found to have mass median diameters in the size range beyond 2 μm, as expected for soil-derived elements. Most of the elements with abnormally high enrichments, in contrast—among them Cu, Zn, Ag, As, Sb, Sn, In, Pb, Cd, Se, S, and Hg—are concentrated in the submicrometer size range. This kind of distribution suggests that highly enriched elements have passed through the gas phase at some point in their history and entered the aerosol phase by condensation. As discussed previously, volcanoes and combustion are likely sources of volatile elements. Another possibility is a biological enrichment. Wood (1974) and Ridley et al. (1977) have pointed out that trace metals such as As, Se, S, Pd, Tl, Pb, Pt, Au, Sn, and Hg serve as acceptors of methyl groups in microbiological processes. This leads to the emission of metal alkyls into the atmosphere. The gas-phase oxidation of these compounds then forms metal oxides, which attach to the atmospheric aerosol. The significance of the process is unknown, however, and further studies are required.

We conclude this section with a brief summary. The inorganic fractions of both marine and continental aerosols have unique compositions differing considerably from those of the underlying source materials. Over the oceans the sea-salt aerosol is modified by condensation processes as well as by the presence of the tropospheric background aerosol. Thus, one finds a deficit of chloride and bromide, and a surplus of sulfate, ammonium, and nitrate, in addition to certain elements of crustal origin. The continental aerosol contains large amounts of sulfate, ammonium, nitrate, and possibly chloride, all resulting from gas-to-particle conversion. In contrast to sea salt, which essentially represents the composition of sea water, the elemental composition of the soil-derived fraction of the continental aerosol differs appreciably from that of crustal rock, average soil or shale. The relative abundances of the major elements differ individually by factors of about three. A number of minor elements are enriched by several orders of magnitude. The behavior is globally widespread, but an unambiguous interpretation is not at hand.

7.5.2 THE ORGANIC FRACTION

Solvent extraction combined with chromatographic techniques generally is used to study organic compounds associated with the atmospheric aerosol. Combustion to CO_2 is additionally used to determine total organic carbon in the samples. This includes elemental carbon present as soot. Due to the limited efficiency of solvent extraction and because many compounds showing up in the chromatograms have not been identified, our knowledge of the composition of the organic fraction of the aerosol is rather incomplete. Up to now, several hundred different compounds have been identified. Although clues as to their origin are slowly emerging, we still know very little about their significance to atmospheric chemistry. Accordingly, little more can be done than to present a survey of the observations.

Table 7-17 compares average concentrations for ether-extractable organic matter in the aerosol with those of total particulate matter at several locations. The efficiency of ether extraction is roughly 50%. The organic compounds are separated into five broad groups: aliphatic, aromatic, and polar neutral compounds, organic acids, and organic bases. The locations may be classified as urban, rural and remote continental, and maritime. While the absolute concentrations decrease as one goes from urban to remote continental and/or maritime regions, the relative compositions of the five organic fractions remain remarkably uniform. The neutral components contribute about 56%, acids 36%, and bases 8% to the total ether-extractable material. Current knowledge is inadequate to determine whether this uniformity is due to a common origin of particulate organic matter or processes in the atmosphere which produce or modify organic particulate matter.

Table 7-18 is added to demonstrate that the abundance of total organic carbon in the marine aerosol is globally very uniform with an average concentration of 0.42 $\mu g/m^3$. Since the residence time of aerosol particles in the troposphere is less than 1 week, Eichmann et al. (1980) argued that such a uniform distribution cannot be maintained solely from continental sources, and that it should be at least supplemented by marine sources or by gas-to-particle conversion processes involving hydrocarbon precursors. A uniform distribution would be in harmony, however, with the concept of a uniform background aerosol, whatever its source. Hoffmann and Duce (1977) and Chesselet et al. (1981) have shown that the size distribution of organic carbon in the marine aerosol does indeed favor the submicrometer size range, which suggests the importance of gas-to-particle conversion. Chesselet et al. (1981) have studied not only the size distribution but the $^{13}C/^{12}C$ isotope ratio as well. The ratios for the smallest particles was close to that for carbon of continental origin ($\delta^{13}C = 26 \pm 2‰$); that for the largest particles was lower ($\delta^{13}C = 21 \pm 2‰$), indicating that these particles were emitted directly from the ocean, presumably together with the formation of sea salt. Chesselet et al. (1981) concluded that about 80% of particulate organic matter in the marine aerosol originates from continental sources. As an average, this high a contribution appears to be excessive in view of the fact that most gaseous precursors that one might think of, primarily the heavier hydrocarbons, are rapidly attacked by OH radicals and have a

Table 7-17. *Average Concentrations ($\mu g/m^3$) of Total Particulate Matter (TPM), Ether-Extractable Organic Matter (EEOM), and of Groups of Neutral Compounds, Organic Acids, and Organic Bases, from Measurements at Various Locations (Hahn, 1980; Eichmann et al., 1980)*

| Location | TPM | EEOM | Neutral compounds | | | Organic acids | Organic bases |
			Aliphatic	Aromatic	Polar		
Mainz, Germany	150	27.1	6.80	3.10	4.90	10	2.0
Deuselbach, Germany	12	2.0	0.53	0.26	0.35	0.7	0.2
Jungfraujoch, Switzerland	17.6	0.9	0.26	0.12	0.20	0.2	0.03
North Atlantic (shipboard)	16.7	0.8	0.20	0.16	0.10	0.23	0.03
Loop Head, Ireland	20.5	0.73		0.38[a]		0.25	0.10
Cape Grim, Tasmania	31	0.5		0.23[a]		0.18	0.08

[a] Sum of neutral compounds.

Table 7-18. *Concentrations of Organic Carbon Associated with the Marine Aerosol*[a]

Location	Sample number	Concentration ($\mu g/m^3$)	Authors
Bermuda	8	0.29 (0.15–0.47)	Hoffman and Duce (1974)
Bermuda	8	0.37 (0.15–0.78)	Hoffman and Duce (1977)
Tropical North Atlantic	5	0.59 (0.33–0.93)	Ketseridis *et al.* (1976)
Hawaii	7	0.39 (0.36–0.43)	Hoffman and Duce (1977)
Eastern Tropical North Pacific	3	0.49 (0.22–0.74)	Barger and Garrett (1976)
Samoa	9	0.22 (0.13–0.41)	Hoffman and Duce (1977)
Eastern Tropical South Pacific	9	0.32 (0.07–0.53)	Barger and Garrett (1976)
Loop Head, Ireland	6[b]	0.57 (0.28–0.86)	Eichmann *et al.* (1979)
Cape Grim, Tasmania	6[b]	0.53 (av.)	Eichmann *et al.* (1980)

[a] Taken partly from the compilation of Duce (1978).

[b] Samples taken in pure marine air.

lifetime in the atmosphere of not more than 4–5 days. The data of Chesselet *et al.* (1981) were derived from samples taken in the Sargasso Sea and at the Enewetak Atoll. Both sites, especially Enewetak, are somewhat influenced by long-range transport of continental aerosols (see below).

Another feature by which organic compounds from terrestrial and marine sources may be distinguished is the relative abundance of compounds containing odd and even numbers of carbon atoms. In the *n*-alkane fraction released from land-based vegetation, there exists a distinct preference for odd carbon numbers in the C_{19}–C_{35} size range, whereas *n*-alkanes from marine biota do not exhibit any carbon number preference. Petroleum-derived hydrocarbons from oil spills and other anthropogenic sources also do not show any carbon number preference. If this feature is detected, it is a sure sign for the presence of material originating from terrestrial vegetation. Eichmann *et al.* (1979, 1980) have studied *n*-alkanes in pure marine air masses at Loop Head, Ireland, and Cape Grim, Tasmania. The results indicate some odd-to-even carbon preference in the aerosol phase, but not in the gas phase (see Fig. 6-7). The relative contribution of terrestrial material for the north Atlantic was estimated as 10–25%. A much higher contribution was found by Simoneit (1977) near the West African coast in particles larger than 1 μm. These evidently had their origin on the continent. Marty and Saliot (1982) explored the size distribution of *n*-alkanes in the equatorial Atlantic and found an increase of the odd-to-even carbon number index in particles larger than 1 μm. This suggests that over the Atlantic Ocean it is primarily the larger particles that have their origin on the continents. The conclusion contrasts with that derived by Chesselet *et al.* (1981) for the Pacific Ocean, derived from the $^{13}C/^{12}C$ isotope ratios, but it agrees with the data of Simoneit (1977).

Gagosian *et al.* (1982) also have studied the $C_{21}-C_{36}$ *n*-alkanes in addition to aliphatic alcohols and acids in aerosols collected at the Enewetak Atoll. Again, the carbon preference numbers indicated a considerable contribution of continental organic matter. In this case it was shown that the temporal trend of land-derived organic material correlated well with other continental indicators such as the elements Al and ^{210}Pb. The Asian continent, particularly its desert areas, is known to release a dust plume into the air over the Pacific Ocean. The plume travels with the westerlies at the 700-mbar (~ 3 km) pressure level in a direction that is seasonally variable due to changes in the large-scale wind patterns. The phenomenon is similar to the Saharan dust plume crossing the Atlantic Ocean with the trade winds. From air-mass trajectory analyses, it appears that Enewetak Atoll was in the range of the Asian dust plume at the time the aerosol samples were taken.

Finally, it should be noted that *n*-alkanes in the marine atmosphere occur mainly in the gas phase. According to the data of Eichmann *et al.* (1979, 1980), not more than 5% of *n*-alkanes with carbon numbers less than C_{28} is associated with the marine aerosol. This observation is due to the low abundances of the *n*-alkanes. Their vapor pressures in marine air range from 10^{-13} to 10^{-12} bar, so that they are smaller than the saturation vapor pressures.

Table 7-19 gives a condensed summary of organic compounds that have been identified in rural continental and marine aerosols. Table 7-20 shows in a similar manner the most abundant compounds in urban aerosols. We shall not discuss the urban situation and refer to the reviews of Daisey (1980) and Simoneit and Mazurek (1981). Typical aerosol components in smog-laden urban areas appear to be dicarboxylic acids and nitrate esters, which are formed as by-products of photochemical reactions involving gas-phase hydrocarbons and nitrogen oxides of nitric acid. In rural areas, dicarboxylic acids are minor components, and nitrate esters have not yet been detected, although undoubtedly they are present. Polycyclic aromatic hydrocarbons such as naphthalene and the higher homologues have received much attention because some of these, especially the five-ring benzo[a]pyrene, are carcinogens. These substances are formed as combustion products. In the atmosphere they are subject to direct photodegradation. They are nevertheless stable enough to undergo long-range transport and are present on a global scale, even though they make up only a minor fraction of total particulate organic matter.

A fairly complete characterization of organic compounds in the lipid, that is, the solvent-soluble fraction of the rural continental aerosol, has been achieved by Simoneit and Mazurek (1982; Simoneit, 1984). They analyzed the extractable lipid material for hydrocarbons, acids, and alcohols, searching in addition for typical molecular markers. Aerosols were collected at various sites in the western United States. The results indicated that the material was a composite derived from two major sources: waxes from vascular plants, and petroleum residues originating from road traffic. Epicuticular plant waxes are abundant on the leaves of many land plants (Kolattukudy, 1976). They frequently occur as microcrystals or hollow tubes with micrometer to submicrometer dimensions (Hall and Donaldson, 1963; Baker and Parsons, 1971) and play a physiological role in water retention (Hall and Jones, 1961). Plant waxes also enter into soils due to the incorporation of senescent and decaying vegetation as well as by the erosion of leaf waxes on healthy plants due to natural weathering. Thus, since soils are subject to wind erosion, the wax particles become resuspended in the air together with soil particles, and this explains their presence in the dust plumes carried from the continents over the oceans.

Apart from the n-alkanes discussed above, epicuticular wax hydrocarbons contain sesqui- and diterpenoids. These compounds are based on the structural skeletons of cadinane and abietane, respectively, which are shown in Fig. 7-24. The sesquiterpenoids recovered by Simoneit and Mazurek (1982) in the rural aerosol were calamenene, tetrahydrocadalane, and cadalene. These compounds presumably are degradation products of cadinane derivatives (various isomers of cadinenes and cadinols), which are ubiquitous in essential oils of many higher plants (Simonsen and Barton, 1961). The major diterpenoid hydrocarbons observed in the aerosol samples were dehydroabietane, dehydroabietin, and retene. The main source of abietane derivatives are coniferous resins. The parent compounds dehydrate fairly rapidly to yield the more stable hydrocarbons found in the aerosols. These may then serve as markers for hydrocarbons arising from vegetation, in addition to the odd-to-even carbon number preference in the n-alkanes.

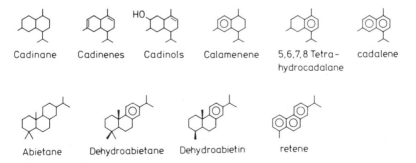

Cadinane   Cadinenes   Cadinols   Calamenene   5,6,7,8 Tetra-   cadalene
                                                    hydrocadalane

Abietane   Dehydroabietane   Dehydroabietin   retene

Fig. 7-24. Molecular markers for hydrocarbons originating from plant waxes (Simoneit and Mazurek, 1982).

Table 7-19. *Identification of Organic Compounds Associated with Continental and Marine Aerosols*

Type of compound		Location	Concentration (ng/m^3)	Authors
		Continental		
Alkanes	C_{21}–C_{25}	Ohio River Valley, Indiana	0.2–2	Barkenbus *et al.* (1983)
Hydrocarbons[a]	C_{15}–C_{35}	Lake Tahoe, California	53–158	Simoneit and Mazurek (1982)
Polycyclic aromatics		National Park locations in Oregon and California	0.01–0.16	Simoneit (1984)
Nitropolyaromates		30 km West of Copenhagen	~0.01	Nielson and Seitz (1984)
Phthalate esters		Point Barrow, Alaska	0.1–20	Weschler (1981)
Formaldehyde		Deuselbach, Germany	39 ± 26	Klippel and Warneck (1980)
Aliphatic alcohols	C_{10}–C_{34}	Lake Tahoe, California	198–524	Simoneit and Mazurek (1982)
Carboxylic acids	C_{12}–C_{32}	Lake Tahoe, California	140–391	Simoneit and Mazurek (1982)
	C_{14}–C_{26}	Ohio River Valley, Indiana	67	Barkenbus *et al.* (1983)
	C_{16}–C_{18}	Ohio River Valley, Indiana	5–24	Barkenbus *et al.* (1983)
Dicarboxylic acids	Succinic	Ohio River Valley, Indiana	10–20	Barkenbus *et al.* (1983)
	Glutaric	Ohio River Valley, Indiana	20–50	Barkenbus *et al.* (1983)
	Malonic	Ohio River Valley, Indiana	3	Barkenbus *et al.* (1983)

Marine

Alkanes	$C_{15}-C_{35}$	Atlantic Ocean	0.1–12	Simoneit (1977)
	$nC_{10}-nC_{30}$	Atlantic Ocean	14 (av.)	Hahn (1980), Ketseridis and Jaenicke (1978)
	$nC_{10}-nC_{30}$	Indian Ocean (Cape Grim)	4.7 (av.)	Eichmann et al. (1980)
	$nC_{15}-nC_{37}$	Tropical Atlantic	6–13	Marty and Saliot (1982)
	$C_{13}-C_{32}$	Enewetak Atoll	0.07–0.24	Gagosian et al. (1982)
Aromatics	Xylenes	Loop Head, Ireland	10.6 (av.)	Eichmann et al. (1979)
	Xylenes	Cape Grim, Tasmania	0.9 (av.)	Eichmann et al. (1980)
Polycyclic aromatics		Tropical Atlantic	0.1–0.2	Marty et al. (1984)
Formaldehyde		Loop Head, Ireland	4.9 ± 2.4	Klippel and Warneck (1980)
Aliphatic alcohols	$C_{10}-C_{34}$	Atlantic Ocean	0.001–30	Simoneit (1977)
		Atlantic Ocean	0.01–10	Simoneit and Mazurek (1982)
	$C_{13}-C_{32}$	Enewetak Atoll	0.07–0.24	Gagosian et al. (1982)
Carboxylic acids	$C_{13}-C_{32}$	Enewetak Atoll	0.04–0.38	Gagosian et al. (1982)
	$C_{12}-C_{32}$	Atlantic Ocean	0.001–30	Simoneit and Mazurek (1982)

[a] Includes branched and cyclic compounds.

357

Table 7-20. Organic Compounds Associated with Urban Aerosols[a]

Type of compound		Concentration (ng/m³)	Location	Authors
Alkanes	C_{18}–C_{50}	1,000–4,000	217 U.S. urban stations	McMullen et al. (1970)
Alkenes		2,000	217 U.S. urban stations	McMullen et al. (1970)
Alkylbenzenes		80–680	West Covina, California	Cronn et al. (1972)
Naphthalenes		40–500	Pasadena, California	Schueltzle et al. (1975)
Polycyclic aromatics		0.09–0.18	Los Angeles, California	Simoneit (1984)
Nitropolycyclic aromatics		0.08–0.07	Southern California	Pitts et al. (1985)
Aromatic esters		29–132	Antwerp, Belgium	Cautreels et al. (1977)
Aromatic acids		90–380	Pasadena, California	Schuetzle et al. (1975)
Aliphatic alcohols	C_{10}–C_{34}	1,360–2,016	Urban Los Angeles, California	Simoneit and Mazurek (1982)
Monocarboxylic acids	C_{10}–C_{32}	217–308	Urban Los Angeles, California	Simoneit and Mazurek (1982)
		36.5	Antwerp, Belgium	Cautreels et al. (1977)
Dicarboxylic acids	C_3–C_7	40–1,350	Pasadena, California	Schuetzle et al. (1975)
Organic nitrates		40–4,010	Pasadena, California	Schuetzle et al. (1975)
Alkylhalides		20–320	Pasadena, California	Schuetzle et al. (1975)
Arylhalides		0.5–3	Pasadena, California	Schuetzle et al. (1975)
Chlorophenols		5.7–7.8	Antwerp, Belgium	Cautreels et al. (1977)

[a] Adapted from Daisey (1980).

The petroleum-derived component associated with the rural aerosol consists of n-alkanes without any carbon number predominance, and many complex branched and cyclic hydrocarbons that cannot be resolved by gas chromatography, so that they appear as a hump in the C_{15}-C_{30} size range. Although the hump itself is indicative of the presence of petroleum residues, one may additionally use pristane and phytane as molecular markers. These quadruply branched C_{19} and C_{20} hydrocarbons are diagenetic products of phytol and are not primary constituents of terrestrial biota (Didyk *et al.*, 1978). They occur in crude petroleum as well as in lubricating oils and auto and diesel engine exhausts, but not in gasoline. Simoneit (1984) also discussed other potential indicator compounds for petroleum residues. He could show that at many nonurban sampling sites in California and Oregon, the petroleum-derived fraction equaled or even exceeded the natural background of biogenic hydrocarbons in the aerosol.

The acids that Simoneit and Mazurek (1982) and others found to be associated with continental aerosols were predominantly straight-chain fatty acids in the range C_{10}-C_{30}. A strong even-to-odd carbon number preference indicated a biological origin. The fatty acid distribution in waxes from composite plant samples did not compare well with those in the aerosol, however. Fatty acids in plant waxes, occurring mainly as esters, fall in the C_{22}-C_{32} range of carbon numbers, with maxima between C_{26} and C_{30}, whereas the aerosol constituents exhibit the principal maximum at C_{16} (palmitic acid), with a secondary maximum at C_{22} or C_{24}. Similar distributions have been observed by Ketseridis and Jaenicke (1978) and Gagosian *et al.* (1982) in marine aerosols, by van Vaeck *et al.* (1979) and Matsumoto and Hanya (1980) in continental aerosols, and by Meyers and Hites (1982) in Indiana rainwaters. The origin of this distribution is obscure. Simoneit and Mazurek (1982) suggest that the homologues below C_{20} are derived from microbial sources, either from microbiota within the atmosphere, or from plant waxes that became inhabited by microorganisms and were degraded. Soil microbiota may additionally contribute to the acids in the aerosol.

Apart from saturated fatty acids, Simoneit and Mazurek (1982) observed low concentrations of unsaturated fatty acids (range C_{14}-C_{17}); α-hydroxy fatty acids (range C_{10}-C_{24}) that are known components of grass wax; dicarboxylic acids (range C_{10}-C_{24}) that probably arise from the direct biodegradation of hydroxy fatty acids; and diterpenoidal acids occurring as diagenetic products of diterpenoids from coniferous resins.

Finally, Simoneit and Mazurek (1982) described n-fatty alcohols as constituents of rural continental aerosols. They ranged from C_{10} to C_{34} with a strong even-to-odd carbon number preference. The distribution generally favored the higher homologues, with a maximum in the range C_{26}-C_{30}. This

is characteristic of vascular plant waxes, so that, again, leaf waxes appear
to be the major source of the material. The fraction below C_{20}, which is
not present in plant waxes, may arise from microbial activity. Molecular
markers for alcohols are the phytosterols. Their distribution in the aerosol
is similar to that of the phytosterol content of many plants.

It was noted earlier (Table 7-12) that according to the study of Winkler
(1974), a good deal of particulate organic matter must be water-soluble.
Few of the compounds discussed above fall into this category, however.
This suggests that a considerable fraction of organic materials associated
with the aerosol has not yet been identified. Additional work is necessary
to resolve this problem.

7.6 Global Distribution, Physical Removal, and Residence Time of the Tropospheric Aerosol

As for other constituents of the atmosphere, it is possible to set up a
mass budget of the aerosol and to calculate its residence time. The main
problem is to characterize the global distribution of particulate matter in
order to determine its total mass in the troposphere. One may then apply
the emission estimates of Table 7-11 to calculate the tropospheric residence
time τ_A with the help of Eq. (4-11). This approach will be discussed in the
first part of this section. Subsequently, we consider an independent method
for estimating the residence time, which results from the use of radioactive
tracers. Finally, the removal of aerosol particles by sedimentation and
impactation at the Earth surface will be discussed.

7.6.1 VERTICAL AEROSOL DISTRIBUTION AND GLOBAL BUDGET

A summary of our knowledge about the vertical distribution of aerosol
particles over the continents of the northern hemisphere is shown in Fig.
7-25. The left-hand side displays the behavior of Aitken particles as determined with condensation nuclei counters; the right-hand side shows the
distribution of large particles as determined with optical techniques. In each
case the range of observations is shown and a single altitude profile is
included for comparison. Apart from balloon soundings, surprisingly few
measurements of large particles exist for the middle troposphere. The
concentration of large particles may be assumed to represent the mass
concentration, whereas that for the Aitken particles is more indicative of
the production of new particles. The number densities of both types of
particles decline rapidly with height in the lower troposphere, with a scale
height of roughly 1 km. In the upper troposphere the decline corresponds
to that of the air density—that is, the mass mixing ratio is approximately

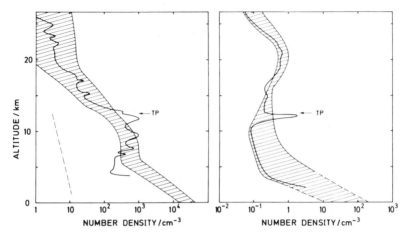

Fig. 7-25. Vertical distribution of the aerosol over the continents. Left: Profile of condensation nuclei, range of results obtained by Junge (1961, 1963), Weickmann (1957), Hoppel *et al.* (1973), Cadle and Langer (1975), Käselau *et al.* (1974), and Rosen and Hofmann (1977, 1978). An example for a single profile is taken from Rosen and Hofmann (1977). The location of the tropopause for this flight is indicated. The dashed line gives the profile for a constant mixing ratio in the troposphere. Right: Vertical distribution of large particles, range of results obtained by Hofmann *et al.* (1975) and Patterson *et al.* (1980) with optical techniques, and by Blifford and Ringer (1969) with impactors; for earlier results see Junge (1963). An example for a single profile is taken from Hofmann *et al.* (1975).

constant. In the stratosphere, the concentration of Aitken particles declines further with increasing altitude, while that for the large particles increases again, reaching a peak near 20 km altitude, which is due to the production of sulfuric acid as discussed in Section 3.4.4 (the Junge sulfate layer).

The observation that the number densities of large and Aitken particles in the troposphere decline with height in a similar manner suggests that the size distribution is approximately preserved. Size distributions of aerosols as a function of altitude have been studied by Blifford and Ringer (1969) with single-stage impactors, and by Patterson *et al.* (1980) with an optical technique. The results indicate that the shape of the size distribution is indeed preserved in the lower troposphere. With increasing height above 5 km the slope toward smaller radii decreases somewhat so that the distribution appears flatter. If the size distribution for particles greater than 0.1 μm in radius is at least approximately independent of altitude, one may combine ground-level concentrations with the altitude profile of Fig. 7-25 to derive the column density of aerosol mass in the continental troposphere. The vertical distribution may be visualized as composed of two contributions: a uniform background evident in the upper troposphere, and a continental aerosol decreasing from the ground on up with a scale height of 1 km. The background aerosol would also be present in the marine environment. Here, the lower troposphere receives a second contribution from sea salt. Its vertical distribution may be gleaned from Fig. 7-14. In this manner it is possible to construct a simple model for the global distribution of the tropospheric aerosol.

The model used here is based on the ground-level concentrations given in Table 7-1. The vertical distribution of continental and marine aerosols are illustrated in Fig. 7-26. The decline of concentrations with height follows from the adopted scale heights and is expressed as

continental $(0 < z < z_T)$: $c_A = c_1^0 \exp(-z/h_1) + c_2^0 \exp(-z/h_2)$

maritime $(0 < z < 600$ m$)$: $c_A = c_3^0 + c_2^0 \exp(-z/h_2)$

$(600$ m $< z < z_T)$: $c_A = c_3^0 \exp[-(z-0.6)/h_3]$

$$+ c_2^0 \exp[-(z-0.6)/h_2] \qquad (7\text{-}29)$$

with $c_1^0 = 45$ μg/m^3, $c_2^0 = 1$ μg/m^3, $c_3^0 = 10$ μg/m^3, $h_1 = 1$ km, $h_2 = 9.1$ km, and $h_3 = 0.9$ km.

Assuming an average tropopause height of 12 km, one obtains by integration over height the following approximate column densities.

continental: $W_{AC} = c_1^0 h_1 + c_2^0 h_2 [1 - \exp(-z_T/h_2)] = 5.17 \times 10^{-5}$ kg/m^2

maritime: $W_{AS} = 600 c_3^0 + c_3^0 h_3 + c_2^0 h_2 [1 - \exp(-z_T/h_2)]$

$$= 2.1 \times 10^{-5} \quad \text{kg/m}^2$$

The total burden of particulate matter in the troposphere is then obtained by applying the surface areas $A_{cont} = 1.49 \times 10^{14}$ m^2 and $A_{sea} = 3.61 \times 10^{14}$ m^2, with the result

$$G_A = W_{AC} A_{cont} + W_{AS} A_{sea} = 1.52 \times 10^{10} \quad \text{kg}$$

In a similar manner one calculates the mass content in the troposphere of sea salt alone as

$$G_{sea\,salt} = 5.4 \times 10^9 \quad \text{kg}$$

In this calculation, the fraction of sea salt transported from the ocean towards the continents is neglected.

The estimates in Table 7-11 indicate a global source strength of particles of $(2.3\text{–}3.9) \times 10^{12}$ kg/yr or $(0.63\text{–}1.0) \times 10^{10}$ kg/day. The values include a production rate of sea salt of 10^{12} kg/yr or 2.74×10^9 kg/day. The corresponding tropospheric residence times are

$$\tau_A = G_A / Q_A = 1.5\text{–}2.5 \quad \text{days}$$

$$\tau_{sea\,salt} = G_{sea\,salt} / Q_{sea\,salt} = 2 \quad \text{days}$$

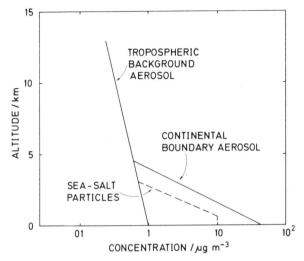

Fig. 7-26. Model for the vertical distribution of particulate matter in the troposphere. The model assumes a superposition of the tropospheric background aerosol with boundary layer aerosol over the continents, and with sea-salt aerosol over the oceans.

Table 7-11 admits the possibility of a five times larger rate of sea-salt production. If this rate is adopted, the residence times for the total aerosol and for sea salt alone would be reduced to 0.7 and 0.4 days, respectively.

One should not expect this procedure to provide more than order-of-magnitude values. The emission rates in Table 7-11 are crude estimates, and the tropospheric mass content of particles predicted by the model distributions must also be considered approximate. The further discussion below and that in Section 8.3 will show, however, that other methods of determining τ_A lead to fairly similar results, which confirm an aerosol residence time of a few days. Note that the residence times of the total aerosol and that of sea salt are almost identical, despite differences in the mass–size distributions, with sea salt favoring the giant particle size range. This suggests that the processes responsible for the removal of particulate matter do not discriminate much between large and giant particles.

7.6.2 RESIDENCE TIMES OF AEROSOL PARTICLES

Independent of budget considerations, the residence time of particles can be estimated by means of radioisotopes as tracers, which become attached to aerosol particles and are removed with them as they are scavenged by precipitation or undergo dry fallout. One of the earliest estimates of this kind was made by Stewart *et al.* (1956) from the decay of fission products following their dispersal in the northern hemisphere after the nuclear weapons tests in Nevada in 1951. The results indicated a residence time of about 1 month, and this value became widely accepted. Stewart *et al.* (1956) had assumed that the debris from the detonations remained confined to the troposphere, but Martell and Moore (1974) showed that the radioactive material from three of the four events had, in fact, penetrated the tropopause. Accordingly, the value of 4 weeks referred more likely to the lower level of the stratosphere than to the troposphere, so that a residence time of 4 weeks should be considered an upper limit to that in the troposphere.

Schemes that one may apply to deduce aerosol residence times from various radioactive elements have been reviewed by Junge (1963), Martell and Moore (1974), and Turekian *et al.* (1977). The published data admit residence times in the range 4–72 days, but crowd into two groups of values averaging 6 and 35 days, respectively. From the evidence available to him, Junge (1963) concluded that the higher value was appropriate to the troposphere as a whole and that the lower values were applicable only to the boundary layer near the Earth surface. Martell and Moore (1974), after having critically reviewed older and newer data, came to the opposite conclusion, namely, that the high values are due to the contribution of stratospheric aerosols, apart from misinterpretations of some data, while the lower values represent the true tropospheric residence time essentially independent of altitude.

Evidence for the second viewpoint comes from measurements of longer-lived radionucleides within the radium decay sequence, specifically bismuth-210 and lead-210. The major routes for nuclei conversion within the radium decay scheme are shown in Fig. 7-27. The direct decay product of radium-226, an alpha-emitter, is radon-222, which escapes the Earth surface. Only the continents are a source; the contribution from the oceans is negligible. Since the half-life time of radon-222 is only 3.8 days, its distribution in the troposphere is rather uneven. Over the continents the mixing ratio declines with increasing altitude (see Fig. 1-9). Over the oceans, the vertical gradient is reversed, as the oceans act as a sink and the zonal circulation keeps supplying material from the middle and upper troposphere. The immediate

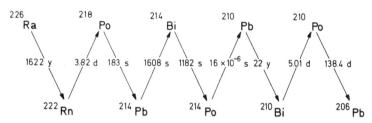

Fig. 7-27. Major routes within the radium decay scheme. Half-life times are shown in years (y), days (d), or seconds (s).

decay products of radon-222 are all short-lived. The first longer-lived species is lead-210 which has a half-life time of 22 years. The subsequent products are bismuth-210 and polonium-210 with half-life times of 5 and 138 days, respectively. If one assumes that all the radium-daughter activities in the atmosphere arise from the decay of radon-222, and that their production and removal are in a steady state, one has

$$dn_2/dt = n_1\lambda_1 - n_2(\lambda_2 + \tau_R^{-1}) = 0 \qquad (7\text{-}30)$$

where n_1 and n_2 are the number densities and λ_1 and λ_2 the radioactive decay constants of the parent and daughter isotopes, respectively. Since ^{210}Pb, ^{210}Bi, and ^{210}Po become attached to atmospheric particles, the removal term τ_R^{-1} is essentially identical with that for the tropospheric aerosol and $\tau_R \approx \tau_A$. There are three ratios of parent to daughter isotopes that may be utilized for the determination of τ_A: $^{222}Rn/^{210}Pb$, $^{210}Pb/^{210}Bi$, and $^{210}Po/^{210}Pb$. By successive application of Eq. (7-30) one obtains for the corresponding residence times

$^{222}Rn/^{210}Pb$: $\quad \tau_R = (\lambda_{Rn} n_{Rn}/n_{Pb} - \lambda_{Pn})^{-1}$

$^{210}Pb/^{210}Bi$: $\quad \tau_R = (\lambda_{Pb} n_{Pb}/n_{Bi} - \lambda_{Bi})^{-1} \qquad (7\text{-}31)$

$^{210}Pb/^{210}Po$: $\quad \tau_R = [-b + (b^2 - 4a)^{1/2}]/2a$

where $\quad a = \lambda_{Bi}(\lambda_{Po} n_{Pb}/n_{Po})$ $\quad$ and $\quad b = (\lambda_{Po} + \lambda_{Bi})$

From what has been said above, the model applies only to purely continental conditions, and the interference by marine air must be avoided. The measurements of Poet et al. (1972) and Moore et al. (1973) were conducted in the central United States. For the first two ratios of radionuclides, the results indicated residence times in the ranges 1.8–9.6 and 3.7–9.2 days, respectively, without any evidence for a systematic variation with altitude (in the troposphere). Averaged values for the residence times are 3.4 and 7.5 days, respectively. The $^{210}Pb/^{210}Po$ ratios, in contrast, gave tropospheric residence times falling into the range 12–49 days, and the vertical profile of ^{210}Po also was found to differ appreciably from that expected from the decay of radon 222. Previous measurements of $^{210}Pb/^{210}Po$ ratios by Lehmann and Sittkus (1959), Burton and Stewart (1960), Pierson et al. (1966), and Marenco and Fontan (1972) had given similar results. The explanation for the discrepancy appears to arise from sources of ^{210}Pb, ^{210}Bi, and ^{210}Po other than radioactive decay within the atmosphere. Soil particles are the most likely contributors, since a part of the tropospheric aerosol originates at the earth surface. In addition, Vilenskiy (1970) discovered that the top layer of the ground surface is enriched with lead-210. Coal burning and forest fires presumably are additional sources of radionuclides. Marenco and Fontan (1972) have given a list of various possibilities. Because of the relatively long lifetime of polonium-210, the surface source affects the $^{210}Pb/^{210}Po$ ratio most strongly, making this nucleide pair unsuitable for an estimate of the aerosol residence time. Poet et al. (1972), for example, showed that up to 85% of polonium-210 in the atmosphere is of a terrestrial origin. The ratios $^{210}Bi/^{210}Pb$ and $^{222}Rn/^{210}Pb$ are subject to quite moderate changes, so that correction factors may be applied. If the terrestrial component is taken into account, the effective residence time derived from $^{210}Bi/^{210}Pb$ is lowered, whereas that derived from $^{222}Rn/^{210}Pb$ is raised. The apparent residence times obtained from these nucleide ratios, 7.5 and 3.4 days, respectively, thus bracket the true tropospheric residence time of particulate matter, which must be approximately 5 days. A similar value had been deduced much earlier by Blifford et al. (1952) and by Haxel and Schumann (1955) from the ratio of short-lived to long-lived radon daughters in surface air. Due to several shortcomings discussed by Junge (1963), the method was not considered reliable, however.

A residence time of 5 days confirms the order of magnitude derived in the preceding section from budget considerations. The aerosol residence time thus is appreciably shorter than the time of vertical transport due to eddy diffusion, which takes approximately 4 weeks to bridge the distance between surface and tropopause. The disparity has led many authors to suggest that the residence time in successive layers of the troposphere increases with altitude (see, for example, Marenco and Fontan, 1973). The data of Moore et al. (1973) demonstrate, however, that the residence times deduced from both ratios, $^{210}Bi/^{210}Pb$ and $^{222}Rn/^{210}Pb$, are essentially independent of altitude between 2 km and the tropopause. It is a different question whether the residence time value applies to the aerosol as a whole or to particles in a specific size range. Since the decay of radon is essentially a gas-to-particle conversion process, one expects the radon decay products to attach primarily to particles in the accumulation mode. Accordingly, it is first of all the large particles whose tropospheric residence time is 5 days. How far this value extends to other size ranges remains to be discussed.

Since the aerosol residence time is shorter than the time constant for vertical transport by eddy diffusion, an efficient process is needed for the removal of particles from the troposphere. Scavenging by precipitation is believed to be the principal process (see Section 8.3). In this regard, it is interesting to note that the ratio of $^{210}Bi/^{210}Pb$ in rainwater, first investigated by Fry and Menon (1962) and subsequently by Poet et al. (1972), leads to values for the aerosol residence time of 6–7 days, which compare well with those obtained for dry aerosols. The good agreement demonstrates that precipitation is, in fact, important in removing particles from the tropospheric air space. If this is so, the removal rate and the local residence time for aerosol particles will depend to some extent on the average precipitation rate. The measurements of Poet *et al.* (1972) and Moore *et al.* (1973) were made at middle latitudes, where the precipitation rate has a relative maximum. Higher precipitation rates are found near the equator. Here, the residence time may be even shorter. Hence, it should be recognized that a residence time of 5 days applies to the midlatitudes of the northern hemisphere, and different values may occur at other latitudes.

The preceding considerations referred to bulk material of particulate matter. We turn now to discuss time constants for the removal of individual particles as a function of particle size. Jaenicke (1978c, 1980) has studied this aspect in some detail, and Fig. 7-28 summarizes his findings. Three processes come into play.

1. Aitken particles are removed mainly by coagulation with other particles. The corresponding time constant is obtained from the decline of

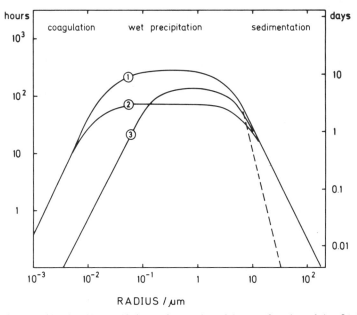

Fig. 7-28. Combined residence lifetimes of aerosol particles as a function of size. [Adapted from Jaenicke (1978c, 1980).] Important removal processes, active in various size ranges, are indicated. Coagulation and sedimentation time constants were calculated; the time constant for wet removal is the residence time derived from ^{210}Bi/^{210}Pb and ^{222}Rn/^{210}Pb ratios (Martell and Moore, 1974). Curves 1 and 2 represent the background aerosol for τ_{wet} equal to 12 and 3 days, respectively. Curve 3 represents the continental aerosol with $\tau_{wet} = 6$ days. The dashed line is calculated from a simple model for sedimentation equilibrium, as described in Section 7.6.3.

concentration to $1/e$ of the initial value in the absence of other sources. Since coagulation is a volume process, the time constant is equivalent to the particle lifetime. As discussed in Section 7.3.1, the rate of coagulation depends on the mobility of Aitken particles as well as on the number density of the entire aerosol population. Figure 7-28 shows that the lifetime of Aitken particles increases essentially with the particle radius squared. The number density of background aerosol is almost two orders of magnitude smaller than that of the continental aerosol, so that the lifetime of background Aitken particles is increased accordingly.

2. The main removal process for particles in the 0.1–10 μm size range is wet precipitation. Recall from Section 7.3.2 that aerosol particles serve as cloud condensation nuclei. The largest particles are readily activated, but with decreasing particle radius the barrier for cloud drop formation increases, and eventually it becomes too high. The lower size limit lies near 0.1 μm, but it depends somewhat on the total particle number density. For continental aerosols the cutoff probably occurs somewhere near 0.2 μm; for the marine background aerosol the cutoff lies somewhat below 0.1 μm. At any rate, it is clear that the range covers giant particles and a good deal of large particles. Combined these particles comprise most of the aerosol mass. In Fig. 7-28 the residence times deduced above are applied, and they establish an upper limit.

3. For particles greater than 10 μm in radius, sedimentation becomes increasingly important as a removal process. The solid curve shown in Fig. 7-28 was calculated by Jaenicke (1978c) from the time for sedimentation of particles starting at 1.5 km altitude, which represents an average height for the exponential falloff of number density with a scale height of 1.5 km. This approach must be questioned, however. Although the assumption of a scale height of 1.5 km is realistic for the bulk of the continental aerosol, it cannot be applied to particles larger than 10 μm because their residence times become sufficiently small to favor a sedimentation equilibrium, such that each particle size group adjusts to its own scale height. This concept will be discussed further below. Residence times resulting from sedimentation equilibrium are shown in Fig. 7-28 by the dashed line.

The residence time of the aerosol has important implications with regard to the transport and distribution of substances associated with particulate matter. For the bulk of the aerosol, the residence time is sufficient to permit particles to spread around the globe within the mean zonal flow, but in latitudinal direction the average distance of transport must be less, and an interhemispheric exchange of aerosol particles can have only local significance in the vicinity of the interhemispheric tropical convergence zone. Particulate matter deposited onto the continents will be subject to resuspension into the air, whereby the material may spread over larger distances after all. Mineral dust deposited onto the surface of the oceans is irretrievably lost from the atmosphere and contributes to the formation of deep-sea sediments. According to Griffin *et al.* (1968), more than 50% of the carbon-free mass of ocean sediments consist of clay minerals, which except in areas of river influence are predominantly of aeolian origin. The Sahara desert is a well-known source of mineral dust. A large portion of it is carried westward with the trade winds and crosses the Atlantic Ocean. Schütz (1980) has summarized the present knowledge on the dust flow. As Saharan air masses leave the continent, they are undercut by cooler air moving in from the northeast, and an inversion layer is established that impedes the vertical exchange of air and preserves the dust flow at levels above the trade inversion. In this manner Sahara dust is transported across the Atlantic Ocean into the Caribbean Sea, although its concentration is steadily diluted by downward mixing and precipitation.

7.6.3 SEDIMENTATION AND DRY DEPOSITION

The sedimentation of particles—that is, their downward motion due to gravitational settling—follows Stokes's law. Turbulent diffusion represents an opposing force, and sedimentation equilibrium is established when both forces cancel. Assuming a constant production rate of particles at the earth surface for each size group and a balance of upward and downward fluxes in the atmosphere leads to the following equation:

$$-K_z n_M \frac{d}{dz}[n(r)/n_M] - v_s(r)n(r) = 0 \qquad (7\text{-}32)$$

Here, n_M is the number density of air molecules, $n(r)$ that of particles with radius r, and $v(r)$ is the sedimentation velocity. For an isothermal atmosphere, $n_M = n_M^0 \exp(-z/H)$, with H being the scale height. For simplicity we shall neglect the variation of K_z with height in the boundary layer and assume a constant $K_z = 20 \text{ m}^2/\text{s}$. The above equation then has the solution

$$n(r) = n_0(r) \exp[-(v_s(r)/K_z + 1/H)z]$$
$$= n_0(r) \exp(-z/h) \qquad (7\text{-}33)$$

The scale height of the exponential distribution of $n(r)$,

$$h = H/(1 + v_s H/K_z)$$

decreases with increasing sedimentation velocity. This quantity is given by

$$v_s(r) = b(r)gV_p(r)(\rho_p - \rho_{air}) \approx b(r)gV_p(r)\rho_p \qquad (7\text{-}34)$$

where $b(r)$ is the mobility of a particle, g is the acceleration due to gravity, $V_p(r)$ is the volume of the particle, and ρ_p is its density, against which the density of air can be neglected. Inserting $V_p(r) = 4\pi r^3/3$ and $b(r)$ from Eq. (7-6) in the approximate form for larger particles $b(r) = (6\pi\eta r)^{-1}$, one obtains

$$v_s(r) = \tfrac{2}{9}(g\rho_p/\eta)r^2 = (2.38 \times 10^{-4})r^2 \quad \text{m/s}$$

Here, $\eta = 1.83 \times 10^{-5}$ kg/m s denotes the viscosity of air, r is inserted in units of micrometers, and a particle density of 2×10^3 kg/m^3 is assumed. The column mass content in the troposphere for particles with radius r is given by

$$W_p(r) = V_p\rho_p \int_0^{z_T} n(r)\, dz = V_p\rho_p h n_0(r)(1 - e^{-z/h})$$

$$\approx V_p\rho_p h n_0(r) \qquad (7\text{-}35)$$

The approximation is valid for particles with radii greater than 10 μm. The corresponding mass flux returning to the earth surface due to sedimentation is given by

$$F_p(r) = V_p\rho_p n_0(r)v_s(r) \qquad (7\text{-}36)$$

The ratio of the last two equations is equivalent to the local residence time of sedimenting particles

$$\tau_p(r) = W_p/F_p = h/v_s(r) = \frac{H/v_s(r)}{[1 + v_s(r)H/K_z]} \qquad (7\text{-}37)$$

Again, r must be inserted in units of micrometers. For particles with radii greater than 10 μm, one has $v_s(r)H/K_z \gg 1$, so that Eq. (7-37) simplifies to

$$\tau_p(r) = K_z/v_s^2(r) \qquad (7\text{-}38)$$

The resulting residence times, which are shown in Fig. 7-28 by the dashed line, should still be considered upper limits in view of the fact that values for K_z in the planetary boundary layer generally are smaller than 20 m^2/s. Junge (1957) and Fabian and Junge (1970) have shown how one can, in principle, incorporate the variation of K_z in the boundary layer. This refinement will not be discussed, however, because meteorological conditions in the boundary layer are too variable for a generalized model to be applicable.

In Section 1-6 we have discussed the concept of dry deposition of gases that are absorbed at the Earth surface. The loss by absorption sets up a gradient of the mixing ratio or concentration, which maintains a downward flux of material due to turbulent diffusion. This flux was expressed as the product of a deposition velocity and the concentration of the substance at a reference height in a manner similar to Eq. (7-36). The similarity of Eqs. (7-36) and (1-31) show that sedimentation is a dry deposition process and that for sedimenting particles the deposition velocity is identical with the sedimentation velocity. For particles in sedimentation equilibrium, the concentration gradient is directed upward because a uniform source was assumed to be active at the ground surface. The gradient reverses, however, as soon as the source is shut off. This raises the question of to what extent particles smaller than 10 μm that are not effectively removed by sedimentation can undergo dry deposition. Inasmuch as the particles are sticky, they are removed from the atmosphere by coming into contact with vegetation and other surface elements. Due to turbulent mixing, a downward flux is then established, which may be expressed in terms of a deposition velocity similar to that of sedimenting particles. Thus we expect also particles smaller than 10 μm to undergo dry deposition.

Figure 7-29 shows deposition velocities as a function of particle size. They were determined by laboratory and field experiments as reviewed by McMahon and Denison (1979) and by Sehmel (1980) and refer to grass

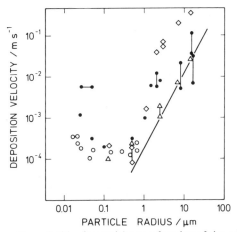

Fig. 7-29. Dry deposition velocities for particles as a function of size according to field and laboratory measurements of Chamberlain (1953, 1966a), Möller and Schumann (1970), Sehmel and Sutter (1974), Clough (1975), Little and Wiffen (1977), and Wesley et al. (1977). Filled symbols are for deposition onto grass and open symbols are for deposition onto water surface. The solid line indicates the sedimentation velocity. Wind speed: $\triangle$, $u \approx 2$ m/s; $\bigcirc$, $u \approx 8$ m/s; $\diamondsuit$, $u = 14$ m/s.

and water surfaces. For particles with radii greater than 1 μm, the deposition rate is given essentially by the sedimentation velocity. At higher wind speeds, however, the deposition rate is enhanced due to the increase in atmospheric turbulence. For submicrometer particles the deposition velocity attains a relative minimum of some 10^{-4} m/s in the 0.1–1.0 μm size range, whereas for Aitken particles the deposition velocity rises again with decreasing particle size. The last effect is caused by the increase in particle mobility, which leads to an increase in the frequency of collisions with surface elements.

For comparison with the data in Fig. 7-28, it is instructive to calculate tropospheric residence times resulting exclusively from dry deposition. We confine the calculation to the size range 0.05–1.0 μm, where the deposition velocity is roughly size-independent. For continental aerosols this size region contains approximately one-half of the total mass of all aerosol particles. The residence time is then given by the tropospheric column content of such particles divided by their flux to the ground,

$$\tau_p \approx \frac{1}{2}\left(\frac{W_{AC}}{F_p}\right) = \frac{1}{2}\left[\frac{W_{AC}}{(c_1^0/2)v_g}\right] = 86 \quad \text{days}$$

the result is obtained by inserting the values $W_{AC} = 5.17 \times 10^{-5}$ kg/m^2 and $c_1^0 = 45$ μg/m^3 taken from Section 7.6.1, and $v_d = 1.5 \times 10^{-4}$ m/s taken from Fig. 7-29. These data lead to a residence time of 86 days, which is about 20 times the value derived from radon product tracers. Thus, it appears that dry deposition cannot be the major removal process for particles in the 0.1–1.0 μm size range. The discussion in Section 8.3 will show that nucleation scavenging followed by wet precipitation provides a more effective removal process for large particles. Dry deposition for 10-μm-sized particles will be competitive with wet precipitation, but in this case the process is confined to the boundary layer because eddy diffusion from the middle to the lower troposphere becomes rate-determining.

It should be noted that the data in Fig. 7-29 refer exclusively to surfaces with small roughness length (water and grass). High-growing vegetation may be more effective in collecting aerosol particles, in that stems and leaves of bushes and trees act as filters, intercepting particles by impaction. Sehmel and Hodgson (1976) have shown theoretically how the minimum for the deposition velocity, apparent in Fig. 7-29, is raised and shifted toward larger sizes by an increase in the roughness height of the vegetation. Field data that quantify the effect are rare. Höfken et al. (1981) have studied size-discriminated aerosol particles outside and underneath the canopies of beech and spruce forests and found support for the theoretical prediction of Sehmel and Hodgson (1976). The results of Höfken et al. (1981) indicate that deposition velocities for submicrometer-sized particles are raised appreciably to values approaching 10^{-2} m/s. This is sufficient to make dry deposition to forested areas a process that cannot be ignored, even though it would be confined to the boundary layer region. The effect is predicted to be particularly significant for ammonium sulfate, because this constituent is concentrated in the submicrometer size range.

8 | Chemistry of Clouds and Precipitation

The condensation of water vapor and its precipitation from the atmosphere in the form of rain, snow, sleet, or hail are important not only for the water cycle, but also because they bring to the earth surface other atmospheric constituents, primarily those substances that have a pronounced affinity toward water in the condensed state. Cloud and precipitation elements may incorporate both aerosol particles and gases. The uptake mechanisms are discussed in this chapter, together with the inorganic composition of cloud and rain water that they determine. These processes are, in principle, well understood. Another subject requiring discussion is the occurrence of chemical reactions in the liquid phase of clouds. The oxidation of SO_2 dissolved in cloud water is considered especially important. As a result of laboratory studies, the conversion of SO_2 to sulfate is now known to proceed by several reaction pathways in aqueous solution.

8.1 The Water Cycle

The column density of water in the troposphere is determined largely by the vertical distribution of the vapor. Liquid droplets and ice crystals represent only a minor fraction of the total abundance of H_2O. Even in dense clouds the mass of water in the vapor phase predominates over that in the condensed state. Clouds are instrumental, of course, in the removal of water from the atmosphere by precipitation.

The maximum burden of water vapor in the air is given by the local saturation vapor pressure p_s. This quantity is a strong function of temperature (see Table A-2). An air parcel rising from the ground surface on up experiences a decrease in temperature and a corresponding lowering of p_s. Condensation sets in when the saturation level is reached. Above this level the air becomes continuously drier as more and more water condenses out and is left behind. Conversely, sinking dry air eventually picks up water vapor as it mixes with moister air at lower altitudes. At the same time, a subsiding air parcel encounters rising temperatures, causing condensed water to evaporate. This mechanism of convective transport coupled with the requirement of material balance thus demands that at altitudes above the saturation temperature level the water vapor pressure adjusts to approximately one-half the saturation vapor pressure, or a relative humidity of 50%.

This prediction can be checked with the help of radiosonde data. The instrumentation normally is not suitable to measure the low abundance of water in the upper troposphere and stratosphere, so that the data are restricted to altitudes below 7 km. Oort and Rasmussen (1971) have compiled zonally averaged, mean monthly specific humidities (H_2O mass mixing ratios) as a function of height for the northern hemisphere. Average water vapor pressures computed from their values are shown in Fig. 8-1 as a function of temperature. If, as a precaution, one uses only data for altitudes

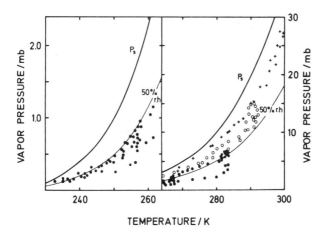

Fig. 8-1. Zonally averaged partial pressures of water vapor for January and July calculated from specific humidities presented by Oort and Rassmusson (1971) and Newell *et al.* (1972) for latitudes between 10°S and 75°N. Data at pressure levels of 400, 500, and 700 mbar are shown by dots, data at 850 mbar by open circles, and data at 1000 mbar by crosses.

above the 850-mbar level, the points are found to scatter reasonably well around the 50% relative humidity curve. Averaged H_2O vapor pressures at the surface level are included in Fig. 8-1 for comparison. They correspond to relative humidities of 75–80%. Such high values are typical for surface air over the oceans. Generally, the relative humidity is quite variable even at altitudes above the 850-mbar level. In addition to the expected statistical scattter, there are regional influences such as the meridional Hadley circulation, which causes the subtropical latitudes to be drier than the more humid tropics at practically all altitude levels.

The radiosonde data also provide the total amount of precipitable water in the atmosphere. Zonal mean values are shown in Fig. 8-2. The data are based mainly on the maps of Bannon and Steele (1960). More recent data presented by Newell et al. (1972) are in good agreement. The amount of precipitable water decreases from the equator to the poles due to the poleward decrease of temperature in the troposphere.

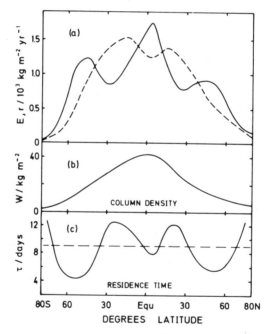

Fig. 8-2. (a) Zonally averaged, mean annual rates of evaporation $\bar{E}$ and precipitation $\bar{R}$ (solid line) versus latitude as given by Sellers (1965). (b) Mean annual column densities W of precipitable water vapor, from data presented by Sellers (1965) and Newell et al. (1972). (c) Mean residence times of water vapor, $\tau = W/\bar{R}$, calculated from (a) and (b). The broken line gives the global mean value $\bar{\tau} = 9$ days.

Water enters into the atmosphere by evaporation from the ocean surface, lakes and rivers, vegetation, and the uppermost layer of soils. On the continents, the local balance of surface water is determined by evaporation E, precipitation R, and losses due to the runoff of water toward rivers and underground reservoirs, which carry it further into the oceans. A similar balance applies to the oceans except that here the runoff from the continents represents a gain rather than a loss. To the benefit of the continents, the oceans evaporate more water than they receive by precipitation, although on a global scale both must balance.

Figure 8-2 depicts zonally averaged, mean annual rates of evaporation and precipitation as a function of latitude. Precipitation rates peak at the equator and in the middle latitudes. The first peak results from the convergence and updraft of moist air advected with the Hadley circulation. The secondary maxima are caused by the high frequency of cyclone activity in the westerlies. The subtropical high-pressure regions are much drier because of subsiding dry air in the downward branch of the Hadley cell. Here, and in contrast to the situation in other latitude zones, the rate of evaporation exceeds that of precipitation. To balance the zonal water budget requires a transport of moisture from the subtropical source regions toward lower and higher latitudes. The poleward flux is maintained by eddy transport and the gradient of precipitable water with latitude. In the tropics the flow of water vapor follows the direction of the Hadley circulation. The transport is most effective in the lowest strata of the troposphere, where the highest H_2O mixing ratios are found.

A closer inspection of the global water budget shows that the evaporation rates are higher in the southern compared with the northern hemisphere, whereas the integrated precipitation rates are nearly equal in both hemispheres. The budget is balanced by a flux of water vapor across the equator. The difference in evaporation rates is due, on one hand, to the greater ratio of ocean to land areas in the southern hemisphere and, on the other hand, to the greater extent of subtropical desert areas in the northern hemisphere where the evaporation rates are much diminished. Transport of water vapor across the equator takes place in both directions, depending on the seasonal position of the Hadley cells and the intertropical convergence zone. The strongest transequatorial H_2O flux occurs in northward direction and is accomplished by the southern Hadley circulation during the period April–November. During the remaining months of the year the H_2O flux is directed southward but with much lower intensity.

Newell *et al.* (1972) have analyzed the Hadley motion in some detail, and they estimated a mean annual flux of water vapor across the equator of 1.4×10^4 Pg yr (1 Pg = 10^{12} kg). A similar flux is obtained from the differences in evaporation and precipitation rates integrated over all latitude zones of each hemisphere. From the data presented in Fig. 8-2 one finds 1.6×10^4 Pg/yr. These values may be compared with the total mass of water precipitated annually in each hemisphere, which is 2×10^5 Pg/yr. The net transfer of water vapor across the equator thus amounts to about 5% of the total annual water budget in each hemisphere.

Finally, we use the data of Fig. 8-2 to estimate from the ratio of the H_2O column density and the precipitation rate the seasonally and zonally averaged residence time of water vapor in the atmosphere. Its distribution with latitude is shown in the lower section of Fig. 8-2. The residence times are found to range from 4.5 to 12.5 days, with an average of about 9 days, which is shown by the dashed line. The results are approximate because they neglect the meridional transport of water vapor from net source to net sink regions. While it seems reasonable that the residence time has a maximum in the dry subtropics, it is from these regions that up to 50% of the water actually being precipitated is transferred toward adjacent latitude belts. The true residence time is lowered accordingly, and the maxima near 25° latitude probably disappear if appropriate corrections are made. The increase of the residence time toward the poles must be considered real, however, since the high latitude zones represent sink rather than source regions. Therefore, we estimate H_2O residence times in the troposphere of about 8 days in the 0–30° latitude regions, 5 days in the 45–65° latitude belts, and greater than 12 days in the polar regions.

8.2 Cloud and Rain Formation

Before we discuss the chemical composition of cloud and rain waters, it will be useful to assemble some background information on the processes leading to the formation of clouds and precipitation. The subject has been treated comprehensively in the books of Fletcher (1962), Mason (1971), and Pruppacher and Klett (1980), which should be consulted for further details.

According to the discussion in Section 7.3.2, cloud drops are generated by the condensation of water vapor onto aerosol particles when in a rising air parcel the relative humidity transcends the saturation level due to adiabatic cooling. The uplift of most air may occur by local convection, in frontal systems, in large cyclones, or along the windward slope of mountains. Aerosol particles are important in two ways. On one hand, they serve as condensation nuclei enabling clouds to form at fairly low degrees of H_2O supersaturation. On the other hand, they contain water-soluble material such as inorganic salts, which contribute significantly to the ionic composition of cloud water. Although with rising relative humidity all particles acquire a certain share of liquid water, it is only the larger particles with radii greater than about 0.1 μm that become activated and grow into cloud droplets. The largest and most soluble particles are activated first. Some of the smaller and thus more numerous particles become activated later, as the air rises toward greater heights and the supersaturation increases. The supersaturation reaches a maximum when the rate of condensation equals the rate at which condensable moisture is made available by adiabatic cooling. The number density of cloud drops is determined at this point. Subsequently, the supersaturation decreases, although condensation still continues. The growth rate of cloud drops and the liquid water content of cloud-filled air are highly correlated with the vertical velocity of uprising air. The maximum of achievable liquid water content is rarely reached, however, because of the entrainment of drier air at the cloud's periphery. This effect is most pronounced in clouds of small width. Generally, the liquid water content increases with height above the cloud base and attains a maximum somewhere in the upper half of the cloud.

Table 8-1 summarizes for several kinds of clouds specific parameters of interest, such as the number density of cloud elements, the range of radii encountered, and the liquid water content. The process of condensation leads to a fairly narrow size distribution of cloud drops centered near 5 μm radius. This feature arises from the fact that droplet growth is mediated by molecular diffusive transport of water vapor. The growth rate is inversely proportional to the droplet's radius. Accordingly, the rate slows down as the drop size increases. Condensation alone is too inefficient a process, therefore, to enable the production of large cloud drops during the lifetime of a cloud. The larger drops are thought to form from smaller ones by collisions and coalescence, in a manner resembling the coagulation of particles treated in Section 7.3.1, except that the collision probabilities are modified by hydrodynamic and gravitational effects. The usual concept employed to describe collisional growth of cloud drops is the accretion of small drops by larger ones in the assembly, due to their higher than average terminal fall velocity (sedimentation coagulation). The gravitational settling of large cloud elements, which causes them to sweep up smaller droplets in their path, is partly counterbalanced by the uprising air motion within the cloud. Since the settling velocity and consequently also the accretion rate increase with the size of the collector drop, its growth accelerates with time. This makes deep clouds more effective in the production of large cloud drops compared with the rather shallow fair weather cumuli. Actual drop size spectra observed in clouds support this view. For examples, see Pruppacher and Klett (1980).

Another important aspect of cloud formation is the generation of ice particles at temperatures below the freezing point of water. The spontaneous glaciation of liquid water drops in clouds is a fairly rare event unless the

Table 8-1. *Typical Drop Sizes and Liquid Water Contents for Fogs and Several Types of Clouds*[a]

| Cloud type | | Drop number density (cm^{-3}) | Drop radius | | | Liquid water content (g/m^3) | Vertical extent (m) |
			Range (μm)	Most frequent (μm)	Average (μm)		
Fog		2000	0.5–30	0.8	—	0.1–0.5	—
Fair-weather cumulus							
(a) Continental	Nonprecipitating	300	2–30	5	5	0.1–0.4	750
(b) Maritime		60	2–25	12	15	0.4–0.5	
Cumulus congestus (continental)		100	2–70	5	20	0.5–1.0	1200–2000
Cumulonimbus (continental)	Precipitating	75	2–100	5	25	2.0	5000

[a] According to data assembled by Mason (1971) and Pruppacher and Klett (1980).

temperature approaches $-20°C$. This value corresponds to an altitude of about 6 km. In the middle troposphere, therefore, supercooled liquid water clouds are a common phenomenon. The probability of ice formation increases with decreasing temperature, so that at 250 K all clouds contain a certain population of ice particles. Mixed clouds consisting of supercooled water drops and ice particles are unstable because the H_2O equilibrium vapor pressure over ice is lower than that over liquid water at the same temperature. The ice crystals thus grow by sublimation at the expense of the liquid phase. The initial generation of ice particles takes place in much the same way as the formation of a liquid water drop, in that a solid particle must serve as a nucleus regardless of whether ice formation proceeds by the freezing of a supercooled liquid drop or by the condensation of water molecules onto the nucleus directly from the vapor phase. Compared with cloud condensation nuclei, a much smaller fraction of particles among the total aerosol population is capable of serving as ice nuclei. It is this feature that makes the occurrence of supercooling so common in clouds. The exact physicochemical nature of ice nuclei has not yet been specified. Insoluble clay particles of the kaolinite type, volcanic ash, and decaying plant leaves have been found particularly effective. In addition to growing by vapor deposition, ice particles may also collide with and accrete supercooled water drops, which then freeze onto them. The process is called riming. In the course of time, riming produces graupel or, in the extreme case, hailstones. Similar to the mechanism of liquid drop growth by coalescence, the ice particles must reach a minimum critical size before riming becomes an effective growth process, but thereafter the growth rate increases rapidly with size.

Most clouds do not lead to precipitation. Those cloud elements that grow by coalescence to a size sufficient for sedimentation into the region below the cloud base then experience relative humidities less than 100%, causing the drops to evaporate. A minimum size is required if a water drop is to survive the sojourn to the earth surface without evaporating completely. Figure 8-3 indicates how much the size of a falling rain drop is reduced by evaporation as it traverses a stagnant air layer of up to 2 km depth. Only drops having an initial radius greater than about 500 μm will reach the ground. Observed size spectra of rain drops extend from perhaps 100 μm in radius (drizzle) up to 4–5 mm. The number frequency decreases approximately exponentially with size because the larger drops suffer collisional breakup (List, 1977).

Two mechanisms are thought to be operative in the formation of rain: the Wegener–Bergeron–Findeisen process, and the warm rain process. The first of these is germane to high-reaching continental clouds and starts with the growth of ice particles in the upper part of the cloud. Under favorable

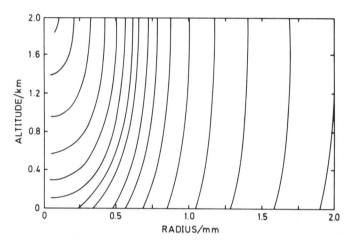

Fig. 8-3. Decrease in the size of rain drops by evaporation during their fall time below clouds. [From Kortzeborn and Abraham (1970), with permission.]

conditions, it will take of the order of 30 min for such ice particles to grow to a size sufficient for removal by sedimentation. On the way toward the ground they enter warmer layers of the atmosphere and melt. Wegener (1911), Bergeron (1935), and Findeisen (1939) have been the principal proponents for this process of cold rain formation, so that it now bears their names. It cannot be the only mechanism, however, since in the tropics and over the oceans one has observed precipitating warm clouds whose tops definitely had not reached beyond the 273 K temperature level. In such clouds the coalescence of liquid cloud elements is the only mechanism that can explain the formation of millimeter-sized raindrops. In this regard, it is significant that the drop size spectrum in marine cumulus clouds features lower number densities and larger drops, on average, then continental clouds (cf. Table 8-1). This feature appears to result from the comparatively low number density of condensation nuclei over the oceans coupled with the role that giant seasalt particles undoubtedly assume in the process of cloud formation. Thus, larger drops with radii of about 25 μm are formed earlier than in continental clouds. This would facilitate the growth of such drops to the millimeter size range during the lifetime of the cloud.

For global considerations of cloud chemistry, it would be desirable to have data on the spatial distribution and the average lifetime of clouds in the troposphere. Such data, unfortunately, are not generally available. De Bary and Möller (1960) have evaluated aircraft observations of cloud coverage over Germany to derive frequency distributions of clouds with height. Their results suggest that at midlatitudes the clouds are concentrated in the 1-4 km altitude regime and that they fill about 10-20% of the entire tropospheric air space. These data should not be extrapolated toward other latitudes, however, and it is even doubtful whether they apply to the interior of the Eurasian continent. Junge (1964), in an unpublished report, estimated small convective clouds to have a lifetime due to reevaporation of 0.5-1 h, whereas that of higher-reaching deep clouds would be 2-3 h. The nuclei set free by dissipating clouds are, in the course of time, subject to renewed condensation and eventually pass through a series of such condensation-evaporation cycles before they are irretrievably removed from the atmosphere by precipitation. According to estimates by Junge (1964), a cloud condensation nucleus undergoes about 10 condensation-evaporation cycles, on average, during the time of its residence in the troposphere.

The number of cycles and the time spent in the liquid phase may be significant to the chemistry of the tropospheric aerosol. Various chemical reactions of gases dissolved in cloud water, primarily the oxidation of SO_2 toward sulfuric acid, act to produce new material of low volatility that remains associated with the particle generated from the cloud drop after its evaporation. Aqueous reactions followed by dissipation of cloud drops represent a gas-to-particle conversion process, which modifies the chemical composition of the aerosol. Evidence for the occurrence of this process has been obtained by Radke and Hobbs (1969), Dinger et al. (1970), Saxena et al. (1970), and Hegg et al. (1980). They found that cloud condensation nuclei activated at a given supersaturation are often more numerous in the air of dissipating clouds than in the ambient air not involved in cloud formation. Easter and Hobbs (1974) have formulated a model specifically applicable to orographic wave clouds in which they put forth an explanation of the phenomenon. The authors assumed that aerosol particles requiring initially a fairly high degree of supersaturation for their activation receive a coating of sulfate due to the in-cloud oxidation of SO_2. Subsequently, after drying up, the particles are more soluble than before and require a lower supersaturation when they are activated again. Hegg and Hobbs (1981, 1982) also have studied the production of particulate sulfate in orographic wave clouds, using aircraft to expose filters upwind and downwind of such clouds. The analysis showed a significant sulfate yield in 16 of 28 observations.

8.3 The Incorporation of Particulate Matter into Cloud and Rain Water

Hydrometeors acquire trace components by scavenging processes occurring within clouds as well as below clouds. In the older literature, these are distinguished as rainout and washout, respectively. The term washout, however, has also been used to describe precipitation scavenging more generally. Slinn (1974) consequently advocated the less ambiguous, alternative expressions, in-cloud scavenging and below-cloud scavenging.

8.3.1 NUCLEATION SCAVENGING

Owing to their function as cloud condensation nuclei, aerosol particles are naturally embedded into cloud drops during cloud formation. Maximum values of supersaturation are about 0.1% for stratiform clouds and 0.3–1% for cumuliform clouds. In continental clouds with droplet number densities of 100–300 per cubic centimeter, essentially all particles with radii greater than 0.2 μm are expected to undergo nucleation scavenging (cf. Fig. 7-6). This fraction comprises roughly 60–80% of total aerosol mass. In marine clouds the situation is even more favorable because of the smaller number density of aerosol particles over the oceans. Once the cloud has formed, the nonactivated fraction of the aerosol population undergoes thermal coagulation with other aerosol particles and with cloud drops. Coagulation removes mainly the smallest and most mobile particles in the Aitken range. Their incorporation into cloud drops adds little extra mass, however. The coagulation of particles with radii around 0.1 μm is fairly slow and the least efficient. Gravitational scavenging of particles by sedimenting water drops also discriminates against particles in the 0.1 μm size range (see further below), so that these particles remain present as an interstitial aerosol.

Let c_A (in μg/m^3) $= c_A^0 \exp(-z_{cb}/h)$ denote the aerosol concentration at the cloud condensation level z_{cb} (cloud base). Here c_A may refer either to an individual constituent of the aerosol or to its entire mass concentration. The aqueous concentration c_w resulting from in-cloud scavenging can then be estimated from a relation first formulated by Junge (1963),

$$c_w = c_A \varepsilon_A / L = c_A (\varepsilon_n + \varepsilon_B + \varepsilon_c)/L \quad \text{mg/kg} \tag{8-1}$$

where ε_A (with $0 < \varepsilon_A < 1$) is the mass fraction of material incorporated into cloud water, and L (in g/m^3) is the average liquid water content of the cloud. Although the overall scavenging efficiency ε_A combines the effects of nucleation scavenging ε_n, attachment to cloud drops by Brownian motion ε_B, and collisional capture ε_c, the first term is the dominant one. We assume values of $\varepsilon_A \approx 0.75$ for continental clouds in background regions, and $\varepsilon_A \approx 0.95$ for marine clouds. For precipitating clouds in the middle latitudes, the normal range of the liquid water content is 1–3 g/m^3. For nonprecipitating clouds, the values shown in Table 8-1 may be used.

Two recent aircraft studies using sulfate as a tracer have provided some observational data on the efficiency of particle scavenging in nonprecipitating clouds ($L = 0.15$–0.5). Hegg and Hobbs (1983) corrected for a possible contribution to sulfate from in-cloud oxidation of SO_2 and found efficiencies in the range 0–174% with an average value of 0.68 ± 0.52 in a series of 12 flights. Saxena and Hendler (1983) reported a range of 9–140% with an average efficiency of 0.56 ± 0.52 from four flights without applying any corrections for SO_2 oxidation. These values are consistent with the preceding estimates, especially if one takes into account the size distribution for sulfate, which favors submicrometer-sized particles. The large scatter in the data is unexpected and may be due to technical difficulties.

Precipitation removes from the atmosphere particulate matter incorporated into cloud water. The associated average mass flux is

$$F = 10^{-3} c_{\mathrm{w}} \bar{R} = 10^{-3} c_A \varepsilon_A \bar{R} / L \qquad (8\text{-}2)$$

where $\bar{R}$ (in kg H_2O/m^2 yr) is the average annual precipitation rate. Locally, the occurrence, rate, and duration of rainfall are distributed with time in a stochastic manner. It is necessary, therefore, to average over a larger area in order to have precipitation somewhere in that region, and a sufficiently long period of time. In the middle latitude belt, $\bar{R} \approx 800$ kg H_2O/m^2 yr. The regional residence time of particulate matter being removed from the troposphere by rainout is defined by the ratio of the column density W to the mass flux F. For the vertical distribution of the aerosol we may use the model presented in Section 7.6. In the lower atmosphere over the continents, the particle concentration declines with height quasi-exponentially with a scale height of $h \approx 1$ km. The cloud condensation level also is located at altitudes near 1 km. Finally, setting $L = 2$ for raining clouds, one obtains for the residence time

$$\tau_{\mathrm{rainout}} = \frac{W}{F} = \frac{c_1^0 h + c_2^0 H[1 - \exp(-z_\mathrm{T}/H)]}{\bar{R} \varepsilon_A c_1^0 \exp(-z_{\mathrm{cb}}/h) / L}$$

$$= \frac{5.2 \times 10^{-5} \times 2}{(800)(0.75)(1.7 \times 10^{-5})} \approx 1 \times 10^{-2} \quad \text{yr or 3.7 days} \qquad (8\text{-}3)$$

This value has the same magnitude as the residence time for the bulk aerosol derived in Section 7.6 in a different manner. The congruence of results shows that in-cloud scavenging and precipitation is the dominant process for the removal of particulate matter from the troposphere.

8.3.2   BELOW-CLOUD SCAVENGING OF PARTICLES

Falling precipitation elements incorporate particles by collisional capture. For water drops, the usual assumption that particles and drops merge upon contact has been experimentally confirmed by Weber (1969). Consider then a single rain drop of radius r_1 moving with velocity v. It would sweep out a cylinder with cross section πr_1^2 and collect the particles contained therein,

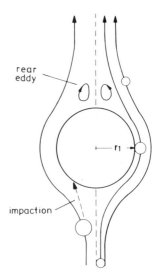

Fig. 8-4. Schematic representation of the air flow around a falling sphere.

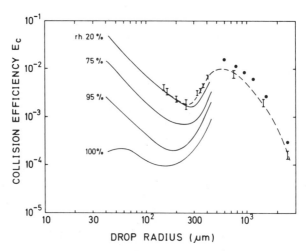

Fig. 8-5. Collision efficiency E_c for the capture of particles ($r = 0.25$ μm) by raindrops as a function of drop radius r_1 and ambient relative humidity. Solid curves were calculated by Grover *et al.* (1977) on the basis of hydrodynamic theory; bars show experimental results of Wang and Pruppacher (1977), and points results of Kerker and Hampel (1974) and of Lai *et al.* (1978). The latter data were, in part, corrected for terminal drop velocity and particle size. The maximum near $r_1 = 500$ μm is due to rear capture of particles in the standing wake eddy for drop sizes 200–600 μm; the fall-off of E_c for larger drop sizes indicates increasing wake eddy shedding.

were it not for the air flow around the nearly spherical drop, which causes the particles to follow the hydrodynamic streamlines. The situation is depicted schematically in Fig. 8-4 to indicate how the hydrodynamic flow pattern modifies the capture cross section. Formally, the effect is taken into account by an efficiency factor $E_c(r_1, r_2)$, which is a function of the radii of both collision partners. From Fig. 8-4, it would seem that E_c is generally smaller than unity, but some experimental results indicate values $E_c > 1$ under certain conditions. Effects that increase the normal collision efficiency include drop oscillations and electric charges. An important phenomenon is the standing eddy developing in the lee of falling water drops with radii in the range 200-600 µm. Particles escaping frontal collisions may then become trapped in the eddy and undergo rear capture. Figure 8-5 shows experimental and theoretical evidence for this effect. For drop sizes larger than 600 µm, the collision efficiency declines as turbulence in the drop's wake causes the eddies to break loose at an increasing rate. Rear capture then becomes less and less significant.

Figure 8-6 shows how the collision efficiency varies with the size of aerosol particles. Three size regions must be distinguished. Small particles with radii below 0.05 µm attach to water drops by Brownian motion, whereas giant particles with radii above 1 µm undergo inertial impaction. Greenfield (1957) first noted the gap between the two regimes. A later investigation by

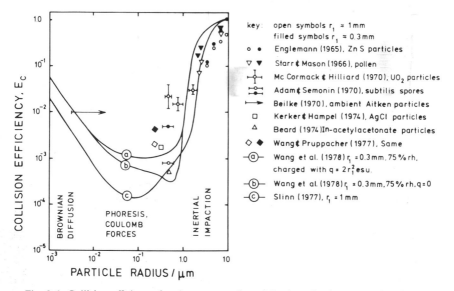

key: open symbols $r_1 \approx 1\,mm$
      filled symbols $r_1 \approx 0.3\,mm$

o •   Englemann (1965), Zn S particles
▽ ▼   Starr & Mason (1966), pollen
⊢+⊣ Mc Cormack & Hilliard (1970), UO$_2$ particles
⊢•⊣ Adam & Semonin (1970), subtilis spores
⊢→   Beilke (1970), ambient Aitken particles
□   Kerker & Hampel (1974), AgCl particles
△   Beard (1974) In-acetylacetonate particles
◇ ◆   Wang & Pruppacher (1977), Same
–(a)– Wang et al. (1978) $r_1 = 0.3\,mm$, 75% rh, charged with $q = 2r_1^2$ esu.
–(b)– Wang et al. (1978) $r_1 = 0.3\,mm$, 75% rh, $q = 0$
–(c)– Slinn (1977), $r_1 = 1\,mm$

Fig. 8-6. Collision efficiency for the capture of particles by rain drops as a function of particle radius. Solid curves indicate calculations, and points represent laboratory results. Three size regimes may be distinguished depending on the dominant type of capture process.

Slinn and Hales (1971) revealed the influence of phoretic forces, which partially closes the Greenfield gap. Most important is thermophoresis, which arises from the heat flux toward a water drop being cooled by evaporation when the ambient relative humidity is less than 100%. The effect is apparent in Fig. 8-5 from the strong increase in the collection efficiency upon lowering the relative humidity from 100 to 20%. Coulomb forces also become important if water drops and aerosol particles both carry electric charges. Curves a and b in Fig. 8-6 represent results from numerical computations by Wang *et al.* (1978) for charged and uncharged drops, respectively. Calculations for large drops are quite approximate because it is not known how to include the phenomenon of wake eddy shedding. Slinn (1976, 1977) has derived a semi-empirical expression for the collision efficiency based in part on the theoretical work of Zimin (1964) and Beard and Grover (1974). Slinn's results are included in Fig. 8-6. The figure also shows a number of laboratory results for comparison with the theoretical predictions. By and large, the data agree with the calculations.

Sood and Jackson (1970) have measured the collection efficiency for various aerosols by falling ice particles with nominal diameters in the range 1–10 mm. The general behavior of the capture efficiency with the size of aerosol particles was found to be similar to that for water drops. Quantitative differences arise from the different shape of the ice crystals and the corresponding changes in hydrodynamic flow patterns.

In order to evaluate the scavenging rate, one has to integrate over all possible collisions. Let $dN_1/dr_1 = f_1(r_1)$ represent the space density size distribution of rain drops (the drop size spectrum), and consider aerosol particles with radii between r_2 and $r_2 + dr_2$. The fraction of such particles that are removed from the air space in unit time then is given by

$$\Lambda(r_2) = \int \pi r_1^2 E_c(r_1, r_2) v(r_1) f_1(r_1) \, dr_1 \qquad (8\text{-}4)$$

(Chamberlain, 1960; Engelmann, 1968). The quantity $\Lambda(r_2)$ is called the washout coefficient. It depends on r_2 in much the same fashion as E_c. Figure 8-7 shows sample calculations for two drop size spectra with maxima at radii of 0.2 and 1 mm, respectively, according to Zimin (1964) and Slinn and Hales (1971). For a given particle size, the washout coefficient is a constant only if the rain drop spectrum does not change with time. This ideal situation is rarely met in nature, due to the usual variation of the rainfall rate. If this assumption is nevertheless made, the contribution of below-cloud scavenging to the total concentration of particulate matter in rainwater is

$$c_w = 10^{-3} c_A^0 \bar{\Lambda} h [1 - \exp(-z_{cb}/h)]/R^* \quad \text{mg/kg} \qquad (8\text{-}5)$$

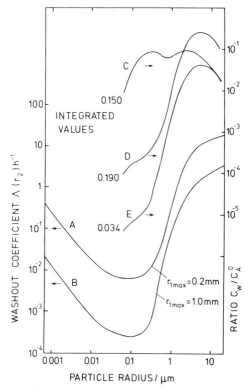

Fig. 8-7. Washout coefficients according to Slinn and Hales (1971) are shown in curves A and B (left-hand scale). They are based on rain drop size spectra of Zimin (1964) with $r_{1\,max} = 0.2$ and 1 mm, respectively, and a precipitation rate of 10 mm/h (10 kg/m² h). Curve C represents the first term and curves D and E the second term in the bracket of Eq. (8-6) in nonintegrated form (right-hand scale applies). These latter three curves are based on the mass–size distribution for the rural continental aerosol in Fig. 7-3. Curve C was calculated with $\varepsilon_A(r_2) = 1$ for $r_2 > 0.5\ \mu m$ and $\varepsilon_A < 1$ for $r_2 < 0.5\ \mu m$, decreasing linearly toward zero at $r_2 = 0.06\ \mu m$. This leads to $\bar{\varepsilon}_A = 0.8$. Curves D and E were obtained by using the washout coefficients of curves A and B, respectively. Note that below-cloud scavenging (curves D and E) affect only giant particles, whereas nucleation scavenging (curve C) incorporates also submicrometer particles.

with

$$\bar{\Lambda} = \int \Lambda(r_2) f_m(r_2)\, d\,(\log r_2)$$

where h is the scale height for the vertical distribution of the aerosol up to cloud base at the level z_{cb}; R^* is the rainfall rate (in kg H₂O/m² s), and $f_m(r_2)$ represents the normalized aerosol mass size distribution, which is assumed to be independent of height. The total concentration of particulate matter in rainwater is then given by the sum of Eqs. (8-1) and (8-5):

$$c_w = c_A^0 \{\varepsilon_A \exp(-z_{cb}/h)/L + 10^{-3}\bar{\Lambda}h[1 - \exp(-z_{cb}/h)]/R^*\} \quad (8\text{-}6)$$

This concentration is proportional to c_A^0. The relative importance of the two terms in the brackets is indicated in Fig. 8-7 by curves A–C, which are based on the mass–size distribution for the rural continental aerosol of Fig. 7-3. The comparison shows that below-cloud scavenging affects primarily particles in the 2–15 μm size range, whereas nucleation scavenging includes all particles with radii above 0.2 μm. Below-cloud scavenging thus enhances such elements in precipitation that are associated with the mineral fraction of the aerosol. Elements residing in the accumulation size range are removed from the atmosphere almost exclusively by in-cloud scavenging. The integrated values for the two terms appearing between the brackets in Eq. (8-6) are 0.15 and 0.23, respectively, for a rain drop size spectrum maximizing at $r_1 = 1$ mm. Both scavenging processes evidently contribute equivalent amounts of particulate matter to its total concentration in rainwater. In the alternative case with $r_{1\,max} = 0.2$ mm the second term in Eq. (8-6) has the value 0.04, so that in this case below-cloud scavenging is less significant compared with the in-cloud scavenging process.

8.4   The Scavenging of Gases by Cloud and Rain Drops

Cloud and rain drops absorb gases to the extent that these are soluble in water. The resulting aqueous solution is said to be saturated when gas and liquid phases are in thermodynamic equilibrium. Uptake (and release) rates are governed by molecular diffusion, and since diffusion coefficients for common gases and water vapor have similar magnitudes, both are incorporated into cloud drops at nearly the same rate. Accordingly, one expects most gases to equilibrate quickly with the liquid phase while the cloud forms. Rain drops, however, falling in the region below clouds experience gas concentrations different from those in clouds, so that the aqueous solution must adjust to the new conditions. The following discussion deals first with equilibria in clouds, then with time scales for the adjustment to equilibrium when conditions change.

8.4.1   Gas–Liquid Equilibria in Clouds

For ideal solutions in which solute interactions can be neglected, the mole fraction x_s of a substance dissolved in water and its equilibrium vapor pressure p in the gas phase above the solution are related by Henry's law,

$$p = \mathbf{H}x_s = \mathbf{H}\nu_s/(\nu_w + \sum \nu_i) \approx \mathbf{H}\nu_s/\nu_w \quad (8\text{-}7)$$

where H is a coefficient that is independent of concentration but varies with temperature approximately as $\exp(-A/T)$, ν_s denotes the number of moles of the solute of interest, and $\nu_w + \Sigma\,\nu_i$ is the total number of moles of liquid water and all solutes present. Textbooks usually treat only binary mixtures, but Henry's law remains valid for multicomponent systems as indicated. Cloud water represents a solution sufficiently dilute to make Henry's law applicable. The mole fraction then approximates to ν_s/ν_w and is proportional to concentration. Frequently, Henry's law is written in the forms

$$p = \mathbf{H}_c c_s \quad \text{or} \quad [s] = K_\mathbf{H}\,p \qquad (8\text{-}7a)$$

Here, c_s denotes the concentration in mol/kg (molality scale), and $[s]$ is the concentration in mol/liter (molarity scale). Both units are related in that $[s] = \rho_w c_s$ where $\rho_w \approx 1$ kg/dm^3 is the density of water. Its variation with temperature causes the molarity scale to depend on temperature, whereas the molality scale does not. In the temperature range 0–25°C, however, the density of water differs from unity by less than 0.3%, so that $[s] = c_s$ with reasonable accuracy. Most Henry coefficients are less well known. From the definitions in Eqs. (8-7) and (8-8), the coefficients involved are related by

$$\mathbf{H} = (10^3/M_w)\mathbf{H}_c = (10^3/M_w)(\rho_w/K_\mathbf{H}) \qquad (8\text{-}8)$$

with M_w denoting the molecular weight of water. Wilhelm et al. (1977) have tabulated Henry coefficients for many gases, expressing their temperature dependencies in a suitable parametric form. A few coefficients of interest are shown in Table 8-2 for a temperature of 283 K, which is taken to refer to the cloud condensation level. The corresponding elevation is roughly 1 km and the pressure 900 mbar. There are no measurements of Henry coefficients for supercooled aqueous solutions that would be applicable to the middle troposphere, but values may be estimated by extrapolating the existing data for the temperature dependence above the freezing point to temperatures below the freezing point of water.

In order to determine the gas–liquid partitioning of a water-soluble atmospheric component in a cloud, consider a fixed volume V_c of cloud-filled air containing ν_0 number of moles of the substance in question. Of this, ν_s moles reside in solution and ν_g moles in the gas phase. We make use of the ideal gas law [Eq. (1-1)] and set

$$L = \nu_w M_w/V_c \quad \text{g/m}^3 \qquad (8\text{-}9)$$

to represent the liquid water content of the cloud. The following relation

Table 8-2. *Henry's Law Coefficients for Several Atmospheric Gases, In-Cloud Scavenging Efficiencies ε_g, Concentrations c_s of Dissolved Substances in Cloud Water for Initial Gas-Phase Mixing Ratios m_0, and Residence Times τ for Rainout[a]*

Constituent	H (bar)[b]	ε_g L = 0.1	ε_g L = 1.0	m_0	c_s (mol/kg), L = 2	τ, L = 2	References
N_2	6.67 (4)	2.0 (−9)	2.0 (−8)	0.78	5.8 (−4)	4.0 (5) yr	Wilhelm et al. (1977)
O_2	3.30 (4)	4.0 (−9)	4.0 (−8)	0.21	3.2 (−4)	2.0 (5) yr	Wilhelm et al. (1977)
CH_4	2.91 (4)	4.5 (−9)	4.5 (−8)	1.5 (−6)	2.5 (−9)	6.7 (5) yr	Wilhelm et al. (1977)
H_2	6.34 (4)	2.1 (−9)	2.1 (−8)	5 (−7)	3.9 (−10)	3.6 (5) yr	Wilhelm et al. (1977)
CO	4.49 (4)	2.9 (−9)	2.9 (−8)	5 (−8)	5.5 (−11)	2.6 (5) yr	Wilhelm et al. (1977)
N_2O	1.43 (3)	9.2 (−8)	9.2 (−7)	3 (−7)	1 (−8)	8.6 (3) yr	Wilhelm et al. (1977)
O_3	3.09 (3)	4.3 (−8)	4.3 (−7)	4 (−8)	6.4 (−10)	1.9 (4) yr	Wilhelm et al. (1977)
C_2H_2	9.59 (2)	1.4 (−7)	1.4 (−6)	3 (−10)	1.6 (−11)	5.5 (3) yr	Wilhelm et al. (1977)
C_2H_4	8.28 (3)	1.6 (−8)	1.6 (−7)	2 (−10)	1.2 (−12)	4.7 (4) yr	Wilhelm et al. (1977)
C_2H_6	1.89 (4)	6.9 (−9)	6.9 (−8)	1 (−9)	2.6 (−12)	1.1 (5) yr	Wilhelm et al. (1977)
CH_3Cl	3.19 (2)	4.1 (−7)	4.1 (−6)	5 (−10)	7.8 (−11)	1.8 (3) yr	Wilhelm et al. (1977)
NO	2.20 (4)	5.9 (−9)	5.9 (−8)	5 (−10)	1.1 (−12)	1.3 (5) yr	Wilhelm et al. (1977)
NO_2	7.93 (3)	1.6 (−8)	1.6 (−7)	1 (−9)	6.3 (−12)	4.5 (4) yr	Lee and Schwartz (1981)
COS	1.49 (3)	8.7 (−8)	8.7 (−7)	5 (−10)	1.7 (−11)	8.5 (3) yr	Wilhelm et al. (1977)
H_2S	3.68 (2)	3.5 (−7)	3.5 (−6)	1 (−10)	1.3 (−11)	2.1 (3) yr	Wilhelm et al. (1977)
$HCHO$	3.85 (−3)	3.3 (−2)	2.5 (−1)	2 (−10)	1.5 (−6)	14 days	Walker (1975)
H_2O_2	2.29 (−4)	3.6 (−1)	8.5 (−1)	1 (−9)	1.7 (−5)	6.3 days	Martin and Damschen (1981)

[a] Rainout data for a liquid water content $L = 2\,g/m^3$. Powers of 10 are given in parentheses.
[b] $T = 283$ K, except for NO_2, 295 K.

then holds:

$$\nu_0 = \nu_s + \nu_g = \nu_s + pV_c/R_gT = \nu_s + \frac{V_cH\nu_s}{R_gT\nu_w}$$

$$= \nu_s(1 + HM_w/R_gTL) \tag{8-10}$$

The fraction of the substance that is incorporated into solution is given by

$$\varepsilon_g = \nu_s/\nu_0 = 1/(1 + HM_w/R_gTL) \tag{8-11}$$

The corresponding concentration in the aqueous phase is

$$c_s = \frac{10^3\nu_s}{\nu_wM_w} = \frac{10^3\varepsilon_g\nu_0}{LV_c} = \frac{10^3\varepsilon_gm_0p_{air}}{R_gTL} \tag{8-12}$$

Here, p_{air} is the atmospheric pressure at the condensation level, and m_0 is the mixing ratio of the substance in the gas phase before condensation took place. For $\varepsilon_g \ll 1$, one has $HM_w/R_gTL \gg 1$, and the concentration simplifies to

$$c_s = 10^3m_0p_{air}/HM_w \tag{8-12a}$$

The factor 10^3 in both equations results from the conversion of kilograms into grams. Table 8-2 shows scavenging efficiencies ε_g for several gases and vapors. The values are small for most gases, but become significant for some important atmospheric constituents such as H_2O_2 and formaldehyde. Concentrations are entered in Table 8-2 for $L = 2$. The concentrations depend on the product ε_gm_0, so that nitrogen and oxygen occur with fairly high concentrations even though the scavenging efficiencies are small. Note that oxygen is somewhat more soluble than nitrogen. Thus, the O_2/N_2 ratio in solution becomes 0.54, compared with 0.27 in the gas phase.

The above equations may be used to derive the residence time of a water-soluble gaseous constituent in the troposphere due to rainout. For simplicity we assume that the cloud–water concentration c_s is approximately preserved during precipitation—that is, we ignore below-cloud processes such as the adjustment of temperature in and evaporation of water from falling rain drops. The average mass flux for the substance being removed by precipitation is

$$F_s = 10^{-3}c_sM_s\bar{R} \tag{8-13}$$

where M_s is the molecular weight of the solute and $\bar{R}$ is the average annual precipitation rate within the region considered. The argument is the same as that used in Section 8.3.1 for the removal of particulate matter from the atmosphere. The residence time for the substance is obtained from the ratio of its column density in the troposphere to the mass flux as

$$\tau_{\text{rainout}} = \frac{W_s}{F_s} = \frac{(M_s/M_{\text{air}})\bar{m}W_T}{10^{-3}c_s M_s \bar{R}}$$

$$= \frac{\bar{m}W_T R_g TL}{m_0 \varepsilon_g M_{\text{air}} p_{\text{air}} \bar{R}}$$

$$= (1.5 \times 10^{-2})\frac{\bar{m}}{m_0}\left(\frac{1}{\varepsilon_g}\right) \quad \text{yr} \qquad (8\text{-}14)$$

where W_T is the column density of air and $\bar{m}$ is the height-averaged mixing ratio of the substance in the troposphere. Averaged over one hemisphere we have approximately $W_T = 8.3 \times 10^3 \text{ kg/m}^2$, $\bar{R} = 10^3 \text{ kg/m}^2 \text{ yr}$, $M_{\text{air}} = 29 \text{ g/mol}$, and we adopt $L = 2 \text{ g/m}^3$ to derive the numerical factor shown. For constituents that are fairly evenly distributed with height, one has $\bar{m} = m_0$ so that the residence time becomes independent of the ratio $\bar{m}/m_0$. In this case it depends only on the fraction of the substance that is dissolved in cloud water.

Values for the residence time due to rainout based on this provision are included in Table 8-2 to indicate the orders of magnitude obtained. For slightly soluble gases the time scales are too long to have any significance. Permanent gases, moreover, are not retained by the ground surface and are soon released again to the atmosphere. A residence time of 1 yr or less requires in-cloud scavenging efficiencies of at least 2%. This condition is met for HCHO and H_2O_2. A scavenging efficiency of unity leads to a residence time of 5.5 days. This should not be taken as an absolute lower limit because clouds then act as efficient sinks and $\bar{m}/m_0$ will be smaller than unity.

In the considerations leading to Eqs. (8-3) and (8-14) it was convenient to use a precipitation rate averaged over a larger region. Locally, there may be dry periods longer than a week, that is, longer than the average residence time for those trace gases whose in-cloud scavenging efficiency is near unity. In such cases, considerable fluctuations of residence times and mixing ratios must occur. Rhode and Grandell (1972) have dealt with this aspect.

A number of atmospheric gases and vapors interact with water to form ions upon dissolution. Most important among them are SO_2, CO_2, NH_3, HCl, and HNO_3. Their reactions with water are reversible, and equilibrium conditions are again set up in clouds. Table 8-3 summarizes the reactions involved and the associated equilibrium constants. Rate coefficients for the individual forward and reverse reactions are included in Table 8-3 as far as they are known. These reactions are very fast, leading to time constants of the order of microseconds or less in establishing the equilibria. As Table 8-3 shows, the dissolution of reactive gases proceeds in two steps. The first

Table 8-3. *Equilibrium Constants for Reactive Gases and the Ions Involved*[a]

Number	Reaction	Coefficient	K_{298}	K_{283}	Units[b]	H^c	$k_f{}^d$	$k_r{}^d$
w	$H_2O \rightleftharpoons OH^- + H^+$	K_w	1.0(−14)	3.0(−15)	M^2		1.3(−3)	1.3(11)
H1	$CO_{2g} + H_2O \rightleftharpoons CO_2 \cdot H_2O$	K_{H1}	3.4(−2)	5.2(−2)	M/bar	1.07(3)		
A1	$CO_2 \cdot H_2O \rightleftharpoons H^+ + HCO_3^-$	K_1	4.6(−7)	3.6(−7)	M		0.04	5.6(4)
A2	$HCO_3^- \rightleftharpoons H^+ + CO_3^{2-}$	K_2	4.5(−11)	3.3(−11)	M		~2.5	~5(10)
H2	$SO_{2g} + H_2O \rightleftharpoons SO_2 \cdot H_2O$	K_{H2}	1.3	2.2	M/bar	2.52(1)		
A3	$SO_2 \cdot H_2O \rightleftharpoons H^+ + HSO_3^-$	K_3	1.3(−2)	2.4(−2)	M		3.4(6)	2.0(8)
A4	$HSO_3^- \rightleftharpoons H^+ + SO_3^{2-}$	K_4	6.4(−8)	7.8(−8)	M		3.0(3)	5.0(10)
H3	$NH_{3g} + H_2O \rightleftharpoons NH_3 \cdot H_2O$	K_{H3}	5.8(1)	1.2(2)	M/bar	4.77(−1)	6.0(5)	3.4(10)
A5	$NH_3 \cdot H_2O \rightleftharpoons NH_4^+ + OH^-$	K_5	1.8(−5)	1.7(−5)	M			
H4	$HNO_{2g} \rightleftharpoons HNO_{2aq}$	K_{H4}	4.9(1)	1.1(2)	M/bar	4.93(−1)	3(5)	5(10)
A6	$HNO_{2aq} \rightleftharpoons H^+ + NO_2^-$	K_6	6.0(−4)	4.4(−4)	M		8.6(6)	~5(10)
H5	$HCOOH_g \rightleftharpoons HCOOH_{aq}$	K_{H5}	3.8(3)	1.0(4)	M/bar	5.50(−3)		
A7	$HCOOH_{aq} \rightleftharpoons H^+ + HCO_2^-$	K_7	1.8(−4)	1.8(−4)	M			
H6	$HCl_g + H_2O \rightleftharpoons H^+ + Cl^-$	K_{H6}	2.9(6)	1.4(7)	M^2/bar			
H7	$HNO_{3g} + H_2O \rightleftharpoons H^+ + NO_3^-$	K_{H7}	3.2(6)	1.5(7)	M^2/bar			
A8	$HSO_4^- \rightleftharpoons H^+ + SO_4^{2-}$	K_8	~1(−2)	—	M			
A9	$HCHO_{aq} + HSO_3^- \rightleftharpoons CH_2(OH)SO_3^-$	K_9	3.6(6)	1.2(7)	M^{-1}			

[a] Calculated mostly from thermochemical data (Wagman et al., 1982). Data for HCl are from Fritz and Fuget (1956); data for HNO₃ from Schwartz and White (1981); rate coefficients were taken from the compilation of Graedel and Weschler (1981); equilibrium constant for reaction A9 from Deister et al. (1986). Powers of 10 are shown in parentheses.
[b] M = mol/liter.
[c] For $T = 283$ K.
[d] In units of mol/liter s.

involves the hydration of the substance, and the second the formation of ions from the hydrate. A third step exists for diprotic acids, which may donate another proton to the solution. Consider CO_2 as an example. The equilibrium relations then are

$$K_H P_{CO_2} = [CO_2 \cdot H_2O]$$

$$K_1[CO_2 \cdot H_2O] = [H^+][HCO_3^-] \tag{8-15}$$

$$K_2[HCO_3^-] = [H^+][CO_3^{2-}]$$

Here we assume again that cloud water represents an ideal solution. For nonideal systems the concentrations would have to be replaced by the corresponding activities $a_s = \gamma_s[s]$, where the activity coefficient is a correction factor accounting for the interactions among solute molecules (Robinson and Stokes, 1970). Departures from ideality become significant for concentrations greater than 10^{-2} mol/liter, that is, for evaporating cloud drops. In Table 8-3 and above, the hydrate of CO_2 is written in the form of an adduct, because it is difficult to distinguish analytically between a CO_2 molecule enclosed in a solvent cage and carbonic acid, H_2CO_3. Both are present in aqueous solutions and may undergo ion pair formation. Accordingly, K_1 is a composite equilibrium constant, which referws to the sum of CO_{2aq} and H_2CO_3 (Stumm and Morgan, 1981). A similar situation exists for SO_2. Note also that hydrogen ions in aqueous solution occur always in the form of H_3O^+ and that the notation H^+ is merely a convention.

The total solute concentration of carbonate species following from the dissolution of CO_2 in water is

$$[s] = [CO_2 \cdot H_2O] + [HCO_3^-] + [CO_3^{2-}]$$

$$= [CO_2 \cdot H_2O]\left\{1 + \frac{K_1}{[H^+]} + \frac{K_1 K_2}{[H^+]^2}\right\}$$

$$= K_{H1} p_{CO_2}/f_{CO_2} \tag{8-16}$$

where f_{CO_2} is a function solely of hydrogen ion concentration, identical to the reciprocal of the expression in the curved brackets. Since

$$[s] = \rho_w(10^3/M_w)x_s$$

one obtains by combining Eqs. (8-7) and (8-16) Henry's law in the modified form

$$p_{CO_2} = f_{CO_2} H_{CO_2} x_s = H^*_{CO_2} x_s \tag{8-17}$$

Similar results are derived for other weak acids and for ammonia (Warneck, 1986). In each case one can define a factor f_a, which expresses the dependence on proton concentration, and a modified Henry's law coefficient

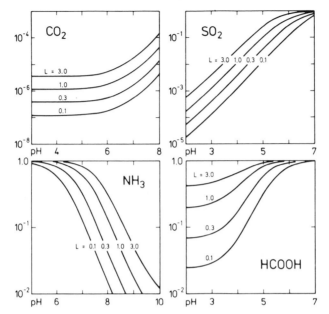

Fig. 8-8. Gas–liquid scavenging efficiencies ε_g for CO_2, SO_2, NH_3, and HCOOH in clouds with liquid water content L (in g/m^3) as a function of the pH of cloud water. [From Warneck (1986), with permission.]

$H^* = f_s H$. This coefficient must then be inserted into Eq. (8-11) to determine the degree of gas–liquid partitioning of the substance in clouds. Figure 8-8 shows ε_g for several gases of interest as a function of pH. The acid-forming gases are poorly absorbed by cloud water when the pH values are low. Ammonia shows the opposite behavior.

The procedure just outlined fails for the strong acids HCl and HNO_3 because it is difficult to measure in the laboratory K_H and the dissociation constant separately. Thus, only their product is known. Accordingly, one has

$$K_{HX} = K_{diss} K_H = [H^+][X^-]/p_{HX}$$

where X refers to either Cl or NO_3. The concentration of X^- in solution is

$$[s] = [X^-] = K_{HX} p_{HX}/[H^+]$$

and the modified Henry coefficient becomes

$$H^*_{HX} = 10^3 \rho_w [H^+]/K_{HX} M_w \qquad (8\text{-}18)$$

Table 8-4 lists in a fashion similar to Table 8-2 modified Henry coefficients, the resulting in-cloud scavenging efficiencies, concentrations in the aqueous

Table 8-4. *Modified Henry's Law Coefficients for Reactive Gases, In-Cloud Scavenging Efficiencies ε_g, Concentrations c_s in the Aqueous Phase for Initial Gas-Phase Mixing Ratios m_0, Residence Times τ for Rainout for a Liquid Water Content of Rain Clouds of $L = 2\,\mathrm{gm}^{-3}$ with Powers of 10 Shown in Parentheses*

Constituent	pH	H^* (bar)	ε_g $L=0.1$	ε_g $L=1.0$	m_0	c_s (mol/kg) $(L=2.0)$	τ $(L=2.0)$
CO_2	≤ 5	1.0 (3)	1.2 (−7)	1.2 (−6)	3.3 (−4)	1.5 (−5)	6 (3) yr
SO_2	4.5	3.3 (−2)	3.9 (−3)	3.8 (−2)	1 (−9)	1.4 (−5)	40 days
HNO_2	4.5	3.4 (−2)	3.8 (−3)	3.7 (−2)	5 (−10)	6.8 (−7)	76 days
HCOOH	4.5	8.3 (−4)	1.4 (−1)	6.1 (−1)	5 (−10)	7.3 (−6)	7.2 days
NH_3	4.5	8.4 (−6)	9.4 (−1)	9.9 (−1)	1 (−9)	1.9 (−5)	5.5 days
HCl	≥ 1	≤ 4 (−7)	~1.0	~1.0	5 (−10)	9.6 (−6)	5.5 days
HNO_3	≥ 1	≤ 4 (−7)	~1.0	~1.0	5 (−10)	9.6 (−6)	5.5 days

phase, and the associated residence times for the substances due to rainout. For $pH = 4.5$, a typical value for cloud water, NH_3, HCl, and HNO_3 are essentially completely scavenged from the gas phase. With regard to HCl and HNO_3, the liquid water content of the system would have to decrease appreciably before the vapors are revolatilized from the solution. In fact, it is necessary to go all the way back to the aerosol stage ($L \approx 10^{-5}$) in order to return the acids to the vapor phase.

The above treatment of reactive gases assumes that cloud water represents a uniform solution. This is unfortunately not true. The pH of an individual cloud drop is determined by the dissolution of electrolytes contained in the particle having served as the condensation nucleus, by the incorporation of reactive gases such as SO_2, NH_3, HNO_3, etc., and by the size of the water drop. The pH of individual cloud elements accordingly varies over a wide range, but the distribution function is not known. It is not possible, therefore, to derive appropriate averages, and our description of the scavenging of reactive gases remains approximate.

Simple Henry's law considerations also ignore interactions between solutes. Such reactions do occur, however, and as long as they are reversible they may be incorporated into the equilibrium concept of the in-cloud scavenging of gases. The most important process of this nature according to current knowledge (Warneck *et al.*, 1978; Jacob and Hoffmann, 1983; Richards *et al.*, 1983; Munger *et al.*, 1984; Deister *et al.*, 1986) is the formation of hydroxymethane sulfonate, an adduct resulting from the interaction of hydrated formaldehyde with bisulfite and/or sulfite:

$$H_2C(OH)_2 + HSO_3^-, SO_3^{2-} \rightleftharpoons CH_2(OH)SO_3^- + H_2O, OH^-$$

The rate of approach to equilibrium is faster in the alkaline pH range, since SO_3^{2-} reacts with formaldehyde more rapidly than HSO_3^- (Skrabal and

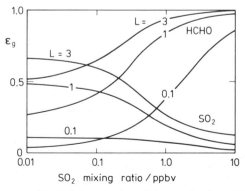

Fig. 8-9. Enhancement of in-cloud scavenging efficiencies for formaldehyde and SO_2 due to their interaction in aqueous solution.

Skrabal, 1936; Boyce and Hoffmann, 1984), but the stability of the adduct is greatest in the pH region 3-6 where HSO_3^- is dominant. The interaction between formaldehyde and bisulfite leads to total concentrations of both compounds in solution that are much greater than those given by Henry's law alone. Warneck (1986) has used an iteration procedure for calculating scavenging efficiencies for the $SO_2/HCHO$ system, and some results are shown in Fig. 8-9 for an initial formaldehyde mixing ratio of 0.2 ppbv. Depending on which of the two components is present in excess, the scavenging efficiency for either SO_2 or HCHO is considerably enhanced. In the marine atmosphere, for example, the residence time of SO_2 due to rainout is by a factor of four lower than in the absence of formaldehyde. This is important for the marine sulfur cycle (see Section 10.4.2).

8.4.2 TIME CONSTANTS FOR THE ADJUSTMENT TO EQUILIBRIUM

Any perturbation of the solution equilibrium described by Henry's law sets up a flux of material tending to reequilibrate the system according to

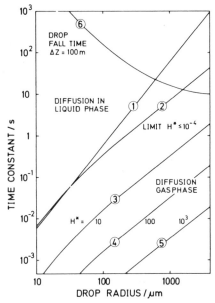

Fig. 8-10. Time constants for the approach to the generalized Henry's law gas–solution equilibrium for cloud drops and raindrops with radius r_1 ($T = 293$ K assumed). The numbering refers to the following cases. (1) Mixing by diffusion inside the drop; the same curve is obtained for diffusive transport in the gas phase for a highly soluble gas and a gas–liquid volume ratio of 10^6; (2) Case (1) calculated with D_g^* replacing D_g. (3) Steady-state approximation applicable to slightly soluble gases, $H^* = 10$. (4), (5) The same for $H^* = 100$ and $H^* = 10^3$, respectively. (6) Free-fall time for a drop over a distance of 100 m.

the new external conditions. The mass flux is determined by molecular diffusion, both inside and outside the liquid water drop, by the transport resistance at the gas-liquid interface, and by the rate of hydrolysis and ion formation. Schwartz and Freiberg (1981; see also Schwartz, 1986) have given a thorough analysis of the overall problem. Table 8-5 presents expressions for the time constants associated with the individual processes. Some numerical values are entered in the fourth column of the table, and Fig. 8-10 illustrates the dependence on drop size for diffusion-limited flow.

Hydrolysis and ion-dissociation equilibria are attained very swiftly in most cases, so that these processes are not rate-determining. The time for the approach to solution equilibrium at the gas-liquid interface is governed by the finite rate at which gas molecules strike the drop's surface. A sticking coefficient $\xi \leq 1$ mut be introduced to account for reflective collisions that do not lead to absorption by the liqud phase. Experimental work reviewed by Danckwerts (1970) has shown that the sticking coefficient usually is close to unity except in the presence of surfactants. As Table 8-5 shows, the time constant for the adjustment to phase equilibrium at the gas-liquid interface depends on the reciprocal of H^* squared. This leads to time scales ranging over many orders of magnitude. Slightly soluble gases such as CO_2 are quite rapidly accommodated, whereas for those gases that favor the aqueous phase, such as NH_3 or HNO_3, the time scales for the adjustment become appreciable.

The characteristic time for diffusion inside a cloud or rain drop follows from the mathematical solution of Eq. (1-19). Diffusion coefficients in the aqueous phase are of the order of $D_L = 1.8 \times 10^{-9}$ m^2/s. The resulting time constants for the approach to uniform concentrations are shown in Fig. 8-10. For large drops the equilibration takes longer than the fall time over a distance of 100 m. Large drops, however, also develop an internal circulation due to frictional drag that enhances mass transport by mixing. The time constant for diffusion inside a water drop thus represents an upper limit to the true mixing time.

Coefficients for diffusion in the gas phase are about four orders of magnitude greater than for the liquid phase. An additional enhancement is due to turbulence arising in the air stream surrounding a falling water drop. The latter effect can be taken into account by a semi-empirical factor first employed by Frössling (1938) in treating the evaporation of water from rain drops. The effective diffusion coefficient in the vicinity of the drop thus is

$$D_g^* = D_g(1 + 0.4\,\mathrm{Re}^{1/2}\mathrm{Sc}^{1/3}) \tag{8-19}$$

where D_g is the gas-phase diffusion coefficient in the absence of turbulence, $\mathrm{Re} = r_1 v / \eta^*$ is the Reynolds number, $\mathrm{Sc} = \eta^* / D_g$ is the Schmidt number,

Table 8-5. *Dissolution of Atmospheric Gases in Cloud and Rain Drops: Time Constants for the Adjustment to Gas–Liquid Equilibrium*[a]

Process	Parameters	Expression for time constant	Numerical values for time constant(s)	References
Diffusion inside spherical drops	D_L	$r_1^2/\pi^2 D_L$	See Fig. 8-10 curve 1 for $D_L = 1.8 \times 10^{-9}\ \mathrm{m^2/s}$	Postma (1980), Crank (1975)
Approach to ion-dissociation equilibrium and hydrolysis	$k_f, k_r, [\mathrm{H^+}]$ [anion] = $[\mathrm{X^-}]$	$\dfrac{1}{k_f + k_r([\mathrm{H^+}] + [\mathrm{X^-}])}$	$SO_2, 3 \times 10^{-7}$; $CO_2, 10^{-7}$; $NH_3, 10^{-6}$ (pH = 5)	Eigen *et al.* (1964)
Adjustment of phase equilibrium at the gas–liquid interface for diffusion-limited flow inside the drop	$D_L, \bar{c}, \mathbf{H^*}$	$D_L \left(\dfrac{4 R_g T \rho_w}{\bar{c} \mathbf{H^*} M_w} \right)^2$	$SO_2, 5 \times 10^{-3}$; $CO_2, 5 \times 10^{-13}$, $H_2O_2, 5.8$; $HNO_3, 7 \times 10^7$	Danckwerts (1970)
Gas-phase diffusion, drop internally well-mixed, constant concentration at large distances from the drop	$D_g^*, \mathbf{H^*}, r_1$	$\dfrac{10^3 \rho_w R_g T r_1^2}{3 D_g^* \mathbf{H^*} M_g}$	See Fig. 8-10, curves 3, 4, 5	Hales (1972)
Gas-phase diffusion, highly soluble gases and vapors, resulting in a substantial reduction of gas concentrations far from the drop	D_g, L, r_1	$\dfrac{r_1^2 (V_g/V_L)^{2/3}}{\pi^2 D_g}$	See Fig. 8-10, curves 1 and 2	—

[a] r_1, drop radius; D_L and D_g, diffusion coefficients for the aqueous and the gas phase, respectively; D_g^*, gaseous diffusion coefficient corrected for turbulent flow (see text); k_f and k_r, forward and reverse rate coefficients for the equilibrium reactions in Table 8-3; $\bar{c}$, gas-phase molecular velocity; ξ, accommodation coefficient; $\mathbf{H^*}$, modified Henry's law coefficient; R_g, gas constant; T, temperature; ρ_w, density of water; M_g, molecular weight of gas molecule; V_g/V_L, gas/liquid volume ratio.

v is the terminal fall velocity of the water drop, and $\eta^* = \eta/\rho_{\text{air}}$ is the kinematic viscosity of air. The Frössling factor becomes noticeably different from unity for water drops whose radii exceed 30 μm. Rain drops 1 mm in size create enough turbulence to enhance molecular diffusion by a factor of ten.

For slightly soluble gases, the concentration far away from the water drop remains essentially constant, and the concentration gradient in the vicinity of the drop's surface follows the concentration change in the aqueous phase. If the water drop is assumed to be internally well mixed, Eq. (1-20) is applicable, and one obtains the mass flux of solute molecules in the form

$$\frac{d[s]}{dt} = 4\pi D_g^* r_1 (n_\infty - n_r)/N_A V_L = (4 D_g^* r_1/R_g T)(p_\infty - p_r)$$

$$= (3 D_g^* H^* M_w/10^3 \, \rho_w R_g T r_1^2)([s]_\infty - [s]) \qquad (8\text{-}20)$$

where the subscripts r and ∞, respectively, indicate the conditions near the drops surface and far away from it, $V_L = 4\pi r_1^3/3$ is the volume of the drop, and $[s]_\infty$ denotes the equilibrium concentration in the aqueous phase. The time constant for the approach to equilibrium is the reciprocal of the factor appearing on the right-hand side. Figure 8-10 shows the dependence on drop size for several values of H* with $D_g = 1.5 \times 10^{-5} \text{ m}^2/\text{s}$.

The treatment is not applicable to gases whose generalized Henry coefficients have values much less than unity, because gas-phase depletion then leads to a decrease of concentrations far away from the drop so that the assumptions made in deriving Eq. (8-20) are invalidated. It is unfortunate that solutions to this more general problem are not available, since many important gases like SO_2, NH_3, or HNO_3 fall into this category. In order to provide a coarse estimate for the time constant associated with the scavenging of strongly soluble substances, note that in this case cloud or rain drops tend to soak up almost all the material initially present in the gas phase. Accordingly, the equilibrium concentration in the aqueous phase is determined essentially by the gas-to-liquid volume ratio. For precipitating clouds as well as for steady rainfalls this volume ratio is of the order of $V_g/V_L = 10^6$. If one considers the gaseous substances around a liquid water drop of average size to reside in a spherical shell, the distance over which diffusive transport must be accomplished is $\bar{r}_1(V_g/V_L)^{1/3}$, on average. The time constant for the approach to equilibrium for highly soluble gases may then be estimated by analogy with diffusion inside a spheric water drop. The corresponding expression is shown in the last line of Table 8-5. For a volume ratio of 10^6, the distance is $10^2 \bar{r}_1$, and since $D_g = 10^4 D_L$, the time constant has a similar magnitude to that for diffusion inside the drop. Curve

2 in Fig. 8-10 shows the dependence on drop size if the diffusion coefficient contains the Frössling factor.

The data in Table 8-5 and Fig. 8-10 lead to the following conclusions. Small water drops adjust to Henry's law equilibria within a fraction of a second, which suggests that liquid water clouds are essentially always near equilibrium. The situation is less favorable for larger raindrops. If the drops are internally well mixed, the adjustment to changing external conditions is rapid for slightly soluble gases. Their concentration in the aqueous phase is then given by Eq. (8-12a). Comparatively long time scales are involved in the equilibration of strongly soluble gases and vapors. In this case the resistances to mass transport in the gas phase and at the gas–liquid interface are both appreciable and a fall distance of 100 m is inadequate to bring the system into equilibrium. In addition, due to the inhomogeneous distribution of trace gases like SO_2, NH_3, etc., a rain drop falling in the below-cloud region experiences rising gas-phase concentrations so that it cannot attain equilibrium. Rain drops then scavenge only a fraction of the material that they are potentially capable of absorbing. Prolonged rainfalls, however, will eventually be effective in depleting the gas phase concentrations.

8.5 Inorganic Composition and pH of Cloud and Rain Water

The major ions that are commonly determined in precipitation are the anions sulfate (SO_4^{2-}), chloride (Cl^-), and nitrate (NO_3^-), and the cations of the alkali and alkaline earth metals (Na^+, K^+, Mg^{2+}, Ca^{2+}) in addition to ammonium (NH_4^+). The hydrogen ion concentration [H^+] is inferred from measurements of the pH by virtue of the relation $pH = -\log_{10}(a_{H^+})$ where $a_{H^+} = \gamma[H^+]$ is the proton activity. For sufficiently dilute solutions such as those represented by cloud and rain waters, the activity coefficient γ is close to unity.

Table 8-6 presents an overview on the concentrations of the major ions in rainwater observed at various locations. Table 8-7 provides some information on cloud and fog waters. In maritime regions seasalt is an important source of cloud condensation nuclei, and it undergoes effective below-cloud scavenging as well. Sodium chloride accordingly contributes the largest fraction of all ions in rainwater. Some of the other ions usually are somewhat enriched in comparison with their relative abundances in seasalt. The enrichment of potassium and calcium is due to the admixture of aerosol from continental sources, and that of sulfate arises from the oxidation of gaseous precursors such as dimethyl sulfide of SO_2. This excess sulfate is associated almost exclusively with submicrometer-sized particles (see Section 7.5.1).

Table 8-6. *Inorganic Ion Composition (µmol/liter) of Rainwater at Various Locations*

	Continental						Maritime		
	North Sweden[a] 1969	Belgium[a] 1969	European, U.S.S.R.[b] 1961–1964	Hubbard Brook, New Hampshire[c] 1963–1974	California[d] 1978–1979	San Carlos, Venezuela[e] 1979–1980	Cape Grim, Tasmania[f] 1977–1981	Amsterdam Island,[e] 1979–1980	Bermuda,[e] 1979–1980
SO_4^{2-}	21	63	82.0	29.8±1.2	19.5	1.6±1.25	79.1	26.3±22.4	24.4±23.8
Cl^-	11	55	60.0	14.4±2.5	28	4.3±3.4	1349	406±467	264±337
NO_3^-	5	36	17.7	23.1±1.7	31	3.5±3.6	5.0	2.7±2.1	7.9±9.1
HCO_3^-	21	—	71.7	—	—	—	—	—	—
HCO_3^{-}[g]	—	0.15	1.73	0.077	0.15	0.33	5.6	0.315	0.27
NH_4^+	6	25	52.8	12.1±0.7	21	17.0±10	2.0	5.1±7.0	4.8±7.1
Na^+	13	42	65.2	5.4±0.6	24	2.7±2.0	1297	334±371	221±282
K^+	5	6	17.3	1.9±0.5	2	1.1±1.7	32.3	7.2±7.8	6.5±8.0
Ca^{2+}	16	33	38.1	4.3±0.6	3.5	0.25±0.35	42.9	7.7±8.6	7.2±7.1
Mg^{2+}	5	15	52.5	1.8±0.4	3.5	0.4±0.45	122	36.4±40	24.6±30.4
H^+	—	38	3.3	73.9±3.0	39	17±10	1.02	18.2±12.7	21.1±26.9
pH (av.)		4.42	5.48	4.13	4.41	4.77	5.99	4.74	4.68
Sum of anions[h]	79	217	313	97.1	98	11	1512	461	321
Sum of cations[h]	66	207	320	105	93	39	1662	453	273
Number of samples	180	180	n.g.[j]	n.g.	n.g.	14	56	26	67
Annual precip.[i]	360	648	—	1310	697	4000	1120	1120	1131

[a] Granat (1972).
[b] Petrenchuk and Selezneva (1970).
[c] Likens et al. (1977).
[d] Liljestrand and Morgan (1980), Morgan (1982).
[e] Galloway et al. (1982).
[f] Ayers (1982).
[g] Calculated from pH.
[h] Sum of charges in µequiv./liter.
[i] Average annual amount of precipitation in mm.
[j] n.g., Not given.

Table 8-7. Concentrations (μmol/liter) of Major Inorganic Ions in Cloud and Fog Waters

	European U.S.S.R.[a]		Southern England,[b] nonprecipitating	Whiteface Mt., New York,[c] intercepted clouds	Central Germany,[d] mountain fogs	Pasadena, California,[c] ground fogs
	Frontal precipitating	Nonprecipitating				
SO_4^{2-}	29	117	40	26–70	387	240–472
Cl^-	22.2	76.8	94	1.7–3.1	205	480–730
NO_3^-	3.2	16.1	18.6	140–215	450	1220–3250
HCO_3^-	11.5	16.4	–	–	–	–
HCO_3^{-}[e]	1.1	0.06	0.14–91	0.023–0.045	0.72	0.4–0.005
NH_4^+	28.8	11.1	22.1	32–89	710	1290–2380
Na^+	17	29.8	95.2	2.3–11	295	320–500
K^+	5.1	20.4	12.5	13–20	85	33–53
Ca^{2+}	10	29.3	33.2	5–10	110	70–265
Mg^{2+}	12.3	40.0	12.3	1.1–3.1	–	45–160
H^+	5.0	94.4	0.06–40	126–251	12.6	14–1200
pH	5.3	4.02	4.4–7.2	3.6–3.9	5.1	4.85–2.92
Sum of anions[f]	95	327	193	194–358	1429	2180–4924
Sum of cations[f]	100	294	220	185–397	1323	1887–4983
Number of samples	125	194	23	Not given	19	4

[a] Petrenchuk and Selezneva (1970).
[b] Oddie (1962).
[c] Munger et al. (1983).
[d] Mrose (1966).
[e] Calculated from pH.
[f] Sum of charges in μequiv./liter.

On the continents the presence of seasalt in precipitation diminishes with the distance from the coast, by about 80% during the first 10 km and then more slowly as indicated by the chloride content of rainwater. The influence of seasalt is strongest in regions with frequent advections of marine air masses, but it is still evident hundreds of kilometers inland, as the data for the Soviet Union in Table 8-6 attest. Junge (1963) has discussed the geographical distribution of chloride in precipitation for Europe and the United States, and Hingston and Gailitis (1976) and Hutton (1976) have presented similar data for Australia. A contribution of gaseous HCl to chloride in rainwater is known to exist, but its quantification has remained difficult.

The major fraction of potassium, magnesium, and calcium in continental rainwaters is due to the mineral component of aerosol particles originating from soils. The contribution of seasalt to these elements is small, whereas for sodium it can be significant. Nitrate and ammonium enter cloud water as constituents of the atmospheric aerosol as well as by gas–liquid scavenging of HNO_3 and NH_3. The relative proportions to which gases and particles contribute are unknown, however. Continental sulfate and ammonium are linked in that ammonia partially neutralizes sulfuric acid formed in the oxidation of SO_2. The complete neutralization would produce $(NH_4)_2SO_4$ with a molar ratio of SO_4^{2-}/NH_4^+ of one-half, while the formation of ammonium bisulfate, NH_4HSO_4, would correspond to a ratio of unity. The molar SO_4^{2-}/NH_4^+ ratio in individual rainfalls is extremely variable. Even monthly averages scatter widely. The data in Table 8-6 indicate that in the northern hemisphere the sulfate/ammonium ratio usually exceeds the values expected if either NH_4HSO_4 or $(NH_4)_2SO_4$ was the precursor. The supply of ammonia evidently is insufficient to balance the flux of excess sulfur from anthropogenic emissions. In the southern hemisphere the situation is different. The measurements at San Carlos, Venezuela (Table 8-6), and other data compiled in Table 10-15 suggest that sulfate is a relatively minor anion in continental rainwaters of the southern hemisphere.

The total ion concentration at San Carlos is much lower than elsewhere. This feature is due to the very high rate of precipitation at that tropical rainforest site. Although there is considerable scatter in the data from all stations, the concentrations usually are inversely correlated with the precipitation rate. Likens *et al.* (1984), for example, have presented evidence for this relation at Hubbard Brook, New Hampshire. At many locations there exists also a seasonal variation of ion concentration due to a variation in the precipitation rate. In convective showers, the development of the rainfall intensity frequently goes through a maximum, while at the same time ion concentrations go through a minimum (Gatz and Dingle, 1971; Kins, 1982). This behavior contrasts with frontal precipitation events, where the initial high concentrations diminish with time toward a steady level that is fairly independent of rainfall rate. The latter effect is explained by the evaporation of raindrops during the initial stages of precipitation, and by a gradual cleansing of the atmosphere from scavengable material due to washout.

Table 8-7 shows that ion concentrations in cloud water are fairly similar to those observed in rainwater. In fog waters, however, they are much higher. This phenomenon must be caused by a higher concentration of particulate matter in the ground-level atmosphere as compared to cloud levels according to Eq. (8-1), since the liquid water content L does not differ enough between fog and cloud systems.

The major cations and anions are not always well balanced. In Table 8-6, the imbalance is especially striking in the data from San Carlos, which show a deficit of anions. A simple calculation on the basis of Eq. (8-12a) reveals that the concentrations of bicarbonate resulting from atmospheric CO_2 are below $1 \mu M$ as long as the pH of the solution is smaller than 5.3. Bicarbonate thus cannot account for the deficit. Galloway et al. (1982), Likens et al. (1983), and Keene et al. (1983) have called attention to the importance of weak organic acids, primarily formic and acetic acid. Both appear to contribute appreciably also to the hydrogen ion concentration. At Hubbard Brook, carboxylic acids make up about 16% by weight of total dissolved organic matter. The average molar concentration of these acids as reported by Likens et al. (1983) is $4.3 \mu M$ which represents but a small fraction of total anion activity. At Amsterdam Island, according to the data of Galloway and Gaudy (1984) the sum of the molar concentrations of formic and acetic acid averages $3.7 \mu M$. At San Carlos, their concentrations are somewhat higher so that their contribution to total anion activity becomes important. The origin of organic acids in rainwater is not yet established. There is some indication for a seasonal variation, which suggests vegetation as a source.

Other minor anions are the halides F^-, Br^-, and I^-, in addition to bisulfite, HSO_3^- (and sulfite, SO_3^{2-}), the latter resulting from the incorporation of gaseous SO_2 into cloud and rainwater. Richards et al. (1983) have explored Californian cloud waters with regard to dissolved SO_2 and found the aqueous concentrations much higher than expected from the known Henry's law and acid dissociation constants. High concentrations of formaldehyde were also observed, suggesting that both compounds occurred combined in the form of hydroxymethane sulfonate as discussed in Section 8.4.1. The mixing ratio of HCHO in unpolluted regions of the lower atmosphere is about 0.2 ppbv. The resulting concentrations of $CH_2(OH)SO_3^-$ are 1–$3 \mu mol/liter$, which is similar to the concentration of formaldehyde in rainwater (Klippel and Warneck, 1978; Thompson, 1980; Zafiriou et al., 1980). The concentrations found by Richards et al. (1983) were an order of magnitude higher, presumably due to local pollution.

Finally, we discuss hydrogen ion concentrations and pH. Tables 8-6 and 8-7 indicate that cloud water and rainwaters are moderately acidic, on average—that is, the pH values are less than 7. Hydrogen ions derive from the dissociation of acids, which thus serve as proton donors, whereas bases act as proton acceptors. The associated ion equilibria are summarized in Table 8-3. Strong acids like HNO_3 are fully dissociated even at H^+ concentrations approaching 1 mol/liter. For weak acids, such as CO_{2aq}, the degree of dissociation depends on concentration, but generally remains less than unity. Cloud water and rainwater are mixtures of strong and weak acids, which are but partially neutralized by alkalis. The proton activity and pH in individual cloud and rain drops, just as in bulk solution, adjust to a value governed by the condition of charge balance between positive and negative ions in combination with the various ion dissociation and Henry's law equilibria involved. The resulting set of equations enables one to calculate the H^+ concentration of the solution, provided the concentrations of the different acids and bases in the system are known. For atmospheric conditions the input parameters unfortunately are incompletely defined, making it necessary to use simplifying assumptions. One type of assumption appears to hold for polluted regions where the oxidation of SO_2 and NO_2 is a major source of the strong acids H_2SO_4 and HNO_3, respectively. If these acids dominate the weak ones, the proton concentration is approximately determined by the excess of total anion over total cation equivalents. Cogbill and Likens (1974) applied this principle to compare predicted with observed H^+ concentrations in rainwaters of the northeastern United States. They found a statistically significant linear correlation indicating that the strong acids were indeed dominant. A second type of assumption is often made for maritime conditions. Here, seasalt is the major contributor to the ionic composition of atmospheric waters. Since in aqueous solutions of seasalt strong acids and alkalis are almost perfectly balanced, the hydrogen-ion concentration is determined by the weak acids, primarily dissolved carbon dioxide. For a CO_2 mixing ratio of 330 ppmv, the pH of the solution in equilibrium with the ground level atmosphere is about 5.6, depending somewhat on pressure and temperature. In the past, this value has been considered a reference level for the remote atmosphere. It must be recognized, however, that the presence of even small amounts of gaseous ammonia, sulfur dioxide, and/or organic acids may change the pH considerably from that due to CO_2 alone. And indeed, the data in Table 8-6 for marine rainwaters clearly show average pH values distinctly different from pH = 5.6. The data from Cape Grim in Tasmania are slightly higher, whereas the pH values at Amsterdam Island and at Bermuda are lower by almost one unit. Bermuda is episodically subject to pollution from the North American continent (Jickells *et al.*, 1982), but Amsterdam Island is not thus affected.

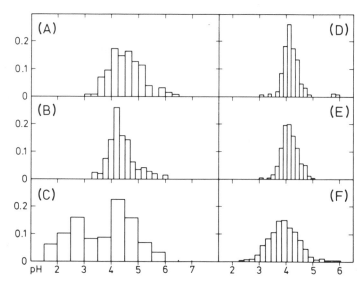

Fig. 8-11. Frequency distribution of pH values observed in hydrometeors. (A) Ernst (1938), 47 bulk rain samples collected in Bad Reinerz, combined with 80 samples showing a similar distribution collected in Oberschreiberhau (both locations in upper Silesia, central Europe), 1937-1938. (B) Mrose (1966), 206 bulk rain samples collected at Dresden-Wahnsdorf (German Democratic Republic), 1957-1964. (C) Esmen and Fergus (1976), about 200 individual rain drops collected during a single rainstorm in Delaware, 1974. (D, E) Likens *et al.* (1984), weekly samples of bulk rainwater collected at Hubbard Brook Experimental Forest Station, New Hampshire, D 1964-1968, E 1975-1979. (F) Falconer and Falconer (1980), 824 measurements of cloud water collected continuously at Whiteface Mountain, New York, 1977.

Similar to other ions, proton concentration and pH fluctuate widely among individual rainwater samples. Figure 8-11 shows frequency distributions of pH values observed at several locations. The data from Bad Reinerz are mainly of historic interest as they were among the first ever obtained (Ernst, 1938). The measurements at Dresden, 200 km further to the west, were made 20-25 yr later, but they show a similar, perhaps slightly narrower and thus more peaked distribution (Mrose, 1966). The data from Hubbard Brook Experimental Forest in New Hampshire feature an even narrower distribution compared with that at Dresden. In particular, the tail at values beyond pH = 6 is nearly absent. Likens *et al.* (1984) consider it significant that the isolated occurrence of rainfalls with pH values around 6 during 1964-1968 have disappeared a decade later. The average proton concentration and pH have remained essentially the same, however. In the same period the annual average of the concentration of sulfate declined whereas that of nitrate increased.

Figure 8-11 includes the distribution of pH in cloud water collected at the summit of Whiteface Mountain in upper New York State. The distribution is broader than that of rainwater at Hubbard Brook, and it extends toward values of pH < 3. The average pH in nonprecipitating clouds in 1977 was 3.55. The onset of precipitation increased the pH abruptly due to the mixing of cloud and rain drops at the collector surface. On average, the pH is by about half a unit higher in precipitating clouds compared with nonprecipitating ones. Esmen and Fergus (1976) have explored the pH of individual raindrops. Figure 8-11 shows one of the distributions observed during four separate rainstorms. All the data display two maxima, one in the pH region 2.5–3.0, the other one in the region 4–5. The total range covered extends from pH = 2 to pH = 8, although high pH values are rather infrequent. The observations of Esmen and Fergus do not give any information on the extent to which the pH of individual rain drops is correlated with their size, but Georgii and Wötzel (1970) have presented evidence for several ions other than H^+ showing that drop size and ion concentration are inversely related. Thus, the smallest drops are probably more acidic than the larger ones. This would explain why the collection of rainwater in bulk not only narrows the pH distribution but removes also the left-hand tail of pH values less than 3. Bulk collection represents a volume-averaging process that discriminates against small drops, as these contribute the least volume. Small drops merging with larger ones undergo a dilution that causes the lowest and the highest pH values to disappear. It is possible that the comparatively narrow distributions at Hubbard Brook are due to the weekly collection procedure. The other data shown in Fig. 8-11 were obtained by sampling individual rainfalls (or clouds). It will be important to study the relation between pH and drop size and the averaging process during bulk collection in more detail in the future.

Some remarks should be added regarding trends in precipitation acidity. In Europe, where a network of sampling stations has been in existence since the early 1950s, Odén (1968, 1976) first called attention to a decrease in the annual average pH of precipitation at a number of Scandinavian stations coupled with a simultaneous rise of sulfate concentrations, which he suggested to arise from the increased amounts of anthropogenic SO_2 being converged to H_2SO_4. According to Likens (1976; Likens and Bormann, 1974, 1975; Likens et al., 1979) a similar development has occurred in the United States. His analysis is based on a smaller and less continuous data set, however. Kallend et al. (1983) have recently reexamined the European network data with regard to acidity trends. Of 120 sampling sites, 29 show a statistically significant trend of increasing hydrogen ion concentration, typically by a factor of 3 to 4 within 20 yr. At the same time the concentrations of nitrate increased substantially, whereas the rise of sulfate was less conspicuous. The rise in acidity has not been steady, however, as expected if due to the increase in SO_2 emissions. Rather, the higher proton concentrations, where they do occur, appear as more frequent intermittent high monthly values. The reasons for this behavior are not obvious, and the problem requires further study.

8.6 Chemical Reactions in Cloud and Fog Waters

The notion that chemical reactions in the aqueous phase may be important to atmospheric chemistry dates back at least 30 yr, when Junge and Ryan (1958) called attention to the great potential of cloud water for the oxidation of dissolved SO_2 by heavy-metal catalysis. At that time the process appeared to be the only viable oxidation mechanism for atmospheric SO_2. Later, when the concept of OH radical reactions gained ground, the gas-phase oxidation of SO_2 by OH was recognized to be equally important. The recent revival of interest in aqueous phase reactions is connected with efforts to achieve a better understanding of the origins of rainwater acidity. An oxidation of NO_2 to nitric acid also takes place in cloud water. Contrary to previous ideas, however, this process was recently shown to have little influence on atmospheric reactions of NO_2.

8.6.1 THE OXIDATION OF SO_2 IN AQUEOUS SOLUTION

As discussed in Section 8.4.1, the dissolution of SO_2 in water yields SO_{2aq}, HSO_3^-, and SO_3^{2-}, in addition to undissociated H_2SO_3, which is indistinguishable from SO_{2aq}. The sum of these species is designated S(IV). The oxidation of S(IV) occurs by ozone, H_2O_2, and by oxygen in the presence of a catalyst. Rate expressions and coefficients are summarized in Table 8-8, which is discussed below.

The recognition that ozone is an important oxidant in clouds is due to Penkett (1972). The reaction with S(IV) may be written

$$O_3 + SO_{2aq}, HSO_3^-, SO_3^{2-} \rightarrow O_2 + H_2SO_4, HSO_4^-, SO_4^{2-}$$

Ozone reacts with each of the S(IV) species, albeit with different velocities. The resulting rate expression is

$$-\frac{d[S(IV)]}{dt} = k_{10}[O_{3aq}][S(IV)]$$

$$= [O_{3aq}](k_a[SO_{2aq}] + k_b[HSO_3^-] + k_c[SO_3^{2-}])$$

The dependence on proton concentration shown in Table 8-8 is obtained by virtue of the equilibria 3 and 4 of Table 8-3. The reaction rate is greatest at high pH because the most rapid reaction occurs between ozone and

Table 8-8. *Rate Expressions for the Oxidation of Sulfur(IV) Compounds in Aqueous Solution*[a]

	Reaction	Rate expression $R = -d[S(IV)]/dt$ (mol/liter s)	Rate coefficients[b]	References
A10	$S(IV) + O_3 \rightarrow SO_4^{2-} + O_2$	$R_{10} = k_{10}[O_3][S(IV)]$ $k_{10} = \dfrac{k_a + k_b K_3/[H^+] + k_c K_3 K_4/[H^+]^2}{1 + (K_3/[H^+])(1 + K_4/[H^+])}$	$k_a = 2.0 \times 10^4$ $k_b = 3.2 \times 10^5$ $k_c = 1.0 \times 10^9$	Penkett et al. (1979b), Maahs (1983), Martin (1984), Hoigné et al. (1985)
A11	$S(IV) + H_2O_2 \rightarrow SO_4^{2-} + H_2O$	$R_{11} = k_{11}[H_2O_2][S(IV)]$ $k_{11} = \dfrac{k_{df}[H^+](k_{dr}/k_e + [H^+])^{-1}}{1 + [H^+]/K_3 + K_4/[H^+]}$	$k_{df} = 5.6 \times 10^6$ $k_{dr}/k_e = 0.1$	Penkett et al. (1979b), Martin and Damschen (1981), Kunen et al. (1983), McArdle and Hoffmann (1983)
A12	$S(IV) + \tfrac{1}{2}O_2 \xrightarrow{Fe^{3+}} SO_4^{2-}$	$R_{12} = k_{12}[Fe^{3+}][S(IV)]/[H^+]$ $R_{12a} = k_{12a}[Fe^{3+}][S(IV)]^2$	$k_{12} \approx 0.2$ pH 0–4 $k_{12a} = 10^5 - 10^7$ pH 5–8	Brimblecombe and Spedding (1974a,b), Fuzzi (1978), Martin (1984)
A13	$S(IV) + \tfrac{1}{2}O_2 \xrightarrow{Mn^{2+}} SO_4^{2-}$	$R_{13} = k_{13}[Mn^{2+}][S(IV)]$ $k_{13} = k_f/(1 + [H^+]/K_3 + K_4/[H^+])$	$k_f = 5 \times 10^3$ pH 3–8	Ibusuki and Barnes (1984), Martin (1984)
A14	$S(IV) + \tfrac{1}{2}O_2 \xrightarrow{Fe^{3+},Mn^{2+}} SO_4^{2-}$	$R_{14} = R_{13} + R_{12}\left\{1 + \dfrac{1.7 \times 10^3 [Mn^{2+}]^{1.5}}{6.31 \times 10^{-6} + [Fe^{3+}]}\right\}$		Martin (1984)

[a] Number of reactions sequential to those in Table 8-3.
[b] At about 25°C, in units of liter/mol s (or liter²/mol² s where applicable).

sulfite. Although the stoichiometry of the reaction is suggestive of a simple O-atom transfer, the mechanism must be more complicated, since Esphenson and Taube (1965) showed by isotope labeling that two oxygen atoms rather than one are incorporated into product sulfate.

Penkett et al. (1979b) first suggested H_2O_2 to be an important oxidant for S(IV) in cloud water. According to the proposal of Hoffmann and Edwards (1975), the reaction proceeds with bisulfite in two steps:

$$HSO_3^- + H_2O_2 \underset{k_{dr}}{\overset{k_{df}}{\rightleftharpoons}} A^- + H_2O$$

$$A^- + H^+ \xrightarrow{k_e} H_2SO_4$$

where the intermediate A^- presumably has the structure of the peroxy-monosulfurous anion and the rate-limiting step is the acid-catalyzed rearrangement of A^-. This mechanism leads to the rate expression

$$-\frac{d[S(IV)]}{dt} = k_{11}[H_2O_2][S(IV)]$$

$$= \frac{k_{df}[HSO_3^-][H_2O_2][H^+]}{(k_{dr}/k_e + [H^+])}$$

Values for k_{df} and k_{dr}/k_e are entered in Table 8-8. For $[H^+] < 10^{-2}$ the overall reaction rate is directly proportional to $[H^+]$. The rate thus decreases with rising pH, a behavior opposite to that for the reaction of S(IV) with ozone.

The oxidation of S(IV) by oxygen may be represented schematically by

$$SO_3^{2-} + \tfrac{1}{2}O_2 \rightarrow SO_4^{2-}$$

The reaction was first studied by Titoff (1903), who showed that it is catalyzed by the ions of transition metals such as copper, iron, or manganese, and that it is difficult to eliminate traces of such metals from aqueous solutions. The claim of later investigators, such as Fuller and Christ (1941), to have observed the uncatalyzed oxidation of S(IV) must, therefore, be viewed with suspicion. In fact, the reaction can be stopped by small amounts of metal-complexing agents (Tsunogai, 1971a; Huss et al., 1978). These observations strongly suggest that the uncatalyzed reaction, if it exists, is very slow. Hegg and Hobbs (1978) have reviewed results of laboratory experiments in which an uncatalyzed reaction was assumed to take place. In all cases the rate was proportional to the concentration of S(IV) and practically independent of that of O_2, but the first-order rate coefficients reported by individual authors scattered over many orders of magnitude.

Controversial results also were obtained for the pH dependence. We shall not discuss the data in detail, because in cloud systems the catalyzed oxidation of S(IV) is more important.

In terms of abundance and effectiveness, iron and manganese are the dominant catalysts in cloud water and rainwater. Concentrations of iron in rainwater fall into the range 0.002–0.8 mg/kg (Penkett *et al.*, 1979a; Rohbock, 1982; Tanaka *et al.*, 1980; Weschler and Graedel, 1982), whereas those of manganese are an order of magnitude lower. Roughly 50% of both metals are dissolved in the aqueous phase, and the remainder is associated with insoluble matter. A portion of the material may be complexed with organic compounds, but neither the fraction nor its significance are known.

Hegg and Hobbs (1978) and Martin (1984) have reviewed empirical rate laws derived from laboratory studies of S(IV) oxidation in the presence of iron and manganese. The rate expressions reported by many investigators are independent of oxygen concentration as long as it is present in large excess. Laboratory data for the ferric iron-catalyzed reaction are consistent only in the pH range 0–4, where the rate of SO_4^{2-} formation increases linearly with the concentrations of iron and S(IV), and inversely with that of hydrogen ions. Brimblecombe and Spedding (1974a) and Fuzzi (1978) have found that raising the pH to values greater than 5 changes the rate expression from first- to second-order in S(IV). For sulfur concentrations of about 10^{-4} mol/liter as used by Fuzzi (1978), the oxidation rate goes through a maximum in the vicinity of pH = 4 and then drops off toward higher pH values. Brimblecombe and Spedding observed a maximum in the pH range 5–6. Ferric ions exist in equilibrium with hydroxylated complexes $Fe(OH)_n^{3-n}$ with $n = 1, 2, 3$. The preponderant compound at pH > 4 is $Fe(OH)_3$. Its solubility is low, however, so that it usually occurs as a colloidal precipitate. At high pH values the reaction rate is less sensitive to pH than in the region where pH < 4, but the rate is still proportional to the iron(III) concentration, which suggests that colloidal $Fe(OH)_3$ is involved in the catalysis. Ferrous iron is less effective.

Rate laws reported for the manganese-catalyzed oxidation of S(IV) fall into two categories, depending on the concentration range. For high Mn(II) and S(IV) concentrations ($> 10^{-4}$ mol/liter), the rate is nearly independent of S(IV) concentration and exhibits second-order behavior with regard to manganese (Huss *et al.*, 1982). These conditions are not germane to the atmosphere. Ibusuki and Barnes (1984) have recently explored the reaction at Mn(II) concentrations less than 10 μmol/liter, which are more relevant to the atmosphere. They found rates proportional to the concentrations of Mn(II) and S(IV) at pH = 4.5, and a pH dependence over the range 3–8 that is indicative of HSO_3^- being the reactive species. Martin (1984) has presented some supporting data for pH values below 3. The rate expression thus reads

$$-\frac{d[S(IV)]}{dt} - k_{13}[Mn^{2+}][HSO_3^-]$$

The same rate law was observed by Cheng *et al.* (1971) in a study of S(IV) oxidation in suspended solution drops. Mn(II) resembles trivalent iron in that it becomes hydroxylated as the pH is raised, but $Mn(OH)_2$ is much more soluble than $Fe(OH)_3$, which makes the system simpler to investigate.

At pH = 4, iron and manganese appear to be about equally effective as catalysts. Penkett *et al.* (1979a), however, in studying the oxidation of S(IV) in actual rainwater samples, found the rate of sulfate formation to correlate well with Mn(II), whereas the correlation with Fe(III) was extremely poor even though iron generally was present in excess. In part, this behavior may be explained by the occurrence of iron in the insoluble state, but Penkett *et al.* as well as Barrie and Georgii (1976) also noted a synergism between iron and manganese, in the sense that when both metals are present the S(IV) oxidation rate was greater than that expected from the sum of the rates observed for each catalyst separately. Martin (1984) made a more detailed study of this effect and showed that it may be quantified by an empirical factor to be added in the rate expression for iron catalysis. Table 8-8 includes his expression for the mixed system. Synergism will be important only for Mn(II) concentrations exceeding 5 μmol/liter, which are rare in cloud water and rainwater.

The reaction mechanisms responsible for the metal-catalyzed oxidation reactions are largely speculative. Huie and Peterson (1983), Hoffmann and Boyce (1983), and Hoffmann and Jacob (1984) have reviewed the various possibilities. Two types of mechanisms are frequently invoked. One is the incorporation of sulfur species as ligands into the coordination sphere of the metal ion, followed by a slow reaction with oxygen. The other type of mechanism consists of a chain reaction involving the sulfite free radical, SO_3^-, which is generated by charge transfer between sulfite or bisulfite and the metal ion. Both mechanisms probably are applicable, depending on the concentration range and the kind of transition metal involved.

Finally, we combine the data of Table 8-8 with gas–liquid scavenging efficiencies to derive SO_2 to SO_4^{2-} conversion rates in cloud-filled air. For this purpose the water drops are assumed to have a uniform pH and transport limitations are ignored. The reaction rate in the liquid phase is denoted by

$$-\frac{d[S(IV)]}{dt} = k[X][S(IV)]$$

where $k[X]$ is the product of the overall rate coefficient and the concentration of oxidant or metal catalyst in solution. By virtue of Eqs. (8-10) and (8-12) one obtains the relative loss rate of SO_2 contained in both gas-phase and aqueous solution as

$$-\frac{dv_s/dt}{v_0} = \frac{10^{-3}R_g TL}{\rho_w m_0 p_{air}}\left(-\frac{d[S(IV)]}{dt}\right) = k[X]\varepsilon_g(SO_2) \qquad (8\text{-}21)$$

in units of s^{-1}. A factor of 3.6×10^5 converts these units into %/h. Figure 8-12 compares relative SO_2 loss rates caused by the processes listed in Table 8-8 in a hypothetical cloud with a liquid water content of $L = 0.3$. In the pH range 2–5, hydrogen peroxide is the dominant oxidant, mainly because it is so effectively absorbed into the aqueous phase. In the alkaline pH region ozone becomes more important than H_2O_2, but cloud drops with pH > 6 are rare. For the most frequent values of pH, around 4, the rate of SO_2 oxidation by ozone is less rapid than that by metal catalysis. The last process, therefore, will be significant in systems where H_2O_2 concentrations are low. Figure 8-12 shows that the addition of iron in concentrations 10 times those of Mn(II) accelerates the oxidation rate about fivefold at pH = 4, whereas in the regions below pH = 3 and above pH = 5 the effect of iron is almost imperceptible. Figure 8-12 includes rates for two other aqueous SO_2

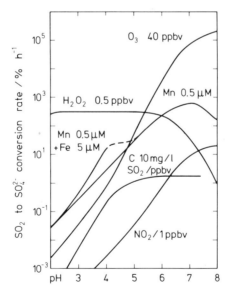

Fig. 8-12. Aqueous oxidation of SO_2 in clouds as a function of pH; relative rates for reactions of dissolved SO_2 with ozone, H_2O_2, NO_2, and oxygen, the latter catalyzed by manganese/iron, or solid carbon for the concentrations indicated. Calculations assume SO_2 scavenging efficiencies for a liquid water content $L = 0.3$ g/m^3 (see Fig. 8-8), and mass transport limitations are neglected. This restriction probably exists for the reaction with H_2O_2.

oxidation processes discussed by Martin (1984), namely, the oxidation by solid carbon particles (Brodzinsky et al., 1980) and by NO_2 (Schwartz and White, 1983). These reactions are compartively ineffective. A comparison of aqueous with gas-phase SO_2 oxidation is made in Section 10.3.3 (see Table 10-10).

8.6.2. REACTIONS OF NITROGEN COMPOUNDS

Table 8-9 lists processes leading to the oxidation of NO_2 to nitrous or nitric acid in aqueous solution. The significance of these reactions in clouds may be assessed by comparing their rates with that for the oxidation of NO_2 by ozone, which is the major dark reaction in the gas phase. Table 8-9 includes the latter process for comparison.

The hydrolysis of NO_2 proceeds via the intermediates N_2O_4 and N_2O_3. Their concentrations are governed by rapid equilibria with the precursors, so that the reactions of N_2O_4 and N_2O_3 with water determine the rate of NO_2 hydrolysis. Rate expressions are shown in the second column of Table 8-9. The associated relative NO_2 loss rates are obtained in a similar manner as loss rates for SO_2 if one uses Eq. (8-11) in an appropriately modified form. The results depend on the partial pressures for NO_2 and NO. Loss rates for typical mixing ratios and conditions near cloud base are shown in column 3 of Table 8-9. In addition to reacting with water, NO_2 has also been reported to interact with sulfites (Takeuchi et al., 1977). The reaction yields HNO_2 and sulfate as products. The mechanistic details of the process have not yet been delineated, but Schwartz and White (1983) have deduced a rate expression that may serve to indicate the significance of the reaction in clouds. With regard to the conversion of SO_2 to sulfate, it may be gleaned from Fig. 8-12 that NO_2 is less important than other oxidants of S(IV). With regard to losses of NO_2, however, the reaction with S(IV) appears to override losses due to hydrolysis.

The gas-phase oxidation of NO_2 by ozone produces first NO_3, which then attaches to another NO_2 molecule to form N_2O_5. Its subsequent redissociation leads to an equilibrium between both compounds. At 283 K, roughly 50% of NO_3 is present as N_2O_5. In the gas phase the reaction of N_2O_5 with water is immeasurably slow, whereas in aqueous solution it is rapid. Neither the rate coefficient for reaction A20 nor the Henry's law coefficient for N_2O_5 is known, but the values probably are similar to those for N_2O_4. With this assumption and adopting $H(N_2O_4) = 5$ bar as evaluated by Schwartz and White (1981, 1983), one finds that the loss of N_2O_5 in cloud drops is faster than the formation of NO_3. Accordingly, the reaction between NO_2 and ozone will be the rate-determining step in the formation of HNO_3. Table 8-9 shows that the resulting NO_2 loss rate is many orders

Table 8-9. Rate Expressions $R = -d[NO_2]/dt$ and Rate Coefficients k for Reactions of NO_2 in aqueous solution, and the Associated Relative Loss Rates for NO_2 in Clouds with a Liquid Water Content $L = 0.3$ g/m³, $T = 283$ K[a]

Number	Reaction	Rate expression and coefficients (in units of mol, liter, s)	Loss rate (%/h)	References
A15	$NO_2 + NO_2 \rightleftharpoons N_2O_4$	$K_{15} = 6.5 \times 10^4$ $k_f = 4.5 \times 10^8$; $k_r = 6.9 \times 10^3$	—	Grätzel et al. (1969)
A16	$N_2O_4 + H_2O \rightarrow HNO_2 + HNO_3$	$R_{16} = 2k_{16}K_{15}[NO_2]^2$ $k_{16} = 1 \times 10^3$	1.5×10^{-5} $m_0(NO_2) = 1$ ppbv	Grätzel et al. (1969) Treinin and Hayon (1970)
A17	$NO_2 + NO \rightleftharpoons N_2O_3$	$K_{17} = 1.37 \times 10^4$ $k_f = 1.1 \times 10^9$; $k_r = 8 \times 10^4$	—	Grätzel et al. (1970)
A18	$N_2O_3 + H_2O \rightarrow 2HNO_2$	$R_{18} = k_{18}K_{17}[NO_2][NO]$ $k_{18} = 5.3 \times 10^2$	1.5×10^{-7} $m_0(NO) = 0.5$ ppbv	Grätzel et al. (1970) Treinin and Hayon (1970)
A19	$2NO_2 + S(IV) \rightarrow 2HNO_2 + SO_4^{2-}$	$R_{19} = (k_{19a}[HSO_3^-] + k_{19b}[SO_3^{2-}])[NO_2]$ $k_{19a} = 3 \times 10^5$; $k_{19b} = 1 \times 10^7$	2.7×10^{-3} $m_0(SO_2) = 1$ ppbv pH = 4	Takeuchi et al. (1977) Schwartz and White (1983)
G	$O_3 + NO_2 \rightarrow NO_3 + O_2$	$k_{62} = 2 \times 10^{-17}$ [b]		
G	$NO_3 + NO_2 \rightleftharpoons N_2O_5$	$R = 2k_{62}n(NO_2)n(O_3)$	13	See Table A-5
A20	$N_2O_5 + H_2O \rightarrow 2HNO_3$	$k_{20} \approx k_{16}$; $H(N_2O_5) \approx 55$ bar (assumed)	$m_0(O_3) = 40$ ppbv	

[a] Mass transport limitations are neglected.
[b] In units of cm³/molecule s.

Table 8-10. *Rate Expressions* $R = -d[N(III)]/dt$ *and Rate Coefficients* k *for Reactions of* HNO_2 *or* NO_2^- *in Aqueous Solution, and Relative Loss Rate of* HNO_2 *at pH = 4 in Clouds with a Liquid Water Content* $L = 0.3$ g/m^3, $T = 283$ K^a

Number	Reaction	Rate expression and coefficients (in units of mol, liter, s)	Loss rate (%/h)	References
A21	$NO_2^- + 2HSO_3^- \rightarrow HON(SO_3)_2^{2-} + OH^-$	$R_{21} = k_{21}[H^+][NO_2^-][HSO_3^-]$ $k_{21} = 4.8 \times 10^3$, pH 4–7	2.8×10^{-4} $m_0(SO_2) = 1$ ppbv	Oblath et al. (1981)
A22	$2HNO_2 + 2S(IV) \rightarrow 2SO_4^{2-} + N_2O + H_2O$	$R_{22} = 2k_{22}[H^+]^{1/2}[N(III)][S(IV)]$ $k_{22} = 1.42 \times 10^2$, pH 0–3	—	Martin (1984)
A23	$NO_2^- + O_3 \rightarrow NO_3^- + O_2$	$R_{23} = k_{23}[NO_2^-][O_3]$ $k_{23} = 3.5 \times 10^5$	2.7×10^{-1} $m_0(O_3) = 40$ ppbv	Penkett (1972), Garland et al. (1980), Damschen and Martin (1983), Hoigné et al. (1985)
A24	$HNO_2 + H_2O_2 \rightarrow HOONO + H_2O$	$R_{24} = 24[H^+][H_2O_2][HNO_2]$ $k_{24} = 4.6 \times 10^3$, pH 0–4	5.1×10^{-3} $m_0(H_2O_2) = 0.5$ ppbv	Damschen and Martin (1983)
A25	$HOONO \rightarrow HNO_3$	$R_{25} = (k_{25a} + k_{25b}[H^+])[HOONO]$ $k_{25a} = 5.27 \times 10^{-2}$; $k_{25b} = 0.596$		Benton and Moore (1970)

^{a} Mass transport limitations are neglected.

of magnitude greater than the rates of the other reactions listed. The latter processes thus are comparatively unimportant. The reason for this behavior is the low solubility of NO_2 in water and the correspondingly small fraction of NO_2 in the liquid phase of clouds.

Table 8-10 is added to show rate expressions for reactions of HNO_2 and NO_2^- in aqueous solution. The sum of both species is usually denoted by $N(III)$. The reaction of $N(III)$ with bisulfite was discussed by Martin (1984). The reaction is acid-catalyzed, whereby its rate increases with decreasing pH. A change in the order of the dependence on proton concentration occurs near $pH = 3.5$. This value is close to the switchover point between NO_2^- and HNO_2 in the equilibrium of both species. Above $pH = 4$ the reaction product is hydroxylamine disulfonate. Below $pH = 3$ the major products are N_2O and sulfate, because hydroxylamine disulfonate hydrolyzes to hydroxylamine, which then reacts further with nitrous acid to give N_2O. Although some interesting chemistry may result from these reactions in clouds, the rates are too slow to gain much significance. At $pH = 4$, the overriding aqueous reaction is that of NO_2^- with dissolved ozone, which oxidizes nitrite to nitrate. At lower pH values the reaction between nitrous acid and hydrogen peroxide comes to prominence, provided sufficient H_2O_2 is present. The incipient product in this case is pernitrous acid, which is unstable, however, and eventually isomerizes toward nitric acid.

Chapter

9 | Nitrogen Compounds in the Troposphere

The better-known among a great number of nitrogen compounds in the atmosphere are ammonia, several nitrogen oxides, and nitric acid. These gases are discussed in the present chapter. NH_3, N_2O, and to some extent NO and NO_2 as well are natural constituents of air arising as an outflow from the biosphere. Although atmospheric chemistry views the biosphere merely as a source (or sink) of trace gases, it will be helpful for the subsequent discussions to give initially a summary of the biochemical processes that are responsible for the release of nitrogeneous volatiles from soil and aquatic environs. The individual pathways of elemental nitrogen in the biosphere and in the atmosphere represent portions of a larger network of fluxes involving all geochemical reservoirs. This aspect will be considered in Section 12.3 from a different viewpoint.

9.1 Biochemical Processes

The following description of the biological nitrogen cycle emphasizes the production of volatile compounds that can escape to the atmosphere. Delwiche (1970) has presented a popular account of the nitrogen cycle. A compilation of review articles edited by Clark and Rosswall (1981) contains many more details and numerous literature citations. Textbooks on microbiology provide additional information.

Figure 9-1 shows pathways of biological nitrogen utilization for the soil-plant ecosystem. With appropriate modifications, the scheme may also be applied to aquatic ecosystems. Nitrogen enters the biosphere largely by way of bacterial nitrogen fixation, a process by which N_2 is reduced and incorporated directly into living biomass. A limited number of bacteria are fitted with the special enzyme system necessary for this task. Examples include phototrophic Cyanobacteria in natural waters, heterotrophs like *Clostridia* and *Azobacter* in the soil, and symbionts associated with plants, such as *Rhizobia* living in the root nodules of legumes. The details of nitrogen fixation were reviewed by Burns and Hardy (1975). It is an energetically costly process, and the biosphere as a whole tends to preserve its store of fixed nitrogen by extensively recirculating it within a given ecosystem. The term fixed nitrogen is used to describe nitrogen contained in chemical compounds that all plants and microorganisms can utilize. Figure 9-1 serves as a guide for the biochemical transformations of the various compounds representing fixed nitrogen.

Dead organic matter undergoes decomposition by a host of different bacteria. Thereby, organic nitrogen is mineralized to ammonium unless it is assimilated. NH_4^+ and NH_3 are in aqueous equilibrium, and if environmental factors like pH, temperature, buffer capacity of the soil, etc. are favorable, ammonia can be released to the atmosphere. Another important source of ammonia is urea arising from animal excreta.

Under aerobic conditions, a number of specialized bacteria derive their energy needs from the oxidation of ammonia to nitrite and then further to nitrate. This process is called nitrification. Focht and Verstraete (1977),

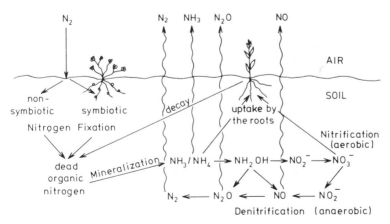

Fig. 9-1. The biological nitrogen cycle in the atmosphere-soil system. Contributions of nitrogen due to rainwater, dry deposition, and fauna are not shown.

Belser (1979), and Schmidt (1982), among others, have reviewed the various aspects of nitrification. In soils, the most common nitrifying bacteria are *Nitrosomonas*, which converts ammonium to nitrite, and *Nitrobacter*, which oxidizes nitrite to nitrate. As chemoautotrophs, both species feature nitrification rates several thousand times faster than the less populous heterotrophic nitrifiers. As Fig. 9-1 shows, hydroxylamine is the only identifiable intermediate product in the oxidation chain. Other intermediates have been postulated, specifically $(HNO)_2$, to explain the formation of N_2O as a side product, but a real need for such an intermediate is not obvious. NH_2OH can be oxidized to NO_2^- directly, and N_2O can be formed by the interaction of NH_2OH with NO_2^-. Since the process is enzyme-controlled, the true intermediate would be enzyme-bound anyway. More enigmatic are the observations of Makarov (1969), Kim (1973), Smith and Chalk (1979), and Lipschultz *et al.* (1981), who found NO and/or NO_2 as products of nitrification. These oxides may arise as side products from the oxidation of NH_2OH to NO_2^-, or they may be formed afterward due to the chemical instability of nitrous acid, for example, by the reaction $2HNO_2 \rightarrow NO + NO_2 + H_2O$. There is evidence for both pathways being operative. Calculations of thermochemical equilibria by Van Cleemput and Baert (1976) show that aerobic conditions favor NO_2 as decomposition product, and anaerobic conditions lead to NO, N_2O, and N_2. Bollag *et al.* (1973), in an experimental study, found that in acidic soils (pH = 5) nitrite mainly decomposes abiotically to NO and NO_2, whereas in alkaline soils its consumption is dominated by biochemical processes. The rate is highest for anaerobic conditions. In this case denitrification occurs and N_2 and N_2O are the major products (see further below). On the other hand, Lipschultz *et al.* (1981) experimented with *Nitrosomonas* in pure culture (pH = 7.5). They demonstrated that NO and N_2O evolve concurrently with NO_2^- the latter accumulating at a constant or slightly accelerating rate. These results essentially preclude NO_2^- as a precursor of NO and N_2O and suggest that they are direct products of nitrification. About 1% of nitrogen enters this route.

Nitrate is the major end-product of nitrification. Most plants (except certain bog species) can utilize nitrate as well as ammonium to satisfy their nitrogen needs. The process involves uptake through the roots, assimilatory reduction of NO_3^- to NH_3 and metabolic conversion into organic nitrogen compounds. In this manner, fixed nitrogen is circulated through the ecosystem. A certain fraction of nitrate is leached from the soil together with soluble cations and enters the ground waters. Yet another fraction undergoes bacterial reductions toward N_2. This process, which is termed denitrification, is the last one to be discussed here.

Focht and Verstraete (1977), Knowles (1981), and Firestone (1982) have provided reviews of biological denitrification. As Fig. 9-1 shows, the reduction of NO_3^- toward N_2 proceeds via NO_2^-, NO, and N_2O as intermediates. There is reasonable agreement that N_2O, in addition to NO_2^-, is an obligatory intermediate, whereas the role of NO is less clear. Dentrification requires a habitat that is best characterized as being almost but not completely anaerobic. Almost all denitrifying bacteria are aerobic species turning toward oxides of nitrogen as a source of oxygen when the level of O_2 falls to low values. In soils, pore moisture is the major factor regulating the diffusive supply of oxygen, and anaerobic conditions are quite generally produced by water-logging. In aerated soils, anaerobic microniches often are established in the interior of soil crumbs where the supply of oxygen is reduced due to consumption in the outer layers. Thus, aerobic and anerobic domains may exist side by side. The probability for finding anaerobic conditions increases, of course, as one goes deeper into the soil.

A great many genera of bacteria have the capacity to denitrify, and it is difficult to identify those organisms that, are functionally the more important. As Gamble *et al.* (1977) stated, the most commonly isolated bacteria are *Pseudomonas* and *Alcaligenes*, but this may simply reflect an inherent selectivity of the cultural media used for growth, which cannot simulate the complexity of the soil environment. Since most denitrifiers are chemoheterotrophs, the availability of organic carbon compounds in soil as an energy source is an important factor controlling the rate at which denitrification takes place. Another critical parameter is soil pH. Dentrification proceeds optimally in the pH range 6–8, whereas in naturally acidic soils it is inhibited.

The dominant product from denitrification is N_2. This process, therefore, represents a reversal of nitrogen fixation, returning fixed nitrogen from the biosphere to the pool of atmospheric N_2. The release of N_2O also constitutes a loss of fixed nitrogen, as the main fate of N_2O in the atmosphere is photodecomposition by ultraviolet (UV) radiation with N_2 and O atoms as products. The release of NO or NO_2, in contrast, does not lead to a loss of fixed nitrogen, because these compounds are oxidized further to HNO_3, which is then returned to the earth surface by wet and dry deposition.

Owing to the complexity of soil biochemistry, the emissions of nitrogeneous gases to the atmosphere are influenced by many parameters, and the emission rates are highly variable. This variability makes it difficult to derive representative averages for global source estimates. An important parameter is temperature, because the rates of biochemical reactions rise exponentially with temperature, at least in the range 288–308 K. Soil temperature undergoes seasonal and diurnal variations, causing the emission rates to follow similar cycles. The observation of diurnal and/or seasonal oscillations in the concentration of a trace gas measured in continental surface air, with maxima occurring in the late afternoon and/or in the summer months, provides a sure indicator for the importance of soil biology as a source of the trace gas studied.

9.2 Ammonia, NH_3

The French scientist DeSaussure is credited with having discovered atmospheric ammonia in rainwater in 1804. Until recently, interest in atmospheric NH_3 was focused on its role as a source of fixed nitrogen to soil and plants. In the 1950s it was recognized, primarily through the work of Junge (1954, 1956), that ammonia is *the* single trace gas capable of neutralizing the acids produced by the oxidation of SO_2 and NO_2. The ammonium salts of sulfuric and nitric acid thus formed become part of the atmospheric aerosol. In Chapter 7 it was shown that acids and ammonia enter particulate matter from the gas phase whereby ammonium is concentrated in the submicrometer range of particle sizes. Giant particles generated by wind force from the Earth's surface contain practically no ammonium.

For a complete understanding of atmospheric ammonia, it is necessary to differentiate between gaseous NH_3 and particulate NH_4^+. In this regard, most of the existing data are unsatisfactory, mainly due to difficulties in separating the two fractions. The older measurements, summarized by Eriksson (1952), did not distinguish gaseous from particle-bound NH_3, so that they provide at best the sum of both. It is now common to place a filter upstream of NH_3 sampling devices in order to remove particulate NH_4^+. During long sampling times, ammonium-rich particles may interact with calcareous soil-derived particles on the filter, causing the release of NH_3 and an overestimate of the gaseous NH_3 concentration. Ferm (1979) and Bos (1980) have described so called denuder tubes, in which NH_3 is sampled by diffusion to the acid-coated walls of the tube, whereas particles pass through it to be collected on a back-up filter. Denuders unquestionably are superior to other sampling devices in differentiating between NH_3 and NH_4^+, but their sampling rates are low. Wet-chemical, colorimetric techniques, often Berthelot's indophenol blue method, are then used to determine the amounts of NH_3 and NH_4^+ collected. The detection of atmospheric NH_3 by long-path infrared absorption using the sun as background source has not been accomplished until very recently (Murcray et al., 1978). A spectral resolution better than 0.1 cm^{-1} is required for the identification of NH_3. This explains the failure to detect it in earlier spectra of lesser resolution (Migeotte and Chapman, 1949; Kaplan, 1973).

9.2.1 DISTRIBUTION IN THE TROPOSPHERE

Table 9-1 compiles results from ground-level measurements of gaseous and particulate ammonia, made in Europe, North America, and over the oceans during the past decade. On the European continent, total $NH_3 + NH_4^+$ concentrations average about 6 μg/ m³ with standard deviations indicating moderate fluctuations. The results agree reasonably well with determinations of gaseous NH_3 made in the 1950s and summarized by Junge (1963). Fewer data are available for the American continent, particularly with respect to simultaneous measurements of NH_3 and NH_4^+, but the total concentrations are fairly similar to those in Europe. Most data reveal seasonal variations with maximum values occurring in the summer. Detailed studies of the annual cycle were performed by Goethel 1980 in a rural community 40 km east of Frankfurt (West Germany), and by Tjepkema et al. (1981) at a forest site near Petersham in central Massachusetts. Several authors have noted a close correlation of NH_3 mixing ratio with temperature, which suggests that soil microorganisms contribute much to atmospheric ammonia. Goethel (1980) and Bos (1980) further reported diurnal cycles during summerly fair weather periods, with NH_3 mixing ratios attaining a maximum at midday and falling to lower values at night. Again, this behavior is explicable in terms of soil temperature.

By means of small aircraft, Georgii and Müller (1974) and Georgii and Lenhard (1978) have explored the vertical distribution of NH_3 and NH_4^+ in the tropospheric boundary layer over Europe up to 3 km altitude. Figure 9-2 shows that both species decrease with height nearly exponentially, the

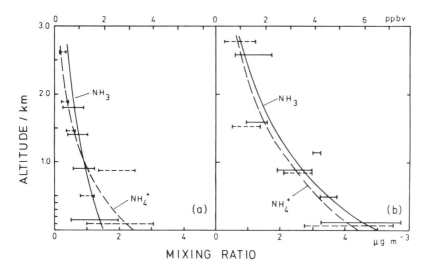

Fig. 9-2. Vertical distribution of gaseous ammonia and particulate ammonium over Europe according to data of Lenhard (1977) and Georgii and Lenhard (1978). The lower scale gives mixing ratios in units of μg/m³ (NPT); the upper scale gives mixing ratios for NH_3 only in units of ppbv. (a) Winter, (b) summer.

Table 9-1. *Surface Abundances of Gaseous NH_3 and Particulate NH_4^+ in Europe, North America, and over the Oceans[a]*

Authors	Location	Season	NH_3 ($\mu g/m^3$)	NH_4^+ ($\mu g/m^3$)	NH_3/NH_4^+ ($\mu g/\mu g$)	Remarks
			Continental data			
Breeding *et al.* (1973)	Mississipi Valley	Autumn	0.8–7.7	—	—	The report does not
	Glasgow, Illinois	Spring	3.0 ± 0.46			mention use of a prefilter
Georgii and Müller (1974)	Western Germany	Winter	6 ± 4	—	—	Extrapolated from averaged vertical profiles derived
		Summer	17 ± 10	—	—	by aircraft measurements
Georgii and Lenhard (1978)	Western Germany	Winter	1.7	2.6	0.6	
		Summer	5.5	6.0	0.9	
	Frankfurt/Main		—	—	0.4	
	Island of Sylt		—	—	2.5	
	Jungfraujoch (Swiss Alps)	Summer	0.2	—	—	3600 km Altitude
Ferm (1979)	Near Gothenburg, Sweden	Autumn	0.21	1.11	0.33	Denuder tubes
Lenhard and Gravenhorst (1980)	Western Germany	Winter	1.9 ± 0.6	2.6 ± 0.9	0.73	100 m Above ground
		Summer	4.8 ± 0.8	4.9 ± 0.8	1.0	

Reference	Location	Season				Notes
Tjepkema et al. (1980)	Petersham, Massachusetts	Winter	0.017 ± 0.012	0.97 ± 0.37	0.019	Denuder tubes, annual variation recorded for period of 1 yr
		Summer	0.13 ± 0.05	2.1 ± 1.0	0.067	
Goethel (1980)	Schlüchtern, Central Germany	Winter	2.0 ± 1.1	3.5 ± 1.8	0.57	Seasonal variation given
		Summer	3.5 ± 1.8	3.7 ± 1.9	0.95	
Bos (1980)	Terschelling, Netherlands	October	3.7 ± 1.9	—	—	Denuder tubes
Cadle et al. (1982)	Warren, Michigan	Winter	0.8	—	—	
		Spring	1.3	—	—	
	Commerce City, Colorado	Winter	3.4	—	—	
			Maritime data			
Tsunogai (1970)	North Pacific		1.15	0.52	2.2	
	Tropical Pacific		0.56	0.23	2.4	
	South Pacific		0.13	0.50	0.26	
Ayers and Gras (1980)	Cape Grim, Tasmania		0.06 ± 0.03	—	—	
Georgii and Gravenhorst (1977)	Central Atlantic		0.2	—	—	
	Sargasso Sea		6	1.5	4	
Gras (1983)	Antarctic		0.016	0.05	0.32	Denuder tubes

[a] To convert to volume mixing ratios: 1 $\mu g/m^3$ NH_3 = 1.3 ppbv.

decrease being equivalent to a scale height of 1-2 km depending on season. At the Jungfraujoch, a mountain station in the Swiss Alps 3600 m above sea level, Georgii and Lenhard found $m(NH_3) = 0.26$ ppbv, confirming the results obtained by aircraft. The abundance of NH_3 in the upper troposphere is expected to be still smaller. Column abundances estimated from infrared solar spectra by Kaplan (1973) for the air over France and by Murcray *et al.* (1978) over Colorado suggest average tropospheric mixing ratios of <0.08 and 0.5 ppbv, respectively. Hoell *et al.* (1980) used an infrared heterodyne radiometer technique to estimate the distribution of NH_3 over Hampton, Virginia. The inferred mixing ratios were about 1 ppbv at 12 km, compared with 10 ppbv at the ground level in March 1979, while in August of the same year they found 0.4 ppbv compared with 1.2 ppbv. Their distributions (not the ground-level values) are at odds with the other data and should be treated with caution until further confirmation by independent techniques.

   Table 9-1 includes the few data from maritime regions. The exploratory study by Tsunogai (1971b) is open to criticism with regard to the sampling procedure used, but it established the general behavior of marine NH_3: the mixing ratio declines with increasing distance from the shoreline until it reaches a background value of the order of 0.1 ppbv. The background level determined by Ayers and Gras (1980) for the Indian Ocean is 0.074 ppbv. These data indicate that the continents are the major source regions and that the sea acts as a receptor of ammonia (and ammonium) of continental origin. It is still not entirely clear whether the low background reflects a residual of continental NH_3 or a Henry's law equilibrium with NH_3 and NH_4^+ dissolved in seawater. The ammonium content of the ocean surface waters is biologically controlled. It varies over the range 0.05-2.0 mmol/m^3, with values from the open ocean tending toward the lower end of the range. The early calculations of Buch (1920) indicated an equilibrium mixing ratio for NH_3 in marine air of 0.5 ppbv or less. The discussion was recently revived by Georgii and Gravenhorst (1977). They assumed an average $NH_3 + NH_4^+$ concentration in seawater of 0.3 mmol/m^3 and calculated equilibrium mixing ratios in the range 0.03-0.4 ppbv for surface water temperatures varying from 5 to 25°C. Ayers and Gras (1980) also discussed the NH_3 mixing ratios at Cape Grim (Tasmania) in terms of an equilibrium with the ammonium content of the ocean. The calculations refer only to gaseous NH_3 and do not include the conversion to particulate NH_4^+ and its return to the sea. Owing to the temperature dependence of the equilibrium, one would expect the tropics to favor higher atmospheric background mixing ratios of NH_3 than the polar regions. The observational data are insufficient to confirm this concept, however. Moreover, the temperature effect may be masked by fluctuations of NH_3 resulting from variations in biological activity. In the Caribbean and the Sargasso Seas, Georgii and Gravenhorst (1977) found NH$_3$ mixing ratios over 10 times greater than in the Central Atlantic, which they cautiously interpret to arise from favorably high local concentrations of dissolved NH$_4^+$. If so, the Sargasso Sea would have to be identified as a true marine source region of atmospheric ammonia. The sixth column of Table 9-1 lists concentration ratios of NH$_3$ to particulate NH$_4^+$. Most of these are smaller than unity indicating a preference of atmospheric ammonia for the particulate phase. Junge (1963) had previously concluded that the gas phase dominates. The change in outlook can be traced to improvements in sampling techniques and a better discrimination between gaseous and particulate ammonia. The NH$_3$/NH$_4^+$ ratio should be determined by the rate at which ammonia is tied to aerosol particles following the production of sulfuric and nitric acid, relative to the rates of ammonia supply and its removal from the atmosphere by precipitation. The quantitative aspects of this relation remain to be investigated.

The observational data provide a coarse estimate of the tropospheric mass content of nitrogen present as NH$_3$ and NH$_4^+$, if we adopt a procedure similar to that used in Section 7.6.1. for aerosols. We must assume that the concentrations found in Europe and the United States can be extrapolated

Table 9-2. *Mass Content of NH$_3$ and NH$_4^+$ Nitrogen in the Troposphere and Turnover Rate Resulting from Known Residence Times τ^a*

	Region	A $(10^{12}\,m^2)$	c_0 $(\mu g/m^3)$	h (km)	Tropospheric mass content (Tg N)	τ (days)	Flux (Tg N/yr)
NH$_3$	Continents	149	2.5	1.5	0.39		
	Oceans	361	0.14	1.5	0.05		
	Background	510	0.06	∞	0.16		
	$G_T(NH_3)$				0.60	5.7	38.4
NH$_4^+$	Continents	149	3.5	1.5	0.52		
	Oceans	361	0.05	1.5	0.02		
	Background	510	0.06	∞	0.16		
	$G_T(NH_4^+)$				0.70	5.0	51.1
$G_T(NH_3)+G_T(NH_4^+)$					1.30	Total flux	89.5

^{a} The mass content is estimated from a model distribution assuming a uniform background mixing ratio superimposed by a boundary layer component. The vertical concentration profile is $c = c_0 \exp[-(1/h + 1/H)z]$ with $H = 9.1$ km, and the average tropopause level is 11 km. The mass content is the product of integrated column density and surface area A.

to other continents as well. To describe the concentrations in the upper troposphere, we take a background mixing ratio of $0.12 \, \mu g \, m^3$, subdivided equally between NH_3 and NH_4^+ and appropriately weighted with the atmospheric scale height. The model parameters and the results are summarized in Table 9-2. Ammonia is removed from the atmosphere mainly by wet precipitation. By using the residence times for the rainout of aerosol particles and NH_3, derived in Chapters 7 and 8, one also obtains an estimate for the flux of ammonia through the atmosphere. The data in Table 9-2 suggest a turnover rate of about 80 Tg N/yr. The following evaluation of the NH_3 budget yields a total flux that is about a factor of two smaller. The mass content of NH_3 and NH_4^+ in the troposphere thus may be an overestimate.

9.2.2   SOURCES AND SINKS OF ATMOSPHERIC NH_3

Table 9-3 lists processes contributing to the tropospheric budget of NH_3, and estimates for the associated global fluxes derived by several authors. The sources are discussed first. They may be classified as combustion, bacterial decomposition of animal excreta, and emanation from soils.

The nitrogen content of coal is 1–2% by mass, present mainly as organic nitrogen, of which a large fraction is released as NH_3 upon heating. Coking of coal once provided the principal source of industrial ammonia before the Haber–Bosch synthesis replaced it. Wood contains about 0.5% nitrogen; leaves and newly grown twigs contain more, typically 1–2%. Miner (1969) compiled NH_3 emission factors for the combustion of various materials: 1 g/kg for coal, $0.12 \, kg/m^3$ for fuel oil, $10 \, mg/m^3$ for natural gas, 1.2 g/kg for wood, and 0.15 g/kg for NH_3 released from forest fires (see also National Research Council Subcommittee on Ammonia, 1979). The value for coal refers to domestic use and must be taken as an upper limit, because high-temperature combustion units oxidize ammonia further to NO. According to United Nations (1978) statistics, the world consumption of hard coal is about 2,000 Tg/yr, of which 400 Tg/yr is coked and must be deducted, and that of lignite is 750 Tg/yr. Combining these figures gives an NH_3 release rate from coal combustion of 2 Tg N/yr. The combustion of fuel oil and natural gas produces much less and can be neglected. NH_3 emissions from automobiles also are small. The estimate in Table 9-3 is based on an emission factor of 30 mg NH_3/km (Harkins and Nicksic, 1967) and a world car population of 275 million. Biomass burning appears more important. Seiler and Crutzen (1980) estimated that the burning of plant matter in slash and shift agriculture, deforestation, savanna and bushland clearings, wild forest fires, and the use of wood as fuel consumes 6,800 Tg of biomass annually. Using the emission factors of Miner (1969) for forest fires and wood as fuel, one calculates that biomass burning releases NH_3 at a rate of 2.1–8.1 Tg/yr.

Table 9.3. *Estimates for the Strengths of the Sources and Sinks of Ammonia (Tg N/yr) in the Troposphere According to Various Authors*

Process	Söderlund and Swensson (1976)	Dawson (1977)	Böttger et al. (1978)	Stedman and Shetter (1983)	Crutzen (1983)	Present estimates or adopted[a]
Sources						
Coal combustion	4–12	—	0.03	<2	—	≤2
Automobiles	—	—	0.2–0.3	—	—	0.2[a]
Biomass burning	—	—	—	—	60	2–8
Domestic animals	20–35[b]	—	20–30	23	20	22
Wild animals	2–6	—	—	3	—	4[a]
Human excrements	[b]	—	—	1.5	—	3
Soil emission	—	38	1	(51)[c]	—	15
Fertilizer losses	—	—	1.2–2.4	3.5	3	3
Sum of sources	26–53	—	22–34	83	—	54
Sinks						
Wet precipitation on the continents	20–80	35	15±7	50[d]	—	30[a]
over the oceans	8–26	—	6±6	10	—	8[a]
Dry deposition (land)	69–151	—	[e]	14	—	10
Reaction with OH	3–8	—	3	9	—	1
Sum of sinks	100–265	—	24±13	83	—	49

[a] Adopted as representative from the work of the authors listed.
[b] Human excrements included in the figure for domestic animals.
[c] Not a valid estimate; adopted to balance the budget.
[d] Obtained by extrapolation, not by integration over latitude belts.
[e] Böttget et al. (1978) consider dry deposition already included in wet precipitation.

As discussed by Eriksson (1952) and Georgii (1963), coal combustion seems to have been regarded as a prominent source of ammonia for many years, at least in the cities. In 1970, Healey et al. showed that this source is dwarfed by NH_3 released from animal urine. The process involved is the bacterial decomposition of urea to NH_3 and CO_2. Healy et al. (1970) listed urea excretion rates for domestic animals as cattle 140, swine 60, sheep 45, and poultry 3, in units of gram urea per animal and day. From this information and world population statistics on domestic animals (FAO, 1981) one can extrapolate to a urea production in the developed countries of 43 Tg/yr, with contributions from cattle of about 51%, pigs 17%, sheep 20%, and poultry 8%. Animals in the developing world generally are less well fed, so that the urea production per head is lower, perhaps half as much. The corresponding global production rate amounts to 39 Tg/yr, to which cattle and buffaloes contribute 56%, pigs 13%, sheep 12%, goats 9%, and poultry 5%. Total urea production sums to 82 Tg/yr or 37 Tg N/yr. Healy et al. (1970) assumed 10% of urea nitrogen to be volatilized as NH_3, but most field experiments, especially those of Watson and Lapins (1969) and Denmead et al. (1974), indicate higher percentages ranging up to 50%. If 30% is taken as representative, the NH_3 release rate is 11 Tg N/yr. To this must be added the contribution of feces. Their nitrogen content is similar to that of urine according to the data assembled by Böttger et al. (1978), although it is not as easily volatilized by bacteria. Nevertheless, assuming a similar percentage to be liberated as ammonia doubles the total source from animal excretions to 22 Tg N/yr. The world domestic animal population has risen only slightly during the past decade.

The human body discharges about 30 g of urea daily. Healy et al. (1970) thought that most of it ends up in the sewers before it is mineralized. The assumption probably holds true for the cities and many smaller communities in the developed countries, but should not be generalized. Some NH_3 may even escape from sewer systems. If one-half of the ammonia from human urea is released to the atmosphere, a world population of 3.6 billion represents an NH_3 source of 2.7 Tg N/yr.

Söderlund and Swensson (1976) estimated the contribution to atmospheric ammonia from wild animals as 2–6 Tg N/yr. The value is based on the ratio of urea excretion to plant matter ingested by herbivores, 6 mg/g as given by Loehr and Hart (1970), on an estimated 3–10% consumption by wild animals of the annual new growth of green-plant material (primary productivity), which totals about 10^5 Tg/yr, and on the 10% figure adopted by Healy et al. (1970) for the conversion of urea to ammonia.

Whereas the various estimates for NH_3 production from animals are reasonably consistent, it is evident from Table 9-3 that the rate of emissions from soils is essentially unknown. A scarcity of field observations exists mainly for natural soils. Emissions of NH_3 and other nitrogen compounds from N-fertilized soils have received much study by agronomists interested in keeping such losses small. The investigations have included fields dressed with ammonium salts, urea, manure, and sewage sludge, and dairy fields and cattle feed lots. The subject has been reviewed by Allison (1955, 1966) and more recently by Freney et al. (1981). Field losses reported range from 5 to 50%. Since rainfall minimize losses, the low values presumably are most representative. A 5-10% loss combined with a world fertilizer production of 55 Tg N/yr results in emissions of 3 Tg N/yr.

The above-mentioned studies are relevant also to natural soils in that they have identified important parameters influencing the NH_3 exhalation rate. Most important are temperature, pH, and moisture content of the soil. Ammonia is pictured to interact with soil water according to the equilibrium $NH_3 + H_2O \rightleftharpoons NH_4^+ + OH^-$. Increasing the water content of the soil shifts the equilibrium toward the right causing NH_3 to be more tightly bound. Increasing the pH—that is, the OH^- concentration—shifts the equilibrium toward the left, causing NH_3 to be released. The magnitude of the equilibrium constant favors the escape of NH_3 from soils with $pH > 6$. More acidic soils are not expected to emanate NH_3 except in the autumn when topped with decaying plant litter. Raising the temperature of the soil enhances the rate at which NH_3 is made available by microbial activity, and it increases the partial pressure of NH_3 in the soil due to the evaporation of water.

Dawson (1977) made a valiant attempt to determine the global rate of NH_3 emissions from soils by means of a model incorporating the various physicochemical and microbiological processes involved. The value obtained by him, 47 Tg/yr or 38 Tg N/yr, must be considered an overestimate, however, because the model ignores the interaction of NH_3 with the plant canopy above the soil. Porter et al. (1972) and Hutchinson et al. (1972) first demonstrated that growing plants assimilate ammonia. Denmead et al. (1976) observed plant canopy absorption of NH_3 emitted from the soil (or decaying organic matter) underneath a clover–grass pasture. Goethel (1980) studied the NH_3 flux from twin plots, one of which had the plant cover removed, and found emissions twice as high from the bare plot compared with the covered one.

Table 9-4 lists the few measurements of NH_3 emissions from natural, unfertilized soils. The fluxes reported by Denmead et al. (1976) are unreasonably high in view of the low values of soil pH. Goethel's (1980) measurements are the only ones covering the period of a whole year. His average exhalation rate agrees quite well with the value given by Dawson (1977) for the same latitude band if correction is made for the difference in soil pH and absorption by the plant cover. It thus appears that Dawson's global rate,

Table 9-4. NH$_3$ Fluxes from Natural Soils, Determined with the Flux Box Method unless Otherwise Noted [a]

Authors	Soil parameters	Observed flux μg N/m² h	Remarks
Hooker et al. (1973)	Tilled soil (Duroc type), silt-loam, pH = 7.3	6–14	Derived from concentration gradient in the air above plant cover
Denmead et al. (1976)	Ungrazed pasture, soil pH = 6, clover and rye grass cover	200	
Georgii and Lenhard (1978)	Various soils, pH 6–8	1–17	Soils with pH = 4 showed no production, seasonal variation with higher summer values
Marshall and Debell (1980)	Clay soil, pH = 5, underneath Douglas fir stand; cover plants were clipped	0–430 av. 100	Control in fertilizer experiments, semi-open plots
Goethel (1980)	Various soils, pH 4–7, covered with weeds, in orchards, pastures, and forests	0–7	Measurements mainly during midday; acidic soils gave no emissions, but sometimes were sinks; snow cover also absorbed ammonia
Goethel (1980)	Brown soil, pH 6.4, grass cover underneath fruit trees	~3 av.	Intermittent measurements over a whole year; positive fluxes Mar.–Nov., negative fluxes Dec.–Feb.; maximum fluxes in late fall during decomposition of plant litter

[a] Losses inside the box are corrected for.

reduced by a factor of 2-3, is a reasonable estimate. Böttger *et al.* (1978) simply extrapolated the data of Hooker *et al.* (1973) and Georgii and Lenhard (1978) to the whole world. Their result of 1 Tg N/yr sets an absolute lower limit.

The various sources of ammonia sum to a total of about 55 Tg N/yr. The predominant fate of NH_3 in the atmosphere is conversion to NH_4^+ and its return to the ground surface by wet precipitation. The annual deposition is determined by the concentration of NH_4^+ in rainwater and the cumulative amount of rainfall. Both parameters are not entirely independent of each other, and they fluctuate considerably at any given location, so that extended measurement series are needed in order to derive the average deposition rate. Data from such measurements are abundant from the European and North American continents; elsewhere they are scarce. Eriksson (1952a), Steinhardt (1973) and Böttger *et al.* (1978) have compiled many of the available data. The first authors included in their tables observations from the second half of the last century; Böttger *et al.* considered only the period 1950-1977. More recent data are available from the stations cooperating with the Background Air Pollution Monitoring Network established by the World Meteorological Organization. Regional maps for the deposition of NH_4^+ were first prepared by Angström and Högberg (1952) and Emanuelson *et al.* (1954) for Sweden. Junge (1963) subsequently discussed similar maps for a number of constituents of rainwater for Western Europe and the United States. Figure 9-3 illustrates the global distributions of NH_4^+ concentrations in rainwater and of NH_4^+ deposition rates according to Böttger *et al.* (1978). Both quantities are not distributed in the same way but the general behavior is similar in that the deposition is concentrated on the continents and decreases toward the oceans. Thus, NH_3 is removed from the atmosphere essentially in the regions of its origin, and this is in accordance with the short tropospheric residence time of NH_3 of only a few days.

The global deposition rate of the continents is obtained by integrating over the land area within selected latitude belts, extrapolating the data wherever necessary, and then summing over all latitudes. Table 9-3 shows results from several such studies. Figure 9-4 presents the distribution with latitude for the deposition rates deduced by Böttger *et al.* (1978) from their own compilation of data, and by Dawson (1977) from the tabulation of Eriksson (1952). The distribution of NH_3 sources given by the same authors is included to show that relative to each other both distributions are comparable. Dawson considered only emissions from soils, whereas the main source according to Böttger *et al.* is animals. Since both sources feature similar distributions, it is not possible to determine from deposition measurements which of the two sources is dominant.

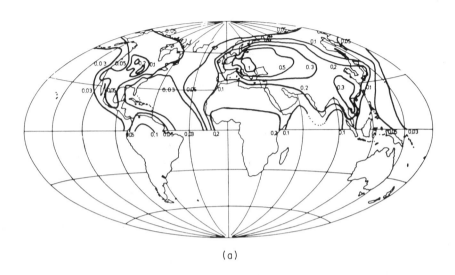

(a)

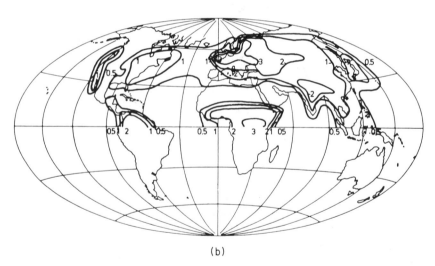

(b)

Fig. 9-3. Global distribution of wet NH_4^+ deposition according to Böttger *et al.* (1978), derived from measurements during the period 1950–1977. (a) The NH_4^+ nitrogen content of rainwater in ppmw. (b) The deposition rate in units of $100\,mg/m^2\,yr$. Reproduced with permission of Kernforschungsanlage Jülich.

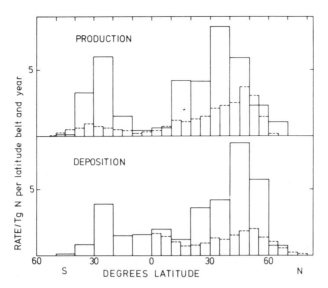

Fig. 9-4. Distribution with latitude of continental NH₃ emissions and wet NH₄⁺ deposition as estimated by Dawson (1977) (———) and Böttger *et al.* (1978) (- - -). The last authors did not treat deposition data for the southern hemisphere.

The discrepancy between the absolute deposition rates obtained by Söderlund and Swensson (1976) and Dawson (1977), on one hand, and by Böttger *et al.* (1978) on the other, raises concern about the data base. Since the compilation of Eriksson (1952) used by the first group of authors includes many data from the past century, whereas that of Böttger *et al.* covers only recent decades, one might suspect the deposition rate to have decreased with time. Brimblecome and Stedman (1982) demonstrated, however, that from 1850 to 1980 the annual deposition of NH₄⁺ at several places in Europe and in North America has remained fairly constant. The wet deposition of NO₃⁻, in contrast, has risen remarkably during the same period. Accordingly, the earlier data should be given equal weight and the global NH₄⁺ removal rate calculated by Böttger *et al.* (1978) may be too low. A rate of 30 Tg N/yr is adopted here.

The three values shown in Table 9-3 for the deposition of NH₄⁺ over the oceans are crude estimates due to the sparsity of measurements and uncertainties concerning the amount of precipitation in maritime regions. A value of 8 Tg N/yr corresponds to an average deposition rate of 22 mg N/m²/yr or about one-tenth of that occurring on land.

Additional losses must be assigned to dry deposition of gaseous NH₃ and particulate NH₄⁺, and to reaction of NH₃ with OH radicals. The large rate that Söderlund and Swensson (1976) estimated for dry deposition is questionable. Gravenhorst and Böttger (1980) pointed out that because in the past many rain samplers were left open continuously during wet and dry periods, dry deposition is already included in the collection of material. This argument applies to dry deposition to flat terrain, but it does not include the interception of both gaseous NH_3 and particulate NH_4^+ by the leaves and stems of plants. In view of the canopy losses described by Denmead *et al.* (1976), interception must be important, and the high-reaching foilage of trees is expected to be most effective in this regard.

A great number of studies summarized by Höfken *et al.* (1981) have shown that the concentrations of many trace substances in rainwater are higher underneath the forest canopy than in clearings at the same location. The effect is mainly caused by dry-deposited material being washed off the foliage. From their own measurements covering the period February–October 1980, Höfken *et al.* found enrichments for NH_4^+ averaging 1.7 for beech and 2.7 for spruce trees. Part of the enrichment is caused by the eduction or leaching of ammonium from the foliage. The relative fraction of NH_4^+ resulting from leaching increases with the amount of precipitation and may reach 40% of the total. The results of Höfken *et al.* (1981) also show that particulate NH_4^+ makes an important contribution to total inter-cepted ammonia. This contrasts with the earlier opinion of Söderlund and Swensson (1976), who thought dry deposition of gaseous NH_3 to be most effective. If these results are representative for the whole world, the total deposition onto woodlands should be 70% greater than wet deposition alone. Since forests and brushwood cover roughly 50% of the continental surface, losses due to interception are expected to augment wet deposition by one-third, or 10 Tg N/yr if wet deposition is 30 Tg N/yr. The sum of these loss processes approximately balances the sources of ammonia.

Finally, we consider the reaction of ammonia with OH radicals:

$$NH_3 + OH \rightarrow NH_2 + H_2O$$

The rate coefficient for this reaction is endowed with a moderate tempera-ture dependence (see Table A-4) so that the value for $T = 283$ K, $k = 1.4 \times 10^{-13}$ cm/molecule s, may be taken as representative for the whole troposphere. If an average tropospheric OH concentration of 5×10^5 molecules/cm^3 is adopted, the lifetime of NH_3 due to the reaction is

$$\tau = 1/k_{120}\, n(OH) = 1.43 \times 10^7 \quad s \quad or \quad 165 \quad days$$

This time constant is 30 times greater than that associated with rainout of NH_3. As a sink process, the reaction with OH is relatively unimportant. The absolute sink strength can be estimated from the global mass of gaseous NH_3 in Table 9-2. A rate of 1 Tg N/yr is obtained.

The fate of the NH_2 radical in the atmosphere is still uncertain. The reaction of NH_2 with oxygen is very slow (Lesclaux and De Missy, 1977; Cheskis and Sarkisov, 1979). The laboratory results of Hack *et al.* (1982) suggest than an NH_2-O_2 addition complex is formed that may be revert to the initial reactants but does not yield new products. The reactions of NH_2 with O_3, NO, and NO_2 proceed rapidly with rate coefficients at 300 K of 1.8×10^{-13}, 2.3×10^{-11}, and 2.0×10^{-11}, respectively, according to the data of Hack *et al.* (1981) and Kurasawa and Lesclaux (1979, 1980). The reactions with NO and NO_2 are favored relative to that with ozone unless the NO_x mixing ratio falls below 220 pptv. This situation exists over the oceans and in the upper troposphere. The reaction with nitric oxide

$$NH_2 + NO \rightarrow N_2 + H_2O$$

is well known to reduce NO to molecular nitrogen, thus providing a sink for NO_x as well as for NH_3. The reaction with NO_2 has been postulated by Kurasawa and Lesclaux (1979) to proceed in a similar manner

$$NH_2 + NO_2 \rightarrow N_2O + H_2O$$

because other laboratory work has shown N_2O to be a product of the overall reaction mechanism. It should be noted, however, that the formation of N_2O may also result from the reaction sequence

$$NH_2 + NO_2 \rightarrow 2HNO$$

$$2HNO \rightarrow H_2O + N_2O$$

In the atmosphere, HNO would presumably transfer the odd hydrogen atom to other receptors, such as O_2, rather than react with itself. Until the products of this reaction are firmly established, one must treat the conversion of NH_2 to N_2O as one of several conceivable reaction pathways.

9.3 Nitrous Oxide, N₂O

Atmospheric N_2O was discovered in 1938 by Adel via infrared absorption features in the solar spectrum. For the next 30 yr, N_2O aroused little interest, presumably because it is neither a hazardous pollutant nor does it display any particular chemical activity. In fact, there are no gas-phase reactions that remove it from the troposphere as far as we know. In the stratosphere (see Chapter 3), N_2O undergoes photodecomposition, and it reacts with $O(^1D)$. The second reaction is the major source of higher nitrogen oxides in that region, and since these reduce ozone catalytically via chain reactions, N_2O is an important agent in controlling the stratospheric ozone balance. The recognition of this relationship by Crutzen (1970, 1971) and McElroy and McConnell (1971) stimulated many researchers toward taking a closer look at the behavior and the budget of N_2O in the troposphere.

The older data and concepts concerning tropospheric N_2O have been reviewed by Hahn and Junge (1977) and by Pierotti and Rasmussen (1977). The results of several recent measurements of N_2O mixing ratios in various parts of the world are collected in Table 9-5. The preferred analytical technique currently is gas chromatography using the electron-capture detector. All recent observations show that the distribution of N_2O in the troposphere is uniform and that the short-term variability is small. Relative fluctuations reported generally amount to 1% or less. The absolute value of the mixing ratio is 310 ± 10 ppbv to a good approximation. The values given by individual investigators differ somewhat, due to unavoidable calibration errors. Rasmussen and Pierotti (1978) conducted an interlaboratory calibration test, with samples averaging 328.7 ± 1.9 ppbv according to their determination. Nine of 10 samples sent out were reported back to have N_2O mixing ratios in the range 312–333 ppbv, with a mean of 323.5 ± 8.7; one result fell below 300 ppbv. A more recent interlaboratory comparison by Rasmussen and Khalil (1981b) faired little better.

The earlier studies, particular those of Goody (1969) and Schütz et al. (1970) indicated a much larger variability (about ± 40 ppbv), which in the light of the recent data is likely to have resulted from instrumental effects. The insufficient sensitivity of thermal conductivity detectors then in use required a preconcentration procedure, which gave rise to larger errors. The difficulties of instrument calibration, influence of temperature, etc. also were underrated at that time.

On the basis of the older data and using the inverse relationship between residence time and variance of mixing ratio (see page 135), Hahn and Junge (1977) estimated the residence time of N_2O in the troposphere to range from 4 to 12 yr, whereas Singh et al. (1979a), Pierotti and Rasmussen (1977), and Weiss (1981) concluded from the much smaller variations observed by them that the residence time is 20–30 yr or longer. Actually, such conclusions should be accepted with caution. Although, on account of its great stability, N_2O appears to be long-lived in the troposphere, it was pointed out by Junge (1974) that the concept of a decreasing variability of mixing ratio with increasing residence time requires an uneven distribution of the sources and sinks of the trace gas being considered. In the case of N_2O, the surface sources are spread fairly evenly over the entire globe, which leads one to question whether the concept is really applicable. And since ground-level measurements by necessity are made in source regions, there is a tendency to observe fluctuations of the source strength rather than those of background mixing ratios even if instrumental effects are negligible.

Table 9-5. *Some Recent Measurements of N_2O Mixing Ratios in the Troposphere[a]*

Investigators	Dectector[b]	Locations	Years		Average mixing ratio (ppbv)
Yoshinari (1976)	HID	Atlantic Ocean	1972		328 ± 5
		Caribbean			
Pierotti and Rasmussen (1977)	ECD	NE Pacific Ocean	1976	$31°N-14°S$	332 ± 9
		Pullman, Washington	1976		328.5 ± 0.7
		Flights between New Zealand and Alaska	1976	$80°N-59°S$	330 ± 5
Tyson et al. (1978)	ECD	Flights between New Zealand and Alaska	1976	Northern hemisphere	306 ± 21
				Southern hemisphere	314 ± 39
Cicerone et al. (1978)	ECD	Ann Arbor, Michigan	1976–1977		329.5 ± 3.3
Goldan et al. (1978)	ECD	Various locations, mainly Boulder, Colorado	1976–1977	Corrected[c]	325.8 ± 1.2
					304.5 ± 1.2
Singh et al. (1979a)	ECD	Various locations around the world	1975–1978	Northern hemisphere	311 ± 2.3
				Southern hemisphere	311 ± 2.8
Roy (1979)	ECD	Cape Grim, Tasmania	1977		338 ± 9
		Various places, southern hemisphere	1977–1978		337 ± 3
Weiss (1981)	USD, ECD	Various locations around the world	1976–1980	$70°N-90°S$	300.2 ± 0.1
		South pole	1978		299.2 ± 0.2

[a] For other measurements prior to 1976 see the compilation of Pierotti and Rasmussen (1977).

[b] Gas chromatographic detectors: HID, helium-ionization detector; ECD, electron-capture detector; USD, ultrasonic phase-shift detector in conjunction with nondispersive infrared analysis.

[c] Calibration error was corrected by Golden et al. (1981).

Table 9-6. N_2O Emission Rates from Soils Determined by the Chamber Technique[a]

Authors	Soil type and cover	Observation period	Emission (kg N/ha yr)	Remarks
Ryden et al. (1978)	Haploxeroll cropped to celery	76 h in June 1977	82.5	Fertilized with 120 kg N/ha of $(NH_4)_2SO_4$
Freney et al. (1978)	Grass-clover, soil type not given	2 h around noon for a period of 5 months	0.5	"Typical" rates, unfertilized, dry after wetting with 5 mm H_2O
McKenney et al. (1978)	Sandy loam planted to corn	4 days in June 1977	0.4	Unfertilized
			1.0	Fertilized, 112 kg N/ha
	Planted to tobacco	20 days in July 1977	1.6	Unfertilized
			7.6	Fertilized, 112 kg N/ha
Denmead et al. (1979)	Sandy loam, grass sward	5 months from winter to summer	0.18	Dry conditions
			0.6–5.5	Moist, diurnal variation
Breitenbeck et al. (1980)	Fallow Harps soil	96 days	1.25	No fertilizer
			1.36	125 kg N/ha $CaNO_3$
			1.91	125 kg N/ha urea
			2.12	125 kg N/ha $(NH_4)_2SO_4$
Roy (1979)	Rye grass and clover; soil type not given	Not given	0.17–9.6	Field grazed by sheep
Bremner et al. (1980)	Various Iowa soils used for corn and soybean production	12 months April–April	0.34–1.97 av. 1.2	No fertilization, seasonal variation, maximum in June and July

444

Reference	Soil	Time period	Emission[a]	Treatment
Conrad and Seiler (1980a)	Aeolian sand, sparse coverage, nonagricultural	72 days, Sept.–Oct. 1978	~0.2	Untreated, low in Sept., high in Aug.
Mosier and Hutchinson (1981)	Nunn clay loam planted to corn	4 months, May–Sept.	7.5	Fertilized with 200 kg N/ha NH$_3$
Bremer et al. (1981)	Webster, Canisteo, and Harps soil planted to corn in 1978 to soybeans in 1979	139 days following June 22, 1979	4.4–6.5 32–51	Unfertilized 250 kg N/ha NH$_3$ applied
Ryden	Wickham clay loam sown to perennial rye grass	Aug.–Dec. (1980) Mar. 1980–Mar. 1981	−0.7[b] 3.25	Unfertilized 250 kg N/ha NH$_4$NO$_3$
Seiler and Conrad (1981)	Loess, Loess loam Aeolian sand	Intermittently over a 3-year period	0.04–0.26 0.17–1.1	Unfertilized Unfertilized
Duxbury et al. (1982)	Mineral soils, weed grasses	May 1979–May 1981	0.9–1.7	Unfertilized
	Covered with alfalfa		2.3–4.2	Unfertilized
	Forested		1	Unfertilized
	Organic soils, bare and fallow		59–165	Unfertilized
	St. Augustine grass or sugar cane		7–97	Unfertilized

[a] Results are prorated to annual emission values; 1 hectare = 10^4 m^2.
[b] Negative sign means uptake of N$_2$O.

445

Weiss (1981) claimed to have detected an excess of 0.7 ppbv N_2O in the northern hemisphere as opposed to the southern, but the data are marginal. More substantial is his finding of an increase of the N_2O mixing ratio by about 0.2% per year during the period 1976–1980. Goldan et al. (1980) deduced a rise of about 1% yr from measurements in the upper troposphere, and Rasmussen et al. (1981) reported an annual rise of 0.4% at the South Pole. Weiss attributed the effect to the growth of fossil-fuel combustion. The true cause remains to be established, however.

The principal sources of N_2O are microbial processes in soils and natural waters. Not too long ago it was thought that N_2O production is associated primarily with denitrification, despite long-standing laboratory evidence recently reviewed by Bremner and Blackmer (1981), showing that nitrifying bacteria also give rise to N_2O. In soil systems, the importance of nitrification has now been convincingly demonstrated by the observation of much higher N_2O emission rates in the case of ammonium salt fertilization as opposed to nitrate. Table 9-6 summarizes N_2O exhalation rates for a variety of soils, both fertilized and unfertilized. All studies listed employed chambers installed on the ground with their interior exposed to the soil surface, and the flux was determined from the rise of the N_2O level inside the chamber. Formerly, it was believed that N_2O originates in the more anaerobic deeper strata of the soil, that it diffuses toward the soil surface, that diffusion sets up a vertical gradient of N_2O concentrations in the soil, and that one can calculate the flux of N_2O from the gradient and a suitably chosen diffusion coefficient (Albrecht et al., 1970; Burford and Stefanson, 1973). This notion must be abandoned. The field observations of Seiler and Conrad (1981) clearly show surface emission rates and vertical gradients to be entirely uncorrelated. Surface emissions apparently arise from bacterial nitrification in the uppermost layer of the soil, whereas the vertical gradient is set up by the local balance of N_2O production and loss, the latter being due to denitrification in anaerobic niches (Rosswall, 1981). Figure 1-13 illustrated that N_2O does not accumulate indefinitely in the observation chamber but reaches a steady-state of about 400 ppbv after some time. The same value obtains if one starts the experiment with N_2O mixing ratios quite different from atmospheric, higher or lower. The particular soil studied thus serves as a source when the N_2O level in the chamber is lower than 400 ppbv, and as a sink when it is higher. Although other types of soil harboring different populations of microorganisms are expected to attain other steady-state N_2O mixing ratios, all the results in Table 9-6 except one show soils to be a source of N_2O. Generally, therefore, the steady-state mixing ratio must be higher than that in the atmosphere.

The N_2O flux increases with rising temperature and/or moisture content of the top soil. This leads to diurnal and seasonal patterns of the flux, with maxima at midday and during the summer, and to a fairly drastic enhancement of the release rate when the top soil is wetted by rainfall or irrigation. From diurnal variations of the flux, Conrad *et al.* (1983) quantified the temperature dependence in the form of an Arrhenius expression. The apparent activation energy was quite variable, however. The dependence on soil moisture was also studied and may be expressed by a third-order power law.

Few of the studies included in Table 9-6 were extended over periods long enough to provide annual averages, and none was conducted outside the temperate latitude zone of the northern hemisphere, so that extrapolation of these data to a global scale involves large uncertainties. About 12% of the continental area is covered with perpetual ice, lakes, and streams, and 14% is deserts (see Table 11-5). Vegetation occupies an area of 100×10^{12} m^2. The emission rates in Table 9-6 for unfertilized, plant-covered soils range from 0.2 to 2 kg N/ha yr. From the data of Bremner *et al.* (1980), the annual average may be taken as 1 kg N/ha yr. If these values are representative for the world, the global source strength for N₂O released from soils is 2–20 Tg N/yr, with a mean of 10 Tg N/yr.

Fertilization enhances the N₂O output. As the use of nitrogen fertilizers is expected to increase in the future in order to supply a growing world population with food, concern has been voiced by Crutzen (1974b, 1976) and McElroy (1975; McElroy *et al.* 1977a) and others about the prospect that the level of N₂O in the troposphere, its flux into the stratosphere, and the associated ozone destruction rate might rise to the point where the UV shielding capacity of the ozone layer is impaired. The problem may be a serious one but is, at the present time, difficult to assess, because the budget of tropospheric N₂O is still inadequately defined. The field measurements indicate that fertilization increases the N₂O emission rate only for a limited period of time, whereupon the flux falls back to normal values. By integrating the excess flux over the period of enhancement one can determine the percentage of fertilizer nitrogen that is lost as N₂O. Table 9-7 lists a number of such measurements. The lowest value is found for nitrate fertilizer, and the highest when anhydrous (liquid) ammonia is applied. If one assumes an average release rate of 0.2%, the current fertilizer production of 55 Tg N/yr (for 1980, United Nations Statistical Yearbook) corresponds to a global emission rate of 0.1 Tg N/yr. This amount is sufficiently small compared with natural emissions to be of no immediate concern. To be correct, however, one should also consider the fate of fixed nitrogen removed with the crop. The consumption of agricultural products by domestic animals and humans eventually leads to additional N₂O emissions from the microbial utilization of nitrogen appearing in the excreta.

The world ocean represents another important source of atmospheric nitrous oxide. Numerous measurements have shown that the surface waters of the Atlantic and Pacific Oceans are somewhat enriched with N₂O relative

Table 9-7. *Fraction (Average Percent) of Fertilizer Nitrogen Released from Soils as N_2O*

Authors	Amount applied (kg N/ha)	Time interval of observation (days)	Type of fertilizer			
			NH_4^+	NO_3^-	Urea	NH_3 [a]
Breitenbeck et al. (1980)	125	Up to 96	0.11	0.02	0.08	—
Bremner et al. (1981)	250	139	—	—	—	5.4
Conrad and Seiler (1980a)	100	72	0.09	0.01	—	—
Conrad et al. (1983)	100	20	0.14	0.03	—	—
Mosier and Hutchinson (1981)	200	86	—	—	—	1.3

[a] "Anhydrous" ammonia.

to the concentrations expected for an equilibrium distribution between air and seawater. The data were reviewed by Hahn (1981). The highest concentrations, up to five times the equilibrium value, were found in the upwelling waters of the tropics (Elkins et al., 1978). The areal extent of these regions is small, however. Elsewhere, the saturation values fall in the range 90–180% with an average of 125%. The corresponding flux of N_2O into the atmosphere can be calculated from the stagnant-film model (see Section 1.6). With the data on film thickness and diffusivity given by Broecker and Peng (1974) and Peng et al. (1979), $\Delta z \approx 35$ μm and $D_L = 1.8 \times 10^{-9}$ m²/s (at 15°C), respectively, one obtains an average flux of 2.4 ng N/m² s. If it is allowed to extrapolate the measurements to the whole ocean area, the global flux of N_2O from the marine sources sums to 26 Tg N/yr. Other investigators have derived similar values: Singh et al. (1979a) 13–19, Pierotti and Rasmussen (1980) 38, Hahn (1981) 28 (8–70) Tg N/yr.

Since large areas of the oceans are as yet unexplored, it is possible that the source strength of marine N_2O is overestimated. Some authors (e.g., McElroy et al., 1976) have suggested that parts of the ocean may act as a sink for atmospheric N_2O. Conceptually, the most reasonable mechanism for N_2O removal is downward transport into the deep sea by subsiding cold waters near the poles. According to estimates of Munk (1966) and Gordon (1975) for the Antarctic and of Broecker (1979) for the North Atlantic, the rate of subsidence totals about 50×10^6 m³/s. At 0°C, using the solubility data of Weiss and Price (1980), the concentration of N_2O in seawater in equilibrium with that in air is 0.7 mg/m³. Thus, less than 1 Tg of N_2O nitrogen per year is removed from the atmosphere in this manner. This is a small fraction of the production rate.

Much higher concentrations of N_2O than at the surface are found in deeper ocean waters. It is still unknown how much of the N_2O escaping to the atmosphere is produced near the surface and how much is brought upward from deeper levels by diffusive transport. The distribution of N_2O with depth is shown in Fig. 9-5. N_2O concentrations generally maximize in the zone where O_2 has a minimum, at 500–1000 m depths. In this region O_2 is used up by the bacterial oxidation of organic matter, and since here N_2O and nitrate are positively correlated, Cohen and Gordon (1979) suggested N_2O to arise primarily from nitrification. The turnover rate of fixed nitrogen between the pools of organic nitrogen in the biota and inorganic nitrogen (mainly NO_3^-) dissolved in the sea can be derived from the net rate of carbon fixation by photosynthesis (35 Pg/yr, see page 552) and the average C:N ratio in plankton (5.7 by weight; Redfield et al., 1963) as 6,000 Tg/yr. If the fraction of NH_4^+ turned into N_2O in seawater is similar to that in soil systems, which is about 0.1% according to Table 9-7, one calculates an N_2O production rate of 6 Tg N/yr from nitrification alone.

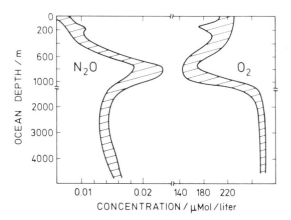

Fig. 9-5. Vertical profiles of N_2O and O_2 concentrations in the Atlantic Ocean (Gulf Stream area and Sargasso Sea) in 1971/1972, according to Yoshinari (1976).

From similar considerations Cohen and Gordon (1979) found production rates of 4–10 Tg N/yr.

Denitrification as a source of marine N_2O must not be ignored, however. The incubation experiments of Gundersen *et al.* (1972) with seawater samples taken from various depths have shown that nitrification and denitrification occur simultaneously in the same strata. In the open ocean the O_2 concentration never falls to values so low as to establish truly anaerobic conditions. Thus, compared with soil systems there is a lesser need for microorganisms to utilize N_2O as a source of oxygen. Accordingly, one expects a higher yield of N_2O relative to N_2 for denitrification in the ocean. In the steady state, the total rate of denitrification should equal that of biological nitrogen fixation augmented by the input of fixed nitrogen from continental sources. A very rough estimate based on the data in Table 12-8 suggest a total addition of fixed nitrogen of about 150 Tg N/yr. After deducting the N_2O production rate due to nitrification from the rate of N_2O emission to the atmosphere, one finds that 5–20% of nitrogen generated by denitrification emerges as N_2O. This fractions is several hundred times greater than that indicated by Table 9-7 for soil systems.

Tables 9-8 summarizes the budget of N_2O in the troposphere. In addition to emanations from soils and seawater, N_2O is released from polluted rivers and estuaries, as a product in the burning of agricultural wastes and biomass, and in the combustion of fossil fuels. Yet another source process of uncertain magnitude is the reaction $NH_2 + NO_2 \rightarrow N_2O + H_2O$ associated with the oxidation of ammonia by OH radicals.

Table 9-8. *Tropospheric Budget of Nitrous Oxide*

	Tg N/yr	Comment or reference
Sources		
Emanation from soils	10 (2–20)	Low values in Table 9-6 extrapolated
Release from the ocean	26 (12–38)	Hahn (1981) and references cited therein
Polluted rivers and estuaries	~1	Kaplan et al. (1978)
Biomass burning	2 (<8)	Crutzen et al. (1979), Crutzen (1983)
Fossil-fuel combustion	1.6	Weiss and Craig (1976)
Artificial fertilizer	0.1	Based on a fertilizer production of 55 Tg N/yr
Total sources	40.7 (16.7–68.7)	
Sinks		
Flux into the stratosphere	10 (8–15)	McElroy et al. (1976), Schmeltekopf et al. (1977), Johnston et al. (1979)

Destruction in the stratosphere provides the only well-known sink for atmospheric N_2O. Sources and sink are approximately in balance if one accepts the lower limit value for the former and the upper limit value for the latter. In this case, the residence time of N_2O in the troposphere is 85 yr. The data in Table 9-8, however, suggest a global source strength greater than 15 Tg N/yr. If correct, additional loss processes must be operative. A turnover rate of 40 Tg N/yr corresponds to a residence time of 30 yr. The value is still consistent with the observed high stability of N_2O in the troposphere.

The quest for additional sinks has met with little success. Bates and Hays (1967) had thought that N_2O photolysis occurs even in the troposphere, but a careful reinvestigation of the long-wavelength tail of the N_2O absorption spectrum by Johnston and Selwyn (1975) established beyond doubt that photoabsorption in the gas phase is insignificant at wavelengths above 300 nm. The ground-level photodissociation rate, as measured by Stedman et al. (1976), is less than $10^{-11}\,s^{-1}$. Rebbert and Ausloos (1978) have shown, however, that N_2O does undergo photodecomposition at wavelengths greater than 300 nm when it is adsorbed on silica or dry desert-type sands. Heat treatment enhances the effect, whereas the presence of moisture reduces it. Schütz et al. (1970), Junge et al. (1971), and Pierotti et al. (1978) all observed a depression of the N_2O mixing ratio in air masses originating from the Saharan Desert. Thus it is possible that hot, dusty deserts serve as sink areas for tropospheric N_2O. The magnitude of the effect unfortunately is hard to quantify. Other processes that have been investigated and were found unimportant include the reactions of N_2O with OH radicals studied by Biermann et al. (1976) and reactions with negative ions studied by Fehsenfeld and Ferguson (1976). Finally, Seiler et al. (1978a) showed that growing plants are neither sources nor sinks of atmospheric N_2O. The possibility that certain soils, primarily swampy anaerobic soils, may act as sinks cannot be entirely discounted, but the data dealing with this aspect are inadequate.

9.4 Nitrogen Dioxide, NO_2, and Related Compounds

The higher nitrogen oxides are very reactive, and for this reason they hold a prominent position in tropospheric chemistry. Six oxides are potentially important. They fall into two groups: NO, NO_2, NO_3, and N_2O_3, N_2O_4, and N_2O_5. The last three compounds arise from the corresponding ones in the first group by association with NO_2. The dinitrogen oxides are thermally unstable, however, which gives rise to association-dissociation equilibria. Table 9-9 lists equilibrium constants for these reactions at high and low termperatures and product/reactant ratios for a range of NO_2

Table 9-9. *Equilibrium Constants and Relative Abundances for Higher Nitrogen Oxides and Peroxides Originating as Addition Products from* NO_2 [a]

Reaction	T (k)	K_{equ} (bar⁻¹)	Ratio $p(x \cdot NO_2)/p(x)$ as function of $p(NO_2)$ in bar					Reference
			(−7)	(−8)	(−9)	(−10)	(−11)	
$NO + NO_2 \rightleftharpoons N_2O_3$	300	0.47	4.7 (−8)	4.7 (−9)	4.7 (−10)	4.7 (−11)	4.7 (−12)	(a)
$NO_2 + NO_2 \rightleftharpoons N_2O_4$	200	1.53 (3)	1.5 (−4)	1.5 (−5)	1.5 (−6)	1.5 (−7)	1.5 (−8)	(a)
	300	5.86	5.9 (−7)	5.9 (−8)	5.9 (−9)	5.9 (−10)	5.9 (−11)	(a)
$NO_3 + NO_2 \rightleftharpoons N_2O_5$	200	5.53 (5)	5.5 (−2)	5.5 (−3)	5.5 (−4)	5.5 (−5)	5.5 (−6)	(a)
	300	4.52 (8)	4.5 (1)	4.5	4.5 (−1)	4.5 (−2)	4.5 (−3)	(b)
$HO_2 + NO_2 \rightleftharpoons HO_2NO_2$	250	9.36 (11)	9.4 (4)	9.4 (3)	9.4 (2)	9.4 (1)	9.4	(b)
	300	3.14 (8)	3.1 (1)	3.1	3.1 (−1)	3.1 (−2)	3.1 (−3)	(c)
$CH_3O_2 + NO_2 \rightleftharpoons CH_3O_2NO_2$	250	5.86 (11)	5.9 (4)	5.9 (3)	5.9 (2)	5.9 (1)	5.9	(c)
	300	5.14 (7)	5.1	5.1 (−1)	5.1 (−2)	5.1 (−3)	5.1 (−4)	(d)
$CH_3(CO)O_2 + NO_2 \rightleftharpoons PAN$	250	1.07 (11)	1.1 (4)	1.1 (3)	1.1 (2)	1.1 (1)	1.1	(d)
	300	2.64 (11)	2.6 (4)	2.6 (3)	2.6 (2)	2.6 (1)	2.6	(e)
	250	2.29 (15)	2.3 (8)	2.3 (7)	2.3 (6)	2.3 (5)	2.3 (4)	(e)

[a] Powers of 10 in parentheses. References: (a) JANAF (Stull and Prophet, 1971); (b) Graham and Johnston (1978); (c) Graham *et al.* (1978); (d) DeMore *et al.* (1982), after Baldwin (1982) and Bahta *et al.* (1982); (e) estimated from rate coefficients for forward and reverse reactions given by Baulch *et al.* (1982). The equilibrium constant at 263.3 K was determined by Bahta *et al.* as 7.6(9) bar⁻¹. This value agrees with the other data.

partial pressures to show that the abundances of N_2O_3 and N_2O_4 are negligible in comparison with NO and NO_2, respectively, whereas N_2O_5 predominates over NO_3 when temperatures are low and/or NO_2 concentrations are high. Accordingly, we may ignore N_2O_3 and N_2O_4, but not N_2O_5.

Table 9-9 includes similar data for the association products of NO_2 with peroxy radicals. Peroxyacetyl nitrate (PAN) is thermally the most stable of these adducts. It is important in all regions of the troposphere. Pernitric acid and methylperoxy nitrate are less stable but may become significant in the upper troposphere where temperatures are low. Model calculations by Logan *et al.* (1981) suggest that as much as 50% of NO_2 may be present as $HOONO_2$ at higher altitudes. Observational data are lacking, however, so that the real contribution of these addition products to the total reservoir of NO_2 in the troposphere remains to be established.

The oxidation of NO_2 eventually leads to the formation of nitric acid and aerosol nitrate, which are deposited at the earth surface. The relevant oxidation pathways are indicated in Fig. 9-6. The following discussion deals first with observations of reaction intermediates; then with tropospheric abundances of NO_2, PAN, and HNO_3/ aerosol nitrate, and finally with the budget of nitrogen oxides and their oxidation products in the troposphere.

9.4.1 NITROGEN OXIDE CHEMISTRY

Figure 9-6 summarizes our current understanding of the chemical reactions involving nitrogen oxides in the troposphere. Photolytically induced

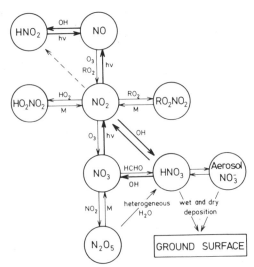

Fig. 9-6. Oxidation scheme for nitrogen oxides and related compounds. Photochemical processes are indicated by bold arrows.

Table 9-10. *Reactions Involving Nitrogen Oxides and Their Oxidation Products*[a]

Reaction	Reactant	m	n[b]	k[c], j[d]	τ (s)
$O_3 + NO \rightarrow NO_2 + O_2$	O_3	30 ppbv	7.5 (11)	1.8 (−14)	7.5 (1)
$RO_2 + NO \rightarrow NO_2 + RO$	RO_2	0.1 ppbv	2.5 (9)	8 (−12)	5 (1)
$OH + NO \rightarrow HNO_2$	OH	—	2 (6)	5 (−12)	1 (5)
$HNO_2 + h\nu \rightarrow OH + NO$	$h\nu$	—	—	8.4 (−4)	1.2 (3)
$NO_2 + h\nu \rightarrow NO + O$	$h\nu$	—	—	3.5 (−3)	2.8 (2)
$O_3 + NO_2 \rightarrow NO_3 + O_2$	O_3	30 ppbv	7.5 (11)	3.2 (−17)	4.2 (4)
$NO_3 + h\nu \rightarrow NO_2 + O$	$h\nu$	—	—	1.6 (−1)	6.2
$NO_3 + HCHO \rightarrow HNO_3 + HCO$	HCHO	1 ppbv	2.5 (10)	6 (−15)	6.7 (3)
$NO_3 + NO_2 \rightarrow N_2O_5$	NO_2	6 ppbv	1.5 (11)	7 (−13)	9.5
$N_2O_5 + h\nu \rightarrow NO_3 + NO_2$	$h\nu$	—	—	6 (−6)	1.7 (5)
$N_2O_5 + H_2O \rightarrow 2HNO_3$	H_2O	2.5%	6 (17)	<2 (−21)	>8.3 (2)
$N_2O_5 + H_2O \xrightarrow{\text{hetero}} 2HNO_3$	Rural aerosol	—	—	—	1.2 (2)
$OH + NO_2 \rightarrow HNO_3$	OH	—	2 (6)	1.3 (−11)	3.8 (4)
$OH + HNO_3 \rightarrow H_2O + NO_3$	OH	—	2 (6)	1.5 (−13)	3.3 (6)
$HNO_3 + h\nu \rightarrow OH + NO_2$	$h\nu$	—	—	1 (−7)	1 (7)
HNO_3, wet deposition	—	—	—	—	4.7 (5)[e]
HNO_3, dry deposition	—	—	—	—	3.3 (5)[e]
$2NO_2 + H_2O \rightarrow HNO_3 + HNO_2$	H_2O	2.5%	6 (17)	8 (−38)	7 (7)[f]
$NO + NO_2 + H_2O \rightarrow 2HNO_2$	H_2O	2.5%	6 (17)	4.4 (−40)	2.5 (10)[f]
	NO_2	6 ppbv	1.5 (11)		

[a] The associated time constants are obtained from the respective rate or photodissociation coefficients and reactant concentrations for ground-level rural atmospheric conditions. Data are based in part on a compilation of Ehhalt and Drummond (1982). Orders of magnitude are shown in parentheses.
[b] In units of molecules/cm^3.
[c] In units of cm^3/molecules.
[d] In units of s^{-1}.
[e] Assumes uniform vertical distribution.
[f] Half-lifetime.

pathways are indicated by bold arrows. These processes are active only during the day, whereas the others occur at all times. Table 9-10 is added to show time constants associated with the individual reaction steps.

The photochemical interconversion between NO and NO$_2$ was discussed in Section 5-2. During the day NO$_2$ undergoes photodissociation, forming NO and O atoms that quickly attach to molecular oxygen, producing ozone. Back-reactions of NO with ozone and with peroxy radicals generated from hydrocarbons establish a steady state between NO and NO$_2$. In the absence of local sources, their molar ratio should be given by the steady-state equation

$$m(NO_2)/m(NO) = [k_{60}n(O_3) + \sum k_x n_x(RO_2)]/j(NO_2)$$

where the summation sign denotes the combined effect of all peroxy radicals including HO_2. Ratios slightly greater than unity are expected in midlatitudes for clear sky conditions if $k_{60}n(O_3) \gg k_x n_x(RO_2)$. This assumption follows from model calculations, which, however, do not generally include hydrocarbons other than methane. The first measurements of NO_2/NO ratios designed to check on the above relationship were made by O'Brien (1974) and Stedman and Jackson (1975) in polluted city air containing high ozone concentrations. The observed ratios were in reasonable agreement with those predicted from simultaneous measurements of $j(NO_2)$ and $n(O_3)$, and there was no need to include RO_2 reactions. Subsequent measurements under clean air conditions revealed NO_2/NO ratios exceeding unity by a considerable margin. Drummond (1977) found values some three times higher than expected in the clean air of Wyoming; Ritter et al. (1979) observed a similar discrepancy at a rural site in Michigan; and Kelly et al. (1980) reported ratios of 4-10 at Niwot Ridge, a mountain site west of Boulder, Colorado. Helas and Warneck (1981), who did not investigate the photostationary state, found NO_2/NO ratios of 10 or greater in marine air masses of the northern Atlantic, and Stedman and McEwan (1983) made similar observations at a mountain site in New Zealand. It thus appears that either the photostationary state is seriously perturbed or that the measurement techniques are, in that they underestimate NO and /or overestimate NO_2. If the effect were caused by reactions of NO with RO_2 radicals, their daytime number densities would have to reach $(2-5) \times 10^9$ molecules/cm^3. Such high values are not only in excess of chemical model predictions, but they would also lead to unacceptably high ozone production rates except in regions with very low NO_2 mixing ratios. Mihelcic et al. (1985) have, nevertheless, observed RO_2 concentrations of the indicated order of magnitude in ground-level air using a cryogenic trapping technique followed by electron paramagnetic resonance detection. The photostationary state between NO and NO_2 thus remains an open problem that requires further study.

As indicated in Fig. 9-6, NO_2 and NO are also involved in the formation of nitrous acid, HNO_2. The reaction of NO with OH is relatively unimportant. Moreover, HNO_2 is rapidly photolyzed by sunlight, whereby OH and NO are reestablished. There exists, however, a nighttime process of HNO_2 formation, which was first observed by Perner and Platt (1979) in polluted air (see Fig. 5-3). The reaction most likely involves NO_2, and for this reason it is indicated in Fig. 9-6 by the dashed arrow, although the true reaction path is as yet unknown. Kessler and Platt (1984) have studied a number of combustion sources with regard to HNO_2 emissions. They concur with Pitts et al. (1984a) that the most effective source is the gasoline engine. Its exhaust gases contain up to 0.15% HNO_2 relative to NO, which is the major nitrogen oxide released from this source. In the atmosphere NO is largely converted to NO_2, but the ratio of HNO_2 to NO_2 observed in air ranges up to 5%, so that automobiles can account for only a fraction of nocturnal HNO_2. The reactions

$$NO + NO_2 + H_2O \rightarrow 2HNO_2$$

$$2NO_2 + H_2O \rightarrow HNO_2 + HNO_3$$

do, in principle, provide pathways for the formation of HNO_2, although in the gas phase the reactions are slow, as Table 9-10 shows. Kessler et al. (1982) thus were led to suggest that HNO_2 arises from heterogeneous processes, that is, by reactions of NO_2 with aerosol particles, fog droplets, or materials at the earth surface. Owing to the poor solubility of NO_2 and NO in water, however, the above reactions are not very effective even in clouds (cf. Table 8-9). More rapid may be the reaction of NO_2 with sulfite ions resulting from the dissolution of SO_2 in liquid water. Yet even this reaction is too slow to explain the observed rate of HNO_2 formation in the atmosphere. In this connection, it should be noted that nitrous acid appears also in smog chambers, where it has been shown to arise from a dark reaction of NO_2 with water adsorbed on the walls of such chambers (Sakamaki et al. 1983; Kessler and Platt, 1984; Pitts et al., 1984b). The concentration of NO_2 in the experiments usually was higher than in ambient air. Nevertheless, the rate of the reaction was lineally dependent on NO_2 concentration, in contrast to expectation if HNO_2 arose from the reaction $2NO + H_2O$. In addition, NO was found to evolve concurrently with HNO_2. While the mechanism responsible for this behavior has remained obscure, it is likely that the production of HNO_2 in the atmosphere is based on a similar process.

The oxidation of NO_2 to nitric acid occurs in two ways, either directly by reaction with OH radicals, or indirectly by reaction with ozone. The second reaction, which produces NO_3 as an intermediate, is a thousand times slower that the equivalent reaction of ozone with NO. Moreover, NO_3 is photolyzed at a faster rate than NO_2, so that during the day the reaction chain leading from NO_2 via NO_3 to N_2O_5 and then further to nitric acid is interrupted. The situation changes at dusk. Noxon et al. (1980) and Platt et al. (1980a, 1981) have detected the NO_3 radical by optical absorption and have followed the buildup of NO_3 concentrations during the night. Figure 9-7 gives two examples for such observations. If there were no losses, the concentration of NO_3 would have to increase steadily until dawn, when the rising sun causes its destruction. Such a behavior may be approximated by the data in Fig. 9-7a. Usually, however, the concentration of NO_3 goes through a maximum and then drops off, as in Fig. 9-7b. This behavior indicates that loss reactions are important. Since NO_3 and N_2O_5 come

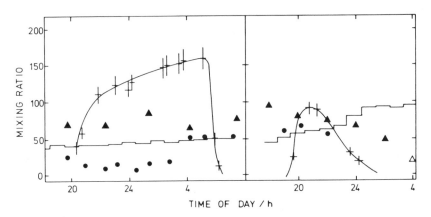

Fig. 9-7. Nighttime behavior of NO$_3$ in the ground-level air near Deuselbach, Western Germany. Shown are mixing ratios of O$_3$(Δ, ppb), NO$_2$ (●, 0.1 ppb), NO$_3$ (+, ppt), and % relative humidity (stepped curve). [Redrawn from data of Platt *et al.* (1981).]

quickly to an equilibrium, they are not independent of each other and one must consider losses of both species.

NO$_3$ is known to react with aldehydes, alkenes, terpenes, and substituted benzenes at moderate rates (Morris and Niki, 1974; Japar and Niki, 1975; Bandow *et al.* 1980; Atkinson *et al*, 1984, b, 1985; Cantrell *et al.*, 1985). The significance of these reactions to atmospheric chemistry remains to be clarified, however. The laboratory studies show that aldehydes undergo hydrogen abstraction, whereas alkenes react with NO$_3$ by addition. Formaldehyde is one of the more abundant constituents of air, and its reaction with NO$_3$ is entered in Fig. 9-6 for illustrations. The gas-phase reaction of NO$_3$ with water is endothermic and slow. Even the reaction of NO$_3$ with SO$_2$, which is exothermic, is negligibly slow (Calvert and Stockwell, 1984). The reaction NO$_3$ + NO → 2NO$_2$ is rapid and important in polluted air. Under pristine conditions, daytime NO is quickly converted to NO$_2$ after the sun sets, so that the reaction of NO$_3$ with NO should become negligible, too, at night.

Stockwell and Calvert (1983) have argued that the nighttime formation of NO$_3$ and its reaction with formaldehyde

$$O_3 + NO \rightarrow NO_3 + O_2$$
$$NO_3 + HCHO \rightarrow HCO + HNO_3$$
$$HCO + O_2 \rightarrow CO + HO_2$$

constitute a nocturnal source of HO$_2$ radicals and that their interaction with

NO_2 might lead to the formation of nitrous acid via

$$HO_2 + NO_2 \rightarrow HNO_2 + O_2 \qquad k = 3 \times 10^{-15}$$

$$HO_2 + NO_2 \rightleftharpoons HOONO_2 \qquad k_f = 2 \times 10^{-12}$$

The process would have to compete with the formation of pernitric acid, but since this compound redissociates, a certain share of HNO_2 would be produced nonetheless. Recent observations by Pitts *et al.* (1984c) have ascertained that HNO_2 is formed at night even in the absence of ozone (and NO_3), so that this process cannot explain the extent of nocturnal HNO_2 formation discussed above.

Nitrogen pentoxide, N_2O_5, which is in equilibrium with NO_3, is the anhydride of nitric acid, and the hydration of N_2O_5 undoubtedly is a major loss process for NO_3 and N_2O_5 combined. Again it has been found (by Morris and Niki, 1973) that the gas-phase reaction of N_2O_5 with water vapor is negligible. However, the reaction proceeds readily with water adsorbed on the walls of reaction vessels. In the atmosphere the process presumably occurs in a similar manner, with liquid water attached to particulate matter. There is some observational evidence to support this idea. Platt *et al.* (1981) have noted a dependence of NO_3 behavior on relative humidity. High relative humidities appear to suppress NO_3. In a number of observations, such as that shown in Fig. 9-7b, the night started out with low relative humidities and NO_3 rose as expected. Later, the relative humidity increased to values high enough to cause fog formation, whereupon the NO_3 concentration dropped below the detection limit. The main effect of raising the relative humidity is to increase the water content of aerosol particles. Condensed water thus appears to provide an important agent for the removal of N_2O_5 and NO_3. The reaction product is nitric acid.

In the troposphere, HNO_3 represents a stable terminal product of NO_2 oxidation. Table 9-10 shows that rates for the removal of nitric acid by photodissociation and reaction with OH are small compared with loss rates due to wet and dry deposition. A portion of HNO_3 becomes attached to the atmospheric aerosol, where it is neutralized by ammonia and/or soil and seasalt cations. A significant fraction of HNO_3 remains in the gas phase, however. The stability of aerosol nitrate is determined in part by the pH of the particles, which may be sufficiently low to revolatilize nitric acid. This process establishes a quasi-equilibrium between gas and particulate phases. In Fig. 9-6 this relation is indicated by the double arrow. The equilibrium depends also on the water content of the aerosol. Tang (1980) calculated equilibrium HNO_3 and NH_3 pressures for model aerosol containing mixtures of NH_4HSO_4 and NH_4NO_3 and found a strong decrease of HNO_3 volatility when the relative humidity was increased from 85 to 99%.

Stelson et al. (1979, 1982) directed their attention to the opposite end of the relative humidity scale and suggested from the knowledge of partial pressures above crystalline NH_4NO_3 that this material tends to evaporate when the relative humidity falls to values below the deliquescence point of NH_4NO_3 (62%). Accordingly, the partitioning of nitrate between gas phase and particulate matter will be quite variable.

Nitrogen dioxide is converted to nitric acid by OH radicals during the day and by reaction with ozone during the night. The combined rates of these two process determine the lifetime of NO_2 in the troposphere. From their observations, Platt et al. (1981) estimated the time constant for the nightime destruction of $NO_3 + N_2O_5$ as 40 min or less. This is short enough to make the reaction of NO_2 with ozone the rate determining step in nightime losses of NO_2. Its average lifetime then is given by

$$\bar{\tau}_{NO_2} = 1/[f_{dark}k_{62}\bar{n}(O_3) + k_{115}\bar{n}(OH)]$$

where f_{dark} denotes the fraction of darkness during a full day, and $\bar{n}(O_3)$ and $\bar{n}(OH)$ are the daily averages for the number densities of ozone and of OH radicals, respectively. For near-surface conditions in midlatitudes, $\bar{n}(O_3) \approx 6 \times 10^{11}$ and $\bar{n}(OH) \approx 5 \times 10^5$ molecules/cm^3. With $f_{dark} = \frac{1}{3}$ for summer conditions one finds $\bar{\tau} = 1.2$ days. In the summer both reactions contribute about equally to the total loss of NO_2. During winter $\bar{n}(OH)$ decreases whereas the dark period increases, so that then the reaction with ozone is the dominant loss process. The average lifetime remains almost unchanged. There are considerable diurnal variations, of course. The midday lifetime in summer may be as short as 5 h, if OH number densities reach peak values of 4×10^6 molecules/cm^3.

In a number of studies, field measurements have been utilized to derive NO_2 lifetimes. For this purpose it is necessary to compare the behavior of NO_2 with that of a tracer of lower reactivity. Chang et al. (1979) examined data from Los Angeles and from St. Louis in this manner and found NO_2 lifetimes of 1–2 days by two independent methods. Spicer (1980, 1982) made aircraft measurements in the urban plumes of Phoenix, Arizona, Boston, Massachusetts, and Philadelphia, Pennsylvania. He found daylight lifetimes of more than 1 day in the first study and of 4–8 h in the other two studies. His measurements confirmed that NO_2 is converted to HNO_3, particulate nitrate, and peroxyacetyl nitrate. The field studies thus support the estimate given above.

9.4.2 TROPOSPHERIC DISTRIBUTION OF NO_x

Owing to the photochemical coupling between nitric oxide, NO, and nitrogen dioxide, NO_2, the sum of their mixing ratios is usually treated as a single variable, which is designated NO_x. The concentrations of NO_3 and N_2O_5 do not reach those of NO_x, so that they can be ignored in this context. Although a large fraction of NO_x is originally injected into the atmosphere in the form of NO, high NO/NO_2 ratios are found only in the source areas, that is, in the cities. Here, the abundance of NO_x may reach 100 ppbv. Outside the source regions, NO reacts rapidly with ozone and NO_2 becomes the dominant component of NO_x. Nitric oxide can now be measured at a level of a few pptv by analyzers based on the chemiluminescence of the $NO + O_3$ reaction. The determination of NO_2 by the same technique requires a reduction of NO_2 to NO prior to analysis. This may be done by hot catalytic converters (e.g., molybdenum), $FeSO_4$ crystals at ambient temperatures, or photodissociation of NO_2. Unfortunately, each conversion technique may add NO from the breakup of other nitrogen compounds such as PAN. Even HNO_3 causes an interference if hot catalytic converters are used. In addition, the chemiluminescence analyzers are plagued with a variable background chemiluminescence that appears to be generated from reactions of ozone with impurities in the gas supplies or at the walls of the reactor. Wet-chemical colorimetric analysis of NO_2, such as the diazotization procedures of Saltzman (1954) and Jacobs and Hocheiser (1958), also are subject to interferences—for example, by ozone—and they lack the sensitivity needed for measurements in the sub-ppbv range of mixing ratios. Long-path optical absorption in the near-ultraviolet spectral region offers yet another method for measuring NO_2, but again it is necessary to take into account underlying spectra of other interfering compounds.

Despite their shortcomings, the indicated techniques furnish at least approximate values for NO_x mixing ratios. Due to the short residence time of NO_x in the troposphere, the mixing ratios show large fluctuations anyway. Longer series of observations generally are necessary for averaging. Table 9-11 summarizes ground-based measurements in so-called rural areas, that is, in areas not directly influenced by anthropogenic sources. Flatland and valley sites in populous regions feature NO_x mixing ratios of the order of 3 ppbv in summer, and 10 ppbv in winter. A seasonal variation exhibiting a winter maximum has now been observed at several sites in Europe and the United States. Figure 9-8 compares the annual variation of NO_2 with that of SO_2. Shown are monthly mean values obtained at the environmental station at Deuselbach, Germany, a rural community situated in hilly terrain. The data show a close correlation between NO_2 and SO_2. Part of the winter maximum may be due to an increase in the burning of fossil fuels for residential heating. Another part may result from the greater frequency of low-lying inversion layers in winter as compared to summer conditions, causing trace gases to be dissipated more slowly from the boundary layer to the free troposphere. The data discussed in Chapter 10 reveal, however,

Table 9-11. *Summary of Ground-Based Measurements of NO_x at Rural and Mountain-Top Sites[a]*

Authors	Location and measurement period	Mixing ratio (ppbv)	Remarks
Galbally (1975)	Aspendale, Australia, suburban	13 av. NO_2	WC, values decreasing with wind speed
	Cape Otway	2 av. NO_2	
Drummond (1977)	Wyoming	0.12–0.41	CL, hot carbon
Ritter et al. (1979)	Rural Michigan, 10 days in June 1977	0.3–0.5	CL, $FeSO_4$
Kelly et al. (1980)	Niwot Ridge, Colorado, 2910 m a.s.l.		CL, $FeSO_4$
	7 days in April 1979	0.25 av.	
	10 days in January 1979	0.21 av.	
Harrison and McCartney (1980)	Rural Northwest England, May–Sept. 1977	10 ± 8.2	CL, hot molybdenum
Martin and Barber (1981)	Bottesford, England, 20 km east of Nottingham, 1978/1979	10 av. summer 21 av. winter	CL, seasonal variation, large contribution by NO
Bollinger et al. (1984), Logan (1983)	Niwot Ridge, Colorado, 2910 m s.a.l.	0.30 av. summer 0.24 av. winter	CL, photolysis cell; high variability due to advection of polluted air from Denver
Pratt et al. (1983)	La Moure County, North Dakota, 1977–1981	3.44 av. NO 1.79 av. NO_2	CL, hot molybdenum, NO is higher than NO_2 (this is unusual)
	Wright County, Minnesota, 1977–1981	3.17 av. NO 5.94 av. NO_2	
Stedman and McEwan (1983)	Mt. John, New Zealand, 1030 m a.s.l., March–April 1981	0.127 av.	CL, $FeSO_4$

462

Reference	Location	Value	Method/notes[a]
Shaw and Paur (1983)	Ohio River Valley, May 1980–August 1981	4.5–16 NO_2 8.2 av. 0.4–12.5 NO 1.9 av.	CL, seasonal variation with winter maximum; $NO/NO_x = 0.22$, on average; NO shows great variability
Umweltbundesamt (1980, 1982)	Deuselbach, Germany, 1968–1982, 15-y average	3.88 av. summer 7.67 av. winter	WC, 24-h sampling, NO_2 only, seasonal variation
	Mt. Schauinsland, Germany	1.82 av. summer 2.28 av. winter	
Broll et al. (1984)	Deuselbach, Germany, 1980/81	4.5 av. summer 16.3 av. winter	CL, $FeSO_4$, seasonal variation
Meszaros and Horvath (1984)	Central Hungary, 1974–1983	3.16 av.	WC, seasonal variation with winter
Johnston and McKenzie (1984)	Rural New Zealand, 29 Nighttime measurements	0.02–1.36 NO_2	LPA (9.2 km), 420–450 nm, low values in windy nights, high values underneath inversion layers
Sjödin and Grennfelt (1984)	Sweden, 1981–1983, Ekeröd (agricultural area)	2.05 av. summer 4.63 av. winter	WC, NO_2 only, seasonal variation noted
	Rydakungsgard (agricultural)	1.12 av. summer 1.90 av. winter	
	Velen (forest)	0.68 av. summer 1.78 av. winter	
	Bredkalen (forest)	0.41 av. summer 0.88 av. winter	

[a] Code for measurement techniques: WC, wet chemical; CL, chemiluminescence ($NO + O_3$), material for NO_2 to NO conversion indicated; LPA long-path optical absorption.

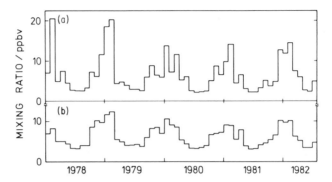

Fig. 9-8. Annual variation of mean monthly (a) SO_2 and (b) NO_2 mixing ratios in Deuselbach, a rural community of Western Germany. [Data taken from regular reports of the German Environmental Administration (Umweltbundesamt).]

that a large part of the seasonal cycle for SO_2 arises from the greater photochemical oxidation rate in the summer. For NO_2, the known oxidation processes do not predict a similar variation in the rate of its conversion to HNO_3, so that the origin of the seasonal cycle of NO_2 is less well understood. This problem obviously requires further study.

Table 9-11 includes data from several mountain sites. Due to local circulation patterns, these sites receive alternately uprising air from the

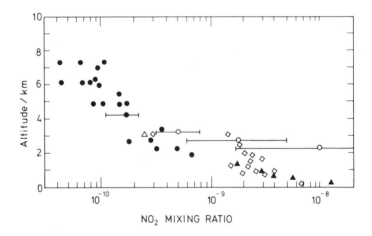

NO_2 MIXING RATIO

Fig. 9-9. Vertical distribution of NO_x in the continental troposphere, aircraft observations. Filled circles, over Wheatland, Wyoming, open circles, over Denver, Colorado (Kley *et al.*, 1981); triangles, single flight over southern Germany in May 1964, diamonds, several flights over Germany in the winter of 1962/1963 (Georgii and Jost, 1964).

valleys and clean air from aloft. The few existing data point toward a decrease of NO_x mixing ratios with elevation. At Niwot Ridge, Colorado, and at Mt. John, New Zealand, the values are reduced to 100–200 pptv. Still lower values are expected in the free troposphere. Figure 9-9 shows results from aircraft explorations over the continents. Again, the data base is small, but the general behavior of NO_x as a function of altitude is made evident. In the lower troposphere the NO_x mixing ratios decrease in accordance with a scale height of about 1 km. In the upper troposphere there appears to exist a natural background of NO_x of some ten pptv. Such a background is expected owing to the small influx of nitrogen oxides and HNO_3 from the stratosphere combined with a contribution to NO_x from the photolytic dissociation of HNO_3 (Kley et al., 1981; Logan et al., 1981). The last process is slow, but the lifetime of NO_2 with regard to its reconversion toward HNO_3 increases in the upper troposphere compared with the lower, because the concentration of OH radicals declines with altitude up to the tropopause, and the abundance of liquid water that might convert N_2O_5 to HNO_3 decreases as well.

Table 9-12 adds further observations of NO_x in marine air masses and in the free troposphere. Measurements onboard ships and at coastal stations are made difficult due to self-contamination, and the original data must be carefully screened to eliminate conceivable effects of pollution. With these precautions it is then possible to show that the purest maritime air masses contain NO_x mixing ratios less than 100 pptv. Broll et al. (1984) have

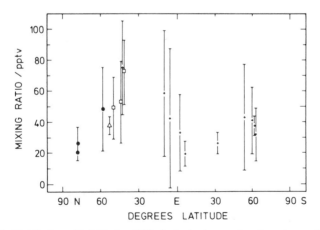

Fig. 9-10. Distribution with latitude of NO_x in the background surface air over the Atlantic Ocean. Derived from measurements on board of ships for clean air mass conditions. Data for Loop Head at the Irish west coast are included for comparison. ●, Meteor—Spitzbergen; △, Loop Head; □, Meteor—Dakar; +, Meteor—Antarctic. [Adapted from Broll et al. (1984).]

Table 9-12. *Measurements of NO_x in the Remote Marine and Free Troposphere[a]*

Authors	Location	Mixing ratio (pptv)	Remarks
	Surface-bound observations in the marine atmosphere		
Cox (1977)	Adrigole, Ireland, July–November, 1974	340 av. NO_2	CL, hot stainless steel, background value
McFarland et al. (1979)	Tropical Pacific Ocean noontime values	~4 NO	CL, photolysis cell; diurnal variation, maximum at noon
Galbally and Roy (1981)	Cape Grim, Tasmania, July–August 1979	<50–300	CL, hot molybdenum
Helas and Warneck (1981)	Loop Head, Ireland, June 1979	100–200 91%ile 87 ± 47 NO_2 37 ± 6 nightime	CL, $FeSO_4$; diurnal variation of NO_2 noted, data trajectory selected for clean marine air
Platt and Perner (1980)	Loop Head, Ireland	90–300 NO_2	LPA, pure marine air
Liu et al. (1983)	Tropical Pacific Ocean, 6.5°N–3.7°S, noon ±3 h	2.8–5.7 NO	CL, photolysis cell, diurnal variation, maximum at noon
Broll et al. (1984)	Atlantic Ocean, 60°N–60°S	20–70 NO_2	CL, $FeSO_4$; see Fig. 9-10
	Observations in the free troposphere		
Schiff et al. (1979)	Pacific Ocean, 20°N-37°S, 6 km altitude	≦40 NO	CL, perturbed by instrumental background, aircraft
Roy et al. (1980)	Australia, 8–10 km altitude	100–300 NO	CL, aircraft
Kley et al. (1981)	Wyoming, 4–7 km altitude	40–200 NO_2	CL, photolysis cell, see Fig. 9-9, aircraft
Noxon (1981, 1983)	Mauna Loa, Hawaii, 2.5–3.5 km altitude	30 ± 10 NO_2	LPA against the setting sun
Drummond et al. (1985)	Southern Germany, 2–6 km altitude	2–19 NO	CL, aircraft
	North Atlantic and South America		
	60°S–67°N, 2–7 km altitude	0–40 NO	
	60°S–67°N, 8–12 km altitude	0–180 NO	

[a] Code for measurement techniques as in Table 9-11.

explored the latitudinal distribution of (nighttime) NO_2 in the surface air over the Atlantic Ocean. The results are shown in Fig. 9-10. The appearance of a broad maximum in the northern hemisphere near 40° latitude suggests that anthropogenic NO_x from the North American continent spreads with the westerlies into the nearly source-free regions of the marine atmosphere, and that the remnants are still noticeable west of Europe. If this interpretation is correct, the North Atlantic Ocean must be considered polluted by NO_x. The lowest NO_2 mixing ratios, with values in the range of 20-30 pptv, were encountered near the equator and in the polar regions. These values presumably indicate the natural background of NO_x in the remote troposphere. Noxon (1983) has used optical spectroscopy to determine the NO_2 mixing ratio in the free troposphere accessible from Mauna Loa, Hawaii (3970 m above sea level). He found values of 30 ± 10 pptv, which compare well with those in the air above the equatorial Atlantic. The data thus give the impression that the troposphere contains a fairly uniform background of NO_2 of about 30 pptv.

Daytime mixing ratios of NO in marine surface air are much lower than those of NO_2. McFarland et al. (1979) and Liu et al. (1983) reported values of about 4 pptv over the central Pacific. Table 9-12 further includes the results of several aircraft studies directed at measurements of NO in the free troposphere. Generally the NO mixing ratios remain low until one reaches the 8-10 km altitude regime. Here, they increase by at least an order of magnitude. This behavior is most readily made apparent by the recent measurements of Drummond et al. (1985), but an explanation has not yet been offered. Further measurements are needed, specifically, simultaneous and reliable measurements of NO and NO_2, if the behavior of NO_x in the free troposphere is to be better understood.

In summary, our knowledge of NO_x in the troposphere is clearly inadequate. But the data presented above provide an approximate indication for the tropospheric distribution. Superimposed on a background mixing ratio of about 30 pptv (mainly NO_2) is a contribution of NO_x originating from continental sources. Starting with high values in the source regions, the mixing ratios decline with elevation above the ground and with distance from the coast until they merge with the background values. The situation in the upper troposphere is insufficiently explored.

9.4.3 PEROXYACETYL NITRATE, PAN

As discussed previously, PAN arises as a secondary product in the oxidation of hydrocarbons in the presence of nitrogen oxides. PAN and its higher homologues were first discovered by Stephens et al. (1956) in laboratory studies designed to understand the formation of photochemical smog, before Scott *et al.* (1957) actually demonstrated the occurrence of PAN in Los Angeles air. The precise structural formula of the substance was initially unknown until Stephens *et al.* (1961) established it as that of peroxyacetyl nitrate. In photochemical smog, PAN is one of the substances responsible for eye irritation and plant damage (Taylor, 1969). The customary technique for measuring PAN in ambient air is gas chromatography using the electron-capture detector. Difficulties still exist in calibrating such instruments, in view of the thermal instability of the compound.

Temple and Taylor (1983) have summarized ground-based observations of PAN in the air of Europe, Japan, and the United States. Grosjean (1984) presented supplemental data for the Los Angeles basin. This smog-affected area is burdened with the highest PAN mixing ratios in the world. In downtown Los Angeles, for example, values as high as 65 ppbv are not uncommon (Lonneman *et al.*, 1976). Diurnal averages rarely exceed 10 ppbv, however. The majority of all measurements were done in an exploratory manner, and few data cover periods long enough to reveal annual trends. Two exceptions are the time series from Riverside, California (Grosjean, 1984) and from Harwell, England (Brice *et al.*, 1984). The former location is a well-known smog receptor site east of Los Angeles; the latter may be described as a rural setting. At both sites, as well as at others, PAN frequently correlates with ozone and correspondingly undergoes a diurnal variation with a maximum in the early afternoon. The correlation is similar to that observed in smog chambers (cf. Fig. 5-1). It results from the fact that both ozone and PAN are products of the NO_x-catalyzed oxidation of hydrocarbons. Indeed, PAN has been suggested to represent a more useful indicator for photochemical air pollution than ozone, because PAN has no large natural sources (Nieboer and Van Ham, 1976; Penkett *et al.*, 1977).

Figure 9-11 shows the annual variation of monthly mean PAN mixing ratios at Riverside and Harwell. In the first cases two independent time series exist, one based on 24-h measurements, the other limited to daylight hours. The annual averages are 3.4 and 4.4 ppbv, respectively. The 30% difference presumably reflects the diurnal variation, but the seasonal distribution of the former set of data appears to be more uniform than the latter, which peaks in the second half of the year. Both data sets agree in that the lowest monthly mean value occurs at the turn of the year. At Harwell, the PAN mixing ratios are an order of magnitude smaller than in California, averaging to 0.36 ppbv annually. In addition, the seasonal variation is more pronounced, showing a definite maximum in later summer and a minimum in winter.

Unusually high PAN mixing ratios were observed at Harwell in February and April of 1981 in comparison with the same months of previous years. Figure 9-11 shows that the monthly means for February and April are raised

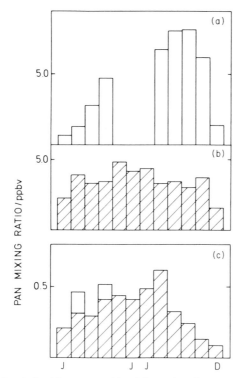

Fig. 9-11. Annual variation of mean monthly mixing ratios of PAN at Riverside, California (upper frame) and at Harwell, England (lower frame). (a) Values shown combine data from 1967/1968 and 1980; measurements were limited to daytime (Temple and Taylor, 1983). (b) 24-h Measurements in 1975/1976 (Grosjean, 1984). (c) Values shown combine data from 1974/1975 and 1980/1981 (Brice *et al.*, 1984. The high values shown for February and April result from high mixing ratios observed during these months in 1981. The lower values (hatched bars) are obtained when the 1981 data are omitted.

considerably if the high 1981 values are included in the total average. In both months the high values for PAN were accompanied by peak concentrations of aerosol sulfate and nitrate, signaling the advection of polluted air. A more detailed analysis of the February data revealed that the high values were caused primarily by two pollution episodes featuring PAN mixing ratios of nearly 3 ppbv. The meteorological situations during these periods were characterized by low-lying temperature inversions and the advection of air masses from the European mainland. Evidently, the formation and transport of PAN in polluted air masses can be important even during the cold season of the year.

While these observations strengthen the view that PAN is mainly associated with photochemical air pollution, it has recently been shown that there exists also a natural background of PAN in the troposphere. Singh and Hanst (1981) have argued that such a background should arise from the oxidation of ethane and propane, which are the most abundant nonmethane hydrocarbons in the troposphere. Their reactions with OH radicals lead first to the formation of acetaldehyde and acetone, respectively, which then undergo either photodecomposition or further reactions with OH radicals to produce acetylperoxy and finally PAN. The reaction pathways are shown in Fig. 9-12. The photolysis of acetyldehyde is the only process that does not lead to acetylperoxy radicals, but this pathway is comparatively unimportant. Singh and Hanst (1981) used a simple model to estimate PAN mixing ratios in the free troposphere, ranging from 20 to 360 pptv with a maximum at altitudes near 10 km.

Recent aircraft explorations have confirmed the ubiquity of PAN in the atmosphere (Singh and Salas, 1983; Meyrahn et al., 1984, 1986). Since the individual measurements indicate a greater variability, the data are here combined to illustrate the general trend. The results are shown in Fig. 9-13. Over the European continent the PAN mixing ratios decrease with altitude, starting from fairly high levels. This behavior indicates the importance of ground-level sources, giving rise to an upward flux of PAN toward greater heights, from where it will be globally distributed. The gradient further suggests a partial destruction of PAN during vertical transport by turbulent mixing. In the marine atmosphere, in contrast, the PAN mixing ratios show no significant dependence on altitude. In one of their four flights above the northeastern Pacific, Singh and Salas (1983) encountered much higher than

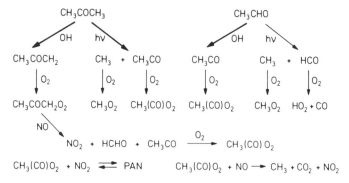

Fig. 9-12. Routes to the formation of PAN from acetone (left) and acetaldehyde (right). In addition to formation by association of acetylperoxy radicals with NO_2, there is a loss of PAN via the reaction of acetylperoxy radicals with NO as shown by the second reaction of the last line. Important pathways are accentuated.

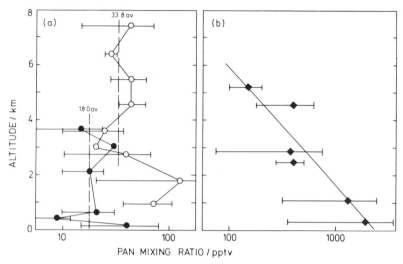

Fig. 9-13. Vertical distribution of PAN in the atmosphere. (a) Over the oceans from data of Singh and Salas (1983) for the Pacific (○), and of Meyrahn *et al.* (1986) for the Atlantic (●). (b) Over the central European continent, southern Germany (Meyrahn *et al.*, 1984).

average PAN levels. These data appear to have been influenced by land contact, and they were omitted in deriving the averages shown in Fig. 9-13. The resulting average mixing ratio in the middle and upper troposphere is about 34 pptv. The data of Meyrahn *et al.* (1986), which were obtained in the region of the Azores, yield an average of 18 pptv. Thus, PAN is present in the remote troposphere with mixing ratios in the range 15–50 pptv. This agrees approximately with predictions, even though the predicted rise with increasing altitude has not been substantiated. By comparison with the background level of NO₂ (about 30 pptv) it becomes apparent that a significant fraction, perhaps as much as 50%, is stored in the form of PAN, so that both must be considered together.

9.4.4 NITRIC ACID, HNO₃, AND PARTICULATE NITRATE, NO₃⁻

The recognition that gaseous nitric acid can coexist in the atmosphere with nitrate bound to aerosol particles has led to a reappraisal of filter techniques for the sampling of particulate nitrate. By means of Teflon membrane filters, which are immune to HNO_3 and other nitrogen compounds, it is now possible to collect aerosol nitrate without interference, and to absorb gaseous HNO_3 separately on an impregnated back-up filter. Alternatively, one may use a wall-coated diffusion-denuder tube for the absorption of gaseous HNO_3 and collect nitrate-containing particles on a back-up filter. Neither technique prevents losses of particulate nitrate as HNO_3 under conditions of low relative humidity, or by the interaction with acidic sulfate particles if the filters are overloaded. Joseph and Spicer (1978) and Kelly *et al.* (1979) have described modified chemiluminescence analyzers for the determination of ambient HNO_3 mixing ratios from the difference of signals derived from $NO_x + HNO_3$ on one hand, and NO_x after the removal of HNO_3, on the other hand. Appel *et al.* (1981), Forrest *et al.* (1982), Spicer *et al.* (1982) and Anlauf *et al.* (1985) have conducted field intercomparison tests for these methods with generally good agreements.

Order of magnitude levels for particulate nitrate in various parts of the world were shown in Table 7-13. Here, we confine the discussion to simultaneous measurements of gaseous HNO_3 and aerosol nitrate. Table 9-13 summarizes some recent data. By and large, both species are present in the atmosphere with comparable concentrations. Most of the ground-based measurements indicate a moderate excess of particulate nitrate, so that the HNO_3/NO_3^- mass ratio is less than unity. In the free troposphere the relation is reversed, however. Nitric acid has a very high dry deposition velocity of $(2-3) \times 10^{-2}$ m/s (Huebert and Robert, 1985), which is at least an order of magnitude greater than that for aerosol particles. There can be no doubt that in the ground-level atmosphere the HNO_3/NO_3^- ratio will be strongly influenced by losses of HNO_3 due to dry deposition at the earth surface.

Several investigators have noted a pronounced diurnal variation of HNO_3 featuring a maximum at midday and a minimum at night. Particulate nitrate often shows an opposite behavior, with an increase during the night and a decrease during the day. The diurnal cycle of HNO_3 is anticorrelated with that of relative humidity, which normally rises during the night due to the decline in temperature. This relation is illustrated in Fig. 9-14a. Connected with the nocturnal rise in relative humidity is a growth in the liquid water content of particulate matter (see Fig. 7-7), which, in turn, increases the absorbing capacity of aerosol particles with regard to gaseous HNO_3. Thus it appears that at least part of the diurnal variation results from the condensation of HNO_3 onto aerosol particles at night, followed by a partial reevaporation during the day when the relative humidity declines. A second possible explanation for the diurnal cycle of HNO_3 may lie in the difference of reaction mechanisms responsible for the oxidation of NO_2 to HNO_3, night and day (cf. Fig. 9-6). The nighttime formation of HNO_3 involves the interaction of N_2O_5 with liquid water already associated with aerosol particles, whereas the daytime oxidation process occurs by the reaction of NO_2 with OH radicals, which produces HNO_3 in the gas phase. The oxidation of NO_2 during the day is paralleled by a similar oxidation of SO_2 to H_2SO_4, but HNO_3 is much more volatile than H_2SO_4, with the consequence that

Authors	Location and period of measurement	HNO₃ NO₃⁻ ($\mu g/m^3$)	HNO₃ ppbv	HNO₃ $\mu g/m^3$	HNO₃/NO₃⁻ ($\mu g/\mu g$)	Remarks
	Ground-based measurements					
Forrest et al. (1982)	Claremont, California, Aug./Sept. 1979, urban	12.5	4.4	12.4	1.0	Average for 9 days
Grosjean (1983)	Claremont, California, Sept./Oct. 1980, pollution episode	8.4	4.0	11.4	1.35	Diurnal variation
Shaw et al. (1982)	Research Triangle Park, North Carolina, semi-rural, June/Aug. 1980	0.66	0.54	1.53	2.3	Average for 11 days
Cadle et al. (1982)	Abbeville, Louisiana, rural, June/Aug. 1979	0.9	0.72	2.0	2.1	Average values, diurnal variation noted at Abbeville
	Luray, Virginia, July/Aug. 1980, rural	0.44	0.36	1.0	2.1	
Cadle (1985)	Warren, Michigan, suburban, June 1981–June 1982	3.06	0.96	2.69	0.88	Averages for summer
		2.32	0.41	1.16	0.50	Fall
		2.78	0.63	1.78	0.64	Winter
		1.67	0.38	1.07	0.64	Spring
Meixner et al. (1985)	Jülich, Germany, regionally polluted, 2 years, 1982/1983	7.68	0.39	1.08	0.14	Average for winter
		4.60	1.02	2.88	0.62	Average for summer
Huebert (1980)	Pacific Ocean, 7°N–9°S	0.22	0.038	0.11	0.48	Average
	Aircraft measurements					
Huebert and Lazrus (1980a)	Boundary layer, continental, 30–50°N	0.84	0.31	0.87	1.03 ⎫	0–3 km Altitude average values
	Boundary layer, marine, 0–30°N	0.40	0.07	0.2	0.49 ⎬	
	Boundary layer, marine, 0–55°S	0.17	0.07	0.2	1.17 ⎭	
	Free troposphere					
	Continental, 30–50°N	0.18	0.16	0.45	2.5 ⎫	3–8 km Altitude average values
	Marine, 55°S–30°N	0.13	0.11	0.31	2.3 ⎭	

the condensation of sulfuric acid is preferred over that of nitric acid, even though both acids are partially neutralized by ammonia. The acidification of aerosol particles by sulfuric acid sets up a temporary barrier to the condensation of HNO_3 until, at the end of the day, the production of sulfuric acid subsides and the continuing absorption of ammonia neutralizes the particles. Stelson *et al.* (1979, 1982) and particularly Richards (1983) have presented extensive discussions of these processes.

Two recent studies (Cadle, 1985; Meixner *et al.*, 1985) have shown that in addition to diurnal variations, the abundance of nitric acid undergoes an annual cycle with a maximum in summer and a minimum in winter and early spring. Figure 9-14b presents a synopsis of the observations of Meixner *et al.* (1985) in a regionally polluted, semirural area of Germany. The seasonal variation probably arises from the same processes responsible for the diurnal cycle, namely, the variation in temperature and relative humidity, and the relative importance of the two principal NO_2 oxidation mechanisms. No pronounced seasonal variation is expected from the nightime oxidation scheme. The daytime oxidation by OH radicals, however, increases in importance during the summer months, because the production rate for OH radicals maximizes during that season.

Huebert and Lazrus (1980a) have performed aircraft measurements of HNO_3 and particulate nitrate over the North American continent and over large parts of the Pacific Ocean. In the free troposphere the data indicate

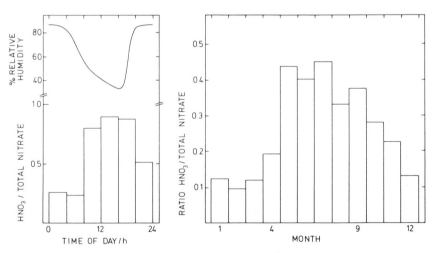

Fig. 9-14. Diurnal and annual behavior of gaseous nitrate in ground level air relative to total nitrate (gas plus particulate nitrate). Left: Averaged diurnal variation of HNO_3 and relative humidity at Claremont, California (Grosjean, 1983). Right: Monthly mean values of HNO_3 in Jülch, Germany, for the period 1982/1983 (Meixner *et al.*, 1985).

a rather uniform horizontal distribution, with little difference between the northern and the southern hemispheres, as well as between continental and maritime regions. Seventy percent of total nitrate, on average, is present in the gas phase. This contrasts with the situation in the boundary layer, where the percentage is smaller. Over the oceans the vertical distribution of nitric acid also is fairly uniform with but a slight indication for a downward gradient. Huebert and Lazrus have considered the influx of nitric acid from the stratosphere as a source of tropospheric HNO_3 but found other sources, such as the production (and oxidation) of NO_x from lightning discharges, to be more important. A relatively small downward gradient would nevertheless be needed to dispose of the input of HNO_3 from the stratosphere.

The tropospheric background of gaseous and particulate nitrate combined amounts to about 0.5 $\mu g/m^3$ STP. This corresponds to a gas-phase mixing ratio of 180 pptv if all the nitrate were present in the form of HNO_3. It is then easy to see that the abundance of HNO_3 plus NO_3^- in the troposphere is a factor of six greater than that of background NO_x. The ratio is in harmony with the concept that NO_x must first be converted to HNO_3 before it is removed from the troposphere. In this case, the relative abundance is given by the ratio of the residence times for both classes of compounds. According to the discussion in Chapter 8, the residence times for the wet removal of gaseous and particulate nitrate are approximately equal, with values of about 5 days. whereas the time constant for the conversion of NO_2 to nitrate may be set as 1 day. Thus, we expect a relative abundance of nitrate to NO_x of five. These arguments ignore losses of HNO_3 by dry deposition, which would lower the residence time of nitrate somewhat, and they neglect the fraction of NO_2 that is stored in the form of PAN and other peroxy nitrates. It is not yet possible to set up a detailed budget for NO_x and its reaction products in the remote troposphere, but one can define a minimum flux of nitrogen associated with NO_x by combining the nitrogen content of background nitrate with its residence time of 5 days. A uniform concentration of nitrate of 0.5 $\mu g/m^3$ represents a tropospheric nitrogen content of 0.37 Tg N. The value must be a lower limit because it does not yet include the amount of nitrate–nitrogen contained in the continental boundary layer. The corresponding flux is $0.37(365/5) = 26$ Tg N/yr, which averages to a deposition rate of 0.05 g N/m^2 yr. This flux agrees within a factor of two with that observed for nitrate in precipitation at remote marine locations.

9.4.5 SOURCES AND SINKS OF NO$_x$ IN THE TROPOSPHERE

The principal routes to the production of NO_x are combustion processes, nitrification and denitrification in soils, and lightning discharges. The major

Table 9-14. *Global Budget of* NO_x *in the Troposphere According to Estimates of Ehhalt and Drummond* (1982) *and of Logan* (1983)

Type of source or sink	Global flux (Tg N/yr)	
	Ehhalt and Drummond	Logan
Production		
Fossil-fuel combustion	13.5 (8.2–18.5)	19.9 (14–28)
Biomass burning	11.5 (5.6–16.4)	12.0 (4–24)
Release from soils	5.5 (1–10)	8.0 (4–16)
Lightning discharges	5.0 (2–8)	8.0 (2–20)
NH_3 oxidation	3.1 (1.2–4.9)	Uncertain (0–10)
Ocean surface (biologic)	—	1
High-flying aircraft	0.3 (0.2–0.4)	—
Stratosphere	0.6 (0.3–0.9)	0.5
Total production	39 (19–59)	48.4 (25–99)
Losses		
Wet deposition of NO_3^-, land	17 (10–24)	19 (8–30)
Wet deposition of NO_3^-, oceans	8 (2–14)	8 (4–12)
Wet deposition, combined	24 (15–33)	27 (12.–42)
Dry deposition of NO_x	—	16 (11–22)
Total loss	24 (15–40)	43 (23–64)

removal mechanism is oxidation to HNO_3 followed by wet and dry deposition. Recent efforts by Böttger *et al.* (1978), Ehhalt and Drummond (1982), and Logan (1983) to quantify the sources and the sinks have led to a global budget in which the flux of NO_x into the troposphere and the rate of nitrate deposition are approximately balanced. Table 9-14 presents a summary of the individual processes and the associated flux rates.

The annual production of NO_x from the combustion of fossil fuels is estimated from empirical emission factors for various combustion processes combined with the worldwide consumption of coal, oil and natural gas. Logan (1983) has given a convenient tabular summary of emission factors. Due to unavoidable fluctuations, the emission factors must be considered uncertain by about 30%. Table 9-15 provides a breakdown of global emission estimates for NO_x according to fuel consumption rates. The estimates of Logan (1983) actually were more detailed than Table 9-15 shows. She found a larger production rate from the combustion of fossil fuels than did Ehhalt and Drummond (1982), with the largest difference showing up in NO_x from transportation. The former estimate was based on the amount of fuel used, whereas the latter relied on the world population of automobiles. During the recent decade there have been attempts to reduce emissions of NO_x

Table 9-15. Global NO_x Emissions from the Burning of Fossil Fuels and Biomass[a]

Source type	Annual consumption (E&D)	(L)	(S&C)	Emission factors (E&D)	(L)	Global source strength (Tg N/yr) (E&D)	(L)	(S&C)
Fossil fuels								
Hard coal	2150 Tg ⎫	2696 Tg	—	1.0–2.8 g/kg	2.7 g/kg	3.9 (1.9–5.8)	6.4	—
Lignite	810 Tg ⎬			0.9–2.7 g/kg		1.6 (0.8–2.3)	—	—
Light fuel oil	300 Tg ⎫	1.39 Tm³	—	1.5–3.0 g/kg	2.2 g/m³	0.7 (0.5–0.9)	3.1	—
Heavy fuel oil	470 Tg ⎬			1.5–3.1 g/kg		1.1 (0.7–1.5)	—	—
Natural gas	1.04 Tg	1.2 Tm³[b]	—	0.6–3.0 g/kg	1.9 g/m³[b]	1.9 (0.6–3.1)	2.3	—
Industrial sources[c]	—			—		—	1.2	—
Automobiles	$(4.1–5.4) \times 10^{12}$ km	1.0 Tm³	—	0.9–1.2 g/km	8.0 g/m³	4.3 (3.7–6.4)	8.0	—
Total						13.5 (8.2–18.5)	19.9	—
Biomass burning								
Savanna	$(6–14) \times 10^{3}$ Tg	2000 Tg	1200 Tg	1.0 g/kg	1.7 g/kg	3.1 (1.8–4.3)	3.4	2.1
Forest clearings	$(2.7–6.7) \times 10^{3}$ Tg	4100 Tg	2700 Tg	1.0–1.6 g/kg	2.0 g/kg	2.1 (0.8–3.4)	8.2	4.7
Fuel wood	—	850 Tg	1100 Tg	—	0.5 g/kg	2.0 (1–3)	0.4	0.5
Agricultural waste	—	15 Tg	1900 Tg	—	1.6 g/kg	4.0 (2–6)	0.02	3.3
Total						11.2 (5.6–16.4)	12.0	10.6

[a] Estimates according to Ehhalt and Drummond (1982) (E&D) and Logan (1983) (L). Emission factors refer to grams nitrogen per unit of fuel consumed. For biomass-burning, Seiler and Crutzen (1980) (S&C) have given annual consumption rates differing somewhat from those of the other authors. The data of S&C and the resulting NO_x production rates are included for comparison.
[b] Petroleum refining and manufacture of nitric acid and cement; global emissions were obtained by scaling U.S. emissions for each industrial process.

from motor vehicles by means of control devices. Logan (1983) pointed out that in the United States, where the control is now mandatory, these measures have had little effect on the total emissions.

Table 9-15 includes a breakdown of estimates for the release of NO_x from the burning of biomass. In natural fires and the burning of wood, agricultural wastes, etc., the temperatures are rarely high enough to cause the oxidation of nitrogen in the air. The NO_x emission, therefore, originates mainly from the nitrogen content of the fuel. Logan (1983) reviewed a number of experimental determinations of emission factors that indicate that the yields are highest for grass and agricultural refuse fires (1.3 g N/kg fuel), less for prescribed forest fires (0.6 g N/kg), and still lower for the burning of fuel wood in stoves and fireplaces (0.4 g N/kg). The values roughly reflect the differences in the nitrogen contents of the materials burned. The nitrogen content of grass, leaves, and twigs is approximately 1% by mass, whereas fuel wood contains only about 0.2%. Not all of the nitrogen is oxidized to NO_x. Experiments performed by Clements and McMahon (1980) suggest an average conversion efficiency of 25%. By combining this figure with the nitrogen contents of plant materials, one obtains emission factors in reasonable agreement with those determined more directly.

Biomass burning is associated mainly with the agricultural practices in the tropics, which include plant, slash, and shift practices as well as a natural or intentional burning of savanna vegetation at the end of the dry season. Forest wildfires and the use of wood as a fuel make a lesser contribution. Table 9-15 includes a breakdown for biomass burning due to Seiler and Crutzen (1980), together with appropriate emission factors. Comparison shows that Ehhalt and Drummond (1982) overestimate fuel wood as a source, while Logan's estimate for the production of NO_x from the clearance of forests and woodlands is higher than the other data would indicate. Yet the total production rates are quite consistent. Crutzen *et al.* (1979) have derived a separate estimate that came to 14 Tg N/yr.

The biochemical release of NO_x from soils is poorly understood, and the flux estimates of Table 9-14 must be viewed with caution. Both rely on the observations of Galbally and Roy (1978), who used the flux box method in conjunction with the chemiluminescence detection of NO_x. They found average fluxes of 5.7 and 12.6 μg N/m^2 h on ungrazed and grazed pastures, respectively. NO was the main product. More recent measurements of Slemr and Seiler (1984) indicate that the release of NO_x from soils depends critically on the temperature and the moisture content of the soil, which makes it difficult to extrapolate the results to a global scale. Slemr and Seiler found an average release rate of 20 μg N/m^2 h for uncovered natural soils, about evenly divided between NO and NO_2. A grass coverage reduced the escape flux, whereas fertilization enhanced it. Ammonium fertilizers were about five times more effective than nitrate fertilizers. This suggests that nitrification as a source of NO_x is more important than denitrification. According to Slemr and Seiler (1984), a global flux of 10 Tg N/yr represents an upper limit to the release of NO_x from soils.

It should be added that the oceans are a negligible source of NO_x. Zafiriou and McFarland (1981) observed a supersaturation of seawater with regard to NO in regions with relatively high concentrations of nitrite due to upwelling conditions. Here, the excess NO must arise from the photo-decomposition of nitrite by sunlight. Logan (1983) estimated a local source strength of 1.3×10^{12} molecules/m² s under these conditions. Linear extrapolation results in a global flux of 0.35 Tg N/yr.

Thunderstrom electricity has been considered a major source of NO_x ever since Liebig (1827) proposed it as a natural mechanism for the fixation of atmospheric nitrogen. Several authors have searched for an increase of nitrate in thunderstorm precipitation, in comparison with normal rainwater, as evidence for the production of NO_x in lightning discharges. The results reported by Hutchinson (1954), Viemeister (1960), Visser (1961, 1964), Reiter (1970), and Gambell and Fisher (1964) were mainly negative or, at best, inconclusive. The most pronounced effect is expected in the equatorial region, where thunderstorm activity is higher than elsewhere. Yet Visser (1961), who worked in Kampala, Uganda, and made daily observations of NO_3^- in rainwater over a period of a year failed to detect any correlation between NO_3^- concentration and the number of lightning flashes. In retrospect it appears that a positive correlation is unlikely due to the time delay between NO_x production and HNO_3 formation, which must precede the incorporation of nitrate into cloud and rain water. Inherent in all studies was the assumption that NO_x reacts speedily with cloud water to form HNO_3. This viewpoint is made untenable by the data of Table 8-9, which indicate that the reaction $2NO_2 + H_2O \rightarrow HNO_3 + HNO_2$ is slow at low concentrations of NO_2.

Electrical discharges in air generate NO_x by thermal dissociation of N_2 due to ohmic heating inside the discharge channel and shockwave heating of the surroundings. Laboratory studies by Chameides et al. (1977) and Levine et al. (1981) indicate a yield of NO_x of 6×10^{16} molecules per joule of spent energy. Great uncertainties exist, however, about the total energy deposited by lightning in the atmosphere. Noxon (1976, 1978) first studied the increase of NO_x in the air during a thunderstorm. By relating it to the frequency of cloud-to-ground flashes, he computed a production rate of 10^{26} molecules of NO_x per lightning flash. These results provide the basis for estimating the global production of NO_x from thunderstorm electricity, and Table 9-16 compiles a variety of such estimates.

Table 9-16. Estimates of the Production of NO_x by Lightning[a]

Authors	NO_x production rate		Energy deposition		Global rate (Tg N/yr)
	10^{16} Molecules/J	10^{25} Molecules per flash or stroke	10^{-4} J/m^2 s	Strokes/s	
Laboratory experiments					
Chameides et al. (1977)	3–7	—	16	—	20–40
Levine et al (1981)	5±2	—	1–10	—	2–18
Peyrous and Lapeyre (1982)	1.6–2.6	—	16	—	9–15
Theoretical model calculations					
Tuck (1976)	—	1.1	—	500	4
Griffing (1977)	—	12–20	—	100	9–15
Chameides (1979)	8–17	—	16	—	47–90
Hill et al. (1980)	—	6	—	100	4
Atmospheric observations					
Noxon (1976), Dawson (1980)	—	10	—	100	7
Drapcho et al. (1983)	—	40	—	100	30

[a] Adapted from Levine (1984).

The global frequency of lightning flashes is now much better known than previously, due to satellite observations (Orville and Spencer, 1979; Orville, 1981; Turman and Edgar, 1982). The data suggest a frequency of $40-120 \text{ s}^{-1}$, with a distribution maximizing in the tropics. The major problem is the conversion of the observed counts at dawn and dusk to diurnal averages. Each lightning flash consists of a number of forward and return strokes. Most of the estimates for NO_x production in Table 9-16 are based on a global flash frequency of 100 s^{-1} and an average of five strokes per flash. Hill (1979) and Dawson (1980) have pointed out that most of the flash energy is dissipated in the first stroke, so that a total frequency of 100 s^{-1} should be more appropriate than 500 s^{-1}.

A comparison of the various sources of NO_x in Table 9-14 shows that the anthropogenic contribution is preponderant. The global production of NO_x by lightning discharges is potentially the largest natural source. Its magnitude may be equivalent to that of human-made emissions, but a sufficiently precise quantification is presently not possible. The input of HNO_3 from the stratosphere and the oxidation of ammonia to NO_x are comparatively minor sources.

As discussed in Chapter 8, wet precipitation provides an efficient mechanism for the removal of gaseous and particulate nitrate from the atmosphere. The addition of nitrate (and ammonium) by rainwater to the plant-soil ecosystem constitutes an important source of fixed nitrogen to the terrestrial biosphere, and until 1930 practically all studies of nitrate in rainwater were concerned with the input of fixed nitrogen into agricultural soils. Erikson (1952a), Steinhardt (1973), and Böttger et al. (1978) have compiled many of the available data. Despite the wealth of information it remains difficult to derive a global average for the deposition of nitrate, because of an uneven global coverage of the data, unfavorably short measurement periods at many locations, and inadequate collection and handling techniques for rainwater samples in many cases. In addition, the concentration of nitrate in rainwater has increased in those parts of the world where the utilization of fossil fuels has led to a rise in the emissions of NO_x, primarily Western Europe and the United States. The problems are similar to those encountered in determining the wet deposition rate for ammonium (Section 9.2.2) and sulfate (Section 10.3.5).

In view of these difficulties, it is somewhat surprising to find that the two independent estimates given in Table 9-14 for the wet deposition of nitrate are essentially identical. Ehhalt and Drummond (1982) relied on the detailed evaluation of data by Böttger et al. (1978). Their analysis emphasized measurements from the period 1950-1977. They prepared a world map for NO_3^- deposition rates, which were then integrated along 5° latitude belts. The geographic distribution of NO_3^- deposition is shown in Fig. 9-15. Logan

Chapter

10 | Sulfur Compounds in the Atmosphere

10.1 Introductory Remarks

Large quantities of sulfur dioxide enter the atmosphere each year from anthropogenic sources, mainly the combustion of fossil fuels and the smelting of metals. SO_2 indisputably ranks as a prominent pollutant, and it is understandable that research of the past 30 yr dealing with atmospheric sulfur has concentrated on such problems as the dispersal of SO_2 from power stations and urban centers, its conversion to sulfuric acid, the formation of sulfate aerosols, and the deposition of sulfate and SO_2 at the ground surface.

The existence of natural sources of atmospheric sulfur has always been acknowledged. Known source processes include volcanic emissions of SO_2 (and some H_2S), aeolian generation of particulate sulfate (e.g., in the form of seasalt), and the emanation of reduced sulfur compounds from the biosphere. The classical representative of the last group is H_2S, which is thought to be produced most copiously in anaerobic marshlands and estuaries. In the recent decade a number of additional sulfides were discovered in the atmosphere, namely, carbonyl sulfide, COS (Hanst *et al.*, 1975), carbon disulfide, CS_2 (Sandalls and Penkett, 1977), dimethyl sulfide, CH_3SCH_3 (Maroulis and Bandy, 1977), and methyl mercaptan, CH_3SH, and dimethyl disulfide, $CH_3S_2CH_3$ (Hill *et al.*, 1978). Various other mercaptans have been identified by Burnett (1969), Stephens (1971), and others over cattle feed lots, where they arise from rotting manure. These discoveries have shifted the emphasis away from H_2S and have provided a much better understanding of biogenic sulfur emissions.

The geochemical cycle of sulfur involves mobilization of the element by the weathering of continental rocks, followed by river transport to the oceans (mainly in the form of dissolved sulfate) and deposition in marine sediments. Tectonic uplift of sedimentary deposits eventually replaces the weathered material on the continents, thus completing the cycle (cf. Fig. 11-3). It is a curious observation that even after correcting for anthropogenic contributions, the rivers carry more sulfate to the oceans than the weathering of rocks can supply. A recent estimate by Ivanov et al. (1983) for the natural flux of sulfate to the oceans is 100 Tg S/yr, whereas the weathering of rocks is estimated to feed into the rivers an amount of 33–42 Tg S/yr (Berner, 1971; Granat et al., 1976; Lein, 1983). Volcanic emissions are not capable of closing the gap. While the problem has remained unresolved, it has suggested to many authors the existence of a marine source of volatile sulfur whose aerial transport toward the continents might balance the account. Traditionally, the role of the volatile has been assigned to H_2S, with the implicit assumption that H_2S is formed by bacterial sulfate reduction in marine muds. Yet, according to Östlund and Alexander (1963) H_2S is quickly oxidized in aerated waters, so that little of it should escape from the ocean to the atmosphere except from shallow coastal waters. In 1972 Lovelock et al. reported the presence of dimethyl sulfide (DMS) in seawater, and they suggested that, because of its greater stability toward aqueous oxidation, DMS is a more important volatile than H_2S. The concept gained ground from the studies of Maroulis and Bandy (1977), who first detected DMS in marine air, of Nguyen et al. (1978), who demonstrated its widespread occurrence in the world oceans, and of Barnard et al. (1982), who showed DMS to be ubiquitous in the air over large parts of the Atlantic Ocean. Thus, the nature of marine volatile sulfur is now well established. In the atmosphere, DMS is rapidly oxidized and sulfur is quickly returned to the ocean surface.

Yet another problem of atmospheric sulfur that has recently been brought closer to solution is the origin of the stratospheric sulfate aerosol layer discovered by Junge et al. (1961). This phenomenon has been discussed in Section 3.4.4.

10.2 Reduced Sulfur Compounds

According to present knowledge, hydrogen sulfide, dimethyl sulfide, carbonyl sulfide, and carbon disulfide are the most important reduced sulfur

Table 10-1. *Observed Mixing Ratios of Sulfides in the Atmosphere*

Authors	Location	Mixing ratios Range (pptv)	Average pptv	Average ng S/m³	Remarks
	Hydrogen sulfide				
Breeding *et al.* (1973)	Jacksonville, Illinois	120–140	130	186	Rural area
	Bowling Green, Missouri	290–340	320	458	Rural area
	Athensville, Illinois	80–150	100	143	Rural area
Slatt *et al.* (1978)	Miami, Florida	41–91	58±23	83±33	Marine air
		121–829	336±258	481±369	Continental air considered polluted
	Barbados	0–45	7±8	10±11.5	Marine air
	Sal Island, Cape Verdes	0–12	4±4	5.7±5.7	Marine air
Jaeschke *et al.* (1978)	West Germany, Kleiner Feldberg (820 m a.s.l.)	230	230	330	March 1976, mean daytime value
	Wetlands, Rotes Moor		53	76	Mean nighttime value
	Upper Bavaria, meadows		718	1026	
	Sylt Island, tidal flats		33	47	
			950	1360	
Delmas *et al.* (1980)	France, Toulouse, urban	0–122	65	94	During winter months
	Toulouse, residential	14.5–49.5	29	41	In May
		36.2–151	76	110	In June
	Loire Valley, near Landes	27.5–121	56	80	In December
	Rural area	23.3–159	63	91	In April
	In a pine forest	38.5–51.2	44	63	In April
	Pic du Midi (2980 m a.s.l.)	1.3–12.6	5.1	7.3	Free atmosphere
	Ivory Coast, tropical forest		4253	6082	Anoxic soil regions
			459	656	Aerobic soils

Reference	Location				Comments
Jaeschke et al. (1980)	West Germany, Kleiner Feldberg (820 m a.s.l.)		293	419	February mean, 1977
			80.3	115	August mean, 1977
Jaeschke and Hermann (1981)	Frankfurt/Main, city		145	207	Annual mean, 1977
			164	235	October 1976, mean daytime value
			810	1158	
			187	267	Mean nightime value
Delmas and Servant (1982)	Florida, various locations, swampy regions	66–336			
Hermann and Jaeschke (1984)	Gulf of Guinea	5.5–45.4	14	20	Marine air
	North and Central Atlantic Ocean	10–110	26	38	Marine air
Dimethyl sulfide					
Maroulis and Bandy (1977)	Wallops Island, Virginia	40–110	58 ± 2.1	83 ± 3	Diurnal cycle, maximum values at night
Barnard et al. (1982)	Cape Henry, Virginia	30	—		Mainly marine air
Andreae et al. (1983)	Atlantic Ocean	0.7–31	4.3	6.1	No diurnal variation
	Gulf of Mexico		6.8	9.7	
Andreae and Raemdonck (1983)	Remote Pacific Ocean	42–70	60 ± 24	86 ± 35	Diurnal variation with nightime maximum
Bürgermeister (1984)	Peru Shelf	2.2–349	14.6 ± 6.8	21 ± 9.8	
	Frankfurt, West Germany, city		32 ± 55	46 ± 79	Seasonal variation, maximum in winter,
	Kleiner Feldberg (820 m a.s.l.)	2.5–20	6.5	9.2	Highest values under inversion layers
	Wiesbaden, city	9–440	106	151	Anthropogenic pollution by paper industry

(continued)

Table 10-1. (*Continued*)

Authors	Location	Range (pptv)	Average pptv	Average ng S/m³	Remarks
			Carbon disulfide		
Sandalls and Penkett (1977)	Harwell, England		190±110	543±314	Relatively clean air
Maroulis and Bandy (1980)	Philadelphia, Pennsylvania		37±10	106±29	Considered polluted
			190±114	543±326	Marine air
Bandy et al. (1981)	Wallops Island, Virginia		39±8	111±23	
	Free troposphere, U.S.A.		≤3	—	5 km Altitude, aircraft flights
Jones et al. (1983)	Harwell, England		15	42.8	Average in clean air derived by comparison with F11 data
			Carbonyl sulfide		
Sandalls and Penkett (1977)	Harwell, England		510±50	730±71	
Maroulis et al. (1977)	Philadelphia, Pennsylvania		434±56	620±80	No significant contribution by pollution
	Wyoming		454±28	650±40	
	Oklahoma		510±35	730±50	
Torres et al. (1980)	Free troposphere		509±63	728±90	Evenly distributed between continental and marine locations, northern and southern hemisphere
	Boundary layer		518±74	740±71	
Inn et al. (1979, 1981)	Stratosphere, lower (15.2 km)	281–524		—	Declines with altitude
	Upper (31 km)	14–18		—	See Fig. 3–13
Rasmussen et al. (1982a)	Surface layer above ocean		530±20	758±28	

[a] Values refer to the ground level unless otherwise noted.

compounds in the atmosphere. We consider, in turn, their atmospheric abundances, reactions and biogenic sources.

10.2.1 ATMOSPHERIC MIXING RATIOS

In spite of recent improvements in measurement techniques, the actual number of observational data on sulfides is still small, and our knowledge is correspondingly patchy. The few measurements that exist are summarized in Table 10-1. Data obtained for H_2S prior to 1970, such as those discussed by Junge (1963), are not included because the analytical methods of that time were of doubtful integrity and the values reported often are incompatibly high compared with recent data. Most measurements refer to the ground level. All sulfides except COS react rapidly with OH radicals and their atmospheric lifetimes are short, on the order of days (Table 10-2). As a consequence, the mixing ratios fluctuate considerably and decrease strongly with distance from sources and with elevation above ground.

Observations on H_2S may be summarized as follows. On the continents, in regions not directly influenced by pollution, H_2S mixing ratios range from 30 to about 100 pptv. Higher values are encountered in the vicinity of cities or larger settlements, thus indicating the importance of anthropogenic sources. Higher values also are observed over anaerobic swamps, salt marshes, tidal flats, and similar regions of high biological productivity. Jaeschke et al. (1978, 1980) have investigated the vertical distribution of H_2S over various parts of Germany by means of aircraft ascents. The mixing ratio decreases with altitude with scale heights from 700 to 2000 m, depending on meteorological conditions and the strength of the underlying sources. The strongest gradients occurred over industrially polluted areas. Table 10-1 further shows that H_2S is present also in marine air masses. Surprisingly, it is observed not only at coastal stations but even over the open ocean. Here, the biochemical sources must be located in near-surface waters, as H_2S from the deeper strata would have no chance of escaping to the atmosphere.

Dimethyl sulfide in the air was first observed by Maroulis and Bandy (1977) at a coastal site of the eastern United States. The data showed little dependence on wind direction, implying the existence of both continental and marine sources. In view of the short atmospheric residence time of DMS of less than a day, it is also possible, however, that local, estuarine sources predominated. Subsequent studies by Barnard et al. (1982) and Andreae and Raemdonck (1983) of DMS over the open ocean indicated mixing ratios ranging from 0 to 180 pptv. The highest values are found in regions of upwelling waters near the equator, where biological productivity is high. Elsewhere the mixing ratios average to about 6 pptv. Even the high values are two orders of magnitude below the mixing ratios predicted from the concentrations of DMS observed in surface seawater if both were in equilibrium in accordance with Henry's law. The concentration difference leads to a steady flux of DMS from the ocean into the atmosphere. The flux rate can be estimated from the stagnant-film model discussed in Section 1.6. Emission rates of the order of 200 μg S/m day have been reported. Several authors have extrapolated their results to the global scale. Nguyen et al. (1978) obtained 27 Tg S/yr, Barnard et al. (1982) 34–56 Tg S/yr, and Andreae and Raemdonck (1983) 38.5 Tg S/yr, uncertain by 30%. The latter result is based on the largest set of data from various parts of the world ocean.

Maroulis and Bandy (1977) and Andreae and Raemdonck (1983) reported diurnal variations of atmospheric DMS mixing ratios featuring a buildup overnight and a decline during the day. Such a behavior was predicted by Graedel (1979) on account of reaction with OH radicals in the sunlit atmosphere. In ocean regions with low DMS mixing ratios, diurnality appears to be absent. In these regions, seawater concentrations of DMS are not significantly lower than those in high-productivity zones and the calculated sea-to-air fluxes have similar magnitudes. The difference in the atmospheric behavior is puzzling. Low mixing ratios and absence of diurnality may be explained by the region occurrence of an additional removal process of a nonphotochemical type, which overrides destruction of DMS by OH radicals in such a region. Andreae and Raemdonck (1983) speculated that the advection of continental air masses might be responsible for the additional sink.

Continental mixing ratios of DMS were explored by Bürgermeister (1984) in the Rhine–Main area of West Germany. In the city of Frankfurt, he found the highest values (900 pptv) on winter days when a well-developed inversion layer restricted the vertical exchange of air. High values were also observed in the city of Wiesbaden in an industrial district, whereas in the outskirts of the city the concentrations were markedly lower. Anthropogenic sources evidently predominate in that region. In the hills north of Frankfurt the DMS mixing ratio is reduced to 6 pptv, on average, but the anthropogenic influence is still noticeable by an enhancement in air masses arriving from the southwest (from Wiesbaden). It is not possible to decide whether 6 pptv represent the natural background of DMS in that region, or simply a reduction of mixing ratio with altitude due to the higher elevation of the Kleiner Feldberg site.

Carbon disulfide is another compound whose abundance in the atmosphere is poorly known. Again, the few measurements that exist indicate an appreciable anthropogenic contribution. Jones et al. (1983) analyzed the frequency distribution of CS_2 mixing ratios at Harwell, England, and concluded, by comparing it with similar data for $CFCl_3$, that the average CS_2

level in clean air is about 15 pptv. Maroulis and Bandy (1980) correlated their data obtained at Philadelphia and at Wallops Island with wind direction and found mixing ratios of about 37 pptv in relatively clean continental or marine air masses. The reactivity of CS_2 toward OH radicals is comparable to that of H_2S, and equivalent scale heights are expected. Indeed, the low CS_2 mixing ratios observed by Bandy et al. (1981) in aircraft flights at 5 km altitude suggest a scale height less than 2 km. Lovelock (1974b) detected the presence of CS_2 in seawater but at concentrations much less than that of dimethyl sulfide, indicating that the flux of CS_2 into the atmosphere is relatively unimportant. Measurements of CS_2 in the air over the open ocean have not been reported.

Carbonyl sulfide contrasts with the other reduced sulfur compounds in that it is evenly distributed throughout the troposphere, with a mixing ratio near 500 pptv. Thus, COS is the most abundant sulfide in the atmosphere. The uniform distribution points toward a residence time in the troposphere of at least 1 yr. The variability of the mixing ratio, expressed in Table 10-1 by the standard deviation, may be used in conjunction with Junge's (1974) relation (see page 135) to estimate a residence time of about 2 yr. Actually, the value is a lower limit since part of the variance is already inherent in the gas-chromatographic measurement technique. In the stratosphere, the COS mixing ratio declines due to photodecomposition by ultraviolet radiation. By fitting to their observational data points a one-dimensional diffusion profile, Inn et al. (1981) determined the upward flux of COS into the stratosphere as 10^{11} molecules/m^2 s, or 0.16 Tg/yr on a global scale. The mass content of COS in the troposphere amounts to 4.4 Tg. Combining both figures gives a residence time of COS in the troposphere of 28 yr due to stratospheric losses alone. The next section will show that reactions of COS with OH radicals and O atoms are slow. The associated residence time is at least 44 yr. Losses to the stratosphere and by tropospheric reactions combine to give a residence time of 17 yr, and this value sets an upper limit.

Johnson (1981) proposed that hydrolysis of COS in seawater provides another sink, and he estimated a destruction rate equivalent to a residence time of 11 yr. Rasmussen et al. (1982a), in contrast, found surface water of the Pacific Ocean to be slightly supersaturated with COS and concluded that the ocean serves as a source of COS. By extrapolating their results, they estimated a global marine source strength of 0.3 ± 0.7 Tg S/yr. The flux is sufficient to balance the known sinks, but the error margin is appreciable and the marine source probably is small compared with COS emissions from continental soils, which are discussed in Section 10.2.3. If so, additional sinks must exist and will have to be identified. Turco et al. (1981a) suggested that the tail of the absorption spectrum of COS extends to wavelengths longer than 300 nm, so that COS photodecomposition may occur to some extent in the troposphere. The significance of this process remains to be determined.

10.2.2 ATMOSPHERIC REACTIONS AND LIFETIMES

The oxidation of sulfides in the atmosphere is initiated mainly by reaction with OH radicals. Table 10-2 lists the pertinent rate coefficients for two conditions: in air at atmospheric pressure, and in the absence of oxygen. The data show that O_2 accelerates the reactions in several cases, most notably that of CS_2. This effect is explained by the incipient formation of an OH addition complex, which then has three options: redissociation toward the original reactants, decomposition to new products, or interaction with O_2 to form a different set of products. The individual reaction pathways are discussed further below.

Table 10-2. *Rate Coefficients at 298 K for Reactions of OH Radicals, O Atoms, and Ozone with Various Sulfides*[a]

	Rate coefficient, $k_{298}(10^{-12} \text{ cm}^3/\text{molecule s})$				
Compound	OH, O_2 present	OH, O_2 absent	O_3	O	Lifetime in the troposphere
H_2S	$5.0^b, 5.2^c$	$5.4^c, 5.3^e$	$2 \times 10^{-8 \, h}$	$0.027i^{\,j}$	4.4 days
CH_3SH	$90^b, 130^c$	$3.3^c, 36^e$		1.9^k	0.3 days
CH_3SCH_3	$9.1^b, 45^c$	$1.8^c, 4.3^e$	$1 \times 10^{-4 \, i}$	$63^k, 48^l$	0.6 days
CH_3SSCH_3	223^b	211^e			0.1 days
CS_2	$0.43^b, 2.0^d$	$0.5^d, 0.07^f$		5.5^j	12 days
COS	$<0.04^b$	$0.007^f, 6 \times 10^{-4 \, g}$	0.014^j		44 yr

[a] Tropospheric lifetimes refer to reaction with OH radicals assuming $n(\text{OH}) = 5 \times 10^5$ molecules/cm^3; for COS, the reaction with O atoms is included assuming $n(\text{O}) = 3 \times 10^4$ atoms/cm^3.
[b] Cox and Sheppard (1980).
[c] Barnes *et al.* (1984).
[d] Barnes *et al.* (1983).
[e] Wine *et al.* (1981).
[f] Atkinson *et al.* (1978).
[g] Leu and Smith (1981).
[h] Becker *et al.* (1975).
[i] Martinez and Herron (1978).
[j] Baulch *et al.* (1980).
[k] Slagle *et al.* (1976).
[l] Lee *et al.* (1976).

Sulfides react also with oxygen atoms and with ozone. The corresponding rate coefficients are included in Table 10-2 as far as they are known. Laboratory investigations of ozone reactions are complicated by the observation of fractional orders, autocatalytic behavior, and chemiluminescence, all features indicating the occurrence of radical chain processes triggered by a slow, possibly endothermic intiation step. Becker *et al.* (1975) reported for the direct reaction of O_3 with H_2S a rate constant of less than 2×10^{-20} cm^3/molecule s, and Martinez and Herron (1978) determined for the bimolecular reaction of O_3 with CH_3SCH_3 a rate coefficient of $<1 \times 10^{-16}$ cm^3/molecule s. In the atmosphere, ozone would begin to compete with OH for H_2S and CH_3SCH_3 if the ozone reactions had rate coefficients greater than about 3×10^{-18} and 3×10^{-17}, respectively. The laboratory evidence suggests, therefore, that the reactions with ozone are relatively unimportant, and we shall not consider them in detail.

The reactions of sulfides with atomic oxygen are quite rapid, as Table 10-2 shows, with rate constants comparable to those of OH reactions in many cases. Note, however, that the average number densities of O atoms in the troposphere are in the range 10^4-10^5 atoms/cm^3—that is, an order of magnitude lower than those for OH. Accordingly, the latter is the preferred reagent. COS is an exception because it reacts with OH quite slowly. Table 10-2 includes approximate atmospheric lifetimes of sulfides based on their reactions with OH radicals. Values of the order of days indicate a rapid destruction of sulfides in the troposphere. The major exception is again COS, which is more stable.

Oxidation mechanisms will be discussed next, and we begin with H_2S. The absence of a rate-accelerating effect of oxygen, and the identification of HS as a product in the work of Leu and Smith (1982) demonstrated that the reaction

$$OH + H_2S \rightarrow H_2O + HS$$

proceeds mainly by abstraction. The subsequent fate of the HS radical in the atmosphere is somewhat uncertain, since its reaction with O_2 is slow (see Table 10-3), and conceivable alternatives have yet to be studied. HS may be considered the sulfur analog of OH, although the reactions with carbon monoxide and with methane are disallowed because they are endothermic. McElroy *et al.* (1980) speculated that HS reacts with HCHO and H_2O_2 by hydrogen abstraction, for example,

$$HS + HCHO \rightarrow H_2S + HCO$$

Both reactions are exothermic. In this manner, H_2S would be reconstituted and its lifetime raised to a value greater than that given in Table 10-2. Since, however, the tropospheric mixing ratios of HCHO and H_2O_2 are smaller

with OH proceeds via an addition complex. Table 10-4 outlines the sequence of reactions expected to follow the formation of the DMS–OH complex. The first step is the further addition of an oxygen molecule, and the second one is the decomposition of the resulting O_2–DMS–OH complex. A number of conceivable decomposition pathways are indicated. It is seen that methanesulfonic acid arises as a direct product after the elimination of CH_3. The methyl radical then is further oxidized to formaldehyde. The formation of CH_3S from the O_2–DMS–OH complex is also possible, but it would require the simultaneous generation of CH_3OH, and since methanol was not observed in the experiments, the process cannot be significant. The O_2–DMS–OH complex may also disintegrate with the ejection of H_2O. The resulting $CH_3SCH_2O_2$ product presumably is unstable, because further breakup leads to a considerable energy release, as Table 10-4 shows. The ultimate products arising from H_2O elimination are expected to be formaldehyde and SO_2, but the detailed mechanism has not been worked out. Finally, it should be noted that the $CH_3SCH_2O_2$ product formed by OH addition may not be identical with that obtained from the OH abstraction mechanism. The latter, therefore, must still be involved to explain the occurrence of CH_3SNO in the experiments. According to the yields reported by Niki et al. (1983), H-atom abstraction may contribute up to 21% to the total reaction.

Table 10-4. *Some Possible Reactions Following from the Formation of an Addition Complex between OH Radicals and Dimethyl Sulfide*[a]

Reaction	Heat of reaction (kJ/mol)
$OH + CH_3SCH_3 \rightarrow CH_3S(OH)CH_3$	
$CH_3S(OH)CH_3 + O_2 \rightarrow \left[CH_3S\overset{OH}{\underset{OO}{-}}CH_3 \right]^*$	
$\left[CH_3S\overset{OH}{\underset{OO}{-}}CH_3 \right]^* \rightarrow CH_3\overset{O}{\underset{O}{S}}OH + CH_3$	
$\rightarrow CH_3SCH_2O_2 + H_2O$	$\leqslant -3$
$\rightarrow CH_3SO + CH_2O + H_2O$	-415.2
$\rightarrow CH_2SO + CH_3O + H_2O$	-275.7
$\rightarrow CH_3 + SO + CH_2O + H_2O$	-199.8
$\rightarrow CH_3\overset{O}{S}CH_3 + HO_2$	-150.4
$\rightarrow CH_3S + CH_3OH + O_2$	-59.2

[a] The reaction energies are based on data reviewed by Benson (1978).

The major product from the OH + DMS reaction is methanesulfonic acid, and the reaction scheme shown in Table 10-4 offers a simple route to its formation. CH_3SO_3H is a condensable product that in the atmosphere becomes associated with aerosol particles. Panter and Penzhorn (1980) have demonstrated its presence in continental aerosols, while Saltzman et al. (1983) found it to occur also in the marine aerosol. Methanesulfonic acid is unstable in aqueous solution. Once associated with the aerosol, it becomes rapidly oxidized to sulfuric acid.

Finally, we discuss the oxidation of CS_2 and COS. Jones et al. (1982, 1983) and Barnes et al. (1983) have studied the OH-induced oxidation of CS_2 in air at atmospheric pressure and observed as products SO_2 and COS in equal amounts. The rate-accelerating effect of O_2 then suggests the following simple oxidation mechanism:

$$OH + CS_2 \rightleftharpoons SCSOH$$
$$SCSOH + O_2 \rightarrow OCS + SO_2 + H$$

The reaction of O_2 with the CS_2–OH addition complex is sufficiently exothermic to be able to generate H atoms directly. Alternatively, one may also consider the formation of an HSO_2 intermediate, which would then react further with O_2 to form SO_2 and HO_2. Since in the atmosphere hydrogen atoms attach to O_2 anyway, the products from both schemes are the same.

The reaction of OH with COS is again thought to proceed via an OH–COS addition complex, but in the absence of a rate-accelerating effect of O_2, the main fate of the complex is reversion to the original reactants. The actual forward reaction yielding new products is a minor pathway showing bimolecular behavior. By means of mass spectrometry, Leu and Smith (1981) identified HS as one of the products, so that the reaction takes the course

$$OH + COS \rightarrow CO_2 + HS$$

All other conceivable product channels are endothermic and can be ignored. The further reactions of HS follow the scheme of H_2S oxidation outlined in Table 10-3. The reaction of COS with atomic oxygen, which supplements the OH + COS reaction, leads to the formation of SO and CO. Neither process is very effective, however, in removing COS from the troposphere.

In the stratosphere, at altitudes above about 20 km, COS is exposed to ultraviolet radiation of wavelengths shorter than 250 nm. In this spectral region, COS has its first absorption band and is subject to photodissociation. The photofragments are CO and sulfur atoms. As dicsussed by Okabe (1978), a major portion of the sulfur atoms are excited to the metastable ^1D state, but the excess energy is rapidly removed by collisions with molecular nitrogen (Little et al., 1972). Sulfur atoms in the ^3P ground state, and presumably in the excited ^1D state as well, react readily with oxygen and are then oxidized to SO_2. The reaction sequence following from the photodecomposition of COS is included in Table 10-3, which summarizes the H_2S, CS_2, and COS oxidation mechanisms.

10.2.3   BIOGENIC EMISSIONS

In this section we consider primarily the release of sulfides from soils. Sulfur is a component of the essential amino acids methionine, cysteine, and cystine, which the living cell must either synthesize (plants, microorganisms) or procure with the food supply (animals). In the course of biological utilization, sulfur enters into a complex chain of oxidation–reduction reactions whereby the element is circulated through various biospheric reservoirs of organic and inorganic sulfur compounds. Below, we give first a brief description of the biochemical sulfur cycle, then a survey of results from field studies attempting to quantify emission rates.

The sulfur cycle combines three principal pathways of biological sulfur utilization. (1) Organisms capable of synthesizing S-containing amino acids do so by the intracellular reduction of sulfate absorbed from the environment. Although the reduction leads to H_2S as an intermediate, virtually all of it is assimilated and none is released (in contrast to dissimilatory sulfate reduction described below). (2) Dead organic matter is subject to microbial degradation. The process liberates to the environment sulfur mainly in the form of H_2S and organic sulfides. It is during this stage that volatile sulfides may escape from the soil to the atmosphere. In aerobic media, sulfides are chemically unstable, and a variety of microorganisms become engaged in reoxidizing sulfides toward sulfate, whereby the cycle is completed. The best known and presumably the most important among the sulfur oxidizers are thiobacilli. They exploit the oxidation process as an energy source. (3) In anaerobic environments the oxidation of sulfur is suppressed and sulfides accumulate. In this case, sulfate and other sulfur oxides provide a source of oxygen for anaerobic respiration, and they undergo dissimilatory reduction toward H_2S. Only a small family of strict anaerobes (the desulfuricants) is able to reduce sulfate, whereas many other microorganisms utilize oxygen contained in sulfites or thiosulfate. Dissimilatory sulfate reduction is most important in marine muds and saline soils with high concentrations of sulfate. In normal soils, the fraction of sulfur present as sulfate is less than 10% of the total, since sulfur occurs mostly in organic compounds, either carbon-bound or in the form of sulfate esters. In anaerobic habitats, ferric iron (Fe^{3+}) is reduced to the ferrous form (Fe^{2+}) before H_2S appears, so that the latter is precipitated as FeS. This reaction is considered the major route to inorganic sulfur fixation in anaerobic domains. In aerobic strata, iron sulfide is unstable and undergoes reoxidation. Further details on the

Table 10-5. *Biochemical Origin of Volatile Sulfides Produced in Soils by the Microbial Degradation of Organic Matter under Aerobic and Anaerobic Conditions*[a]

Volatile	Biochemical precursors
H_2S [b]	Proteins, polypeptides, cystine, cysteine, gluthathione
CH_3SH	Methionine, methionine sulfoxide, methionine sulfone, S-methyl cysteine
CH_3SCH_3	Methionine, methionine sulfoxide, methionine sulfone, S-methyl cysteine, homocysteine
CH_3SSCH_3	Methionine, methionine sulfoxide, methionine sulfone, S-methyl cysteine
CS_2	Cysteine, cystine, homocysteine, lanthionine, djenkolic acid
COS	Lanthionine, djenkolic acid

[a] From Bremner and Steele (1978).

[b] H_2S can also be formed by anaerobic dissimilatory sulfate reduction, as indicated in the text.

sulfur cycle, especially on the microorganisms involved, can be found in the texts of Alexander (1977) and Zinder and Brock (1978).

Information about the biochemical precursors of volatile sulfur compounds has been obtained from artificial culture studies, reviewed by Kadota and Ishida (1972), and from incubation studies of natural and amended soils, reviewed by Bremner and Steele (1978). Table 10-5 summarizes our knowledge about the origin of sulfides, which have been observed to emanate from soils as degradation products under aerobic as well as anaerobic conditions. As indicated above, H_2S is formed not only by the degradation of organic compounds, but also by anaerobic reduction of sulfate.

Similar biochemical decomposition reactions are expected to take place in aquatic environs. Here, however, dimethyl sulfide dominates the product mixture, because of its association with the growth of algae, which constitute the largest portion of aquatic biomass. The main precursor of DMS in algae, according to Challenger (1951), is dimethyl propiothetin, $(CH_3)_2S^+CH_2CH_2COO^-$, a ternary sulfonium compound that is thought to form methionine. Although dimethyl propiothetin is known to undergo enzymatic cleavage to DMS in the living algal cell, little DMS appears to be released directly. Many observations suggest that DMS in aquatic environments derives largely from the bacterial decomposition of dimethyl propiothetin leaked from aged cells. For example, as discussed by Bremner and Steele (1978), very high concentrations of DMS have been observed during the bacterial putrefaction of algae following algal blooms.

Table 10-6 presents results from field studies attempting to quantify the flux of H_2S evolving from soils. Most of the observations refer to saline marshlands and anaerobic swamps where H_2S emissions are appreciable but whose areal extents probably are not globally significant. An unexpected discovery is the unusual diurnal behavior of H_2S in such source regions.

Table 10-6. *Hydrogen Sulfide Emissions from Tidal Regions, Marshes, Swamps, and Common Soils*

Authors	Location and/or soil type	Flux (g S/m² yr)		Remarks
		Average	Range	
Hansen *et al.* (1978)	Shallow coastal lagoons, two sites	17.6	0–84	Release occurs mainly at night
		442	0–1880	
Jaeschke *et al.* (1978)	Tidal flats	0.042	—	
	Swamps	0.035	—	
	Various aerobic soils	None	—	
Hill *et al.* (1978)	Salt marsh, Long Island	0.55	0.03–0.92	
Aneja *et al.* (1979)	Salt marsh, two sites	0.19 (±0.03)	—	Includes COS, greater emission of DMS, temperature dependence
		0.04 (±0.01)	—	No DMS found
	Mudflats, two sites	0.41 (±0.05)	—	
		0.12 (±0.05)	—	
Delmas *et al.* (1980)	Various soils in France			
	Toulouse, grass sward	0.044	0–0.082	Temperature 3–13°C
		0.058	0.043–0.088	15–20°C
		0.23	0.15–0.35	20–25°C
	Loire valley	0.018	0.0023–0.03	3–9°C
	Landes	0.038	0–0.069	9–12°C
	Pine forest	0.022	0.013–0.034	9–12°C
	Ivory Coast, littoral forest	0.3	0.1–2.5	
Adams *et al.* (1980)	Soils of the eastern U.S.A.			Simultaneous emission of other sulfides is common, see Table 10-7
	Saline marshes		0.003–1940	
	Nonsaline swamps		0.11–0.198	
	Histosols (peat, muck)		0.002–0.478	
	Dry mineral soils		0.001–2.46	
Goldberg *et al.* (1981)	Saltwater marsh, Wallops Island, Virginia	3–16	—	One day in July, strong buildup during night
Aneja *et al.* (1981)	Freshwater marsh	0.6	—	

The highest fluxes are observed at night and during the early morning hours, when according to the interpretation of Hansen *et al.* (1978) the oxygen concentration in stagnant waters falls to low values. During the day, oxygen is produced by algal photosynthesis, H_2S is subject to bacterial oxidation, and the H_2S flux decreases. The H_2S mixing ratio above high-production regions show a similar diurnal variation (see, e.g., Goldberg *et al.*, 1981) because micrometeorological conditions favor the accumulation of H_2S during the night and its dilution by enhanced vertical mixing during the day.

Reports on H_2S emissions from common mineral soils are not very consistent and in part contradictory. For example, Bloomfield (1969) and Siman and Jansson (1976) detected the evolution of H_2S from water-logged soils amended with sulfate and incubated under nitrogen atmosphere, whereas Banwart and Bremner (1976), who studied 25 soils from Iowa, failed to detect H_2S under any conditions, even the most favorable ones. The latter authors ascribe the lack of H_2S emission to sorption by the soils, possibly accompanied by fixation as FeS. The field measurements of Jaeschke *et al.* (1978, 1980) indicate that normally aerobic soils absorb H_2S rather than emit it. Farwell *et al.* (1979), who studied agriclutural, forest, and marsh soils, also found only the last type to emit measurable quantities of H_2S. Delmas *et al.* (1980), by contrast, found H_2S to evolve from various soils in France, whose classification was not given but that cannot have differed much from those studied by Jaeschke *et al.* (1978).

The most comprehensive set of measurements of reduced sulfur compounds emanating from soils is that of Adams *et al.* (1979, 1981a,b). They separated H_2S, DMS, DMDS (dimethyl disulfide), COS, CS_2, and CH_3S by means of advanced gas chromatography and studied a great number of soils of various types from the eastern United States. The data show that H_2S evolves in many cases, not only from saline or swampy soils, but that it is not necessarily the dominant volatile. In the gaseous compositions published, H_2S was absent only in 8 out of 30 cases, and then either DMS or CS_2 was the major sulfur compound. Table 10-7 shows for four major groups of soils the average distribution among the sulfur compounds released and its variability. The most abundant gases are H_2S, DMS, COS, and CS_2, whereas the other sulfur compounds are comparatively insignificant.

Temperature is an important parameter controlling the sulfur emission rate, as Table 10-6 indicates for H_2S. Adams *et al.* (1980, 1981a, b) have studied the temperature dependence of total sulfur emissions in some detail and found them to correlate equally well with ambient temperature as with soil temperature. The former was then used as a parameter, and linear regression analysis was employed to group the data according to soil order. By forcing the individual regression lines through a common intercept at region come primarily from the Florida peninsula, where the percentage of wetlands and coastal marshlands is higher than elsewhere in the study region. Adams *et al.* (1981a) attributed the effect in part to the increase in tropical biomass, and by extrapolating their data toward the equator they estimated a tropical sulfur flux of $0.5-1.0 \text{ g S}/\text{m}^2$ yr for the 5°N latitude band. Since this agrees approximately with the results of Delmas *et al.* (1980) from the littoral forests of the Ivory Coast (Table 10.6), Adams *et al.* (1981a) felt confident that their model describes the situation in the tropics correctly; by summing over all continental land areas, they derived a global sulfur flux from all soils of 64 Tg S/yr. The value must be considered an overestimate, however, for several reasons.

The tropics undoubtedly contribute the largest share to the global sulfur flux, because the highest mean temperatures and the most rapid biospheric turnover occur in that region. But it would be wrong to assume that the entire land area within the 25°N–25°S latitude belt is covered with soils similar to those found in Florida and in the littoral forests of the Ivory Coast. In fact, Florida soil types are rather inconspicuous on Food and Agricultural Organization (FAO, 1977) world soil maps of Africa, Australia, and South America. If the data from Florida are disregarded as atypical, a linear extrapolation of the remaining curve in Fig. 10-1 toward lower latitudes leads one to estimate a global sulfur flux of about 6 Tg S/yr, that is, only one-tenth of that proposed by Adams *et al.* (1981a). The difference reflects the large uncertainty involved in extrapolating their data.

The second reason for rejecting the high sulfur emissions projected by these authors is the amount of sulfate deposited annually by tropical precipitation (see Table 10-15). The observations indicate a wet deposition rate of $0.1-0.2 \text{ g S}/\text{m}^2$ yr, which, if it represents the return flux of sulfides emanating from soils, corresponds to an integrated emission rate of 5–10 Tg S/yr from the $50 \times 10^6 \text{ km}^2$ of land mass residing in the 25°N–25°S latitude belt. The extratropical regions would contribute less than 3 Tg S/yr, so that the resulting global sulfur flux is again much smaller than that estimated by Adams *et al.* (1981a).

The most compelling argument against a high emission rate is based on the budget of COS in the atmosphere. According to previous considerations, the atmospheric residence time of COS cannot be less than 2 yr. Since there are 4.4 Tg of COS in the troposphere harboring 2.35 Tg S, the flux of sulfur associated with COS is, at the most, 1.2 Tg S/yr. We shall assume that it is exclusively due to soil emissions and that other sources can be neglected. One part of COS is emitted directly, whereas another part arises from the oxidation of CS_2 in the air. Fifty percent of the sulfur in CS_2 is converted to COS. Table 10-7 gives the fractions that COS and CS_2 contribute to the total quantity of sulfur released by all sulfides from various soil types. These data may be combined with the turnover rate of atmospheric COS to estimate the total sulfur flux. The highest flux is obtained if one assumes that all soils behave like saline marshes. In this case, the percentage of $COS + \frac{1}{2}CS_2$ is $2.3 + 6.5 = 9.8$ and the total sulfur flux amounts to $1.2/0.098 = 12$ Tg S/yr. The lowest flux is obtained for mineral soils as the dominant sources. In that case, one finds a total emission rate of 4.1 Tg S/yr. A soil composition similar to that of the eastern United States with 7% coastal wetlands, 18.7% high-organic inland soils, and 74.3% dry mineral soils leads to a global sulfur flux of about 5 Tg S/yr. All values are rather uncertain, not only because the measured compositions of soil-derived sulfides vary considerably, but also because we have used a COS turnover rate in the atmosphere near the upper limit. Nevertheless, the range of values is congruent with that derived from the deposition of sulfate by wet precipitation. Taken together, therefore, the various arguments point toward a global emission of sulfides from soils of 8 ± 5 Tg S/yr, which is a far cry from the 64 Tg S/yr proposed by Adams *et al.* (1981a).

10.3 Sulfur Dioxide, SO_2, and Particulate Sulfate, SO_4^{2-}

Sulfur dioxide enters the atmosphere via direct emissions from volcanic and anthropogenic sources as well as by the oxidation of organic sulfides as described in Section 10.2.2. Owing to a rapid further oxidation to sulfuric acid, SO_2 also accounts for a good deal of sulfate associated with the atmospheric aerosol. Both components, therefore, are appropriately discussed together. In the following sections we consider the magnitude of human-made and natural emissions, the rate at which SO_2 is converted to particulate sulfate, the distribution of SO_2 and SO_4^{-2} in the troposphere, and the removal of both components by wet and dry deposition. The observational data are then used to construct, in Section 10.4, regional and global sulfur budgets.

10.3.1 ANTHROPOGENIC EMISSIONS

A breakdown of processes causing the release of SO_2 from anthropogenic sources is shown in Table 10-8. The combustion of coal now contributes 60% to all such emissions, that of petroleum and its products 28%, and the smelting of nonferrous ores as well as miscellaneous industrial processes take up the remainder. Individual emission estimates are derived as usual by combining statistical production data with emission factors. Fossil fuels contain sulfur primarily in the form of organic sulfur compounds. Combustion turns them into SO_2, which is vented with the flue gases. The ashes retain very little sulfur, so that emission factors for coals correspond closely are expected to release sulfur mainly as SO_2. Upon entering the atmosphere both compounds are oxidized as hot volcanic gases react with oxygen; H_2S is partially converted to SO_2, and SO_2 to sulfuric acid.

A first estimate for the global rate of sulfur emissions from volcanoes was made by Kellog et al. (1972). Their assessment was based on the total volume of lava and ashes ejected during the time period from 1500 to 1914 according to Sapper (1927), the estimate by MacDonald (1955) of the weight of the gases evolved from fresh lava on Hawaii, and the analysis of Shepherd (1938), who found 10% by weight of the gases in the form of SO_2. The average flux of volcanic SO_2 derived in this manner is 0.75 Tg S/yr. The result actually is a lower limit to the real value, because it considers only SO_2 released during eruptions, whereas the much longer quiescent periods of volcanic activity are not taken into account. The same deficiency is inherent in later attempts by Friend (1973) and Cadle (1975) to improve on the estimate of Kellog et al. (1972). In the meantime, more direct

Table 10-9. *Measured or Estimated Rates of SO_2 Emissions from Volcanoes*[a]

Authors	Volcano	Emission rate (10^3 kg SO_2/day)	Category
Italy			
Haulet et al. (1977)	Etna	3,740	Eruptive (I)
Zettwoog and Haulet (1979)	Etna	3,325	Preeruptive
Zettwoog and Haulet (1979)	Etna	1,130	Posteruptive
Malinconico (1979)	Etna	1,000	Intraeruptive
Jaeschke et al. (1982)	Etna	142	Posteruptive
Jaeschke et al. (1982)	Etna	955/1,600	Preeruptive
Nicaragua			
Stoiber and Jepson (1973)	Telica	20	Posteruptive
Stoiber and Jepson (1973)	Momotombo	50	Posteruptive
Stoiber and Jepson (1973)	San Cristobal	360	Posteruptive
Stoiber and Jepson (1973)	Masaya	180	Preeruptive
Taylor and Stoiber (1973)	Cerro Negro	120[b]	Eruptive (I)
Taylor and Stoiber (1973)	Cerro Negro	2,000[b]	Eruptive (II–IV)
Guatemala			
Stoiber and Jepson (1973)	Pacaya	260	Eruptive (I)
Stoiber and Bratton (1978)	Pacaya	300	Posteruptive
Stoiber and Jepson (1973)	Santiaguito	420	Eruptive
Stoiber and Jepson (1973)	Santiaguito	300/1,500	Eruptive (IV)
Rose et al. (1973)	Fuego	10,000[b]	Eruptive (II–IV)
Stoiber and Jepson (1973)	Fuego	40	Posteruptive
Crafford (1975)	Fuego	55,000[c]	Eruptive (V)
Crafford (1975)	Fuego	423	Intraeruptive
Stoiber and Bratton (1978)	Fuego	300/1,500	Eruptive (I)

Table 10-9. (*Continued*)

Authors	Volcano	Emission rate (10^3 kg SO_2/day)	Category
Hawaii			
Naughton *et al.* (1975)	Kilauea	$1,280^d$	Eruptive (I–IV)
Naughton *et al.* (1975)	Kilauea	280	Undetermined
Stoiber and Malone (1975)	Mauna Ulu	30	Extraeruptive
Stoiber and Malone (1975)	Sulfur Banks	7	Extraeruptive
Alaska			
Stith *et al.* (1978)	St. Augustine	$86,400^c$	Eruptive (V)
Stith *et al.* (1978)	St. Augustine	432^c	Posteruptive
Stith *et al.* (1978)	St. Augustine	8,640	Eruptive (II–IV)
Stith *et al.* (1978)	St. Augustine	86	Posteruptive
Stith *et al.* (1978)	Mt. Martin	3	Extraeruptive
Washington State			
Hobbs *et al.* (1981)	St. Helens	10/50	Preeruptive
Casadevall *et al.* (1981)	St. Helens	1,000/1,900	Preeruptive
and Hobbs *et al.* (1981)	St. Helens	1,300/2,400	Eruptive (V)
Casadeval *et al.* (1981)	St. Helens	900	Intraeruptive
Casadeval *et al.* (1981)	St. Helens	600/3,600	Posteruptive
Japan and U.S.S.R.			
Okita (1971)	Mihara	345	Posteruptive
Okita and Shimozuru (1975)	Asama	142	Posteruptive
Okita and Shimozuru (1975)	Asama	780	Posteruptive
Crafford (1975)	Karimski	173	Undetermined

[a] Classification of activity: eruptive (intensity), intensity scale after Tsuya (1955) in terms of total volume of ejected material I ($=10^{-5}$ km^3)–IX ($=10^2$ km^3); preeruptive, intensification of activity before an eruption; intraeruptive, phase of repose between two paroxysmal eruptions; posteruptive, permanent fuming or fumarolic activity; extraeruptive, exclusive fumarolic and solfataric activity. [From Berresheim and Jaeschke (1983).]
[b] Calculated from the SO_4^{2-} content of ash particles.
[c] Estimated value.
[d] Calculated from the SO_2 content of lava.

information has been obtained on volcanic SO_2 emissions by means of aircraft samplings of the sulfur plume, and by remote measurement techniques. Table 10-9 presents a compilation of observational data prepared by Berresheim and Jaeschke (1983). To determine from these data average emission rates requires a correlation with the intensity of volcanic activity at the time of observation. Berresheim and Jaeschke have performed the analysis and thereby achieved a more realistic assessment of the situation than was hitherto available.

Five categories of volcanic activities were distinguished, namely, eruptive, preeruptive, intraeruptive, posteruptive, and extraeruptive phases. The first category was further subdivided into nine groups of differing intensities. The definitions are given in the footnote to Table 10-9. The authors then considered the worldwide frequency of eruptions in the period between 1961 and 1979, appropriately classified by the strength of individual eruptions. An average number of 55 volcanoes undergo eruptions each year. This is about one-tenth of all active volcanoes currently in existence. By summing over the entire amount of SO_2 released within each intensity class, Berresheim and Jaeschke found an average emission rate of 0.5 Tg S/yr for the time period considered. The value differs little from that of Kellog *et al.* (1972). A much greater total emission rate was obtained, however, for the steadier, noneruptive activities. Among some 520 live volcanoes, there are about 365 that may be classified as posteruptive, and an additional 100 in the extraeruptive phase. Volcanoes in the latter category may be neglected to a first approximation becuase they emit only a small fraction of sulfur released by the others. The preeruptive and intraeruptive phases are relatively short and may also be disregarded.

According to Table 10-9, the average SO_2 flux for a single volcano in the posteruptive activity class is roughly 3×10^5 kg SO_2/day. The use of this value leads to a global sulfur emission rate of 20 Tg S/yr. Berresheim and Jaeschke (1983) adopted a more cautious attitude. On the basis of MacDonald's (1972) description of 516 active volcanoes, they assigned a source strength of the magnitude suggested by the average of the data in Table 10-9 to not more than 90 volcanoes. The remaining 275 volcanoes with posteruptive activity were taken to produce only 5×10^4 kg SO_2/day each. The global sulfur flux from nonerupting volcanoes is then reduced to 7 Tg S/yr. The rate is still an order of magnitude greater than that associated with erupting volcanoes. Berresheim and Jaeschke also used their own measurements of H_2S in the plume of Mt. Etna to estimate the rate of H_2S emissions from all volcanoes with post- and extraeruptive activities as about 1 Tg S/yr. Cadle (1980) has considered COS and CS_2 as effluents and found their contributions negligible by comparison.

Although total volcanic sulfur emissions may reach or even exceed 20 Tg S/yr, there exists support for a lower average release rate from the geochemical data discussed in Chapter 12. The continuous degassing of the Earth's interior since the formation of the planet 4.5 billion yr ago has by now virtually depleted the upper mantle of juvenile volatiles. As acidic gases, CO_2 and SO_2 are responsible for the chemical weathering of virgin rocks, with the result that CO_2 is transformed into (bi)carbonate and SO_2 into sulfate. In this form, carbon and sulfur are washed into the oceans and eventually come to rest in marine sediments. Cycles of sedimentary rocks are discussed in Section 11.3.1 in conjunction with the geochemistry of carbon dioxide (see also Fig. 11-3). It is clear that the losses of carbon and sulfur from the upper mantle, caused by volcanic emissions are to continue. The replacement is accomplished by the subduction of sedimentary rocks in those geological zones, where marine and continental crustal plates meet and by their counteracting motion force crustal material downward into regions of the upper mantle. Assuming steady-state conditions, one thus expects modern volcanoes to emit the volatile elements carbon and sulfur in roughly the same proportions as they occur in average sedimentary rock. Table 12-5 shows that the mass ratio of sulfur and carbon residing in the sediments is $1.2/6.2 = 0.19$, whereas Table 11-8 suggests that the mass flux of carbon associated with volcanic emissions or the subduction of sedimentary rocks into the upper mantle is 23 Tg C/yr. Taken together, both values lead to a mass flux of sulfur subducted into the upper mantle and reemitted by volcanoes of 4.5 Tg S/yr. The rate agrees surprisingly well with the estimate derived from direct observations of volcanic sulfur emissions.

10.3.3 CHEMICAL CONVERSION OF SO_2 TO PARTICULATE SULFATE

Pathways for the oxidation of SO_2 in the atmosphere are complex and have not yet been fully delineated. The reactions occur in the gas phase, in fog and cloud droplets, and on the surface of aerosol particles. Some of the reaction mechanisms were discussed in Sections 7.4.3 and 8.6. Here, we are primarily interested in the overall oxidation rate and the associated lifetime of SO_2 with respect to its conversion to H_2SO_4. Information on this point is available from laboratory and field observations.

Table 10-10 summarizes processes that have been studied in the laboratory, including the inferred atmospheric conversion rates and time constants. The direct gas-phase photooxidation of SO_2 due to its absorption of solar radiation is slow because the excitation energy is mainly lost in collisions with air molecules rather than utilized to convert SO_2 to SO_3. More important are oxidation reactions initiated by radicals. The best known of these is the reaction of SO_2 with OH. The mechanism leads first to the formation of an adduct, then by its reaction with oxygen to SO_3, which finally reacts with water to produce H_2SO_4. Peroxy radicals such as HO_2, CH_3O_2, etc. appear to be potentially less effective in oxidizing SO_2. An exception are the Criegee intermediates produced in the reaction of ozone with alkenes. Unfortunately, it is difficult to assess their contribution to SO_2 oxidation, for rate coefficients are unavailable. Several other gas-phase reactions that have been studied are not listed in the table because they are insignificant in the atmopshere. Among them are reactions of SO_2 with ozone, NO_3, CH_3O, and electronically excited O_2 molecules.

In the presence of clouds and fogs, SO_2 dissolves to some extent in liquid water, where it forms HSO_3^- and SO_3^{2-}. These ions are subject to oxidation

Table 10-10. *Summary of SO_2 Oxidation Processes Studied in the Laboratory: Application to Atmospheric Conditions and Estimates of Rates and SO_2 Lifetimes*

Process	Rate (%/h)	Time constant (days)	Remarks
Direct photooxidation[a]			
$SO_2 + h\nu \rightarrow SO_2^* \xrightarrow{O_2} SO_3 + O$	<0.04	>100	SO_2^* involves both singlet and triplet states. Quenching by N_2 makes this process unimportant in the atmosphere (Calvert et al., 1978).
Reactions involving radicals			
$SO_2 + OH \rightarrow HOSO_2 \xrightarrow{O_2} SO_3 + HO_2$	0.35	12	Conversion rate is based on $n(OH) = 1 \times 10^6$ molecules/cm³. Rate is subject to diurnal, seasonal, and latitudinal variation.
$SO_2 + HO_2 \rightarrow$ products	<0.04	>100	Conversion rate is based on $n(HO_2) = 2.5 \times 10^8$ molecules/cm³ and the rate coefficient given by Payne et al. (1973). This is an upper limit (Calvert et al., 1978).
$SO_2 + RO_2 \rightarrow$ products	<0.4	>11	Conversion rate is based on $n(RO_2) = 7 \times 10^7$ molecules/cm³ assuming CH_3O_2 to be the principal species. The rate coefficient used, $k = 10^{-14}$ cm³/molecule s, is an upper limit value (Calvert and Stockwell, 1984).
$SO_2 + R\dot{C}HOO\cdot \rightarrow RCHO + SO_3$	>0.1	<40	Rate coefficients for the individual reactions are unavailable, but the process is considered important (Calvert and Stockwell, 1984).

512

Aqueous reactions in clouds^{b,c}

Reaction			Comments
$SO_3^{2-} + \frac{1}{2}O_2 \rightarrow SO_4^{2-}$	$(3.5\times10^{-4})^c$ 7×10^{-5}	$(1.2\times10^4)^c$ 6×10^4	This reaction is promoted by catalysts, and the existence of an uncatalyzed oxidation is uncertain. We adopt the low rate coefficient of Beilke et al. (1975) and Brimblecombe and Spedding (1974b), $k = 0.001\ s^{-1}$ at pH = 5.
$S(IV) + \frac{1}{2}O_2 \xrightarrow{Mn^{2+},\,Fe^{3+}} SO_4^{2-}$	$(6)^c$ 1.2	$(0.7)^c$ 3.5	Rate expression according to Table 8-8 for synergistic catalysis by manganese and iron, adopting $[Mn] = 5\times10^{-7}\ M$ and $[Fe] = 5\times10^{-6}\ M$.
$S(IV) + O_3 \rightarrow SO_4^{2-} + O_2$	$(0.3)^c$ 0.06	$(15)^c$ 75	Rate expression from Table 8-8, $p(O_3) = 4\times10^{-8}$ bar.
$S(IV) + H_2O_2 \rightarrow SO_4^{2-} + H_2O$	$(21)^c$ 4.1	$(0.2)^c$ 1.0	Rate expression from Table 8-8, $[H_2O_2] = 1.5\times10^{-5}$ corresponding to $m_0(H_2O_2) = 100$ pptv before cloud condensation. The concentration of H_2O_2 in cloud water is highly uncertain.
Oxidation of SO_2 on particles			
Fly-ash particles in power-plant plumes	22	0.2	Initial oxidation rate $0.37\ m^3/g_{aerosol}$ min (Dlugi and Jordan, 1982); assumed concentration of particles $10\ mg/m^3$.
Natural aerosol of same composition	0.1	40	Prorated to a particle concentration of $50\ \mu g/m^3$.

^{a} See Tables 2-4 and 2-6.

^{b} Rate estimates are based on clouds with a liquid water content of $0.1\ g/m^3$, $T = 283\ K$, pH = 4, and a fraction of SO_2 in the aqueous phase according to Eq. (8-11). Mass transport limitations and rate changes due to acidification by product sulfuric acid are neglected.

^{c} The values in parentheses refer to cloud-filled air; the other values assume that clouds fill only about 20% of the entire air space of the lower troposphere.

513

Table 10-11. SO_2 to SO_4^{2-} Conversion Rates Deduced from Field Observations

Authors	Rate (%/h) Range	Rate (%/h) Average	Time constant (days)	Remarks
	Observations in power-plant plumes			
Cantrell and Whitby (1978)	0.41–4.9	1.5	2.9	Coal-fired power station, Labadie, Missouri;
Husar et al. (1978)	0.1–4.8	1.5	2.7	July and August 1974, 1976, diurnal variation of conversion rate noted.
Hegg and Hobbs (1980)	0–5.7	1.1	3.6	Data from three coal-fired power stations, Midwestern U.S.A., June 1978; review of previous data.
Gillani et al. (1981)	0.3–3.1	1.4	3.0	Coal-fired power station, Labadie, Missouri, July 1976.
	0.1–5.8	2.–	2.0	Coal-fired power station, Cumberland, Tennessee, August 1978.
Forest et al. (1981)	0.2–7.0	2.2	1.9	Coal-fired power station, Cumberland, Tennessee, August 1978.
Wilson (1981), night	0.1–1.3	0.5	8.3	Analysis of 230 measurements in the plumes
Wilson (1981), whole day	0–8.4	1.5	2.8	of 9 power plants and 2 metal smelters; plume age greater than 1 h; strong dependence on solar radiation dose.
Meagher et al. (1983), winter	—	0.15	28	Averages from measurements in the plumes
Meagher et al. (1983), summer	—	1.30	3.2	of five coal-fired power plants; seasonal variation described.
Gillani and Wilson (1983), dry	0.2–2.1	0.97	4.3	Analysis of data from three coal-fired power
Gillani and Wilson (1983), wet	0.3–8.0	1.77	2.3	plant plumes under dry and wet conditions.

Urban plume and regional budget studies

Reference				Remarks
Eliassen and Saltbones (1975)	0.3–1.7 / 2.6–14[a]	0.8 / 8.1[a]	5 / 0.5[a]	Comparison of trajectory transport calculations with observations at six stations in Western Europe, based on an areal emission inventory.
Breeding et al. (1976)	5.3–32	16.4	0.3	Ground-based measurements in the plume of St. Louis, Missouri.
Mészáros et al. (1977)	7.2–35	—	—	Ground-based measurements in a suburban area of Budapest, Hungary, correlation analysis with wind speed, temperature, data grouping.
Alkezweeny (1978)	8–11.5	9.8	0.4	Urban plume of St. Louis, Missouri; aircraft observations on three days in August 1975.
Elshout et al. (1978)	0.6–4.4	1.7	2.4	Ground-based measurements in Arnhem, The Netherlands, of polluted air originating in the Ruhr Valley; four selected days in winter 1972.
Forrest et al. (1979)	0–4	2	2	Urban plume of St. Louis, Missouri, followed by manned balloon on a day in June 1976.
Miller and Alkezweeny (1980)	1–9	4	1	Urban plume of Milwaukee, Wisconsin, over Lake Michigan, on 2 days in August 1976, 1 day in July 1977.
McMurry and Wilson (1983), dry	0–5	—	—	Urban plume of Columbus, Ohio, ground-based measurements in July and August 1980.
McMurry and Wilson (1983), r.h. > 75%	≤12			
Nordlund (1983)	—	8.5[a]	0.5[a]	Seasonal average; from budget considerations of measurements at many stations in Western Europe.

[a] Total removal rate includes dry deposition of SO_2 to the ground surface.

517

that losses of SO_2 due to dry deposition to the ground are unavoidable. The effect should show up most markedly in the data from ground-based observations, whereas aircraft measurements should be less influenced. The two regional budget studies included in Table 10-11 clearly indicate that in polluted regions dry deposition can be more significant in removing SO_2 from the atmosphere than conversion to sulfate.

Dry deposition, however, cannot account for the great variation in the SO_2 to SO_4^{2-} conversion rate revealed by some of the aircraft studies. Specifically, the measurements of Miller and Alkezweeny (1980) in the urban plume of Milwaukee, Wisconsin, indicated average rates of 9% and 0.7%, respectively, on two consecutive days of nearly identical meteorological conditions. Hydrocarbon mixing ratios were measured along with several other pollutants. While total nonmethane hydrocarbon reactivities toward OH radicals differed little on both days, alkenes were considerably more abundant on the first day when the SO_2 oxidation rate was high compared with the second day when it was low. Qualitatively, the data appeared to confirm the importance of SO_2 oxidation by Criegee intermediates produced in the oxidation of alkenes. Miller and Alkezweeny (1980) thought that the dramatic difference in the SO_2 to SO_4^{2-} conversion rates was caused by the differing quality of the upwind air masses. On the first day the air moved into the city from the southwest. High ozone and sulfate levels were typical of well-aged polluted air (with Chicago as a possible point of origin). On the second day, in contrast, the wind direction was from the northwest and ozone and sulfate concentrations were low. This leads one to conclude that the SO_2 oxidation rate is much enhanced by the presence of oxidants or oxidant precursors carried along with polluted air.

In general, and despite the uncertainties noted, the observational data of Table 10-11 confirm the predictions of the laboratory studies. The combination of all data suggests an SO_2 oxidation rate of 2–4%/h and a corresponding lifetime of SO_2 of 2–4 days.

10.3.4 DISTRIBUTION OF SO_2 AND SO_4^{2-} IN THE TROPOSPHERE

The combustion of fossil fuels is the dominant source of sulfur dioxide in the industrialized regions of the world. Before 1950, coal provided the major fuel for energy generation, steel production, and domestic applications. The comparatively high sulfur content of many coals gave rise to urban SO_2 levels often exceeding 100 ppbv (Meetham et al., 1964; Jacobs, 1959). In England the former practice of heating homes by burning coal in open fireplaces further burdened the atmosphere with high concentrations of carbonaceous particles. The development of radiation fogs under stagnant and humid weather conditions (usually during the winter months), if persisting for several days, then led to incidents of the infamous London-type smog, in which smoke, fog, SO$_2$, and H$_2$SO$_4$ combined to cause bronchial irritation, vomiting, and in some cases death due to heart failure. The worst episode on record occurred in London during 4 days in December 1952, when SO$_2$ and smoke concentrations rose to levels well above 1 mg/m^3. During this episode the mortality reached twice the normal rate (Wilkins, 1954).

After 1950, coal for domestic heating was gradually replaced by light fuel oil and natural gas; both are low-sulfur fuels. Many new electric power plants were constructed outside the cities. They were fitted with taller stacks discharging the effluents higher into the atmosphere, thus spreading SO$_2$ more over the countryside. Altshuller (1980) and Husar and Patterson (1980) have discussed this development for the United States. Between 1963 and 1973, urban SO$_2$ mixing ratios in the northeastern United States declined by a factor of five to about 10 ppbv, while nonurban SO$_2$ values remained fairly constant at 3.5 to 4 ppbv. A similar though less extensive reduction, starting from lower levels, occurred in Midwestern cities in spite of rising SO$_2$ emissions from new power stations in that region. On the British Isles, the average urban SO$_2$ concentration decreased by a factor of two from 1960 to 1975 (Department of the Environment, 1983). The average annual SO$_2$ level in London now is 23 ppbv. In Frankfurt, West Germany, SO$_2$ mixing ratios declined to 32 ppbv during the same period (Georgii, 1983).

The distribution of SO$_2$ on the European continent was first estimated by DeBary and Junge (1963), who used data from a network of monitoring statsons set up in the 1950s at nonurban sites. Figure 10-3 shows the results of a more recent and more ambitious study (Ottar, 1978), which combined aircraft observations, measurements at 70 ground stations, an emissions inventory, and various advection models. The distribution is similar to that derived earlier by DeBary and Junge. The highest SO$_2$ levels appear in and around industrial centers. Over the ocean the mixing ratios fall off to quite low values.

In North America, the rural background of SO$_2$ seems to have received less attention than urban SO$_2$. Altshuller (1976) listed only six nonurban SO$_2$ monitoring stations, all located in the northeast, as opposed to 48 urban measurement sites. The annual average SO$_2$ level at the six stations during the period 1968–1972 was about 3.5 ppbv. In the late 1970s a concerted effort was made in the Sulfate Regional Experiment to determine SO$_2$ and particulate sulfate in a larger region east of the Mississipi River. The program involved 54 ground stations and supplemental aircraft observations (Mueller et al., 1980). The distribution of SO$_2$ shown in Fig. 10-4 incorporates these data as well as additional results from six Canadian stations operated in 1978/1979 (Whelpdale and Barrie, 1982). The situation resembles that in

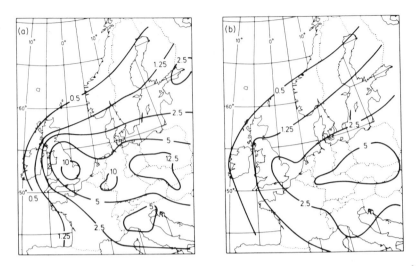

Fig. 10-3. Distribution of (a) SO_2 and (b) particulate sulfate over Europe, in units of $\mu g\,S/m^3$. [Adapted from Ottar (1978).]

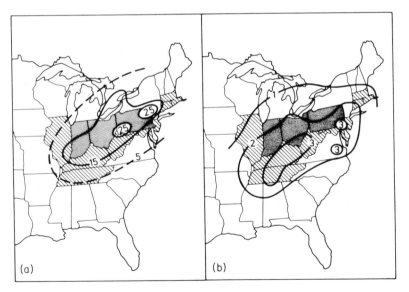

Fig. 10-4. Distribution of (a) SO_2 and (b) particulate sulfate in the northeastern U.S.A., in units of $\mu g\,S/m^3$. [Adapted from Mueller *et al.* (1980).] SO_2 emission densities in various states are indicated by heavy stipples, 20–30 g/m^2 yr; light stipples, 10–20 g/m^2 yr; no stipples, less than 10 g/m^2 yr (from Husar and Patterson, 1980).

Europe with SO_2 levels peaking in the industrialized regions. High concentrations also occur along the Ohio River valley where numerous power plants are located (Benkovitz, 1982).

Figures 10-3 and 10-4 include the distributions of particulate sulfate. On both continents the SO_4^{2-} distribution is similar to, though somewhat broader than, that of SO_2 indicating that particulate sulfate arises primarily from the oxidation of SO_2 and that the conversion is rapid relative to long-distance transport. Further support for SO_2 being the main precursor of SO_4^{2-} derives from the mass distribution of sulfate with particle size of the continental aerosol (see Fig. 7-20). According to the discussions in Chapter 7, the process of gas-to-particle conversion channels material primarily into the submicron size range, wheras wind-generated particles such as seasalt or mineral dust appear in the size range greater than 1 μm radius. The continental aerosol carries its burden of sulfate mostly in the submicrometer size fraction. Exceptions are found in coastal areas where the influence of seasalt SO_4^{2-} shows up in the coarse particle mode, and in regions where mineral deposits of sulfate are exposed to aeolian weathering (Ryaboshapko, 1983).

At the monitoring stations in Europe and North America where a longer record of data exists, the average SO_2 mixing ratio generally follows an annual cycle with high values in winter and low ones in summer. The trend showed up already in the early data treated by DeBary and Junge (1963). Figure 9-8 presented a record from a more recent measurement series at a rural station in Germany. Martin and Barber (1981) have observed a similar behavior in Bottesford in central England. Figure 10-5 is added here to show the seasonal variation of both SO_2 and SO_4^{2-} at three stations in the Ohio River valley, as reported by Shaw and Paur (1983). The sum of both components exhibits a winter maximum, as one would expect from the increased consumption of fossil fuels during the cold season of the year. The concentration of sulfate, however, varies in a manner opposite to that of SO_2, going through a minimum in winter and a maximum in summer. The existence of such a seasonal cycle for sulfate itself had been noted earlier by Hidy et al. (1978) and by Husar and Patterson (1980) from observations at several rural as well as urban sites. The summer maximum for SO_4^{2-} (and the simultaneous minimum for SO_2) clearly indicates an increase in the rate of SO_2 oxidation during the warmer and brighter period of the year, presumably as a result of enhanced photochemical activity, specifically the rise of OH concentrations during the summer months. The data in Fig. 10-5 supplement nicely the results of Wilson (1981) and Meagher et al. (1983), which had demonstrated diurnal and seasonal cycles regarding the rate of SO_2 to SO_4^{2-} conversion in power plant plumes.

The continental background of SO_2 in regions not directly influenced by anthropogenic emissions has remained largely unexplored. Breeding et al.

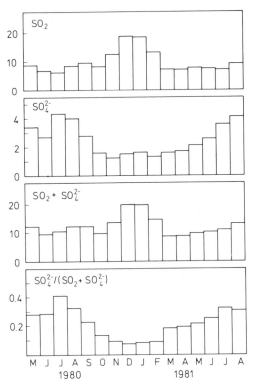

Fig. 10-5. Annual variation of monthly means for gaseous SO_2, particulate SO_4^{2-}, their sums, and ratios, in the Ohio River valley, in units of $\mu g\,S/m^3$. Averages are plotted, from three stations located in the western, central, and eastern valley regions. [Data from Shaw and Paur (1983).]

(1973) and Ryaboshapko (1983) have summarized the few data in existence. For example, Georgii (1970) measured 0.75 ppbv in Colorado; Lodge *et al.* (1974) found 0.1–1.3 ppbv in the American tropics; Delmas *et al.* (1978) observed 3.5–10 ppbv in the tropical rain forests of West Africa, but 10 times lower values in the savanna; and Fischer *et al.* (1968) reported a range of 0.3–1.4 ppbv for Antartica. The latter values are questionable, however, in view of the much lower SO_2 mixing ratios found by Nguyen *et al.* (1974) and by Bonsang *et al.* (1980) over the Subantarctic Ocean (see Table 10-13). Additional data quoted by Ryaboshapko (1983) came from the Caucasus, where SO_2 mixing ratios at 2000 m elevation ranged from 0.12 to 0.22 ppbv, and from Borovoe, Kasakhstan, where the range was 0.1–0.28 ppbv. All these measurements were taken within a few weeks at the most. Since SO_2

exhibits a considerable short-term variability, none of these results provide more than an order-of-magnitude estimate.

The vertical distribution of SO_2 in the troposphere has been explored by means of aircraft ascents, mainly through the efforts of Georgii and Meixner (1980), and Meixner (1984). Figure 10-6 shows the results for a number of individual flights. The SO_2 mixing ratio decreases with height in the first few kilometers above the ground and then reaches an almost constant level in the upper troposphere. Individual altitude profiles are strongly influenced by the difference in winter and summer mixing ratios at the ground surface, by the vertical stability of the atmosphere, by the presence of inversion layers, and by other meteorological factors. The selection of data in Fig. 10-6 suggests an average scale height of 1250 ± 500 m in the boundary layer. In this region SO_2 is lost by reactions with OH and other radicals, and by oxidation in clouds. A simple one-dimensional eddy diffusion model incorporating a constant sink term, but no volume sources, is compatible with the observed scale height if the lifetime of SO_2 with respect to oxidation is about 4 days. In the upper troposphere, the concentration of OH is much smaller than that in the boundary layer, and the scavenging of SO_2 by clouds also is less significant. Accordingly, one expects the lifetime of SO_2 to be considerably extended, and this may be the reason for the almost constant SO_2 background.

Over the Atlantic Ocean west of Europe, the SO_2 mixing ratio is essentially independent of altitude, as Fig. 10-6b shows. The values are similar to those in the continental upper troposphere, indicating the presence of a fairly uniform background of SO_2 in the entire troposphere. This idea is supported

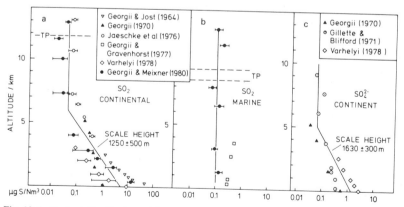

Fig. 10-6. Vertical distribution of (a, c) SO_2 and particulate sulfate in the atmosphere over the continents, and of (b) SO_2 over the Atlantic Ocean. Tropopause levels are indicated.

by the measurements of Maroulis *et al.* (1980), who probed the troposphere horizontally and found practically no latitudinal dependence in the 5-6 km altitude band over the Pacific Ocean. The numerical data are shown in Table 10-12. The average SO_2 mixing ratio in the free troposphere over North America seems to be higher than that over Europe, but the flight level of 5-6 km chosen by Maroulis *et al.* (1980) is not high enough to preclude perturbations by the meteorology of the boundary layer. Evidence for such perturbations may be seen in the greater variability of their data compared with the others.

Maroulis *et al.* (1980) suggested that the background of SO_2 in the troposphere arises from the oxidation of reduced sulfur compounds. COS is a particularly likely candidate because of its uniform tropospheric distribution. The rate coefficients used by Maroulis *et al.* have undergone revision, however. The newer data in Table 10-2 lead to SO_2 production rates resulting from the oxidation of COS, which are insignificant compared with the rate of global transport of SO_2 by vertical and horizontal mixing. The oxidation of other reduced sulfur compounds, such as H_2S, CS_2, or CH_3SCH_3, is so rapid that they are lost predominantly in the boundary layer. The fraction that escapes destruction and enters into the upper troposphere by turbulent mixing must be quite small. Chatfield and Crutzen (1984) proposed a rapid transfer of reduced sulfur compounds in the strong updraft regions of the tropics, the so-called hot towers. Sulfur dioxide, of course, would be subject to the same transport mechanism. Chatfield and Crutzen used their model mainly to explain the higher SO_2 mixing ratios reported by Maroulis *et al.* (1980) in the 5-6 km altitude regime as opposed to that in the marine boundary layer (see Table 10-13). In view of the overlapping variances of both data sets, however, the difference does not appear to be statistically very significant. Presently, the most plausible explanation for the uniform background of SO_2 is its extended lifetime in the uper troposphere coupled with horizontal transport. In the marine atmosphere, the scheme works only because the ocean acts as an indirect source of SO_2, which compensates the capacity of seawater to scavenge SO_2 by absorption.

Figure 10-6c shows the vertical distribution of particulate sulfate observed over the continents. The data indicate again a decrease in concentration with increasing height above the ground toward a nearly constant value of about 70 ng S/st m^3 in the upper troposphere. Normalizing the concentration to standard air density facilitates comparison with the mixing ratios for SO_2. The SO_4^{2-} concentration declines with altitude more slowly, corresponding to an average scale height of 1630 ± 300 m. The difference in the two scale heights must be ascribed to the conversion of SO_2 to SO_4^{2-} during vertical transport. In polluted regions the ground-level concentration of SO_4^{2-} usually is smaller than that of SO_2, whereas in the upper troposphere

Table 10-12. Sulfur Dioxide and Particulate Sulfate in the Remote Troposphere (outside the Boundary Layer)

Authors	Location	Mixing ratio		Remarks
		pptv	ng S/std m³	
	Sulfur dioxide			
Maroulis et al. (1980)	Continental U.S.A.	160 ± 100	228 ± 143	5-6 km Altitude
	Pacific Ocean, 57°S–37°N	85 ± 28	121 ± 40	No significant interhemispheric gradient
Georgii and Meixner (1980)	Bay of Biscayne	95 ± 28	136 ± 40	Average from one flight, 1–12 km altitude
Meixner (1984)	Continental Europe	44 ± 15	63 ± 21	Average from 3 flights above 4 km altitude
	Particulate sulfate			
Gillette and Blifford (1971)	Scottsbluff, Nebraska		78 ± 11	Altitude 4–9 km
	Death Valley, California		94 ± 10	Altitude 4–9 km
	Pacific Ocean, west of U.S.A.		44 ± 26	Altitude 4–9 km
Adams et al. (1977)	Chacaltaya Mountain, Bolivia		118 ± 41	5,200 m a.s.l.
Maenhaut and Zoller (1977)	South Pole		49 ± 10	2,800 m a.s.l.
Lezberg et al. (1979)	Continental U.S.A.		65 ± 51	Near tropopause
			155 ± 51	Above tropopause
Huebert and Lazrus (1980a,b)	Continental U.S.A.		130 av.	5-6 km Altitude
	Pacific Ocean, 57°S–37°N		78 av.	5-6 km Altitude
Cunningham and Zoller (1981)	South Pole, winter		29 ± 10	2,800 m a.s.l.
	South Pole, summer		76 ± 24	

Table 10-13. *Sulfur Dioxide Mixing Ratios in the Remote Marine Troposphere (Surface Air and Boundary Layer)*

Authors	Location	Mixing ratio		Remarks
		pptv	ng S/m³ STP	
Nguyen et al. (1974)	Subantarctic Pacific	89.6 ± 83.1	128 ± 119	Some continental influence
	South Indian Ocean	71.4 ± 75.6	102 ± 108	(Australia)
	Antarctic	65.4 ± 17.3	93 ± 25	
Prahm et al. (1976)	Faroe Islands	65 ± 18	92 ± 25	Clean air masses by trajectory selection
Ryaboshapko (1983)	Central Pacific Ocean	24.5 ± 1.5	35 ± 2	Trajectory selection, no land contact
Bonsang et al. (1980)	South Indian Ocean	36.3 ± 26	51 ± 37	
	Central Atlantic Ocean	22.6 ± 10.9	32 ± 15	
	Pacific Ocean, 57°S–43°N	54 ± 19	77 ± 27	0–2 km Altitude
Maroulis et al. (1980)	Gulf of Guinea	28.7 ± 10.4	40 ± 15	
Delmas and Servant (1982)	Atlantic Ocean, 30°S–43°N	54.2 ± 42.6	77 ± 61	No significant latitudinal dependence
Ockelmann (1982)	Arctic, 70–80°N	23.5 ± 17.6	34 ± 25	
Nguyen et al. (1983)	North Atlantic Ocean	87.5 ± 47	125 ± 67	
	Central Atlantic Ocean	51.5 ± 68.4	74 ± 98	
	South Indian Ocean	33.2 ± 38.8	47 ± 55	127 Data points
Herrmann and Jaeschke (1984)	Atlantic Ocean, 30–45°N	27 ± 8	38 ± 11	Trajectory selected, no land contact

Table 10-14. *Particulate Sulfate in the Remote Marine Troposphere, Surface Air, and Boundary Layer*

Authors	Location	Sulfate ($\mu g/m^3$ STP)		Remarks
		Range	Average	
Nguyen et al. (1974)	North Atlantic Ocean	0.43–0.18	0.78 total	
	South Pacific Ocean	0.17–0.90	0.42 Total	
Mészáros and Vissy (1974)	South Atlantic Ocean	—	0.32 Excess	
Prahm et al. (1976)	North Atlantic Ocean	0.12–0.36	0.25 total	Trajectory analysis indicates pure marine air
		0.03–0.23	0.14 Excess	
Gravenhorst (1975)	North Atlantic Ocean	0.10–0.29	0.23 Total	Data selected to exclude continental influence
	Meteor cruise 23	0.10–0.23	0.17 Excess	
	Meteor cruise 32	0.20–0.70	0.44 Total	
		0.12–0.37	0.27 Excess	
Bonsang et al. (1980)	Central Atlantic	0.42–0.64	0.54 Total	
		0.08–0.29	0.24 Excess	
	South Indian Ocean	0.28–8.40	1.89 Total	
		0.13–1.54	1.03 Excess	
Huebert and Lazrus (1980)	Pacific Ocean	0.08–1.30	0.37 Total	0–2 km Altitude
Heintzenberg et al. (1981)	Spitzbergen	0.50–0.84	0.65 Total	Marine air masses
Maenhaut et al. (1981)	Samoa	—	0.063 Excess	Particles <1 μm
Maenhaut et al. (1983)	Central Pacific, 95–150°W	—	0.046 Excess	2-Month average (particles <1 μm)
Horvath et al. (1981)	Indian Ocean	0.07–0.84	0.57 Total	
		0–0.62	0.29 Excess	

527

both concentrations are approximately the same. A crossover of concentrations at an intermediate altitude is sometimes observed (Georgii, 1970; Varhelyi, 1978). The scale height for SO_4^{2-} in the boundary layer also is larger than that for total aerosol mass, which is of the order of 1 km according to Fig. 7-25. With increasing altitude, therefore, the aerosol becomes enriched with sulfate.

Table 10-12 includes concentrations of particulate sulfate in the free troposphere at elevations above 4 km. Although the data come from various parts of the world, the concentrations are fairly uniform, ranging from 20 to 130 ng S/m^3 STP. Thus, sulfate is an important constituent of the tropospheric background aerosol. If, as in Section 7.6, we assume a mixing ratio for the background aerosol of 1 $\mu g/m^3$ STP, sulfate is found to contribute roughly 25% by mass.

Finally, we present in Tables 10-13 and 10-14 data for SO_2 and particulate sulfate at the ocean surface. The majority of measurements were made onboard of ships. In near-coastal areas the atmosphere often is burdened with SO_2 and/or SO_4^{2-} of continental origin. The standard procedure for eliminating such perturbations from the data is the selection according to air mass trajectories having had no land contact for at least 3–4 days. Bonsang et al. (1980) have used radon-222 as an indicator for continental air. This element emanates from the Earth's crust and has a half-life of 3.8 days, which is similar to that of SO_2 in the lower atmosphere. The elimination of sulfate takes longer, however. Bonsang et al. estimate a minimum time of 10 days. Particularly the North Atlantic Ocean is affected by terrigeneous sulfate carried along with the Saharan dust plume.

Mixing ratios of SO_2 in pure marine air fall into the range 20–80 pptv independent of location. Büchen and Georgii (1971) had reported higher values in the North Atlantic Ocean compared with its southern part, but the more recent data of Ocklemann (1982) have not confirmed this trend and show instead a fairly even distribution of SO_2 between the two hemispheres. The uniformity of mixing ratios is also evident in the aircraft observations of Maroulis et al. (1980).

Laboratory studies of Spedding (1972) and of Beilke and Lamb (1974) have revealed that the absorption capacity of seawater for SO_2 is sufficient to make the ocean surface an almost perfect sink for atmospheric SO_2. Liss and Slater (1974) estimated the resulting deposition velocity to be 0.0044 m/s, whereas Spedding (1972) derived a value of 0.014 m/s. The associated tropospheric residence time of SO_2, assuming a uniform vertical distribution, ranges from 6 to 17 days. The absence of significant horizontal (and vertical) gradients in the SO_2 mixing ratio thus makes it clear that the marine background cannot be due solely to residual SO_2 from continental sources. The uniform occurrence of SO_2 over the oceans is particularly striking in view of the imbalance of anthropogenic emissions between the northern and southern hemispheres.

A widespread marine source of SO$_2$ is required to explain these observations. In Section 10.2 it was shown that the oceans release hydrogen sulfide and dimethyl sulfide, and that both are rapidly oxidized in the atmosphere by reaction with OH radicals. The processes convert hydrogen sulfide fully to SO$_2$, whereas dimethyl sulfide yields primarily methanesulfonic acid, and SO$_2$ accounts for only 25% of all products. Let us see whether the oxidation of these compounds suffices to explain the SO$_2$ mixing ratios observed in marine air. For this purpose we assume steady-state conditions and use the lifetimes for H$_2$S and DMS given in Table 10-2. The mixing ratio for SO$_2$ at the ocean surface then is

$$m(\text{SO}_2) = \tau_{\text{SO}_2}\left[\frac{m(\text{H}_2\text{S})}{\tau_{\text{H}_2\text{S}}} + 0.25\frac{m(\text{DMS})}{\tau_{\text{DMS}}}\right]$$

According to Table 10-1 the mixing ratios for marine H$_2$S are in the range 7–14 pptv, and those for dimethyl sulfide range from 5 to 60 pptv. If the lifetime for SO$_2$ is about 4 days, one obtains

$$m(\text{SO}_2) = 4\left[\frac{(7-14)}{4.4} + 0.25\frac{(5-60)}{0.6}\right] = 15\text{--}96 \quad \text{pptv}$$

The calculated SO$_2$ mixing ratios fall into the same range of values as the measurements summarized in Table 10-13. The results thus demonstrate that the oxidation of reduced sulfur compounds does indeed explain the natural background of SO$_2$ existing in the marine atmosphere. Additional sources are not required. Bonsang et al. (1980) and Nguyen et al. (1983) have furnished further evidence for the biogenic origin of marine SO$_2$ by noting a close correlation between the SO$_2$ mixing ratio and primary biological productivity. As an indicator for the latter quantity they used the concentration of chlorophyll a in the surface waters of the sea. The highest SO$_2$ mixing ratios were found in regions of upwelling waters west of the Peruvian coast, where according to the observations of Andreae and Raemdonck (1983) the concentrations of dimethyl sulfide also are very high (cf. Table 10-1). At least qualitatively, therefore, the origin of SO$_2$ in marine air appears to be well established.

Table 10-14 shows a number of data for particulate sulfate over the open ocean under conditions eliminating continental effects as far as possible. As noted earlier, the marine aerosol contains sulfate from seasalt, as well as sulfate arising from the oxidation of sulfur compounds in the gas phase. The latter fraction is called excess sulfate. As described in Section 7.5, the excess can be determined either by a size fractionation of the aerosol or from the excess of sulfate to sodium mass ratio over that found in seawater. The data in Table 10-14 suggest that excess sulfate contributes 40–75% to total sulfate, The concentrations of both types of sulfate are quite variable, however. The concentration of seasalt aerosol depends on the wind force, whereas the flux of gaseous sulfur compounds varies with biological productivity. Bonsang *et al.* (1980) have studied the latter aspect and found that excess sulfate is linearly correlated with SO_2, whose concentration, in turn, is coupled to the strength of biological activity. Bonsang *et al.* assumed excess sulfate to be formed exclusively by the oxidation of SO_2, but this viewpoint mast be amended if the biogenic sulfur emissions consist mainly of dimethyl sulfide. According to the laboratory data discussed in Section 10.2.2, the dominant product resulting from the oxidation of dimethyl sulfide in the air is methanesulfonic acid rather than SO_2, a compound expected to condense onto preexisting aerosol particles. Saltzman *et al.* (1983) recently confirmed the presence of methanesulfonic acid in marine aerosol samples taken at island and coastal stations in the Pacific and Indian Oceans and in the Gulf of Mexico. The relative abundance of the compound was only 2–11% of excess sulfate, however. Even lesser amounts were found in marine rainwater samples. Saltzman *et al.* (1983) consequently suggested that methanesulfonic acid in the aerosol phase undergoes rapid further oxidation to sulfuric acid. We are thus led to the conclusion that a large share of excess sulfate, probably more than 50%, is formed not via SO_2 but along other routes in the dimethyl sulfide oxidation mechanism.

10.3.5 WET AND DRY DEPOSITION OF SULFATE

Sulfate in rainwater has been under scrutiny for over a century. Eriksson (1952b, 1960, 1966), Steinhardt (1973), Granat *et al.* (1976), and Ryaboshapko (1983) have discussed many of the individual observations in various parts of the world. The majority of studies addressed ecological problems such as the input of sulfur and other elements into agricultural soils. Smith (1872) was probably the first to note the relation between rainwater chemistry and air pollution, but his work was not immediately followed up. In the 1950s, when interest in the fate of air pollutants revived, networks of monitoring stations in Europen and in the United States established for the first time the regional distribution patterns for the concentrations of sulfate and other ions in precipitation. The results were reviewed by Junge (1963). Fifteen years later, Granat (1978) discussed sulfate in precipitation again. He had available a record of European network data spanning two decades. The period is long enough to search for trends and, indeed, the deposition of sulfate was found to have increased with time, although not to the extent expected from the simultaneous rise in anthropogenic SO_2 emissions.

Since 1972, the World Meteorological Organization collects precipitation data from about 110 stations in 72 countries. Georgii (1982) has analyzed a portion of the data with regard to SO_4^{2-}, NO_3^-, and pH. Figure 10-7 shows the global distribution of sulfate in precipitation for 1979. Although at any

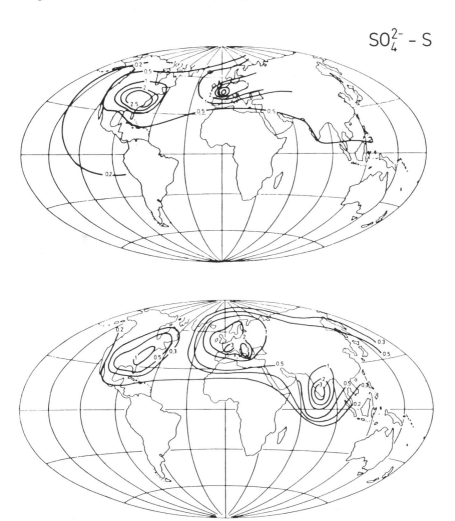

Fig. 10-7. Sulfate in rain water. Upper part: Global average distribution of concentration in units of mg S/liter. Lower part: Deposition rate in units of g S/m² yr. [From Georgii (1982), with permission.]

given location the SO_4^{2-} concentration undergoes large variations, the smoothed distribution of annual averages changes little from year to year. Concentrations and deposition rates maximize in the eastern United States and in Western Europe. The amount being deposited depends on the concentration in rainwater and on the precipitation rate. The impact of the latter is illustrated by the high rate of SO_4^{2-} deposition over India. It is not yet clear how much of the deposition there is due to marine sulfate carried inland with the monsoon rains. The contribution may be significant.

The data in Fig. 10-7 do not cover the southern hemisphere. Granat *et al.* (1976) discussed some sporadic measurements in Africa and in Australia. The results are shown in Table 10-15, together with more recent data obtained by Stallard and Edmond (1981) and by Galloway *et al.* (1982) in South America. Most of the deposition rates reported are of the order of 0.1–0.2 g S/m^2 yr. Although the data are selective, they are consistent with an extrapolation of values shown in Fig. 10-7 and presumably represent the natural background. The total deposition of SO_4^{2-} on all continents can be estimated by summation within individual latitude bands. This gives 55–60 Tg S/yr. The value should be corrected downward, however, because precipitation collectors at most stations are left open continuously during sampling periods, so that dry deposition of aerosol particles and scavengeable gases adds to the total quantity of material accumulating. Georgii *et al.* (1983) have used automated dry only/wet only samplers in a 2-yr study at a dozen stations in Germany in order to determine the relative contribution of dry-deposited material. Sulfate collected during dry periods amounted to 9–29% of the total, depending on location. The average for all stations was 20%. If these results are typical, the global rate of wet sulfate deposition would reduce to about 45 Tg S/yr. Even this figure may be too high because it disregards the deposition of sulfur dioxide—either wet or dry—that is subsequently oxidized to sulfate in aqueous solution.

More significant than dry deposition onto flat terrain probably is the interception of aerosol particles by high-growing vegetation. The arguments are the same as those presented in Section 9.2.2 for particulate ammonium. Höfken *et al.* (1981) have studied the filtering action of forests in some detail. They found, for example, that the concentration of particulate sulfate is up to 35% lower in the air underneath the canopy than above it. Rainfalls rinse the dry-deposited material off the foliage, so that rainwater collected below the canopy is enriched with sulfate and other trace substances compared with rainwater collected in forest clearings. From the observed enrichments, Höfken *et al.* calculated ratios of dry to wet deposition rates for sulfate of 0.9 for beech and 4.1 for spruce trees. The corresponding ratios for ammonium were 0.4 and 0.8, respectively. Since in the polluted regions of Western Europen sulfate and ammonium are closely associated,

Table 10-15. *Sulfate in Precipitation on the Continents of the Southern Hemisphere*

Authors	Location	Concentration (mg S/liter)	Deposition rate (g S/m² yr)	Remarks
Africa				
Hesse (1957)	Kiyuhu	—	0.1	
Visser (1961)	Kampala, Uganda	1.8	2.28	$\bar{R} = 1300$ mm, 10 out of 59 samples less than 0.1 mg/l
Eriksson (1966)	Zaire	—	0.47	Average from 3 sites
Bromfield (1974)	Northern Nigeria	0.13	0.114	Average from 11 sites
Australia				
Hingston (1958)		—	0.13–0.7	As reported by Eriksson (1960)
Galloway et al. (1982)	Katherine	0.1	0.092	$\bar{R} = 916$ mm
South America				
Stallard and Edmond (1981)	Amazon region	0.16	—	Average from 8 sites
Galloway et al. (1982)	San Carlos, Venezuela	0.05	0.19	$\bar{R} = 3914$ mm

10.4 Tropospheric Sulfur Budgets

The circulation of sulfur through the atmosphere has so far eluded a satisfactory description. The difficulties arise from the very inhomogeneous concentration patterns of many sulfur compounds in the atmosphere, the problems of quantifying dry and wet deposition rates, and the uncertainties connected with biogenic source strengths. With regard to the latter aspect, some clarifying progress has now been made. In view of the short residence times of all sulfur compounds except COS, the local behavior of atmospheric sulfur can be described only by regional budgets. These may then serve as building blocks in the construction of the global sulfur cycle. Regional budgets have been discussed by Rhode (in Granat *et al.*, 1976) and Mészarós *et al.* (1978) for Western Europe, and by Galloway and Whelpdale (1980) for the northeast United States. Ryaboshapko (1983) presented estimates for the atmospheric sulfur balance in polluted, clean continental, dusty continental, and marine environments, and Kritz (1982) has worked out a local budget for the marine atmosphere. The following discussion deals first with the flux pattern for sulfur in the polluted atmosphere, then with the unperturbed marine atmosphere, and finally with the global sulfur budget.

10.4.1 THE REGIONALLY POLLUTED CONTINENTAL ATMOSPHERE

Figure 10-8 presents a flux diagram for the disposal of sulfur from anthropogenic sources in the continental atmosphere of the northern hemisphere. Biogenic and volcanic emissions are comparatively minor and are ignored. More substantial may be aeolian sources of particular sulfate, but these are also neglected. The troposphere is subdivided into urban, regionally polluted, and remote regions. The corresponding ground-level concentrations are indicated in Fig. 10-8, in addition to the scale heights in the lower atmosphere, which were estimated from the vertical profiles shown in Fig. 10-6. The flux scheme starts with the injection of SO_2 and smaller amounts of particulate SO_4^{2-} into the urban atmosphere. This is an oversimplification, of course. Transport then carries SO_2 and SO_4^{2-} into the regionally polluted atmosphere, and from there into remote regions of the continents. In each region, SO_2 is partially converted to SO_4^{2-}, and both compounds are removed by wet and dry deposition. Wet deposition rates for SO_2 make allowance for the enhancement resulting from the interaction of bisulfite with formaldehyde in clouds (see Section 8.4.1). The dry deposition velocity is about 8 mm/s, according to Table 1-12. The rate of wet deposition of sulfate is determined primarily by the residence time of aerosol particles due to rainout, approximately 5 days. In the polluted atmosphere the rate should be slightly higher because of below-cloud scavenging of particles by falling rain drops. The rate of dry deposition of particulate sulfate is

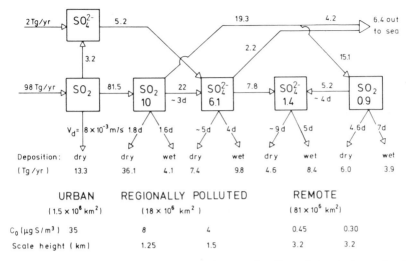

Fig. 10-8. Flux diagram for the disposal of anthropogenic sulfur emissions in the continental troposphere of the northern hemisphere. Fluxes are given in units of Tg S/yr. Numbers on boxes indicate column densities in units of mg S/m². They were derived from the adopted ground-level concentrations and scale heights below 5.5 km. Above this level, mixing ratios are assumed constant with $m(SO_2) = 65$ ng S/m³ and $m(SO_4^{2-}) = 70$ ng S/m³. Chemical conversion of SO_2 to SO_4^{2-} occurs only in the lower troposphere. Dry deposition velocities are 8 mm/s for SO_2, 0.15 mm/s for SO_4^{2-} on flat terrain, and 5 mm/s for SO_4^{2-} interception by forests, which are assumed to occupy 50% of the continental area.

uncertain. The estimate in Fig. 10-8 is based on the process of particle interception by forest canopies.

One conclusion derived from the box model of Fig. 10-8 is the great importance of dry deposition of SO_2, which takes up about 50% of the budget in urban and polluted regions. Rhode (in Granat *et al.*, 1976), Garland (1977, 1978), Mészarós *et al.* (1978), and Eliassen (1978) have reached similar conclusions in conjunction with considerations of the sulfur budget of Western Europe. Their estimated percentages for the contribution of dry deposition were 22-45, 45, 43, and 55, respectively. These authors also estimated losses of sulfur from the European territory by advection to other regions as 21-43, 37, 9, and 17% of the total sulfur budget. In Fig. 10-8, the flux of sulfur due to transport from the polluted to the remote atmosphere amounts to 22.9% of the total input. Another 6.4% is carried out to sea. While a flux of 6.4 Tg S/yr represents a relatively small fraction of anthropogenic sulfur, it is substantial compared with marine emissions from the biosphere, which are of the order of 10-15 Tg S/yr in the northern hemisphere.

The box model of Fig. 10-8 further predicts a wet deposition rate for SO_2 and sulfate in the polluted regions amounting to 0.77 g S/m^2 yr, or 14% of total anthropogenic sulfur. The former value compares well with estimates for Europe; the latter is much lower. The authors listed above reported average wet deposition rates of 0.97, 0.67, 1.1, and 0.62 g S/m^2 yr. The corresponding percentages of their total budgets are 32, 19, 48, and 27, respectively. The fluxes were in most cases derived from actual measurements of sulfate in rainwater collected at various stations of European networks. The reasons for the disparate relative contributions of wet sulfur deposition to the total budgets are not obvious. It is of interest, however, to note that dry deposition of particulate sulfate in the model presented here is almost as significant as the wet deposition of sulfate.

10.4.2 THE CIRCULATION OF SULFUR IN THE UNPERTURBED MARINE ATMOSPHERE

The transformation of reduced sulfur compounds to SO_2 and excess sulfate in the marine atmosphere can be quantified by means of a steady-state, local circulation model. An appropriate flow diagram is shown in Fig. 10-9. The necessary elements for constructing this model have been assembled by Bonsang et al. (1980) and by Kritz (1982). The column densities for SO_2 and excess particulate sulfate are based on the observed concentrations presented in Section 10.3.4, and that for H_2S is estimated from the data in Table 10-1. The flux of dimethyl sulfide emanating from the ocean surface was discussed in Section 10.2.1. The lifetime of H_2S is given in Table 10-2, and the pathways for the oxidation of dimethyl sulfide are apportioned according to the laboratory data of Niki et al. (1983) and Hatakeyama and Akimoto (1983). Methanesulfonic acid and SO_2 account for 75% of sulfur-bearing products. The fate of the remaining 25% is not precisely established. Therefore, it is indicated in Fig. 10-9 as "unknown." Methanesulfonic acid is assumed to be associated entirely with aerosol particles and to undergo oxidation to sulfate in the condensed phase. The individual fluxes follow from the rates for the deposition of SO_2 and excess sulfate to the ocean surface. A dry deposition velocity for SO_2 of 5 mm/s is adopted according to Liss and Slater (1974). The rate for the wet deposition of SO_2 is based on the interaction of bisulfite with formaldehyde in clouds, which leads to a residence time of 9 days due to rainout. Excess sulfate is assumed to have an atmospheric residence time equal to that of the normal background aerosol.

The oxidation of H_2S and DMS and the conversion of SO_2 to SO_4^{2-} occur primarily in the boundary layer. SO_2 in the free troposphere is comparatively stable. The flux of H_2S is small relative to that of DMS, so that H_2S

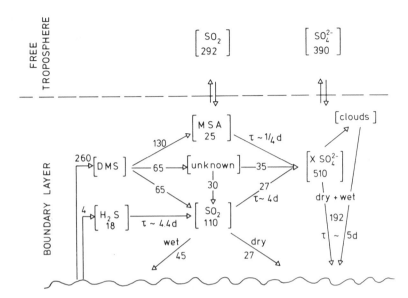

Fig. 10-9. Flux diagram for sulfur in the unperturbed marine atmosphere. Fluxes are given in units of μg S/m² day. Numbers in boxes indicate column densities in units of μg S/m². DMS, Dimethyl sulfide; MSA, methane sulfonic acid (associated with the aerosol). The mixing ratio of SO_2 is 60 ng S/m³, independent of altitude. The mixing ratio of SO_4^{2-} is 280 ng S/m³ in the boundary layer and 80 ng S/m³ in the free troposphere. Contrary to the model of Kritz (1982), the fluxes are confined to the boundary layer. There exists no significant net flux into or out of the free troposphere. The dry deposition velocity for SO_2 is 5 mm/s.

contributes little to the formation of SO_2 and excess sulfate. The time constant chosen for the conversion of SO_2 to SO_4^{2-} is similar to that in the continental atmosphere. In order to balance the fluxes of sulfur through the reservoir of SO_2, it is then necessary that a certain portion of the product designated "unknown" becomes oxidized to SO_2. In the flux scheme of Fig. 10-9, roughly one-half of the unknown product is converted to SO_2, but the fraction depends on the rate at which SO_2 is further oxidized to particulate sulfate. Owing to the deposition of SO_2, about 27% of the total sulfur flux is diverted from reaching the reservoir of excess sulfate.

The turnover of sulfur in the boundary layer is much faster than the exchange of air with the upper troposphere. Vertical mixing, however, is still faster than interhemispheric air exchange, so that the sulfur cycles in the two hemispheres are essentially decoupled. In the model depicted in Fig. 10-9, SO_2 and SO_4^{2-} are allowed to communicate freely with the upper troposphere, but the exchange is not considered to result in significant net fluxes in either upward or downward direction. Kritz (1982) has postulated nonnegligible downward fluxes for both SO_2 and SO_4^{2-}. These would have to be maintained by horizontal transport in the upper troposphere. As discussed previously, there is no evidence for a significant horizontal gradient of the SO_2 mixing ratio in the upper troposphere. For excess SO_4^{2-}, a comparison of the data in Tables 10-12 and 10-14 indicates higher mixing ratios near the ocean surface compared with the upper troposphere, so that an upward rather than a downward flux due to turbulent mixing exists. The upward flux is needed to bring SO_4^{2-} to the cloud level, from where its removal by rainout takes place. In the free troposphere, the fraction of SO_2 and SO_4^{2-} that has penetrated the cloud layer intermingels with SO_2 and SO_4^{2-} of continental origin. Both marine and continental sources thus contribute to the general background of sulfur in the troposphere.

10.4.3 THE GLOBAL SULFUR BUDGET

To provide a historic perspective, we present in Table 10-17 an overview on atmospheric sulfur budgets derived by various authors. The early investigators had no solid information on biogenic sources. Their magnitude was inferred from the imbalance between estimated deposition rates and emissions, which were considered relatively well known, primarily SO_2 from fossil fuel combustion and sulfate in seasalt. As more observational data became available, the deposition rates were revised downward and the inferred volatile emissions from the biosphere decreased accordingly. Granat et al. (1976) were the first to adopt an estimate (by Hitchcock, 1975) for sulfur emissions from soils, but the marine contribution to excess sulfate still had to be determined by difference. It is noteworthy that the outcome of 27 Tg S/yr compares well with recent estimates obtained more directly from measurements (26–50 Tg S/yr). In spite of the agreement, the budget presented by Granat et al. (1976) cannot be accepted without revision because of the reassessment by Cullis and Hirschler (1980) of human-made sources. Ryaboshapko (1983), who incorporated the change in his budget, included also a greater production rate for sulfur in seaspray than Eriksson's (1960) original estimate. The need for a larger flux of seasalt is confirmed by the data discussed in Section 10.3.5. Möller's (1984) interest was mainly to expose the great uncertainties inherent in all models that estimate the contribution from biogenic sulfur sources by balancing the global budget.

Table 10-17 includes a global atmospheric sulfur budget based on the emission estimates discussed in this chapter and the flux diagrams shown in Figs. 10-8 and 10-9. The marine budget of 36 Tg S/yr supplied by the biosphere must be augmented by about 6.8 Tg S/yr from anthropogenic sources. In addition, about one-half of the sulfur from volcanic emissions

Table 10-17. *Global Atmospheric Sulfur Fluxes (Tg S/yr) as Proposed by Several Authors*

Type of source or sink	Eriksson (1960)	Robinson and Robbins (1970b)	Kellog et al. (1972)	Friend (1973)	Granat et al. (1976)	Ryaboshapko (1983)	Möller (1984)	Present
Gaseous emissions								
Anthropogenic emissions	39	73	50	65	65	101 (84–118)[b]	75	103[b]
Biogenic emissions, total	267	98	90	106	32	41 (8–77)	70	43
From oceans	190[a]	30[a]	—	48[a]	27[a]	24 (3–45)	35	36
From soils	77	68	—	58	5	17 (5–32)	35	7
Volcanic gases	—	—	1.5	2	3	28 (14–42)	2	7
Particulate sulfate emissions								
Seaspray	44	44	44	44	44	140 (77–203)[c]	175	150
Mineral dust	—	—	—	—	0.2	20 (10–30)	—	?
Anthropogenic SO$_4^{2-}$	—	—	—	—	—	12 (10–14)	—	3
Wet and dry deposition								
SO$_2$ over the oceans	70	25	[d]	25	16[e]	11 (5–17)	30	15
SO$_2$ over the continents	77	26	15	15	44[e]	17 (7–27)	—	71
Excess SO$_4^{2-}$ over oceans	102	31	29	31	17[e]	107 (70–145)	117	28
Excess SO$_4^{2-}$ onto continents	57	90	96	102	23[e]	67 (44–92)	—	42
Seasalt over oceans	40	40	44	40	41	125 (69–180)	157	135
Seasalt onto continents	4	4	—	4	3	15 (8–23)	18	15
Sum of fluxes								
Total budget	350	215	185	217	144	342 (203–484)	322	306
Flux from continent to ocean	—	26	—	8	17	100 (50–150)	—	11
Flux from ocean to continents	15	4	—	4	15	20 (10–30)	—	18

[a] Resulting as a balance in the cycle.
[b] Includes 3 (1–5) Tg S/yr of reduced sulfur.
[c] From the balance of fluxes in the marine atmosphere.
[d] Included in excess SO$_4^{2-}$.
[e] Not originally presented in this form, but calculated from the data given.

is assumed to be directed toward maritime regions. Rates for the deposition of SO_2 and its conversion to SO_4^{2-} are apportioned according to the flux distribution in Fig. 10-9. The resulting deposition of SO_2 is 15 Tg S/yr, and the rate of excess SO_4^{2-} formation over the oceans is 31 Tg S/yr. Since marine excess sulfate is associated with seasalt aerosol, of which 10% is deposited on land, 3 Tg S/yr of excess sulfate likewise is deposited on the continents. This leaves 28 Tg S/yr to return to the oceans. The strength of biogenic emissions from soils excludes carbonyl sulfide, which is long-lived and enters into a separate cycle. In the present context, the contribution of about 1 Tg S/yr of COS may be neglected. According to the data in Table 10-9 and the distribution of continental soil types, one estimates that 25% of the emissions that do not lead to COS consist of dimethyl sulfide. One-half of DMS is oxidized to methanesulfonic acid and then to sulfate; the other half produces mainly sulfur dioxide. H_2S and 50% of CS_2 are converted directly to SO_2. The source strength of biogenic sulfur from soils thus leads to 1 Tg S/yr of sulfate via methanesulfonic acid and 6 Tg S/yr of SO_2. Its conversion to sulfate and the deposition of both SO_2 and SO_4^{2-} is assumed to follow the flux pattern for the remote continental atmosphere shown in Fig. 10-8. This results in a total deposition of SO_2 on the continents, including that from anthropogenic and volcanic sources, of 71 Tg S/yr. It must be increased by 3 Tg S/yr from marine excess sulfate transported toward the continents.

The present budget differs from that of Ryaboshapko (1983) mainly by the greater deposition rate for SO_2 over the continents, and a correspondingly smaller fraction of continental sulfur transported toward the oceans. The dominant factor causing the difference is the extent of dry deposition of SO_2 in the urban and regionally polluted atmosphere. Ryaboshapko (1983) assumed for the polluted regions of the world an area of $12 \times 10^6 \, km^2$ and derives a flux for the dry deposition of SO_2 of 12 ± 7 Tg S/yr. The present estimate, in comparison, is based on an area of $19.5 \times 10^6 \, km^2$, in which roughly 50 Tg S/yr of SO_2 undergoes dry deposition. If dry deposition of SO_2 undergoes dry deposition. If dry deposition of SO_2 were overestimated, more SO_2 would have to leave the continental and enter the marine atmosphere. A global transport model based on the actual average wind field may clarify the situation.

11 Geochemistry of Carbon Dioxide

11.1 Introduction

Atmospheric carbon dioxide is chemically quite inert except at high altitudes, where it is subject to photodissociation. At the Earth's surface, CO_2 occurs dissolved in the oceans and in the surface waters of the continents. Here, it partakes in several geochemically important reactions, such as the weathering of rocks and the formation of limestone deposits. Another important process involving CO_2 is its assimilation by plants. Carbon is a key element of life, and atmospheric CO_2 provides the principal source of it. The consequence of all interactions is a complex system of carbon fluxes connecting a number of well-differentiated geochemical reservoirs: the atmosphere, the biosphere, the oceans, and the sediments of the Earth's crust. Reservoirs and fluxes combined describe the geochemical carbon cycle. The atmosphere cannot be isolated from the rest of the system, because the abundance of CO_2 in air is governed by the behavior of the other reservoirs to which the atmosphere is coupled and by the associated exchange processes. To elucidate what controls atmospheric CO_2 thus requires a detailed discussion of the major reservoirs and their interactions.

The current great interest in carbon dioxide derives from the observed rise of its concentration in the atmosphere and a growing concern about the prospect of climatic changes if the trend continues. It was Callendar

(1938) who first noted the increase since the turn of the century, but systematic measurements of the trend were not begun until 1957. Since then, a unique record of atmospheric CO_2 mixing ratios has been obtained at Mauna Loa, Hawaii, and somewhat less extensively at the South Pole (see Fig. 1-2). These data document the annual increase of CO_2 over a period of at least 20 yr. Following Callendar, the effect has generally been attributed to the combustion of fossil fuels by humans. Only recently has it been recognized by Bolin (1977) and others that the destruction of forests accompanying the expansion of arable land areas during the past 100 yr must have liberated additional nonnegligible quantities of CO_2 to the atmosphere.

As an infrared-active molecule, CO_2 assumes a significant role in determining the heat balance of the atmosphere. By intercepting thermal radiation emitted from the Earth's surface, CO_2 raises the temperature in the troposphere (the so-called greenhouse effect), while at the same time it serves as a cooling agent in the upper atmosphere by radiating heat away toward space. At the turn of the century, Arrhenius (1896, 1903) estimated that the surface temperature would rise by 9°C if the abundance of atmospheric CO_2 were tripled. Recent studies, all carried out with the aid of high-speed electronic computers, have revealed the enormous complexity of quantifying the effect. Thus, Plass (1956) provided the first realistic model for the 15-μm absorption band of CO_2; Gebhart (1967) included the absorption of solar radiation in the near-infrared; Möller (1963) took into account the overlap of H_2O and CO_2 absorption and the feedback by the increase of water evaporation due to the rise in surface temperature (a very significant effect, since water vapor is the main infrared absorber), and Manabe and Weatherald (1967) considered the convective readjustment of the troposphere accompanying the change in heat fluxes. The most comprehensive effort to date is that of Manabe and Weatherald (1975, 1980), who developed a three-dimensional circulation model incorporating radiative, convective and advective heat transport as well as the water-vapor balance resulting from evaporation and precipitation. The authors found a temperature rise of 2.9 K, on average, when the present CO_2 level is doubled, with a stronger warming in the polar regions. Similar one-dimensional radiative–convective models reviewed by Schneider (1975) and by Ramanathan and Coakley (1978) gave a temperature increase of 2–3 K, depending on assumptions concerning the cloud-top behavior.

11.2 The Major Carbon Reservoirs

In describing the natural carbon cycle, it will be useful to commence with a characterization of the major reservoirs. Table 11-1 presents an overview.

Table 11-1. *Geochemical Carbon Reservoirs*; *Mass Contents in Pg C* (10^{12} *kg*)

Reservoir	Mass content of carbon	Remarks and references
Atmosphere		
Present level	7.0×10^2	$m(CO_2) = 330$ ppmv
Preindustrial	6.15×10^2	Assuming $m(CO_2) = 290$ ppmv
Oceans		
Total dissolved CO_2	3.74×10^4	Bolin *et al.* (1981); average concentration, see Table 11-2
Dissolved CO_2 in mixed layer	6.70×10^2	Depth of mixed layer 75 m; Bolin *et al.* (1981)
Living biomass carbon	3	Mainly plankton; Mopper and Degens (1979)
Dissolved organic carbon	1.0×10^3	Average concentration 0.7 g/m^3; Williams (1975)
Sediments		
Carbonates, continental and shelf	2.7×10^7	Carbonates: total of 5×10^7 as given by Garrels and Perry (1974) subdivided according to Hunt (1972); for accounting see Table 11-3
Carbonates, oceanic	2.3×10^7	
Organic carbon, continent and shelf	1.0×10^7	
Organic carbon, oceanic	0.2×10^7	
Biosphere		
Terrestrial biomass carbon	6.5×10^2	Includes deadwood and plant litter; Ajtay *et al.* (1979)
Soil organic carbon	2×10^3	Somewhat uncertain; estimate from Ajtay *et al.* (1979)
Oceanic organic carbon	1×10^3	From value above for oceans, dissolved organic carbon

We shall have to differentiate between organic (or reduced) carbon and inorganic (or oxidized) carbon. The former includes living biomass as well as biological decay products such as plant debris, soil humus, and metamorphic organic compounds. The latter refers to CO_2, bicarbonate, and carbonate.

11.2.1 CARBON DIOXIDE IN THE ATMOSPHERE

In 1978, the mixing ratio of CO_2 was 329 ppmv, on average, and it has been rising further since. The distribution of CO_2 in the atmosphere is fairly uniform, so that we can immediately calculate the total atmospheric content as 2570 Pg (1 Pg $= 10^{15}$ g). The corresponding mass of carbon is 700 Pg. The contribution due to other carbon-containing compounds can be neglected.

The preindustrial level of atmospheric CO_2—that is, the level that must have existed in the years before 1860—was 290 ppmv or less. The precise value is somewhat in doubt. Information on this matter can be obtained in several ways. One is to assess the reliability of measurements made during the second part of the last century; another involves a back-extrapolation of CO_2 values based on an estimate of total fossil fuel consumption coupled with auxiliary assumptions on the fraction that has remained in the atmosphere; and a third method relies on CO_2 preserved in ancient ice samples from glaciers, etc. The first method was applied by Callendar (1958) and by Bray (1959). Their analyses gave 290-295 ppmv. The second approach makes use of the data assembled by Keeling (1973a) and Rotty (1981, 1983) for the integrated amount of CO_2 released from combustion and cement manufacture since 1860. The cumulative input by the end of 1977 was 150 Pg. Bacastow and Keeling (1981) have also shown that the increase of atmospheric CO_2 for the period 1958-1978 corresponds to 54% of the cumulative emissions from fossil fuel combustion during the same period. If one assumes that this fraction has stayed constant for the last hundred years, one obtains the preindustrial carbon content of the atmosphere by subtraction: $700 - (0.54 \times 150) = 619$ Pg, which is equivalent to 292 ppmv. This calculation does not take into account the CO_2 released by forest clearings and must be considered an upper limit. Neftel *et al.* (1983) have studied gas inclusions in Greenland ice cores and report a value of 271 ± 9 ppmv for the CO_2 level of the atmosphere 600 yr ago.

Most ground-based measurements of CO_2 in air reveal diurnal and seasonal variations due to the assimilation and respiration of CO_2 by land biota [see, for example, Figs. 1-1 and 1-2, and Fig. 5 in Junge (1963)]. The diurnal variations are rapidly dampened with height above the ground, whereas the seasonal variations permeate the entire troposphere and disappear not before one transcends the tropopause. Bischof (1977, 1981) has collected an extensive set of data on tropospheric CO_2 variations from measurements by aircraft in the northern hemisphere. Part of the data has been analyzed by Bolin and Bischof (1970), who found them consistent with our notions about the large-scale mixing processes in the atmosphere, in particular with a vertical exchange time of about 1 month. In the southern hemisphere, the amplitude of the seasonal variation is smaller than in the northern hemisphere and the oscillations exhibit a phase shift by 6 months. In qualitative terms, both observations are readily understood by the decoupling of the two hemispheres, by the smaller land area available for the support of assimilating land biota in the southern hemisphere, and by the displacement of the seasonal cycle by half a year. The quantitative aspects will be discussed in Section 11.3.3.

11.2.2 THE OCEANS

The total volume of the world ocean is 1.35×10^{18} m^3. The value is obtained by combining the areal extent of the oceans with the mean depth of 3730 m, derived by Menard and Smith (1966) from hypsometric charts of all ocean basins.

The vertical structure of the ocean features a decrease of temperature with increasing depth and a correspondingly stable stratification. A shallow surface layer of 50–100 m thickness is vertically well mixed due to agitation by wind force. This portion represents only a small subvolume, but it is of crucial importance to the exchange of CO_2 with the atmosphere. Compared with the bulk of the ocean, the mixed layer responds quickly to changes in the atmosphere and it must be treated as a separate reservoir. The depth of the mixed layer is variable. We adopt the recommendation of Bolin *et al.* (1981) and use a value of 75 m. This is a seasonal average obtained by Bathen (1972) from measurements in the Pacific Ocean.

Intermediate between the mixed layer and the deep sea, at depths between 100 and 1000 m, lies the main thermocline, a region where mixing is imperfect and exchange with the atmosphere is slow but nevertheless faster than in the denser and cooler waters of the deep ocean. At high latitudes, a mixed layer often cannot be discerned and the surface waters mix directly with the deeper strata. Circulation in the main body of the ocean is accomplished by the downward motion of cold surface waters near the poles and a slow updrift at low latitudes. Estimated rates of downflow are 30×10^6 m^3/s as given by Munk (1966) and Gordon (1975) for the Antarctic, and $(10–30) \times 10^6$ m^3/s as derived by Broecker (1979) for the North Atlantic. The resulting turnover time of deep ocean waters is 700–1000 yr.

The total amount of carbon dioxide dissolved in the ocean is known from measurements at various locations in the three main ocean basins and the Antarctic. Takahashi *et al.* (1980, 1981a,b) have provided summaries of recent data. Measurement techniques involved infrared gas analysis and automated potentiometric titration by the Gran (1952) procedure, the latter in conjunction with the determination of total alkalinity. Both methods gave concordant results. Inconsistencies noted by Broecker and Takahashi (1978) in the data from the Pacific and Indian Oceans were later traced to the nonideal behavior of the glass electrodes used in the titration, and appropriate corrections were applied. Table 11-2 lists mean values of total CO_2 thus obtained for three depth ranges in seven areas of the world ocean. Figure 11-1 is added to show the average distribution with depth together with other data of interest. The average concentration of CO_2 in the surface layer is 2.0 mmol/kg, or 2.05 mol/m^3 if one adopts a value of 1025 kg/m^3 for the

Table 11-2. *Arithmetic Means of Total CO_2 Concentrations ($\mu mol/kg$) in Seawater in Three Depth Ranges of Seven Ocean Areas, and Averages for the World Ocean (Takahashi et al., 1981a)[a]*

Depth range	North Atlantic	South Atlantic	North Pacific	South Pacific	North Indian	South Indian	Antarctic	Average
0–50 m	1944±25	1961±34	1985±77	1971±12	1933±13	1936±25	2172±49	2002
50–1200 m	2124±96	2142±94	2227±140	2146±112	2211±83	2160±109	2230±47	2182
>1200 m	2181±21	2212±38	2356±32	2309±31	2330±16	2298±17	2263±12	2288
Percent area	11.9	9.7	21.9	21.4	3.4	13.6	18.2	—
Whole ocean	(Volume-weighted arithmetic mean)							2,254

[a] The data are normalized to a mean salinity of the world ocean of 34.78‰.

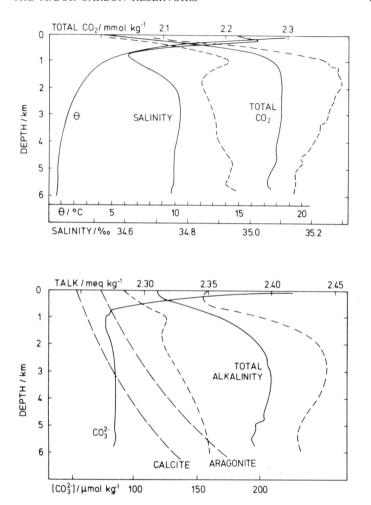

Fig. 11-1. Global mean conditions in the world ocean for the following quantities: potential temperature θ (i.e., temperature corrected for adiabatic heating), salinity, concentrations of total CO_2 and CO_3^{2-}, and total alkalinity. [Adapted from Takahashi *et al.* (1981a).] Dashed curves indicate the spread of total CO_2 and alkalinity; long-dashed curves show the critical dissolution regions for calcite and aragonite according to Broecker and Takahashi (1978).

density of seawater. The total mass of inorganic carbon in the mixed layer then amounts to 670 Pg, which is approximately the same as that residing in the atmosphere. The average concentration of dissolved CO_2 for the entire ocean of 2.25 mmol/kg leads to a total mass content of 37,400 Pg for inorganic carbon.

The dissolution of CO_2 in seawater involves a set of ion equilibria, which can be represented by the sequence

$$CO_{2gas} \underset{H_2O}{\overset{K_H}{\rightleftharpoons}} CO_{2aq} \underset{-H^+}{\overset{K_1}{\rightleftharpoons}} HCO_3^- \underset{-H^+}{\overset{K_2}{\rightleftharpoons}} CO_3^{2-} \underset{Ca^{2+}}{\overset{K_s}{\rightleftharpoons}} CaCO_{3s}$$

The first three relations were discussed in Section 8.4 in conjunction with the dissolution of CO_2 in cloud water. The last reaction is specific to conditions existing in the oceans and leads to solid calcium carbonate as a precipitate. For a concentrated salt solution such as seawater, the effects of complexing and other ionic interactions cannot be neglected and ion activities must be inserted in the equilibrium relations [Eq. (8-15)] rather than concentrations. Thus, HCO_3^- is found partially complexed with Na^+, and a major portion of CO_3^{2-} occurs complexed to Mg^{2+}. A review of the data has been given by Stumm and Brauner (1975). A rigorous description of all ion equilibria, however, is not required to quantify the carbon dioxide system. For practical purposes it suffices to incorporate the various effects into the equilibrium constants and to work with apparent values of K_H, K_1, and K_2. The relative concentrations of CO_{2aq}, HCO_3^-, and CO_3^{2-} resulting from Eq. (8-15) then refer to the stoichiometric concentrations of the free ions plus those bound to other ionic species. The effective equilibrium constants depend on the salinity (salt content) of seawater in addition to temperature, and to some extent also on pressure. Edmond and Gieskes (1970), Disteche (1974), and Skirrow (1975) have reviewed the data and tabulated them. The salinity of seawater is close to 35‰ and the pH value lies in the vicinity of pH = 8. One then finds that about 0.6% of dissolved carbon dioxide is present as CO_{2aq}, 90% as bicarbonate, and 9% as carbonate ions. The amount of suspended $CaCO_3$ is small by comparison.

From equilibrium constants and observations of total CO_2, temperature, pressure, salinity, and alkalinity, Takahashi et al. (1981a) calculated the concentration of CO_3^{2-} as a function of depth at various locations. Averaged values are shown in Fig. 11-1. It is apparent that the concentration of CO_3^{2-} decreases with depth, although that of total CO_2 increases. The behavior is due mainly to the decrease in temperature. Figure 11-1 shows also the critical dissolution curves proposed by Broecker and Takahashi (1978) for the two principal crystalline forms of $CaCO_3$ in the oceans: calcite and aragonite. The solubility of $CaCO_3$ increases with depth because of the decrease of temperature and the increase of pressure. Near the surface, CO_3^{2-} is found to be supersaturated relative to calcium carbonate, whereas at greater depths undersaturation occurs.

Our knowledge of carbonate sedimentation has been reviewed by Cloud (1965), Bathurst (1975), and Holland (1978). Despite the apparent supersaturation of CO_3^{2-} in the upper strata of the sea, there is little evidence for the direct, inorganic chemical precipitation of $CaCO_3$ except in warm and shallow waters of the tropics. Today as well as during the recent history of the Earth, the predominant mode of $CaCO_3$ production takes place by shell-forming organisms such as protozoa (foraminifers), algae (coccolithophores), mollusks, corals, etc., all residing in the sunlit surface waters. The majority of carbonate sediments are formed in the nutrient-rich shelf regions together with deposits of detritus of continental origin advected by the rivers. In the open sea, the calcareous remains of organisms, while settling toward the ocean floor, enter eventually into the region of undersaturation and become subject to dissolution. The rate of dissolution increases with depth until at a certain critical level, called the carbonate compensation depth, the rate of dissolution equals the influx of solid from above. Below this level calcium carbonate disappears rapidly. According to Gieskes (1974), as much as 80% of $CaCO_3$ precipitated redissolves in the deep ocean. Only those skeletal remains that have undergone transformations in the intestines of predators and that as fecal pellets are shielded by an organic coating can partially escape dissolution and reach deeper strata. From sedimentary evidence assembled by Pytkowicz (1970) and Bathurst (1975), the carbonate compensation level has been shown to occur at depths near 5000 m, although as a kinetic boundary its position must be quite variable.

Organic carbon in the ocean has been the subject of detailed reviews by Menzel (1974), Williams (1975), Parsons (1975), and Mopper and Degens (1979). It is customary to distinguish particulate and dissolved organic carbon on the basis of filtration with a 0.45-μm filter. For both fractions the concentrations fluctuate widely in varius parts of the ocean. In addition, there exist discrepancies between different analytical techniques, so that estimates of the average content of organic carbon in seawater involve large uncertainties. Concentrations reported for waters below the mixed layer range from zero to 120 mg/m^3 for particulate and 0.3–1.7 g/m^3 for dissolved organic carbon. Following Williams (1975), we adopt average concentrations of 20 mg/m^3 and 0.7 g/m^3, respectively, which give total contents of 30 and 10^3 Pg for the whole ocean. Living biomass is concentrated in plankton residing in the near-surface waters. Its mass has been estimated to amount to about 3 Pg, and that of bacteria is an order of magnitude lower.

The dominant source of organic carbon in seawater is the photosynthetic fixation of CO_2 by unicellular algae (phytoplankton) in the photic zone. Their growth by cell division is rapid, but the population is kept in balance by grazing species (zooplankton). DeVooys (1979) has discussed the state of the art for determining the rate of primary production of marine biomass. Recent estimates, all based on the take-up of radiocarbon, fall into the range of 23–80 Pg/yr with a probable value of about 35 Pg/yr. The contribution of kelps, rockweeds, and other macrophytes that grow in coastal regions and feature high local production rates is not globally significant.

Dissolved organic carbon appears to arise mainly from phytoplankton, both directly, through exudation, and indirectly, by the decay of dead cells. The chemical composition of the exudate is complex, and a full characterization is still wanting. Amino acids, carbohydrates, fatty acids, hydrocarbons, and aromatic compounds have been identified. A good deal of the material is scavenged by bacteria and other heterotrophs. The remainder appears to undergo polymerization, since the molecular weight has been found to increase with depth from values around 1500 in the mixed layer to more than 10,000 at depths of 5000 m (Mopper and Degens, 1979). By means of radiocarbon dating, Williams et al. (1969) estimated the average age of dissolved organic carbon in deep waters to be 3400 yr. The result must be considered a minimum age owing to the possibility of contamination by ^{14}C generated by atomic bomb tests. Nevertheless, the radiocarbon age provides an upper limit of 0.3 Pg/yr for the flux of dissolved organic carbon required to maintain a steady-state reservoir of 10^3 Pg in the deep ocean. The rate corresponds to about 0.7% of total primary production. Removal processes presumably include metabolic oxidation and conversion to particulate organic carbon, but the details of organic carbon recycling are uncertain. An even smaller fraction of primary production enters the sediments and becomes trapped there. Estimates for this fraction range from 10^{-4} to 4×10^{-3} (Garrels and Perry, 1974; Walker, 1974; Mopper and Degens, 1979), depending on location. Low values apply to deep sea sediments; higher values are associated with continental shelf regions. The rate of carbon deposition is difficult to measure, because again, much of the organic material entering the sediment–water interface is subject to microbial utilization and decays.

11.2.3   CARBON IN SEDIMENTARY ROCKS

The bulk of the earth's crust consists of igneous rocks representing primary material of magmatic origin. Exposure to the atmosphere and the action of surface waters causes a small fraction of crustal rocks to undergo weathering and erosion. The debris is washed into the continental and oceanic basins, where deposition, compaction, and diagenesis convert this secondary material into sedimentary rocks.

The total mass of the sediments that has been generated from igneous rocks since the formation of our planet amounts to approximately $G_{sed} = 2.4 \times 10^{21}$ kg, excluding volcanic intercalations, or about 8% of the total mass of the crust. The estimate can be derived in two ways. The direct

Table 11-3. *Distribution of Carbon in Sedimentary Rocks, Excluding Volcanogenics*

Sedimentary rock type	Mass (10^19 kg)	Distribution (%)	Carbonate carbon			Organic carbon		
			Weight percent	Mass (10^19 kg)	Distribution (%)	Weight percent	Mass (10^19 kg)	Distribution (%)
Continental shelf and slope[a]								
Clays and shales	83	59	0.84	0.70	21	0.99	0.82	83
Carbonates	25	18	9.40	2.35	69	0.33	0.08	8
Sandstones	32	23	1.05	0.35	10	0.28	0.09	9
Sum	140			3.40			0.99	
Oceanic[a]								
Clays and shales	34	40	0.68	0.23	8	0.22	0.07	33
Carbonates	35	42	7.90	2.76	91	0.28	0.10	48
Sandstones	15	18	0.28	0.04	1	0.26	0.04	19
Sum	84			3.03			0.21	
Total sediments[a]								
Clays and shales	117	52	0.79	0.93	14	0.76	0.89	74
Carbonates	60	27	8.50	5.11	80	0.30	0.18	15
Sandstones	47	21	0.83	0.39	6	0.27	0.13	11
Sum	224			6.43			1.20	
Total sediments[b]								
Clays and shales	178	75	0.42	0.75	15	0.76	1.35	88
Carbonates	36	15	11.4	4.10	81	0.30	0.11	7
Sandstones	26	11	0.71	0.18	4	0.27	0.07	5
Sum	240			5.03			1.53	

[a] From Hunt (1972); inorganic carbon contents originally from Ronov and Yaroshevskiy (1969, 1971).
[b] Using data given by Garrels and MacKenzie (1971); inorganic carbon contents originally from Clarke (1924), organic carbon contents taken from Hunt (1972).

method makes use of the observed spatial extent of continental sedimentary shields combined with seismic evidence about the depths of the deposits and deep drillings on the continents and in the oceans. Thus, Ronov and Yaroshevskiy (1969) obtained a mass of 2.24×10^{21} kg, excluding volcanogenics. An indirect estimate results from geochemical mass balances based on the redistribution of elements during the transformation of igneous to sedimentary rocks by weathering. In this manner, Garrels and MacKenzie (1971) calculate a mass of 2.34×10^{21} kg, whereas Li (1972) estimates 2.4×10^{21} kg.

The carbon content of igneous rocks (~ 100 ppm) is minor compared with that of the sediments. Table 11-3 shows representative data for the distribution of the three major sedimentary rock types: limestones, shales, and sandstones, and the weight percentages of carbon in each. Garrels and MacKenzie (1971) emphasized the difficulties encountered in estimating the relative proportions of limestones, shales, and sandstones for the totality of all sediments because of uncertainties about the individual contributions from Phanerozoic and Precambrian sediments. If we accept the mass ratios of $15:74:11$ that these authors derived from geochemical mass balances, we find 2.1% by weight of carbonate·carbon, on average, for all three sedimentary rock types. The resulting combined mass of inorganic carbon in the sediments is $0.021\,G_{sed} = 5 \times 10^{19}$ kg C. Using the data of Ronov and Yaroshevskiy (1969), Hunt (1972) gave a detailed breakdown of inorganic and organic carbon by sediment types. His summary is included in Table 11-3. The subdivision into oceanic and continental plus shelf sediments gives masses of 0.8×10^{21} and 1.4×10^{21} kg, respectively, with inorganic carbon contents of 3×10^{19} and 3.4×10^{19} kg. The combined total is 6.4×10^{19} kg C. It is interesting to note that Hunt's distribution of inorganic carbon between limestones, shales and sandstones, which is $80:14:6$, compares well with that computed from the data of Garrels and MacKenzie (1971), despite the differences in the ratios of sedimentary rock types adopted for the calculation.

For presentation in Table 11-1 we have selected a value of 5×10^{19} kg for the total mass of inorganic carbon in sediments. Here we have the largest pool of carbon, harboring an amount a thousand times greater than that in the atmosphere–ocean system. If all the carbonates now buried in the sediments were volatilized and the CO_2 were released to the atmosphere, the pressure would rise to 38 bar and the composition of the atmosphere would resemble that presently existing on the planet Venus. As will be considered in more detail in Chapter 12, it is now well accepted that the atmospheres of the terrestrial planets in the solar system originated from thermal outgassing of virgin planetary matter, water vapor and CO_2 providing the principal exhalation products. The difference in evolutionary trends between Venus and Earth arose from the more favorable conditions for the condensation of water on Earth, a prerequisite for the formation of an ocean and the deposition of carbonates.

As was pointed out earlier, the sediments also inherit a small fraction of reduced carbon as a residue of biological activity. Once trapped, the organic compounds eventually undergo stabilization by polymerization reactions and are diagenetically converted to a product called kerogen (Durand, 1980), a geopolymer defined by its insolubility in the usual organic solvents as opposed to soluble bitumens. The formation of kerogen must be viewed as a continuous disproportionation process going hand in hand with an increase in the degree of aromaticity and a progressive elimination of lighter organic compounds. Kerogen is essentially immobile, whereas the lighter fraction tends to migrate and may accumulate to form economically exploitable deposits.

Evidence for the biological origin of reduced carbon in sedimentary rocks comes from two sources. One is the existence of chemofossils, that is, characteristic remnants of biologically important compounds more resistant to chemical degradation than others. Notable examples are the isoprenoids pristane (2,4,6,10-tetramethyl pentadecane) and phytane (2,4,6,10-tetramethyl hexadecane), which arise from the decay of chlorophyll (Eglinton and Calvin, 1967; Didyk et al., 1978; McKirdy and Hahn, 1982; Hahn,

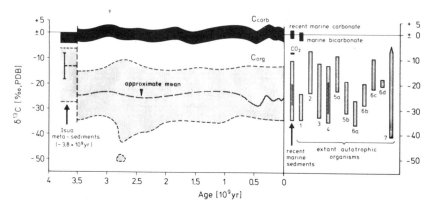

Fig. 11-2. Left: Geological record for the ^{13}C contents (in permille PDB standard) of carbonates (C_{carb}) and organic carbon (C_{org}) in the sediments. The spread of values is indicated. Right: Carbon-13 content in atmospheric CO_2, marine carbonates, organic material of modern sediments, and various organisms: (1) C_3 plants, (2) C_4 plants, (3) CAM (crassulacean acid metabolism) plants, (4) eukaryotic algae (black bar, range for marine plankton), (5) cyanobacteria from (a) natural communities and (b) culture experiments, (6) several nonoxygenic photosynthetic bacteria, and (7) methanogenic bacteria. [From Schidlowski (1984), with permission.]

1982). The second kind of evidence is obtained from the $^{13}C/^{12}C$ isotope ratio of sedimentary reduced carbon. Schidlowski (1983a) and Schidlowski *et al.* (1983) have reviewed this aspect in some detail. The various assimilatory pathways by which carbon enters the biosphere always favor the light ^{12}C isotope over the heavier ^{13}C carbon, so that the latter is somewhat depleted in organic compounds compared with the inorganic carbon reservoir. The $^{13}C/^{12}C$ ratio in the contemporary biosphere is 2–3% smaller than that of oceanic bicarbonate, which represents the largest pool of inorganic carbon accessible to the biosphere. In the sediments, the isotopic compositions of both organic and inorganic carbon are preserved with only minor alterations. A synopsis of $^{13}C/^{12}C$ data is given in Fig. 11-2 in the conventional notation expressing the isotope ratio of a sample by its permille deviation from that of a standard. The left-hand side shows the spread of ^{13}C values in sedimentary carbonates (C_{carb}) and reduced carbon (C_{org}) over geological times, the record dating back almost 3.8 billion yr into the past. The right-hand side gives recent ^{13}C values observed in various autotrophic organisms and the marine, atmospheric, and sedimentary environments. Figure 11-2 demonstrates that the $^{13}C/^{12}C$ ratio of carbonates is tied to the zero permille line, whereas that of organic carbon is clearly displaced toward negative values scattering around 25‰. The large spread must be taken to reflect variations in the isotope effect by individual organisms and environmental factors.

The constancy of the isotope shift is remarkable. It implies that biogenic organic carbon has been incorporated into the sediments since the earliest Precambrian when marine bacteria and blue–green algae were the sole agents of biological carbon fixation. This interpretation is supported by the morphological fossil record, which if one includes the microfossils and stromatolites (see page 623) extends back to 3.5 Gyr ago. The possibility of an abiotic formation of reduced carbon in the sediments has been considered, of course. Schidlowski *et al.* (1983) pointed out, however, that the magnitude of the observed iostope shift is hard to match by abiotic reactions, and a convincing reaction scheme yielding an effect of the required size has yet to be found.

Since organic and inorganic carbon in the sediments derive from a common source, namely, primordial carbon dioxide, the negative ^{13}C isotope shift of C_{org} must be compensated by a corresponding positive isotope shift of inorganic carbon relative to the $^{13}C/^{12}C$ ratio associated with geochemically undifferentiated primary carbon. This condition simply follows from the requirement of mass balance for ^{12}C and ^{13}C. Expressed in quantitative terms, one obtains an equation relating the ratio of organic to inorganic carbon in the sediments to the isotope shifts in both fractions. Using the conventional notation

$$\delta^{13}C = 10^3[(^{13}C/^{12}C)_{sample}/(^{13}C/^{12}C)_{stand} - 1]$$

in permille and setting $x = C_{org}/(C_{org} + C_{carb})$, the mass balance equation can be cast in the form

$$\delta^{13}C_{prim} = (1 - x)\delta^{13}C_{carb} + x\,\delta^{13}C_{org} \qquad (11\text{-}1)$$

a result which the reader may verify. Here, $\delta^{13}C_{prim}$ denotes the isotopic composition of the primordial carbon input to the system, commonly put as being close to $-5‰$. With $\delta^{13}C_{carb} = \pm0‰$ and $\delta^{13}C_{org} = -25‰$ one finds by solving the above equation that $x \approx 0.2$. The value implies a partitioning of total sedimentary carbon between C_{org} and C_{carb} in the ratio $C_{org}/C_{carb} = x/(1 - x) = 1:4$. The mass of reduced carbon in sedimentary rocks amounts to one-quarter of that of carbonate carbon, or 1.2×10^{19} kg. If both fractions are added, one obtains for the total mass of sedimentary carbon the value 6.2×10^{19} kg.

Reliable information on the content of reduced carbon in sedimentary rocks can also be derived from direct measurements. These have been reviewed and evaluated by Ronov and Yaroshevskiy (1969), Hunt (1972), and Schidlowski (1982). Table 11-3 presents distributions among sedimentary rock types as given by Hunt (1972). Continental clays and shales are found to contain the highest weight percentages of organic carbon with an average of about 1%, whereas the averages for carbonates and sandstones are about 0.2%. Marine sediments contain between 0.1 and 0.3%, except those in proximity to the continents. A reliable estimate of the average reduced carbon content in all sediments depends again on the proper selection of the frequency distribution of limestones, shales, and sandstones. Fortunately, errors are less severe here than for the carbonates, because shales not only represent the largest mass of sedimentary rocks but also contain the highest fraction of reduced carbon. The detailed breakdown of Table 11-3 leads to an average content of organic carbon in the sediments of 0.54–0.62%. The corresponding mass of reduced carbon in all sediments is $(1.3-1.5) \times 10^{19}$ kg. Only about 1% of it occurs in economically exploitable amounts as coal or petroleum.

The last result is remarkably close to that obtained in an entirely independent manner from the $^{13}C/^{12}C$ ratio. It may also be combined with the range of estimates for the total mass of inorganic carbon, $(5.0-6.4) \times 10^{19}$ kg, to derive for the ratio C_{carb}/C_{org} a value of 4 ± 0.8, in good agreement with that deduced from the ^{13}C and ^{12}C mass balance. In Table 11-1 a ratio of four was adopted and the mass of organic carbon was adjusted accordingly.

11.2.4 THE TERRESTRIAL BIOSPHERE

The last reservoir of carbon to be discussed is organic matter accrued by the terrestrial biosphere. Here again, we have to distinguish several subreservoirs. One comprises the living biomass of plants (and animals), another leaf litter, deadwood, and other debris, and a third soil humus. The following account is based largely on the detailed review of Ajtay et al. (1979). As in the marine environment, it is the photosynthetic process that provides the source of all organic matter in the terrestrial biosphere. Net primary productivity is defined as the rate of storage of organic material in plant tissue resulting from the uptake of CO_2 in excess of that released again by respiration. Net primary productivity frequently is given in terms of dry weight of organic matter accumulated per year, and a suitable conversion factor must be used to determine the mass of carbon thus fixed. The carbon content of biomass is variable but usually higher than that of pure hexose. For living biomass, a conversion factor of 0.45 is commonly employed. Plant litter and humus require the application of somewhat higher values, 0.5 and 0.6, respectively.

Methods for the assessment of terrestrial primary productivity and biomass were reviewed in Lieth and Whittaker (1975). The techniques include harvests from small sample plots, dimensional analysis and census of forest stands, growth relations between different plant tissues (for example, leaf or twig dry weight versus stem dry weight), and gas-exchange measurements. Primary productivity and biomass depend heavily on the type of vegetation considered, and on environmental factors such as water and nutrient supply, temperature, duration of the photoproductive period, etc. Estimation of the world's total productivity and biomass thus involves a classification of the biosphere into ecosystem types and a detailed accounting of land areas occupied combined with data characterizing the desired quantities for each individual ecosystem. Of the numerous estimates reported in the literature as summarized by Whittaker and Likens (1973), only the most recent ones are based on a sufficiently detailed differentiation of vegetation types to be considered reliable.

Table 11-4 compares two detailed estimates of global net primary productivity and biomass. The compilation of Whittaker and Likens (1973, 1975) mainly uses the classification and area assignment developed by Lieth (1975) and, as the authors emphasize, reflects the situation existing in 1950. The more recent assessment by Ajtay et al. (1979) followed the same principles but made use of newer data and a slightly more extensive subdivision according to ecosystem types. Concerning primary productivity, both estimates are in reasonable accord and are consistent also with earlier estimates of Olson (1970) and SCEP (1970). These data suggest a global productivity

Table 11-4. Comparison of Two Estimates for Terrestrial Living Biomass and Net Primary Production Rates

Ecosystem type	Land area[a] (10^12 m^2)	Mean biomass[a] (kg/m^2)	Mass of carbon[a] (Pg)	Land area[b] (10^12 m^2)	Mean biomass[b] (kg/m^2)	Mass of carbon[b] (Pg)	Net primary productivity[a] (kg/m^2 yr)	Net primary productivity[a] (Pg/yr)	Net primary productivity[b] (kg/m^2 yr)	Net primary productivity[b] (Pg/yr)
Tropical rain forests	17.0	45	344	10	42	189	2.2	16.8	2.3	10.4
Tropical seasonal forests	7.5	35	117	4.8	25	54	1.6	5.4	1.6	3.4
Evergreen forests	5.0	35	79	3	30	41	1.3	2.9	1.5	2.0
Deciduous forests	7.0	30	95	3	28	38	1.2	3.8	1.3	1.8
Boreal forests	12.0	20	108	10.5	22	105	0.8	4.3	0.93	4.4
Woodlands and shrubs	8.5	6	22	4.5	12	24	0.7	2.7	1.1	2.2
Savannah	15.0	4	27	22.5	6.5	66	0.9	6.1	1.75	17.7
Temperate grassland	9.0	1.6	6	12.5	1.6	9	0.6	2.4	0.78	4.4
Tundra and alpines	8.0	0.6	2	9.5	1.4	6	0.14	0.5	0.22	0.9
Semidesert shrubs	18.0	0.7	6	21	0.8	7	0.09	0.7	0.14	1.3
Cultivated land	14.0	1	6	16	0.5	3	0.65	4.1	3.6	3.3
Swamps and marshes	2.0	15	14	2	13	12	3.0	2.7	0.94	6.8
Miscellaneous	26	0.08	1	30	0.4	6	0.4	0.4	0.10	1.3
Sums	149		827	149		560		52.8		59.9

[a] From Whittaker and Likens (1975).
[b] From Ajtay et al. (1979).

of about 55 Pg C/yr. The Russian school (Bazilevich *et al.*, 1971; Rodin *et al.*, 1975) has derived a higher value of 77 Pg C/yr. It is based on a more detailed accounting but appears to overestimate the productivities in some ecosystems and thus should be considered an upper limit. All results indicate that the net primary productivity on land exceeds that in the oceans in spite of the smaller area available to terrestrial life.

Turning to living biomass, the two estimates in Table 11-4 agree that forests harbor more than 80% of the world's biomass. In contrast to primary productivity, however, a disparity exists for total biomass. Inspection of the individual entries shows that the difference arises mainly from the smaller area of forest stands adopted by Ajtay *et al.* (1979) compared with that of Whittaker and Likens (1975). Specifically, the area of tropical rain forests is smaller, and it is compensated by a greater area of savannah and grass lands. The reason for the difference is not obvious. Part of the discrepancy undoubtedly results from uncertainties inherent in the estimates, but another part may signify the decline of the biomass since 1950 due to deforestation. A great deal of forest clearing certainly has taken place, particularly in the tropics. If the difference of 270 Pg carbon between the two estimates were assigned entirely to the reduction of tropical forests by humans, the biomass would have decreased by 30% during the last 30 yr. This is clearly an exaggeration. Independent estimates by Bolin (1977) and by Woodwell *et al.* (1978) put the annual net loss of biomass carbon due to deforestation at 1.0 and 5.8 Pg C, respectively, which if applied linearly would amount to a loss of biomass of between 30 and 174 Pg C over a period of 30 yr. The divergent numbers show how uncertain such estimates presently are. We come back to this problem in Section 11.3.3.

Various other assessments of global biomass should be cited for comparison with those derived in Table 11-4. Bowen (1966) gave a value of 518 Pg C, Bolin (1970) 450 Pg C, Baes *et al.* (1976) 680 Pg C, and Bazilevich *et al.* (1971) 1080 Pg C. The latter estimate refers to a reconstructed plant cover of the earth without corrections for agricultural areas and forest cuts, assuming an optimum vegetation for each of the bioclimatic zones considered. Bazilevich *et al.* were mainly interested in the potential resources, and their value must be regarded as an upper limit. Terrestrial animals represent only a small fraction of total living biomass and, according to the estimates of Bowen (1966) and Whittaker and Likens (1973), may be neglected in the present context.

The preceding estimates do not yet include deadwood and plant litter and must be corrected accordingly. Data on dead phytomass still attached to living plants are scarce and extrapolation is difficult. Ajtay *et al.* (1979) collected a number of data for above-ground dead biomass in various ecosystems, from which they concluded that total standing dead material equals roughly 5% of living biomass, on average.

Table 11-5. Rates of Plant Litter Fall, Mass of Plant Litter, and Organic Soil Carbon in Different Ecosystem Types[a]

Ecosystem type	Surface area[b] (10⁶ km²)	Litter fall[c] (kg/m² yr)	Total (Pg C/yr)	Litter mass[c] (kg/m²)	Total (Pg C)	Organic soil carbon (kg/m²)	Total (Pg)
Tropical rain forests	10	1.85	8.3	0.65	3.3	8	80
Tropical seasonal forests	4.5	1.30	2.6	0.85	1.9	9	40.5
Mangroves	0.3	0.6	0.1	10.0	1.5	8	2.4
Temperate forests	7.5	0.86	2.8	2.5	9.4	12	90
Boreal forests	9	0.59	2.4	3.5	15.8	15	135
Temperate woodlands	2	1.22	1.1	2.5	2.5	12	24
Chaparal, maquis, brushland	2.5	1.0	1.1	0.5	0.6	12	30
Savannah grassland	19	1.5	12.8	0.35	3.3	12	228
Savannah woodland	3.5	0.8	1.3	0.35	0.6	10	36
Temperate grassland, dry	7.5	0.55	1.9	0.5	1.3	30	225
Temperate grassland, wet	5	0.9	2.0	0.325	1.2	14	70
Tundra, high arctic	5.1	0.105	0.23	0.36	0.9	6.5	33.2
Scrub tundra	4.4	0.2	0.4	5.0	11.0	20	88
Desert, semidesert	21	0.125	1.2	0.1	1.1	8	168
Extreme desert	9	0.015	0.06	0.015	0.1	2.5	22.5
Swamps, marshes, bog and peatland	3.5	0.6	0.95	2.5	4.4	64	225
Cultivated land	16	0.425	3.1	0.05	0.4	8	128
Human areas	2	0.3	0.2	0.3	0.2	5	10
Sums			42.6		59.5		1636

[a] Slightly condensed from Ajtay et al. (1979). Conversion factors for carbon in dry organic matter: litter fall 0.45, litter mass 0.5.

[b] Note: 17.5×10^6 km² is perpetual ice, lake and streams, etc., which are not included.

[c] Dry mass.

Dead material shed by plants is called litter. Reiners (1973) made an attempt to estimate total litter fall using four different procedures and derived values ranging from 37 to 64 Pg C/yr. The results are based largely on the estimates provided by Whittaker and Likens (1973) for the areal extent, primary productivity, and mean biomass of different ecosystems. Ajtay *et al.* (1979) repeated Reiner's calculations using their own data and obtained for the annual litter fall values between 37 and 49 Pg C/yr. One of their detailed estimates is presented in Table 11-5. These results demonstrate that the global primary production of the biosphere of about 55 Pg C/yr turns mainly into plant litter. Ajtay *et al.* further estimate a consumption by herbivores of about 6 Pg C/yr, including consumption by domestic livestock. This leaves only about 10% of net primary productivity to be incorporated annually into living biomass.

Observational data for the mass of litter concern mainly forest ecosystems but are scarce for savannah and grass lands and estimates are correspondingly tenuous. The data of Ajtay *et al.* (1979) in Table 11-5 suggest a value of 60 Pg C for the global mass of carbon contained in plant litter of all types. The amount represents about 10% of carbon in living biomass. If we add the 5% assumed to reside in standing dead material, we obtain a total mass of 650 Pg for carbon in the biosphere, both living and dead. This value is entered in Table 11-1.

Finally, we must consider organic matter in the soil. Again it is necessary to extrapolate from field samples and world maps of soil types to obtain an estimate for the global mass of carbon present in soil as humus. Soil carbon is particularly difficult to assess because it generally decreases exponentially with depth, and it is not surprising that estimates vary widely. The early results of Waksman (1938), Bolin (1970), and Baes *et al.* (1976) range from 400 to 1080 Pg C. More recent studies produced higher values. Bohn (1976) and subsequently Ajtay *et al.* (1979) presented detailed data based on UNESCO world soil maps and determined the global mass of soil carbon as 2950 and 2070 Pg, respectively. Schlesinger (1977) noted the close relationship between humus accumulation and temperature and wetness as environmental factors. From his data, and using the humus content of soils according to ecosystem types, Ajtay *et al.* prepared another detailed account from which they derived an independent estimate of 1640 Pg C as shown in Table 11-5. The value chosen here and entered in Table 11-1 lies midway between the extremes of the recent estimates. The reader should consult the original studies for further details.

11.3 The Global Carbon Cycles

The carbon reservoirs described in the preceding sections are coupled to each other by a variety of exchange processes. The resulting total network

Table 11-6. *Overview on Carbon Cycles Involving Atmospheric CO_2, and the Associated Turnover Times in the Atmosphere*

Type of cycle	Time constant (yr)
Geochemical rock cycles (see Table 11-8 for specification)	$(2.4-30) \times 10^3$
Exchange with the ocean	
Mixed layer (Table 11-9)	4-10
Deep ocean	20
Exchange with the biosphere (see Table 11-10 for specification)	
Short-term storage	15
Long-term storage	75
Soil humus	200

of fluxes is complex, but by considering the nature of the interactions the fluxes can be arranged into a system of closed cycles which can be dealt with individually. Table 11-6 gives an overview on the relevant flux cycles. Associated with each cycle is a turnover time of carbon in the atmospheric reservoir. This quantity provides a useful criterion for classification. There are two main groups. One arises from geochemical processes and features time constants of several thousand years. On this time scale, the atmosphere and the ocean tend to be in equilibrium. The other set of cycles describes the exchange of CO_2 with the biosphere and with the ocean. Here, the characteristic times are of the order of 10 yr. The great difference in the time scales shows that the two groups of cycles are essentially decoupled and can be treated separately.

11.3.1 THE GEOCHEMICAL CYCLES

The slow carbon cycles are intimately connected with the geochemical cycles of sedimentary rocks, which involves all crustal elements. Although our interest is focused on carbon, we must discuss some related elements but can do so only superficially. For aspects not covered, especially those of mineralogy, reference is made to the geochemistry texts of Krauskopf (1979) and Garrels and MacKenzie (1971).

The salient features of the geochemical (rock) cycle are shown in Fig. 11-3 in the form of a simplified box model. The driving forces are volcanism, tectonics, and weathering reactions. Volcanic exhalations supply carbon dioxide from the earth's mantle to the atmosphere, and weathering causes the erosion and chemical breakdown of rocks by reactions involving CO_2 dissolved in terrestrial surface waters. With regard to carbon we can distinguish three types of reactions: (1) the weathering of silicates, (2) the

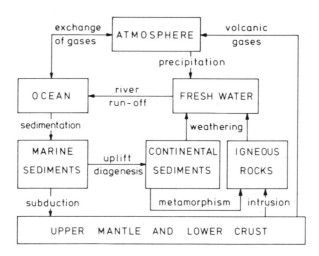

Fig. 11-3. Geochemical cycles involving igneous and sedimentary rocks. The cycles are driven by the weathering action of atmospheric CO_2 dissolved in terrestrial surface waters, by the supply of CO_2 to the atmosphere from volcanic exhalations, and by the tectonic displacement of crustal material.

weathering of carbonates, and (3) the oxidation of organic carbon in sedimentary rocks following their exposure to atmospheric oxygen. The inorganic reactions will be treated first.

Feldspars comprise the major fraction of igneous rocks. They consist of aluminosilicates of sodium, potassium, and calcium. Their weathering may be exemplified by the following generic reactions:

$$2KAlSi_3O_8 + 2CO_2 + 3H_2O \rightleftharpoons Al_2Si_2O_5(OH)_4 + 4SiO_2 + 2K^+ + 2HCO_3^-$$

K-feldspar   kaolinite   quartz

$$CaAl_2Si_2O_8 + 2CO_2 + 3H_2O \rightleftharpoons Al_2Si_2O_5(OH)_4 + Ca^{2+} + 2HCO_3^-$$

Anorthite   kaolinite

These represent summations of several individual processes converting igneous rock into clay minerals, quartz, and dissolved salts. Following a suggestion of Sillén (1963), who noted that the reactions resemble an acid–base titration, most authors (e.g., Siever, 1968; Pytkowicz, 1975) now present them in the form

$$2KAlSi_3O_8 + 2H^+ + 2A^- + H_2O \rightleftharpoons Al_2Si_2O_5(OH)_4 + 4SiO_2 + 2K^+ + A^-$$

It is then apparent that the characteristic feature of aluminosilicate weathering is the fixation of hydrogen ions in the place of cations that are released. The source of the hydrogen ions has been left unspecified and A^- may be any suitable anion. As we have seen in Chapter 8, the acidity of rainwater is determined by acids stronger than H_2CO_3. Once these are consumed, however, H_2CO_3 takes over. In soils, moreover, the abundance of CO_2 usually is greater than in the atmosphere due to biological decay processes. Analyses of ground waters beneath granitic and basaltic rocks are shown in Table 11-7. HCO_3^- is the major anion, demonstrating the significance of CO_2 as weathering agent. A good deal of sedimentary rocks also consists of silicates, which are weathered in much the same way as igneous rock silicates. Of particular importance to the carbon cycle are sedimentary limestones. They occur mainly as calcite ($CaCO_3$), with a smaller fraction of dolomite [$CaMg(CO_3)_2$]. Both are poorly soluble in water, whereas the corresponding bicarbonates dissolve easily. The weathering of carbonates may be written

$$CaCO_3 + H^+ + HCO_3^- \rightleftharpoons Ca^{2+} + 2HCO_3^-$$

$$CaMg(CO_3)_2 + 2H^+ + 2HCO_3^- \rightleftharpoons Ca^{2+} + Mg^{2+} + 4HCO_3^-$$

Here, the hydrogen ions are used to mobilize carbon as bicarbonate and, as in the case of silicate weathering, an equivalent amount of CO_2 is used up to establish charge balance.

The weathering products from both igneous and sedimentary rocks eventually enter the rivers and are swept into the ocean. The river load contains suspended particulate material in addition to dissolved salts. The relative proportions are indicated in Table 11-7. Silicon is transported mainly in the quartz and clay-mineral particles of the suspended load. Only a small fraction appears dissolved as silicic acid. The alkali and earth alkaline metals are abundant in both fractions. Sodium and chloride are not exclusively weathering products but are in part recycled via the atmosphere as seasalt. Sulfate and magnesium are less affected in this way, but a significant part of sulfate appears to be of anthropogenic origin. Holland (1978) has given a thorough discussion of the provenances of individual river constituents.

In the ocean, the material advected by the rivers participates in the formation of new sediments whereby the geochemical cycle is closed. The suspended load is redeposited in the continental shelf regions. The dissolved elements are temporarily stored in the ocean reservoir before they, too, are incorporated into sedimentary deposits. In the long run, the river influx must be balanced by an appropriate sedimentary output for each element considered. A certain degree of accumulation is nevertheless evident from the higher concentrations of the principal ions in seawater as compared to those in river water (Table 11-7, columns 6 and 7). Residence times in the ocean can be computed by means of Eq. (4-11) from the ratio of the reservoir's mass content to the observed river influx. The residence times

Table 11-7. *Comparison of Components in Natural Waters*[a]

| Constituent | Molar distribution (%) of dissolved constituents | | | Concentrations (mg/kg) | | | |
| | Ground water | | Rivers | River load | | Ocean, dissolved | Residence time in the ocean (10^6 yr) |
	Granitic	Basaltic		Dissolved	Suspended		
SiO_2	14.2	12.5	9.1	13.1	244	62	0.014
Al	—	—	—	—	35	0.002	—
Fe	0.2	0.06	0.5	0.7	25	0.002	0.0001
Mg^{2+}	6.3	10.4	7.0	4.1	5.3	1,294	10[c]
Ca^{2+}	16.1	12.3	15.6	15	11	412	0.8
Na^+	10.3	8.9	11.4	6.3	7.6	10,760	95[c]
K^+	1.6	1.6	2.4	2.3	4.2	399	5.0
HCO_3^-	43.1	44.4	39.9	58.4	—	145	0.075
SO_4^{2-}	3.9	1.9	4.9	11.2	—	2,712	8[c]
Cl^-	4.2	7.9	9.2	7.8	—	19,350	225[c]
Average total concentration (mg/kg)	221	334	120	119	332[b]		

[a] Dissolved constituents in groundwaters beneath granitic and basaltic rocks, from White *et al.* (1963); dissolved constituents in river waters, world average (Livingstone, 1963); suspended load of rivers, composition from Garrels and MacKenzie (1972) combined with estimate of world river efflux, 18 Pg/yr (Holeman, 1968); dissolved constituents in seawater, from Table 7-15; and residence time of elements in the ocean based on river input, taking the water flux to be 4.6×10^4 Pg/yr (Holland, 1978).

[b] Total load adopted is 400 mg/kg. The metals usually are reported as oxides so that the difference is due to bound oxygen.

[c] After correction for contribution by seasalt assuming two-thirds of Cl^- in the river load to derive from the fallout of seasalt. Note that the aerial transport of aerosol particles from the continents to the oceans (Table 7-11) is negligible compared with that by the suspended load discharged annually via the rivers.

are shown in the last column of Table 11-7. The values must be considered order-of-magnitude estimates because of the implicit assumption that the current continental runoff rates have persisted throughout geological times. The longest residence times are associated with sodium and chloride, but even for these elements the values are much shorter than the age of the Earth, indicating that they have been recycled many times.

The fate of alkali and alkaline earth elements in the ocean has intrigued geochemists for some time. Clearly identified removal mechanisms are the deposition of $CaCO_3$, as described earlier, which provides the major loss process for calcium and inorganic carbon, and the formation of salt deposits by evaporation of seawater in closed-off estuaries. Although evaporites form a relatively minor fraction of all sediments, it appears that chloride is lost mainly by this route and sodium to an appreciable extent.

The removal processes for potassium and magnesium have remained obscure. Sillén (1961), who considered seawater to be in equilibrium with the underlying oceanic sediments, was impressed by the potential ion exchange capacity of marine clay minerals. Indeed, in geochemical mass balances most potassium and much magnesium are recovered in shales. Following Sillén's proposal, a number of workers (e.g., MacKenzie and Garrels, 1966; Siever, 1968; Pytkowicz, 1975) have advocated the reconstitution of silicates in the ocean's sediments by reactions such as

$$5Al_2Si_2O_5(OH)_4 + 4SiO_2 + 2K^+ \rightleftharpoons 2KAl_5Si_7O_2(OH)_4 + 2H^+ + 5H_2O$$

Kaolinite   quartz   "illite"

which result in a release of hydrogen ions and thus represent a reversal of weathering. The hydrogen ions would eventually combine with bicarbonate ions and allow a return of CO_2 to the atmosphere. MacKenzie and Garrels (1966) constructed a set of reactions for the purpose of mass balance considerations, which predict that reverse weathering removes magnesium largely as chlorite, potassium as illite, and sodium partly as montmorillonite (all are clay minerals). The transformations must be viewed strictly as model reactions, however,. Drever (1974) and Holland (1978) have reviewed the observational evidence for this scheme and found little support for it. Especially magnesium presents a problem. As discussed by Bathurst (1975) and Holland (1978), the direct deposition of dolomite in the ocean is difficult despite an apparent supersaturation of magnesium, and most dolomite formations must have resulted from postdepositional alterations of calcium carbonates. Holland has pointed out that the reaction of seawater with basalts of the midoceanic ridges should not be neglected, since it has been shown to fix magnesium in favor of potassium and calcium, which are liberated. Although the process helps to improve the magnesium balance, it raises problems with quantifying the potassium and calcium inputs.

When we now consider the impact of all these processes on the inorganic carbon budget, the following three cycles can be discerned in Fig. 11-3:

1. One cycle of carbon that can be recognized involves the mobilization of carbonate during the weathering of limestones, transferral of Ca^{2+} and HCO_3^- to the ocean, and the redeposition as calcium carbonate. The cycle is closed by the diagenesis of the precipitate into carbonate rocks followed by tectonic uplift. The fate of magnesium from dolomites is presently not obvious. On a molar basis, the Mg/Ca ratio in all limestones is 0.25. If magnesium is used to replace calcium in suboceanic rocks, the cycle would be closed because the calcium released would also form carbonate deposits. Otherwise the cycle is not completely closed.

2. In spite of many uncertainties still inherent in the concept of reverse weathering, it appears reasonable to assume to a first approximation that the amount of CO_2 tied up as bicarbonate during the weathering of silicates is released again to the atmosphere–ocean system when the cations involved are removed into oceanic sediments. The corresponding flow of inorganic carbon represents an approximately closed cycle from the atmosphere via the rivers into the ocean, and back again to the atmosphere. The carbon flux within this cycle is augmented by the amount of atmospheric CO_2 used up in the weathering of limestones, as an equivalent quantity is set free upon the deposition of $CaCO_3$ in the ocean.

3. An exception to the above scheme must be seen in the weathering of calcium-bearing silicates in igneous rocks. In this case, only half of the CO_2 consumed in the mobilization of calcium is returned to the atmosphere when calcium is removed from the ocean; the other half enters the $CaCO_3$ reservoir of the sediments. The resulting loss of CO_2 from the atmosphere must be balanced by volcanic exhalations if stationary conditions are to prevail. It is probable that in the long run the cycle is closed by the subduction of carbonate sediments into region of the upper mantle due to plate-tectonic activity (cf. Fig. 11-3). The subduction zones are located in coastal regions where marine and continental plates meet.

The magnitudes of the fluxes associated with the three cycles can be estimates from the calcium and magnesium data in Table 11-7 and auxiliary information. According to Blatt and Jones (1975), roughly 25% of the continental surface is covered with igneous rocks, the remainder exposes sedimentary rocks. Garrels and MacKenzie (1971) reported the CaO content of igneous rocks to be 4.9% by weight, that of limestones 43%, shales 1.5%, and sandstones 3.2%. Using the distribution of these rocks in continental sediments given in Table 11-2 we find that the ratio of calcium from silicates in igneous rocks to that from sedimentary carbonates is about 12:88. The contribution to dissolved calcium in river waters stemming from evaporite gypsum has been estimated by Garrels and MacKenzie (1972) and Holland (1978) to be 6% and 9%, respectively. The correspondingly adjusted concentration of Ca^{2+} in river water, 14 mg/kg, leads to a calcium flux of $[(14 \times 10^{-3}/40)\text{mol/kg}][(4.6 \times 10^{16})\text{kg H}_2\text{O yr}] = 16.1$ Tmol/yr. About 12% of it is associated with the weathering of silicates. The remaining fraction, which is assigned to carbonate rocks, must be increased by about 25% to account for the contribution from magnesium carbonates. Roughly one-half of the magnesium found dissolved in river water is released in this manner. The resulting carbon fluxes in cycles (1) and (3) are 17.8 and 1.9 Tmol/yr, respectively, or 213 and 23 Tg C/yr. The average bicarbonate content of river effluents of 58 mg/kg corresponds to a total carbon flux of 44 Tmol/yr or 530 Tg C/yr. The excess of 295 Tg C/yr over the sum of the fluxes in cycles (1) and (3) must be attributed to the carbon flux in cycle (2).

Table 11-8 summarizes the results. Again it should be emphasized that the fluxes derived are current ones and that the uncertainties are considerable, even though Garrels and MacKenzie (1972), Garrels and Perry (1974), and Holland (1978), all using similar procedures, have obtained comparable results. To indicate the range of values, the estimates of the first authors are included in Table 11-8. The fluxes may also be used to calculate the residence times of carbon in the participating reservoirs, and Table 11-8 includes these data. The longest residence times are associated with the sediments on account of the high carbon content of this reservoir. The values of a few hundred million years indicate that carbonates have turned over at least 10 times since the first deposits occurred after the planet earth had formed 4.5 billion yr ago.

The inorganic carbon cycles must be supplemented by yet another cycle, cycle (4), resulting from the weathering of reduced carbon in sedimentary rocks. The geological record indicates that, if one disregards occasional fluctuations, the content of kerogen in sedimentary rocks has stayed remarkably constant throughout geological times (Schidlowski, 1982). This observation implies the existence of a stationary state. Since sedimentary rocks have been recycled several times and since, further, fresh sediments are always endowed with organic detritus from the marine biosphere, one would expect sedimentary organic carbon to accumulate with time, were it not removed in the exterior branch of the rock cycle. For this reason it is generally assumed that reduced carbon is oxidized to CO_2 when being exposed to air, even though kerogen is recognized to be chemically rather inert. The mechanism of oxidation is unknown, and it may be mediated by bacteria (Shneour, 1966). Unfortunately, the fate of fossil carbon cannot

The production of organic carbon by photosynthesis represents in effect a separation of carbon from oxygen in CO_2, and the burial of organic carbon in the sediments accordingly leaves an equivalent amount of oxygen behind. The mass of oxygen in the atmosphere is understood to have been generated in this fashion. This aspect will be followed up in Chapter 12. In the closed carbon cycle, the annual production of O_2 resulting from carbon burial is balanced by a corresponding loss due to the weathering of reduced carbon in sedimentary rocks.

11.3.2   EXCHANGE OF CO_2 BETWEEN ATMOSPHERE AND OCEAN

The fact that the Pacific, the Atlantic, and the Indian Oceans are separate entities with mixing characteristics of their own makes it difficult to model the world ocean in a realistic manner. The simplest approach, and the one that will be followed here, is to subdivide the ocean into two reservoirs: the windmixed layer and the deep sea. Figure 11-4 shows how they are connected to the atmosphere. The exchange of CO_2 takes place mainly via the mixed layer. The polar currents subsiding from the surface directly into the deep ocean are comparatively small and will be ignored, even though Siegenthaler (1983) showed that they are not entirely negligible. Keeling (1973b) has given a thorough mathematical analysis of the reservoir system

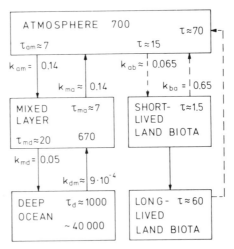

Fig. 11-4. Exchange of CO_2 between atmosphere and ocean in a two-box ocean model. Reservoir contents are shown in units of Pg C ($= 10^{12}$ kg C), exchange coefficients in yr^{-1}, and residence times in years. For comparison, the exchange of atmospheric carbon with the biosphere is shown in parallel. For simplicity the biosphere is represented by the two reservoirs of short-lived photosynthetic active components and the long-lived structural material. A more detailed breakdown of the biospheric reservoir is shown in Fig. 11-6.

shown in Fig. 11-4. The main deficiency of the two-box ocean model is its neglect of the transition zone of the main thermocline connecting the mixed layer with the deep sea. More recently, Oeschger *et al.* (1975) developed a diffusion model that better accounts for the transport of material in the domain of the thermocline. The short-comings of the two-box model are being felt most severely when one considers variations on a time scale short compared with the turnover time of the deep ocean, because the usual assumptions of a well-mixed reservoir can then not be maintained. Keeling (1973b) pointed out, however, that the box-model approach requires only that the outgoing fluxes are proportional to the total content of the reservoir whereby the usefulness of box models is considerably extended.

We consider first the exchange of CO_2 between the atmosphere and mixed layer of the ocean. The majority of studies concerned with CO_2 exchange have made use of radiocarbon data. Carbon-14 is formed naturally in the atmosphere by the interaction of cosmic ray–produced neutrons with nitrogen. The global ^{14}C production rate is about 3×10^{26} atoms/yr (Lingenfelter, 1963; Lingenfelter and Ramaty, 1970). Temporarily, ^{14}C is stored in the atmosphere, the biosphere, and the mixed layer of the ocean, before it is transferred to the deep sea, where it decays with a half-life time of 5730 yr. Nuclear weapons tests beginning in 1954 and culminating in the early 1960s caused additional short-term inputs, resulting in a drastic increase of atmospheric ^{14}C levels. The history of excess ^{14}C in the troposphere is depicted in Fig. 11-5. A significant fraction of bomb ^{14}C was injected into the stratosphere. The annual oscillations appearing

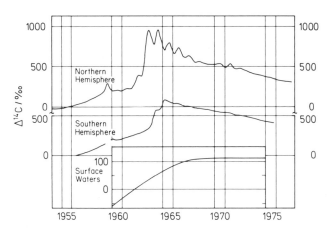

Fig. 11-5. History of ^{14}C activity in the troposphere and in the surface waters of the ocean (insert). [Adapted from the compilations of Tans (1981).] The data are given as permille deviation from the National Bureau of Standards oxalic acid standard.

superimposed on the decline of ^{14}C activity following the 1963 test moratorium must be attributed to the seasonal variation of air exchange across the tropopause. The tests of this period were conducted in the northern hemisphere. The response of the southern hemisphere lags behind by about 1 yr on account of the equatorial barrier to eddy transport imposed by the interhemispherical tropical convergence zone. Figure 11-5 also shows the rise of excess ^{14} concentrations in the surface waters of the ocean. Here, a new steady state is reached about 4 yr after the last major injection into the atmosphere.

Exchange rates between the troposphere and the mixed layer of the ocean can be derived from the ^{14}C data in several ways: (1) by considering the steady state flux of natural ^{14}C from the atmosphere through the mixed layer into the deep ocean; (2) from the inventory of excess bomb ^{14}C in the atmosphere and the ocean; and (3) by evaluating the change in the natural $^{14}CO_2/^{12}CO_2$ ratio due to dilution by the addition of inactive CO_2 from the combustion of fossil fuels to the atmospheric reservoir (Suess effect). Independently, one may also use the stagnant film model as described in Section 1-6 to estimate the CO_2 exchange rate. Table 11-9 gives a summary of results obtained by various investigators as well as the equivalent residence times of CO_2 in the atmosphere. They average to $\tau_{am} \approx 7$ yr. Below,

Table 11-9. *Transfer of CO_2 from the Atmosphere to the Mixed Layer of the Ocean*[a]

Authors	Method[b]	k_{am} (yr^{-1})	τ_{am} (yr)
Münnich and Roether (1967)	(a)	0.18	5.4
Bien and Suess (1967)	(a)	0.04	25
Nydal (1968)	(a)	0.1–0.2	5–10
Young and Fairhall (1968)	(a)	0.25	4
Machta (1971)	(a)	0.46	2.15
Stuiver (1981)	(a)	0.14	6.8
Revelle and Suess (1957)	(b)	0.1	10
Fergusson (1958)	(b)	0.14–0.5	2–7
Craig (1957)	(c)	0.1–0.25	7 ± 3
Broecker (1963)	(c)	0.1–0.14	7–10
Oeschger et al. (1975)	(c)	0.137	7.3
Peng et al. (1979)	(d)	0.114	8.8

[a] Compilation of transfer coefficients k_{am} and corresponding residence times of CO_2 in the atmosphere deduced by a number of investigators.

[b] Methods: (a) transfer of excess bomb ^{14}C to ocean and biosphere; (b) change of atmospheric ^{14}C activity due to dilution by the addition of fossil fuel CO_2 (Suess effect); (c) steady state of natural ^{14}C exchange and decay in the ocean; (d) application of stagnant-film model independent of ^{14}C.

we shall illustrate the procedure employed in deriving k_{am} and k_{md} by the first of the methods indicated above.

The mass contents of ^{12}C in the atmosphere, the mixed layer, and the deep ocean are abbreviated by G_a, G_m, and G_d, respectively, and the corresponding quantities for radiocarbon are indicated by an asterisk. Fluxes are assumed to be proportional to the mass content of the three reservoirs; the notation of the exchange coefficients is shown in Fig. 11-4. The measurements provide activities expressed as $^{14}C/^{12}C$ mole ratios. Their average values in the three reservoirs are denoted by

$$R_a = fG_a^*/G_a \qquad R_m = fG_m^*/G_m \qquad R_d = fG_d^*/G_d \qquad (11\text{-}2)$$

where the factor $f = 12/14$ is the mass ratio of the two isotopes. It will later be seen to cancel in the flux balance equations. In thermodynamic equilibrium, the $^{14}C/^{12}C$ ratios of total dissolved CO_2 in seawater and that in air are not identical because the equilibrium constants for solubility and ionization differ slightly for both isotopes. The resulting fractionation has been determined in the laboratory for $^{13}C/^{12}C$ and amounts to 8 permille as discussed by Siegenthaler and Münnich (1981). The fractionation for $^{14}C/^{12}C$ will be twice as large, so that

$$\frac{k_{am}^*}{k_{ma}^*} = \alpha^* \frac{k_{am}}{k_{ma}} \qquad \text{with} \quad \alpha^* = 1.016 \qquad (11\text{-}3)$$

In the steady state, the radioactive decay of ^{14}C in the ocean must be balanced by the net uptake from the atmosphere,

$$k_{am}^* G_a^* - k_{ma}^* G_m^* = \lambda(G_m^* + G_d^*) \approx \lambda G_d^* \qquad (11\text{-}4)$$

Here, $\lambda = (1/8267)\ \text{yr}^{-1}$ is the ^{14}C decay constant. For ordinary CO_2, equilibrium conditions can be assumed:

$$k_{am} G_a = k_{ma} G_m \qquad (11\text{-}5)$$

Dividing Eq. (11-4) through by G_a and converting to activities yields

$$k_{am}^* R_a - k_{ma}^* R_m(G_m/G_a) = R_d(G_d/G_a) \qquad (11\text{-}6)$$

which with the help of the preceding equations can be cast in the form

$$k_{am} \approx k_{am}^* = \frac{\lambda(R_d/R_a)}{(1 - R_m/\alpha^* R_a)}\left(\frac{G_d}{G_a}\right) \qquad (11\text{-}7)$$

The decay of ^{14}C in the sediments is nearly balanced by the river input of ^{14}C. Both quantities amount to only a few percent of that present in the ocean and can be neglected. From measurements conducted during the period preceding the major weapons tests, Broecker (1963) and more recently Oeschger *et al.* (1975) estimated $R_d/R_a = 0.85 \pm 0.05$ and $R_m/\alpha^* R_a = 0.95 \pm 0.015$. The mass of carbon in the deep sea is $G_d \approx$ 38,000 Pg if organic carbon is included, and that of carbon dioxide in the atmosphere was about 650 Pg during the period considered. The insertion of these data into Eq. (11-7) yields $k_{am} = 0.133 \pm 0.047$ yr^{-1}. The equivalent atmospheric residence time, $\tau_{am} = 1/k_{am} = 7.5$ yr, is long compared with the time for tropospheric mixing. The uncertainty in the value arises primarily from the fact that the denominator of Eq. (11-7) represents a difference of two almost equal numbers, and for the same reason one must include the seemingly small correction for isotope fractionation. The coefficient for the reverse flux, k_{ma}, is determined by the equilibrium relation of Eq. (11-5). It turns out that $k_{ma} \approx k_{am}$, since the mass of CO_2 dissolved in the mixed layer of the ocean is essentially the same as that in the atmosphere.

Consider now the exchange of inorganic carbon between the mixed layer and the deep ocean. The steady-state conditions for the latter reservoir are

$$k_{md} G_m^* = (k_{dm} + \lambda) G_d^*$$
$$k_{md} G_m = k_{dm} G_d \tag{11-8}$$

In this case the isotope fractionation is expected to be negligible and the asterisk of the rate coefficients for the exchange of radiocarbon can be dropped. Elimination of k_{dm} and conversion to activities gives

$$k_{md} R_m = k_{md} R_d + \lambda R_d (G_d/G_m) \tag{11-9}$$

which after dividing through by R_a can be rearranged to yield

$$k_{md} = \frac{\lambda (G_d/G_m)(R_d/R_a)}{R_m/R_a - R_d/R_a} \tag{11-10}$$

The numerical values given above for the quantities appearing in this expression lead to $k_{md} = 0.05$ yr^{-1} or a residence time of inorganic carbon in the mixed layer of 20 yr. The range of uncertainty is similar to that for k_{am} and amounts to $\pm 50\%$. In contrast to k_{am}, the values obtained for k_{md} and τ_{md} depend on the assumed thickness of the mixed layer, so that the results reported in the literature differ somewhat. Oeschger *et al.* (1975) give $\tau_{md} = 22.7$ yr; Houtermans *et al.* (1973) obtained a value of the order of 30 yr. We can check on the validity of the above result by calculating the equilibrium residence time of CO_2 dissolved in the deep ocean, which is independent of G_m:

$$\tau_{dm} = 1/k_{dm} = \tau_{md} G_d/G_m = 1100 \quad \text{yr}$$

The value is in satisfactory agreement with the turnover time of 700–1000 yr resulting from the circulation of deep ocean waters.

Further insight into the behavior of the ocean–atmosphere system can be gained by considering a sudden perturbation of atmospheric ^{14}C (Houtermans *et al.*, 1973) and the subsequent adjustment to new steady state conditions. We set $G_a^*(t) = \bar{G}_a^* - \Delta G_a^*$ and $G_m^*(t) = \bar{G}_m^* - \Delta G_m^*$ where $\bar{G}_a^*$ and $\bar{G}_m^*$ represent the inventories of ^{14}C in the atmosphere and mixed layer, respectively, when the new steady state is reached, and ΔG_a^* and ΔG_m^* are the deviations from these values at time t. The equations governing the temporal change following an input pulse then reduce to

$$\frac{d\,\Delta G_a^*}{dt} = k_{am}\,\Delta G_a^* - k_{ma}\,\Delta G_m^* \tag{11-11a}$$

$$\frac{d\,\Delta G_m^*}{dt} = (k_{am} - k_{dm})\,\Delta G_a^* - (k_{ma} + k_{md} + k_{dm})\,\Delta G_m^* \tag{11-11b}$$

If we confine the interest to a time scale of 50–100 yr, the radioactive decay of ^{14}C can be neglected to a first approximation. Also neglected for simplification is the influence of the terrestrial biosphere, although its omission is not really justified (in contrast to treating the steady state) because it acts as a temporary sink for ^{14}C on the time scale of a few decades.

If the initial conditions are chosen to represent a sudden increase ΔG_{a0}^* of ^{14}C in the atmosphere (for example, due to the atomic weapon tests), the solutions to the above system of differential equations are

$$\frac{\Delta G_a^*}{G_{a0}^*} = A \exp(-\lambda_1 t) + B \exp(-\lambda_2 t) \tag{11-12a}$$

$$\frac{\Delta G_m^*}{G_{a0}^*} = C[\exp(-\lambda_2 t) - \exp(-\lambda_1 t)] \tag{11-12b}$$

where A, B, and C are appropriate constants with which we shall not be concerned, and λ_1 and λ_2 are the reciprocal time constants for the relaxation of the system to the new steady state. They are given by

$$\tau_{1,2}^{-1} = \lambda_{1,2} = \frac{a+d}{2} \pm \tfrac{1}{2}[(a-d)^2 + 4bc]^{1/2} \tag{11-13}$$

with $a = k_{am}$, $b = k_{ma}$, $c = k_{am} - k_{dm}$, and $d = k_{am} + k_{md} + k_{dm}$, and by inserting for the exchange coefficients the values discussed previously one finds

$$\tau_1 = \frac{1}{\lambda_1} = 3.3 \quad \text{yr} \qquad \tau_2 = \frac{1}{\lambda_2} = 42 \quad \text{yr}$$

The first relaxation time is associated with the equilibration of excess ^{14}C between the atmosphere and the mixed layer of the ocean; the second relaxation time describes the much slower rate of filling up the deep sea reservoir. The relaxation time for the first of these processes agrees with the time constant observed for the rise of ^{14}C in the mixed layer after the last major injection of ^{14}C into the atmosphere by nuclear weapons tests (see Fig. 11-5 for comparison). Note that equilibration with the mixed layer is faster than one would expect from the atmospheric residence time under steady-state conditions. After a time span of 10 yr or more, the atmosphere and the mixed layer approach steady state and they may be combined into a single reservoir. The relaxation is then determined by the second process. The excess flux of ^{14}C into the deep ocean now is

$$k_{md} \, \Delta G_m \approx \frac{1}{\tau_2} (\Delta G_m^* + \Delta G_a^*) = \frac{1}{\tau_2} \left(1 + \frac{k_{ma}^*}{k_{am}^*} \right) \Delta G_m^* \qquad (11\text{-}14)$$

or

$$\tau_2 = \frac{k_{am} + k_{ma}}{k_{am}} \, \tau_{md}$$

The fact that the relaxation time τ_2 is about twice the residence time τ_{md} is seen to be a consequence of treating the atmosphere and the mixed layer together as one reservoir. The decline of ^{14}C activity in the atmosphere that has been observed in recent years is more rapid than a relaxation time of 42 yr would predict. The small additional injections of ^{14}C since 1963 can be ignored in this context, but not the influence of the biosphere, which acts as another absorbing reservoir on the time scale of interest, and therefore leads to a faster decay of atmospheric ^{14}C than the present treatment, neglecting the biosphere, can offer.

Compared with excess $^{14}CO_2$, the addition of fossil fuel-derived $^{12}CO_2$ to the atmosphere–ocean system forms a much more massive perturbation, which affects also the ion equilibria in the mixed layer. According to Eq. (8-15), the absorption capacity for CO_2 of seawater depends on the hydrogen ion concentration. This quantity is determined largely by the charge balance equation

$$A + [H^+] = [HCO_3^-] + 2[CO_3^{2-}] + [H_2BO_3^-] + [OH^-]$$

where A stands for the net charge of all other ions, both positive and negative, that do not vary significantly with changes of the CO_2 content. Using the substitution procedure in Eq. (8-16), the equilibrium relations between CO_{2aq}, HCO_3^-, and CO_3^{2-} on one hand, and H_3BO_3 and $H_2BO_3^-$ on the other, modify the above equation to

$$A = \frac{K_H K_1}{[H^+]} \left(1 + \frac{2K_2}{[H^+]} \right) P_{CO_2} + \frac{[H_3BO_3] + [H_2BO_3^-]}{1 + [H^+]/K_B} + \frac{K_w}{[H^+]} - [H^+] \qquad (11\text{-}15)$$

where K_H, K_1, and K_2 are the apparent equilibrium constants for the dissolution of CO_2 in seawater and K_B and K_w refer to the dissociation equilibria of boric acid and water, respectively. This expression shows that the hydrogen ion concentration is a function of the partial pressure of CO_2 in the atmosphere. Solving for $[H^+]$ in terms of p_{CO_2} and inserting the result into Eq. (8-16) yields a nonlinear relation between p_{CO_2} and the total concentration $\Sigma C = [CO_2]_{aq} + [HCO_3^-] + [CO_3^-]$ of inorganic carbon in seawater. A rise of p_{CO_2} increases $[H^+]$, thus lowering the capacity of seawater for the additional absorption of CO_2. In other words, the surface layer of the ocean appears to evade the uptake of CO_2. This relation can be expressed in terms of a buffer factor ζ defined by

$$\frac{\delta p_{CO_2}}{p_{CO_2}^0} = \zeta \frac{\delta \Sigma C}{C^0} \quad \text{or} \quad \frac{\delta G_a}{G_a^0} = \zeta \frac{\delta G_m}{G_m^0} \tag{11-16}$$

where δ denotes an infinitesimal change in pressure or concentration and the superscript zero indicates preindustrial conditions. The buffer factor can be calculated from the known equilibrium constants, and it has also been measured at various points on the sea surface. Both procedures gave concordant results, as Sundquist *et al.* (1979), among others, have shown. The values for ζ range from 8 to 14 depending on the temperature. The most common choice for the average ocean is $\zeta \approx 10$ for current conditions. The buffer factor is not independent of ΣC, however, and it is expected to increase with the further rise of atmospheric CO_2 in the future [see Keeling (1973b) for details].

The buffer factor has the consequence that the exchange coefficient k_{ma} associated with the dissolution equilibrium between atmosphere and mixed layer must be replaced by ζk_{ma} when the equilibrium is perturbed. The uptake capacity of the mixed layer is reduced to one-tenth of its equilibrium value, and the relaxation time for the transfer of excess CO_2 toward the deep ocean, given by Eq. (11-14) for a pulse input, is raised to $\tau_2 \approx 220$ yr. This is an important result. It shows that it takes several centuries to drain from the atmosphere the excess of CO_2 injected by the combustion of fossil fuels. It makes little difference that combustion must be represented by a continuous source function, because any continuous function can be expressed by a time series of pulses. In the box-diffusion model of the ocean discussed by Siegenthaler and Oeschger (1978), the response to a pulse input leads to a nonexponential decay of atmospheric CO_2, which after equilibration with the mixed layer is somewhat faster than that in the two-box model treated here, but the time scales are still roughly the same.

11.3.3 THE INTERACTION BETWEEN ATMOSPHERE AND BIOSPHERE

The biological carbon cycle in the ocean is essentially short-circuited. Despite the high turnover rate it has little effect on the atmosphere, because marine organisms and detritus contain only a minute fraction of total carbon in the ocean. Accordingly, we focus our attention onto the terrestrial biosphere. Ocean–atmosphere exchange models usually treat the terrestrial biosphere as a delay line in which carbon is temporarily stored before it is returned to the atmosphere from where it was taken (see Fig. 11-4). It was pointed out earlier, however, that the biosphere as a whole does not represent a uniform reservoir with a well-defined residence time, and a subdivision into several compartments, each with its own characteristic behavior, is necessary. How many such compartments are required to provide a realistic description of the biosphere is a matter of current research, and a consensus on the most appropriate subdivision has not yet been reached. One possible scheme is shown in Fig. 11-6. It follows naturally from our previous discussion and distinguishes four important subreservoirs on which information is available: leaves and other assimilating parts of plants; wood contained in roots, stems, branches and standing deadwood; leaf and wood litter; and soil humus. In this model, the various ecosystems are lumped together. More precisely, one should treat each ecosystem separately and then sum equivalent reservoir contents and fluxes. For steady-state conditions or annual averages, the results of both procedures will be identical. The more general case of time-dependent fluxes, for which the results are expected to differ, has not yet been studied in sufficient detail.

The fluxes in Fig. 11-6 are annual averages estimated by the methods discussed in Section 11.2.4. If stationary-state conditions are assumed, the fluxes may be taken as constant. The residence times in each reservoir are readily evaluated and the results are summarized in Table 11-10. Several

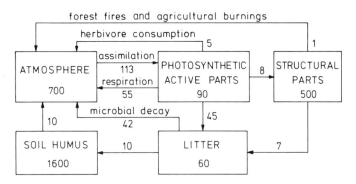

Fig. 11-6. Four-reservoir model of organic carbon in the terrestrial biosphere. Reservoir contents in units of Pg C ($=10^{12}$ kg C) and fluxes in units of Pg C/yr follow from the data assembled in Section 11.2.4. The flux due to biomass burning (Seiler and Crutzen, 1980) excludes deforestation. Emanuel *et al.* (1981) have employed a similar compartmental subdivision but somewhat different fluxes.

Table 11-10. *Carbon Reservoirs of the Terrestrial Biosphere, Fluxes and Residence Times*[a]

Reservoir	Carbon content (Pg)	Net flux (Pg C/yr)	Residence time (yr)
Atmosphere, short-term storage	700	47	15
Atmosphere, long-term storage		8–10	70–87
Assimilating parts of biosphere	90	58	1.5
Structural parts including standing deadwood	500	8	63
Stem, branch, twig and leaf litter	60	52	1.2
Soil humus	2000	10	200

[a] For interactions and connections with the atmosphere see Fig. 11-6. Individual data were gleaned from the summary of Ajtay *et al.* (1979).

subcycles of carbon through the system can be identified. One is the direct exchange of CO_2 with the atmosphere, due to assimilation during daylight and respiration at all times. About one-half of the gross primary production appears as plant tissue and constitutes net primary production. Most of this material is returned as CO_2 to the atmosphere when leaves and other assimilating parts wither, turn to litter, and are decomposed. This pathway forms the second subcycle. The ensuing residence times in both participating compartments are of the order of 1 yr. A third subcycle is provided by the small fraction of organic carbon that is converted to wood, whereas a fourth subcycle involves the fraction of plant litter that enters the soil and becomes humus. This fraction is also eventually metabolized to CO_2 and is expired to the atmosphere. Wood and soil humus are the major reservoirs of organic carbon, featuring residence times of the order of 63 and 200 yr, respectively. Herbivore consumption followed by metabolization of green plant tissue and biomass burnings represent comparatively minor pathways for the return of CO_2 to the atmosphere.

In considering the effect of the biosphere on atmospheric CO_2 we must distinguish three time scales. The first is given by the diurnal variation of primary productivity. It may be locally significant but averages out over the whole troposphere. The next time scale is determined by the growth and decay of foliage and other assimilating parts of plants in the second subcycle described above. The corresponding residence time of CO_2 in the atmosphere is 15 yr. This is sufficiently short to produce a seasonal variation of atmospheric CO_2. The third time scale follows from the slower flux of carbon through the major biospheric reservoirs (wood and soil humus) and leads to an atmospheric residence time of 70–90 yr.

The seasonal oscillation of CO_2 mixing ratios in the troposphere has been investigated by means of eddy diffusion models and is well understood. Junge and Czeplak (1968) made a first attempt to relate the variations to the temporal imbalance between biospheric net uptake (by photosynthesis) and release (by decaying plant litter), using a simple one-dimensional north–south transport model. Machta (1971) and more recently Pearman and Hyson (1981) worked with two-dimensional models. The input data required are the eddy diffusion coefficients as a function of latitude and height, and the biospheric source functions. Seasonal variations due to anthropogenic activities, such as home heating, and temperature variations of ocean surface waters are small enough to be negligible. The early studies employed the net primary productivity estimates of Lieth (1965), which have since been improved, together with an estimate for the fraction of organic carbon already suffering decay during the growth period. Net uptake and release are assumed to balance out over a complete annual cycle. In the middle and high latitudes the growth season is the warm period of the year; in subtropical regions it is the rainy season. In the tropics, where temperature and solar intensities vary little over the year, growth and decay are nearly always balanced and the seasonal effects are nil.

Although the calculations of Junge and Czeplak (1968) and of Machta (1971) gave results in reasonable accord with observations, there were still differences with respect to amplitude and phase. Pearman and Hyson (1981) adopted an inverse procedure. By matching calculated to measured variations of atmospheric CO_2, they sought the biospheric source function that best reproduces the observational data. Figure 11-7a shows for the two measurement stations Mauna Loa and South Pole the level of agreement that can be reached. Figure 11-7b shows the corresponding seasonal variations of the biospheric source for three latitude bands of equal area in the northern hemisphere. These curves may be interpreted as a superposition of two curves, one describing the release of CO_2 by a continuous decay of plant material and the other representing the net uptake due to photosynthesis during the growth period. The area underneath each curve gives the amplitude of the imbalance within the chosen latitude band. The resulting distribution with latitude is shown in Fig. 11-7c. The largest amplitudes are found in midlatitudes and, as expected on account of the distribution of land masses, in the northern hemisphere. Summation over the entire globe yields a peak-to-peak amplitude for the CO_2 imbalance of 15 Pg C/yr, which corresponds roughly to 25% of global net primary productivity.

Finally, we consider time periods much longer than 1 yr. Two questions may be raised: what is the response of the biosphere to changes in the

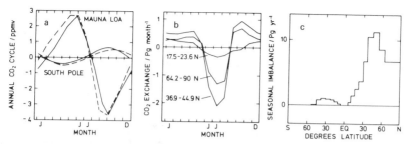

Fig. 11-7. Exchange of atmospheric CO_2 with the terrestrial biosphere. Left: (solid line) Seasonal variation of atmospheric CO_2 at two stations (northern and southern hemisphere) compared with (dashed line) model calculations by Pearman and Hyson (1981). Center: Rate of CO_2 exchange between biosphere and atmosphere for three latitude bands in the northern hemisphere derived from a best fit of the model to observational data. Right: Integrated annual CO_2 net exchange rate versus latitude.

atmospheric CO_2 level, and to what extent has the reduction of biomass due to the deforestation by humans contributed to the rise of CO_2 in the atmosphere? These are difficult problems of current concern to which satisfactory answers are still being sought, and the following comments can do little more than to indicate the difficulties.

The first problem was reviewed by Lemon (1977). The dependence of net primary productivity on the CO_2 mixing ratio in air has been studied for a number of plant species, and Fig. 11-8 illustrates the observed behavior

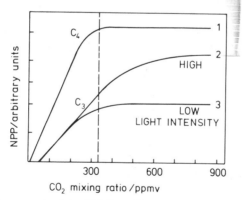

Fig. 11-8. Net primary productivity (NPP): typified response of plants to changes in the atmospheric CO_2 concentration. [Adapted from Berry (1975).] C_3 and C_4 plants differ in their photosynthetic mechanism. The former generate a three-carbon carboxylic acid (phosphoglyceric acid) as the first identifiable photosynthetic product; the latter form first a four-carbon dicarboxylic acid (oxaloacetic acid); see Devlin and Barker (1971) for details. Examples for C_4 plants are tropical grasses such as maize, sorghum, and sugar cane. All forest species are C_3 plants.

for two different types of plants. The initial response is linear, but saturation sets in for higher CO_2 concentrations. Similar saturation curves are found for the dependence of net primary productivity on light intensity and annual precipitation. The saturation effect appears to be due to the closure of the stomata (leaf openings), which regulate the intercellular CO_2 level and the water transpiration rate. As Fig. 11-8 shows, C_4 plants are more efficient than C_3 plants in that they reach saturation at lower CO_2 partial pressures. At the present level of CO_2 in the atmosphere, C_4 plants operate already near saturation, whereas C_3 plants offer a potential for increasing the net primary productivity with rising concentration of CO_2, provided other factors such as nutrient and water supply are not rate-limiting. In models of the long-term exchange of CO_2 between biosphere and atmosphere, reviewed by Bolin et al. (1981), the net flux of carbon entering the biosphere is expressed in the form

$$F_{ab} = F^0 f_a(G_{LB}/G_{LB}^0) f_b(G_A/G_A^0) \qquad (11\text{-}17)$$

where G_{LB} denotes the carbon content of living biomass, G_A that of the atmosphere, and the superscript zero refers to preindustrial steady-state conditions. The term F_{ab} is a nonlinear function of the size of the biosphere, whereas the return flux F_{ba} is taken to be proportional to G_{LB}. The function f_b is chosen to model in some fashion the shape of curve 2 in Fig. 11-8. The function f_a must also be a limiting function, since the biosphere cannot grow indefinitely in view of limitations imposed by continental area and the supply of nutrients and water. Unfortunately, neither the shape of f_a nor the potential size of the biosphere are adequately defined, so that the above relation remains speculative. Moreover, greenhouse studies have shown that the faster uptake of CO_2 at higher mixing ratios may pertain only to the growth period and need not continue when plants reach maturity. Final biomass is not necessarily increased with added CO_2, but the life cycle of a plant may be shortened.

There is some evidence from mass balance considerations that the biosphere has, in fact, acted as a sink for excess CO_2. As discussed previously, the buildup of CO_2 in the atmosphere during the 20-yr period 1958–1978 corresponds to 54% of that released by the combustion of fossil fuels over the same period. The calculations of Siegenthaler and Oeschger (1978) and Peng et al. (1979), which are based on the fairly reliable oceanic box-diffusion model, show that another 35% has entered the oceans. This leaves about 11% unaccounted for, and it seems reasonable to assign the remaining fraction to an uptake by the biosphere.

The conclusion that the biosphere has acted as a sink for excess CO_2 contrasts with the suggestion first made by Bolin (1977) that the decimation of forests due to human activities has led to the release of CO_2 in amounts

Table 11-11. *Estimates of CO_2 Released from the Biosphere due to Human Influences[a]*

Authors	Annual input (Pg C/yr)	Cumulative input[d] (Pg C)	Preindustrial $m(CO_2)$ (ppmv)
Bolin (1977)	1.0 ± 0.6[b]	70 ± 30 (1800–1975)[b]	275
J. A. Adams et al. (1977)	0.4–4[b]	—	—
Stuiver (1978)	1.2[c]	120 (1850–1950)[c]	268
Woodwell et al. (1978)			
Vegetation only	5.8 (1.5–13)[b]	—	—
Including detritus and humus	7.8 (2–18)		
Wong (1978)	1.9[b]	—	—
Wagener (1978)	—	170 (1800–1935)[c]	
Freyer (1978)	—	70 (1860–1974)[c]	—
Siegenthaler et al. (1978)	—	133–195 (1860–1974)[c]	
Hampicke (1979)	1.5–4.5[b]	—	—
Seiler and Crutzen (1980)	0 ± 2[b]	—	—
Moore et al. (1981)	2.2–4.7[b]	—	—
Peng et al. (1983)	1.2[c]	240 (1850–1975)[c]	243

[a] For comparison, the present rate of fossil fuel combustion is about 6 Pg C/yr and the total CO_2 release from fossil fuels since the beginning of industrialization is about 156 Pg C.

[b] Based on biomass change data.

[c] Based on $^{13}C/^{12}C$ data from tree rings.

[d] Dates in parentheses.

comparable to that from fossil fuel combustion. Although there can be no doubt that the annual clearing of biomass (mostly in the tropics) is significant, the mass of CO_2 thus liberated is rather uncertain, with estimates ranging from 0.4 to 18 Pg C/yr, as summarized in Table 11-11. But even the smaller of these values are difficult to reconcile with the results from the ocean models, so that a controversy developed regarding the role of the biosphere in the long-term budget of atmospheric CO_2. Seiler and Crutzen (1980) have made a careful analysis of the processes involved in biomass burning. They concluded that only about one-half of the cleared biomass is actually burned and that a considerable fraction of the carbon thus freed is fixed again by the regrowth of vegetation in the cleared areas, by reforestation in temperate latitudes, and by other effects. According to Seiler and Crutzen (1980), the uncertainties in estimating the CO_2 release are such that they would allow the biosphere to have acted either as a source or as a sink with magnitude up to 2 Pg C/yr. The mass of carbon released from fossil-fuel combustion in the period 1958–1978 is 74.7 Pg (Rotty, 1981). If the biosphere had consumed 11% of it, the annual uptake rate would have averaged to 0.4 Pg C/yr. The amount is well within the range of uncertainty indicated. These comments concern only the recent decades, of course, for which a record of atmospheric CO_2 exists, not the land use due to colonization in the past century.

Another approach to the problem, by which one had hoped to settle the controversy, is to use the $^{13}C/^{12}C$ ratio in tree rings as an indicator for the net transfer of carbon between the biosphere and the atmosphere. The idea basically is as follows. Due to the kinetic isotope effect associated with photosynthesis, the $^{13}C/^{12}C$ ratio in plant matter is about 18‰ smaller than that of CO_2 in air (see Fig. 11-2). The isotope shift is essentially preserved in fossil carbon, so that the combustion of larger amounts of biomass and/or fossil carbon temporarily reduces the atmospheric $^{13}C/^{12}C$ value until the exchange with the much larger reservoir of inorganic carbon in the ocean slowly corrects the perturbation. The history of atmospheric ^{13}C is recorded in tree rings, which can be accurately dated by dendrochronology. The available measurements have been reviewed by Stuiver (1978, 1982) and by Peng et al. (1983). Many of the trees studied show indeed a decrease of the $^{13}C/^{12}C$ ratio since the middle of the last century, although the data exhibit much scatter due to environmental factors. In order to differentiate between the contributions of biomass and fossil-fuel burning, Stuiver (1978) corrected the data by means of the Suess effect, that is, the dilution of the natural $^{14}C/^{12}C$ ratio by the injection of fossil fuel–derived CO_2 into the atmosphere. The corrected data provided no substantial evidence for any net change in the $^{13}C/^{12}C$ ratio between 1945 and the present. Prior to 1945, however, the change was appreciable, indicating a decrease of the biomass since 1850. Other authors, notably Freyer (1978), have come to similar conclusions, as Table 11-11 attests.

Unfortunately, the effect is hard to quantify. The major problem is the natural variability of the ^{13}C tree-ring data caused by local factors such as irradiance, relative humidity, temperature, vicinity to anthropogenic pollution sources, etc. To minimize environmental influences, Peng et al. (1983) selected nine trees from sites far from urban regions, trees that were not water-stressed and that had experienced consistent temperatures during the growth season. From the individual ^{13}C data, a composite curve was constructed whose deconvolution yielded a time history for the CO_2 input from the biosphere. The release of CO_2 from the use of fossil fuels was taken from the compilation of Rotty (1981), and the uptake of CO_2 by the ocean was modeled in the fashion of Oeschger et al. (1975). The results of the study are shown in Fig. 11-9 and in Table 11-11. The cumulative input into the atmosphere of CO_2 resulting from human manipulation of the biosphere was calculated to be 240 Pg, about twice that of other estimates and higher than the concurrent input due to fossil fuel combustion. The preindustrial CO_2 mixing ratio in air deduced from the data is 243 ppmv, almost 30 ppmv

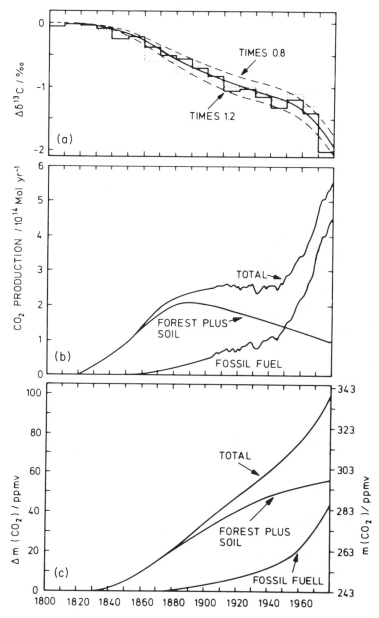

Fig. 11-9. (a) History of atmospheric ^{13}C recorded in tree rings. (b) Deconvolution of tree ring data to determine the history of CO_2 released to the atmosphere. Deducting from the resulting curve the fossil fuel-derived part gives the contribution arising from deforestation and soil manipulations. (c) The corresponding rise of CO_2 mixing ratio in the atmosphere. [Adapted from Peng *et al.* (1983), with permission.]

lower than the value obtained by Neftel *et al.* (1983) from air enclosures conserved in the Greenland ice shield. This is a serious disparity, which heightens the controversy about the role of the biosphere as a source of excess CO_2, rather than resolving it.

It will be clear that the results of such models depend critically on the choice of the ^{13}C data available in tree rings. Some trees feature $^{13}C/^{12}C$ ratios declining much less than the curve adopted by Peng *et al.* (1983), just sufficient to accommodate the release of CO_2 from fossil fuel consumption. Stuiver (1982) reports the ^{13}C record for two sequoia trees that grew 2 km apart, yet the two records to not correlate well with each other, presumably due to local effects. Stuiver, therefore, cautions against the use of composite curves until the origin of the variability in the ^{13}C data is better understood. Accordingly, the results in Fig. 11-9 should be taken as one example for the treatment of such data, possibly one tending to the extreme, but still tentative at present.

11.4 Summary

The diversity of arguments in the preceding sections calls for a summation, which we shall focus onto the most essential points in the discussion, namely, what controls the abundance of CO_2 in the atmosphere, and how does the system respond to perturbations? The conclusions can be summarized as follows.

With regard to CO_2, the atmosphere (and the biosphere as well) represents only a small appendix to the much larger ocean reservoir. In the absence of perturbations and on a time scale sufficiently long to guarantee a well-mixed ocean, the partial pressure of CO_2 in the atmosphere will be governed by solution equilibrium. On a much shorter time scale, only the upper layer of the ocean is equilibrated with atmospheric CO_2, so that deviations from equilibrium with the deep ocean are expected. The inorganic carbon content of the ocean is determined by the balance between the weathering of rocks and the sedimentation of calcium carbonate. Table 11-7 shows that the influx of fresh carbon arises mainly from the weathering of limestones. The current rate of biological $CaCO_3$ formation is about five times greater than that of carbonate deposition. The mismatch appears to result from the dissolution of $CaCO_3$ below the carbonate compensation level. Broecker (1973) consequently suggested that it is this level that controls the global $CaCO_3$ deposition rate. The role of the carbonate compensation level is still somewhat obscure, but, in principle, it may adjust to the amount of CO_2 dissolved in the ocean and thus provide a feedback mechanism, which by regulating the $CaCO_3$ deposition tends to keep the carbon content of the ocean nearly constant. According to Table 11-7, the time constant for

CO_3^{2-} removal is 1.7×10^5 yr. It could be reduced by a factor of five under favorable conditions, namely, if the biological $CaCO_3$ production rate stayed constant and all of the $CaCO_3$ formed was also deposited. The geochemical CO_2 cycle clearly is coupled to that of calcium. For present conditions, however, calcium is not the controlling factor, because its concentration in moles per kilogram is about five times greater in the ocean than that of dissolved CO_2, and the time constant for changes is correspondingly larger.

Consider now a perturbation of the system, specifically the addition of CO_2 to the atmosphere. Equilibration with the mixed layer of the ocean is achieved quickly within a few years, although the evasion factor prevents a favorable partitioning of CO_2 between both reservoirs and more than 90% of the added CO_2 stays initially in the atmosphere. The excess is eventually transferred into the deep sea. Complete equilibrium is achieved slowly and requires a time of the order of 1000 yr. In the interim period, the biosphere may utilize some of the excess CO_2 and store it temporarily as organic carbon in biomass and soil humus. The residence time is longest in soil humus (see Table 11-10), amounting to some hundred years, on average. This is still short compared with the time needed to attain complete equilibrium between atmosphere and ocean, so that it is the latter process that determines the long-term behavior of the system.

The exploitable fuel reserves have been estimated as 7×10^3 Pg of carbon (Zimen et al., 1973, 1977). The amount corresponds to nearly 19% of that currently residing in the ocean–atmosphere reservoir. If all of the fossil fuel were burned and the resulting CO_2 transferred into the ocean, the evasion factor would rise from 10 to 15 and the mixing ratio of CO_2 in the atmosphere, at equilibrium, would increase by a factor of 2.7 to about 800 ppmv. Rather dramatic climatic changes are then expected to take place. Much higher mixing ratios would prevail temporarily before equilibrium is reached, depending on the rate at which fossil fuel reserves are consumed. Siegenthaler and Oeschger (1978) have investigated several conceivable scenarios by means of model calculations. In the worst case, if most of the fuel reserves are utilized during the next century, the mixing ratio of CO_2 in the atmosphere would rise to 3000 ppmv for some time. Levels that high will have very deleterious consequences for life on earth, and the prospect of such a development should be cause for serious concern, more so than any other effect of local or global pollution.

The slow geochemical cycles begin to take hold long after the excess CO_2 has been transferred into the deep sea. On account of the inorganic chemistry involved, one would expect the increase in atmospheric CO_2 to cause an enhancement in the weathering rate. In reality, this concept is hard to prove because of the unknown contribution to weathering by CO_2 produced biogenically in soils. It might well be that the weathering rate is largely biologically controlled. In the oceans, however, the deposition of $CaCO_3$ appears to be controlled by an abiotic process, even though the formation of $CaCO_3$ is biologically mediated. Hence one expects an increase of inorganic carbon to enhance the deposition rate somewhat. Again, it is difficult to make quantitative predictions, but the time scale for the removal is not expected to change much so that the excess of inorganic carbon will remain in the ocean for a hundred thousand years to come.

Chapter

12 The Evolution of the Atmosphere

Two groups of atmospheric gases remain to be discussed: the noble gases, and the major constituents of air, nitrogen and oxygen. The great stability of their abundances indicates residence times in the atmosphere much longer than those for other constituents, and in discussing these gases we are invariably led to consider the evolution of the atmosphere over a time span reaching back to the formation of Earth some 4.5 billion yr ago.

Our knowledge about the primitive atmosphere that developed on Earth after its creation is quite fragmentary, thus leaving much room for speculation. Nevertheless, there is sufficient evidence to show that the early atmosphere was a secondary feature, produced by thermal outgassing of virgin planetary matter, in contrast to a primordial atmosphere composed of gases originally present in the solar nebula. The evidence comes mainly from the depletion of the rare gases on Earth relative to their solar abundances, and from the consumption of excess volatiles during the formation of the sediments. These topics will be discussed in the next two sections before we consider the behavior of nitrogen and oxygen.

12.1 The Noble Gases

Mixing ratios of the noble (or rare) gases in air are given in Table 12-1. Included are the corresponding mass contents in the atmosphere and the

Table 12-1. *Mixing Ratios, Total Mass Contents, and Isotopic Compositions of Stable Noble Gases in the Atmosphere*

Constituent	Mixing ratio (ppmv)	G_A (Pg)	Isotopic composition (%)		
He	5.24	3.71	^{4}He($\sim$100)	^{3}He(1.25×10^{-4})	
Ne	18.2	65.0	^{20}Ne(90.5)	^{21}Ne(0.268)	^{22}Ne(9.23)
Ar	9340	6.6×10^4	^{36}Ar(0.337)	^{38}Ar(0.063)	^{40}Ar(99.6)
Kr	1.14	16.9	^{78}Kr(0.354)	^{80}Kr(2.270)	^{82}Kr(11.56)
			^{83}Kr(11.55)	^{84}Kr(56.90)	^{86}Kr(17.37)
Xe	0.087	2.0	^{124}Xe(0.096)	^{126}Xe(0.09)	^{129}Xe(1.919)
			^{129}Xe(26.44)	^{130}Xe(4.08)	^{131}Xe(26.89)
			^{132}Xe(26.89)	^{134}Xe(10.44)	^{136}Xe(8.87)

observed isotopic distributions. The fractions dissolved in the ocean are comparatively small, but those present in portions of Earth's crust or mantle are not necessarily negligible. Due to their chemical inertness, the noble gases are not expected to enter into chemical cycles, so that they must either have accumulated in the atmosphere as a result of terrestrial exhalation or represent a remnant of a primordial atmosphere. We shall see that the latter possibility can be rejected. A third possibility, the capture of solar wind particles by the planet, also turns out to be unimportant. By combining the observed extraterrestrial proton flux with known solar rare gas abundances, Axford (1970) and Banks and Kockarts (1973) have shown that the accretion rate is too small by several orders of magnitude to account for the rare-gas content of the atmosphere.

Helium and ^{40}Ar are recognized to have originated as products from radioactive decay processes in the earth's crust and mantle. Thus, ^{40}Ar derives from K-shell electron capture in ^{40}K, and ^{4}He is generated by nuclear disintegrations within the uranium and thorium decay series. These gases, therefore, have definitely accumulated. Argon is entirely retained by Earth's gravitational field, whereas helium has a tendency to escape slowly to interplanetary space so that it does not accumulate to the same degree as argon. The rates of production and the total amounts of argon and helium that have been generated since the formation of Earth can, in principle, be calculated from the terrestrial abundances of the precursor elements. Unfortunately, the abundances are not well known and we must be satisfied with estimates. Ganapathy and Anders (1974) estimated the average potassium content of the earth as 170 ppmw. The fraction present as ^{40}K is 1.2×10^{-4}, its total radioactive decay constant is 5.54×10^{-10} yr^{-1} and about 10% of the product is ^{40}Ar (Steiger and Jäger, 1977). With the mass of

Earth (5.98×10^{24} kg) one calculates a total ^{40}Ar production over the last 4.5 Gyr of 1.4×10^5 Pg. The amount is about twice that presently residing in the atmosphere. If Earth had the average composition of chondritic meteorites (an incorrect assumption), the total production of argon would have been five times higher (Turekian, 1964), and this value sets an upper limit. The lower limit is given by the amount in the atmosphere. Although the uncertainties are large, it appears that perhaps one-half of radiogenic argon has been exhaled and that a comparable fraction remains in Earth's crust and mantle.

Similar considerations apply to helium. Zartman *et al.* (1961) and Wasserburg *et al.* (1963) have studied helium and argon in natural gases, including geothermal and bedrock gases of varied chemical compositions. In most samples, the radiogenic He/Ar ratios fell into the range 6–25, which compares well with production rates calculated from abundances of uranium, thorium, and potassium in average igneous and sedimentary rocks. The helium content of the gases varied by a factor of 2000; hence, it is not unreasonable to conclude that the degassing of helium and argon occurs at roughly equivalent rates.

The observed ^{4}He/^{40}Ar production ratio of about 10, on average, corresponds to a mass ratio of about unity. And yet, as Table 12-1 shows, the ratio in the atmosphere is 1/1800. Apparently, most of the helium has escaped the gravitational field of Earth. We can derive an estimate for the residence time of helium in the atmosphere in the following way. The average release rate for argon over geological times to reach the current atmospheric abundance has been $6.6 \times 10^{16}/4.5 \times 10^9 = 1.4 \times 10^7$ kg/yr. The current production rate of ^{40}Ar is $\exp(-\lambda \Delta t) = 0.082$ times the rate 4.5 Gyr ago, where $\lambda = 5.54 \times 10^{-10}$ is the ^{40}K decay constant, and the average production rate over the period Δt is

$$[1 - \exp(-\lambda \Delta t)]/\lambda \Delta t \exp(-\lambda \Delta t) = 4.45 \qquad (12\text{-}1)$$

times the present value. If one accepts the observed average mass ratio, ^{4}He/^{40}Ar ≈ 1, one obtains for the current helium release rate and the He residence time in the atmosphere the values

$$Q(\text{He}) = 1.4 \times 10^7/4.45 = 3.3 \times 10^6 \quad \text{kg/yr}$$

$$\tau(\text{He}) = G(\text{He})/Q(\text{He}) = 3.7 \times 10^{12}/3.3 \times 10^6 \approx 10^6 \quad \text{yr}$$

The total production rate of helium cannot be much higher. As was pointed out by Nicolet (1957), the heat flow of Earth places a limit on the rate of helium production. If the heat originates entirely from the decay of uranium and thorium, an average heat flow of 3×10^{-3} J/m^2 s would correspond to a helium production rate of 3.5×10^6 kg/yr. Clearly, these values are quite uncertain, but at least they do not depend on assumptions concerning the uranium and thorium contents of Earth.

The mechanism of the thermal escape of a gaseous constituent from Earth's gravitational field was first described by Jeans (1916) on the basis of kinetic theory. Recent developments have been reviewed by Hunten (1973) and by J. C. G. Walker (1977). The loss of gas particles to interplanetary space takes place in the outermost region of the atmosphere, in the exosphere. Here, the gas densities are reduced to values so low that collisions between gas particles become rather improbable. For such conditions, an upward-moving particle will be lost when, on its last collision at altitudes near the exobase, it has acquired sufficient kinetic energy to overcome the backpull of Earth's gravitational field. Otherwise the particle will return to the exobase in a ballistic orbit. The kinetic energy required for escape is $\frac{1}{2}mv^2 > mg(r)r$ where $r = r_0 + z$ is the distance from the center and r_0 the radius of Earth, and $g(r) = g_0(1 + z/r_0)^{-2}$ the acceleration due to gravity at the altitude $z = z - r_0$. The minimum velocity for escape then is

$$v_e = [2g(r_c)r_c]^{1/2} \tag{12-2}$$

where r_c denotes the critical altitude level, essentially the base of the exosphere. At this level, the particles are assumed to have a Maxwellian distribution of velocities, although it is recognized that the high-energy tail of the distribution is modified by the loss of fast particles. The error incurred is considered tolerable. After weighting the number density n_c at the critical level with the Maxwellian distribution function and integrating over all directions and velocities greater than v_e, one obtains an expression for the escape flux:

$$F_e = n_c \left(\frac{R_g T}{2M} \right)^{1/2} \left[1 + \frac{Mg(r_c)r_c}{R_g T} \right] \exp \left[\frac{Mg(r_c)r_c}{R_g T} \right] \tag{12-3}$$

The overriding factor is the exponential term. On Earth, only hydrogen and helium have atomic masses small enough and velocities high enough to overcome the gravitational field at temperatures existing in the thermosphere (750–1500 K). For hydrogen, the escape flux is sufficiently high to cause a supply problem so that, in effect, the escape rate is transport-limited (Hunten, 1973). This is not the case for helium.

The exobase is determined essentially by the average collision cross section $\sigma_c(2 \times 10^{-14} \text{ cm}^2)$. The critical altitude is given with sufficient accuracy by the condition

$$\int_{r_c}^{\infty} \sigma_c n_T(r) \, dr = \sigma_c n_T(r_c) H(r_c) = 1 \tag{12-4}$$

where n_T is the total number density and H is the scale height. Both depend

Table 12-2. *Conditions at the Base of the Exosphere, Helium Escape Fluxes, and Annual Helium Loss Rates as Function of Temperature[a]*

		$T(K)$					
Parameter	Units	750	1000	1250	1500	1750	2000
z_c	km	400	450	550	650	700	800
n_{He}	atoms/cm^3	2.0 (6)	2.4 (6)	1.6 (6)	1.3 (6)	1.4 (6)	1.2 (6)
n_T	particles/cm^3	3.6 (7)	5.2 (7)	3.4 (7)	2.3 (7)	2.9 (7)	2.3 (7)
$g(r)$	m/s^2	8.68	8.55	8.31	8.07	7.96	7.74
F_e(He)	atoms/m^2 s	1.6	2.6 (4)	5.7 (6)	2.2 (8)	3.3 (9)	2.2 (10)
F_e(He)	kg/yr	1.9 (−4)	3.2	7.2 (2)	2.8 (4)	4.4 (5)	3.0 (6)

[a] Partly using data compiled by Banks and Kockarts (1973). Numbers in parentheses indicate powers of 10.

on temperature, which varies with the time of the day (weakly) and with solar activity (strongly). Table 12-2 shows as a function of temperature the altitude of the exobase and other parameters required to evaluate the escape flux. The total number density at the critical level is dominated by atomic oxygen, and helium is the next important constituent. These conditions follow from the fact that at altitudes above 120 km O_2 is largely dissociated, mixing no longer prevails, and molecular diffusion is the dominant transport mechanism. Each constituent then establishes its own scale height. Table 12-2 includes the resulting helium escape fluxes. Even with temperatures on the high side one finds that the loss rate falls short of that required to accomodate the release of helium to the atmosphere from terrestrial sources. MacDonald (1963, 1964) considered the variations in the escape flux over a complete solar cycle and, taking the average, found the loss rate too low by a factor of 30. Although the resulting atmospheric residence time of helium ($\sim 6 \times 10^7$ yr) is still short compared with the age of Earth, it is obvious that Jeans escape does not balance the helium budget.

A variety of suggestions has been made to overcome the difficulty [see Hunten (1973) for details]. The most reasonable one appears to be the proposition by Axford (1968) that helium leaves Earth's atmosphere mainly in the ionized state. Ion flow would be most effective in the polar regions where the magnetic field lines are open, but there is some doubt whether the ionization rate is adequate. This question has not yet been resolved, so that our understanding of the atmospheric helium budget remains unsatis-factory.

We now turn from the radiogenic components to consider the residual rare gases. Brown (1949) and Suess (1949) independently noted that the noble gases on earth are severely underrepresented compared to their

Table 12-3. *Abundances of the Rare Gases in the Sun, in Earth's Atmosphere, and in a Number of Meteorites, Relative to Silicon* $(X_{Si} = 10^6)^a$

Constituent	$(X/X_{Si})_{solar}$	$(X/X_{Si})_{earth}$ [b]	DF_{earth}	$DF_{meteorite}$ [c]
He	3.1×10^9	—	—	7.9–10.9
Ne	1.6×10^7	1.0×10^{-4}	11.20	8.0–11.4
$^{36}Ar + ^{38}Ar$	2.4×10^5	2.1×10^{-4}	9.05	7.3–8.9
Kr	5.1×10^1	6.7×10^{-6}	6.90	5.7–7.7
Xe	3.6	4.9×10^{-7}	6.85	4.2–6.7

[a] Data from Signer and Suess (1963). The deficiency factor is defined as $DF = -\log_{10}(X/X_{Si})_{sample}/(X/X_{Si})_{solar}$.
[b] Assumes a terrestrial abundance of silicon of 14% by weight.
[c] Assumes an average silicon content of meteorites of 20% by weight.

abundances in the solar system. Signer and Suess (1963) later demonstrated similar deficiencies for the primordial noble gas components in certain meteorities. Table 12-3 lists relative abundances of neon, $^{36}Ar + ^{38}Ar$, krypton, and xenon in the sun, on Earth, and in meteorites. Silicon is used as the reference element. The left-hand side of Table 12-3 shows that the terrestrial noble gases are indeed significantly depleted by many orders of magnitude. Similar deficiencies occur for several other volatile elements, for example, hydrogen and nitrogen (Urey, 1952). In discussing relative abundances of elements in the aerosol, we have found it convenient to introduce (on page 325) a fractionation factor, which in view of the observed enrichments relative to crustal abundances we had called an enrichment factor. In the present context, the deficiency of volatile elements on Earth can be treated in the same manner. Because of the large range of values, it is useful to work with the negative logarithm of the fractionation factor and to define the deficiency by

$$DF = -\log_{10}[(X/Si)_{sample}/(X/Si)_{solar}]$$

Values derived from the noble gases on Earth are entered in column three of Table 12-3. It is then easy to see that the deficiency increases toward the lighter noble gases. If this feature is interpreted to have resulted from Jeans escape, one finds that the mass of the body from which the escape took place must have been 100 times smaller than that of the present Earth, provided unreasonably high temperatures are excluded. The result may be used to support the argument that any primordial atmosphere that Earth might have had was removed before Earth had agglomerated to its present size.

Noble gases in meteorites have been studied after releasing them from the solid by heating. They have at least four sources: radiogenesis of ^4He

and ^{40}Ar; spallation processes induced by interaction with cosmic rays; implantation of solar wind particles; and the incorporation of primordial gases during the formation of solid matter by condensation from the solar nebula. The last component is the one of interest here. It occurs primarily in dark grains imbedded in material of a lighter texture, so that it can be isolated and studied separately. Deficiencies for this fraction, taken from the compilation of Signer and Suess (1963), are entered in the last column of Table 12-3 to demonstrate that their magnitudes are similar to those for the terrestrial noble gases. Several of the meteorites investigated have potassium–argon ages close to 4.5 Gyr and must have arisen concurrently with the material that formed the earth. Yet their primordial noble gas abundances correlate well with the terrestrial ones, thus indicating a common origin of both.

Concepts about the origin of Earth and the other planets have been reviewed by Arrhenius *et al.* (1974), Grossman and Larimer (1974), Alfven and Arrhenius (1976), J. C. G. Walker (1977), and Ringwood (1979). It is now fairly well accepted that the planets formed at moderate temperatures by the accretion of dust grains and larger aggregates thereof (planetesimals), produced by condensation from the plasma phase surrounding the developing sun. Certain classes of meteorites, especially carbonaceous chondrites, are thought to represent residues of early condensates. And although by composition they should not necessarily be identified with the material that had formed Earth, they provide first-hand information about the products resulting from primordial condensation. On account of their small size, meteorites cannot have held an atmosphere of their own, and they must have acquired their primordial noble gas content during the initial condensation stage. The characteristic distribution of noble gas abundances appears to be the result of fractionation associated either with the parent gas phase or with the specific processes by which rare-gas atoms were implanted into the growing crystals. The fact that the abundances of primordial noble gases in meteorites and of noble gases on Earth are closely related suggests that the Earth inherited its store of noble gases in the same manner as meteorites—that is, the planitesimals must have served as host bodies during the accretion stage of planet formation. The gases were subsequently released together with other volatiles when the material became heated. Within the framework of this concept it is not necessary to assume that the earth even possessed a primordial atmosphere other than that derived by outgassing. The thermal history of Earth, the nature of the degassing products, and the probable composition of the resulting early atmosphere will be discussed in the next section.

The time scale of accretion is not well established but must have been shorter than $\sim 5 \times 10^8$ yr. From studies of lunar craters it can be inferred that the period of extensive impactation of accreting bodies ended about 3.9 Gyr ago (Neukum, 1977). The age of the oldest known sediments from the Isua formation is 3.7 Gyr (Moorbath *et al.*, 1973). The rocks contain well-rounded pebbles, indicating the presence of liquid water at that time, in addition to carbonates and other evidence for weathering processes similar to those being active today.

12.2 The Primitive Atmosphere

Thermal exhalation of volatiles from Earth's interior is a process occurring today in conjunction with volcanism, heating being provided by radioactive decay. Since the level of radioactivity is declining steadily, it is virtually certain that the current degassing rate is only a small fraction of that 4.5 Gyr ago. During Earth's accretion, additional heat was supplied by impactation, and conceivably by the gravitational settling of metallic iron to the core. The last point is a matter of controversy [See Ringwood (1979) for details]. If Earth had accreted homogeneously, core formation would have occurred comparatively late and the accompanying sudden energy release would have been sufficient for a complete melting of the planet (Birch, 1965; Hanks and Anderson, 1969). Such a massive heating event leads to fundamental difficulties, as we shall see below, and must be rejected. Alternatively, there are plausible arguments for an inhomogeneous accretion process. During the cooling of the solar nebula, high-temperature refractories would have condensed out first, giving rise to metal-rich protoplanets, which then collected low-temperature condensates at a later stage. According to this model, the formation of a core resulted from the specific sequence in which the material accumulated, not from the later separation of materials having different densities, so that the above-mentioned difficulties are circumvented. As discussed by Ringwood, however, there are observations in the geochemical domain that appear to be in conflict with inhomogeneous accretion, and for this reason the model has not found a universal acceptance.

In the present context, the details of accretion are not a matter of concern as long as excessive surface temperatures are avoided. The principal argument against a massive heating of Earth late during its development is the accompanying catastrophic release of most if not all volatiles contained in virgin planetary matter. Water vapor must have been an important degassing product in view of the existence of an ocean on Earth. The release of water in amounts comparable to those now present in the seas combined with elevated temperatures would have produced a thick steam atmosphere nearly opaque to infrared radiation. As Rasool and DeBergh (1970) first demonstrated for the planet Venus, such conditions would have led to an

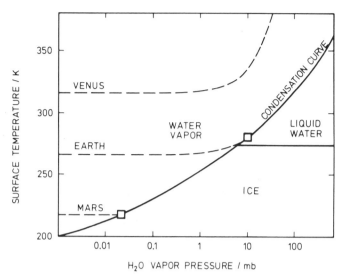

Fig. 12-1. Surface temperatures expected on the three planets Venus, Earth, and Mars as a function of the water vapor pressure. An increase in the vapor pressure increases the retention of infrared radiation in the atmosphere, raising the temperature via the greenhouse effect. Overlaid is the phase diagram of water. On Earth and Mars the starting (radiation equilibrium) temperatures are low enough for water to condense out when the temperature intersects the condensation curve. On Venus, the temperature rises more rapidly and runs away. [Adapted from Walker (1977), originally modeled by Rasool and DeBergh (1970).]

irreversible greenhouse effect, causing temperatures to remain high. Figure 12-1 shows from the planets Venus, Earth, and Mars the temperatures expected due to infrared absorption by water vapor as a function of its surface partial pressure. The dashed curves begin on the left at temperatures for radiative equilibrium in the absence of any absorbing atmosphere. Increasing the water vapor pressure causes the temperature to rise until, for Mars and Earth, the temperature curve intersects the saturation pressure curve, whereupon the rise is halted. If the initial temperature is too high, as is the case for Venus, the saturation curve cannot be reached and the temperature runs away. These results argue against average surface temperatures greatly exceeding 300 K on the primitive Earth, since without doubt the temperature must have been favorable enough for water to condense. Figure 12-1 suggests that the principal factor responsible for such conditions is the optimum distance from the sun. Venus, being much closer, has temperatures that are too high, whereas the more distant planet Mars features temperatures too low to maintain liquid water for long before it freezes.

If it is permissible to assume that the inner planets Venus, Earth, and Mars have accreted from roughly the same type of meteoritic material, one

Table 12-4. *Data for the Planets Venus, Earth, and Mars. Physical Parameters and the Main Constituents of Atmospheres*[a]

Parameter		Venus	Earth	Mars	
Mass of planet (kg)		4.88 (24)	5.98 (24)	6.42 (23)	
Acceleration of gravity (m/s^2)		8.88	9.81	3.73	
Radius (km)		6053	6371.3	3380	
Surface area (m^2)		4.6 (14)	5.1 (14)	1.44 (14)	
Surface temperature (K)		730	288	218	
Surface pressure (bar)		91	1	0.007	
Mass of atmosphere (kg)		4.78 (20)	5.1 (18)	2.5 (16)	
Composition	CO_2	96	0.03	95.3	
of the atmosphere	N_2	3.4	78.08	2.7	
(in percent)	O_2	6.9 (−3)	20.9	0.13	
	H_2O	0.1–0.5	2	0.03	
	^{40}Ar	(2–7) (−3)	0.93	1.6	
Ratios: mass	CO_2	9.4 (−5)	3.8 (−5)[b]	4.0 (−8)[c]	>4 (−8)[d]
of a volatile	N_2	2.1 (−6)	8.0 (−7)[b]	6.8 (−10)[c]	4 (−8)[d]
to mass of	H_2O	(1–5) (−7)	2.8 (−4)[b]	5 (−12)[c]	>5 (−6)[d]
the planet	^{40}Ar	(2–7) (−9)	1.1 (−8)	5.6 (−10)	5.6 (−10)

[a] From the compilations of Owen *et al.* (1977), Oyama *et al.* (1979) and Pollack and Yung (1980). Orders of magnitude are indicated in parentheses.

[b] From Table 12-5; includes CO_2 in carbonates, N_2 in shales, and H_2O in the ocean.

[c] Actual values.

[d] Estimates including material in near-surface reservoirs or lost to space, according to McElroy *et al.* (1977b) and Pollack and Black (1979). Venus and Earth have about the same size, and the amounts of volatiles outgassed are roughly equivalent. Mars has outgassed to a lesser extent, as judged from the ^{40}Ar content. The fate of H_2O on Venus is still undetermined.

expects their degassing products and atmospheres to have similar chemical compositions provided just a simple accumulation took place and further changes were kept minimal. Table 12-4 compares abundances of the major constituents in the atmospheres of the three planets. It shows that the atmospheres of Venus and Mars are indeed similar in that CO_2 is the major constituent. The much lower abundance of CO_2 in the terrestrial atmosphere must be attributed to its removal by sediment formation, which, in turn, follows from the presence of liquid water on Earth but not on the other planets. If all the CO_2 now buried in the sediments were brought into the atmosphere, CO_2 would rank as the dominant constituent also on Earth. Hence it appears that CO_2 was a major exhalation product on all three planets, in addition to water vapor. Yet, it is equally apparent that the three atmospheres have undergone evolutionary changes. Venus has lost most of the water initially present, and the Earth has developed a considerable amount of oxygen that the other planets are lacking.

Magmatic gases released from volcanoes today contain water vapor and carbon dioxide as the major components, with smaller contributions of SO_2/H_2S, HCl, CO, H_2, and N_2. A large share of them probably has undergone extensive recycling so that they should not be considered juvenile degassing products. It is nevertheless possible to derive approximately the average composition of juvenile gases by tracing them to the geochemical reservoirs in which they now reside and by setting up an inventory of the accumulated masses. Geochemists have long recognized that water in the hydrosphere, CO_2 buried as carbonate in the sediments, chloride and sulfate dissolved in seawater, and nitrogen contained in the atmosphere are far too abundant to have originated from rock weathering alone and, hence, must have been supplied as thermal exhalation products. Rubey (1951), who was one of the first to prepare a quantitative account of them, appropriately termed them "excess" volatiles. Table 12-5 shows an inventory for the most important volatile elements in addition to water, fashioned after that presented by Rubey. More recent data are used in Table 12-5, but the results are very similar to the earlier ones and the conclusions remain the same. Li (1972) employed a somewhat different method, using mass balance relations for the weathering reactions in conjunction with ^{18}O, ^{13}C, and ^{34}S isotope data in order to fix the cumulative masses of individual volatiles. His results are entered in the last row of the table to show that concerning the total amounts of volatiles very little divergence of opinion exists. The individual entries in Table 12-5 indicate that, indeed, water resides mainly in the oceans, carbon in the sediments, and nitrogen in the atmosphere, whereas chlorine and sulfur are about evenly divided between oceans and sediments. In the latter domain, Cl and S occur mainly in evaporites, sulfur being additionally present in reduced form in shales.

Table 12-6 compares the relative abundances of excess volatiles with compositions of gases escaping from volcanoes and fumaroles, and gases that can be extracted from solidified lava. Only a few selected data are given. Volcanic gases are notoriously difficult to study, because contamination with air is almost unavoidable and chemical reactions may change the composition during the time between collection and analysis. Moreover, all measurements reveal a marked variability in composition, which is partly due to variations in temperature and the accompanying changes in chemical equilibria. These difficulties notwithstanding, it can be seen from Table 12-6 and has been emphasized also by Rubey (1951) that the excess volatiles occur in relative proportions comparable to those found for the same elements in volcanic emanations. This implies either that the degassing conditions during the early period of Earth's history were roughly the same as today, or that current volcanic exhalations feed largely on volatiles being recycled from the store of sedimentary deposits. Both conclusions probably

Table 12-5. *Estimates for the Masses (kg) of Water and Other Volatiles Residing in the Atmosphere, the Ocean, and the Sediments, with Contributions from the Weathering of Igneous Rocks and "Excess" (Magmatic) Exhalations*[a]

Source	H_2O	C	CO_2	N_2	Cl	S
Atmosphere	Negligible	7.0 (14)	2.5 (15)	3.9 (18)	Negligible	Negligible
Hydrosphere	1.4 (21)	3.4 (16)	1.4 (17)	2.2 (16)	2.7 (19)	1.2 (18)
Sedimentary rocks	3.3 (20)[b]	6.2 (19)[c]	2.3 (20)	9.0 (17)[d]	2.2 (19)[e]	1.2 (19)[f]
Total volatiles	1.7 (21)	6.2 (19)	2.3 (20)	4.8 (18)	4.9 (19)	1.3 (19)
Supplied by weathering of igneous rocks[g]	3.0 (19)	6.9 (17)	2.5 (18)	5.9 (16)[d]	6.5 (17)	8.8 (17)
Excess volatiles	1.7 (21)	6.2 (19)	2.2 (20)	4.2 (18)	4.8 (19)	1.2 (19)
Excess volatiles according to Li (1972)[h]	1.7 (21)	5.6 (19)	2.0 (20)	3.9 (18)	4.2 (19)	1.3 (19)

[a] Orders of magnitude are shown in parentheses.
[b] Garrels and MacKenzie (1971).
[c] From Table 11-1.
[d] Wlotzka (1972) corrected to a total sediment mass of 2.4 (21) kg.
[e] Sediment distributions of Hunt (1972), Table 11-3, assuming 3% of evaporites.
[f] Hosler and Kaplan (1966).
[g] 2.4×10^{21} kg derives from the weathering of 2.15×10^{21} kg of igneous rock. Element abundance taken from Horn and Adams (1966).
[h] The estimates of Li (1972) were derived in part from isotope balances combined with mass balances of rock weathering.

Table 12-6. *Composition of Volcanic Gases (% by Volume) Compared with the Composition of Excess Volatiles*

Location	H_2O	CO_2	N_2	Cl	S	H_2	CO	References
Hawaii, composite[a]	79.3	11.6	1.3	0.05	6.9	0.6	0.4	Eaton and Murata (1960)
Kilauea, Hawaii[b]	97.3	2.3	—	—	0.43	0.23	—	Heald *et al.* (1963)
Surtsey, Iceland[c]	86	6.0	0.07	0.4	2.7	4.7	0.4	Sigvaldason and Elisson (1968)
Erta'Ale, Ethiopia[d]	79.4	10.0	0.18	0.4	7.0	1.5	0.5	Giggenbach and LeGuern (1976)
Lassen Peak, California, gas extracted from lava	93.7	2.1	0.6	0.3	0.9	0.4	0.6	Shepherd (1925)
Mauna Loa, Hawaii, lava	73.2	15.3	5.2	0.2	0.2	4.4	1.4	Shepherd (1938)
Excess volatiles	93.6	4.6	0.2	1.2	0.4	—	—	Last two lines from Table 12-5

[a] Composite from data reported by Shepherd (1938), Jagger (1940), and Naughton and Terada (1954).
[b] Corrected for the presence of air.
[c] Average of the three least contaminated samples.
[d] Average of 18 samples.

603

are applicable. The oxidation state of volcanic gases is mainly determined by the Fe^{3+}/Fe^{2+} ratio in the magmas from which the gases evolve. It is interesting to note that Precambrian basalts contain Fe_2O_3 and FeO in essentially the same ratio as recent primary rocks (Eaton and Murata, 1960; Holland, 1978), indicating that the oxidation state has not changed significantly with time.

Free oxygen is not a volcanic exhalation product. The partial pressure of oxygen above basaltic melts is about 10^{-8} bar at 1500 K (Katsura and Nagashima, 1974), and it decreases toward lower temperatures (Heald et al., 1963). The O_2 concentration is low enough to cause the appearance in volcanic gases of H_2 and CO (see Table 12-6), which result from the high-temperature dissociation of H_2O and CO_2, respectively. The fractions of H_2 and CO are found to agree roughly with those expected from thermochemical equilibria with their precursors, provided oxygen partial pressures of 10^{-7} bar or lower are inserted in the calculations (Holland, 1978; Matsuo, 1978).

In the primitive atmosphere, oxygen was essentially absent. The prevalence of anoxic conditions during the first 2.5 Gyr on Earth's history is documented by the preservation in ancient detritals of minerals like uranite and pyrite (Schidlowski, 1966, 1970), which are unstable in oxidizing environments. Authors addressing the problem of the origin of life (e.g., Rutten, 1971; Miller and Orgel, 1974) emphasize that the initial development of life demanded a reducing habitat. Volcanic gases having a modern composition presumably would have provided a sufficiently reducing environment, but many workers in the field have adopted the more radical suggestion of Urey (1952) and Miller and Urey (1959) of a primitive atmosphere consisting mainly of methane and ammonia. Their principal arguments are not concerned with outgassing conditions, and we shall deal with them further below.

The possibility exists, of course, that primitive volcanic gases were more reducing than those being released now, but the admissible margin is rather narrow. Suppose for the sake of argument that the 5×10^{19} kg of inorganic carbon residing in the sediments had been issued originally as CO. Its subsequent oxidation to CO_2 would have required 6.7×10^{19} kg of oxygen. As we shall see later, the mass of oxygen derived from photochemical processes is approximately 3.4×10^{19} kg, that is, only one-half of that required. Since further amounts of oxygen were needed to oxidize other elements such as sulfur and bivalent iron, no more than a small fraction of volatile carbon can have occurred as CO, and most of it must have been released as CO_2. Even greater difficulties would have to be met had we assumed that carbon had emanated predominantly in the form of methane. We conclude that not only the relative amounts but also the oxidation state of the primitive degassing products were similar to the volcanic gases of today.

With the assumption that the reservoir of releasable volatiles is limited, Li (1972) suggested for the degassing rate and the accumulation of volatiles G_s at time t the functions

$$dG_s/dt = G_{s0}\lambda \exp(-\lambda t)$$
$$G_s = G_{s0}[1 - \exp(-\lambda t)]$$

(12-5)

respectively, where G_{s0} is the available total amount of an individual volatile. The degassing rate is directly related to the accumulation rate of the sediments, and Li employed the age frequency distribution of ancient sedimentary deposits, with appropriate corrections for losses due to recycling, to estimate the rate constant as $\lambda = 1.16 \times 10^{-9}$ yr^{-1}. This value lies between two extremes: a rapid degassing shortly after the formation of the earth ($\lambda \to \infty$), and a very slow, almost linear, degassing up to the present time ($\lambda \le 2 \times 10^{-11}$ yr^{-1}). The latter case is unrealistic, however. Accepting Li's estimate, we find that two-thirds of the volatiles had accumulated within the first 10^9 yr after the earth's creation. Today, the degassing of juvenile components is essentially complete and 99% of volcanogenic gases represents recycled material.

Within the framework of this model, the ocean formed early during Earth's history, in agreement with geological evidence. CO_2 would have dissolved mainly in the ocean and formed carbonates by virtue of the weathering of igneous rocks. In all probability, the abundance of CO_2 in the atmosphere has never differed greatly from that of today. HCl as well as sulfur compounds (SO_2 and/or H_2S) would also have dissolved in the ocean, whereas nitrogen, hydrogen, and carbon monoxide are much less soluble and must have at least temporarily accumulated in the atmosphere. Nitrogen will be discussed in the next section. Carbon monoxide is not of much interest, because on one hand it is supplied at a higher rate by the photolysis of CO_2, and on the other hand it is rapidly converted back to CO_2 by reacting with OH radicals derived from the photolysis of water vapor, whereby the steady-state abundance is kept small (Kasting et al., 1979). Hydrogen is considered important and requires comment.

If it is assumed that the ratio of hydrogen to water vapor in juvenile gases is the same as in modern volcanic gases (~1%), one estimates from Table 12-5 a potential for the outgassing of hydrogen of $G_{H_2} = 1.9 \times 10^{18}$ kg. The amount is a factor of 6.7 greater than that of nitrogen. The result should not be taken to imply that the early atmosphere was dominated by hydrogen, because Jeans escape provides a leak that keeps the abundance of H_2 below that of nitrogen at all times. As Hunten (1973) and Hunten and Strobel

(1974) have shown, the rate of hydrogen escape in the present atmosphere is controlled by eddy diffusion in the homosphere. This is a consequence of the high temperatures in the exosphere, causing Jeans escape to be faster than its supply by upward eddy diffusion. Since the concentration of CO_2 in the primitive atmosphere was held low by dissolution in the ocean, the outer atmosphere lacked an efficient infrared emitter and the temperature at the exobase presumably was nearly as high as it is today. Transport limitation exists so long as the temperature at the exobase exeeds 600 K. Accordingly, it appears that the escape rate was always determined by transport through the homosphere.

In order to estimate the abundance of hydrogen in the primitive atmosphere, we assume steady-state conditions and set the escape flux equal to that resulting from thermal degassing:

$$F_{esc} \approx \frac{m(H_2) K_z n_T(z_{hp})}{H} = a(G_2/M_{H_2})\lambda \exp(-\lambda t) \tag{12-6}$$

The left-hand side represents the approximation given by Hunten (1973) for the eddy flux in the vicinity of the homopause. The term K_z is the eddy diffusion coefficient, H the scale height, and $n_T(z_{hp})$ the total number density at the altitude of the homopause. We assume that the transition from a well-mixed to a diffusion-controlled atmosphere occurs at the same number density as in today's atmosphere and take $n_T(z_{hp}) = 2.5 \times 10^{19}$ m^{-3} to be a constant. On the right-hand side of Eq. (12-6) the degassing rate must be multiplied with a factor $a = N_L/\gamma A_{earth} = 38.1$ (molecules yr/mol m^2 s) to convert to common flux units. By integrating the one-dimensional diffusion equation [Eq. (1-12)] from the ground on up to height z_{hp} for a constant flux F_{esc}, one finds the mixing ratio at the homopause, $m(H_2) \approx \frac{1}{2}m_0(H_2)$ to be about one half of that at ground level. The dry hydrogen mixing ratio is given by the pressure ratio

$$m_0(H_2) = p_0(H_2)/[p_0(H_2) + p_0(N_2) + p_0(CO_2)] \tag{12-7}$$

with $p_0(N_2) = 0.78 [1 - \exp(-\lambda t)]$, and $p_0(CO_2) = 3 \times 10^{-4}$. Combining this with Eq. (12-5) yields for the hydrogen pressure

$$p_0(H_2) = \frac{A \exp(-\lambda t)[p_0(N_2) + p_0(CO_2)]}{[1 - A \exp(-\lambda t)]} \text{ bar} \tag{12-8}$$

where we have set $A = 2aH(G_{H_2}/M_{H_2})\lambda/K_z n_T(z_{hp})$.

Figure 12-2 shows the course of hydrogen partial pressure and mixing ratio in the primitive atmosphere during the first 3 billion yr after Earth's formation. The critical parameter is K_z. We have here adopted the value appropriate for present conditions at the homopause, $K_z = 10^2$ m^2/s, which

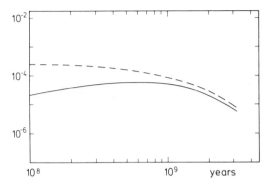

Fig. 12-2. Partial pressure (in bar, solid line) and mixing ratio (dashed line) of hydrogen in the primitive atmosphere on Earth according to Eqs. (12-7) and (12-8). $K_z = 10^2\ m^2/s$, $H = 9\ km$, $A = 3 \times 10^{-4}$. The eddy diffusion coefficient corresponds to present conditions at the homopause. If K_z in the primitive atmosphere had been smaller by a certain factor, the hydrogen partial pressure would have been greater by the same factor and vice versa.

leads to partial pressures of H_2 of the order of 5×10^{-5} bar. One cannot exclude the possibility that K_z was lower, in which case the hydrogen pressure would have been greater by a corresponding factor. Even for $K_z = 1\ m^2/s$, however, hydrogen is still a minor constituent relative to nitrogen. It is important to realize that the present treatment breaks down when oxygen accumulates in the atmosphere, because hydrogen is then oxidized to water. The abundance of hydrogen for those conditions is determined by the rate of oxidation rather than the escape rate. As we shall see in Section 12.4, the transition to oxidizing conditions occurred about 10^9 yr before present.

Hydrogen is discussed here in some detail, because Miller and Urey (1959) suggested that its presence in the primitive atmosphere led to the formation of methane on account of the reaction

$$CO_2 + 4H_2 \rightleftarrows CH_4 + 2H_2O$$

The reaction is interesting in that its equilibrium favors the left-hand side when temperatures are elevated, whereas it favors the right-hand side when the temperature is reduced to 300 K or lower. At 300 K, the thermochemical data (Stull and Prophet, 1971) yield an equilibrium constant

$$K_{equ} = p(CH_4)p^2(H_2O)/p(CO_2)p^4(H_2) = 5.2 \times 10^{19}\quad bar^{-2} \quad (12\text{-}9)$$

Taking $p(H_2O) = 3 \times 10^{-2}$, $p(CO_2) = 3 \times 10^{-4}$, $p(H_2) = 5 \times 10^{-5}$, we find an equilibrium pressure for methane of 100 bar. The calculation ignores the fact that the supply of hydrogen is limited and the methane pressure cannot exceed the initial pressure of hydrogen, but the result demonstrates the tendency of the reaction to convert all the available hydrogen to methane. At 400 K, in contrast, the equilibrium constant is 2.7×10^{12} bar^{-2}, and by using the same data one finds an equilibrium CH_4 pressure of 5×10^{-6} bar, which is much less than that of hydrogen.

The decisive question is whether the conversion of H_2 to methane can proceed at an adequate rate. At higher temperatures, chemical equilibria generally are established quickly, but, as we have seen, the reaction then favors H_2 and CO_2 as reaction products. At low temperatures where the reaction would generate methane, it is immeasurably slow. Under laboratory conditions, hydrogen and carbon dioxide can coexist for many years without showing any signs of reaction. Miller and Urey (1959) argued that the exposure to energetic radiation in the primitive atmosphere would cause the mixture to react nonetheless within geological times. The problem thus reduces to one of reaction kinetics. A detailed study of it seems to be lacking, but a number of facts should be noted. If CO_2 is to be reduced to methane, the C-O bond must be broken. This is one of the tightest bonds known in chemistry, leading, for example, to a dissociation energy for CO of 1,072 kJ/mol. Ultraviolet radiation with wavelengths less than 110 nm might supply the necessary energy, although only in the outer atmosphere as the lower atmosphere is shielded by CO_2. For such conditions methane, if it were formed, would itself become subject to photodecomposition by ultraviolet radiation (see Table 2-6). It is highly doubtful, therefore, whether methane can build up appreciable concentrations in the upper regions of the primitive atmosphere. Even if this had been accomplished, methane would have to diffuse downward into the lower atmosphere. There it would meet OH radicals derived from the photodissociation of water vapor and it would suffer destruction. Thus, while we cannot refute Miller and Urey's argument outright, it appears that the formation of methane in the manner envisioned by them is rather unlikely.

There is a different route to methane formation, however. Methanogenic bacteria are among the most ancient forms of life. They utilize exactly the above reaction as a source of energy and produce methane from hydrogen and carbon dioxide. After the development of life, therefore, methane must have been a component of the atmosphere. Along this route the abundance of methane would be controlled by the bacterial source strength and photochemical destruction in the atmosphere, rather than by chemical equilibrium with hydrogen. If bacterial methanogenesis was extremely efficient, the entire supply of hydrogen might have been utilized for the production of methane, and its abundance could have been appreciable, but probably not greater than about 10^{-5} bar. The consequences have not yet been explored and remain speculative.

12.3 Nitrogen

According to the concepts discussed in the preceding section, nitrogen evolved by thermal outgassing from Earth's interior and accumulated mainly in the atmosphere. The accumulated mass is 3.9×10^{18} kg. Wlotzka (1972) has reviewed the nitrogen contents of other geochemical reservoirs. As Table 12-5 shows, the sediments contain 9×10^{17} kg of nitrogen, predominantly in the form of organic compounds in shales. This amount represents an outflow from the biosphere and must be counted as part of the total mass of nitrogen that has evolved. Igneous rocks contain about 25 ppmw N as ammonium and 3 ppmw N in elementary form, summing to a total of 7×10^{17} kg for the entire content of igneous rocks in the earth's crust. The amount of juvenile nitrogen remaining in the mantle is unknown. Most of the N_2 found in volcanic gases today presumably derives from the recycling of sediments by subduction.

Owing to its utilization by the biosphere, atmospheric nitrogen participates in two slow cycles. One is the exchange of N_2 between the atmosphere and the biosphere via processes of nitrogen fixation and denitrification; the other cycle involves the burial and recovery of organic nitrogen within the turnover of the sediments. Table 12-7 lists estimates for the inventories of fixed nitrogen in various compartments of the biosphere and in the oceans. It can be seen that this fraction is small compared with that residing in the atmosphere. Few organisms are capable of utilizing atmospheric nitrogen directly. Generally, there is a shortage of supply, forcing plants to rely extensively on nitrogen offered by decaying organic matter. According to Rosswall (1976), 95% of the nitrogen present in terrestrial ecosystems is recycled within the soil and between soil and vegetation. The biochemical pathways operative in converting different nitrogen compounds were treated in Section 9.1. Here, we are interested primarily in the rate of nitrogen fixation as the principal source of nitrogen in the biosphere. There are two routes. In addition to the biochemical reduction of N_2 by bacteria and blue-green algae, we must consider abiotic oxidation to nitrate, which all plants and most microorganisms can use directly. Nitrogen returns from the biosphere to the atmosphere in the form of N_2 and N_2O, which are products of denitrification, whereby the cycle is closed.

Estimates for the fluxes associated with N_2-fixation processes are listed in Table 12-8. The individual entries carry the usual uncertainties. The global rate of biological nitrogen fixation in terrestrial habitats was taken from the compilation of Burns and Hardy (1975). It represents the sum of rates for agricultural systems (44 Tg/yr, mainly from legumes), from grasslands (45 Tg/yr), from forests (40 Tg/yr), and from miscellaneous sources

Table 12-7. *Global Nitrogen Contents in the Biosphere, in Soils and in the Ocean* [a]

Reservoir	Mass (Tg N)	Remarks
Terrestrial		
Land plants	0.75–1.1 (4)	Based on C/N = 75 and Table 11-4
Litter and deadwood	1.5 (3)	Using C/N = 60 and 90 Pg C of litter and deadwood
Animals	2.0 (2)	Delwiche (1970)
Soil humus (organic fraction)	2.0 (5)	Based on C/N = 10 and Table 11-1
Soil, inorganic fraction (mainly fixed, insoluble NH_4^+)	1.6 (5)	Delwiche (1970); the figure is questionable: Söderlund and Svensson (1976) suggest instead 1.6 (4)
Sum	3.7 (5)	
Marine		
Plankton	5.2 (2)	Based on C/N = 5.7 (Redfield *et al.*, 1963) and Table 11-1
Animals	1.7 (2)	Delwiche (1970)
Dead organic matter (particulates)	5.3 (3)	Based on C/N = 5.7 and 30 Pg of particulate carbon (see Section 11.2.2)
Dissolved organic matter	3.7 (5)	Based on C/N = 2.7 (Duursma, 1961) and Table 11-1
Inorganic nitrogen (NO_3^-)	5.7 (5)	Söderlund and Svensson (1976), originally Sverdrup *et al.* (1942) and Emery *et al.* (1955)
Sum	9.4 (5)	
Dissolved N_2	2.2 (4)	From Table 12-5
Total mass of nitrogen	13.3 (5)	

[a] Values in Tg, orders of magnitude shown in parentheses; partly based on data from Chapter 11.

(10 Tg/yr). The total is about three times larger than previous estimates of Hutchinson (1954) and Delwiche (1970). The rate of biological nitrogen fixation in marine environments is less well defined, as the range of estimates given by Söderlund and Svenson (1976) shows. These authors considered also denitrification in the oceans, for which they derived a rate of 91 Tg/yr. Since it should balance the rate of marine nitrogen fixation, both probably amount to about 100 Tg/yr. Terrestrial and marine conversion rates combined transfer about 220 Tg of nitrogen per year from the atmosphere to the biosphere. Supplementary fixation processes are the natural oxidation of N_2 by lightning discharges and by ionizing radiation in the atmosphere, by industrial processes (mainly the production of fertilizers for agriculture), and by the inadvertent oxidation of nitrogen in air during the combustion of fossil fuels. The fluxes in Table 12-8 show that the biological nitrogen fixation is the dominant process, although the application of fertilizers is expected to grow in the future and may eventually reach a similar magnitude.

Table 12-8 also lists residence times of nitrogen in the atmosphere that one calculates from the individual fluxes. Biological nitrogen fixation results in a residence time of 18 million yr, which is much shorter than the age of Earth and its atmosphere. Even the oxidation of N_2 in lightning discharge leads to a residence time that is shorter. Lewis and Randall (1923) and Sillen (1966) pointed out that in the presence of free oxygen and liquid

Table 12-8. *Global Fluxes and Atmospheric Residence Times of Nitrogen Associated with Biological Nitrogen Fixation, Inorganic Atmospheric Oxidation, Agricultural Fertilizer Production, Combustion Process, and the Sedimentary Cycle*

Process	Flux (Tg N/yr)	τ (yr)	Remarks
Biological fixation			
Terrestrial	139		Burns and Hardy (1975), Hardy and Havelka (1975)
Marine	20–120		Söderlund and Svensson (1976)
Combined	159–259	1.8×10^7	
Lightning flashes	8	5×10^8	Table 9-14
Cosmic rays	0.04	9.7×10^{10}	Warneck (1972)
Industrial fertilizer	36	1.1×10^8	From Söderlund and Svensson (1976), for the year 1970
Combustion and biomass burning	30	1.3×10^8	Table 9-14
Sedimentary cycle	4.3	9×10^8	Ratio $N/C = 0.072$; carbon flux from Table 11-6

water, nitrogen is thermodynamically unstable because the equilibrium of the reaction

$$N_2 + 2\tfrac{1}{2}O_2 + H_2O \rightleftharpoons 2H^+ + 2NO_3^-$$

favors the right-hand side. If the entire amount of oxygen in the atmosphere were used to convert nitrogen to nitrate, the mass of atmosphere N_2 would be reduced by 11% to 3.5×10^{18} kg and the difference of 4.1×10^{17} kg would have to reside as nitrate in the ocean (or the sediments). The present amount of nitrate in the oceans is 5.7×10^{14} kg, as Table 12-7 shows. There can be no doubt that biological denitrification and the accompanying release of N_2 to the atmosphere keep the concentration of nitrate at a low level. The external biological cycle comprising nitrogen fixation and denitrification occurs at a rate at least 10 times faster than NO_3^- formation via lightning discharges and, therefore, is the controlling process.

Also entered in Table 12-8 is the flux of nitrogen associated with the cycle of the sediments. It is derived from the flux of organic carbon given in Table 11-6 and the ratio of organic nitrogen to carbon in the sediments. The small rate obtained makes it evident that this cycle is trivial compared with that between the atmosphere and the biosphere. The fate of organic nitrogen exposed to the atmosphere upon weathering of the sediments has not been explored. Presumably it becomes part of organic matter in the soil, is utilized by microorganisms, and so returns to the biosphere from which it originated.

12.4   Oxygen

The prebiotic, primitive atmosphere was practically devoid of oxygen, since it is not a constituent of volcanic gases. How oxygen came to arise and attain its present level is a fascinating problem that will be discussed below.

12.4.1   SOURCES OF OXYGEN

In the primitive atmosphere the only source of oxygen would be the photodissociation of H_2O and CO_2. In the early treatment of these processes by Berkner and Marshall (1964, 1966) and Brinkmann (1969) it was assumed that the photodecomposition of water vapor was confined to the lowest regions of the atmosphere, that the products were hydrogen and oxygen, that hydrogen underwent Jeans escape causing oxygen to accumulate, and that eventually oxygen overtook water vapor as an effective ultraviolet absorber, whereupon H_2O was shielded from further photodestruction. Due to the emphasis placed on water vapor and an insufficient understanding

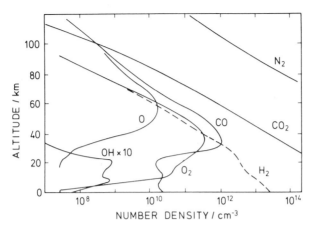

Fig. 12-3. Steady-state concentrations of N_2, CO_2, CO, O_2, H_2, O, and OH in the prebiotic primitive atmosphere, according to model calculations of Kasting *et al.* (1979). These authors assumed a steady flux of H_2 of 10^{12} m^{-2} s^{-1} through the atmosphere, entering it by volcanic exhalations and leaving by Jeans escape.

of the factors controlling the escape of hydrogen, the early studies concentrated on the amount of oxygen required as a radiation shield, but ignored all aspects of reaction kinetics. These are important, however.

Kasting *et al.* (1979) recently clarified the situation by means of a one-dimensional steady-state diffusion model incorporating the essential photochemical reactions. Figure 12-3 shows some of their results. The overriding source of oxygen at all altitudes is the photodecomposition of CO_2. Molecular oxygen is formed by the recombination of oxygen atoms ($z \approx$ 60 km) and by the reaction $O + OH \rightarrow O_2 + H$ ($z < 30$ km), the OH radicals being produced by the photodissociation of H_2O. The steady-state O_2 mixing ratio is a function of altitude. It rises from a value near 10^{-12} at the ground level to almost 10^{-5} at 60 km. The small concentrations of O_2 in the lower atmosphere are partly due to an efficient catalytic chain in which hydrogen and oxygen combine to produce water. Another reason for the low oxygen level is the presence of H_2 due to volcanic outgassing. Kasting *et al.* (1979) assumed an H_2 flux of 7×10^7 kg/yr, which they considered to be the present rate. Walker (1977) derived a rate about three times higher. According to the discussion in Section 12.2 the flux may have been an order of magnitude higher 4 billion yr ago. Such conditions would have reduced the oxygen concentration even further.

In discussing Jeans escape of hydrogen generated by the photodissociation of water vapor, we reiterate that the escape rate is determined by eddy diffusion of hydrogenous compounds in the region below the homopause.

In the lower atmosphere, the decomposition of H_2O by ultraviolet radiation is largely suppressed by back-reactions. The study of Kasting *et al.* (1979) shows that these reactions become ineffective at altitudes above 20 km, and beyond this level the dissociation of H_2O is essentially complete. The supply to these altitudes is determined by the temperature minimum in the middle atmosphere, so that the escape rate of H_2 derived from water vapor should have been similar to that existing today. The present rate of Jeans escape of hydrogen is about 2×10^{12} atoms/m^2 s or 2.6×10^{10} mol/yr (Hunten and Strobel, 1974; Liu and Donahue, 1974). The net loss is somewhat reduced by the accretion of hydrogen from the solar wind but not by much (Walker, 1977). The corresponding rate of oxygen production is 4.2×10^8 kg/yr. If one assumes that the rate has remained unchanged over geological times, the total gain of oxygen in 4.5 billion yr is 1.9×10^{18} kg. This value compares well with that presently residing in the atmosphere and ocean combined (1.2×10^{18} kg). The comparison is misleading, however, because it neglects losses of oxygen due to the oxidation of reduced compounds. In the model of Kasting *et al.* (1979), the oxygen gain is essentially balanced by losses resulting from the oxidation of hydrogen emanating from volcanoes. Additional losses are important, as we shall see below, so that Jeans escape cannot account for the mass of oxygen that must have been generated throughout geological times.

Our current understanding of the evolution of oxygen is founded on the recognition that oxygen is a by-product of the assimilation of CO_2 by green plants and phytoplankton. Recall from Chapter 11 that the annual exchange of carbon between atmosphere and biosphere is almost balanced. The same must hold true for the equivalent exchange of oxygen, since the amount of oxygen generated by photosynthesis is used up again in the conversion of organic matter toward CO_2. However, a tiny fraction of organic carbon is lost each year by its deposition in the sediments, and this process then imposes a similar imbalance on oxygen. The evolution of oxygen thus is intimately tied to that of organic carbon imbedded in the sediments. The photosynthetic process may be represented by the reaction

$$n CO_2 + n H_2O \xrightarrow{h\nu} (CH_2O)_n + n O_2$$

which shows that 1 mol of oxygen is liberated for each mole of carbon that becomes fixed. The relation accordingly determines the mass of oxygen that must have been generated from the mass of carbon buried in the sediments. Table 11-1 showed that the sediments harbor 1.2×10^{19} kg of organic carbon. The corresponding amount of oxygen is 1.2×10^{19} $(32/14) = 3.2 \times 10^{19}$ kg, or 25 times the mass currently residing in the atmosphere. If we add the amount derived from the photodissociation of H_2O and escape of hydrogen, we shall have to account for a total of 3.4×10^{19} kg of oxygen. Since the atmosphere contains so little of it, it is clear that the dominant part of it has been used up in oxidation reactions, and these must be identified in order to establish a mass balance.

12.4.2 THE BUDGET AND CYCLES OF OXYGEN

Oxidizable elements or compounds are made available from two geochemical sources: volcanic emanations yield H_2, CO, and sulfur compounds such as SO_2 and H_2S; and the weathering of igneous rocks supplies bivalent iron in the form of $FeSiO_3$. Other elements occuring in reduced form are of lesser importance. For an estimate of the total quantities of oxidizable material released during geological times, we turn again to the data assembled by Li (1972). Table 12-9 lists the pertinent reactions and attempts to establish a budget for oxygen. Several comments are required. Before oxygen accumulated in the atmosphere, the oxidation of hydrogen exhaled from volcanoes cannot have exceeded the rate of oxygen production from H_2O and CO_2 photolysis. Afterward, all the hydrogen from geothermal processes was oxidized. When the transition took place is not well known, but it probably occurred not earlier than 2 Gyr ago. The estimate in Table 12-9 is obtained by extrapolating the present rate of hydrogen release back over 1.5 Gyr, which gives 2.4×10^{18} kg. Two-thirds of the amount produced by H_2O photodissociation and presumably consumed in reacting with hydrogen is then added. The amount of CO that we assume to have been released with volcanic gases is based on a volume ratio of $CO_2/CO = 37$, in accordance with the thermal equilibrium at 1,500 K (Matsuo, 1978), combined with the total carbon content of sedimentary rocks (Table 11-1). Sulfur also presents a problem in that modern volcanic emissions contain both SO_2 and H_2S, but we do not known the partitioning, neither now or for earlier times. We assume with Li (1972) that H_2S predominated and this assumption maximizes the consumption of O_2.

Only one-half of the sulfur inventory occurs as sulfate. The other half is present as pyrite in the sediments due to the action of sulfate-reducing bacteria living in marine muds. The reduction is made possible by organic carbon. The reaction may be written

$$2Fe_2O_3 + 8SO_4^{2-} + 15CH_2O + 16H^+ \rightarrow 4FeS_2 + 15CO_2 + 23H_2O$$

Here, 3.75 mol of organic matter are consumed for each mole of pyrite deposited, and an equivalent quantity of O_2 is left behind in the atmosphere-ocean system. The excess O_2 is used up again in the oxidation of pyrite when it is brought to Earth's surface by the sedimentary rock cycle. With regard to the oxygen budget, it is important to note that disulfide represents

Table 12-9. *Geochemical Oxygen Budget, Based Mainly on Data Presented by Li (1972) from Mass Balance Relations Describing the Weathering of Igneous Rocks by Primary Magmatic Volatiles*

Material	Reaction	Amount released ($\times 10^{22}$ mol)	Mass of oxygen ($\times 10^{18}$ kg)	Remarks
		Production		
C_{org}	$CO_2 + H_2O \xrightarrow{h\nu} CH_2O + O_2$	10	32.0	Photosynthesis and burial in the sediments; from Table 11-1.
H_2O	$H_2O \xrightarrow{h\nu} \frac{1}{2}O_2 + H_2$	1.2	1.9	Jeans escape of hydrogen, present rate
Sum			33.9	
		Consumption		
H_2	$H_2 + \frac{1}{2}O_2 \rightarrow H_2O$	1.5	3.6	Two-thirds of O_2 from H_2O photodissociation plus losses to H_2 released since the advent of O_2 in the atmosphere
CO	$CO + \frac{1}{2}O_2 \rightarrow CO_2$	1.4	2.2	Assuming CO_2/CO mole ratio of 37 in volcanic exhalations
H_2S	$H_2S + 2O_2 \rightarrow H_2SO_4$	4.2 (total)	13.3	Assuming sulfur entirely released as H_2S. About 50% now occurs as SO_4^{2-} in the ocean and in evaporites, the rest as Fe_2S in shales.
	$2H_2S + \frac{1}{2}O_2 \rightarrow 2H^+ + S_2^{2-} + H_2O$		1.7	
$FeSiO_3$	$2Fe^{2+} + \frac{1}{2}O_2 \rightarrow Fe_2O_3$	12.4	9.9	
Free in the atmosphere		0.4	1.2	Including the amount dissolved in the ocean
Sum			31.9	

a higher oxidation stage of sulfur than sulfide. The substitute reaction entered in Table 12-9 demonstrates that the oxidation of 1 mol of sulfide to disulfide consumes 0.25 mol of oxygen. The mass of sulfur occurring as pyrite in the sediments has been estimted by Holser and Kaplan (1966) as 6.65×10^{18} kg. The corresponding mass of oxygen consumed is 1.7×10^{18} kg, and this value is entered in the table. Finally, we consider the oxidation of bivalent iron. FeO made available by the weathering of igneous rocks is rapidly oxidized to Fe_2O_3 when exposed to air. In a way similar to sulfate, however, Fe_2O_3 can be reduced to Fe^{2+} in anaerobic environments such as the ocean sediments. The reduction is again mediated by organic carbon and leads to the occurrence of bivalent iron in sedimentary rocks in amounts greater than that represented by pyrite. The total mass of FeO that was converted to Fe_2O_3 and remains in that oxidation state can be estimated from the difference in the average contents of FeO and Fe_2O_3 in igneous and sedimentary rocks. Difficulties are encountered because of widely varying iron contents, and representative averages are not easily established. Holland (1978) cites average abundances of FeO in igneous rocks and in the sediments of 3.5 and 1.6 by weight respectively. If we adopt a total mass of 2.4×10^{21} kg for the sediments and assume that the weathering of 1 kg of igneous rock results in 1.12 kg of sedimentary rock (due to the incorporation of volatiles), we find with $M_{FeO} = 72 \times 10^{-3}$ kg/mol that

$$\left(\frac{3.5}{1.12} - 1.6\right)10^{-2}\frac{2.4 \times 10^{21}}{72 \times 10^{-3}} = 5.1 \times 10^{20} \quad \text{mol}$$

of FeO have been oxidized to Fe_2O_3. The value should be considered a lower limit, because basaltic rocks often contain higher weight percentages of FeO than the granitic rocks on which the 3.5% estimates of Holland is based. Li (1972) assumes instead a weight percentage of 5.8 in average igneous rock, so that he derived the higher value of 12.4×10^{20} mol of FeO that was oxidized to Fe_2O_3. This value is entered in Table 12-9 to maximize the consumption of O_2, but it is rather uncertain.

The total mass of oxygen consumed by the reactions listed, added to that contained in the atmosphere, sums to about 3.1×10^{19} kg. The result provides a reasonable balance between production and losses only because we have made an effort to maximize the oxygen consumption. On the whole, the budget is subject to much uncertainty and is as yet unsatisfactory. But the data show that the major reservoirs of oxygen are sulfate in seawater and in evaporites, and Fe_2O_3 in sedimentary rocks. Only 4% of total oxygen resides in the atmosphere. One must appreciate the peculiarity of this distribution. Since oxidative weathering causes a steady drain on O_2, we can understand its presence in the atmosphere only if it is continuously replenished. The subsequent discussion deals with the ensuing cycle of oxygen. We consider first the flux of oxygen through the atmosphere, then the mechanism controlling its abundance, and finally the rise of atmospheric O_2 during the earth's history.

The weathering of rocks leads to a loss of oxygen, mainly due to the oxidation of reduced materials that sedimentary rocks inherit on account of biological activities in the marine environment. The resulting rate of oxygen consumption can be determined from the average contents of C_{org}, FeS_2, and FeO in sedimentary rock and the flux of organic carbon associated with the geochemical rock cycle. The latter quantity was discussed in Section 11.3.1 (see Table 11-8). For completeness, we must also consider O_2 losses incurred by the weathering of igneous rocks. Table 12-10 summarizes the data. Shown are the abundances in sedimentary and igneous rocks of the substances of interest (using values in line with the preceding discussion), the associated material fluxes, and the rates of oxygen consumption that one calculates for a complete oxidation of C_{org} to CO_2, FeO to Fe_2O_3, and S_2^{2-} to SO_4^2. The oxidation of organic carbon is found to contribute most to the total loss, whereas the oxidation of bivalent iron in igneous rocks accounts for only 10%. The total O_2 loss rate is about 270 Tg/yr. Other authors have come up with similar estimates. Holland (1973, 1978) derived a loss rate of 400 Tg/yr, and Walker (1974) obtained 300 Tg/yr. Walker (1977) also pointed out that reduced volatiles exhaled from volcanoes consume oxygen at a rate that is negligible compared with weathering.

The loss of oxygen due to weathering should be balanced by a similar net production rate if the cycle is closed. Walker (1974) has examined rates

Table 12-10. *Oxygen Consumption Rates Associated with the Weathering of Sedimentary and Igneous Rocks*[a]

| Material | Content (wt. %) | Flux rate | | Rate of oxygen consumption Tg/yr |
		Tg/yr	T mol/yr	
Sedimentary rocks				
C_{org}	0.5	60	5.0	160
FeO	1.6	192	2.67	21
S_2^{2-}	0.28	34	0.52	62
Igneous rocks				
FeO	5.8	232	3.22	26
Total rate				269

[a] Based on a flux of organic carbon of 60 Tg/yr (Table 11-6) and average content of 0.5% C_{org} in sedimentary rock and a ratio of 25:75 of igneous to sedimentary rocks being weathered. The amount of oxygen consumed by reduced volatiles exhaled from volcanoes is negligible by comparison (~1%).

for the burial of organic carbon in several aquatic environments. He concluded that the highest rates are found for marine sediments known as "blue muds" occurring on the continental slopes where carbon contents and accumulation rates are both high. Walker estimated global burial rates ranging from 8 to 22 Tmol/yr for such locations. The equivalent O_2 production rate of 250-700 Tg/yr has the correct magnitude to compensate the loss due to weathering. Walker further found that the contribution of carbon burial in deep sea sediments is an order of magnitude smaller despite the larger area covered, and anoxic basins like the Black Sea are even less effective. Globally negligible, too, is the burial of organic carbon in freshwater lakes.

Some of the organic carbon in the continental shelf regions originates from the terrestrial biosphere. Meybeck (1981) estimated that the flux of organic matter washed into the sea by rivers totals 400 Tg/yr, the major share being advected by tropical streams. How much of this material survives to become incorporated into marine sediments is quite uncertain but is generally thought to be small (Holland, 1978). Griffin and Goldberg (1975) have shown that charcoal derived from forest fires also enters marine sediments. The corresponding carbon fluxes are an order of magnitude smaller than those from the marine biosphere.

A turnover rate of about 300 Tg/yr corresponds to a residence time of oxygen in the atmosphere-ocean reservoir of

$$\tau_{atm}(O_2) = 4 \times 10^6 \text{ yr}$$

The weathering of rocks would deplete atmospheric oxygen in a geologically short period of time, were it not replenished by the activity of the biosphere. The latter is the controlling factor, of course, whereas the weathering reactions are a secondary phenomenon. But a residence time of 4 million yr indicates the time scale on which changes in the atmospheric oxygen level are to be expected. Such changes presumably have occurred even after the atmospheric oxygen reservoir was well established.

The above value for $\tau_{atm}(O_2)$ may also be compared with the residence time resulting from the rapid exchange of carbon between the biosphere and CO_2 in the atmosphere-ocean system. For this purpose, we combine the net primary productivities of the marine and the terrestrial biospheres given in Chapter 11 as 35 and 58 Pg/yr, respectively. Correcting for the different molecular weight of carbon and oxygen, we find

$$\tau_{atm}(O_2) = \frac{(1.2 \times 10^6)(12)}{(35+55)(32)} = 5 \times 10^3 \quad \text{yr}$$

The turnover rate of atmospheric oxygen resulting from the interaction with the biosphere is about a thousand times faster than that due to the geochemical cycle. The same result would have been obtained had we compared the pertinent fluxes directly. Yet the residence time associated with the fast cycle of oxygen is already long enough to make seasonal variations of oxygen difficult to detect.

We turn now to the question about the mechanism that must control the abundance of oxygen in the atmosphere. First, it should be noted that despite the rapid turnover, the fast cycle of carbon cannot be the controlling factor because the size of the biosphere is too small for that purpose. The quantity of oxygen required to convert the entire mass of organic carbon in the biosphere to CO_2, including dead organic carbon of the soil and in the oceans, amounts to less than 1% of the mass of oxygen residing in the atmosphere. Accordingly, the control must be exercised by processes associated with the slower geochemical cycle. Broecker (1970, 1971), Holland (1973, 1978), and Walker (1974, 1977) have examined these processes, and the following discussion summarizes their main arguments.

The current high concentration of atmospheric oxygen ensures a fairly rapid oxidation of reduced material in sedimentary rocks upon their exposure to air during weathering. Certainly this is true for bivalent iron and sulfide, which are both soluble but which are not significant as constituents in river waters. Earlier, on page 569, we had discussed the possibility that some of the organic carbon escapes oxidation and undergoes resedimentation. If so, the rate of oxygen consumption is somewhat diminished, but the conclusions that follow would not be substantially altered. For simplicity, we assume with Walker (1974) that oxidation is complete. The rate of oxygen loss is then limited by the rate at which reduced material in rocks is made available by erosion and chemical weathering, and since these processes are independent of the abundance of free oxygen, they cannot stabilize it. The situation clearly was different in the Precambrian, when oxygen started to evolve and its concentration was much lower than it is today. We shall consider the early evolution of oxygen further below and for the moment confine our attention to the modern oxygen cycle.

If, under limiting conditions, weathering reactions are not effective in controlling O_2, we must look to the source of it, namely, the burial of reduced carbon in marine sediments. The efficiency of this process is determined by two opposing factors. One is the supply of organic matter by the biosphere, and the other is the rate at which oxygen dissolved in seawater is made available for the reoxidation of organic material. The details of the oxidation processes operating along the food chain and by microbial decay are as yet inadequately understood. For the purpose of illustration, we may use the following very simple, and undoubtedly imperfect, model. Organic carbon generally meets aerobic waters while traveling from the photosynthetic production region near the sea's surface toward the sediment layer at the bottom of the ocean. Along the way, organic matter is subject to oxidation and the fraction reaching the ocean floor will depend on both the depth of the water layer traversed and the average concentration of oxygen in it. Somewhere near the ocean water–sediment interface, either above or below it, conditions become anoxic, and upon reaching this region, organic carbon is shielded from further oxidation. Subsequent reactions such as fermentation, sulfate reduction, etc. do not change the local oxidation state and have no consequences on the oxygen balance. From this model it comes clear that if the oxygen concentration in the atmosphere–ocean system were lowered, less organic carbon would be oxidized and more of it would become buried. The result would be an increase in the net production of oxygen. Conversely, if the oxygen level in the atmosphere were raised, more oxygen would dissolve in the ocean, more organic carbon would undergo oxidation, and less of it would be incorporated into the sediments. The net production of oxygen would then decrease. In this manner, a feedback mechanism is established that keeps the oxygen concentration at a level sufficiently high for an almost complete oxidation of organic matter. According to this model, the fraction of organic carbon escaping oxidation should be higher in relatively shallow waters as opposed to the deep sea, a feature explaining in part the high rates of carbon burial along the ocean's margins. The model also makes clear that the biosphere regulates primarily oxygen dissolved in the ocean. The role of the atmosphere is that of a buffering reservoir. This is a consequence of the relatively low solubility of oxygen in seawater, which causes the physical partitioning to favor the atmosphere as the main reservoir.

12.4.3 THE RISE OF ATMOSPHERIC OXYGEN

Finally, some notes are in order concerning the evolution of free oxygen toward its present abundance, starting from the very low levels in the primitive atmosphere. This process must be viewed as an integral part of the early history of life and the development of photosynthetic carbon fixation. Our knowledge in this respect is quite fragmentary. Compared with the rich macroscopic fossil record spanning the last 600 million yr (Phanerozoicum), the sediments from the more ancient geological period (Precambrium) contain hardly any evolutionary indicators. The few that exist are summarized in Fig. 12-4 and will be discussed below. Supplemental information comes from the realm of microorganisms, especially anaerobic bacteria, some of which undoubtedly represent very ancient forms of life. Recently, the technique of nucleotide sequence analysis as applied to certain proteins and nucleic acids has made it possible to establish the genealogy of bacteria with some confidence (Schwartz and Dayhoff, 1978; Fox et al., 1980). The results have greatly benefited our understanding of cellular

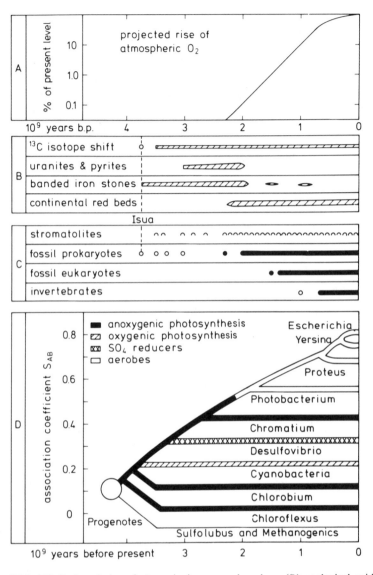

Fig. 12-4. (A) Projected rise of atmospheric oxygen based on (B) geological evidence (Schidlowski, 1978) and (C) fossil remains (Arwamik, 1982; McAlester, 1968; Pflug, 1978; Schidlowski, 1983a,b; Schopf, 1975; Schopf and Dehler, 1976). (D) Evolution of bacteria according to Kandler (1981). A few lines of development are selected for illustration. The phylogenetic tree is based on nucleotide sequences in 16 S ribosomal RNAs (Fox *et al.*, 1980). The association coefficient S_{AB} ($0 < S_{AB} < 1$) is a measure of the overlap of common sequences, with $S_{AB} = 0$ indicating complete difference and $S_{AB} = 1$ full identity. The oldest line is represented by the Archebacteria, which include *Sulfolubus*, living in hot, hydrogen sulfide-rich

evolution. A few descent lines are included in Fig. 12-4 for illustration, although the time scale for the branching points must remain somewhat conjectural for the present.

Phototrophic bacteria hold a prominent place in the hierarchy of all bacteria. The majority of photosynthetic species is not capable of utilizing water as a hydrogen donor, and they must make use of hydrogen derived from organic compounds or H_2S in processes that do not liberate free oxygen. Table 12-11 compares a number of simplified reactions for the fixation of carbon, and the associated energy needs to show that the reaction of CO_2 with water as hydrogen donor demands the highest energy input. One expects this process to have been adapted subsequent to the exploitation of other reactions with lesser energy requirements and, indeed, the phylogenetic tree in Fig. 12-4 bears this out in that cyanobacteria appear later than the green sulfur bacteria *Chloroflexus* and *Chlorobium*, although all three species represent very early forms of life. Among the bacteria, the splitting of water seems to have been mastered only by the cyanobacteria (also known as blue-green algae) and by the recently discovered prokaryotic algal species *Prochloron* (Lewin, 1976). The chloroplasts contained in the more advanced eukaryotic cells of green plants are believed to have descended from cyanobacteria or *Prochloron*-like precursors, which entered into a symbiotic relation with the eukaryotic host cell.

Figure 12-4 will now be discussed in more detail. The record of reduced carbon in sedimentary rocks and the associated $^{13}C/^{12}C$ isotope shift date back to the oldest deposits of 3.5 Gyr ago, and even further if one makes allowance for the metamorphic alterations of the still older Isua formation (see Fig. 11-2). The biological origin of the isotope shift has been convincingly demonstrated, as reviewed by Schidlowski *et al.* (1983). The implication that life had been in full swing 3.5 billion yr ago is supported by several lines of fossil evidence: stromatolites, microfossils, and banded iron formations.

1. Stromatolites are the commonest and most conspicuous fossils (Walter, 1976). They consist of bun-shaped or columnar, laminated structures, usually in limestone deposits; since they are essentially indistinguishable from modern analogs forming today by colonies of blue-green

---

springs, and the methanogenic bacteria; *Chlorobium* and *Chloroflexus* are green sulfur bacteria. Cynobacteria are the only oxygen producers. Time markers for oxygenic photosynthesis are the $^{13}C/^{12}C$ shift, the occurrence of banded iron stones, and stromatolites; the beginning of the $^{34}S/^{32}S$ shift is an indicator for the evolution of sulfate reducers 3.3 Myr ago (Monster *et al.*, 1979); the advent of the continental redbeds is taken to indicate that the Pasteur point was reached, and aerobic bacteria should have appeared at the same time.

Table 12-11. *Free-Energy Requirements for Reactions Reducing Carbon Dioxide to the Carbohydrate Level[a]*

Number	Process	ΔG (kJ/mol)	Examples of bacteria utilizing the reaction
1	$CO_2 + 2H_2 \rightarrow CH_2O + H_2O$	-4.2	Chemoautotrophs (e.g., methanogenics)
2	$CO_2 + 2H_2S \xrightarrow{h\nu} CH_2O + H_2O + 2S$	50.2	Chlorobacteriaceae (e.g., *Chlorobium*)
3	$CO_2 + H_2O + \frac{1}{2}H_2S \xrightarrow{h\nu} CH_2O + \frac{1}{2}H_2SO_4$	117.2	Thiorodoceae (e.g., *Chromatium*)
4	$CO_2 + 2H_2O^* \xrightarrow{h\nu} CH_2O + H_2O + O_2^*$	470.7	Cyanophyaceae and chloroplasts

[a] Reaction 1 is exoergic, whereas the other reactions require an energy input. Photosynthesis using water as hydrogen donor and resulting in the release of free oxygen has the highest energy demand; asterisk indicates oxygen atoms transferred.

algae, stromatolites from the Precambrian generally are taken as evidence for the presence of cyanobacteria during that period, although it should be clear that other species may have produced similar biosedimentary structures. The earliest stromatolites occur in 3.5-Gyr-old rocks (Lowe, 1980; Walter *et al.*, 1980), and they become quite common in the areally more extensive sediments younger than 2.2 Gyr.

2. Microfossils, spheroidal or filamentous objects, can be discerned in stromatolites and nonstromatolitic cherts. On the basis of morphology and size, they are ascribed to unicellular organisms, either bacteria or eukaryotic cells, the latter making their appearance in sediments younger than 1.4 Gyr (Cloud, 1976; Schopf and Oehler, 1976). The biogenicity of microfossils has been firmly established for strata younger than about 2.3 Gyr on grounds of cellular differentiation or similarity to living organisms combined with carbonaceous composition and careful elimination of contamination by younger intrusions. More debatable are the primitive features occurring in 3.5 Gyr-old rocks (Dunlop *et al.*, 1978) and the still tentative microfossils described by Pflug (1978; Pflug and Jaeschke-Boyer, 1979) in the 3.7-Gyr-old metasediments from the Isua formation. They are objects of current controversy (Roedder, 1981; Bridgewater *et al.*, 1981).

3. Banded iron stones are sedimentary structures consisting of alternating layers of iron oxides and iron-poor siliceous material. With few exceptions, their occurrence is confined to sediments older than about 2 billion years. Since the contemporaneous occurrence of detrital uranites and pyrites indicates the absence of free oxygen at that time, soluble ferrous iron was not immediately immobilized as it is today by being oxidized to insoluble ferric iron, and it was washed into the oceans, where it accumulated. The currently accepted explanation for the origin of banded iron formations was formulated by Cloud (1972, 1973), who considers it a biosedimentary structure generated by the action of cyanobacteria or their progenitors. The local production of oxygen due to photosynthesis converted ferrous to ferric hydroxides, which subsequently deposited to form hematite. The presence of ferrous iron in the ancient seas certainly must have kept the oxygen concentration at a low level, permitting oxygen-sensitive organisms to survive. The periodicity of deposition giving rise to the alternating iron-rich and iron-poor microbands is still not well understood. It may have resulted from seasonal variations in the supply of vital nutrients. While this last point leaves many questions open, there exists some support for the biogenicity of banded iron formations from chemical indicators. Fiebiger (1975) studied hydrocarbons in a variety of banded

iron stones and found evidence for photosynthetic activity on account of the presence of phytane and pristane, two stable decomposition products of chlorophyll. If the association of banded iron formation with photosynthetic activity is correct, it suggests that photosynthetic organisms were in existence already 3.7 Gyr ago (banded iron deposits in Isua, West Greenland).

The incipient rise of oxygen in the ancient atmosphere is signaled by the appearance of the first continental redbeds of ferric iron about 2 Gyr ago. This event essentially coincides with the termination of widespread banded iron stone formation, and the pyrite and uranite detritals disappear at the same time. By then, in a long sequence of evolutionary steps, the early organisms must have learned to cope with the toxicity of free oxygen by having developed a variety of anti-oxidant enzymes such as cytochromes, carotenoids, and eventually superoxide dismutase. Before this capability had evolved, the rate of photosynthetic carbon fixation was limited by the rate at which the inorganic oxygen acceptors ferrous iron and hydrogen sulfide were made available by weathering processes and volcanic activity, respectively. The added capacity for tolerating free oxygen provided room for an expansion of primary productivity to higher levels, so far as the nutrient supply permitted it. These events led to an increase in the oxygen production rate and the rise of oxygen in the atmosphere. Now, as oxidative weathering took hold, ferrous iron no longer reached the seas, and bacteria unable to develop defense mechanisms against an aerobic environment had to withdraw into anoxic niches.

At the same time, the respiratory metabolism developed. In the utilization of energy stored in organic compounds, respiration is an order of magnitude superior to fermentation, opening up a new era in the evolution of life. A number of bacteria are facultative anaerobes, reverting to fermentation when the oxygen level in the environment decreases below the so-called Pasteur point, about 1% of the present abundance of oxygen in the atmosphere. The corresponding O_2 concentration thus indicates the minimum level required for respiration. The advent 1.4 Gyr ago of the first eukaryotic cells, which are contingent upon the respiratory matabolic pathway, makes it likely that the atmospheric oxygen level had reached the Pasteur point at that time. The possibility exists, of course, that the first eukaryotes lived in locally oxygen-rich niches and that atmospheric oxygen reached the Pasteur point somewhat later. According to McAlester (1968), the earliest metazoa appeared 700 Myr ago. This was a soft-bodied fauna dependent on diffusive transport of oxygen to the interior cells, so that by that time the oxygen level in the atmosphere almost certainly had risen beyond the Pasteur point.

Life began to inhabit the land surface in the upper Silurian, about 400 Myr ago. During this period and afterward the abundance of oxygen in the atmosphere must have been high enough to form an ozone layer of sufficient optical depth to shield terrestrial life from lethal ultraviolet radiation. The precise functional relationship between oxygen and ozone is uncertain because it depends on trace gases like N_2O and CH_3Cl, whose abundances in earlier times are not known. Ratner and Walker (1972) have studied the development of the ozone layer as a function of oxygen concentration for a pure oxygen atmosphere, whereas the calculations of Hesstvedt *et al.* (1974) allowed for the additional presence of water vapor. Both groups found that an O_2 concentration of 1% of the present level produces enough ozone to shield the earth's surface from radiation of wavelengths below 290 nm, except for a small contribution in the atmospheric window near 210 nm. The latter effect is completely eliminated when the O_2 concentration is raised to 10% of the present value. Since according to the preceding arguments the 1% level had been reached 700 Myr ago at the latest, it appears likely that 400 Myr ago the O_2 concentration was higher and had approached the 10% mark. This value was used to construct the rise curve in Fig. 12-4.

The following 100 Myr saw the first air-breathing lungfishes and primitive amphibians, indicating that the oxygen level had risen still further. Fossil remains of insects, including giant dragonfly-like species featuring a 70-cm wingspan, first become abundant in the late Carboniferous and early Permian strata (about 300 Myr ago; McAlester, 1968). As is well known, the respiratory tracts of insects, the tracheae, conduct oxygen passively by diffusion. This kind of respiratory system is considered to require rather high concentrations of oxygen and suggests that modern O_2 levels had been attained by that time. In fact, Schidlowski (1971) argued that giant insects might have needed an O_2 concentration in excess of that existing now. On the other hand, the continuous evolution of mammalian species from the Triassic on seems to preclude significant excursions of the atmospheric oxygen level during the last 200 million yr.

Although the qualitative nature of the preceding arguments leaves many questions open, there can be little doubt that atmospheric oxygen is a consequence of the biosphere. Indeed, Lovelock (1979; Lovelock and Margulis, 1974) considers oxygen a biological contrivance, specifically developed to further the evolution of life. This viewpoint is not entirely without justification if one recalls that the respiratory metabolism is an order of magnitude more effective than fermentation in exploiting energy stored in organic compounds. Respiration certainly made possible the tremendous diversification of life forms starting with the Phanerozoicum.

12.5   Atmospheric Gases: Cumulative versus Cyclic Behavior

The main objective of the present chapter was to show that the earth's atmosphere developed together with the ocean as a consequence of geochemical processes, primarily by thermal exhalation of volatiles from the planet's interior. To summarize briefly, the rare gases and nitrogen reside mainly in the atmosphere because their solubilities in seawater are small. Carbon dioxide, in contrast, is much more soluble, and it interacted with dissolved calcium to become overwhelmingly trapped in limestone sediments. Oxygen as a biogenic constituent is unique in that it appears to have accumulated in the atmosphere as an excess over the amount required for the oxidation of reduced elements or compounds derived from exhalations or the weathering of igneous rocks. Seemingly, all these gases must be considered as having accumulated. Certainly this is true for the rare gases, but the others are also subject to geochemical and biological cycles, raising the question which of the two processes, accumulation or cycling, actually controls the abundance of these gases in the atmosphere. For CO_2, the discussions of the preceding chapter have shown that cycling exerts the dominant control. For nitrogen and oxygen, the situation is not immediately obvious, so that some additional comments concerning this aspect are in order to conclude the present chapter.

Junge (1972; see also Junge and Warneck, 1979) has studied the problem and proposed a set of criteria that are based on the distribution of an atmospheric constituent among the varous geochemical reservoirs, and that facilitate an assignment to one of the two categories. For a better understanding of the criteria it is instructive to reconsider initially the exchange of material between two coupled reservoirs A and B. In the steady state, the exchange fluxes are equal and one has

$$k_{AB}G_A = k_{BA}G_B \qquad (12\text{-}10)$$

where the exchange coefficients k_{AB} and k_{BA} are the reciprocals of the residence times τ_A and τ_B, respectively. Therefore, the above relation can also be written

$$\tau_A/\tau_B = G_A/G_B$$

Assume now that $G_A \gg G_B$. It is then evident that the residence time τ_A is much greater than τ_B, and the substance being considered must accumulate in reservoir A even if it is initially introduced to reservoir B. Due to the smaller residence time, the content of reservoir B is controlled by that of A via the flux cycle existing between both reservoirs. Two examples that were discussed previously may illustrate this behavior: (1) the amount of water in the ocean (reservoir A) is much greater than that in the atmosphere

Table 12-12. Contents of H_2O, N_2, CO_2 (as Carbon), and O_2 in the Four Principal Geochemical Reservoirs[a]

Constituent	G_{sed}	G_{ocean}	G_{atm}	G_{bio}	G_{mobile}/G_{sed}	G_{atm}/G_{ocean}	G_{atm}/G_{bio}
H_2O	3.3 (8)	1.4 (9)	1.3 (4)	—	4.2	9.0 (−6)	na
N_2	9.0 (5)	2.2 (4)	3.9 (6)	1.3 (3)	4.4	1.7 (2)	3.0 (3)
$CO_2(C_{org}+C_{inorg})$	6.2 (7)	3.8 (4)	7.0 (2)	2.5 (3)	6.6 (−4)	1.8 (−2)	2.8 (−1)
C_{org}	1.2 (7)	b	3.1[c]	3.5 (3)	na[h]	na	8.8 (−4)
O_2	6.5 (7)[d]	1.7 (4)	1.2 (6)	—	na	7.0 (1)	na
C_{org} + other reduced material expressed as O_2	4.8 (7)[e]	b	(8.3)[c]	9.3 (3)	2.5 (−2)[f]	na	1.3 (2)[g]

[a] Values (in Pg) from Tables 11-1, 12-5, 12-8, and 12-10; orders of magnitude shown in parentheses. Ratios of reservoir contents are used as criteria for establishing control over the atmospheric reservoir—accumulation or cycles. $G_{mobile} = G_{atm} + G_{ocean} + G_{bio}$.

[b] The content of C_{org} in the ocean is included in G_{bio}.

[c] Methane.

[d] Counting only the amount of oxygen fixed as sulfate and trivalent iron but including the amount of SO_4^{2-} in the ocean.

[e] Includes FeS_2 and FeO.

[f] Mobile O_2 over C_{org} and other reduced material expressed as O_2.

[g] Atmospheric oxygen over C_{org} expressed as O_2.

[h] na, Not applicable.

(reservoir B), so that the latter is controlled by the former and the water cycle; (2) a similar distribution between ocean and atmosphere exists for CO_2, so that again the ocean determines the atmospheric cycle, provided, of course, that interactions with other geochemical reservoirs are ignored.

For the more general case, we consider the distribution of a substance in the four principal geochemical reservoirs: the sediments, the ocean, the atmosphere, and the biosphere. The total amount of a volatile constituent exhaled during the history of the earth is given by

$$G_{tot} = G_{sed} + G_{ocean} + G_{atm} + G_{bio} = G_{sed} + G_{mobile} \qquad (12\text{-}11)$$

The distribution of H_2O, N_2, CO_2, and O_2 in the four reservoirs are summarized in Table 12-12. From the geochemical point of view we may combine the contents of the ocean, the atmosphere, and the biosphere, and consider this the mobile fraction, in contrast to the amount of substance fixed in the sediments. The ratio G_{mobile}/G_{sed} then tells us whether the substance has accumulated mainly in the sediments or in the other reservoirs. If the ratio is appreciably greater than unity, the constituent has remained mobile; if the ratio is much smaller than unity, the constituent has entered the sediments and the mobile fraction is governed by material exchange with the lithosphere. We can further form the ratios G_{atm}/G_{ocean} and G_{atm}/G_{bio} in order to determine which of the three reservoirs containing the mobile fraction controls the atmospheric reservoir. The three ratios are included in Table 12-12. They provide the desired set of criteria for the categorization of atmospheric constituents. At the same time, they furnish a short-hand summary of our discussions of these constituents and their behavior.

H_2O and N_2 occur mainly in the mobile fraction, as the sediments contain less than one-fifth of G_{tot}. The small value of G_{atm}/G_{ocean} for water indicates accumulation in the ocean reservoir. It controls the content of water vapor in the atmosphere, as discussed above. For nitrogen, the ratios G_{atm}/G_{ocean} and G_{atm}/G_{bio} are both much greater than unity. This demonstrates that nitrogen has, in fact, accumulated in the atmosphere and that the cycles of nitrogen through the biosphere and the sediments are subordinate features. For CO_2, the total of organic and inorganic carbon is used in Table 12-12. This procedure is chosen to enable a comparison of carbon in the atmosphere with that in the biosphere. The contents of carbon in the ocean and in the sediments are dominated by inorganic carbon anyway. For a better comparison, however, we have also entered organic carbon separately. The three ratios derived from the contents of the individual reservoirs are all smaller than unity. Accordingly, atmospheric CO_2 is governed by the other reservoirs. Which of these dominates the control cannot be decided without considering the individual exchange fluxes. The discussion in Chapter 11 has shown that the ocean and the biosphere control the short-term variations, whereas weathering of sedimentary rocks is responsible for the long-term average level. In the line for organic carbon, only the ratio G_{atm}/G_{bio} is relevant. The value is smaller than unity, indicating that, in accordance with expectation, methane and the other organic compounds in the atmosphere are controlled by the biosphere.

Finally, we consider the behavior of oxygen. This is a special case, inasmuch as oxygen bound to sulfate and trivalent iron in the sediments is permanently fixed and cannot be liberated by weathering. As a consequence there is no direct exchange between the mobile fraction and the reservoir of O_2 in sedimentary rocks, so that the ratio G_{mobile}/G_{sed} becomes meaningless. Since the source of oxygen is the disproportionation of CO_2 to give C_{org} and O_2, and the sink is the reoxidation of C_{org} to CO_2, it will be more appropriate to relate the mobile fraction of O_2 to the contents of organic carbon (and other reduced material) in the sediments and in the biosphere. When this procedure is employed we obtain the last line of Table 12-12, where the contents of reduced material in the sediments and in the biosphere are expressed as equivalents of oxygen. The ratios now show that $G_{mobile}/G_{sed} \ll 1$, which agrees with out earlier conclusions in that the cycle of organic carbon through the sediments controls the behavior of oxygen in the atmosphere. At the same time, the ratio G_{atm}/G_{bio} is much larger than unity, and since this is true also for the ratio G_{atm}/G_{ocean}, we find that interactions with the biosphere or with the ocean are not controlling factors. In this sense we must consider free oxygen to have accumulated in the atmosphere. The behavior of oxygen thus shows hybrid features. On one hand, it is controlled by the burial of organic carbon in the sediments due to biological activity; on the other hand it behaves like a cumulative constituent of the atmosphere. The reason presumably is the slowness of the sediment cycle, which causes the residence time of oxygen in the atmosphere to be a few million years.

References

Abel, N., R. Jaenicke, C. Junge, H. Kanter, P. Rodriguez Garcia Prieto, and W. Seiler (1969). Luftchemische Studien am Observatorium Izana (Teneriffa). *Meteorol. Rundsch.* **22**, 158–167.

Ackerman, M., and F. Biaumé (1970). Structure of the Schumann-Runge bands from (0-0) to the (13-0) band. *J. Mol. Spectrosc.* **35**, 73–82.

Ackerman, M. (1971). Ultraviolet solar radiation related to mesospheric processes. In "Mesospheric Models and related Experiments" (G. Fiocco, ed.) Reidel, Dordrecht, The Netherlands. pp. 149–159.

Adam, J. R., and R. G. Semonin (1970). Collection efficiencies of raindrops for submicron particulates. *In* "Precipitation Scavenging 1970" (R. J. Engelmann and W. G. N. Slinn, coordinators). pp. 151–160. U.S. Atomic Energy Commission, Division of Technology Information, Oakridge, Tennessee.

Adams, F., R. Dams, L. Guzman, and J. W. Winchester (1977). Background aerosol composition on Chacaltaya Mountain, Bolivia. *Atmos. Environ.* **11**, 629–634.

Adams, J. A., M. S. M. Mantovani, and L. L. Lundell (1977). Wood versus fossil fuel as a source of excess carbon dioxide in the atmosphere: a preliminary report. *Science* **196**, 54–56.

Adams, D. F., S. O. Farwell, M. R. Pack, and W. L. Bamesberger (1979). Preliminary measurements of biogenic sulfur-containing gas emissions from soils. *J. Air Pollut. Control Assoc.* **29**, 380–382.

Adams, D. F., S. O. Farwell, E. Robinson, and M. R. Pack (1980). Biogenic sulfur emissions in the SURE region. Final Report **EA-1516**, Project **856-1**. Electric Power Research Institute, Palo Alto, California.

Adams, D. F., S. O. Farwell, E. Robinson, M. R. Pack, and W. L. Bamesberger (1981a). Biogenic sulfur source strengths. *Environ. Sci. Technol.* **15**, 1493–1498.

Adams, D. F., S. O. Farwell, M. R. Pack and E. Robinson (1981b). Biogenic gas emissions from soils in Eastern and Southeastern United States. *J. Air Pollut. Control Assoc.* **31**, 1083–1089.

Adel, A. (1938). Further detail in the rock-salt prizmatic solar spectrum. *Astrophys. J.* **88**, 186–188.

Aikin, A. C., J. R. Herman, E. J. Maier, and C. J. McQuillan (1982). Atmospheric chemistry of ethane and ethylene. *J. Geophys. Res.* **87**, 3105–3118.

Aitken, J. (1923). Collected Scientific Papers (C. Knott, ed.). Cambridge University Press, London and New York.

Air Quality Criteria for Photochemical Oxidants (1970). U.S. Department of Health, Education and Welfare, Public Health Service, Environmental Health Service, Natl. Air Poll. Control Administration Publication No. **AP-63**.

Ajtay, G. L., P. Ketner, and P. Duvigneaud (1979). Terrestrial primary production and phytomass. *In* "The Global Carbon Cycle," *SCOPE* **13**, 129–181.

Albrecht, B., C. Junge, and H. Zakosek (1970). Der N_2O Gehalt der Bodenluft in drei Bodenprofilen, *Z. Pflanzenernähr Bodenkd.* **125**, 205–211.

Aldaz, L. (1969). Flux measurements of atmospheric ozone over land and water. *J. Geophys. Res.* **74** 6934–6946.

Alexander, M. (1977). "Soil Microbiology," 2nd ed. Wiley, New York.

Alfvén, H., and G. Arrhenius (1976). Evolution of the Solar System. National Aeronautics and Space Administration, Washington D.C.

Alkezweeny, A. J. (1978). Measurements of aerosol particles and trace gases in METROMEX. *J. Appl. Meteorol.* **17**, 609–614.

Allara, D. L., T. Mill, D. G. Hendry, and F. R. Mayo (1968). Low-temperature gas- and liquid-phase oxidations of isobutane. *Adv. Chem. Ser.* **76**, 40–57.

Allison, F. E. (1955). The enigma of soil nitrogen balance sheets. *Adv. Argron.* **7**, 213–250.

Allison, F. E. (1966). The fate of nitrogen applied to soils, *Adv. Agron.* **18**, 219–258.

Altshuller, A. P., and S. P. McPherson (1963). Spectrometric analysis of aldehydes in the Los Angeles atmosphere. *J. Air Pollut. Control Assoc.* **13**, 109–111.

Altshuller, A. P., and J. J. Bufalini (1965). Photochemical aspects of air pollution: a review. *Photochem. Photobiol.* **4**, 97–146.

Altshuller, A. P., and J. J. Bufalini (1971). Photochemical aspects of air pollution: a review. *Environ. Sci. Technol.* **5**, 39–64.

Altshuller, A. P. (1976). Regional transformation of sulfur dioxide to sulfate in the U.S. *J. Air Pollut. Control Assoc.* **26**, 318–324.

Altshuller, A. P. (1980). Seasonal and episodic trends in sulfate concentrations in the eastern United States. *Environ. Sci. Technol.* **14**, 1337–1349.

Altshuller, A. P. (1983). Review: Natural volatile organic substances and their effect on air quality in the United States. *Atmos. Environ.* **17**, 2131–2165.

Anastasi, C., and I. W. M. Smith (1976). Rate measurements of OH by resonance absorption. Part 5: Rate constants for $OH + NO_2$ $(+M) \rightarrow HNO_3$ $(+M)$ over a wide range of temperature and pressure. *J. Chem. Soc. Faraday Trans.* 2 **72**, 1459–1468.

Anderson, J. G. (1971). Rocket measurement of OH in the mesosphere. *J. Geophys. Res.* **76**, 7820–7824.

Anderson, J. G. (1976). The absolute concentration of OH $(X^2 II)$ in the earth's stratosphere. *Geophys. Res. Lett.* **3**, 165–168.

Anderson, J. G. (1980). Free radicals in the earth's stratosphere: a review of recent results. Proc. NATO Adv. Study Inst. Atmospheric Ozone (M. Nicolet and A. C. Aikin, eds.), **FAA-EE-80-20**, pp. 233–251. U.S. Department of Transportation, Washington D.C.

Anderson, L. G. (1976). Atmospheric chemical kinetics data survey. *Rev. Geophys. Space Phys.* **14**, 151–170.

Andreae, M. O., W. R. Barnard, and J. M. Ammons (1983). The biological production of dimethyl sulfide in the ocean and its role in the global atmospheric sulfur budget. *In* "Environmental Biochemistry", (R. Hallberg, ed.), *Ecol. Bull.* (*Stockholm*) **35**, 167–177.

Andreae, M. O., and H. Raemdonck (1983). Dimethyl sulfide in the surface ocean and the marine atmosphere: a global view. *Science* **221**, 744–747.

Aneja, V. P., J. H. Overton, L. T. Cupitt, J. L. Durban, and W. E. Wilson (1979). Direct measurements of emission rates of some atmospheric biogenic sulfur compounds. *Tellus* **31**, 174–178.

Aneja, V. P., J. H. Overton, and A. P. Aneja (1981). Emission survey of biogenic sulfur flux from terrestrial surfaces. *J. Air Pollut. Control Assoc.* **31**, 256–258.

Angell, J. K., and J. Korshover (1976). Global analysis of recent total ozone fluctuations. *Mon. Weather Rev.* **104**, 63–75.

Angell, J. K., and J. Korshover (1978). Global ozone variations: an update into 1976. *Mon. Weather Rev.* **106**, 725–737.

Angström, A., and L. Högberg (1952). On the content of nitrogen in atmospheric precipitation in Sweden 2. *Tellus* **4**, 271–279.

Amimoto, S. T., A. P. Force, and J. R. Wiesenfeld (1978/1979). Ozone photochemistry: Production and deactivation of O (2^1D_2) following photolysis at 248 nm. *Chem. Phys. Lett.* **60**, 40–43.

Anlauf, K. G., P. Fellin, H. A. Wiebe, H. I. Schiff, G. I. MacKay, R. S. Braman, and R. Gilbert (1985). A comparison of three methods for measurements of atmospheric nitric acid and aerosol nitrate and ammonium. *Atmos. Environ.* **19**, 325–333.

Appel, B. R., E. M. Hoffer, E. L. Kothny, S. M. Wall, M. Haik, and R. L. Knights (1979). Analysis of carbonaceous material in southern California atmospheric aerosols. 2. *Environ. Sci. Technol.* **13**, 98–104.

Appel, B. R., Y. Tokiwa, and M. Haik (1981). Sampling of nitrates in ambient air. *Atmos. Environ.* **15**, 283–289.

Arijs, E., D. Nevejans, P. Frederik, and J. Ingels (1982). Stratospheric negative ion composition measurements, ion abundances and related trace gas detection. *J. Atmos. Terr. Phys.* **44**, 681–694.

Arin, L. M., and P. Warneck (1972). Reaction of ozone with carbon monoxide. *J. Phys. Chem.* **76**, 1514–1516.

Arnold, F., R. Fabian, G. Henschen, and W. Joos (1980). Stratospheric trace gas analysis from ions, H_2O and HNO_3. *Planet. Space Sci.* **28**, 681–685.

Arnts, R. R., and S. A. Meeks (1981). Biogenic hydrocarbon contribution to the ambient air of selected areas. *Atmos. Environ.* **15**, 1643–1651.

Arnts, R. R., W. B. Peterson, R. L. Seila, and B. W. Gay, Jr. (1982). Estimates of alpha-pinene emissions from a loblolly pine forest using an atmospheric diffusion model. *Atmos. Environ.* **16**, 2127–2137.

Arrhenius, S. (1896). On the influence of the carbonic acid in the air upon the temperature of the ground. *Philos. Mag.* **41**, 237–276.

Arrhenius, S. (1903). "Lehrbuch der kosmischen Physik." Hirzel, Leipzig.

Arrhenius, G., B. R. De, and H. Alfvén (1974). Origin of the Ocean. *In* "The Sea" (E. Goldberg, ed.), vol. 5, pp. 839–861. Wiley (Interscience), New York.

Atkinson, R., R. A. Perry, and J. N. Pitts Jr. (1977). Absolute rate constants for the reaction of oxygen (^3P) atoms with *n*-butane and NO (M = Ar) over the temperature range 298–438 K, *Chem. Phys. Lett.* **47**, 197–202.

Atkinson, R., R. A. Perry, and J. N. Pitts, Jr. (1978). Rate constants for the reaction of OH radicals with COS, CS_2 and CH_3SCH_3. *Chem. Phys. Lett.* **54**, 14-18.

Atkinson, R., K. R. Darnall, A. C. Lloyd, C. M. Winer, and J. N. Pitts, Jr. (1979). Kinetics and mechanisms of the reactions of the hydroxyl radical with organic compounds in the gas phase. *Adv. Photochem.* **11**, 375-488.

Atkinson, R., W. P. L. Carter, K. R. Darnall, A. M. Winer, and J. N. Pitts (1980). A smog chamber and modeling study of gas-phase NO_x—air photooxidation of toluene and the cresols. *Int. J. Chem. Kinet.* **12**, 779-836.

Atkinson, R., S. M. Aschmann, W. P. L. Carter, A. M. Winer, and J. N. Pitts (1982a). Alkyl nitrate formation from the NO_x—air photooxidations of C_2-C_8 *n*-alkanes. *J. Phys. Chem.* **86**, 4563-4569.

Atkinson, R., A. M. Winer, and J. N. Pitts, Jr. (1982b). Rate constants for the gas phase reactions of O_3 with the natural hydrocarbons isoprene and α- and β-pinene. *Atmos. Environ.* **16**, 1017-1020.

Atkinson, R., and A. C. Lloyd (1984). Evaluation of kinetic and mechanistic data for modeling of photochemical smog. *J. Phys. Chem. Ref. Data* **13**, 315-444.

Atkinson, R., C. N. Plum, W. P. L. Carter, A. M. Winer, and J. N. Pitts, Jr. (1984a). Rate constants for the gas-phase reactions of nitrate radicals with a series of organics in air at 298 ± 1 K. *J. Phys. Chem.* **88**, 1210-1215.

Atkinson, R., W. P. L. Carter, C. N. Plum, A. M. Winer, and J. N. Pitts, Jr. (1984b). Kinetics of the gas-phase reactions of NO_3 radicals with a series of aromatics at 296 ± 2 K. *Int. J. Chem. Kinet.* **16**, 887-898.

Atkinson, R., and S. M. Aschmann (1984). Rate constants for the reactions of O_3 and OH radicals with a series of alkynes. *Int. J. Chem. Kinet.* **16**, 259-268.

Atkinson, R., S. M. Aschman, A. M. Winer, and J. N. Pitts, Jr. (1984c). Kinetics of the gas-phase reactions of NO_3 radicals with a series of dialkenes, cycloalkenes and monoterpenes at 295 ± 1 K. *Environ. Sci. Technol.* **18**, 370-374.

Atkinson, R., S. M. Aschmann, A. M. Winer, and J. N. Pitts, Jr. (1985). Kinetics and atmospheric implications of the gas-phase reactions of NO_3 radicals with a series of monoterpenes and related organics at 294 ± 2 K. *Environ. Sci. Technol.* **19**, 159-163.

Attmannspacher, W., and H. U. Dütsch (1970). International Ozone sonde intercomparison at the Observatory Hohenpeisenberg. *Ber. Dtsch. Wetterdienstes* **16**, No. 120.

Attmannspacher, W., and R. Hartmannsgruber (1973). On extremely high values of ozone near the ground. *Pure Appl. Geophys.* **106–108** 1091-1096.

Awramik, S. M. (1982). The pre-phanerozoic fossil record. *In* "Mineral Deposits and the Evolution of the Biosphere," Dahlem Konferenzen (H. D. Holland and O. Schidlowski, eds.), pp. 67-82. Springer-Verlag Berlin and New York.

Axford, W. I. (1968) The polar wind and the terrestrial helium budget. *J. Geophys. Res.* **73**, 6855-6859.

Axford, W. I. (1970) On the origin of radiation belt and auroral primary ions. *In* "Particles and Fields in the Magnetosphere" (B. M. McCormac, ed.), pp. 46-59. Reidel Dordrecht, Holland.

Ayers, and G. P., and J. L. Gras (1980). Ammonia gas concentration over the southern ocean. *Nature* **284**, 539-540.

Ayers, G. P. (1982). The chemical composition of precipitation. A southern hemisphere perspective. *In* "Atmospheric Chemistry," Dahlem Konferenzen, (E. D. Goldberg, ed.), pp. 41-56. Springer-Verlag, Berlin.

Bacastow, R. B., and C. D. Keeling (1981). Atmospheric carbon dioxide concentration and observed airborne fraction. *In* "Carbon Cycle Modelling" (B. Bolin, ed.) *SCOPE* **16**, 103-112.

Bach, W. (1976). Global air pollution and climatic change. *Rev. Geophys. Space Phys.* **14**, 429-474.

Baes, C. F., H. E. Goelle, J. S. Olsen, and R. M. Rotty (1976). The global carbon dioxide problem, **ORNL-5194**, Oak Ridge Nat. Lab., Oak Ridge, Tennessee, pp. 1-72.

Bagnold, R. A. (1941). The Physics of Blown Sand and Desert Dunes." Methuen, London.

Bahe, F. C., W. N. Marx and U. Schurath (1979). Determination of the absolute photolysis rate of ozone by sunlight, $O_3 + h\nu \rightarrow O(^1D) + O_2(^1\Delta_g)$ at ground level. *Atmos. Environ.* **13**, 1515-1522.

Bahe, F. C., U. Schurath, and K. H. Becker (1980). The frequency of NO_2 photolysis at ground levels as recorded by a continuous actinometer. *Atmos. Environ.* **14**, 711-718.

Bahta, A., R. Simonaitis, and J. Heicklen (1982). Thermal decomposition kinetics of $CH_3O_2NO_2$. *J. Phys. Chem.* **86**, 1849-1853.

Baker, G., and R. Shaw (1965). Reactions of methoxyl, ethoxyl and t-butoxyl with nitric oxide and with nitrogen dioxide. *J. Chem. Soc.*, 6965-6974.

Baker, E. A., and E. Parsons (1971). Scanning electron microscopy of plant cuticles. *J. Microsc.* **94**, 39-49.

Baker-Blocker, A., T. M. Donahue, and K. H. Mancy (1977). Methane flux from wetland areas. *Tellus* **29**, 245-250.

Baldwin, A. C., J. R. Barker, D. M. Golden, and D. G. Hendry (1977). Photochemical smog rate parameter estimates and computer simulations. *J. Phys. Chem.* **81**, 2483-2492.

Baldwin, A. C. (1982). Thermochemistry of peroxides. *In* "Chemistry of Functional Groups" (S. Patai, ed.). J. Wiley, New York.

Bandow, H., M. Okuda, and H. Akimoto (1980). Mechanism of the gas-phase reaction of C_3H_6 and NO_3 radicals. *J. Phys. Chem.* **84**, 3604-3608.

Bandy, A. R., P. J. Maroulis, L. Shalaby, and L. A. Wilner (1981). Evidence for a short tropospheric residence time for carbon disulfide. *Geophys. Res. Lett.* **8**, 1180-1183.

Banks, P. M., and G. Kockarts (1973). "Aeronomy," Parts A and B. Academic Press, New York.

Bannon, J. K., and L. P. Steele (1960). Average water-vapour content of the air, Geophys. Mem. No. 102, Meteorol. Office Air Ministry, London.

Banwart, W. L., and J. M. Bremner (1976). Volatilization of sulfur from unamended and sulfate-treated soils. *Soil Biol. Biochem.* **8** 19-22.

Barbier, M., D. Joly, A. Saliot, and D. Tourres (1973). Hydrocarbons from sea water. *Deep Sea Res.* **20**, 305-314.

Barger, W. R., and W. D. Garrett (1976). Surface active organic material in the marine atmosphere. *J. Geophys. Res.* **81**, 3151-3157.

Barkenbus, B. D., C. S. Mac Dougall, W. H. Griest, and J. E. Caton (1983). Methology for the extraction and analysis of hydrocarbons and carboxylic acids in atmospheric particulate matter. *Atmos. Environ.* **17**, 1537-1543.

Barker, J. R., S. W. Benson, and D. M. Golden (1977). The decomposition of dimethyl peroxide and the rate constant for $CH_3O + O_2 \rightarrow CH_2O + HO_2$, *Int. J. Chem. Kinet.* **9**, 31-53.

Barnard, W. R., M. O. Andreae, W. E. Watkins, H. Bingemer, and H. W. Georgii (1982). The flux of dimethyl sulfide from the oceans to the atmosphere. *J. Geophys. Res.* **87**, 8787-8793.

Barnes, I., K. H. Becker, E. H. Fink, A. Reimer, F. Zabel, and H. Niki (1983). Rate constant and products of the reaction $CS_2 + OH$ in the presence of O_2. *Int. J. Chem. Kinet.* **15**, 631-645.

Barnes, I., V. Bastian, K. H. Becker, E. H. Fink (1984). Reactions of OH radicals with reduced sulfur compounds under atmospheric conditions. In "Physico-chemical Behaviour of Atmospheric Pollutants" (B. Versino and G. Angeletti, eds.), *Proc. Eur. Symp. 3rd, Varese, Italy*, pp. 149-157. Dordrecht, Holland.

Barrie, L. A., and H. W. Georgii (1976). An experimental investigation of the absorption of sulfur dioxide by water drops containing heavy metal ions. *Atmos. Environ.* **10**, 743-749.

Bartholomew, G. W., and M. Alexander (1979). Microbial metabolism of carbon monoxide in culture and in soil. *Appl. Environ. Microbiol.* **37**, 932-937.

Bartholomew, G. W., and M. Alexander (1982). Microorganisms responsible for the oxidation of carbon monooxide in soil. *Environ. Sci. Technol.* **16**, 300-301.

Bass, A. M., A. E. Ledford, and A. H. Laufer (1976). Extinction coefficients of NO_2 and N_2O_4. *J. Res. Nat. Bur. Stand.*, **80A**, 143-166.

Bates, D. R., and P. B. Hays (1967). Atmospheric nitrous oxide. *Planet. Space Sci.* **15**, 189-197.

Bathen, K. H. (1972). On the seasonal changes in depth of the mixed layer in the North Pacific Ocean. *J. Geophys. Res.* **77**, 7138-7150.

Bathurst, R. G. C. (1975). Carbonate Sediments and Their Diagenesis," 2nd ed. Elsevier, Amsterdam.

Batt, L., R. D. McCulloch, and R. T. Milne (1975). Thermochemical and kinetic studies of alkyl nitrites (RONO), D(RO-NO), the reactions between RO and NO and the decomposition of RO. *Int. J. Chem. Kinet.* **7** (Symp. 1), 441-461.

Batt, L., and R. D. McCulloch (1976). The gas-phase pyrolysis of alkyl nitrites. II. S-Butylnitrite. *Int. J. Chem. Kinet.* **8**, 911-933.

Batt, L., and R. T. Milne (1977). The gas-phase pyrolysis of alkyl nitrites. III. Isopropylnitrite. *Int. J. Chem. Kinet.* **9**, 141-159.

Batt, L. (1979). The gas-phase decomposition of alkoxy radicals. *Int. J. Chem. Kinet.* **11**, 977-993.

Batt, L. (1980) Reactions of alkoxy radicals relevant to atmospheric chemistry. *In: Eur. Symp. Physico-Chemical Behaviour Atmos. Pollutants, 1st* (B. Versino and H. Ott, eds.), pp. 167-184. Commission of the European Communities, Luxembourg.

Bauer, E. (1974). Dispersion of tracers in the atmosphere and ocean: survey and comparison of experimental data. *J. Geophys. Res.* **79**, 789-795.

Bauer, E. (1979). A catalogue of perturbing influences on stratospheric ozone 1955-1975. *J. Geophys. Res.* **84**, 6929-6940.

Bauer, K., W. Seiler, and H. Giehl (1979). CO Production höherer Pflanzen an natürlichen Standorten. *Z. Pflanzenphysiol.* **94**, 219-230.

Bauer, K., R. Conrad, and W. Seiler (1980). Photooxidative production of carbon monoxide by phototrophic microorganisms. *Biochim. Biophys. Acta* **589**, 46-55.

Baulch, D. L., R. A. Cox. R. F. Hampson, Jr., J. A. Kerr, J. Troe, and R. T. Watson (1980). Evaluated kinetic and photochemical data for atmospheric chemistry. *J. Phys. Chem. Ref. Data* **9**, 295-471.

Baulch, D. L., R. A. Cox, P. J. Crutzen, R. F. Hampson, Jr., J. A. Kerr, J. Troe, and R. T. Watson (1982). Evaluated kinetic and photochemical data for atmospheric chemistry, Supplement I CODATA Task Group on Chemical Kinetics. *J. Phys. Chem. Ref. Data* **11**, 327-496.

Baulch, D. L., R. A. Cox, R. F. Hampson, Jr., J. A. Kerr, J. Troe, and R. T. Watson (1984). Evaluated kinetic and photochemical data for atmospheric chemistry: Supplement II CODATA Task Group on Gas Phase Chemical Kinetics. *J. Phys. Chem. Ref. Data* **13**, 1259-1380.

Baum, F. (1972). CO Emissionen aus Hausbrand Feuerstätten. *Staub* **32**, 54-59.

Baumgartner, A., and E. Reichel (1975). "The World Water Balance." R. Oldenburg Verlag, München.

Baxter, M. S., M. J. Stenhouse, and N. Drudarski (1980). Fossil carbon in coastal sediments. *Nature* **287**, 35-36.

Bazilevich, N. I., L. E. Rodin, and N. N. Rozov (1971). Geographical aspects of biological productivity. *Sov. Geogr. Rev. Transl.* **12**, 293-317.

Beard, K. V. (1974) Experimental and numerical collision efficiencies for submicron particles scavenged by small raindrops. *J. Atmos. Sci.* **31**, 1595-1603.

Beard, K. V., and S. N. Grover (1974). Numerical collision efficiencies for small raindrops colliding with micron size particles. *J. Atmos. Sci.* **31**, 543-550.

Becker, R., and W. Doering (1935). Kinetische Behandlung der Keimbildung in übersättigten Dämpfen. *Ann. Phys. (Leipzig)* **24**, 719-752.

Becker, K. H., A. Inocenncio, and U. Schurath (1975). The reaction of ozone with hydrogen sulfide and its organic derivatives. *Int. J. Chem. Kinet.* **7** (*Symp.* No. 1), 205-220.

Begemann, F., and I. Friedman (1968). Isotopic composition of atmospheric hydrogen. *J. Geophys. Res.* **73**, 1139-1147.

Beilke, S. (1970). Untersuchungen über das Auswaschen atmosphärischer Spurenstoffe durch Niederschläge. Doctoral Thesis, Universität Frankfurt-Main.

Beilke, S., and D. Lamb (1974). On the absorption of SO_2 in ocean water. *Tellus* **26**, 268-271.

Beilke, S., D. Lamb, and J. Müller (1975). On the uncatalyzed oxidation of atmospheric SO_2 by oxygen in aqueous systems. *Atmos. Environ.* **9**, 1083-1090.

Belser, L. W. (1979). Population ecology of nitrifying bacteria. *Annu. Rev. Microbiol.* **33**, 309-333.

Benkovitz, C. M. (1982). Compilation of an inventory of anthropogenic emissions in the United States and Canada. *Atmos. Environ.* **16**, 1551-1563.

Benson, S. W. (1978). Thermochemistry and kinetics of sulfur-containing molecules and radicals. *Chem. Rev.* **78**, 23-35.

Benton, D. J., and P. Moore (1970). Kinetics and mechanism of the formation and decay of peroxynitrous acid in perchloric solutions. *J. Chem. Soc. A*, 3179-3182.

Berg, W. W., P. J. Crutzen, F. E. Grahek, S. N. Gitlin, and W. A. Sedlacek (1980). First measurements of total chlorine and bromine in the lower stratosphere. *Geophys. Res. Lett.,* **7**, 937-940.

Bergeron, T. (1935). On the physics of clouds and precipitation. *Proc. 5th, Assem. U.G.G.I., Lisbon* **2**, 156.

Berkner, L. V., and L. C. Marshall (1964). The history of oxygenic concentrations in the earth's atmosphere. *Discuss. Faraday Soc.* **37**, 122-141.

Berkner, L. V., and L. C. Marshall (1966). Limitation on oxygen concentration in a primitive planetary atmosphere. *J. Atmos. Sci.* **23**, 133-143.

Berner, R. A. (1971). Worldwide sulfur pollution of rivers. *J. Geophys. Res.* **76**, 6597-6600.

Berresheim, H., and W. Jaeschke (1983). The contribition of volcanoes to the global atmospheric sulfur budget. *J. Geophys. Res.* **88**, 3732-3740.

Berry, J. A. (1975). Adaptation of photosynthetic processes to stress. *Science* **188**, 644-650.

Besemer, A. C. (1982). Formation of chemical compounds from irradiated mixtures of aromatic hydrocarbons and nitrogen oxides. *Atmos. Environ.* **16**, 1599-1602.

Bidwell, R. G. S., and D. E. Fraser (1972). Carbon monoxide uptake and metabolism by leaves. *Can. J. Bot.* **50**, 1435-1439.

Bien, G., and H. E. Suess (1967). Transfer and exchange of C-14 between the atmospheric and surface waters of the Pacific Ocean. *In* "Radioactive Dating and Methods of Low-level Counting," *Symp. Proc., Monaco, Int. Atomic Energy Agency, Vienna* (**STI/PUB/152**), pp. 105-115.

Biermann, H. W., C. Zetzsch, and F. Stuhl (1976). Rate constant for the reaction of OH with N_2O at 298 K. *Ber. Bunsen Ges. Phys. Chem.* **80**, 909-911.

Bischof, W. (1977). Comparability of CO_2 measurements. *Tellus* **29**, 435-444.

Bischof, W. (1981). The CO_2 content of the upper polar troposphere between 1963–1979. *In* "Carbon Cycle Modelling" (B. Bolin, ed.), *SCOPE* **16**, 113–116.

Birch, F. (1965). Speculations on the earth's thermal history. *Bull. Geol. Soc. Am.* **76**, 133–154.

Black, G. (1984). Reactions of HS with NO and NO_2, *J. Chem. Phys.* **80**, 1103–1107.

Blak, D. R., E. W. Mayer, S. C. Tyler, Y. Makide, D. C. Montagne, and F. S. Rowland (1982). Global increase in atmospheric methane concentrations between 1978–1980. *Geophys. Res. Lett.* **9**, 477–480.

Blaker, A. J. (1979). An atmospheric absorption model for the Schuman–Runge bands of oxygen. *J. Geophys. Res.* **84**, 3272–3282.

Blanchard, D. C., and A. H. Woodcock (1957). Bubble formation and modification in the sea and its meteorological significance. *Tellus* **9**, 145–158.

Blanchard, D. C. (1963). Electrification of the atmosphere. *Prog. Oceanogr.* **1**, 71–202.

Blanchard, D. C., and L. Syzdek (1972). Variations in Aitken and giant nuclei in marine air. *J. Phys. Oceanogr.* **2**, 255–262.

Blanchard, D. C., and A. H. Woodcock (1980). The production, concentration and vertical distribution of the sea-salt aerosol. *Ann. N. Y. Acad. Sci.* **338**, 330–347.

Blatherwick, R. D., A. Goldman, D. G. Murcray, F. J. Murcray, G. R. Cook, and J. W. van Allen (1980). Simultaneous mixing ratios profiles of stratospheric NO and NO_2 as derived from balloon borne infrared solar spectra. *Geophys. Res. Lett.* **7**, 471–473.

Blatt, H., and R. L. Jones (1975). Proportions of exposed igneous, metamorphic and sedimentary rocks. *Bull. Geol. Soc. Am.* **86**, 1085–1088.

Blifford, Jr., I. H., L. B. Lockhart, Jr., and H. B. Rosenstock (1952). On the natural radioactivity in the air. *J. Geophys. Res.* **57**, 499–509.

Blifford, I. H., Jr. and L. D. Ringer (1969). The size and number distribution of aerosols in the continental troposphere. *J. Atmos. Sci.* **26**, 716–726.

Blifford, I. H., J. W. Burgmeier, and C. E. Junge (1974). Modification of aerosol size distributions in the troposphere, Technical Note **NCAR-TN/STR-98**, National Center for Atmospheric Research, Boulder, Colorado.

Bloomfield, C. (1969). Sulfate reduction in water-logged soils. *Soil Sci.* **20**, 207–221.

Bloomfield, P., M. L. Thompson, and S. Zeger (1982). A statistical analysis of Umkehr measurements of 32–46 km ozone. *J. Appl. Meteorol.* **21**, 1828–1837.

Blundell, R. V. W. G. A. Cook, D. E. Hoare, and G. S. Milne (1965). Rates of radical reactions in methane oxidation. *Int. Symp. Combustion, 10th*, pp. 445–452. The Combustion Institute, Pittsburgh, Pennsylvania.

Bodenstein, M. H. (1918). Die Geschwindigkeit der Reaktion zwischen Stickoxyd und Sauerstoff. *Z. Electrochem.* **24**, 183–201.

Bohn, H. L. (1976). Estimate of orgnic carbon in soils. *J. Am. Soc. Soil Sci.* **40**, 468–470.

Bolin, B. (1970) The carbon cycle. *Sci. Am.* **223**, 124–132.

Bolin, B. and W. Bischof (1970). Variation of carbon dioxide content of the atmosphere in the northern hemisphere. *Tellus* **22**, 431–442.

Bolin, B. (1977). Changes of land biota and their importance for the carbon cycle. *Science* **196**, 613–615.

Bolin, B., A. Björkström, C. D. Keeling, R. Bacastow, and U. Siegenthaler (1981). Carbon cycle modelling. *In* "Carbon Cycle Modelling" (B. Bolin, ed.), *SCOPE* **16**, 1–28.

Bollag, J. M., S. Drzymala, and L. T. Kardos (1973). Biological versus chemical nitrite decomposition in soil. *Soil Sci.* **116**, 44–50.

Bollinger, M. J. C. J. Hahn, D. D. Parrish, P. C. Murphy, D. L. Albritton, and F. C. Fehsenfeld (1984). NO_x measurements in clean continental air and analysis of the contributing meteorology, *J. Geophys. Res.* **89**, 9623–9631.

Bolton, N. E., J. A. Carter, J. F. Emery, C. Feldman, W. Fulkerson, L. D. Hulett, and W. S. Lyon (1975). Trace element balance around a coal-fired steam plant. In "Trace Elements in Fuel" (S. P. Babu, ed.), Adv. Chem. Ser. 141, 175–187.

Bonsang, B., B. C. Nguyen, A. Gaudry, and G. Lambert (1980). Sulfate enrichment in marine aerosols owing to biogenic gaseous sulfur compounds. J. Geophys. Res. 85, 7410–7416.

Borchers, R., P. Fabian, and S. A. Penkett (1983). First measurements of the vertical distribution of CCl_4 and CH_3CCl_3 in the stratosphere. Naturwissenschaften. 70, 514–517.

Bos, R. (1980) Automatic measurement of atmospheric ammonia. J. Air Pollut. Control Assoc. 30, 1222–1224.

Böttger, A., D. H. Ehhalt, and G. Gravenhorst (1978). Atmosphärische Kreisläufe von Stick-oxiden und Ammoniak. Ber. Kernforschungsanlage Jülich, Nr. 1558.

Boubel, R. W., E. F. Darley, and E. A. Schuck (1969). Emissions from burning grass stubble and straw. J. Air Pollut. Control Assoc. 19, 497–500.

Boulaud, D., G. Madeleine, D. Vigla, and J. Bricard (1977). Experimental study on the nucleation of water vapor sulfuric acid binary system. J. Chem. Phys. 66, 4854–4860.

Boultron, C. (1980). Respective influence of global pollution and volcanic eruptions on the past variations of the trace metal content of Antarctic snows since 1880's J. Geophys. Res. 85, 7426–7432.

Bowen, H. J. M. (1966). "Trace Elements in Biochemistry." Academic Press, London.

Boyce, S. D., and M. R. Hoffmann (1984). Kinetics and mechanism of the formation of hydroxymethane sulfonic acid at low pH. J. Phys. Chem. 88, 4740–4746.

Braslau, N., and J. V. Dave (1973). Effect of aerosols on the transfer of solar energy through realistic model atmospheres, Parts I and II. J. Appl. Meterorol. 12, 601–619.

Braslawski, S., and J. Heicklen (1976). The gas phase reaction of O_3 with H_2CO. Int. J. Chem. Kinet. 8, 801–808.

Brasseur, G., and M. Nicolet (1973). Chemospheric processes of nitric oxide in the mesosphere and stratosphere. Planet. Space Sci. 21, 939–961.

Brasseur, G., A. De Rudder, and C. Tricot (1985). Stratosphere response to chemical perturbation. Preprint.

Bray, J. R. (1959). An analysis of the possible recent change in atmospheric carbon dioxide concentration. Tellus 11, 220–230.

Breeding, R. J., J. P. Lodge, Jr., J. B. Pate, D. C. Sheesley, H. B. Klonis, B. Fogle, J. A. Anderson, T. R. Englert, P. L. Haagenson, R. B. McBeth, A. L. Morris, R. Pogue, and A. F. Wartburg (1973). Background trace gas concentrations in the Central United States. J. Geophys. Res. 78, 7057–7064.

Breeding, R. J., H. B. Klonis, J. P. Lodge, J. B. Pate, D. C. Sheesley, T. R. Englert, and D. R. Scars (1976). Measurements of atmospheric pollutants in the St. Louis area. Atmos. Environ. 10, 181–194.

Breitenbeck, G. A., A. M. Blackmer and J. M. Bremner (1980). Effect of different nitrogen fertilizers on emission of nitrous oxide from soil. Geophys. Res. Lett. 7, 85–88.

Bremner, J. M., and C. G. Steele (1978). Role of micro organisms in the atmospheric sulfur cycle. Adv. Microb. Ecol. 2, 155–201.

Bremner, J. M., S. G. Robbins, and A. M. Blackmer (1980). Seasonal variability in emission of nitrous oxides from soil. Geophys. Res. Lett. 7, 641–644.

Bremner, J. M., and A. M. Blackmer (1981). Terrestrial nitrification as a source of atmospheric nitrous oxide. In "Denitrification, Nitrification and Atmospheric Nitrous Oxide" (C. C. Delwiche, ed.), pp. 151–170). Wiley, New York.

Bremner, J. M., G. A. Breitenbeck, and A. M. Blackmer (1981). Effect of anhydrous ammonia fertilization on emission of nitrous oxide from soils. J. Environ. Qual. 10, 77–80.

Brewer, A. W. (1949). Evidence for a world circulation provided by the measurements of helium and water vapor distribution in the stratosphere. *Q. J. R. Meteorol. Soc.* **75**, 351-363.

Brewer, A. W., and J. R. Milford (1960). The Oxford-Kew Ozonesonde. *Proc. R. Soc. London* A **256**, 470-495.

Brewer A. W., and C. T. McElroy (1973). Nitrogen dioxide concentrations in the atmosphere. *Nature* **246**, 129-133.

Bricard, J., F. Billard and G. J. Madeleine (1968). Formation and evolution of nuclei of condensation that appear in air initially free of aerosols. *J. Geophys. Res.* **73**, 4487-4496.

Bridgwater, D., J. H. Allaart, J. W. Schopf, C. Klein, M. R. Walter, E. S. Barghoorn, P. Strother, A. H. Knoll, and B. E. Gorman (1981). Microfossil-like objects from the Archaean of Greenland: a cautionary note. *Nature* **289**, 51-53.

Brice, K. A., S. A. Penkett, D. H. F. Atkins. F. J. Sandalls, D. J. Bamber, A. F. Tuck, and G. Vaughan (1984). Atmospheric measurements of peroxyacetyl nitrate (PAN) in rural, south-east England, seasonal variations, winter photochemistry and long-range transport. *Atmos. Environ.* **18**, 2691-2702.

Briggs, J., and W. T. Roach (1963). Aircraft observations near jet streams. *Q. J. R. Meteorol. Soc.* **89**, 225-247.

Brimblecombe, P., and D. J. Spedding (1974a). The reaction of the metal ion catalyzed oxidation in aqueous solution. *Chemosphere* **1**, 29-32.

Brimblecombe, P., and D. J. Spedding (1974b). The catalytic oxidation of micromolar aqueous sulfur dioxide. *Atmos. Environ.* **8**, 937-945.

Brimblecombe, P., and D. H. Stedman (1982). Evidence for a dramatic increase in the contribution of oxides of nitrogen to precipitation acidity. *Nature* **298**, 460-462.

Brinkmann, R. T. (1969). Dissociation of water vapor and evolution of oxygen in the terrestrial atmosphere. *J. Geophys. Res.* **74**, 5355-5368.

Brodzinski, R., S. G. Chang, S. S. Markowitz, and T. Novakov (1980). Kinetics and mechanism for the catalytic oxidation of sulfur dioxide on carbon in aqueous solution. *J. Phys. Chem.* **84**, 3354-3358.

Broecker, W. S. (1963). Radioisotopes and large-scale oceanic mixing. "The Sea: Ideas and Observations on Progress in the Study of the Seas" (M. N. Hill, ed.), Chap. 4, pp. 88-108. Wiley (Interscience), New York.

Broecker, W. S. (1970). A boundary condition on the evolution of atmospheric oxygen. *J. Geophys. Res.* **75**, 3553-3557.

Broecker, W. S., and T. H. Peng (1971). The vertical distribution of radon in the Bomex area. *Earth Planet. Sci. Lett.* **11**, 99-108.

Broecker, W. S. (1971). A kinetic model for the chemical composition of sewater. *Quat. Res.* (*N.Y.*) **1**, 188-207.

Broecker, W. S. (1973). Factors controlling CO_2 content in the oceans and atmosphere. *In* "Carbon and the Biosphere" (G. M. Woodwell and E. V. Pecan, eds.), pp. 32-50. *AEC Symp. Ser.* **30**, NTIS U.S. Dept. Commerce, Springfield, Virginia.

Broecker, W. S., and T. H. Peng (1974). Gas exchange rates between air and sea. *Tellus* **20**, 21-35.

Broecker, W. S., and T. Takahashi (1978). The relationship between lysocline depth and in situ carbonate concentration. *Deep Sea Res.* **25**, 65-95.

Broecker, W. S. (1979). A revised estimate for the radiocarbon age of North Atlantic deep water. *J. Geophys Res.* **84**, 3218-3226.

Broll, A., G. Helas, K. J. Rumpel, and P. Warneck (1984). NO_x background mixing ratios in surface air over Europe and the Atlantic Ocean. *In* "Physico-chemical Behaviour of

Atmospheric Pollutants", (B. Versino and G. Angeletti, (eds.) pp. 390-399. *Eur. Symp. 3rd, Varese, Italy.* Reidel, Dordrecht, Holland.

Bromfield, A. R. (1974). The deposition of sulphur in rainwater in northern Nigeria. *Tellus* **26**, 408-411.

Brosset, C. (1978). Watersoluble sulfur compounds in aerosols. *Atmos. Environ.* **12**, 25-38.

Brown, H. (1949). Rare gases and the formation of the earth's atmosphere. *In* "The Atmospheres of the Earth and Planets" (G. P. Kuiper, ed.), pp. 258-266. University of Chicago Press, Chicago.

Brown, A. P., and K. P. Davis (1973). "Forest Fire, Control and Use." McGraw-Hill, New York.

Brueckner, G. E., J. D. F. Bartoc, O. K. Moe, and M. E. vanHoosier (1976). Absolute solar ultraviolet intensities and their variations with solar activity in the wavelength region 1750-2100 Å, *Astrophys. J.* **209**, 935-944.

Bruyevich, S. V., and E. Z. Kulik (1967). Chemical interaction between atmosphere and the ocean (salt exchange). *Oceanology (Engl. Transl.)* **7**, 279-293.

Buat-Menard, P., and M. Arnold (1978). The heavy metal chemistry of atmospheric particulate matter emitted by the Mount Etna volcanoe. *Geophys. Res. Lett.* **5**, 245-248.

Buat-Menard, P., and R. Chesselet (1979). Variable influence of the atmospheric flux on the trace metal chemistry of oceanic suspended matter. *Earth Planet. Sci. Lett.* **42**, 399-411.

Buat-Menard, P., J. Morelli, and R. Chesselet (1974). Water-soluble elements in atmospheric particulate matter over tropical and equatorial Atlantic. *J. Rech. Atmos.* **8**, 661-673.

Buch, K. (1920). Ammoniakstudien an Meer-und Hafenwasserproben. *Kemistsamfundets Meddelanden* **1920**, 59-60.

Büchen, M., and H. W. Georgii (1971). Ein Beitrag zum atmosphärischen Schwefelhaushalt über dem Atlantik. *Meteor. Forschungsergeb. Reihe B* **7**, 71-77.

Bürgermeister, S. (1984). Messung von Bodenemissionen und atmosphärischen Konzentrationen des Dimethylsulfids. Diploma Thesis, Institute für Geologie und Geophysik der Universität Frankfurt/Main.

Bufalini, J. J., and A. P. Altshuller (1965). Kinetics of vapor-phase hydrocarbon ozone reactions. *Can. J. Chem.* **43**, 2243-2250.

Bufalini, J. J., B. M. Gay, Jr., and K. L. Brubaker (1972). Hydrogen peroxide formation from formaldehyde photo-oxidation and its presence in urban atmosphere. *Environ. Sci. Technol.* **6**, 816-821.

Buijs, H. L., G. L. Vail, G. Tremblay, and D. J. W. Kendall (1980). Simultaneous measurements of the volume mixing ratios of HCl and HF in the stratosphere. *Geophys. Res. Lett.* **7**, 205-208.

Burford, J. R., and R. C. Stefanson (1973). Measurements of gaseous losses of nitrogen from soils. *Soil. Biol. Biochem.* **5**, 133-141.

Burnett, W. E. (1969). Air pollution from animal wastes. *Environ. Sci. Technol.* **3**, 744-749.

Burns, R. C., and R. W. F. Hardy (1975). Nitrogen Fixation in Bacteria and Higher Plants." Springer-Verlag, New York.

Burrows, J., G. S. Tyndall, and G. K. Moortgart (1985). Absorption spectrum of NO_3 and kinetics of reaction of NO_3 with NO_2, Cl and several stable atmospheric species at 298 K. *J. Phys. Chem.* **89**, 4848-4856.

Burton, W. M., and N. G. Stewart (1960). Use of long-lived natural radioactivity as an atmospheric tracer. *Nature* **186**, 584-589.

Bush, Y. A., A. L. Schmeltekopf, F. C. Fehsenfeld, D. L. Albritton, J. R. McAfee, P. D. Goldan, and E. E. Ferguson (1978). Stratospheric measurements of methane at several latitudes. *Geophys. Res. Lett.* **5**, 1027-1029.

Cadle, R. D., and E. R. Allen (1970). Atmospheric photochemistry. *Science* **167**, 243-249.

Cadle, R. D., A. F. Wartburg, W. H. Pollock, B. W. Gandrud, and J. P. Shedlovsky (1973). Trace constituents emitted to the atmosphere by Hawaiian volcanoes. *Chemosphere* **6**, 231–234.

Cadle, R. D. (1973). Particulate matter in the lower atmosphere, *In* "Chemistry of the Lower Atmosphere" (S. I. Rasool, ed.), Chap. 2, pp. 69–120. Plenum Press, New York.

Cadle, R. D. (1975). Volcanic emissions of halides and sulfur compounds to the troposphere and stratosphere. *J. Geophys. Res.* **80**, 1650–1652.

Cadle, R. D., and G. Langer (1975). Stratospheric Aitken particles near the tropopause. *Geophys. Res. Lett.* **2**, 329–332.

Cadle, R. D., F. G. Fernald, and C. L. Frush (1977). Combined use of lidar and numerical diffusion models to estimate the quantity and dispersion of volcanic eruption clouds in the stratosphere: Vulcan Fuego 1974 and Augustine 1976. *J. Geophys. Res.* **82**, 1783–1786.

Cadle, R. D., and E. J. Mroz (1978). Particles in the eruption cloud from St. Augustine Volcano. *Science* **199**, 455–456.

Cadle, R. D. (1980). A comparison of volcanic with other fluxes of atmospheric trace gas constituents. *Rev. Geophys. Space Phys.* **18**, 746–752.

Cadle, S. H., R. J. Countess, and N. A. Kelly (1982). Nitric acid and ammonia in urban and rural locations. *Atmos. Environ.* **16**, 2501–2506.

Cadle, S. H. (1985). Seasonal variational in nitric acid, nitrate, strong aerosol acidity, and ammonia in an urban area. *Atmos. Environ.* **19**, 181–188.

Callendar, G. S. (1938). The artificial production of carbon dioxide and its influence on temperature. *Q. J. R. Meterol. Soc.* **64**, 223–237.

Callendar, G. S. (1958). On the amount of carbon dioxide in the atmosphere. *Tellus* **10**, 243–248.

Calvert, J. G., and J. N. Pitts, Jr. (1966). "Photochemistry." Wiley, New York.

Calvert, J. G., J. A. Kerr, K. L. Demerjian, and R. D. McQuigg (1972). Photolysis of formaldehyde as a hydrogen atom source in the lower atmosphere. *Science* **175**, 751–752.

Calvert, J. G., F. Su, J. W. Bottenheim, and O. P. Strausz (1978). Mechanism for the homogeneous oxidation of sulfur dioxide in the troposphere. *Atmos. Environ.* **12**, 197–226.

Calvert, J. G. (1980). The homogeneous chemistry of formaldehyde generation and destruction within the atmosphere. "NATO Advanced Study Institute on Atmospheric Ozone: Its Variation and Human Influences" (M. Nicolet and A. C. Aikin, **FAA-EE-80-20**, pp. 153–190. U.S. Dept. Transportation, Federal Aviation Administration, Washington, D.C.

Calvert, J. G., and W. R. Stockwell (1984). Mechanism and rates of the gas-phase oxidations of sulfur dioxides and nitrogen oxides in the atmosphere. *In* "SO_2, NO and NO_2 and Oxidation Mechanisms: Atmospheric Considerations" (J. G. Calvert, ed.), Chap. 1, Acid Precipitation Series, Vol. 3, pp. 1–62. Butterworth, Boston.

Campbell, R. (1977). "Microbial Ecology." Halsted Press, New York.

Campbell, I. M. (1977). "Energy and the Atmosphere: a physical–chemical approach." Wiley, London and New York.

Cantrell, B. K., and K. T. Whitby (1978). Aerosol size distributions and aerosol volume formation for a coal-fired power plant plume. *Atmos. Environ.* **12**, 323–333.

Cantrell, C. A., W. R. Stockwell, L. G. Anderson, K. L. Busarow, D. Perner, C. Schmeltekopf, J. G. Calvert, and H. S. Johnston (1985). Kinetic study of the NO_3–CH_2O reaction and its possible role in nighttime tropospheric chemistry. *J. Phys. Chem.* **89**, 139–146.

Carter, W. P. L., K. R. Darnall, A. C. Lloyd, A. M. Winer, and J. N. Pitts, Jr. (1976). Evidence for alkoxy radical isomerization in photooxidations of C_4–C_6 alkenes under simulated atmospheric conditions. *Chem. Phys. Lett.* **42**, 22–27.

Carter, W. P. L., A. C. Lloyd, J. L. Sprung, and J. N. Pitts, Jr. (1979). Computer modeling of smog chamber data: progress in validation of a detailed mechanism for the photooxidation of propene and *n*-butene in photochemical smog. *Int. J. Chem. Kinet.* **11**, 45–101.

Casadevall, T. J., D. A. Johnston, D. M. Harris, W. I. Rose, L. L. Malinconico, R. E. Stoiber, T. J. Bornhorst, S. N. Williams, L. Woodruff, and J. M. Thompson (1981). SO$_2$ emission rates at Mount St. Helens from March 29 through December, 1980. *U.S. Geol. Surv. Profess. Pap.* **1250**, 193–200.

Castellano, E., and H. J. Schumacher (1962a). Die Kinetik des photochemischen Zerfalls von Ozon in rotgelben Licht. *Z. Phys. Chem. N.F.* **34**, 198–212.

Castellano, E., and H. J. Schumacher, (1962b). Photochemical decomposition of ozone in yellow-red light and the mechanism of its thermal decomposition. *J. Chem. Phys.* **36**, 2238.

Castellano, E., and H. J. Schumacher (1972). The kinetics and the mechanism of the photochemical decomposition of ozone with light of 3340 Å wavelength. *Chem. Phys. Lett.* **13**, 625–627.

Castleman, A. W., Jr., H. R. Munkelwitz, and B. Manowitz (1973). Contribution of volcanic sulphur compounds to the stratospheric aerosol layer. *Nature* **244**, 345–346.

Castleman, Jr., A. W., H. R. Munkelwitz, and B. Manowitz (1974). Isotopic studies of the sulfur component of the stratospheric aerosol layer. *Tellus* **26**, 222–234.

Cauer, H. (1935). Bestimmung des Gesamtoxidationswertes des Nitrits, des Ozons und des Gesamtchlorgehaltes in roher und Vergifteter Luft. *Z. Anal. Chem.* **103**, 321–324, 385–416.

Cauer, H. (1951). Some problems of atmospheric chemistry. *In* "Compendium of Meteorology" (T. F. Mahone, ed.), pp. 1126–1136. Am. Met. Soc., Boston.

Cautreels, W. K., K. van Cauwenberghe, and L. A. Guzman (1977). Comparison between the organic fraction of suspended matter at a background and an urban station. *Sci. Total Environ.* **8**, 79–88.

Chaen, M. (1973). Studies of the production of sea-salt particles on the sea surface. *Mem. Fac. Fish. Kagoshima Univ.* **22**, 49–107.

Challenger, F. (1951). Biological methylation. *Adv. Enzymol.* **12**, 429–491.

Chamberlain, A. C. (1953). Aspects of travel and deposition of aerosol and vapor clouds, **AERE-HP/R 1261**. Atomic Energy Research Establishment, Harwell, England.

Chamberlain, A. C. (1960). Aspects of the deposition of radioactive and other gases and particles. *Int. J. Air Pollut.* **3**, 63–88.

Chamberlain, A. C. (1966a). Transport of Lycopodium spores and other small particles to rough surfaces. *Proc. R. Soc. A* **296**, 45–70.

Chamberlain, A. C. (1966b). Transport of gases to and from grass and grass-like surfaces. *Proc. R. Soc. A* **290**, 236–265.

Chameides, W. L., and J. C. G. Walker (1973). A photochemical theory of tropospheric ozone. *J. Geophys. Res.* **78**, 8751–8760.

Chameides, W. L., and J. C. G. Walker (1976). A time-dependent photochemical model for ozone near the ground. *J. Geophys. Res.* **81**, 413–420.

Chameides, W. L., D. H. Stedman, R. R. Dickerson, D. W. Rush, and R. J. Cicerone (1977). NO$_x$ production in lightning. *J. Atmos. Sci.* **34**, 143–149.

Chameides, W. L. (1979). Effect of variable energy input on nitrogen fixation in instantaneous linear discharges. *Nature* **277**, 123–125.

Chameides, W. L., and D. D. Davis (1980). Iodine: its possible role in tropospheric chemistry. *J. Geophys. Res.* **85**, 7383–7398.

Chameides, W. L., and A. Tan (1981). The two-dimensional diagnostic model for tropospheric OH: an uncertainty analysis. *J. Geophys. Res.* **86**, 5209–5223.

Chang, T. Y., J. M. Norbeck, and B. Weinstock (1979). An estimate of the NO$_x$ removal rate in an urban atmosphere. *Environ. Sci. Technol.* **13**, 1534–1537.

Chapman, S. (1930). A theory of upper atmospheric ozone. *Q. J. R. Meteorol. Soc.* **3**, 103–125.

Chapman, S. (1930). On the annual variation of the upper atmospheric ozone. *Philos. Mag.* **10**, 345-352.

Chatfield, R., and H. Harrison (1976). Ozone in the remote troposphere, mixing versus photochemistry. *J. Geophys. Res.* **81**, 421-423.

Chatfield, R., and H. Harrison (1977a). Tropospheric ozone 1: Evidence for higher background values. *J. Geophys. Res.* **82**, 5965-5968.

Chatfield, R., and H. Harrison (1977b). Tropospheric ozone 2: Variations along a meridivual band. *J. Geophys. Res.* **82**, 5969-5976.

Chatfield, R., and P. Crutzen (1984). Sulfur dioxide in remote oceanic air: cloud transport of reactive precursors. *J. Geophys. Res.* **89**, 7111-7132.

Chepil, W. S. (1951). Properties of soils which influence wind erosion, IV. State of dry aggregate soil structure. *Soil Sci.* **72**, 387-401.

Cheskis, S. G., and O. M. Sarkisov (1979). Flash photolysis of ammonia in the presence of oxygen. *Chem. Phys. Lett.* **62**, 72-76.

Chesselet, R., J. Morelli, and P. Buat-Menard (1972a). Some aspects of the geochemistry of marine aerosols. *In* "The changing chemistry of the oceans" (C. D. Dryssen and D. Jagner, eds.) Nobel Symp. **20**, pp. 93-114. Wiley, New York.

Chesselet, R., J. Morelli, and P. Buat-Menard (1972b). Variations in ionic ratios between reference seawater and marine aerosols. *J. Geophys. Res.* **77**, 5116-5131.

Chesselet, R., M. Fontugne, P. Buat-Menard, V. Ezat, and C. E. Lambert (1981). The origin of particulate organic carbon in the marine atmosphere as indicated by its stable carbon isotopic composition. *Geophys. Res. Lett.* **8**, 345-348.

Chou, C. C., J. G. Lo, and F. S. Rowland (1974). Primary processes in the photolysis of water vapor at 174 nm. *J. Chem. Phys.* **60**, 1208-1210.

Chou, C. C., W. S. Smith, H. VeraRuiz, K. Moe, G. Crescentini, M. J. Molina, and F. S. Rowland (1977). The temperature dependence of the ultraviolet absorption cross section of CCl_2F_2 and CCl_3F and their stratospheric significance. *J. Phys. Chem.* **81**, 286-290.

Ciccioli, P., E. Brancaleoni, M. Possanzini, A. Brachetti, and C. Di Palo (1984). Sampling, identification and quantitative determination of biogenic and anthropogenic hydrocarbons in forested areas. *In* "Physicochemical Behaviour of Atmospheric Pollutants, (B. Versino and G. Angeletti, eds.) *Proc. Symp. 3rd, Varese, Italy* pp. 62-73. Reidel, Dordrecht.

Cicerone, R. J., J. D. Shetter, D. H. Stedman, T. J. Kelly and S. C. Liu (1978). Atmospheric N_2O: measurements to determine its source, sinks and variations. *J. Geophys. Res.* **83**, 3042-3050.

Cicerone, R. J. (1979). Atmospheric carbon tetrafluoride. A nearly inert gas. *Science* **206**, 59-61.

Cicerone, R. J., and J. L. McCrumb (1980). Photodissociation of isotopically heavy O_2 as a source of atmospheric O_2. *Geophys. Res. Lett.* **7**, 251-254.

Cicerone, R. J., and J. D. Shetter (1981). Sources of atmospheric methane: measurements in rice paddies and a discussion. *J. Geophys. Res.* **86**, 7203-7209.

Cicerone, R. G., J. D. Shetter, and C. C. Delwiche (1983). Seasonal variation of methane flux from a Californian rice paddy. *J. Geophys. Res.* **88**, 11022-11024.

Clark, J. H., C. B. Moore, and N. S. Nogar (1978). The photochemistry of formaldehyde: absolute quantum yields, radical reactions and NO reactions. *J. Chem. Phys.* **68** 1264-1271.

Clark, F. E., and T. Rosswall, eds. (1981). Terrestrial nitrogen cycles: processes, ecosystem, strategies and management impacts. *Ecol. Bull. (Stockholm)* **33**.

Clarke, F. W. (1924). "The data of geochemistry," 5th ed., pp. 841. *U.S. Geol. Surv. Bull.* **770**. Government Printing Office.

Clements, H. B., and C. K. McMahon (1980). Nitrogen oxides from burning forest fuels examined by thermogravimetry and evolved gas analysis. *Thermochim. Acta* **35**, 133-139.

Cleveland, W. S., B. Kleiner, J. E. McRae, and J. L. Warner (1976). Photochemical air pollution: Transport from the New York City area into Connecticut and Massachussetts. *Science* **191**, 179-181.

Cleveland, W. S., T. E. Graedel, and B. Kleiner (1977). Urban formaldehyde: observed correlation with source emissions and photochemistry. *Atmos. Environ.* **11**, 357-360.

Cloud, P. E. (1965). Carbonate precipitation and dissolution in the marine environment. "Chemical Oceanography" (J. P. Riley and G. Skirrow, eds.), Vol. 2, Chap. 17, pp. 127-158. Academic Press, New York.

Cloud, P. E. (1972). A working model of the primitive atmosphere. *Am. J. Sci.* **272**, 537-548.

Cloud, P. E. (1973). Paleological significance for the banded iron formation. *Econ. Geol.* **68**, 1135-1143.

Cloud, P. E. (1976). Beginnings of biospheric evolution and their biogeochemical consequences. *Paleobiology* **2**, 351-387.

Clough, W. S. (1975). The deposition of particles on moss and grass surfaces. *Atmos. Environ.* **9**, 1113-1119.

Coffey, P. E., and W. N. Stasiuk (1975). Evidence of atmospheric transport of ozone into urban areas. *Environ. Sci. Technol.* **9**, 59-62.

Coffey, T., W. Stasiuk, and V. Mohnen (1977). Ozone in rural and uban areas of New York. In: *Proc. Int. Conf. Photochemical Oxidant Pollution and Its Control*, pp. 89-97. U.S. Environmental Protection Agency, Research Triangle Park, N.C. Publ. No. **EPA-600/3-77-001a**.

Coffey, M. T., W. G. Mankin, and A. Goldman (1981). Simultaneous spectroscopic determination of the latitudinal, seasonal and diurnal variability of stratospheric N_2O, NO, NO_2, and HNO_3. *J. Geophys. Res.* **86**, 7331-7341.

Cogbill, C. V., and G. E. Likens (1974). Acid precipitation in the Northeastern United States. *Water Resour. Res.* **10**, 1133-1137.

Cohen, Y., and L. I. Gordon (1979). Nitrous oxide production in the ocean. *J. Geophys. Res.* **84**, 347-353.

Colbeck, I., and R. M. Harrison (1985). The concentrations of specific C_2-C_6 hydrocarbons in the air of northwest England. *Atmos. Environ.* **19**, 1899-1904.

Collins, N. M., and T. G. Wood (1984). Termites and atmospheric gas production. *Science* **224**, 84-86.

Connell, P., and H. S. Johnston (1979). The thermal dissociation of N_2O_5 in N_2. *Geophys. Res. Lett.* **6**, 553-556.

Conrad, R., and W. Seiler (1980a). Field measurements of the loss of fertilizer nitrogen into the atmosphere as nitrous oxide. *Atmos. Environ.* **14**, 555-558.

Conrad, R., and W. Seiler (1980b). Role of microorganisms in the consumption and production of atmospheric carbon monoxide by soil. *Appl. Environ. Microbiol.* **40**, 437-445.

Conrad, R., and W. Seiler (1980c). Contribution of hydrogen production by biological nitrogen fixation to the global hydrogen budget. *J. Geophys. Res.* **85**, 5493-5498.

Conrad, R., and W. Seiler (1980d). Photooxidative production and microbial consumption of carbon monoxide in seawater. *FEMS Microbiol. Lett.* **9**, 61-64.

Conrad, R., and W. Seiler (1982a). Arid soils as a source of atmospheric carbon monoxide. *Geophys. Res. Lett.* **9**, 1353-1356.

Conrad, R., and W. Seiler (1982b). Utilization of traces of carbon monoxide by aerobic oligotrophic microorganisms in ocean, lake and soil. *Arch. Microbiol.* **132**, 41-46.

Conrad, R., W. Seiler, G. Bunse, and H. Giehl (1982). Carbon monoxide in seawater (Atlantic ocean). *J. Geophys. Res.* **87**, 8839-8852.

Conrad, R., W. Seiler, and G. Bunse (1983). Factors influencing the loss of fertilizer nitrogen into the atmosphere as N_2O. *J. Geophys. Res.* **88**, 6709-6718.

Conrad, R., and W. Seiler (1985a). Influence of temperature, moisture and organic carbon on the flux of H$_2$ and CO between soil and atmosphere: field studies in subtropical regions. *J. Geophys. Res.* **90**, 5699-5709.

Conrad, R., and W. Seiler (1985b). Characteristics of abiological carbon monoxide formation from soil organic matter, humic acids and phenolic compounds. *Environ. Sci. Technol.* **19**, 1165-1169.

Cowan, M. I., A. T. Glen, S. A. Hutchinson, M. E. MacCartney, J. M. Mackintosch, and A. M. Moss (1973). Production of volatile metabolites by the species of fomes. *Trans. Br. Mycol. Soc.* **60**, 347-360.

Cox, R. A. (1972). Quantum yields for the photo oxidation of sulfur dioxide in the first allowed absorption region. *J. Phys. Chem.* **76**, 814-820.

Cox, R. A., and S. A. Penkett (1972). Aerosol formation from sulphur dioxide in the presence of ozone and olefinic hydrocarbons. *J. Chem. Soc. Faraday Trans. 1* **68**, 1735-1753.

Cox, R. A., (1973). Some experimental observations of aerosol formation in the photo-oxidation of sulphur dioxide. *J. Aerosol Sci.* **4**, 473-483.

Cox, R. A. (1974). Photolysis of gaseous nitrous acid. *J. Photochem.* **3**, 175-188.

Cox, R. A., and R. G. Derwent (1976). The ultraviolet absorption spectrum of gaseous nitrous acid. *J. Photochem.* **6**, 23-34.

Cox, R. A., and M. J. Roffey (1977). Thermal decomposition of peroxyacetylnitrate in the presence of nitric oxide. *Environ. Sci. Technol.* **11**, 900-906.

Cox, R. A. (1977). Some measurements of ground level NO, NO$_2$ and O$_3$ concentrations at an unpolluted maritime site. *Tellus* **29**, 356-362.

Cox, R. A., and D. Sheppard (1980). Reactions of OH radicals with gaseous sulfur compounds. *Nature* **284**, 330-331.

Cox, R. A., R. G. Derwent, and M. R. Williams (1980). Atmospheric photooxidation reactions. Rates, reactivity and mechanisms for reactions of organic compounds with hydroxyl radicals. *Environ. Sci. Technol.* **14**, 57-61.

Crafford, T. C. (1975). SO$_2$ emission of the eruption of the 1974 eruption of volcano Fuego, Guatemala. *Bull. Volcanol.* **39**, 536-556.

Craig, R. A. (1950). The observations and photochemistry of atmospheric ozone and their meteorological significance. Meterorol. Monographs I/2 Amer. Meteorol. Soc., Boston.

Craig, H. (1957). The natural distribution of radiocarbon and the exchange time of carbon dioxide between atmosphere and sea. *Tellus* **9**, 1-17.

Craig, H., and C. C. Chou (1982). Methane: the record in polar ice cores. *Geophys. Res. Lett.* **9**, 1221-1224.

Crank, J. (1975). "The Mathematics of Diffusion," 2nd ed. Oxford Univ. Press, London.

Criegee, R. (1957). The course of ozonization of unsaturated compounds. *Rec. Chem. Prog.* **18**, 111-120.

Criegee, R. (1962). Peroxide pathways in ozone reactions. *In* "Peroxide Reaction Mechanisms" (J. O. Edwards, ed.), pp. 29-39. Wiley (Interscience), New York.

Criegee, R. (1975). Mechanism of ozonolysis. *Angew. Chem. Int. Ed. Engl.* **14**, 745-751.

Cronn, D. R., R. J. Charlson, R. L. Knights, A. L. Crittenden, and B. R. Appel (1972). A Survey of the molecular nature of primary and secondary components of particles in urban air by high resolution mass spectroscopy. *Atmos. Environ.* **11**, 929-937.

Cronn, D. R., R. A. Rasmussen, E. Robinson, and D. E. Harsch (1977). Halogenated compound identification and measurement in the troposhere and lower stratosphere. *J. Geophys. Res.* **82**, 5935-5944.

Cronn, D. R., and D. E. Harsch (1980). Smoky Mountain ambient halocarbon and hydrocarbon monitoring 21-26 September 1978. U.S. Environmental protection Agency Contract No. **RO804033-03-2.** Washington State University, Pullman, Washington.

Crutzen, P. J. (1970). The influence of nitrogen oxides on the atmospheric ozone content. *Q. J. R. Meteorol. Soc.* **96**, 320-325.

Crutzen, P. J. (1971). Ozone production rates in an oxygen-hydrogen-nitrogen oxide atmosphere. *J. Geophys. Res.* **76**, 7311-7327.

Crutzen, P. J. (1973). A discussion of the chemistry of some minor constituents in the stratosphere and troposphere. *Pure Appl. Geophys.* **106-108**, 1385-1399.

Crutzen, P. J. (1974a). Photochemical reactions initiated by and influencing ozone in unpolluted tropospheric air. *Tellus* **26**, 47-57.

Crutzen, P. J. (1974b). Estimates of possible variations in total ozone due to natural and human activities. *Ambio* **3**, 201-210.

Crutzen, P. J., I. S. A. Isaksen, and G. C. Reid (1975). Solar photon events: stratospheric sources of nitric oxide. *Science* **189**, 457-459.

Crutzen, P. J. (1976). Upper limits on atmospheric ozone reductions following increased applications of fixed nitrogen to the soil. *Geophys. Res. Lett.* **3**, 169-172.

Crutzen, P. J., and J. Fishman (1977). Average concentrations of OH in the troposphere and the budgets of CH_4, CO, H_2 and CH_3 CCl_3. *Geophys. Res. Lett.* **4**, 321-324.

Crutzen, P. J., L. E. Heidt, J. P. Krasnec, W. H. Pollock, and W. Seiler (1979). Biomass burning as a source of atmospheric gases CO, H_2, N_2O, NO, CH_3CL, and COS. *Nature* **282**, 253-256.

Crutzen, P. J. (1982). The global distribution of hydrogen. "Atmospheric Chemistry" Dahlem Konferenzen (E. D. Goldberg, ed.), pp. 313-328. Springer-Verlag, Berlin.

Crutzen, P. J., and L. T. Gidel (1983). A two-dimensional photochemical model of the atmosphere. 2: The tropospheric budgets of the anthropogenic chlorocarbons, CO, CH_4, CH_3Cl and the effect of various NO_x sources on tropospheric ozone. *J. Geophys. Res.* **88**, 6641-6661.

Crutzen, P. J., and U. Schmailzl (1983). Chemical budgets of the stratosphere. *Planet. Space Sci.* **31**, 1009-1032.

Crutzen, P. J. (1983). Atmospheric interactions—homogeneous gas reactions of C, N and S containing compounds. "The Major Biochemical Cycles and their Interactions" (B. Bolin and R. B. Cook, eds.), *SCOPE* **21**, 67-114.

Cullis, C. F., and M. M. Hirschler (1980). Atmospheric sulfur: natural and man-made sources. *Atmos. Environ.* **14**, 1263-1278.

Cunningham, W. C., and W. H. Zoller (1981). The chemical composition of remote area aerosols. *J. Aerosol Sci.* **12**, 367-384.

Cunnold, D., F. Alyea, and R. Prinn (1978). A methology for determining the atmospheric life time of fluorocarbons. *J. Geophys. Res.* **83**, 5493-5500.

Cunnold. D. M., R. G. Prinn, R. A. Rasmussen, P. G. Simmonds, F. N. Alyea, C. A. Cardelino, A. J. Crawford, P. J. Fraser, and R. D. Rosen (1983a). The atmospheric life time experiment 3. Life time methology and application to three years of $CFCl_3$ data. *J. Geophys. Res.* **88**, 8379-8400.

Cunnold, D. M., R. G. Prinn, R. A. Rasmussen, P. G. Simmonds, F. N. Alyea, C. A. Cardelino, and A. J. Crawford (1983b). The atmospheric lifetime experiment 4. Results for CF_2Cl_2 based on three years of data. *J. Geophys. Res.* **88**, 8401-8414.

Cvetanovic, R. J. (1965) Excited oxygen atoms in the photolysis of N_2O and NO_2 *J. Chem. Phys.* **43**, 1850-1851.

Czeplak, G., and C. Junge (1974). Studies of interhemispheric exchange in the troposphere by a diffusion model. *Adv. Geophys.* **18B**, 57-72.

Daisey, J. M. (1980). Organic compounds in urban aerosols (1980). *Ann. N. Y. Acad. Sci.* **338**, 50-69.

D'Almeida, G., and R. Jaenicke (1981). The size distribution of mineral dust. *J. Aerosol Sci.* 12, 160-162.

Damschen, D. E., and L. R. Martin (1983). Aqueous aerosol oxidation of nitrous acid by O_2, O_3 and H_2O_2. *Atmos. Environ.* 17, 2005-2011.

Danckwerts, P. V. (1970). "Gas-Liquid Reactions." McGraw-Hill, New York.

Danielsen, E. F. (1964). Report on Project Springfield, DASA-1517 Defense Atomic Support Agency, Washington, D. C.

Danielsen, E. F. (1968). Stratospheric-tropospheric exchange based on radioactivity, ozone and potential vorticity. *J. Atmos. Sci.* 25, 502-518.

Danielsen, E. F., R. Bleck, J. Shedlowsky, A. Wartburg, P. Haagenson, and W. Pollock (1970). Observed distribution of radioactivity, ozone and potential vorticity associated with tropopause folding. *J. Geophys. Res.* 75, 2353-2361.

Danielsen, E. F. (1975). The Nature of transport processes in the stratosphere In "The Natural Stratosphere of 1974" (A. J. Grobecker, ed.), pp. 6-12-6-22. CIAP Monograph 1. U.S. Dept of Transportation, Washington, D.C.

Danielsen, E. F., and V. A. Mohnen (1977). Project Dustom report: ozone transport, *in situ* measurements and meteorological analyses of tropopause folding. *J. Geophys. Res.* 82, 5867-5877.

Darley, E. F., F. R. Burleson, E. H. Mateer, J. T. Middleton, and V. P. Osterli (1966). Contribution of burning of agricultural wastes to photochemical air pollution. *J. Air Pollut. Control Assoc.* 11, 685-690.

Darnall, K. R., A. C. Lloyd, A. M. Winer and J. N. Pitts, Jr. (1976a). Reactivity scale for atmospheric hydrocarbons based on reaction with hydroxyl radical. *Environ. Sci. Technol.* 10, 692-696.

Darnall, K. R., W. P. L. Carter, A. M. Winer, A. C. Lloyd, and J. N. Pitts, Jr. (1976b) Importance of RO_2 + NO in alkylnitrate formation from C_4-C_6 alkane photooxidation under simulated atmospheric condition. *J. Phys. Chem.* 80, 1948-1950.

Davidson, B., J. P. Friend, and H. Seitz (1966). Numerical models of diffusion and rainout of stratospheric radioactive materials. *Tellus* 18, 301-315.

Davies, M., and G. H. Thomas (1960). The lattice energy, infrared spectra and possible cyclization of some dicarboxylic acids. *Trans. Faraday Soc.* 56, 185-192.

Davis, D. D., W. S. Heaps, and T. McGee (1976). Direct measurements of natural tropospheric levels of OH via an aircraft-borne tunable dye laser. *Geophys. Res. Lett.* 3, 331-333.

Davis, D. D., W. S. Heaps, D. Philen, M. Rodgers, T. McGee, A. Nelson, and A. J. Moriarty (1979). Airborne laser induced fluorescence system for measuring OH and other trace gases in the parts per quadrillion to parts per trillion range. *Rev. Sci. Instrum.* 50, 1506-1516.

Davis, D. D., M. O. Rodgers, S. D. Fischer, and W. S. Heaps (1981). A theoretical assessment of the O_3/H_2O interference problem in the detection of OH via LIF. *Geophys. Res. Lett.* 8, 73-76.

Davis, D. D., M. O. Rodgers, S. D. Fischer, and K. Asai (1981). An experimental assessment of the O_3/H_2O interference problem in the detection of natural levels of OH via laser induced fluorescence. *Geophys. Res. Lett.* 8, 69-72.

Dawson, G. A. (1977). Atmospheric ammonia from undisturbed land. *J. Geophys. Res.* 82, 3125-3133.

Dawson, G. A. (1980). Nitrogen fixation by lightning. *J. Atmos. Sci.* 37, 174-178.

Day, J. A. (1964). Production of droplets and salt nuclei by the busting of air bubble films. *Q. J. R. Meteorol. Soc.* 90, 72-78.

De Bary, E., and F. Möller (1960). Die mittlere Vertikale Verteilung von Wolken in Abhängigkeit von der Wetterlage. *Ber. Dtsch. Wetterdienstes* 9 (67).

De Bary, E., and C. E. Junge (1963). Distribution of sulfur and chlorine over Europe. *Tellus* **15**, 370-381.

Deister, U., R. Neeb, G. Helas, and P. Warneck (1986). The equilibrium $CH_2(OH)_2 + HSO_3^- = CH_2(OH)SO_3^- + H_2O$ in aqueous solution. *J. Phys. Chem.* **90**, 3213-3217.

Delany, A., A. C. Delany, D. W. Parkin, J. J. Griffin, E. D. Goldberg, and B. E. Reimann (1967). Airborne dust collected at Barbados. *Geochim. Cosmochim. Acta* **31**, 885-909.

Delany, A. C., W. H. Pollock and J. P. Shedlowsky (1973). Tropospheric Aerosol: The relative contribution of marine and continental components. *J. Geophys. Res.* **78**, 6249-6265.

Delmas, R., J. Baudet, and J. Servant (1978). Mise en evidence des sources naturelles de sulfate en milieu tropical humide. *Tellus* **30**, 158-168.

Delmas, R., J. Baudet, J. Servant, and I. Baziard (1980). Emissions and concentrations of hydrogen sulfide in air of the tropical forest of the Ivory Coast and of temperature regions in France. *J. Geophys. Res.* **85**, 4468-4474.

Delmas, R., and J. Servant (1982). The origins of sulfur compounds in the atmosphere of a zone of high productivity (Gulf of Guinea). *J. Geophys. Res.* **87**, 11019-11026.

Delwiche, C. C. (1970). The nitrogen cycle, *Sci. Am.* **223**, 137-146.

DeMore, W. B., J. J. Margitan, M. J. Molina, R. T. Watson, R. F. Hampson, M. J. Kurylo, D. M. Golden, C. J. Howard, and A. R. Ravishankara (1985). Chemical kinetics and photochemical data for use in stratospheric modeling, evaluation number 7. National Aeronautics and Space Administration, California Jet Propulsion Laboratory, Publication **85-37**. Institute of Technology, Pasadena, California.

Demerjian, K. L., J. A. Kerr, and J. G. Calvert (1974). The mechanism of photochemical smog formation. *Adv. Environ. Sci. Technol.* **4**, 1-262.

Denmead, O. T., J. R. Simpson, and J. R. Freney (1974). Ammonia fluxes into the atmosphere from a grazed pasture. *Science* **185**, 609-610.

Denmead, O. T., J. R. Freney, and J. R. Simpson (1976). A closed ammonia cycle within a plant canopy. *Soil Biol. Biochem.* **8**, 161-164.

Denmead, O. T., J. R. Freney, and J. R. Simpson (1979). Studies of nitrous oxide emission from a grass sward. *J. Am. Soc. Soil Sci.* **43**, 726-728.

Department of the Environment (1983). Digest of Environmental Protection and Water Statistics, No. 6. HMSO, London.

Derwent, R. G. (1982). On the comparison of global, hemispheric, one-dimensional model formulations of halocarbon oxidation by OH radicals in the troposphere. *Atmos. Environ.* **16**, 551-561.

Detwiler, C. R., J. D. Garret, J. D. Purcell, and R. Tousey (1961). The intensity distribution in the ultraviolet solar spectrum. *Ann. Geophys. (Paris)* **17**, 263-272.

Devlin, R. M., and A. V. Barker (1971). "Photosynthesis." Van Nostrand, New York.

de Vooys, G. G. N. (1979). Primary production in aquatic environments. *In* 'The Global Carbon Cycle" (B. Bolin, E. T. Degens, S. Kempe, and P. Ketner, eds.), *SCOPE* **13**, 259-292.

Dickerson, R. R., D. H. Stedman, W. L. Chameides, P. J. Crutzen, and J. Fishman (1979). Actinometric measurements and theoretical calculations of $j(O_3)$, the rate of photolysis of ozone to $O(^1D)$. *Geophys. Res. Lett.* **6**, 833-836.

Dickerson, R. R., D. H. Stedman, and A. C. Delaney (1982). Direct measurement of ozone and nitrogen dioxide photolysis rates in the troposphere. *J. Geophys. Res.* **87**, 4933-4946.

Didyk, B. M., B. R. T. Simoneit, S. C. Brassel, and G. Eglinton (1978). Organic geochemical indicators of paleo-environmental conditions of sedimention. *Nature* **272**, 216-222.

Dimitriades, B., M. C. Dodge, J. J. Bufalini, K. L. Demerjian, and A. P. Altshuller (1976). Correspondence. *Environ. Sci. Technol.* **10**, 934-36.

Dimitriades, B., and A. P. Altshuller (1976). *Int. Conf. Oxidant Problems*: analysis of the evidence/View points presented. Part I: Definitions of key issues. *J. Air Pollut. Control Assoc.* **27**, 299–307.

Dinger, J. E., H. B. Howell, and T. A. Wojciechowski (1970). On the source and composition of cloud nuclei in a subsident air mass over the North Atlantic. *J. Atmos. Sci.* **27**, 791–797.

Dingledy, D. P., and J. G. Calvert (1963). A study of the ethyl-oxygen reaction by flash photolysis. *J. Am. Chem. Soc.* **85**, 856–861.

Disteche, A. (1974) The effect of pressure on dissociation constants and its temperature dependence. *In* "The Sea" (E. D. Goldberg, ed.), Vol. 5, Chap. 2, pp. 81–121. Wiley (Interscience), New York.

Ditchburn, R. W., and P. Young (1962). The absorption of molecular oxygen between 1850 and 2500 Å. *J. Atmos. Terr. Phys.* **24**, 127–139.

Dlugi, R., and S. Jordan (1982). Heterogeneous SO_2 oxidation: its contribution to cloud condensation nuclei formation. *Idojaras* (*J. Hung. Meteorol. Serv.*) **86**, 82–88.

Dobson, G. M. B., and D. N. Harrison (1926). Measurements of the amount of ozone in the earth's atmosphere and its relation to other geophysical conditions, Part 1. *Proc. R. Soc., London* **110**, 660–693.

Dobson, G. M. B. (1973). The laminated structure of the ozone in the atmosphere. *Q. J. R. Meteorol. Soc.* **99**, 599–607.

Dodge, M. C., and R. R. Arnts (1979). A new mechanism for the reaction of ozone with olefins. *Int. J. Chem. Kinet.* **11**, 399–410.

Doyle, G. J. (1961). Self-nucleation in the sulfuric acid water system. *J. Chem. Phys.* **35**, 795–799.

Drapcho, D. L., D. Sisterson, and R. Kumar (1983). Nitrogen fixation by lightning activity in a thunderstorm. *Atmos. Environ.* **17**, 729–734.

Drever, J. I. (1974). The magnesium problem. "The Sea" (E. D. Goldberg, ed.), Vol. 5, Chap. 10, 337–357. Wiley (Interscience), New York.

Driscoll, J. N., and P. Warneck (1968). Primary processes in the photolysis of SO_2 at 1849 Å. *J. Phys. Chem.* **72**, 3736–3740.

Drummond, J. R. (1977). Atmospheric measurements of nitric oxide using a chemiluminescence detector. Ph.D. Thesis, University of Wyoming, Laramie.

Drummond, J. R., and R. F. Jarnot (1978) Infrared measurements of stratospheric composition II, simultaneous NO and NO_2 measurements. *Proc. R. Soc. London A* **364**, 237–254.

Drummond, J. W., and A. Volz (1985). A summary of nitric oxide (NO) measurements obtained during STRATOZ III, 0–12 km 67N–60S. Evidence of air pollution in the upper troposphere. *In* "Physio-chemical Behaviour of Atmospheric Pollutants," (F. A. A. M. De Leeuw and N. D. Van Egmond, ed.), pp. 102–107. COST Action 611, *Proc. Workshop Pollutant Cycles Transport-Modelling Field Experiments, Bilthoven, The Netherlands*.

Drummond, J. W., A. Volz, and D. H. Ehhalt (1985). An optimized chemiluminescence detector for tropospheric No measurements. *J. Atmos. Chem.* **2**, 287–306.

Duce, R. A., J. W. Winchester, and T. W. van Nahl (1965). Iodine, Bromine and Chlorine in the Hawaiian marine atmosphere. *J. Geophys. Res.* **70**, 1775–1799.

Duce, R. A., and E. J. Hoffman (1976). Chemical fractionation at the air/sea interface. *Annu. Rev. Earth Planet. Sci.* **4**, 187–228.

Duce, R. A., G. L. Hoffman, B. J. Ray, I. S. Fletcher, G. T. Wallace, J. L. Fasching, S. R. Piotromicz, P. R. Walsh, E. S. Hoffman, J. M. Miller, and J. L. Heffter (1976). Trace metals in the marine atmosphere: sources and fluxes. *In* "Marine Pollutant Transfer" (H. L. Windum and R. A. Duce, eds.). Heath, Lexington, Massachusetts.

Duce, R. A. (1978). Speculations on the budget of particulate and vapor phase non-methane organic carbon in the global troposphere. *Pure Appl. Geophys.* **116**, 244–273.

Duce, R. A., and R. B. Gasosian (1982). The input of atmospheric n-C_{10} to n-C_{30} alkanes to the ocean. *J. Geophys. Res.* **87**, 7192-7200.

Duce, R. A., V. A. Mohnen, P. R. Zimmerman, D. Grosjean, W. Cautreels, R. Chatfield, R. Jaenicke, J. A. Ogren, E. D. Pellizzari, and G. T. Wallace (1983). Organic Material in the global atmosphere. *Rev. Geophys. Space Phys.* **21**, 921-952.

Dufour, L., and R. Defay (1963). "Thermodynamics of Clouds." Academic Press, New York.

Dütsch, H. U. (1968). The photochemistry of stratospheric ozone. *Q. J. R. Met. Soc.* **94**, 483-497.

Dütsch, H. U. (1971). Photochemistry of atmospheric ozone. *Adv. Geophys.* **15**, 219-322.

Dütsch, H. U. (1980). Ozon in der Atmosphäre: Gefährdet die stratosphärische Verschmutzung die Ozonschicht? Neujahrsblatt der Naturforschenden Gesellschaft in Zürich auf das Jahr 1980, *Vierteljahresschr.* **124**, (5), 1-48.

Dunlop, J. S. R., M. D. Muir, V. A. Milne, and D. I. Groves (1978). A new microfossil assemblage from the archaean of Western Australian. *Nature* **274**, 676-678.

Durand, B., ed. (1980). "Kerogen, Insoluble Matter from Sedimentary Rocks." Editions Technip., Paris.

Durbin, W. G., and G. D. White (1961). Measurements of the vertical distribution of atmospheric chloride particles. *Tellus* **13**, 260-275.

Dutkiewicz, V. A., and L. Husain (1979). Determination of stratospheric ozone at ground level using beryllium 7/ozone ratios. *Geophys. Res. Lett.* **6**, 171-174.

Duursma, E. K. (1961). Dissolved organic carbon, nitrogen and phosphorus in the sea. *Neth. J. Sea Res.* **1**, 1-147.

Duxbury, J. M., D. R. Bouldin, R. E. Terry, and R. L. Tate (1982). Emission of nitrous oxide from soils. *Nature* **298**, 462-464.

Eagan, R. C., P. V. Hobbs, and L. F. Radke (1974). Measurements of cloud condensation nuclei and cloud droplet size distributions in the vicinity of forest fires. *J. Appl. Meteorol.* **13**, 553-557.

Easter, R. C., and P. V. Hobbs (1974). The formation of sulfates and the enhancement of cloud condensation nuclei in clouds. *J. Atmos. Sci.* **31**, 1586-1594.

Eaton, J. P., and K. J. Murata (1960). How volcanoes grow. *Science* **132**, 925-938.

Edmond, J. M., and J. M. T. M. Gieskes (1970). On the calculation of the degree of saturation of seawater with respect to calcium carbonate under in situ conditions. *Geochim. Cosmochim. Acta* **34**, 1261-1291.

Egerton, A., G. J. Minkoff, and K. C. Salooja (1956). The slow oxidation of methane. *Proc. R. Soc. London A* **235**, 158-173.

Eggleton, A. E. J. (1969). The chemical composition of atmospheric aerosols on Tees-Side and its relation to visibility. *Atmos. Environ.* **3**, 355-372.

Eglinton, G., and M. Calvin (1967). Chemical fossils. *Sci. Am.* **216**, 32-43.

Ehhalt, D. H., G. Israel, W. Roether, and W. Stich (1963). Tritium and deuterium content of atmospheric hydrogen. *J. Geophys. Res.* **68**, 3747-3751.

Ehhalt, D. H., and L. E. Heidt (1973a). Vertical profiles of CH_4 in the troposphere and stratosphere. *J. Geophys. Res.* **78**, 5265-5271.

Ehhalt, D. H., and L. E. Heidt (1973b). The concentration of molecular H_2 and CH_4 in the stratosphere. *Pure Appl. Geophys.* **106–108**, 1352-1360.

Ehhalt, D. H. (1974). The atmospheric cycle of methane. *Tellus* **26**, 58-70.

Ehhalt, D. H., L. E. Heidt, R. H. Lueb, and N. Roper. (1974) Vertical profiles of CH_4, H_2, CO, N_2O and CO_2 in the stratosphere. *In Proc. Conf. Clim. Impact Assess. Program, 3rd*, (A. J. Broderick and T. M. Hard, eds.), **DOT-TSC-OST-74-15**, U.S. Department of Transportation Washington, D.C.

Ehhalt, D. H., L. E. Heidt, R. H. Lueb, and W. Pollock (1975). The vertical distribution of trace gases in the stratosphere. *Pure Appl. Geophys.* **113**, 389-402.

Ehhalt, D. H., U. Schmidt, and L. E. Heidt (1977). Vertical profiles of molecular hydrogen in the troposphere and stratosphere. *J. Geophys. Res.* **82**, 5907–5910.

Ehhalt, D. H., and U. Schmidt (1978). Sources and sinks of atmospheric methane. *Pure Appl. Geophys.* **116**, 452–464.

Ehhalt, D. H. (1979). Der atmosphärische Kreislauf von Methan. *Naturwissenschaften* **66**, 307–311.

Ehhalt, D. H., and J. W. Drummond (1982). The tropospheric cycle of NO_x. *In* "Chemistry of the Unpolluted and Polluted Troposphere" (H. W. Georgii and W. Jaeschke, eds.), NATO ASI Series, Vol. C96, pp. 219–251. Reidel, Dordrecht, Holland.

Ehhalt, D. H., J. Rudolph, and U. Schmidt (1986). On the importance of light hydrocarbons in multiphase atmospheric systems *In* "Chemistry of Multiphase Atmospheric Systems" (W. Jaeschke, (ed.) pp. 321–350. NATO ASI Series, Vol. G6. Springer-Verlag, Berlin.

Ehmert, A. (1949). Ein einfaches Verfahren zur Bestimmung kleinster Jodkonzentrationen, Jod- und Natriumthiosulfatmengen in Lösungen. *Z. Naturforsch.* **46**, 321–327.

Ehmert, A. (1951). Ein enfaches Verfahren zur absoluten Messung des Ozongehaltes der Luft. *Meteorol Rundsch.* **4**, 64–68.

Eichmann, R., P. Neuling, G. Ketseridis, J. Hahn, R. Jaenicke, and C. Junge (1979). *n*-Alkane studies in the troposphere—I. Gas and particulate concentrations in North Atlantic air. *Atmos. Environ.* **13**, 587–599.

Eichmann, R., G. Ketseridis, G. Schebeske, R. Jaenicke, J. Hahn, P. Warneck, and C. Junge (1980). *N*-Alkane studies in the troposphere—II: Gas and Particulate concentrations in Indian Ocean Air. *Atmos. Environ.* **14**, 695–703.

Eigen, M., W. Kruse, G. Maas, and L. De Maeyer (1964). Rate constants of protolytic reactions in aqueous solution. *Prog. React. Kinet.* **2**, 285–318.

Eliassen, A., and J. Saltbones (1975). Decay and transformation rates of SO_2 as estimated from emissions data, trajectories and measured air concentrations. *Atmos. Environ.* **9**, 425–429.

Eliassen, A. (1978). The OECD Study of Long Range Transport of Air Pollutants: long range transport modelling. *Atmos. Environ.* **12**, 479–487.

Elkins, J. W., S. C. Wofsy, M. B. McElroy, C. E. Kolb, and W. A. Kaplan (1978). Aquatic sources and sinks for nitrous oxide. *Nature* **275**, 602–606.

Elsaesser, H. W. (1974). Water budget of the stratosphere. (A. J. Broderick and T. M. Hard, eds.), *Proc. Conf. Climat. Impact Assess. Program 3rd*, **DOT-TSC-OST-74-15**, U.S. Department of Transportation, Washington, D.C. pp. 273–283.

H. W., Elsaesser, J. E. Harris, D. Kley, and R. Penndorf (1980). Stratospheric H_2O. *Planet. Space Sci.* **28**, 827–835.

Elshout, A. J., J. W. Viljeer, and H. Van Duuren (1978). Sulphates and sulphuric acid in the atmosphere in the years 1971–1976 in the Netherlands. *Atmos. Environ.* **12**, 785–790.

Emanuel, W. R., G. G. Killough, and J. S. Olson (1981). Modelling the circulation of carbon in the world's terrestrial ecosystems. *In* "Carbon Cycle modelling" (B. Bolin, ed.), *SCOPE* **16**, 335–353.

Emanuelson, A., E. Erikson, and H. Egner (1954). Composition of atmospheric precipitation in Sweden. *Tellus* **6**, 261–267.

Emery, K. O., W. L. Orr, and S. C. Rittenberg (1955). Nutrients in the ocean. *In* "Essays in the Honor of Captain Allan Hancock," pp. 299–309. University of Southern California Press, Los Angeles.

Englemann, R. J. (1965). Rain scavenging of zinc sulfide particles. *J. Atmos. Sci.* **22**, 719–729.

Englemann, R. J. (1968). The calculation of precipitation scavenging, in Meteorology and Atomic Energy 1968. (D. H. Slade, ed.), pp. 208–221. U.S. Atomic Energy Commission, Division of Technical Information, Oak Ridge, Tennessee.

Environmental Protection Agency (1976). National Air Quality and Emissions Trend Report, **EPA-450/1-76-002**. U.S. Environmental Protection Agency, Research Triangle Park, North Carolina.

Eriksson, E. (1952a). Composition of atmospheric precipitation. I. Nitrogen compounds. *Tellus* **4**, 215-232.

Eriksson, E. (1952b). Composition of atmospheric precipitation. II. Sulfur, chloride, iodine compounds; bibliography. *Tellus* **4**, 280-303.

Erikson, E. (1957). The chemical composition of Hawaiian rainfall. *Tellus* **9**, 509-520.

Eriksson, E. (1959). The yearly circulation of chloride and sulfur in nature; meteorological, geochemical and pedological implications. Part I. *Tellus* **11**, 375-603.

Eriksson, E. (1960). The yearly circulate of chloride and sulfur in nature: meteorological, geochemical and pedological implications, Part II. *Tellus* **12**, 63-109.

Eriksson, E. (1966). Air and precipitation as sources of nutrients, *In* Handbuch der Pflanzenernährung und Düngung, (H. Linser, ed.), pp. 774-792. Springer-Verlag, Vienna.

Erlenkeuser, H., E. Suess, H. Wilkomm (1974). Industrialization affects heavy metal and carbon isotopes concentrations in recent Baltic Sea sediments. *Geochim. Cosmochim. Acta* **38**, 823-842.

Ernst, W. (1938). Über pH-Wert Messungen von Niederschlägen. *Balneologe* **5**, 545-549.

Esmen, N. A., and R. B. Fergus (1976). Rain water acidity: pH spectrum of individual drops. *Sci. Total Environ.* **6**, 223-226.

Esphenson, J. E., and H. Taube (1965). Tracer experiments with ozone as oxidizing agent in aqueous solution. *Inorg. Chem.* **4**, 704-709.

Eucken, A., and F. Patat (1936). Die Temperaturabhängigkeit der photochemischen Ozonbildung. *Z. Phys. Chem. B* **33**, 459-474.

Evans, W. F. J., H. Fast, J. B. Kerr, C. T. McElroy, R. S. O'Brien, D. I. Wardle, J. C. McConnell, and B. A. Ridley (1978). Stratospheric constituent measurements from project stratoprobe. *Proc. WMO Symp. Geophys. Aspects Consequences Change Composition Stratosphere*, pp. 55-60. WMO Publ. 511, World Meteorological Organisation, Geneva.

Eyre, J. R., and H. K. Roscoe (1977). Radiometric measurements of stratospheric HCl. *Nature* **226**, 243-244.

Fabian, P., W. F. Libby, and C. E. Palmer (1968). Stratospheric residence time and interhemispheric mixing of Strontium 90 from fallout in rain. *J. Geophys. Res.* **73**, 3611-3616.

Fabian, P., and C. E. Junge (1970). Global rate of ozone destruction at the earth's surface. *Arch. Meteorol. Geophys. Bioklimatol. Ser. A* **19**, 161-172.

Fabian, P. (1974). Comment on a photochemical theory of tropospheric ozone by Chameides. *J. Geophys. Res.* **99**, 4124-4125.

Fabian, P., and P. G. Pruchniewicz (1976). Final Report on Project "Troposphärisches Ozon" (Project TROZ), Report No. **MPAE-W-100-76-21**. Max Planck Institute für Aeronomie, Katlenberg, Lindau, Federal Republic of Germany, 1976.

Fabian, P., and P. G. Pruchniewicz (1977). Meridional distribution of ozone in the troposphere and its seasonal variations. *J. Geophys. Res.* **82**, 2063-2073.

Fabian, P., R. Borchers, K. H. Weiler, U. Schmidt, A. Volz, D. H. Ehhalt, W. Seiler, and F. Müller (1979). Simultaneous measured vertical profiles of H_2, CH_4, CO, N_2O, $CFCl_3$, and CF_2Cl_2 in the midlatitude stratosphere and troposphere. *J. Geophys. Res.* **84**, 3149-3154.

Fabian, P., R. Borchers. S. A. Penkett, and N. J. D. Prosser (1981a). Halocarbons in the stratosphere. *Nature* **294**, 733-735.

Fabian, P., R. Borchers, G. Flentje, W. A. Matthews, W. Seiler, H. Giehl, K. Bunse, F. Müller, U. Schmidt, A. Volz, A. Khedim, and F. J. Johnen (1981b). The vertical distribution of stable trace gases at mid-latitudes. *J. Geophys. Res.* **86**, 5179-5184.

Fabian, P. (1986). Halogenated hydrocarbons in the atmosphere. "The Handbook of Environmental Chemistry IV" (O. Huntziger, ed.), pp. 24-51. Springer-Verlag, Berlin and New York.

Fabry, C., and H. Buisson (1921). Etude de l'extremite du spectre solaire. *J. Phys.* (*Orsay, France*) **2**, 197-226.

Fairchild, P. W., and E. K. C. Lee (1978). Relative quantum yields of $O(^1D)$ in ozone photolysis in the region 250-300 nm. *Chem. Phys. Lett.* **60**, 36-39.

Fairchild, C. E., E. J. Stone, and G. M. Lawrence (1978). Photofragment spectroscopy of ozone in the uv region 270-310 nm and at 600 nm. *J. Chem. Phys.* **69**, 3632-3638.

Falconer, P. D., R. Pratt, and V. A. Mohnen (1978). The transport cycle of atmospheric ozone and its measurements from aircraft and at the earth's surface. *In* "Man's Impact on the Troposphere, Lectures in Tropospheric Chemistry" (J. S. Levine and D. R. Schryer, eds.), pp. 109-147. NASA Reference Publication 1022. National Aeronautics and Space Administration, Langley Research Center, Hampton, Virginia.

Falconer, R. E., and P. D. Falconer (1980). Determination of cloud water acidity at a mountain observatory in the Adirondack Mountains of New York State. *J. Geophys. Res.* **85**, 7465-7470.

Falls, A. H., and J. H. Seinfeld (1978). Continued development of a kinetic mechanism for photochemical smog. *Environ. Sci. Technol.* **12**, 1398-1406.

FAO (1977). Soil map of the world, Food and Agricultural Organisation of the United Nations. UNESCO, Paris.

FAO (1981). 1980 FAO Production Year Book Statistics Ser., No. 34. Food and Agriculture Organization of the United Nations, Rome.

Farlow, N. H., D. M. Hayes, and H. Y. Lem (1977). Stratospheric aerosols: Undissolved granules and physical state. *J. Geophys. Res.* **82**, 4921-4929.

Farlow, N. H., K. G. Suetsinger, H. Y. Lem, D. M. Hayes, and B. M. Tooper (1978). Nitrogen-sulfur compounds in stratospheric aerosols. *J. Geophys. Res.* **83**, 6207-6212.

Farley, F. F. (1978). Correspondence. *Environ. Sci. Technol.* **12**, 99-100.

Farman, J. C., B. G. Gardiner, and J. D. Shanklin (1985). Large losses of total ozone in Antartica reveal seasonal ClO_x/NO_x interaction. *Nature* **315**, 207-210.

Farmer, C. B., O. F. Raper, D. Robbins, R. A. Toth, and C. Müller (1980). Simultaneous spectroscopic measurements of stratospheric species: O_3, CH_4, CO, CO_2, N_2O, H_2O, HCl and HF at northern and southern mid-latitudes. *J. Geophys. Res.* **85**, 1621-1632.

Farwell, S. O., A. E. Sherrard, M. R. Pack, and D. F. Adams (1979). Sulfur compounds volatilized from soils at different soil moisture contents. *Soil Biol. Biochem.* **11**, 411-415.

Feely, H. W., H. Seitz, R. J. Lagomarsino, P. E. Biscaye (1966). Transport and fallout of stratospheric radioactive debris. *Tellus* **18**, 316-328.

Fehsenfeld, F. C., and E. E. Ferguson (1976). Reactions of atmospheric negative ions with N_2O. *J. Chem. Phys.* **64**, 1853-1854.

Fergusson, G. J. (1958). Reduction of atmospheric radiocarbon concentrations by fossil fuel carbon dioxide and the mean life of carbon dioxide in the atmosphere. *Proc. R. Soc. London, Ser. A.* **243**, 561-574.

Ferm, M. (1979). Method for determination of atmospheric ammonia. *Atmos. Environ.* **13**, 1385-1393.

Fiebiger, W. (1975). Organische Substanzen in präkambrischen Itabiriten und deren Nebengesteinen. *Geol. Rundsch.* **64**, 641-652.

Findeisen, W. (1939). Zur Frage der Regentropfenbildung in reinen Wasserwolken. *Meteorol. Z.* **56**, 365-368.

Finlayson, B. J., J. N. Pitts, Jr., and H. Akimoto (1972). Production of vibrationally excited OH in chemiluminescent ozone-olefin reactions. *Chem. Phys. Lett.* **12**, 495-498.

Finlayson, B. J., J. N. Pitts, Jr., and R. Atkinson (1974). Low pressure gas-phase ozone-olefin reactions. Chemiluminescence, kinetics and mechanisms. *J. Am. Chem. Soc.* **96**, 5356–5367.

Finlayson, B. J., and J. N. Pitts, Jr. (1976). Photochemistry of the polluted troposphere. *Science* **192**, 111–119.

Firestone, M. K. (1982). Biological dentrification. *Agronomy* **22**, 289–326.

Fischer, W. H., J. P. Lodge, jr., A. F. Wartburg, and J. B. Pate (1968). Estimation of some atmospheric trace gases in Antarctica. *Environ. Sci. Technol.* **2**, 464–466.

Fischer, K., and U. Lüttge (1978). Light-dependent net production of carbon monoxide by plants. *Nature* **275**, 740–741.

Fischer, H., F. Fergg, D. Rabus, and P. Burkert (1985). Stratospheric H_2O and HNO_3 profile derived from solar occultation measurements. *J. Geophys. Res.* **90**, 3831–3843.

Fishman, J., and P. J. Crutzen (1978a). The distribution of the hydroxyl radical in the troposphere. Atmospheric Science Paper No. 284, Dept. of Atmospheric Science, Colorado State University, Fort Collins, Colorado.

Fishman, J., and P. J. Crutzen (1978b). The origin of ozone in the troposphere. *Nature* **274**, 855–858.

Fishman, J., S. Salomon, and P. J. Crutzen (1979). Observational and theoretical evidence in support of a significant in-situ photochemical source of tropospheric ozone. *Tellus* **31**, 432–446.

Fitzgerald, J. W. (1973). Dependence of the supersaturation spectrum of CCN on aerosol size distribution and composition. *J. Atmos. Sci.* **30**, 628–634.

Fitzgerald, J. W. (1974). Effect of aerosol composition on cloud droplet size distribution. A numerical study. *J. Atmos. Sci.* **31**, 1358–1367.

Flanagan, F. J. (1973). 1972 values for international geochemical reference samples. *Geochim. Cosmochim. Acta* **37**, 1189–1200.

Fletcher, N. H. (1962). "Physics of Rain Clouds." Cambridge University Press, London.

Flohn, H. (1961). Meridional transport of particles and standard vector deviation of upper winds. *Geophys. Pura Appl.* **50**, 229–234.

Flood, H. (1934). Tröpfchenbildung in übersättigter Äthylalkohol-Wasserdampfgemischen. *Z. Phys. Chem. A* **170**, 286–294.

Flyckt, D. L. (1979). Seasonal variation in the volatile hydrocarbon emissions from ponderosa pine and red oak. M. S. Thesis, College of Engineering, Washington State University.

Fontanella, J. C., A Girard, L. Gramont, and N. Louisnard (1975). Vertical distribution of NO, NO_2 and HNO_3 as derived from stratospheric absorption infrared spectra. *Appl. Optics* **14**, 825–839.

Focht, D. D., and W. Verstraete (1977). Biochemical ecology of nitrification and denitrification. *Adv. Microb. Ecol.* **1**, 135–214.

Forrest, J., S. E. Schwartz, and L. Newman (1979). Conversion of sulfur dixoide to sulfate during the Da Vinci flights. *Atmos. Environ.* **13**, 157–167.

Forrest, J., R. W. Garber, and L. Newman (1981). Conversion rates in power plant plumes based on a filterpack data: The coal fired Cumberland Plumes. *Atmos. Environ.* **15**, 2273–2282.

Forrest, J., D. J. Spandau, R. L. Tanner, and L. Newman (1982). Determination of atmospheric nitrate and nitric acid employing a diffusion denuder with a filter pack. *Atmos. Environ.* **16**, 1473–1485.

Fowler, D. (1978). Dry deposition of SO_2 on agricultural crop. *Atmos. Environ.* **12**, 369–373.

Fox, G. E., E Stackebrandt, R. B. Hespel, J. Gibson, J. Maniloff, T. A. Dyer, R. S. Wolfe, W. E. Balch, R. S. Tanner, L. J. Magrum, L. B. Zablen, R. Blakemore, R. Gupta, L. Bonen, B. J. Lewis, D. A. Stahl, K. R. Luehrsen, K. N. Chen, and C. R. Woese (1980). The phylogeny of procaryotes. *Science* **209**, 457–463.

Fraser, P. J. B., and G. I. Pearman (1978). Atmospheric halocarbons in the southern hemisphere. *Atmos. Environ.* **12**, 839–844.

Fraser, P. J., M. A. K. Khalil, R. A. Rasmussen, and A. J. Crawford (1981). Trends of atmospheric methane in the southern hemisphere. *Geophys. Res. Lett.* **8**, 1063–1066.

Fraser, P. J., M. A. K. Khalil, R. A. Rasmussen, and L. P. Steele (1984). Tropospheric methane in the mid-latitudes of the southern hemisphere. *J. Atmos. Chem.* **1**, 125–135.

Frederick, J. E. (1976). Solar corpuscle emission and neutral chemistry in the earth's middle atmosphere. *J. Geophys. Res.* **81**, 3179–3185.

Frederick, J. E., and R. D. Hudson (1979a). Predissociation linewidths and oscillator strengths for the (2-0) to (13-0) Schumann Runge bands of O_2. *J. Mol. Spectrosc.* **74**, 247–258.

Frederick, J. E., and R. D. Hudson (1979b). Predissociation of nitric oxide in the mesosphere and stratosphere. *J. Atmos. Sci.* **36**, 737–745.

Freney, J. R., O. T. Denmead, and J. R. Simpson (1978). Soil as a source or sink for atmospheric nitrous oxide. *Nature* **273**, 530–532.

Freney, J. R., J. R. Simpson, and O. T. Denmead (1981). Ammonia volatilization. *In* "Terrestrial Nitrogen Cycles" (F. E. Clark and T. Rosswall, eds.), *Ecol. Bull.* (*Stockholm*) **33**, 291–302.

Frenkel, J. (1955). "Kinetic Theory of Liquids." Dover, New York.

Freyer, H. D. (1978). Prelimary evaluation of past CO_2 increase derived from ^{13}C measurements in tree rings. *In* "Carbon Dioxide Climate and Society" (J. Williams, ed.), pp. 69–78. Pergamon, New York.

Friedlander, S. K. (1977). "Smoke, Dust and Haze. Fundamentals of Aerosol Behaviour." Wiley, New York.

Friend, J. P. (1973). The general sulfur cycle. *In* "Chemistry of the lower atmosphere" (S. I. Rasool, ed.), pp. 177–201. Plenum Press, New York.

Friend, J. P., R. Leifer, and M. Trichon (1973). On the formation of stratospheric aerosols. *J. Atmos. Sci.* **30**, 465–479.

Fritz, J. J., and C. R. Fuget (1956). Vapor pressure of aqueous hydrogen chloride solutions. *Ind. Eng. Chem.* **1**, 10–12.

Frössling, N. (1938). The evaporation of falling drops. *Gerlands Beitr. Geophys.* **52**, 170–216.

Fry, L. M., and K. K. Menon (1962). Determination of the tropospheric residence time of lead-210. *Science* **137**, 994–995.

Fuchs, N. A. (1964). "The Mechanics of Aerosols." Pergamon Press, Oxford.

Fuller, E. C., and R. H. Christ (1941). The rate of oxidation of sulfite ions by oxygen. *J. Am. Chem. Soc.* **63**, 1644–1650.

Fushimi, K., and Y. Miyake (1980). Contents of formaldehyde in the air above the surface of the ocean. *J. Geophys. Res.* **85**, 7533–7536.

Fuzzi, S. (1978). Study of iron (III) catalyzed sulphur dioxide oxidation in aqueous solution over a wide range of pH. *Atmos. Environ.* **12**, 1439–1442.

Gaedtke, H., and J. Troe (1975). Primary processes in the photolysis of NO_2. *Ber. Bunsen Ges. Phys. Chem.* **79**, 184–191.

Gagosian, R. B., O. C. Zafiriou, E. T. Peltzer, and J. B. Alford (1982). Lipids in aerosols from the tropical North Pacific: Temporal variability. *J. Geophys. Res.* **87**, 11133–11144.

Galbally, I. E. (1971). Surface ozone observations at Aspendale, Victoria 1964–70. *Atmos. Environ.* **5**, 15–25.

Galbally, I. E. (1975). Nitrogen oxides (NO_2 and NO_x) in the air of Aspendale and other places in Victoria. *Clean Air* **9**, 12–15. Commonwealth of Australia.

Galbally, I. E. (1976). Man-made carbon tetrachloride in the atmosphere. *Science* **193**, 573–576.

Galbally, I. E., and C. R. Roy (1978). Loss of fixed nitrogen from soils by nitric oxide exhalation. *Nature* **275**, 734–735.

Galbally, I. E., and C. R. Roy (1980). Destruction of ozone at the earth's surface. *Q. J. R. Meteor. Soc.* **106**, 599-620.

Galbally, I. E., and C. R. Roy (1981). Nitrogen oxides, *In* WMO Baseline Air Monitoring Report 1978, Section 3.1.5, pp. 6-9. Australian Government Publ. Service, Canberra.

Galloway, J. N., and D. M. Whelpdale (1980). An atmospheric sulfur budget for Eastern North America. *Atmos. Environ.* **14**, 409-417.

Galloway, J. N., G. E. Likens, W. C. Keene, and J. M. Miller (1982). The composition of precipitation in remote areas of the world. *J. Geophys. Res.* **87**, 8771-8786.

Galloway, J. N., and A. Gaudy (1984). The composition of precipitation at Amsterdam Island, Indian Ocean. *Atmos. Environ.* **18**, 2649-2656.

Gambell, A. W., and D. W. Fisher (1964). Occurrence of sulfate and nitrate in rainfall. *J. Geophys. Res.* **69**, 4203-4210.

Gamble, T. N., M. R. Betlach, and J. M. Tiedje (1977). Numerically dominant denitrifying bacteria from world soils. *Appl. Env. Microbiol.* **33**, 926-939.

Ganapathy, R., and E. Anders (1974). Bulk Composition of the Moon and Earth estimated from meteorites. *Proc. Lunar Science Conf., 5th, 1974*, 254-256. Lunar Science Institute, Houston, Texas.

Ganapathy, R., and D. E. Brownlee (1979). Interplanetary dust: Trace element analysis of individual particles by neutron activation. *Science* **206**, 1075-1077.

Garber, R. W., J. Forrest, and L. Newman (1981). Conversion rates in power plant plumes based on filter pack data: the oil-fired Northport plume. *Atmos. Environ.* **35**, 2283-2292.

Gardner, E. P., R. D. Wijayaratne, and J. G. Calvert (1984). Primary quantum yields of the photodecomposition of acetone in air under tropospheric conditions. *J. Phys. Chem.* **88**, 5069-5676.

Garland, J. A. (1976). Dry deposition of SO_2 and other gases. "Atmosphere–Surface Exchange of Particulate and Gaseous Pollutants" (R. J. Engelmann and G. A. Sehmel, coordinators) *Symp. Proc Richland Washington 1974*, pp. 212-227. **ERDA CONF-740921**, Nat. Tech. Inform. Service, U.S. Dept. of Commerce, Springfield, Virginia.

Garland, J. A. (1977). The dry deposition of sulphur dioxide to land and water surfaces. *Proc. R. Soc. London A* **356**, 245-269.

Garland, J. A., and J. R. Branson (1977). The deposition of sulphur dioxide to pine forest assessed by a radioactive tracer method. *Tellus* **29**, 445-454.

Garland, J. A. (1978). Dry and wet removal of sulphur from the atmosphere. *Atmos. Environ.* **12**, 349-362.

Garland, J. A. (1979). Dry deposition of gaseous pollutants. *Proc. WMO Symp. Long-Range Transport Pollutants Relation General Circulation stratospheric/tropospheric Exchange Processes*, pp. 97-103. World Meteorological Organization, Geneva, Switzerland.

Garland, J. A., A. W. Elzerman, and S. A. Penkett (1980). The mechanism for dry deposition of ozone to seawater surfaces. *J. Geophys. Res.* **85**, 7488-7492.

Garrels, R. M., and F. T. MacKenzie (1971). "Evolution of Sedimentary Rocks." Norton, New York.

Garrels, R. M., and F. T. MacKenzie (1972). A quantitative model for the sedimentary rock cycle. *Marine Chem.* **1**, 27-61.

Garrels, R. M., and E. A. Perry, Jr. (1974). Cycling of carbon, sulfur and oxygen through geologic time. *In* "The Sea" (E. D. Goldberg, ed.), pp. 303-336. Wiley (Interscience), New York.

Gatz, D. F., and A. N. Dingle (1971). Trace substances in rainwater: concentration variations during convective rains and their interpretation. *Tellus* **23**, 14-27.

Gauthier, M., and D. R. Snelling (1971). Mechanism of singlet molecular oxygen formation from photolysis of ozone at 2537 Å. *J. Chem. Phys.* **54**, 4317-4325.

Gebhart, R. (1967). On the significance of shortwave CO_2 absorption in investigations concerning the CO_2 theory of climate change. *Arch. Meteorol. Geophys. Bioklimatol. B* **15**, 52–61.

Georgii, H. W. (1963). Oxides of nitrogen and ammonia in the atmosphere. *J. Geophys. Res.* **68**, 3963–3970.

Georgii, H. W., and D. Jost (1964). Untersuchung über die Verteilung von Spurengasen in der freien Atmosphäre. *Pure Appl. Geophys.* **59**, 217–224.

Georgii, H. W. (1970). Contribution to the atmospheric sulfur budget. *J. Geophys. Res.* **75**, 2365–2371.

Georgii, H. W., and D. Wötzel (1970). On the relation between drop size and concentration of trace elements in rainwater. *J. Geophys. Res.* **75**, 1727–1731.

Georgii, H. W., and W. J. Müller (1974). On the distribution of ammonia in the middle and lower troposphere. *Tellus* **26**, 180–184.

Georgii, H. W., and G. Gravenhorst (1977). The ocean as source or sink of reactive trace gases. *Pure Appl. Geophys.* **115**, 503–511.

Georgii, H. W., and U. Lenhard (1978). Contribution to the atmospheric NH_3 budget. *Pure Appl. Geophys.* **116**, 385–391.

Georgii, H. W., and F. X. Meixner (1980). Measurement of tropospheric and stratospheric SO_2 distribution. *J. Geophys. Res.* **85**, 7433–7438.

Georgii, H. W. (1982). Review of the chemical composition of precipitation as measured by the WMO, BAPMON Global Environmental Monitoring System. World Meteorological Organization, Geneva, Switzerland.

Georgii, H. W., C. Perseke, and E. Rohbock (1983). Wet and dry deposition of acidic and heavy metal components in the Federal Republic of Germany. *In* "Acid Deposition" (S. Beilke and A. J. Elshout, eds.) pp. 142–148. Reidel, Dordrecht, Holland.

Georgii, H. W. (1983). The atmospheric sulfur budget. *In* "Chemistry of the Unpolluted and Polluted Troposphere" (J. W. Georgii and W. Jaeschke, eds.), pp. 295–324. Reidel, Dordrecht.

Gerstle, R. W., and D. A. Kemnitz (1967). Atmospheric emissions from open burning. *J. Air Pollut. Control Assoc.* **17**, 324–327.

Gidel, L. T., and M. A. Shapiro (1980). General circulation model estimates of the net vertical flux of ozone in the lower stratosphere and the implications for the tropospheric ozone budget. *J. Geophys. Res.* **85**, 4049–4058.

Gidel, L. T., P. J. Crutzen, and J. Fischmann (1983). A two dimensional photochemical model of the atmosphere. 1: Chlorocarbon emissions and their effect on stratospheric ozone. *J. Geophys. Res.* **88**, 6622–6640.

Gieskes, J. M. (1974). The alkalinity-total carbon dioxide system in seawater. *In* "The Sea" (E. D. Goldberg, ed.), Vol. 5, Chap. 3, pp. 123–151.

Giggenbach, W. F., and F. LeGuern (1976). The chemistry of magmatic gases from Erta'Ale, Ethiopia. *Geochim. Cosmochim. Acta* **42**, 25–30.

Gillani, N. V., S. Kohli, and W. E. Wilson (1981). Gas-to-particle conversion of sulfur in power plant plumes. I. Parametrization of the conversion rate for dry, moderately polluted ambient conditions. *Atmos. Environ.* **15**, 2293–2313.

Gillani, N. V., and W. E. Wilson (1983). Gas-to-particle conversion of sulfur in power plant plumes. II. Observations of liquid-phase conversion. *Atmos. Environ.* **17**, 1739–1752.

Gillette, D. A., and I. H. Blifford (1971). Composition of tropospheric aerosols as a function of altitude. *J. Atmos. Sci.*, **28**, 1199–1210.

Gillette, D. A. (1974). On the production of soil wind erosion aerosols having the potential for long-range transport. *J. Rech. Atmos.* **8**, 735–744.

Gillette, D. A., and P. A. Goodwin (1974). Microscale transport of sand-sized soil aggregates eroded by wind. *J. Geophys. Res.* **79**, 4080–4089.

Gillette, D. A. and T. R. Walker (1977). Characteristics of airborne particles produced by wind erosion of sandy soil, high plains of West Texas. *Soil Sci.* **123**, 97-110.

Gillette, D. A. (1978). A wind tunnel simulation of the erosion of soil: effect of soil texture, sandblasting, wind speed and soil condition on dust production. *Atmos. Environ.* **12**, 1735-1743.

Gillette, D. A. (1979). Environmental factors affecting dust emission and wind erosion. *In* "Sahara Dust: Mobilization, Transport, Deposition" (C. Morales, ed.), *SCOPE* **14**, 71-91.

Gillette, D. A. (1980). Major contributions of natural primary continental aerosols: source mechanisms. *Ann. N.Y. Acad. Sci.* **338**, 348-358.

Gillette, D. A., J. Adams, C. Endo, and D. Smith (1980). Threshold velocities for input of soil particles into the air by desert soils. *J. Geophys. Res.* **85**, 5621-5630.

Gilmore, F. R. (1965). Potential energy curves for N_2, NO, O_2 and corresponding ions. *J. Quant. Spectrosc. Radiat. Transfer* **5**, 369.

Girard, A., J. Besson, R. Giraudet, and L. Gramont (1978/79). Correlated seasonal and climatic variations of trace constituents in the stratosphere. *Pure Appl. Geophys.* **117**, 381-394.

Girard, A., G. Fergant, L. Gramont, O. Lado-Bordowsky, J. Laurent, S. Le Boiteux, M. P. Lemaitre, and N. Louisnard (1983). Latitudinal distribution of ten stratospheric species deduced from simultaneous spectroscopic measurements. *J. Geophys. Res.* **88**, 5377-5392.

Glasson, W. A., and C. S. Tuesday (1970a). Hydrocarbon reactivities in the atmospheric photooxidation of nitric oxide. *Environ. Sci. Technol.* **4**, 916-924.

Glasson, W. A., and C. S. Tuesday (1970b). Hydrocarbon reactivity and the kinetics of the atmospheric photooxidation of nitric oxide. *J. Air. Pollut. Control Assoc.* **20**, 239-243.

Glavas, S., and S. Toby (1975). Reaction between ozone and hydrogen sulfide. *J. Phys. Chem.* **79**, 779-782.

Gleason, J. F., A. Sinha, and C. J. Howard (1987). Kinetics of the gas-phase reaction $HOSO_2 + O_2 \rightarrow HO_2 + SO_3$. *J. Phys. Chem.* **91**, 719-724.

Glueckauf, E. (1951). The composition of atmospheric air. *In* "Compendium of Meteorology" (T. F. Mahone ed.), pp. 3-12. Am. Meteorol. Soc., Boston.

Gmitro, J. I., and T. Vermeulen (1964). Vapor liquid equilibria for aqueous sulfuric acid. *Am. Inst. Chem. Eng.* **10**, 741.

Goethel, M. (1980). Untersuchung ausgewählter Quellen und Senken des atmosphärischen Ammoniaks mit einem neuen Probenahmeverfahren. Diploma-Thesis, Institut für Meteorologie und Geophysik der Universität Frankfurt/Main. May, 1980.

Goldan, P. D., Y. A. Bush, F. C. Fehsenfeld, D. L. Albritton, P. J. Crutzen, A. L. Schmeltekopf, and E. E. Ferguson (1978). Tropospheric N_2O mixing-ratio measurements. *J. Geophys. Res.* **83**, 935-939.

Goldan, P. D., W. C. Kuster, D. L. Albritton, and A. L. Schmeltekopf (1980). Stratospheric $CFCl_3$, ClF_2Cl_2 and N_2O profile measurements at several latitudes. *J. Geophys. Res.* **85**, 413-423.

Goldan, P. D., W. C. Kuster, A. L. Schmeltekopf, F. C. Fehlsenfeld, and D. L. Albritton (1981). Correction of atmospheric N_2O mixing ratio data. *J. Geophys. Res.* **86**, 5385-5386.

Goldberg, E. D. (1971). Atmospheric dust, the sedimentary cycle and man. *Geophys.* **3**, 117-132.

Goldberg, A. B., P. J. Maroulis, L. A. Wilner, and A. R. Bandy (1981). Study of H_2S emissions from a salt water marsh. *Atmos. Environ.* **15**, 11-18.

Goldman, A., F. G. Fernald, W. J. Williams, and D. G. Murcray (1978). Vertical distribution of NO_2 in the stratosphere as determined from balloon measurements of solar spectra in the 4500-Å region. *Geophys. Res. Lett.* **5**, 257-260.

Goody, R. M. (1969). Time variations in atmospheric N_2O in Eastern Massachusetts. *Planet. Space Sci.* **17**, 1319-3320.

Gordon A. L. (1975). General ocean circulation. *In* "Numerical Models of Ocean Circulation" (R. O. Reid, ed.), pp. 39-53. National Academy of Science, Washington, D.C.

Götz, F. W. P. (1931). Zum Strahlungsklima des Spitzbergensommers: Strahlungs- und Ozonmessungen in der Königsbucht 1929. *Gerlands Beitr. Geophys.* **31**, 119-156.

Götz, F. W. P., A. R. Meetham, and G. M. B. Dobson (1934). The vertical distribution of ozone in the atmosphere. *Proc. R. Soc. A* **145**, 416-446.

Graedel, T. E. (1979). Reduced sulfur emissions from the open oceans. *Geophys. Res. Lett.* **6**, 329-331.

Graedel, T. E., and C. J. Weschler (1981). Chemistry within aqueous aerosols and raindrops. *Rev. Geophys. Space Phys.* **19**, 505-539.

Graham, R. A., and H. S. Johnston (1978). The photochemistry of NO_3 and the kinetics of the N_2O_5-O_3 system. *J. Phys. Chem.* **82**, 254-268.

Graham, R. A., A. M. Winer, and J. N. Pitts, Jr. (1978). Pressure and temperature dependence of the unimolecular decomposition of HO_2NO_2. *J. Chem. Phys.* **68**, 4505-4510.

Gran, G. (1952). Determination of the equivalence point in potentiometric tirations, Part II. *Analyst (London)* **77**, 661-671.

Granat, L. (1972). On the relation between pH and the chemical composition in atmospheric precipitation. *Tellus* **24**, 550-560.

Granat, L., R. O. Hallberg, and H. Rhode (1976). The global sulphur cycle. *In* "Nitrogen, Phosphorus and Sulphur—Global cycles" (B. H. Svensson and R. Söderlund, eds.), *SCOPE Report 7, Ecol. Bull. (Stockholm)* **22**, 89-134.

Granat, L. (1978). Sulfate in precipitation as observed by the European atmospheric chemistry network. *Atmos. Environ.* **12**, 413-424.

Gras, J. L. (1983). Ammonia and ammonium concentrations in the Antarctic atmosphere. *Atmos. Environ.* **17**, 815-818.

Grätzel, M., A. Henglein, J. Lilie, and C. Beck (1969). Pulsradiolytische Untersuchung einiger Elementarprozesse der Oxidation und Reduktion des Nitritions. *Ber. Bunsen Ges. Physik Chem.* **73**, 646-653.

Grätzel, M., S. Taniguchi, and A. Henglein (1970). Pulsradiolytische Untersuchung der NO Oxidation und des Gleichgewichts $N_2O_3 = NO + NO_2$ in wässriger Lösung. *Ber. Bunsen Ges. Phys. Chem.* **74**, 688-492.

Gravenhorst, G. (1975a). Der Sulfatanteil im atmosphärischen Aerosol über dem Nordatlantik. Ber. Institut für Meteorologie Geophys. der Universität Frankfurt/M., No. 30.

Gravenhorst, G. (1975b). The sulfate component in aerosol samples over the North Atlantic. *Meteor. Forschungsergeb. B* **10**, 22-33.

Gravenhorst, G., and A. Böttger (1980). Ammonia in the troposphere. *In* "Physico-chemical Behaviour of Atmospheric Pollutants", *Proc. Eur. Symp., 1st, Ispra. Italy* (B. Versino and H. Ott, eds.), pp. 383-394. Commission of the European Communities, Luxembourg.

Gray, P., R. Shaw, and J. C. J. Thynne (1967). The rate constants of alkoxy radical reactions. *Prog. React. Kinet.* **4**, 65-117.

Greeley, R., J. D. Iverson, J. B. Pollak, N. Udovich, and B. White (1974). Wind tunnel studies of Martian aeolean processes. *Proc. R. Soc. London, Ser. A* **341**, 331-360.

Greenberg, J. P., and P. R. Zimmerman (1984). Nonmethane hydrocarbons in remote tropical, continental and marine atmospheres. *J. Geophys. Res.* **89**, 4767-4778.

Greenberg, J. P., P. R. Zimmerman, L. Heidt, and W. Pollock (1984). Hydrocarbon and carbon monoxide emissions from biomass burning in Brazil. *J. Geophys. Res.* **89**, 1350-1354.

Greenberg, J. P., P. R. Zimmerman, and R. B. Chatfield (1985). Hydrocarbons and carbon monoxide in African Savannah air. *Geophys. Res. Lett.* **12**, 113-116.

Greenfield, S. M. (1957). Rain scavenging of radioactive particulate matter from the atmosphere. *J. Meteorol.* **14**, 115-125.

Gregory, P. H., and J. M. Hirst (1957). The summer air spora at Rothamsted in 1952. *J. Gen. Microbiol.* **17**, 135-152.

Gregory, P. H. (1973). The Microbiology of the Atmosphere, 2nd ed. Leonard Hill, Aylesburg.

Gregory, P. H. (1978). Distribution of airborne pollen and spores and their long distance transport. *Pure Appl. Geophys.* **116**, 309-315.

Greiner, N. R. (1966). Flash photolysis of H_2O vapor in the presence of D_2, Ar, $H_2^{18}O$. *J. Chem. Phys.* **45**, 99-103.

Greiner, N. R. (1967a). Photochemistry of N_2O essential to a simplified vacuum-ultraviolet actinometer. *J. Chem. Phys.* **47**, 4373-4377.

Greiner, N. R. (1967b). Hydroxyl-radical kinetics by kinetic spectroscopy. I. Reactions with H_2, CO and CH_4 at 300 K. *J. Chem. Phys.* **46**, 2795-2799.

Greiner, N. R. (1970). Hydroxyl radical kinetics by kinetic spectroscopy. VI. Reactions with alkanes in the range 300-500 K. *J. Chem. Phys.* **53**, 1070-1076.

Grennfelt, P., C. Bengtson, and L. Scärby (1983). Dry deposition of nitrogen dioxide to Scots Pine needles. *In* "Precipitation Scavenging, Dry Deposition and Resuspension" (H. R. Pruppocher, R. G. Semonin, W. G. N. Slinn, eds.), Vol. 2. pp. 753-761. Elsevier, New York.

Griffin, J. J., H. Windom, and E. D. Goldberg (1968). The distribution of clay minerals in the world ocean. *Deep Sea Res.* **15**, 433-459.

Griffin, J. J., and E. D. Goldberg (1975). The fluxes of elemental carbon in coastal marine sediments. *Limnol. Oceanogr.* **20**, 456-463.

Griffiing, G. W. (1977). Ozone and oxides of nitrogen production during thunderstorms. *J. Geophys. Res.* **82**, 943-950.

Griggs, M. (1968). Absorption coefficients of ozone in the ultraviolet and visible regions. *J. Chem. Phys.* **49**, 857-859.

Grimsrud, E. P., and R. A. Rasmussen (1975). Survey and analysis of halocarbons in the atmosphere by gas chromatography-mass spectrometry. *Atmos. Environ.* **9**, 1014-1017.

Grimsrud, E. P., H. H. Westberg, R. A. Rasmussen (1975). Atmospheric reactivity of mono-terpene hydrocarbons, NO_x photooxidation and ozonolysis. *Int. J. Chem. Kinet.* **7** (Symp. 1), 183-195.

Groblicki, P., and G. J. Nebel (1971). The photochemical formation of aerosols in urban atmospheres. *In* "Chemical Reactions in Urban Atmospheres" (C. S. Tuesday, ed.), pp. 241-263. American Elsevier, New York.

Grosjean, D. (1977). Aerosols. Chap. 3 *In* "Ozone and other Photochemical Oxidants," pp. 45-125. *Comm. Med. Effects Environ. Pollut.* National Academy of Sciences, Washington D.C.

Grosjean, D., and S. K. Friedlander (1980). Formation of organic aerosols from cyclic olefins and diolefins. *In* "The Character and Origin of Smog Aerosols" (G. M. Hidy, P. K. Mueller, D. Grosjean, B. R. Appel and J. J. Weslowski, eds.), *Adv. Environ. Sci. Technol.* **9**, 435-473.

Grosjean, D. (1983). Distribution of atmospheric nitrogeneous pollutants at a Los Angeles area receptor site. *Environ. Sci. Technol.* **17**, 13-19.

Grosjean, D., R. Swanson, and C. Ellis (1983). Carbonyls in Los Angeles air: contribution of direct emmissions. *Sci. Total Environ.* **29**, 65-85.

Grosjean, D. (1984). Worldwide ambient measurements of peroxyacetyl nitrate (PAN) and implications for plant injury. *Atmos. Environ.* **18**, 1489-1491.

Grossman, L., and J. W. Larimer (1974). Early chemical history of the solar system. *Rev. Geophys. Space Phys.* **12**, 71-101.

Grover, S. N., H. R. Pruppacher, and A. E. Hamielec (1977). A numerical determination of the efficiency with which spherical aerosol particles collide with spherical water drops due to inertial impaction and phoretic and electrical forces. *J. Atmos. Sci.* **34**, 1655-1663.

Gudiksen, P. H., A. W. Fairhall, and R. J. Reed (1968). Roles of mean meridional circulation and eddy diffusion in the transport of trace substances in the lower stratosphere. *J. Geophys. Res.* **73**, 4461-4473.

Gundersen, K., C. W. Mountain, D. Taylor, R. Ohye, and J. Shen (1972). Some chemical and microbiological observations in the Pacific Ocean off the Hawaiian Islands. *Limnol. Oceanogr.* **17**, 524-531.

Gutman, D., N. Sanders, and J. E. Butler (1982). Kinetics of the reactions of methoxy and ethoxy radicals with oxygen. *J. Phys. Chem.* **86**, 66-70.

Haaf, W., and R. Jaenicke (1977). Determination of the smallest particle size detectable in condensation nucleus counters by observation of the coagulation of SO_2 photo-oxidation products. *J. Aerosol Sci.* **8**, 447-456.

Haaf, W. (1980). Accurate measurements of aerosol size distribution II, construction of a new plate condensor electric mobility analyser and first results. *J. Aerosol Sci.* **11**, 201-212.

Haaf, W., and R. Jaenicke (1980). Results of improved size distribution measurements in the Aitken range of atmospheric aerosols. *J. Aerosol Sci.* **11**, 321-330.

Hack, W., O. Horie, and H. Gg. Wagner (1981). The rate of the reaction of NH_2 with O_3. *Ber. Bunsen Ges. Phys. Chem.* **85**, 72-78.

Hack, W., O. Horie, and H. Gg. Wager (1982). Determination of the rate of NH_2 with O_2. *J. Phys. Chem.* **86**, 765-771.

Hahn, J., and C. Junge (1977). Atmospheric nitrous oxide: a critical review. *Z. Naturforsch.* **32a**, 190-214.

Hahn, J. (1980). Organic constituents of natural aerosols. *Ann. N.Y. Acad. Sci.* **338**, 359-376.

Hahn, J. (1981). Nitrous oxide in the oceans. *In* "Denitrification, Nitrification and Atmospheric Nitrous Oxide" (C. C. Delwiche, ed.), pp. 191-277. Wiley, New York.

Hahn, J. (1982). Geochemical fossils of a possible archaebacterial origin in ancient sediments. *Zentralbl. Bakteriol. Mikorbiol. Hyg. Abt. 1 Orig. C* **3**, 40-52.

Hales, J. M. (1972). Fundamentals of the theory of gas scavenging by rain. *Atmos. Environ.* **6**, 635-659.

Hall, D. M. and R. L. Jones (1961). Physiological significance of surface wax on leaves. *Nature* **191**, 95-96.

Hall, D. M., and L. A. Donaldson (1963). The ultrastructure of wax deposits on plant leaf surfaces. 1. Growth of wax on leaves of tripolium repens. *Ultrastruct. Res.* **9**, 259-267.

Hameed, S., J. P. Pinto, and R. W. Stewart 1(979). Sensitivity of the predicted $CO-OH-CH_4$ perturbation to tropospheric NO_x concentrations. *J. Geophys. Res.* **84**, 763-767.

Hamill, P. (1975). The time dependent growth of $H_2O-H_2SO_4$ aerosols by heteromolecular condensation. *J. Aerosol Sci.* **6**, 475-484.

Hamill, P., R. P. Turco, O. B. Toon, C. S. Kiang, and R. C. Whitten (1982). On the formation of sulfate aerosol particles in the stratosphere. *J. Aerosol Sci.* **13**, 561-585.

Hampicke, U. (1979). Net transfer of carbon between the land biaota and the atmosphere induced by man. *In* "The Global Carbon cycle" (B. Bolin, E. T. Degens, S. Kempe and P. Ketner, eds.) *SCOPE* **13**, 219-236.

Hampson, R. F. (1980). Chemical kinetic and photochemical data sheets for atmospheric reactions. U.S. Department of Transportation, Rep. No. **FAA-EE-80-17**.

Hänel, G. (1976). The properties of atmospheric aerosol particles as functions of the relative humidity at thermodynamic equilibrium with the surrounding moist air. *Adv. Geophys.* **19**, 73-188.

Hänel, G., and J. Thudium (1977). Mean bulk densities of dry atmospheric aerosol particles: a summary of measured data. *Pure Appl. Geophys.* **115**, 799-803.

Hanks, T. C., and D. L. Anderson (1969). The early thermal history of the earth. *Phys. Earth Planet. Inter.* **2**, 19-29.

Hansen, M. H., K. Ingvorsen, and B. B. Jorgensen (1978). Mechanisms of hydrogen sulfide release from coastal marine sediments to the atmosphere. *Limnol. Oceanogr.* **23**, 68-76.

Hanst, P. L., L. L. Spiller, D. M. Watts, J. W. Spence, and M. F. Miller (1975). Infrared measurement of fluorocarbons, carbontetrachloride, carbonylsulfide and other atmospheric trace gases. *J. Air Pollut. Control Assoc.* **25**, 1220-1226.

Hanst, P. L., J. W. Spence, and O. Edney (1980). Carbon monoxide production in photooxidation of organic molecules in the air. *Atmos. Environm.* **14**, 1077-1088.

Hardy, R. W. F., and U. D. Havelka (1975). Nitrogen fixation research: a key to world food. *Science* **188**, 633-643.

Harker, A. B., W. Ho, and J. J. Ratto (1977). Photodissociation quantum yield of NO_2 in the region 375-420 nm. *Chem. Phys. Lett.* **50**, 394-397.

Harkins, J. H., and S. W. Nicksic (1967). Ammonia in auto exhaust *Environ. Sci. Technol.* **9**, 751-752.

Harries, J. E. (1976). The distribution of water vapor in the stratosphere. *Rev. Geophys. Space Phys.* **14**, 565-575.

Harries, J. E., D. G. Moss, N. R. W. Swann, G. F. Neill, and P. Gildwarg (1976). Simultaneous measurements of H_2O, NO_2 and HNO_3 in the daytime stratosphere from 15 to 35 km. *Nature* **259**, 300-302.

Harrison, R. M., and H. A. McCartney (1980). Ambient air quality at a coastal site in rural north-west England. *Atmos. Environ.* **14**, 233-244.

Hartley, W. N. (1881). On the absorption of solar rays by atmospheric ozone. *J. Chem. Soc.* **39**, 111.

Harvey, R. B., D. H. Stedman, and W. Chameides (1977). Determination of the absolute rate of solar photolysis of NO_2. *J. Air Pollut. Control Assoc.* **27**, 663-666.

Hatakeyama, S., H. Bandow, M. Okuda, and H. Akimoto (1981). Reactions of CH_2OO and $CH_2(^1A_1)$ with H_2O in the gasphase. *J. Phys. Chem.* **85**, 2249-2254.

Hatakeyama, S., M. Okuda, and H. Akimoto (1982). Formation of sulfur dioxide and methanesulfonic acid in the photooxidation of dimethylsulfide in the air. *Geophys. Res. Lett.* **9**, 583-586.

Hatakeyama, S., and H. Akimoto (1983). Reactions of OH radicals with methanethiol, dimethylsulfide and dimethyldisulfide in air. *J. Phys. Chem.* **87**, 2387-2395.

Hatakeyama, S., K. Izumi, and H. Akimoto (1985). Yields of SO_2 and formation of aerosol in the photooxidation of DMS under atmospheric conditions. *Atmos. Environ.* **19**, 135-141.

Haulet, R., P. Zettwoog, and J. C. Sabroux (1977). Sulphur dioxide discharge from Mount Etna. *Nature* **268**, 715-717.

Haxel, U., and G. Schumann (1955). Selbstreinigung der Atmosphäre. *Z. Phys.* **142**, 127-132.

Hayes, W. (1972). Designing wind erosian control systems in the midwest region. RT SCS Agron. Tech. Note **LI-9**, Soil Conservation Service, USDA, Lincoln, Nebraska.

Heald, E. F., J. J. Naughton, and I. L. Barnes (1963). The chemistry of volcanic gases. 2. Use of equilibrium calculations in the interpretation of volcanic gas samples. *J. Geophys. Res.* **68**, 545-557.

Healey, T. V., H. A. C. McKay, A. Pilbeam, and D. Scargill (1970). Ammonia and ammonium sulfate in the troposphere over the United Kingdom. *J. Geophys. Res.* **75**, 2317-2321.

Heaps, W. S., and T. J. McGee (1982). Balloon borne lidar measurements of strospheric hydroxyl radicals. *Appl. Opt.* **21**, 2265-2274.

Heath, D. F. (1973). Space observations of the variability of solar irradiance in the near and far uv. *J. Geophys. Res.* **78**, 2779-2792.

Heath, D. F., and M. P. Thekaekara (1977). The solar spectrum between 1200-3000 Å. *In* "The Solar Output and Its Variation" (O. R. White, ed.), pp. 193-212. Colorado Associated Univ. Press, Boulder, Colorado.

Heaton, W. B., and J. T. Wentworth (1959). Exhaust gas analysis by gas chromatography combined with infrared detection. *Anal. Chem.* **31**, 349-357.

Hegg, D. A., and P. V. Hobbs (1978). Oxidation of sulfur dioxide in aqueous systems with particular reference to the atmosphere. *Atmos. Environ.* **12**, 241-253.

Hegg, D. A., P. V. Hobbs, and L. F. Radke (1980). Observations of the modification of cloud condensation nuclei in wave clouds. *J. Rech. Atmos.* **14**, 217-222.

Hegg, D. A., and P. V. Hobbs (1980). Measurements of gas-to-particle conversion in the plumes from five coal-fired electric power plants. *Atmos. Environ.* **14**, 99-116.

Hegg, D. A., and P. V. Hobbs (1981). Cloud chemistry and the production of sulfates in clouds. *Atmos. Environ.* **15**, 1597-1604.

Hegg, D. A., and P. V. Hobbs (1982). Measurements of sulfate production in natural clouds. *Atmos. Environ.* **16**, 2663-2668.

Hegg, D. A., and P. V. Hobbs (1983). Preliminary measurements on the scavenging of sulfate and nitrate in clouds. *In* "Precipitation Scavenging Dry Deposition, and Resuspension" (H. R. Pruppacher, R. G. Semonin, and W. G. N. Slinn, eds.), pp. 79-85. Elsevier, New York.

Heicklen, J., K. Westberg, and N. Cohen (1971). *In* "Chemical Reactions in Urban Atmospheres" (C. S. Tuesday, ed.), pp. 55-59. American Elsevier, New York.

Heidt, L. E., R. Lueb, W. Pollock, and D. H. Ehhalt (1975). Stratospheric profiles of CCl_3F and CCl_2F_2. *Geophys. Res. Lett.* **2**, 445-447.

Heidt, L. E., J. P. Krasnec, R. A. Lueb, W. H. Pollock, B. E. Henry, and P. J. Crutzen (1980). Latitudinal distribution of CO and CH_4 over the Pacific. *J. Geophys. Res.* **85**, 7329-7336.

Heidt, L. E., and D. H. Ehhalt (1980). Corrections of CH_4 concentrations measured prior to 1974. *Geophys. Res. Lett.* **7**, 1023.

Heintzenberg, J., H. C. Hansson, and H. Lannefors (1981). The chemical composition of arctic haze at Ny-Alesund, Spitzbergen. *Tellus* **33**, 162-171.

Heisler, S. L., and S. K. Friedlander (1977). Gas to particle conversion on photochemical smog: aerosol growth laws and mechanisms for organics. *Atmos. Environ.* **11**, 157-168.

Helas, G., and P. Warneck (1981). Background NO_x mixing ratios in air masses over the North Atlantic Ocean. *J. Geophys. Res.* **86**, 7283-7290.

Hendry, D. G., and R. A. Kenley (1977). Generation of peroxy radicals from peroxynitrate (RO_2NO_2), decomposition of peroxyacylnitrates. *J. Am. Chem. Soc.* **99**, 3198-3199.

Hering, W. S., and H. U. Dütsch (1965). Comparison of chemiluminescent and electrochemical ozone sonde observations. *J. Geophys. Res.* **70**, 5483-5490.

Hering, W. S., and T. R. Borden, Jr. (1967). Ozone sonde observations over North America, Vol. 4. Report **AFCRL 64-30**(IV), Environmental Research Papers No. 279. Air Force Cambridge Research Laboratories, Bedford, Massachussetts.

Herman, J. R., and J. E. Mentall (1983). O_2 absorption cross sections (187-225 nm) from stratospheric solar flux measurements. *J. Geophys. Res.* **87**, 8967-8975.

Herrmann, J., and W. Jaeschke (1984). Measurements of H_2S and SO_2 over the Atlantic Ocean. *J. Atmos. Chem.* **1**, 111-123.

Herron, J. T., and R. E. Huie (1977). Stopped-flow studies of the mechanism of ozone-alkene reactions in the gas phase. Ethylene. *J. Am. Chem. Soc.* **99**, 5430-5435.

Hesse, P. R. (1957). Sulphur and nitrogen changes in forrest soils of East Africa. *Plant Soil* **9**, 86-96.

Hesstvedt, E., S. E. Henriksen, and H. Hjartarson (1974). On the development of an aerobic atmosphere. A model experiment. *Geophys. Norv.* **31**, 1-8.

Hesters, N. E., E. R. Stephens, and O. C. Taylor (1975). Fluorocarbon air pollutants measurements in the lower stratosphere. *Environ. Sci. Technol.* **9**, 875-876.

Hidalgo, H., and P. J. Crutzen (1977). The tropospheric and stratospheric composition perturbed by NO_x emissions of high-altitude aircraft. *J. Geophys. Res.* **82**, 5833-5866.

Hidy, G. M., and J. R. Brock (1970). "The Dynamics of Aerocolloidal Systems." Pergamon Press, Oxford.

Hidy G. M., and J. R. Brock (1971). An assessment of the global sources of tropospheric aerosols. *Proc. Int. Clean Air Congr., 2nd.* (H. M. Englund and W. T. Beery, eds.), pp. 1088-1097. Academic Press, New York.

Hidy, G. M., P. K. Mueller, and E. Y. Tong (1978). Spatial and temporal distributions of airborne sulfate in parts of the United States. *Atmos. Environ.* **12**, 735-752.

Hill, F. B., V. P. Aneja, and R. M. Felder (1978). A technique for measurements of biogenic sulfur emission fluxes. *J. Environ. Sci. Health A* **13**, 199-225.

Hill, R. D. (1979). A survey of lightning energy estimates. *Rev. Geophys. Space Phys.* **17**, 155-164.

Hill, R. D., R. G. Rinker, and H. D. Wilson (1980). Atmospheric nitrogen fixation by lightning. *J. Atmos. Sci.* **37**, 179-192.

Hilsenrath, E., D. F. Heath, and B. M. Schlesinger (1979). Seasonal and interannual variations in total ozone revealed by Nimbus 4 backscattered ultraviolet experiment. *J. Geophys. Res.* **84**, 6969-6979.

Hingston, F. J., and V. Gailitis (1976). The geographic variation of salt precipitated over Western Australia. *Aust. J. Soil Res.* **14**, 319-335.

Hitchcock, D. R., and A. E. Wechsler (1972). Biological cycling of atmospheric trace gases. Final Report NASW-2128, pp. 117-154. Littleton, Cambridge, Massachusetts.

Hitchcock, D. R. (1975). Dimethyl sulfide emissions to the global atmosphere. *Chemosphere* **3**, 137-138.

Hoare, D. E. (1967). The combustion of methane. "Low Temperature Oxidation" (W. Jost, ed.), Chap. 6, pp. 125-167. Gordon and Breach, New York.

Hobbs, P. V., L. F. Radke, and J. L. Stith (1977). Eruptions of the St. Augustine Volcano: airborne measurements and observations. *Science* **195**, 871-873.

Hobbs, P. V., L. F. Radke, and J. L. Stith (1978). Particles in the eruption cloud from St. Augustine Volcano. *Science* **199**, 457-458.

Hobbs, P. V., L. F. Radke, M. W. Eltgroth, and D. A. Hegg (1981). Airborne studies of the emissions from the volcanic eruptions of Mount St. Helens. *Science* **211**, 816-818.

Höfken, K. D., H. W. Georgii, and G. Gravenhorst (1981). Untersuchungen über die Deposition atmosphärischer Spurenstoffe an Buchen- und Fichtenwald, Bericht des Instituts für Meteorologie und Geophysik, Universität Frankfurt am Main, No. 46.

Hoell, J. M., C. N. Harward, and B. S. Williams (1980). Remote infrared heterodyne radiometer measurements of atmospheric ammonia profiles. *Geophys. Res. Lett.* **5**, 313-316.

Hoffman, G. L. (1971). Particulate trace metals in the Hawaiian marine atmosphere. Ph.D. Thesis, University of Hawaii, Honolulu.

Hoffman, G. L., and R. A. Duce (1972). Consideration of the chemical fractionation of alkali and alkaline earth metals in atmospheric particulate matter over the North Atlantic. *J. Geophys. Res.* **77**, 5161-5169.

Hoffman, E. J., and R. A. Duce (1974). The organic carbon content of marine aerosols collected at Bermuda. *J. Geophys. Res.* **79**, 4474-4477.

Hoffman, E. J., G. L. Hoffman, and R. A. Duce (1974). Chemical fractionation of alkali and alkaline earth metals in atmospheric particulate matter on the North Atlantic. *J. Rech. Atmos.* **8**, 676-688.

Hoffmann, M. R., and J. O. Edwards (1975). Kinetics of the oxidation of sulphite by hydrogen peroxide in acidic solution. *J. Phys. Chem.* **79**, 2096-2098.

Hoffman, D. J., J. M. Rosen, T. J. Pepin, and R. G. Pinnick (1975). Stratospheric aerosol measurements. I: Time variation at northern mid-latitudes. *J. Atmos. Sci.* **32**, 1446-1456.

Hoffman, E. J., and R. A. Duce (1977). Organic carbon in marine atmospheric particular matter, concentration and particle size distribution. *Geophys. Res. Lett.* **4**, 449–452.

Hoffmann, M. R., and S. D. Boyce (1983). Catalytic auto oxidation of sulfur dioxide in relationship to atmospheric systems. *Adv. Environ. Sci. Technol.* **12**, 148–189.

Hoffmann, M. R., and D. J. Jacob (1984). Kinetics and mechanisms of the catalytic oxidation of dissolved sulfur dioxide in aqueous solution: an application to nightime fog water chemistry. Chapter 3 *In* "SO$_2$, NO and NO$_2$ Oxidation Mechanisms: Atmospheric Considerations" (J. G. Calvert, ed.), pp. 101–172. Acid Precipitation Series, Vol. 3. Butterworth, Boston.

Hoigné, J., Bader, H., W. R. Haag, and J. Staehlin (1985). Rate constants of reactions of ozone with organic and inorganic compounds in water III, Inorganic compounds and radicals. *Water Res.* **19**, 993–1004.

Holeman, J. N. (1968). The sediment yield of major rivers of the world. *Water Resour. Res.* **4**, 737–747.

Holdren, M. W., H. H. Westberg, and P. R. Zimmerman (1979). Analysis of monoterpene hydrocarbons in rural atmospheres. *J. Geophys. Res.* **84**, 5083–5088.

Holland, H. D. (1973). Ocean water, nutrients and atmospheric oxygen. *In* "Hydrogeochemistry" (E. Ingerson, ed.), *Proc. IAGG Symp. Hydrogeochemistry Biochemistry, Tokyo, 1970, Vol. 1*, pp. 68–81. The Clark Co., Washington D.C.

Holland, H. D. (1978). "The Chemistry of the Atmosphere and Oceans." Wiley, New York.

Holser, W. T., and I.R. Kaplan (1966). Isotope geochemistry of sedimentary sulfactes. *Chem. Geol.* **1**, 93–135.

Hooker, M. L., G. A. Peterson, and D. H. Sander (1973). Ammonia nitrogen losses from simulated plowing of native sods. *Soil Sci. Am. Proc.* **37**, 247–249.

Hoppel, W. A., J. E. Dinger, and R. E. Ruskin (1973). Vertical profile of CCN at various geographic locations. *J. Atmos. Sci.* **30**, 1410–1420.

Horn, M. K. and J. A. S. Adams (1966). Computer-derived geochemical balances and element abundances. *Geochim. Cosmochim. Acta* **30**, 279–297.

Horowitz, A., and J. G. Calvert (1978). Wavelength dependence of the quantum efficiencies of the primary processes in formaldehyde photolysis at 25 °C. *Int. J. Chem. Kinet.* **10**, 805–819.

Horowitz, A., and J. G. Calvert (1982). Wavelength dependence of the primary processes in acetaldehyde photolysis. *J. Phys. Chem.* **86**, 3105–3114.

Horvath, J. J., and C. J. Mason (1978). Nitric oxide mixing ratios near the stratopause measured by a rocket-borne chemiluminescent detector. *Geophys. Res. Lett.* **5**, 1023–1026.

Horvath, L., E. Meszaros, E. Antal, and A. Simon (1981). On the sulfate, chloride and sodium concentrations in maritime air around the Asian continent. *Tellus* **33**, 382–386.

Houtermans, J. C., H. E. Suess, and H. Oeschger (1973). Reservoir models and production rate variations of natural radiocarbon. *J. Geophys. Res.* **78**, 1897–1908.

Howard, J. A. (1971). Absolute rate constants for reactions of oxyl radicals. *Adv. Free Radical Chem.* **4**, 49–173.

Hübler, G., D. H. Ehhalt, H. W. Pätz, D. Perner, U. Platt, J. Schröder, and A. Tönnisson (1982). Determination of ground level OH concentrations by a long path laser absorption technique. *In* "Physico-chemical Behaviour of Atmospheric Pollutants" (B. Versino and H. Oh, eds.), *Proc. Eur. Symp., 2nd, Varese, Italy*, pp. 2–9. Reidel, Dordrecht, Holland.

Hübler, G., D. Perner, U. Platt, A. Tönissen, and D. H. Ehhalt (1984). Ground level OH radical concentration: new measurements by optical absorption. *J. Geophys. Res.* **89**, 1309–1319.

Hubrich, C., C. Zetsch, and F. Stuhl (1977). *Ber. Bunsen Ges. Phys. Chem.* **81**, 437–442.

Hudson, R. D., V. L. Carter, and J. A. Stein (1966). An investigation of the effect of temperature on the Schumann-Runge absorption continuum of oxygen 1580-1950. *J. Geophys. Res.* **71**, 2295-2298.

Hudson, R. D., V. L. Carter, and E. L. Breig (1969). Predissociation in the Schumann-Runge band system of O_2: Laboratory measurements and atmospheric effects. *J. Geophys. Res.* **74**, 4079-4086.

Hudson, R. D., and S. H. Mahle (1972). Photodissociation rates of molecular oxygen in the mesosphere and lower thermosphere. *J. Geophys. Res.* **77**, 2902-2914.

Hudson, D. D., and E. I. Reed (1979). The stratosphere: present and future. NASA Reference Publication 1049. National Aeronautics and Space Administration, Washington, D.C.

Hudson, R. D., chief editor (1982). "The Stratosphere 1981: Theory and Measurements." WMO Global Research and Monitoring Project Report No. 11. World Meterological Organization, Geneva.

Huebert, B. J. (1980). Nitric acid and aerosol nitrate measurements in the equatorial Pacific region. *Geophys. Res. Lett.* **7**, 325-328.

Huebert, B. J., and A. L. Lazrus (1980a). Tropospheric gas-phase and particulate nitrate measurements. *J. Geophys. Res.* **85**, 7322-7328.

Huebert, B. J., and A. L. Lazrus (1980b). Bulk composition of aerosols in the remote troposphere. *J. Geophys. Res.* **85**, 7337-7344.

Huebert, B. J., and C. H. Robert (1985). The dry deposition of nitric acid to grass. *J. Geophys. Res.* **90**, 2085-2090.

Huie, R. E. and N. C. Peterson (1983). Reactions of sulfur (IV) with transition-metal ions in aqueous solutions. *Adv. Environ. Sci. Technol.* **12**, 117-146.

Hull, L. A. (1981). Terpene ozonolysis products. *In* "Atmospheric Biogenic Hydrocarbons" (J. J. Bufalini and R. R. Arnts, eds.), *Ann. Arbor. Science*, 161-186.

Hunt, B. G. (1966a). Photochemistry of ozone in a moist atmosphere. *J. Geophys. Res.* **71**, 1386-1398.

Hunt, B. G. (1966b). The need for a modified photochemical theory of the ozonosphere. *J. Atmos. Sci.* **23**, 2388-2395.

Hunt, J. M. (1972). Distribution of carbon in crust of earth. *Bull. Am. Assoc. Pet. Geol.* **56**, 2273-2277.

Hunten, D. M. (1973). The escape of light gases from planetary atmospheres. *J. Atmos. Sci.* **30**, 1481-1484.

Hunten, D. M., and D. F. Strobel (1974). Production and escape of terrestrial hydrogen. *J. Atmos. Sci.* **31**, 305-317.

Hunten, D. M. (1975). Estimates of stratospheric pollution by an analytical model. *Proc. Natl. Acad. Sci. U.S.A.* **72**, 4711-4715.

Husar, R. B., and K. T. Whitby (1973). Growth mechanism and size spectra of photochemical aerosols. *Environ. Sci. Technol.* **7**, 241-247.

Husar, R. B., D. E. Patterson, J. D. Husar, N. V. Gillani, and W. E. Wilson (1978). Sulfur budget of a power plant plume. *Atmos. Environ.* **12**, 549-568.

Husar, R. B., and D. E. Patterson (1980). Regional scale air pollution sources and effects. *Ann. N.Y. Acad. Sci.* **338**, 399-417.

Huss, A., Jr., P. K. Kim, and C. A. Eckert (1978). On the uncatalyzed oxidation of sulfur (IV) in aqueous solutions. *J. Am. Chem. Soc.* **100**, 6252-6253.

Huss, A., Jr., P. K. Kim, and C. A. Eckert (1982). Oxidation of aqueous sulfur dioxide. 1. Homogeneous manganese (II) and iron (II) catalysis at low pH. *J. Phys. Chem.* **86**, 4224-4228.

Hutchinson, G. E. (1948). Circular causal systems in ecology. *Ann. N.Y. Acad. Sci.* **50**, 221.

Hutchinson, G. E. (1954). The biogeochemistry of the terrestrial atmosphere. *In* "The Earth as a Planet" (G. P. Kuiper, ed.), pp. 371-433. University of Chicago Press, Chicago, Illinois.

Hutchinson, G. L., R. J. Millington, and D. B. Peters (1972). Atmospheric ammonia: absorption by plant leaves. *Science* **175**, 771-772.

Hutton, J. T. (1976). Chloride in rainwater in relation to distance from ocean. *Search* **7**, 207-208.

Ibusuki, T., and H. M. Barnes (1984). Manganese (II) catalyzed sulfur dioxide oxidation in aqueous solution at environmental concentrations. *Atmos. Environ.* **18**, 145-151.

Ingold, K. U. (1969). Peroxy radicals. *Acc. Chem. Res.* **2**, 1-9.

Inn, E. C. Y., and Y. Tanaka (1953). Absorption coefficient of ozone in the ultraviolet and visible regions. *J. Opt. Soc. Am.* **43**, 870-873.

Inn, E. C. Y., J. F. Vedder, and B. J. Tyson (1979). COS in the stratosphere. *Geophys. Res. Lett.* **6**, 191-193.

Inn, E. C. Y., J. F. Vedder, and D. O'Hara (1981). Measurements of stratospheric sulfur constituents. *Geophys. Lett.* **8**, 5-8.

Inn, E. C. Y., N. H. Farlow, P. B. Russell, M. P. McCormick, and W. P. Chu (1982). Observations. *In* "The Stratospheric Aerosol Layer" (R. C. Whitten, ed.), Chap. 2, pp. 15-68. Springer-Verlag, Berlin.

Ingersoll, R. B., R. E. Inman, and W. R. Fisher (1974). Soil's potential as a sink for atmospheric carbon monoxide. *Tellus* **26**, 151-159.

Inman, R. E., R. B. Ingersoll, and E. A. Levy (1971). Soil: a natural sink for carbon monoxide. *Science* **172**, 1229-1231.

Isaksen, I. S. A., K. H. Mitboe, J. Sunde, and P. J. Crutzen (1977). A simplified method to include molecular scattering and reflection in calculations of photon fluxes and photodissociation rates. *Geophys. Norv.* **31**, 11-26.

Isidorov, V. A., I. G. Zenkevich, and B. V. Ioffe (1985). Volatile organic compounds in the atmosphere of forests. *Atmos. Environ.* **19**, 1-8.

Ivanov, M. V., V. A. Grinenko, and A. P. Rabinovich (1983). The sulfur cycle in continental Reservoirs, Part II. Sulfur flux from continents to ocean. *In* "The Global Biogeochemical Sulphur Cycle" (M. V. Ivanov and J. R. Freney, eds.), *Scope* **19**, 331-356.

Jackman, C. H., J. E. Frederick, and R. S. Stolarski (1980). Production of odd nitrogen in the stratosphere and oecosphere. An intercomparison of source strengths. *J. Geophys. Res.* **85**, 7495-7505.

Jacobs, M. B., and S. Hocheiser (1958). Continuous sampling and ultramicrodetermination of nitrogen dioxide in air. *Anal. Chem.* **30**, 426-428.

Jacobs, M. B. (1959). Concentration of Sulfur Containing Pollutants in a Major Urban Area. American Geophysical Union Monograph No. 3, (J. P. Lodge, ed.), *Proc. Symp. Atmos. Chem. Chlorine Sulfur Compounds*, pp. 81-87. Waverly Press, Baltimore.

Jacob, D. J., and M. R. Hoffmann (1983). A dynamic model for the production of H^+, NO_3^- and SO_4^{2-} in urban fog. *J. Geophys. Res.* **88C**, 6611-6621.

Jaenicke, R., and C. E. Junge (1967). Studien zur oberen Grenzgrösse des natürlichen Aerosols. *Beitr. Phys. Atmos.* **40**, 129-143.

Jaenicke, R. (1978a). Aitken particle size distribution in the Atlantic north east trade winds. *Meteor. Forschungsergebnisse Reihe B* **13**, 1-9.

Jaenicke, R. (1978b). Physical properties of atmospheric particulate sulfur compounds. *Atmos. Environ.* **12**, 161-169.

Jaenicke, R. (1978c). Über die Dynamik atmosphärischer Aitkenteilchen. *Ber. Bunsen Ges. Phys. Chem.* **82**, 1198-1202.

Jaenicke, R. (1978d). The role of organic material in atmospheric aerosols. *Pure Appl. Geophys.* **116**, 283-292.

Jaenicke, R., and L. Schütz (1978). Comprehensive study of physical and chemical properties of the surface aerosols in the Cape Verde Island region. *J. Geophys. Res.* **83**, 3585-3599.

Jaenicke, R. (1980). Natural aerosols. *In* "Aerosols: Anthropogenic and Natural, Sources and Transport" (T. J. Kneip and P. J. Lioy, eds.). *Ann. N.Y. Acad. Sci.* **338**, 317-329.

Jaeschke, W., R. Schmitt, and H. W. Georgii (1976). Preliminary results of stratospheric SO_2 measurements. *Geophys. Res. Lett.* **9**, 517-519.

Jaeschke, W., H. W. Georgii, H. Claude, and H. Malewski (1978). Contributions of H_2S to the atmospheric sulfur cycle. *Pure Appl. Geophys.* **116**, 465-475.

Jaeschke, W., H. Claude, and J. Herrmann (1980). Sources and sinks of atmospheric H_2S. *J. Geophys. Res.* **85**, 5639-5644.

Jaeschke, W., and J. Herrmann (1981). Measurements of H_2S in the atmosphere. *Int. J. Environ. Anal. Chem.* **10**, 107-120.

Jaeschke, W., H. Berresheim, and H. W. Georgii (1982). Sulfur emissions from Mount Etna. *J. Geophys. Res.* **87**, 7253-7261.

Jaggar, T. A. (1940). Magmatic gases. *Am. J. Sci.* **238**, 313-353.

Japar, S. M., C. H. Wu, and H. Niki (1974). Rate constants for the reaction of ozone with olefins in the gas phase. *J. Phys. Chem.* **78**, 2318-2320.

Japar, S. M., and H. Niki (1975). Gas-phase reactions of the nitrate radical with olefins. *J. Phys. Chem.* **79**, 1629-1632.

Jeans, J. H. (1954). "The Dynamical Theory of Gases," 4th ed., Dover, New York.

Jeffries, H. E., D. L. Fox, and R. Kamens (1976). Photochemical conversion of NO to NO_2 by hydrocarbons in an outdoor chamber. *J. Air Pollut. Control Assoc.* **26**, 480-484.

Jeffries, H. E., and M. Saeger (1976). Correspondence. *Environ. Sci. Technol.* **10**, 936-937.

Jickells, T., A. Knap, T. Church, J. Galloway, and J. Miller (1982). Acid rain on Bermuda. *Nature* **297**, 55-57.

Johnson, B., and R. C. Cooke (1979). Bubble populations and spectra in coastal waters: a photographic approach. *J. Geophys. Res.* **84**, 3763-3766.

Johnson, J. E. (1981). The life time of carbonyl sulfide in the troposphere. *Geophys. Res. Lett.* **8**, 938-940.

Johnston, H. S., W. A. Bonner, and D. J. Wilson (1957). Carbon isotope effect during oxidation of carbon monoxide with nitrogen dioxide. *J. Chem. Phys.* **26**, 1002-1006.

Johnston, H. S. (1966). "Gas Phase Reaction Rate Theory.' Ronald Press, New York.

Johnston, H. S. (1971). Reduction of stratospheric ozone by nitrogen oxide catalysis from SST exhaust. *Science* **173**, 517-522.

Johnston, H. S. (1972). Laboratory kinetics as an atmospheric science. *In* "Climatic Impact Assessment Program, Proceedings of the Survey Conference" (A. E. Barrington, ed.), pp. 90-112. **DOT-TSC-OST-72-13**. U.S. Department of Transportation.

Johnston, H. S., and G. Whitten (1973). Instantaneous photochemical rates in the global stratosphere. *Pure Appl. Geophys.* **106–108**, 1468-1489.

Johnston, H. S., S. G. Chang, and G. Whitten (1974). Photolysis of nitric acid vapor. *J. Phys. Chem.* **78**, 1-7.

Johnston, H. S. (1975). Global balance in the natural stratosphere. *Rev. Geophys. Space Phys.* **13**, 637-649.

Johnston, H. S., and G. Selwyn (1975). New cross section for the absorption of near ultraviolet radiation by nitrous oxide (N_2O). *Geophys. Res. Lett.* **2**, 549-551.

Johnston, H. S., and G. Whitten (1975). Chemical reactions in the atmosphere as studied by the method of instantaneous rates. *Int. J. Chem. Kinet. Symp.* **1**, 1-26.

Johnston, H. S., and J. Podolske (1978). Interpretations of stratospheric photochemistry. *Rev. Geophys. Space Phys.* **16**, 491-519.

Johnston, H. S., O. Serang, and J. Podolske (1979). Instantaneous global nitrous oxide photochemical rates. *J. Geophys. Res.* **84**, 5077-5082.

Johnston, P. V. and R. L. McKenzie (1984). Long path absorption measurements of tropospheric NO_2 in rural New Zealand. *Geophys. Res. Lett.* **11**, 69-72.

Jones, I. T. N., and R. P. Wayne (1970a). The photolysis of ozone by ultraviolet radiation IV; effect of photolysis wavelength on primary steps. *Proc. R. Soc. London A* **319**, 273-287.

Jones, I. T. N., and R. P. Wayne (1970b). The photolysis of ozone by ultraviolet radiation, V. Photochemical formation of $O_2(^1\Delta_g)$. *Proc. R. Soc. London A* **321**, 409-424.

Jones, I. T. N., and K. D. Bayes (1973). Photolysis of nitrogen dioxide. *J. Chem. Phys.* **59**, 4836-4844.

Jones, B. M. R., J. P. Burrows, R. A. Cox, and S. A. Penkett (1982). OCS Formation in the reaction of OH with CS_2. *Chem. Phys. Lett.* **88**, 372-376.

Jones, B. M. R., R. A. Cox, and S. A. Penkett (1983). Atmospheric chemistry of carbon disulphide. *J. Atmos. Chem.* **1**, 65-86.

Joseph, D. W., and C. W. Spicer (1978). Chemiluminescence method for atmospheric monitoring of nitric acid and nitrogen oxides. *Anal. Chem.* **50**, 1400-1403.

Jost, D. (1974). Aerological studies on the atmospheric sulfur budget. *Tellus* **26**, 206-212.

Junge, C. E. (1952a). Gesetzmässigkeiten in der Grössenverteilung des atmosphärischen Aerosol über dem Kontinent. *Ber. Dtsch. Wetterdienstes, U.S. Zone* **35**, 261-277.

Junge, C. E. (1952b). Das Grössenwachstum der Aitkenkerne. *Ber. Dtsch. Wetterdienstes, U. S. Zone* **38**, 264-267.

Junge, C. (1953). Die Rolle der Aerosole und der gasförmigen Beimengungen der Luft im Spurenhaushalt der Troposphäre. *Tellus* **5**, 1-26.

Junge, C. E. (1954). The chemical composition of atmospheric aerosols, I. Measurements at Round Hill Field Station June-July 1963. *J. Meteorol.* **11**, 223-333.

Junge, C. E. (1955). The size distribution and aging of natural aerosols as determined from electrical and optical data on the atmosphere. *J. Meteorol.* **12**, 13-25.

Junge, C. E. (1956). Recent investigations in air chemistry. *Tellus* **8**, 127-139.

Junge, C. E. (1957). The vertical distribution of aerosols over the ocean. *In* "Artificial Stimulation of Rain" (J. Weickman and W. Smith, eds.). *Proc. Conf. Phys. Cloud Precip. Particles, 1st*, pp. 89-96. Pergamon Press, London.

Junge C. E., and T. G. Ryan (1958). Study of the SO_2 oxidation in solution and its role in atmospheric chemistry. *Q. J. R. Meteorol. Soc.* **84**, 46-55.

Junge, C. E. (1961). Vertical profiles of condensation nuclei in the stratosphere. *J. Meteorol.* **18**, 501-509.

Junge, C. E., C. W. Chagnon, and J. E. Manson (1961). Stratospheric aerosols. *J. Meterol.* **18**, 81-108.

Junge, C. E. (1962). Global ozone budget and exchange between stratosphere and troposphere. *Tellus* **14**, 363-377.

Junge, C. E. (1963). "Air Chemistry and Radioactivity." Academic Press, New York.

Junge, C. E. (1964). The modification of aerosol size distribution in the atmosphere. Final Tech. Report, Contract-Da 91-591-EVC 2979. European Research Office, U.S. Army.

Junge, C. E., and N. Abel (1965). Modification of aerosol size distribution in the atmosphere and development of an ion counter of high sensitivity. Final. Tech. Report No. Da 91-591-EVC-3404. Meteor. Inst., University of Mainz, Germany.

Junge, C. E., and G. Czeplak (1968). Some aspects of the seasonal variation of carbon dioxide and ozone. *Tellus* **20**, 422-434.

Junge, C. E., and E. McLaren (1971). Relationship of cloud nuclei spectra to aerosol size distribution and composition. *J. Atmos. Sci.* **28**, 382-390.

Junge, C. E., B. Bockholt, K. Schütz, and R. Beck (1971). N_2O measurements in air and seawater over the Atlantic. *Meteor Forschungsergeb. B* **6**, 1-11.

Junge, C. E. (1972). The cycle of atmospheric gases—natural and man-made. *Q. J. R. Meteorol. Soc.* **98**, 711-729.

Junge, C. E. (1974). Residence time and variability of tropospheric trace gases. *Tellus* **24**, 477-488.

Junge, C. E. (1979). The importance of mineral dust as an atmospheric constituent. *In* "Sahara Dust: Mobilization, Transport, Deposition" (C. Morales, ed.), *SCOPE* **14**, 49-60.

Junge, C. E., and P. Warneck (1979). Composition of the atmosphere, *In* "Review of Research on Modern Problems in Geochemistry" (F. R. Siegel, ed.), pp. 139-165. UNESCO, Paris.

Kadota, H., and Y. Ishida (1972). Production of volatile sulfur compounds by microorganisms. *Ann. Rev. Microbiol.* **26**, 127-163.

Kadowaki, S. (1976). Size distribution of atmospheric total aerosols, sulfate, ammonium and nitrate particulates in the Nagoya area. *Atmos. Environ.* **10**, 39-43.

Kadowaki, S. (1977). Size distribution and chemical composition of atmospheric particulate nitrate in the Nagoya area. *Atmos. Environ.* **11**, 671-695.

Käselau, K. H., P. Fabian, and H. Röhrs (1974). Measurements of aerosol concentrations up to a height of 27 km. *Pure Appl. Geophys.* **112**, 877-885.

Kaiser, E. W., and C. H. Wu (1977). Measurement of the rate constant of the reaction of nitrous acid with nitric acid. *J. Phys. Chem.* **81**, 187-190.

Kaiser, E. W., and C. H. Wu (1977). A kinetic study of the gas phase formation and decomposition reactions of nitrous acid. *J. Phys. Chem.* **81**, 1701-1708.

Kajimoto, O., and R. J. Cvetanović (1979). Absolute quantum yields of $O(^1D_2)$ in the photolysis of ozone in the Hartley band. *Int. J. Chem. Kinet.* **11**, 605-612.

Kallend, A. S., A. R. W. Marsh, J. H. Pickels, and V. Proctor (1983). Acidity of rain in Europe. *Atmos. Environ.* **17**, 127-137.

Kandler, O. (1981). Achaebakterien und Phylogenie der Organismen. *Naturwissenschaften* **68**, 183-192.

Kaplan, L. D. (1973). Background concentration of photochemically active trace constituents in the stratosphere and upper atmosphere. *Pure Appl. Geophys.* **106–108**, 1342-1345.

Kaplan, W. A., J. W. Elkins, C. E. Kolb, M. B. McElroy, S. Wofsy, and C. R. Duran (1978). Nitrous oxide in fresh water systems. An estimate for the yield of atmospheric N_2O associated with disposal of human waste. *Pure Appl. Geophys.* **116**, 423-438.

Kasting, J. F., S. C. Liu, and T. M. Donahue (1979). Oxygen levels in the primitive atmosphere. *J. Geophys. Res.* **84**, 3097-3107.

Katsura, T., and S. Nagashima (1974). Solubility of sulfur in some magmas at one atmosphere. *Geochim. Cosmochim. Acta* **38**, 517-531.

Keeling, C. D. (1973a). Industrial production of carbon dioxide from fossil fuels and limestone. *Tellus* **27**, 174-198.

Keeling, C. D. (1973b). The carbon dioxide cycle. Reservoir models to depict the exchange of atmospheric carbon dioxide with ocean and land plants. *In* "Chemistry of the Lower Atmosphere" (S. Rasool, ed.), pp. 251-329. Plenum Press, New York.

Keeling, C. D., R. B. Bacastow, and T. P. Whorf (1982). Measurements of the concentration of carbon dioxide at Mauna Loa Observatory, Hawaii. *In* "Carbon Dioxide Review 1982" (W. C. Clark, ed.), pp. 377-385. Oxford University Press, New York.

Keene, W. C., J. N. Galloway, and J. D. Holden, Jr. (1983). Measurement of weak organic acidity in precipitation from remote areas of the world. *J. Geophys. Res.* **88**, 5122-5130.

Kellog, W. W., R. D. Cadle, E. R. Allen, A. L. Lazrus, and E. A. Martell (1972). The sulfur cycle. *Science* **175**, 587-596.

Kelley, J. J. (1973). Surface ozone in the Arctic atmosphere. *Pure Appl. Geophys.* **106–108**, 1106-1115.

Kelly, T. J., D. H. Stedman, and G. L. Kok (1979). Measurements of H_2O_2 and HNO_3 in rural air. *Geophys. Res. Lett.* **6**, 375–378.

Kelly, T. J., D. H. Stedman, J. A. Ritter, and R. B. Harvey (1980). Measurements of oxides of nitrogen and nitric acid in clean air. *J. Geophys. Res.* **85**, 7417–7425.

Kenley, R. A., J. E. Davenport, and D. G. Hendry (1978). Hydroxyl radical reactions in the gas phase. Products and pathways for the reaction of OH with toluene. *J. Phys. Chem.* **82**, 1095–1096.

Kerker, M. and V. Hampel (1974). Scavenging of aerosol particles by a falling water drop and calculation of washout coefficients. *J. Atmos. Sci.* **31**, 1368–1376.

Kerr, J. B., and C. T. McElroy (1976). Measurements of stratospheric nitrogen dioxide from the AES stratospheric balloon program. *Atmosphere* **14**, 166–171.

Kessler, C., D. Perner, and U. Platt (1982). Spectroscopic measurements of nitrous acid and formaldehyde—implications for urban photochemistry. *In* "Physico-chemical Behaviour of Atmospheric Pollutants", (B. Versino and H. Om, eds.), *Proc. Eur. Symp., 2nd, Varese, Italy*, pp. 393–399. Reidel, Dordrecht, Holland.

Kessler, C., and U. Platt (1984). Nitrous acid in polluted air masses—sources and formation pathways. *In* "Physico-chemical Behaviour of Atmospheric Pollutants" (B. Versino and G. Angeletti, eds.). *Proc. Eur. Symp. 3rd, Varese, Italy*, pp. 412–421. Reidel, Dordrecht, Holland.

Ketseridis, G., J. Hahn, R. Jaenicke, and C. Junge (1976). The organic constituents of atmospheric particulate matter. *Atmos. Environ.* **10**, 603–610.

Ketseridis, G., and R. Jaenicke (1978). Organische Beimengungen atmosphärischer Reinluft: ein Beitrag zur Budget-Abschätzung *In* "Organische Verunreinigungen in der Umwelt," (K. Anrand, H. Hässelbarth, E. Lahmann, G. Müller, and W. Niemitz, eds.), pp. 379–390. Erich Schmidt Verlag, Berlin.

Khalil, M. A. K., and R. A. Rasmussen (1981). Increase of $CHClF_2$ in the earth's atmosphere. *Nature* **292**, 823–824.

Khalil, M. A. K., and R. A. Rasmussen (1983). Sources, sinks and seasonal cycles of atmospheric methane. *J. Geophys. Res.* **88**, 5131–5144.

Khalil, M. A. K., and R. A. Rasmussen (1985). Causes of increasing atmospheric methane: Depletion of hydroxyl radicals and the rise of emissions. *Atmos. Environ.* **19**, 397–407.

Kiang, C. S., and D. Stauffer (1973). Chemical nucleation theory for various humidities and pollutants. *Faraday Symp.* **7**, 26–33.

Kiang, C. S., D. Stauffer, V. A. Mohnen, J. Bricard, and D. Viglia (1973). Heteromolecular nucleation theory applied to gas to particle conversion. *Atmos. Environ.* **7**, 1279–1283.

Kim, C. M. (1973). Influence of vegetation types on the intensity of ammonia and nitrogen dioxide liberation from soil. *Soil Biol. Biochem.* **5**, 163–166.

Kins, L. (1982). Temporal variation of chemical composition of rain water during individual precipitation events. *In* "Deposition of Atmospheric Pollutants" (H. W. Georgii and J. Pankrath, eds.), Reidel, Dordrecht, Holland.

Kleindienst, T. E., G. W. Harris, and J. N. Pitts, Jr. (1982). Rates and temperature dependence of the reaction of OH with isoprene, its oxidation products and selected terpenes. *Environ. Sci. Technol.* **16**, 844–846.

Kley, D., J. W. Drummond, M. McFarland, and S. C. Liu (1981). Tropospheric profiles of NO_x. *J. Geophys. Res.* **86**, 3153–3161.

Klippel, W., and P. Warneck (1978). Formaldehyde in rainwater and on the atmospheric aerosol. *Geophys. Res. Lett.* **5**, 177–179.

Klippel, W., and P. Warneck (1980). The formaldehyde content of the atmospheric aerosol. *Atmos. Environ.* **14**, 809–818.

Knowles, R. (1981). Denitrification. *In* "Terrestrial Nitrogen Cycles" (F. E. Clark and T. Rosswell, eds.). *Ecol. Bull.* (*Stockholm*) **33**, 315-329.

Knudsen, M., and S. Weber (1911). Luftwiderstand gegen die langsame Bewegung kleiner Kugeln. *Ann. Phys.* **36**, 981-994.

Ko, M. K. W., N. D. Sze, M. Lifshits, B. McElroy, and J. A. Pyle (1984). The seasonal and latitudinal behaviour of trace gases and O_3 as simulated by a two-dimensional model of the atmosphere. *J. Atmos. Sci.* **41**, 2381-2408.

Ko, M. K. W., K. K. Tung, D. K. Weisenstein, and N. D. Sze (1985). A Zonal-mean model of stratospheric trace transport in isentropic coordinators: numerical simulation for nitrous oxide and nitric acid. *J. Geophys. Res.* **90**, 2313-2329.

Kockarts, G. (1976). Absorption and photodissociation in the Schumann-Runge bands of molecular oxygen in the terrestrial atmosphere. *Planet. Space Sci.* **24**, 589-604.

Köhler, H. (1936). The nucleus in and the growth of hygroscopic droplets. *Trans. Faraday Soc.* **32**, 1152-1162.

Kolattunkudy, P. E., ed. (1976). "Chemistry and Biochemistry of Natural Waxes." Elsevier, Amsterdam.

Kolbig, J. and W. Warmbt (1978). Messungen des bodennahen Ozons in Mirny-Antarktika. *Z. Meteorol.* **28**, 274-277.

Komhyr, W. D. (1969). Electrochemical concentration cells for gas analysis. *Ann. Geophys.* (*Paris*) **25**, 203-210.

Komhyr, W. D., E. W. Barrett, G. Slocum, and H. K. Weickmann (1971). Atmospheric total ozone increase during 1960's. *Nature* **232**, 390-391.

Kopczynski, S. T., W. A. Lonnemann, F. D. Sutterfield, and P. E. Darley (1972). Photochemistry of atmospheric samples in Los Angeles. *Environ. Sci. Technol.* **6**, 342-347.

Kortzeborn, R. N., and F. F. Abraham (1970). Scavenging of aerosols by rain: a numerical study. *In* "Precipitation Scavenging 1970", (R. J. Engelmann and W. G. N. Slinn, coordinators), pp. 433-446. U.S. Atomic Energy Commission, Div. of Tech. Information.

Kossina, E. (1933). Die Erdoberfläche. *In* "Handbook der Geophysik" (B. Gutenberg, ed.), Vol. 2, pp. 869-954. Bontraeger, Berlin.

Koyama, T. (1963). Gaseous metabolism in lake sediments and paddy soils and the production of atmospheric methane and hydrogen. *J. Geophys. Res.* **68**, 3971-3973.

Koyama, T. (1964). Biogeochemical studies on lake sediments and paddy soils and the production of atmospheric methane and hydrogen. *In* "Recent Researches in the Field of Hydrosphere, Atmosphere and Nuclear Geochemistry", (Y. Miyake and T. Koyama, eds.), pp. 143-177. Murucen, Tokyo.

Krall, A. R., and A. Tolbert (1957). A comparison of the light-dependent metabolism of carbon monoxide by barley leaves with that of formaldehyde, formate and carbon dioxide. *Plant. Physiol.* **32**, 321-326.

Krauskopf, K. B. (1979). "Introduction to Geochemistry," 2nd Ed. McGraw-Hill, New York.

Krey, R. W., and B. Krajewski (1970). Comparison of atmospheric transport model calculations with observations of radioactive debris. *J. Geophys. Res.* **75**, 2901-2908.

Krey, P. W., M. Schomberg, and L. Toonkel (1974). Updating stratospheric inventories to January 1973. Rep. **HASL-281**. U.S. Atomic Energy Commission, New York.

Krey, P. W., R. J. Lagomarsino, and L. E. Toonkel (1977). Gaseous halogens in the atmosphere in 1975. *J. Geophys. Res.* **82**, 1753-1766.

Kritz, M. A., and J. Rancher (1980). Circulation of Na, Cl, and Br in the tropical marine atmosphere. *J. Geophys. Res.* **85**, 1633-1639.

Kritz, M. A. (1982). Exchange of sulfur between the free troposphere, marine boundary layer and the sea surface. *J. Geophys. Res.* **87**, 8795-8803.

Krueger, A. J. (1969). Rocket measurements of ozone over Hawaii. *Ann. Geophys.* **25**, 307–311.

Krueger, A. J. (1973). Mean ozone distribution from several series of rocket soundings to 52 km at latitudes from 58°S to 64°N *Pure Appl. Geophys.* **106–108**, 1272–1280.

Kuis, S., R. Simonaitis, and J. Heicklen (1975). Temperature dependence of the photolysis of ozone at 3130 Å, *J. Geophys. Res.* **80**, 1328–1331.

Kunen, S. M., A. L. Lazrus, G. L. Kok, and B. G. Heikes (1983). Aqueous oxidation of SO_2 by hydrogen peroxide. *J. Geophys. Res.* **88**, 3671–3764.

Kurasawa, H., and R. Lesclaux (1979). Kinetics of the reaction of NH_2 with NO_2, *Chem. Phys. Lett.* **66**, 602–607.

Kurasawa, H., and R. Lesclaux (1980). Rate constant for the reaction of NH_2 with ozone in relation to atmospheric processes. *Chem. Phys. Lett.* **72**, 437–442.

Kuwata, K., M. Uebori, Y. Yamasaki, and Y. Kuge (1983). Determination of aliphatic aldehyde in air by liquid chromatography. *Anal. Chem.* **55**, 2013–2016.

Labitzke, K., and J. J. Barnett (1979). Review of climatological information obtained from remote sensing of the stratosphere and mesosphere. *COSPAR Space Res.* **19**, 97–106.

Labitzke, K. (1980). Climatology of the stratosphere and mesosphere. *Philos. Trans. R. Soc., London A* **296**, 7–18.

Lai, K. Y., N. Dayan, and M. Kerker (1978). Scavenging of aerosol particles by a falling water drop. *J. Atmos. Sci.* **35**, 674–682.

Lal, D., and Rama (1966). Characteristics of global tropospheric mixing based man-made ^{14}C, ^{3}Hand ^{90}Sr. *J. Geophys. Res.* **71**, 2865–2874.

Lal, D. and B. Peters (1967). Cosmic ray produced radioactivity one earth. *In* "Encyclopedia of Physics" (S. Flügge, ed.), Vol. 46/2, pp. 551–612. Springer-Verlag, Berlin and New York.

Lal, S., R. Borchers, P. Fabian, and B. C. Krueger (1985). Increasing abundance of $CBrClF_2$ in the atmosphere. *Nature* **316**, 135–136.

Lamb, R. G. (1977). A case study of stratospheric ozone affecting ground-level oxidant concentrations. *J. Appl. Meteorol.* **16**, 780–794.

Lamontagne, R. A., J. Swinnerton, and V. J. Linnenboom (1974). C_1–C_4 hydrocarbons in the North and South Pacific. *Tellus* **26**, 71–77.

Lazrus, A. L., and B. W. Gandrud (1974). Distribution of stratospheric nitric acid vapor. *J. Atmos. Sci.* **31**, 1102–1108.

Lazrus, A. L., and B. W. Gandrud (1977). Stratospheric sulfate at high altitude. *Geophys. Res. Lett.* **4**, 521–522.

Leach, P. W., L. J. Leng, T. A. Bellar, J. E. Sigsby, Jr., and A. P. Altshuller (1964). Effects of HC/NO_x ratios on irradiated auto exhaust. Part II. *J. Air. Pollut. Control. Assoc.* **14**, 176–183.

Lee, L. C., G. Black, R. L. Sharpless, and T. G. Slanger (1980). $O(^1S)$ yield from O_3 photodissociation of 1700–2400 Å. *J. Chem. Phys.* **73**, 256–258.

Lee, J. H., R. B. Timmons, and L. J. Stief (1976). Absolute rate parameters for the reaction of ground state oxygen with dimethyl sulfide and episulfide. *J. Chem. Phys.* **64**, 300–305.

Lee, L. C., T. G. Slanger, G. Black, and R. L. Sharpless (1977). Quantum yields for the production of $O(^1D)$ from photodissociation of O_2 at 1160–1770 Å, *J. Chem. Phys.* **67**, 5602–5606.

Lee, Y. N., and S. E. Schwartz (1981). Reaction kinetics of nitrogen dioxide with liquid water at low partial pressure. *J. Phys. Chem.* **85**, 840–848.

Lehmann, L., and A. Sittkus (1959). Bestimmung von Aerosolverweilzeiten aus dem RaD und RaF Gehalt der atmosphärischen Luft und des Niederschlags. *Naturwissenschaften* **46**, 9–10.

Leifer, R., K. Sommers, and S. F. Guggenheim (1981). Atmospheric trace gas measurements with a new clean air sampling system. *Geophys. Res. Lett.* **8**, 1079–1082.

Leighton, P. A. (1961). "Photochemistry of Air Pollution." Academic Press, New York.

Lein, A. Yu. (1983). The sulfur cycle in the lithosphere. Part II: Cycling. *In* "The Global Biogeochemical Sulphur Cycle" (M. V. Ivanov and J. R. Freney eds.), *SCOPE* **19**, 95-127.

Lemon. E. (1977). The land's response to more carbon dioxide *In* "The Fate of Fossil Fuel CO_2 in the Oceans" (N. R. Andersen and A. Malahoff, eds.). pp. 97-130. Plenum Press, New York.

Lenhard, U. (1977). Messungen von Ammoniak in der unteren Troposphäre und Untersuchung der NH_3 Quellstärke von Böden. Diploma Thesis Institut für Meteorologie und Geophysik der Universität Frankfurt/Main.

Lenhard, U., and G. Gravenhorst (1980). Evaluation of ammonia fluxes into the free atmosphere over Western Germany. *Tellus* **32**, 48-55.

Lepel, E. A., K. M. Stefansson, and W. H. Zoller (1978). The enrichment of volatile elements in the atmosphere by volcanic activity: Augustine Volcano 1976. *J. Geophys. Res.* **83**, 6213-6220.

Lesclaux, R., and M. De Missy (1977). On the reaction of NH_2 radical with oxygen. *Nouv. J. Chim.* **1**, 443-444.

Lettau, H., and W. Schwerdtfeger (1933). Untersuchungen über atmosphärische Turbulenz und Vertikalaustausch vom Freiballon aus. *Meteorol. Z.* **50**, 25-29; (1934). **51**, 249; (1936). **53**, 44.

Leu, M. T. and R. H. Smith (1981). Kinetics of the gas-phase reaction between hydroxyl and carbonyl sulfide over the temperature range 300-517 K. *J. Phys. Chem.* **85**, 2570-2575.

Leu, M. T. and R. H. Smith (1982). Rate constants for the gas-phase reaction between hydroxyl and hydrogen sulfide over the temperature range 228-518 K. *J. Phys. Chem.* **86**, 73-83.

Levine, J. S. (1984). Nitrogen fixation by lightning activity in a thunderstorm. *Atmos. Environ.* **18**, 2272-2280.

Levine, J. S., R. S. Rogowski, G. L. Gregory, W. E. Howell, and J. Fishman (1981). Simultaneous measurements of NO_x, NO and O_3 production in a laboratory discharge: atmospheric implications. *Geophys. Res. Lett.* **8**, 357-360.

Levy, H. (1971). Normal atmosphere: large radical and formaldehyde concentrations predicted. *Science* **173**, 141-143.

Levy, H. (1972). Photochemistry of the lower troposphere. *Planet. Space Sci.* **20**, 919-935.

Levy, H. (1974). Photochemistry of the troposphere. *Adv. Photochem.* **9**, 369-524.

Levy, H., J. D. Mahlman, and W. J. Moxim (1980). A stratospheric source of reactive nitrogen in the unpolluted troposphere. *Geophys. Res. Lett.* **7**, 441-444.

Lewin, R. A. (1976). Prochlorophyta as a proposed new division of algae. *Nature* **261**, 697-698.

Lewis, G. N., and M. Randall (1923). "Thermodynamics and the Energy of Chemical Substances." McGraw-Hill, New York.

Lezberg, E. A., F. M. Hummenik, and D. A. Otterson (1979). Sulfate and nitrate mixing ratios in the vicinity of the tropopause. *Atmos. Environ.* **13**, 1299-1304.

Li, Y. H. (1972). Geochemical mass balance among lithosphere hydrosphere and atmosphere. *Am. J. Sci.* **272**, 119-137.

Liebig, J. (1827). Une note sur la nitrification. *Ann. Chem. Phys.* **35**, 329-333.

Liebl, K. H., and W. Seiler (1976). CO and H_2 destruction of a soil surface, *In* "Microbial Production and Utilization of Gases" (H. G. Schlegel, G. Gottschalk, and N. Pfenning, eds.), pp. 215-229. Akademie der Wissenschaften, Göttingen.

Lieth, H. (1965). Versuch einer kartographischen Darstellung der Produktivität der Pflanzendecke auf der Erde. "Geographisches Taschenbuch," pp. 72-79. Franz Steiner Verlag, Wiesbaden.

Lieth, H. (1975). Primary production of the major vegetation units of the world In "Primary Productivity of the Biosphere" (H. Lieth and R. H. Whittaker, eds.), *Ecol. Stud.* **14**, 203-215. Springer-Verlag, Berlin and New York.

Lieth, H., and R. H. Whittaker, eds. (1975). "Primary Productivity of the Biosphere." *Ecol. Stud.* **14**, Springer-Verlag, Berlin and New York.

Likens, G. E., and F. H. Bormann (1974). Acid rain: a serious regional environmental problem. *Science* **184**, 1176-1179.

Likens, G. E., and F. H. Bormann (1975). Acidity in rainwater: has an explanation been presented? *Science* **188**, 958.

Likens, G. E. (1976). Acid precipitation. *Chem. Eng. News* **54**, 29-44.

Likens, G. E., F. H. Bormann, R. S. Pierce, J. S. Eaton, and N. M. Johnson (1977). "Biochemistry of a Forested Ecosystem." Springer-Verlag, Berlin and New York.

Likens, G. E., R. F. Wright, J. N. Galloway, and T. J. Butler (1979). Acid rain. *Sci. Am.* **241**, 39-47.

Likens, G. E., E. S. Edgerton, and J. N. Galloway (1983). The composition and deposition of organic carbon in precipitation. *Tellus* **35B**, 16-24.

Likens, G. E., F. H. Bormann, R. S. Pierce, J. S. Eaton, and R. E. Munn (1984). Long-term trends in precipitation chemistry at Hubbard Brook, New Hampshire. *Atmos. Environ.* **18**, 2641-2647.

Liljestrand, H. M., and J. J. Morgan (1980). Spatial variations of acid precipitation in Southern California. *Environ. Sci. Technol.* **15**, 333-339.

Lin, C. L., and W. B. DeMore (1973). O(^{1}D) production in ozone photolysis near 3100 Å. *J. Photochem.* **2**, 161-164.

Lingenfelter, R. E. (1963). Production of carbon-14 by cosmic ray neutrons. *Rev. Geophys.* **1**, 35-55.

Lingenfelter, R. E., and R. Ramaty (1970). Astrophysical and geographic variations in C-14 production. *In* "Radiocarbon Variations and absolute Chronology" (I. U. Olsson, ed.), pp. 513-537. Almquist and Wiksell, Stockholm; Wiley (Interscience), New York.

Linnenbom, V. J., J. W. Swinnerton, and R. A. Lamontagne (1973). The ocean as a source for atmospheric carbon monoxide. *J. Geophys. Res.* **78**, 5333-5340.

Linton, R. W., A Loh, and D. F. S. Natusch (1976). Surface predominance of trace elements in airborne particles. *Science* **191**, 852-854.

Lipari, F., J. M. Dasch, and W. F. Scruggs (1984). 2,4-Dinitrophenylhydrazin-coated florisil sampling cartridges for the determination of formaldehyde in air. *Environ. Sci. Technol.* **19**, 70-74.

Lipschultz, F., O. C. Zafariou, S. C. Wofsy, M. B. McElroy, F. W. Valois, and S. W. Watson (1981). Production of NO and N_2O by soil nitrifying bacteria. *Nature* **294**, 641-643.

Liss, P. S. (1973). Processes of gas exchange across an air-water interface. *Deep Sea Res.* **20**, 221-238.

Liss, P. S., and P. G. Slater (1974). Flux of gases across the air-sea interface. *Nature* **247**, 181-194.

List, R. J., and K. Telegadas (1969). Using radioactive tracers to develop a model of the circulation of the stratosphere. *J. Atmos. Sci.* **26**, 1128-1136.

List, R. (1977). The formation of rain. *Trans. R. Soc. Canada* **15**, 333-337.

Little, D. J., A. Dalgleish, and R. J. Donovan (1972). Relative rate data for the reactions of S(3^1D_2) using the NS radical as a spectroscopic marker. *Faraday Discuss. Chem. Soc.* **53**, 211-216.

Little, P., and R. D. Wiffen (1977). Emission and deposition of petrol engine exhaust Pb-I: Deposition of exhaust Pb to plant on soil surfaces. *Atmos. Environ.* **11**, 437-467.

Liu, S. C., and T. M. Donahue (1974). Realistic model of hydrogen constituents in the lower atmosphere and escape flux from the upper atmosphere. *J. Atmos. Sci.* **31**, 2238-2242.

Liu, B. Y. H., and C. S. Kim (1977). On the counting efficiency of condensation nuclei counters. *Atmos. Environ.* **11**, 1097-1110.

Liu, S. C., M. McFarland, D. Kley, O. Zafiriou, and B. J. Huebert (1983). Tropospheric NO and O_3 budgets in the equatorial Pacific. *J. Geophys. Res.* **88**, 1360-1368.

Livingstone, D. A. (1963). Chemical composition of rivers and lakes. *In* "Data of Geochemistry" (M. Fleischer, ed.), 6th Ed., pp. 440-446. U.S. Geological Survey Professional Paper.

Lloyd, A. C., K. R. Darnall, A. M. Winer, and J. N. Pitts, Jr. (1976). Relative rate constants for the reaction of the hydroxyl radical with a series of alkanes, alkenes and aromatic hydrocarbons. *J. Phys. Chem.* **80**, 789-794.

Lodge, J. P., P. A. Machado, J. B. Pate, D. C. Sheesley, and F. A. Wartburg (1974). Atmospheric trace chemistry in the American humid tropics. *Tellus* **26**, 250-253.

Loehr, R. C. and S. A. Hart (1970). Changing practices in agriculture and their effects on the environment. *Crit. Rev. Environ. Control* **1**, 87-92.

Loewenstein, M., W. J. Borucki, H. F. Savage, J. G. Borucki, and R. C. Whitten (1978a). Geographical variations of NO and O_3 in the lower stratosphere. *J. Geophys. Res.* **83**, 1874-1882.

Loewenstein, M., W. J. Starr, and D. G. Murcray (1978b). Stratospheric NO and HNO_3 observations in the northern hemisphere for three seasons. *Geophys. Res. Lett.* **5**, 531-534.

Logan, J. A., M. J. Prather, S. C. Wofsy, and M. B. McElroy (1978). Atmospheric chemistry: response to human influence. *Philos. Trans. R. Soc.* **290**, 187-192.

Logan, J. A., M. J. Prather, S. C. Wofsy, and M. B. McElroy (1981). Tropospheric chemistry: a global perspective. *J. Geophys. Res.* **86**, 7210-7254.

Logan, J. A. (1983). Nitrogen oxides in the troposphere: global and regional budgets. *J. Geophys. Res.* **88**, 10785-10807.

London, J., and S. J. Oltmans (1979). The global distribution of long-term total ozone variations during the period 1957-1975. *Pure Appl. Geophys.* **117**, 346-354.

Lonneman, W. A., S. L. Kopezynski, P. E. Darley, and F. D. Sutterfield (1974). Hydrocarbon composition of urban air pollution. *Environ. Sci. Technol.* **8**, 229-236.

Lonneman, W. A., J. J. Bufalini, and R. L. Seila (1976). PAN and oxidant measurements in ambient atmosphere. *Environ. Sci. Technol.* **10**, 374-380.

Lonneman, W. A., R. L. Seila, and J. J. Bufalini (1978). Ambient air hydrocarbon concentrations in Florida. *Environ. Sci. Technol.* **12**, 459-463.

Louis, J. F. (1975). Mean meridional circulation. *In* "The Natural Stratosphere of 1974" (A. J. Grobecker, ed.), pp. 6-23-6-31. CIAP Monograph 1. U.S. Dept. of Transportation, Washington, D.C.

Louw, C. W., J. F. Richards, and P. K. Faure (1977). The determination of volatile organic compounds in city air by gas chromatography combined with standard addition, selective substraction, infrared spectrometry and mass spectrometry. *Atmos. Environ.* **11**, 703-717.

Lovas, F. J., and R. D. Suenram (1977). Identification of Dioxirane H_2COO in ozone-olefin reactions via microwave spectroscopy. *Chem. Phys. Lett.* **51**, 453-456.

Lovelock, J. E. (1961). Ionisation methods for the analysis of gases and vapors. *Anal. Chem.* **33**, 162-182.

Lovelock, J. E. (1971). Atmospheric fluorine compounds as indicators of air movement. *Nature* **230**, 379.

Lovelock, J. E. (1972). Atmospheric turbidity and CCl_3F concentrations in rural southern England and southern Ireland. *Atmos. Environ.* **6**, 917-925.

Lovelock, J. E., R. J. Maggs, and R. A. Rasmussen (1972). Atmospheric dimethyl sulfide and the natural sulphur cycle. *Nature* **237**, 452-453.

Lovelock, J. E., R. J. Maggs, and R. J. Wade (1973). Halogenated hydrocarbons in and over the atlantic. *Nature* **242**, 194-196.

Lovelock, J. E. (1974a). Atmospheric halocarbons and stratospheric ozone. *Nature* **252**, 292-294.

Lovelock, J. E. (1974b). CS_2 and the natural sulfur cycle. *Nature* **248**, 625-626.

Lovelock, J. E., and L. Margulis (1974). Atmospheric homostasis by and for the biosphere: The Gaia hypothesis. *Tellus* **26**, 2-12.

Lovelock, J. E. (1975). Natural halocarbons in the air and in the sea. *Nature* **256**, 193-194.

Lovelock, J. E. (1979). "Gaia. A New Look at Life on Earth." Oxford Univ. Press, London.

Lovett, R. F. (1978). Quantitative measurement of airborne sea-salt in the North Atlantic. *Tellus* **30**, 358-364.

Lowe, D. R. (1980). Stromatolites: 3400 Myr old from the archean of Western Australia. *Nature* **284**, 441-443.

Lowe, D. C., U. Schmidt, and D. H. Ehalt (1980). A new technique for measuring tropospheric formaldehyde (CH_2O). *Geophys. Res. Lett.* **7**, 825-828.

Lowe, D. C., and U. Schmidt (1983). Formaldehyde (HCHO) measurements in the nonurban atmosphere. *J. Geophys. Res.* **88**, 10844-10858.

Ludwig, J. H., G. B. Morgan, and T. B. McMullen (1970). Trends in urban air quality. *EOS Trans. Am. Geophys. Union* **51**, 468-475.

Lüttge, U., and K. Fischer (1980). Light dependent net carbon monoxide evolution by C_3 and C_4 plants. *Planta* **149**, 59-63.

Lundgren, D. A. (1970). Atmospheric aerosol composition and concentration as a function of particle size and of time. *J. Air. Pollut. Control. Assoc.* **20**, 603-608.

Luther, F. M. (1975). Large-scale eddy diffusion *In* "The Natural Stratosphere of 1974" (C. J. Grobecker, ed.), pp. 6-31-6-38. CIAP Monograph 1. U.S. Dept. of Transportation, Washington D.C.

Luther, F. M., and R. J. Gelinas (1976). Effect of molecular multiple scattering and surface albedo on atmospheric photodissociation rates. *J. Geophys. Res.* **81**, 1125-1132.

Lyles, L. (1977). Wind erosion: processes and effect on soil productivity. *Trans. ASAE* **20**, 880-884.

Maahs, H. G. (1983). Kinetics and mechanism of the oxidation of S(IV) by ozone in aqueous solution with particular reference to SO_2 conversion in non-urban tropospheric clouds. *J. Geophys. Res.* **88**, 10721-12732.

MacDonald, G. A. (1955). Hawaiian volcanoes during 1952, *U. S. Geol. Surv. Bull.* **1021-B**, 15-108.

MacDonald, G. J. F. (1963). The escape of helium from the earth's atmosphere. *Rev. Geophys.* **1**, 305-349.

MacDonald, G. J. F. (1964). The escape of helium from the earth's atmosphere. *In* "The Origin and Evolution of Atmospheres and Oceans" (P. J. Brancazio and A. G. W. Cameron, eds.), pp. 127-182. Wiley, New York.

MacDonald, G. A. (1972). Volcanoes, Prentice-Hall, Englewood Cliffs, New Jersey.

MacIntyre, F. (1968). Bubbles: a boundary-layer "Microtome" for micron thick samples of liquid surface. *J. Phys. Chem.* **72**, 589-592.

MacIntyre, F. (1972). Flow patterns in breaking bubbles. *J. Geophys. Res.* **77**, 5211-5228.

MacKenzie, F. T., and R. M. Garrels (1966). Chemical mass balance between rivers and oceans. *Am. J. Sci.* **264**, 507-525.

Machta, L., and G. Hughes (1970). Atmospheric Oxygen in 1967 to 1970. *Science* **168**, 1582-1584.

Machta, L. (1971). The role of the oceans and biosphere in the carbon cycle, *In* "The Changing Chemistry of the Oceans" (F. Dryssen and D. Jagner, eds.), Nobelsymposium 20, pp. 121–145. Almqvist and Wikrell, Stockholm.

Machta, L. (1974). Global scale atmospheric mixing. *Adv. Geophys.* **18 B**, 33–56.

Madronich, S., D. R. Hastie, B. A. Ridley, and H. I. Schiff (1983). Measurement of the photodissociation coefficient of NO_2 in the Atmosphere: I. Method and surface measurements. *J. Atmos. Chem.* **1**, 3–25.

Maenhaut, W., and W. H. Zoller (1977). Determination of the chemical composition of the south pole aerosol by instrumental neutron activation analysis. *J. Radioanal. Chem.* **37**, 637–650.

Maenhaut, W., W. H. Zoller, R. A. Duce, and G. L. Hoffman (1979). Concentration and size distribution of particulate trace elements in the south polar atmosphere. *J. Geophys. Res.* **84**, 2421–2431.

Maenhaut, W., M. Darzi, and J. W. Winchester (1981). Seawater and non-seawater aerosol compounds in the marine atmosphere of Samoa. *J. Geophys. Res.* **86**, 3187–3193.

Maenhaut, W., H. Raemdonck, A. Selen, R. Van Grieken, and J. W. Winchester (1983). Characterization of the atmospheric aerosol over the eastern equatorial Pacific. *J. Geophys. Res.* **88**, 5353–5364.

Mahlman, J. D., and W. J. Moxim (1978). Trace simulation using a global general circulation model: results from a midlatitude instantaneous source experiment. *J. Atmos. Sci.* **35**, 1340–1374.

Maigné, J. P., P. Y. Turpin, G. Madelaine, and J. Bricard (1974). Nouvelle methode de determination de la granulometrie d'un aerosol au moyen d'une batterie de diffusion. *J. Aerosol Sci.* **5**, 339–355.

Makarov, B. N. (1969). Liberation of nitrogen dioxide from soil. *Sov. Soil Sci.* **1**, 20–25.

Malinconico, L. L. (1979). Fluctuations in SO_2 emission during recent eruptions of Etna. *Nature* **278**, 43–45.

Manabe, S., and R. T. Weatherald (1967). Thermal equilibrium of the atmosphere with a given distribution of relative humidity. *J. Atmos. Sci.* **24**, 241–259.

Manabe, S., and R. T. Weatherald (1975). The effects of doubling the CO_2 concentration on the climate of a general circulation model. *J. Atmos. Sci.* **32**, 3–17.

Manabe, S., and R. T. Weatherald (1980). On the distribution of climate change resulting from an increase in the CO_2 content of the atmosphere. *J. Atmos. Sci.* **37**, 99–118.

Mankin, W. G., and M. T. Coffey (1983). Latitudinal distributions and temporal changes of stratospheric HCl and HF. *J. Geophys. Res.* **88**, 10776–10784.

Mankin, W. G., M. T. Coffey, D. W. T. Griffith, and S. R. Drayson (1979). Spectroscopic measurement of carbonyl sulfide in the stratosphere. *Geophys. Res. Lett.* **6**, 853–856.

Mann, C. O. (1981). EPA data systems used to report anthropogenic hydrocarbon emission data. *In* "Atmospheric Biogenic Hydrocarbons" (J. J. Bufalini and R. R. Arnts, eds.), Vol. 1, "Emissions." pp. 149–178. Ann Arbor Science Publ., Ann Arbor, Michigan.

Marenco, A., and J. Fontan (1972). Etude par simulation sure modeles numerique du temps de sejour des areosols dans la troposphere. *Tellus* **24**, 428–441.

Maroulis, P. J., and A. R. Bandy (1977). Estimate of the contribution of biologically produced dimethyl sulfide to the global sulfur cycle. *Science* **196**, 647–648.

Maroulis, P. J., A. L. Torres, and A. R. Bandy (1977). Atmospheric concentrations of carbonyl sulfide in the Southwestern and Eastern United States. *Geophys. Res. Lett.* **4**, 510–512.

Maroulis, P. J., A. L. Torres, A. B. Goldberg, and A. R. Bandy (1980). Atmospheric SO_2 measurements on Project Gametag. *J. Geophys. Res.* **85**, 7345–7349.

Maroulis, R. J., and A. R. Bandy (1980). Measurements of atmospheric concentrations of CS_2 in the eastern United States. *Geophys. Res. Lett.* **7**, 681–684.

Marple, V. A., and K. Willeke (1976). Inertial impactors: Theory, design and use. *In* "Fine Particles, Aerosol Generation, Measurement, Sampling and Analysis" (B. Y. H. Liu, ed.), pp. 411–446. Academic Press, New York.

Marshall, V. G., and D. S. Debell (1980). Comparison of four methods of measuring volatilization losses of nitrogen following urea fertilization of forest soils. *Can. J. Soil Sci.* **60**, 549–563.

Martell, E. A., and H. E. Moore (1974). Tropospheric aerosol residence times: a critical review. *J. Rech. Atmos.* **8**, 903–910.

Martens, C. S., J. J. Wesolowski, R. C. Harris, and R. Kaifer (1973). Chlorine loss from Puerto Rican and San Francisco Bay area marine aerosols. *J. Geophys. Res.* **78**, 8778–8792.

Martin, A., and F. R. Barber (1981). Sulfur dioxide, oxides of nitrogen and ozone measured continuously for two years at a rural site. *Atmos. Environ.* **15**, 567–578.

Martin, L. R., and D. E. Damschen (1981). Aqueous oxidation of sulfur dioxide by hydrogen peroxide at low pH. *Atmos. Environ.* **15**, 1617–1622.

Martin, L. R. (1984). Kinetic studies of sulfite oxidation in aqueous solution, in : J. G. Calvert (ed.) "SO$_2$, NO and NO$_2$ Oxidation Mechanisms: "Atmospheric Considerations" Chap. 2, pp. 63–100, Acid Precipitation Series, Vol. 3. Butterworth, Boston.

Martinez, R. I., R. E. Huie, and J. T. Herron (1977). Mass spectrometric detection of dioxirane, H$_2$COO and its decomposition products, H$_2$ and CO, from the reaction of ozone with ethylene. *Chem. Phys. Lett.* **51**, 457–459.

Martinez, R. I., and J. T. Herron (1978). Stopped-flow study of the gas-phase reaction of ozone with organic sulfides: Dimethylsulfide. *Int. J. Chem. Kinet.* **10**, 433–452.

Marty, J. C., and A. Saliot (1982). Aerosols in equatorial Atlantic air: *n*-alkanes as a function of particle size. *Nature* **298**, 144–147.

Marty, J. C., M. J. Tissier, and A. Saliot (1984). Gaseous and particulate polycyclic aromatic hydrocarbons (PAH) from the marine atmosphere. *J. Atmos. Environ.* **18**, 2183–2190.

Marx, W., P. B. Monkhouse, and U. Schurath (1984). "Kinetik und Intensität photochemischer Reaktionsschritte in der Atmosphäre." Gesellschaft für Strahlen- und Umweltforschung, Munich.

Mason, B. (1966). "Principles of Geochemistry," 3rd ed. Wiley, New York.

Mason, B. J. (1971). "The Physics of Clouds," 2nd ed., Oxford Univ. Press (Clarendon), London and New York.

Matsumoto, G., and T. Hanya (1980). Organic constituents in atmospheric fallout in the Tokyo area. *Atmos. Environ.* **14**, 1409–1419.

Matsuo, S. (1978). The oxidation state of the primodial atmosphere. *In* "Origin of Life" (H. Noda, ed.), pp. 21–27. Center Acad. Publ. Japan, Tokyo.

Mauersberger, K., R. Finstad, S. Anderson, and P. Robbins (1981). A comparison of ozone measurements. *Geophys. Res. Lett.* **8**, 361–364.

Maynard, J. B., and W. N. Sanders (1969). Determination of the detailed hydrocarbon composition and potential atmosphere reactivity of full-range motor gasolines. *J. Air. Pollut. Control Assoc.* **19**, 505–510.

Mayrsohn, H., and J. H. Crabtree (1976). Source reconciliation of atmospheric hydrocarbons. *Atmos. Environ.* **10**, 137–143.

Mayer, E. W., D. R. Blake, S. C. Tyler, Y. Makide, D. C. Montagne, and F. S. Rowland (1982). Methane: Interhemispheric concentration gradient and atmospheric residence time. *Proc. Natl. Acad. Sci.* **79**, 1366–1370.

McAlester, A. L. (1968). The history of life, Prentice Hall, Englewood Cliffs, New Jersey.

McAuliffe, C. (1966). Solubility in water of paraffin, cycloparaffin, olefin, acetylene, cycloolefin and aromatic hydrocarbons. *J. Phys. Chem.* **70**, 1267–1275.

McArdle, J. V., and M. R. Hoffmann (1983). Kinetics and mechanism of the oxidation of aquated sulfur dioxide by hydrogen peroxide at low pH. *J. Phys. Chem.* **87**, 5627–5429.

McCarthy, R. L., F. A. Bower, and J. P. Jesson (1977). The fluorocarbon–ozone theory. I: Production and release. World production and release of CCl_3F and CCl_2F_2 through 1975. *Atmos. Environ.* **11**, 491–497.

McConnell, J. C., M. B. McElroy, and S. C. Wofsy (1971). Natural sources of atmospheric CO. *Nature* **233**, 187–188.

McCormak, J. D., and R. K. Hilliard (1970). Scavenging of aerosol particles by sprays. *In* "Precipitation Scavenging 1970," (R. J. Englemann and W. G. N. Sninn, coordinators), pp. 187–204. U.S. Atomic Energy Commission, Division of Tech. Information **NTS/CONF 700601**.

McElroy, M. B., and McConnell (1971). Nitrous oxide: a natural source of stratospheric NO. *J. Atmos. Sci.* **28**, 1095–1098.

McElroy, M. B. (1975). Testimony to U.S. House of Representatives Committee on Science and Technology. Subcommittee on the environment and the Atmosphere, U.S. Government Printing Office, Washington D.C., transcript No. 57-788, pp. 196–199; and Testimony U.S. Senate Committee on Aeronautical and Space Sciences, Subcommittee on the Upper Atmosphere, U.S. Government Printing Office, Washington D.C., transcript No. 60-183, pp. 1025–1049.

McElroy, M. B. (1976). Chemical processes in the solar system: a kinetic perspective. *In* "MTP International Review of Science," (D. Herschbach, ed.) Butterworth, London.

McElroy, M. B., J. W. Elkins, S. C. Wofsy, and Y. L. Yung (1976). Sources and sinks of atmospheric N_2O. *Rev. Geophys. Space Phys.* **14**, 143–150.

McElroy, M. B., S. C. Wofsy, and Y. L. Yung (1977a). The nitrogen cycle: perturbations due to man and their impact on atmospheric N_2O and O_3. *Philos. Trans. R. Soc., London* **277**, 159–181.

McElroy, M. B., T. Y. Kong, and Y. L. Yung (1977b). Photochemistry and evolution of Mars' atmosphere. A Viking perspective. *J. Geophys. Res.* **82**, 4379–4388.

McElroy, M. B., S. C. Wofsy, and N. D. Sze (1980). Photochemical sources for atmospheric H_2S. *Atmos. Environ.* **14**, 159–163.

McEwen, D. J. (1966). Automobile exhaust hydrocarbon analysis by gas chromatography. *Anal. Chem.* **38**, 1047–1053.

McFarland, M., D. Kley, J. W. Drummond, A. L. Schmeltekopf, and R. J. Winkler (1979). Nitric oxide measurements in the equatorial Pacific region. *Geophys. Res. Lett.* **6**, 605–608.

McGrath, W. D., and R. G. W. Norrish (1960). Studies of the reactions of excited atoms and molecules produced in the flash photolysis of ozone. *Proc. R. Soc., London* **254**, 317–326.

McKay, C., M. Pandow, and R. Wolfgang (1963). On the chemistry of the natural radiocarbon. *J. Geophys. Res.* **68**, 3929–3931.

McKenney, D. J., D. L. Wade, and W. I. Findlay (1978). Rates of N_2O evolution from N-fertilized soil. *Geophys. Res. Lett.* **5**, 777–780.

McKinnon, I. R., J. G. Mathiesen, and I. R. Wilson (1979). Gas-phase reaction of nitric oxide with nitric acid. *J. Phys. Chem.* **83**, 779–780.

McKirdy, D. M. and J. Hahn (1982). Composition of kerogen and hydrocarbons in precambrian rocks. *In* "Mineral Deposits and the Evolution of the Biosphere" (H. D. Holland and M. Schidlowski, eds.), pp. 123–154. Springer-Verlag, Berlin and New York.

McMahon, C. K., and P. W. Ryan (1976). Some chemical and physical characteristics of emissions from forest fires, Paper No. 76-2.3 presented at *Annu. Meet. Air Pollut. Control Assoc., 69th, Portland, Oregon, 1976.*

McMahon, T. A., and P. J. Denison (1979). Empirical atmospheric deposition parameters—a survey. *Atmos. Environ.* **13**, 571–585.

McMullen, T. B., R. B. Faoro, and G. B. Morgan (1970). Profile of pollutant fractions in non urban suspended particulate matter. *J. Air Pollut. Control Assoc.* **20**, 369-376.

McMurry, P. H., and J. C. Wilson (1983). Droplet phase (heterogeneous) and gas phase (homogeneous) contributions to secondary ambient aerosol formations as function of relative humidity. *J. Geophys. Res.* **88**, 5101-5108.

Meagher, J. F., E. M. Bailey, and M. Luria (1983). The seasonal variation of the atmospheric SO_2 to SO_4 conversion rate. *J. Geophys. Res.* **88**, 1525-1527.

Meetham, A. R., D. W. Bottom, and S. Cayton (1964). "Atmospheric Pollution: Its Origin and Prevention," 3rd Ed., Pergamon Press, Oxford.

Mehlmann, A. (1986). Grössenverteilung des Aerosolnitrats und seine Beziehung zur gasförmigen Salpetersäure, Ph.D. Thesis, Univ. Mainz.

Meinel, A. B. (1950). OH emission bands in the spectrum of the night sky. *Astrophys. J.* **111**, 555-564.

Meixner, F. X. (1984). The vertical sulfur dioxide distribution at the tropopause level. *J. Atmos. Chem.* **2**, 175-189.

Meixner, F. X., K. P. Müller, G. Aheimer, and K. D. Höfken (1985). Measurements of gaseous nitric acid and particulate nitrate. *In* "Physico-chemical Behaviour of Atmospheric Pollutants", (F. A. A. M. De Leeuw and N. D. Van Egmond, eds.) COST Action 611, *Proc. Workshop Pollut. Cycles Transport-Modelling Field Experiments, Bilthoven, The Netherlands*, pp. 103-114.

Menard, H. W., and S. M. Smith (1966). Hypsometry of ocean basin provinces. *J. Geophys. Res.* **71**, 4305-4325.

Mentall, J. E., B. Guenther, and D. Williams (1985). The solar spectral irradiance between 150-200 nm. *J. Geophys. Res.* **90**, 2265-2271.

Menzel, D. W. (1974). Primary productivity, dissolved and particulate organic matter, and the sites of oxidation of organic matter. *In* "The Sea" (E. D. Goldberg, ed.), Vol. 5, Chap. 18, pp. 659-678. Wiley (Interscience), New York.

Menzies, R. T. (1983). A re-evaluation of laser heterodyne radiometer ClO measurements. *Geophys. Res. Lett.* **10**, 729-732.

Mészáros, A., and K. Vissy (1974). Concentration, size distribution and chemical nature of atmospheric aerosol particles in remote ocean areas. *J. Aerosol Sci.* **5**, 101-110.

Mészáros, E., D. J. Moore, and J. P. Lodge, Jr. (1977). Sulfur dioxide–sulfate relationships in Budapest. *Atmos. Environ.* **11**, 345-349.

Mészáros, E., G. Varhelyi, and L. Haszpra (1978). On the atmospheric sulfur budget over Europe. *Atmos. Environ.* **12**, 2273-2277.

Mészáros, E. (1982). On the atmospheric input of sulfur into the ocean. *Tellus* **34**, 277-282.

Mészáros, E., and L. Horvath (1984). Concentration and dry deposition of atmospheric sulfur and nitrogen compounds in Hungary. *Atmos. Environ.* **18**, 1725-1730.

Meteorological Service of Canada (1961-1969). Ozone data for the world, 1960-1968, Dept. of Transportation, Meteorol. Branch, Toronto.

Meybeck, M. (1981). River transport of organic carbon to the ocean. *In* "Carbon Dioxide Effects Research and Assessment Program, Flux of Organic Carbon by Rivers to the Ocean" (G. E. Likens, F. T. MacKenzie, J. E. Richey, J. R. Sedell, K. K. Turekian, and V. I. Pye, eds.), pp. 219-269. U.S. Dept. of Energy, Washington D.C., **CONF-8009140-UCII.**

Meyers, P. A., and R. A. Hites (1982). Extractable organic compounds in midwest rain and snow. *Atmos. Environ.* **16**, 2169-2175.

Meyrahn, H., G. K. Moortgart, and P. Warneck (1982). The photolysis of acetaldehyde under atmospheric conditions. *In* "Atmospheric Trace Constituents" (F. Herbert, ed.), pp. 65-72. Vieweg and Sohn, Braunschweig.

Meyrahn, H., J. Hahn, G. Helas, P. Warneck, and S. A. Penkett (1984). Cryogenic sampling and analysis of peroxyacetyl nitrate in the atmosphere. *In* "Physico-chemical Behaviour of Atmospheric Pollutants" (B. Versino and G. Angeletti, eds.), pp. 39–43. Reidel, Dordrecht, The Netherlands.

Meyrahn, H., J. Hahn, G. Helas, P. Warneck, and S. A. Penkett (1986). Unpublished data.

Meyrahn, H., J. Pauly, W. Schneider, and P. Warneck (1986). Quantum yields for the photodissociation of acetone in air and an estimate for the lifetime of acetone in the lower troposphere. *J. Atmos. Chem.* **4**, 277–291.

Middleton, P., and C. S. Kiang (1978). A kinetic aerosol model for the formation and growth of secondary sulfuric and acid particles. *J. Aerosol Sci.* **9**, 359–385.

Migeotte, M. V. (1948). Spectroscopic evidence of methane in the earth's atmosphere. *Phys. Rev.* **73**, 519–520.

Migeotte, M. V. (1949). The fundamental band of carbon monoxide at 4.7 microns in the solar spectrum. *Phys. Rev.* **75**, 1108–1109.

Migeotte, M. V., and R. M. Chapman (1949). On the question of atmospheric ammonia. *Phys. Rev.* **75**, 1611.

Mihelcic, D., P. Müsgen, and D. H. Ehhalt (1986). Tropospheric measurements of NO_2 and RO_2 by matrix isolation and electron spin resonance. *J. Atmos. Chem.* (in press).

Milford, J. B., and C. I. Davidson (1985). The size of particulate trace elements in the atmosphere—a review. *J. Air. Pollut. Control Assoc.* **35**, 1249–1260.

Mill, T., F. Mayo, H. Richardson, K. Irvin, and D. A. Allara (1972). Gas- and Liquid-Phase Oxidations of *n*-Butane. *J. Am. Chem. Soc.* **94**, 6802–6811.

Mill, T., and G. Montorsi (1973). The liquid phase oxidation of 2,4-dimethylpentane. *Int. J. Chem. Kinet.* **5**, 119–136.

Miller, S. L., and H. L. Urey (1959). Organic compound synthesis on the primitive earth. *Science* **130**, 245–251.

Miller, S. L., and L. E. Orgel (1974). "The Origin of Life on Earth." Prentice Hall, Englewood Cliffs, New Jersey.

Miller, D. F., and C. W. Spicer (1975). Measurement of nitric acid in smog. *J. Air. Pollut. Control Assoc.* **9**, 940–942.

Miller, D. F., and A. J. Alkezweeny (1980). Aerosol formation in urban plumes over Lake Michigan. *Ann. N.Y. Acad. Sci.* **338**, 219–232.

Miller, C., D. L. Filkin, A. J. Owens, J. M. Steed, and J. P. Jesson (1981). A two-dimensional model of stratospheric chemistry and transport. *J. Geophys. Res.* **86**, 12039–12065.

Millikan, R. A. (1923). The general law of fall of a small spherical body through a gas, and its bearing upon the nature of molecular reflections from surfaces. *Phys. Rev.* **22**, 1–23.

Miner, S. (1969). Preliminary air pollution survey of ammonia, a literature review. *Natl. Air Pollut. Control Admin. Publ.* **APTD 69-25**, Raleigh, N.C.: U.S. Dept of Health Education and Welfare.

Mirabel, P., and J. L. Katz (1974). Binary homogeneous nucleation as a mechanism for the formation of aerosols. *J. Chem. Phys.* **60**, 1138–1144.

Miyake, Y., and S. Tsunogai (1963). Evaporation of iodine from the ocean. *J. Geophys. Res.* **68**, 3989–3994.

Möller, F. (1950). Eine Berechnung des horizontalen Grossaustausches über dem Atlantischen Ozean. *Arch. Meterol. Geophys. Bioklimatol. Ser. A* **2**, 73–81.

Möller, F. (1963). On the influence of changes in the CO_2 concentration in air on the radiation balance of the earth's surface and on climate. *J. Geophys. Res.* **68**, 3877–3886.

Möller, U., and G. Schumann (1970). Mechanisms of transport from the atmosphere to the earth surface. *J. Geophys. Res.* **75**, 3013–3019.

Möller, D. (1984). On the global natural sulfur emission. *Atmos. Environ.* **18**, 29–39.

Molina, M. J., and F. S. Rowland (1974). Stratospheric sink for chlorofluoromethanes: chlorine atom catalyzed destruction of ozone. *Nature* **249**, 810-812.

Mohnen, V. A. (1977). The issue of stratospheric ozone intrusion. Atmospheric Sciences Research Center, Publ. No. 428. New York State University, Albany, New York.

Monahan, E. C. (1968). Sea spray as a function of low elevation wind speed. *J. Geophys. Res.* **73**, 1127-1137.

Monohan, E. C. (1971). Oceanic Whitecaps. *J. Phys. Oceanogr.* **1**, 139-144.

Monster, J., P. W. U. Appel, H. G. Thode, M. Schidlowski, C. M. Carmichael, and D. Bridgwater (1979). Sulfur isotope studies in early archaean sediments from Isua, West Greenland: implications for the antiquity of bacterial sulfate reduction. *Geochim. Cosmochim. Acta* **43**, 405-413.

Moorbath, C., R. K. O'Nions, and R. J. Pankhurst (1973). Early archaean age for the Isua iron formation West Greenland. *Nature* **245**, 138-139.

Moore, B., R. D. Boone, J. E. Hobbie, R. A. Houghton, J. M. Melillo, B. J. Peterson, G. R. Shaner, C. J. Vörösmarty, and G. M. Woodwell (1981). A simple model for analysis of the role of terrestrial ecosystem in the global carbon budget. *SCOPE* **16**, 365-385.

Moore, H. E., S. E. Poet, and E. A. Martell (1973). ^{222}R, ^{210}Pb, ^{210}Bi, and ^{210}Po profiles and aerosol residence times versus altitude. *J. Geophys. Res.* **78**, 7065-7075.

Moortgat, G. K., and C. E. Junge (1977). The role of SO_2 oxidation for the background stratospheric sulfate layer in the light of new reaction rate data. *Pure Appl. Geophys.* **115**, 759-774.

Moortgat, G. K., E. Kudszus, and P. Warneck (1977). Temperature dependence of $O(^1D)$ formation in the near UV photolysis of ozone. *J. Chem. Soc., Faraday Trans. 2* **73**, 1216-1221.

Moortgat, G. K., and E. Kudszus (1978). Mathematical expression for the $O(^1D)$ quantum yields from the O_3 photolysis as a function of temperature (230-320 K) and wavelength (295-320 nm). *Geophys. Res. Lett.* **5**, 191-194.

Moortgat, G. K., and P. Warneck (1979). CO and H_2 quantum yield in the photodecomposition of formaldehyde in air. *J. Chem. Phys.* **70**, 3639-3651.

Moortgat, G. K., W. Seiler, and P. Warneck (1983). Photodissociation of HCHO in air: CO and H_2 quantum yields at 220 and 300 K. *J. Chem. Phys.* **78**, 1185-1190.

Mopper, K., and E. Degens (1979). Organic carbon in the ocean, nature and cycling. *SCOPE* **13**, 293-316.

Morgan, J. J. (1982). Factors governing the pH, availability of H^+, and oxidation capacity of rain. *In* "Atmospheric Chemistry" (E. D. Goldberg, ed.), pp. 17-40. Dahlem Konferenzen, Springer-Verlag, Berlin and New York.

Morris, E. D., and H. Niki (1971). Reactivity of hydroxyl radicals with olefins. *J. Phys. Chem.* **75**, 3640-3641.

Morris, E. D., and H. Niki (1973). Reaction of dinitrogen pentoxide with water. *J. Phys. Chem.* **77**, 1929-1932.

Morris, E. D., and H. Niki (1974). Reaction of the nitrate radical with acetaldehyde and propylene. *J. Phys. Chem.* **78**, 1337-1338.

Mosier, A. R., and G. L. Hutchinson (1981). Nitrous oxide emissions from cropped fields. *J. Environ. Qual.* **10**, 169-173.

Mount, G. H., and G. J. Rottman (1981). Solar intensities: The solar spectral irradiance 1200-3184 Å near solar maximum: 15. July 1980. *J. Geophys. Res.* **86**, 9193-9198.

Mount, G. H. and G. J. Rottman (1983). The solar absolute spectral irradiance at 1150-3173 Å: May 17, 1982. *J. Geophys. Res.* **88**, 5403-5410.

Mount, G. H., and G. J. Rottman (1985). Solar absolute spectral irradiance 118-300 nm: July 25, 1983. *J. Geophys. Res.* **90**, 13031-13036.

Moyers, J. L., L. E. Rauweiler, S. B. Hopf, and N. E. Korte (1977). Evaluation of particulate trace species in southwest desert atmosphere. *Environ. Sci. Technol.* **11**, 789-795.

Mrose, H. (1966). Measurements of pH and chemical analysis of rain, snow and fog water. *Tellus* **18**, 266-270.

Mroz, E., and W. H. Zoller (1975). Composition of atmospheric particulate matter from the eruption of Heimaey, Iceland. *Science* **190**, 461-464.

Mroz, E. J., A. L. Lazrus, and J. C. Bonnelli (1977). Direct measurements of stratospheric fluoride. *Geophys. Res. Lett.* **4**, 149-150.

Mueller, T. K., G. M. Hidy, K. Warren, T. F. Lavery, and R. L. Baskett (1980). The occurrence of atmospheric aerosols in the north-eastern United States. *Ann. N.Y. Acad. Sci.* **338**, 463-482.

Münnich, K. O. (1963). Der Kreislauf des Radiokohlenstoffs in der Natur. *Naturwissenschaften* **50**, 211-218.

Münnich, K. O., and W. Roether (1967). Transfer of bomb C-14 and tritium from the atmosphere to the ocean. Internal mixing of the ocean on the basis of tritium and C-14 profiles, *In* "Radioactive Dating and Methods of Low Level Counting", *Symp. Proc., Int. Atom. Agency, Monaco*, pp. 93-104. Vienna (**STI/PUB/152**).

Munger, J. W., D. J. Jacob, J. M. Waldman, and M. R. Hoffmann (1983). Fog water chemistry in an urban atmosphere. *J. Geophys. Res.* **88**, 5109-5121.

Munger, J. W., D. J. Jacob, and M. R. Hoffmann (1984). The occurrence of bisulfite aldehyde addition products in fog and cloud water. *J. Atmos. Chem.* **1**, 335-350.

Munk, W. H. (1966). Abyssal recipes. *Deep Sea Res.* **13**, 707-730.

Murcray, D. G., A. Goldman, W. J. Williams, F. H. Murcray, J. N. Brooks, J. van Allen, R. N. Stocker, J. J. Kosters, D. B. Barker, and D. E. Snider (1974). Recent results of stratospheric trace gas measurements from balloon-borne spectrometers. *Proc. CIAP Conf. 3rd.* (A. J. Broderick and F. M. Hard, eds.), pp. 184-192. **DOT-TSC-OSF-74-15** Department of Transportation, Washington D.C.

Murcray, D. G., A. Goldman, C. M. Bradford, G. R. Cook, J. W. van Allen, F. S. Bonomo, and F. H. Murcray (1978). Identification of the v_2 vibration-rotation band of ammonia in ground level solar spectra. *Geophys. Res. Lett.* **5**, 527-530.

Murcray, D. G., A. Goldman, F. H. Murcray, and W. J. Williams (1979). Stratospheric distribution of $ClONO_2$. *Geophys. Res. Lett.* **6**, 857-859.

Murozumi, M., T. J. Chow, and C. Patterson (1969). Pollutant lead aerosols, terrestial dusts and sea salts in Greenland snow strata. *Geochim. Cosmochim. Acta* **33**, 1247-1294.

Nastrom, G. D. (1977). Vertical and horizontal fluxes of ozone of troposphere from first year of GASP data. *J. Appl. Meteorol.* **16**, 740-744.

National Research Council (1976). "Vapor-phase Organic Pollutants." National Academy of Sciences, Washington, D.C.

National Research Council, Subcommittee on Ammonia (1979). "Ammonia." University Park Press, Baltimore.

Naughton, J. J., and K. Terada (1954). Effect of eruption of Hawaiian volcanoes on the composition and carbon isotope content of associated volcanic and fumarolic gases. *Science* **120**, 580-581.

Naughton, J. J., V. A. Lewis, D. Thomas, and J. B. Finlayson (1975). Fume compositions found at various stages of activity at Kilauea volcano, Hawaii. *J. Geophys. Res.* **80**, 2963-2966.

Neckel, H., and D. Labs (1981). Improved data of solar spectral irradiance from 0.33 to 1.25 μm. *Solar Phys.* **74**, 231-249.

Neely, W. B., and J. H. Ploncka (1978). Estimation of time-averaged hydroxyl radical concentration in the troposphere. *Environ. Sci. Technol.* **12**, 317-321.

Neftel, A., H. Oeschger, J. Schwander, B. Stauffer, and R. Zumbrunn (1983). Ice core sample measurements give atmospheric CO_2 content during the last 40,000 years. *Nature* **295**, 220–223.

Neitzert, V., and W. Seiler (1981). Measurement of formaldehyde in clean air. *Geophys. Res. Lett.* **8**, 79–82.

Neligan, R. E. (1962). Hydrocarbons in the Los Angeles atmosphere. Comparison between hydrocarbons in automobile exhaust and those found in the Los Angeles atmosphere. *Arch. Environ. Health* **5**, 581–591.

Nelson, P. F., S. M. Quigley, and M. Y. Smith (1983). Sources of atmospheric hydrocarbons in Sidney: a quantitative determination using a source reconciliation technique. *Atmos. Environ.* **17**, 439–449.

Neukum, G. (1977). Lunar cratering. *Philos. Trans. R. Soc. London A* **285**, 267–272.

Newell, R. E., J. M. Wallace, and J. R. Mahoney (1966). The general circulation of the atmosphere and its effects on the movement of trace substances: Part 2. *Tellus* **18**, 363–380.

Newell, R., D. G. Vincent, and J. W. Kidson (1969). Interhemispheric mass exchange from meteorological and trace substance observations. *Tellus* **21**, 641–647.

Newell, R. E. (1970). Stratospheric temperature change from the Mt. Agung volcanic eruption of 1963. *J. Atmos. Sci.* **27**, 977–978.

Newell, E., J. W. Kidson, D. G. Vincent, and G. J. Boer (1972). "The General Circulation of the Tropical Atmosphere and Interactions with Extratropical Latitudes," Vol. 1. MIT Press, Cambridge, Massachusetts.

Newell, R., G. Boer, and J. Kidson (1974). An estimate of the interhemispheric transfer of carbon monoxide from tropical general circulation data. *Tellus* **26**, 103–107.

Nguyen, B. C., B. Bonsang, and G. Lambert (1974). The atmospheric concentration of sulfur dioxide and sulfate aerosols over antarctic, subantarctic areas and oceans. *Tellus* **26**, 241–248.

Nguyen, B. C., A. Gaudry, B. Bonsang, and G. Lambert (1978). Reevaluation of the role of dimethyl sulfide and the natural sulphur cycle. *Nature* **237**, 452–453.

Nguyen, B. C., B. Bonsang, and A. Gaudry (1983). The role of the ocean in the global atmospheric sulfur cycle. *J. Geophys. Res.* **88**, 10903–10914.

Nicolet, M. (1954). Dynamic effects in the high atmosphere. *In* "The Earth as a Planet" (G. P. Kuiper, ed.) Chap. 13, pp. 644–712. University of Chicago Press, Chicago, Illinois.

Nicolet, M. (1957). The aeronomic problem of helium. *Ann. Geophys.* (*Paris*) **13**, 1–19.

Nicolet, M. (1971). Aeronomic reactions of hydrogen and ozone. *In* "Mesospheric Models and Related Experiments" (G. Fiocco, ed.), pp. 1–51. Reidel, Dordrecht, The Netherlands.

Nicolet, M. (1975). On the production of nitric oxide by cosmic rays in the mesosphere and stratosphere. *Planet. Space Sci.* **23**, 637–649.

Nicolet, M. (1978). "Etude des Reactions Chimiques de l'Ozone dans la Stratosphere." Institut Royal Meteorologique de Belgique, Brussels.

Nicolet, M., and W. Peetermans (1980). Atmospheric absorption in the O_2–Schumann–Runge band spectral range and photodissociation rates in the stratosphere and mesosphere. *Planet. Space Sci.* **28**, 85–103.

Nicolet, M. (1981). The solar spectral irradiance and its action in the atmospheric photodissociation processes. *Planet. Space Sci.* **29**, 951–974.

Nieboer, H., and J. Van Ham (1976). Peroxyacetyl nitrate (PAN) in relation to ozone and some meteorological parameters at Delft in the Netherlands. *Atmos. Environ.* **10**, 115–120.

Nielson, T., and B. Seitz (1984). Occurrence of Nitro-PAH in the atmosphere in a rural area. *Atmos. Environ.* **18**, 2159–2165.

Niki, H., E. E. Daby, and B. Weinstock (1972). Mechanism of smog reactions. *In* "Photochemical Smog and Ozone Reactions" (R. F. Gould, ed.), *Adv. Chem. Ser.* **113**, 116–176.

Niki, H., P. D. Maker, C. M. Savage, and L. P. Breitenbach (1977). Fourier transform IR spectroscopic observations of propylene ozonide in the gas phase reaction of ozone-*cis*-butene-formaldehyde. *Chem Phys. Lett.* **46**, 327-330.

Niki, H., P. D. Maker, C. M. Savage, and L. P. Breitenbach (1978). Mechanisms for OH radical initiated oxidation of olefin-NO mixtures in ppm concentrations. *J. Phys. Chem.* **82**, 135-137.

Niki, H., P. D. Maker, C. M. Savage, and L. P. Breitenbach (1981). An FRIR study of mechanisms for the HO radical initiated oxidation of C_2H_4 in the presence of NO: detection of glycolaldehyde. *Chem. Phys. Lett.* **80**, 499-503.

Niki, H., P. D. Maker, C. M. Savage, and L. P. Breitenbach (1983). An FRIR study of mechanism for the reaction $OH + CH_3SCH_3$. *Int. J. Chem. Kinet.* **15**, 647-654.

Nojima, K., K. Fukaja, S. Fukni, and S. Kanno (1974). Formation of glyoxals by the photochemical reaction of aromatic hydrocarbons in the presence of nitrogen monoxide. *Chemosphere* **5**, 247-257.

Nordlund, G. (1983). Seasonal averages of net decay rate of SO_2 over northern Europe. *Atmos. Environ.* **17**, 1199-1201.

Noxon, J. F. (1975). Nitrogen dioxide in the stratosphere and troposphere measured by ground-based absorption spectroscopy. *Science* **189**, 547-549.

Noxon, J. F. (1976). Atmospheric nitrogen fixation by lightning. *Geophys. Res. Lett.* **3**, 463-465.

Noxon, J. F. (1978). Tropospheric NO_2. *J. Geophys. Res.* **83**, 3051-3057 (correction: *J. Geophys. Res.* **85**, 4560-4561, 1981).

Noxon, J. F. (1979). Stratospheric NO_2 global behaviour. *J. Geophys. Res.* **84**, 5067-5076.

Noxon, J. F., R. Norton, and G. Marovich (1980). NO_3 in the troposphere. *Geophys. Res. Lett.* **7**, 125-128.

Noxon, J. F. (1981). NO_x in the mid-Pacific troposphere. *Geophys. Res. Lett.* **8**, 1223-1226.

Noxon, J. F. (1983). NO_3 and NO_2 in the mid-Pacific troposphere. *J. Geophys. Res.* **88**, 11017-11021.

Nyberg, A. (1977). On air-borne transport of sulphur over the North Atlantic. *Q. J. R. Meteorol. Soc.* **103**, 607-615.

Nydal, R. (1968). Further investigation on the transfer of radiocarbon in nature. *J. Geophys. Res.* **73**, 3617-3635.

Oana, S., and E. S. Deevy (1960). Carbon 13 in lake waters and its possible bearing on paleolimnology. *Am. J. Sci.* **258A**, 253-272.

Oblath, S. B., S. S. Markowitz, T. Novakov, and S. G. Chang (1981). Kinetics of the formation of hydroxylamine disulfonate by reaction of nitrite with sulfites. *J. Phys. Chem.* **85**, 1017-1021.

O'Brien, R. J. (1974). Photostationary state in photochemical smog studies. *Environ. Sci. Technol.* **8**, 579-583.

Ockelmann, G. (1982). Untersuchung der Verteilung des Atmosphärischen Schwefeldioxids über dem Atlantik und der Arktis. Diploma Thesis, Institut für Meteorologie und Geophysik der Universität Frankfurt/Main.

Oddie, B. C. V. (1962). The chemical composition of precipitation at cloud levels. *Q. J. R. Meteorol. Soc.* **88**, 535-538.

Odén, S. (1968). The acidification of air and precipitation and its consequence on the natural environment. *Swed. Nat. Sci. Res. Counc. Ecol. Comm., Bull.* **1**, p. 68.

Odén, S. (1976). The acidity problem—an outline of concepts. *Water, Air, Soil Pollut.* **6**, 137-166.

Oeschger, H., U. Siegenthaler, U. Schotterer, and A. Gugelmann (1975). A box diffusion model to study the carbon dioxide exchange in nature. *Tellus* **27**, 168-192.

Ogawa, M. (1971). Absorption cross sections of O_2 and CO_2 continua in the Schumann and far-uv regions. *J. Chem. Phys.* **54**, 2550-2556.

Okabe, H. (1978). "Photochemistry of Small Molecules." Wiley, New York.

Okita, T., and D. Shimozuru (1975). Remote sensing measurements of mass flow of sulfur dioxide gas from volcanoes. *Bull. Volcanol. Soc. Jpn.* **19-3**, 153-157.

Oltmans, S. J., and W. D. Komhyr (1976). Surface ozone in Antarctica. *Geophys. Res.* **81**, 5359-5364.

Olson, J. S. (1970). Geographic index of world ecosystem. *In* "Analysis of Temperate Forest Ecosystems" (D. Reichle, ed.), *Ecol. Stud.* **1**, 297-304.

Oort, A. H., and E. M. Rasmusson (1971). Atmospheric circulation statistics, National Oceanic and Atmospheric Administration Prof. Paper 5. U.S. Dept. of Commerce, Rockville, Maryland.

Orr, C., K. Hurd, W. Hendrix, and C. E. Junge (1958a). The behaviour of condensation nuclei under changing humidities. *J. Meteorol.* **15**, 240-242.

Orr, C., K. Hurd, and W. J. Corbett (1958b). Aerosol size and relative humidity. *J. Colloid. Sci.* **13**, 472-482.

Ortgies, G., K. H. Gericke, and F. J. Comes (1980). Is uv laser induced fluorescence a method to monitor tropospheric OH? *Geophys. Res. Lett.* **7**, 905-908.

Orville, R. E. (1981). Global distribution of midnight lightning, September to November 1977. *Mon. Weather Rev.* **109**, 391-395.

Orville, R. E., and D. W. Spencer (1979). Global lightning flash frequency. *Mon. Weather Rev.* **107**, 934-943.

Östlund, H. G., and J. Alexander (1963). Oxidation rate of sulfide in seawater, a preliminary study. *J. Geophys. Res.* **68**, 3997-3997.

Ottar, B. (1978). An assessment of the OECD study on long range transport of air pollutants (LRTAP). *Atmos. Environ.* **32**, 454-465.

Owen, T., K. Biemann, D. R. Rushneck, J. E. Biller, D. W. Howarth, and A. L. Lafleur (1977). The composition of the atmosphere at the surface of Mars. *J. Geophys. Res.* **82**, 4635-4639.

Owens, A. J., C. H. Hales, D. L. Filkin, C. Miller, J. M. Steed, and J. P. Jesson (1985). A coupled one-dimensional radiation-convective chemistry transport model of the atmosphere. 1. Model structure and steady-state perturbation calculations. *J. Geophys. Res.* **90**, 2283-2311.

Oyama, V. I., G. C. Carle, F. Woeller, and J. B. Pollack (1979). Venus lower atmospheric composition. Analysis by gas chromatography. *Science* **203**, 802-805.

Pack, D. H., J. E. Lovelock, G. Cotton, and C. Curthoys (1977). Halocarbon behavior from a longtime series. *Atmos. Environ.* **11**, 329-344.

Paetzold, H. K., and E. Regener (1957). Ozon in der Erdatmosphäre. *In* "Encyclopaedia of Physics" (S. Flügge, ed.), pp. 370-426. Springer-Verlag, Berlin and New York.

Pales, J. C., and C. D. Keeling (1965). The concentration of atmospheric carbon dioxide in Hawaii. *J. Geophys. Res.* **70**, 6053-6076.

Palmer, H. B., and D. J. Seery (1973). Chemistry of pollutant formation in flames. *Annu. Rev. Phys. Chem.* **24**, 235-262.

Paneth, J. A. (1937). The chemical composition of the atmosphere. *Q. J. R. Meteorol. Soc.* **63**, 433-438.

Pannetier, R. (1970). Original use of the radioactive tracer gas Krypton 85 to study the meridian atmospheric flow. *J. Geophys. Res.* **75**, 2985-2989.

Panter, R., and R. D. Penzhorn (1980). Alkyl sulfonic acids in the atmosphere. *Atmos. Environ.* **14**, 149-151.

Paraskepopoulos, G., and R. J. Cvetanović (1969). Competitive reactions of the excited oxygen atoms (O^1D). *J. Am. Chem. Soc.* **91**, 7572-7577.

Park, J. H. (1974). The equivalent mean absorption cross sections of the O_2 Schumann-Runge bands: Application to the H_2O and NO photodissociation rates. *J. Atmos. Sci.* **31**, 1893-1897.

Parkin, D. W., D. R. Phillips, R. A. L. Sullivan, and L. R. Johnson (1972). Airborne dust collections down the Atlantic. *Q. J. R. Meteorol. Soc.* **98**, 789-808.

Parsons, T. R. (1975). Particulate organic carbon in the seas. *In* "Chemical Oceanography," 2nd Ed. (J. P. Riley and G. Skirrow, eds.), pp. 365-383. Vol. 2. Academic Press, London.

Parry, H. D. (1977). Ozone depletion by chlorofluoromethanes? Yet another look. *J. Appl. Meteorol.* **16**, 1137-1148.

Pasquill, F. (1974). "Atmospheric Diffusion," 2nd ed. Ellis Harwood, Chichester.

Pate, C. T., R. Atkinson, and J. N. Pitts (1976). The gas phase reaction of O_3 with a series of aromatic hydrocarbons. *J. Environ. Sci. Health A* **11**, 1-10.

Patterson, R. K., and J. Wagman (1976). Mass and composition of an urban aerosol as a function of particle size for several visibility levels. *Prepr. Pap. Natl. Meet. Div. Environ. Chem., Am. Chem. Soc.* **14**, 102-107.

Patterson, E. M., C. S. Kiang, A. C. Delany, A. F. Wartburg, A. C. D. Leslie, and B. J. Huebert (1980). Global measurements of aerosols in remote continental and marine regions: concentrations, size distributions and optical properties. *J. Geophys. Res.* **85**, 7361-7376.

Payne, W. A., L. J. Stief, and D. D. Davis (1973). A kinetic study of the reaction of HO_2 with SO_2 and NP. *J. Am. Chem. Soc.* **95**, 7614-7619.

Pearman, G. I., and P. Hyson (1981). A global atmospheric diffusion simulation model for atmospheric carbon studies. *In* "Carbon Cycle Modelling" (B. Bolm, ed.), *SCOPE* **16**, 227-240.

Peiser, C. D., C. C. Lizada, and S. F. Yang (1983). Dark metabolism of carbon monoxide in lettuce leaf disks. *Plant Physiol.* **70**, 397-400.

Peirson, R., and I. Cambray (1967). Interhemispheric transfer of debris from nuclear explosions using a simple atmospheric model. *Nature* **216**, 755-758.

Peng, T. H., W. S. Broecker, G. G. Mathieu, and Y. H. Li (1979). Radon evasion rates in the Atlantic and Pacific Oceans as determined during the GEOSECS program. *J. Geophys. Res.* **84**, 2471-2486.

Peng, T. H., W. S. Broecker, H. D. Freyer, and S. Trumbore (1983). A deconvolution of the tree ring based delta/13 record. *J. Geophys. Res.* **88**, 3609-3620.

Penkett, S. A. (1972). Oxidation of SO_2 and other atmospheric gases by ozone in aqueous solution. *Nature* **240**, 105-106.

Penkett, S. A., F. J. Sandalls, and B. M. R. Jones (1977). PAN measurements in England, analytical methods and results. *VDI Ber.* **270**, 47-54.

Penkett, S. A., B. M. R. Jones, and A. E. J. Eggleton (1979a). A study of SO_2 oxidation in stored rainwater samples. *Atmos. Environ.* **13**, 139-147.

Penkett, S. A., B. M. R. Jones, K. A. Brice, and A. E. J. Eggleton (1979b). The importance of atmospheric ozone and hydrogen peroxide in oxidising sulphur dioxide in cloud and rainwater. *Atmos. Environ.* **13**, 323-337.

Penkett, S. A., F. J. F. Atkins, and M. H. Unswork (1979c). Chemical composition of the ambient aerosol in the Sudan Gezire. *Tellus* **31**, 285-327.

Penkett, S. A., R. G. Derwent, P. Fabian, R. Borchers, and U. Schmidt (1980a). Methylchloride in the stratosphere. *Nature* **283**, 58-60.

Penkett, S. A., N. J. D. Prosser, R. A. Rasmussen, and M. A. K. Khalil (1980b). Measurement of $CHFCl_2$ in background tropospheric air. *Nature* **286**, 793-795.

Penkett, S. A., N. J. D. Prosser, R. A. Rasmussen, and M. A. K. Khalil (1981). Atmospheric measurements of CF_4 and other fluorocarbons containing the CF_3 grouping. *J. Geophys. Res.* **86**, 5172-5178.

Perner, D., D. H. Ehhalt, H. W. Pätz, U. Platt, E. P. Röth, and A. Volz (1976). OH-Radicals in the lower troposphere. *Geophys. Res. Lett.* 3, 466-468.

Perner, D., and U. Platt (1979). Detection of nitrous acid in the atmosphere by differential optical absorption. *Geophys. Res. Lett.* 6, 817-920.

Perry, R. A., R. Atkinson, and J. N. Pitts, Jr. (1977a). Kinetics and mechanism of the gas phas reaction of OH-radicals with aromatic hydrocarbons on the temperature range 296-473 K. *J. Phys. Chem.* 81, 296-304.

Perry, R. A., R. Atkinson, and J. N. Pitts, Jr. (1977b). Rate constants for the reactions of OH radicals with dimethyl ether and vinyl methyl ether over the temperature range 299-427 K. *J. Chem. Phys.* 67, 611-614.

Peterson, J. T., and C. E. Junge (1971). Sources of particulate matter in the atmosphere. *In* "Man's Impact on Climate" (W. H. Matthews, W. W. Kellog, and G. D. Robinson, eds.), pp. 310-320. MIT Press, Cambridge, Massachusetts.

Peterson, J. T. (1976). Calculated actinic fluxes (290-700) nm for air pollution photochemistry applications. **EPA-600/4-76-025**, U.S. Environmental Protection Agency, Research Triangle Park, North Carolina.

Petrenchuk, O. P., and E. S. Selezneva (1970). Chemical composition of precipitation in regions of the Soviet Union. *J. Geophys. Res.* 75, 3629-3634.

Pentrenchuk, O. P. (1980). On the budget of sea salts and sulfur in the atmosphere. *J. Geophys. Res.* 85, 7439-7444.

Peyrous, R., and R. M. Lapeyre (1982). Gaseous products created by electrical discharges in the atmosphere and condensation nuclei resulting from gas-phase reactions. *Atmos. Environ.* 16, 959-986.

Pflug, H. D. (1978). Yeast-like microfossils detected in oldest sediments of the earth. *Naturwissenschaften* 65, 611-615.

Pflug, H. D., and H. Jaeschke-Boyer (1979). Combined structural and chemical analysis of 3800 Myr old microfossils. *Nature* 280, 483-486.

Pierotti, D., and R. A. Rasmussen (1977). The atmospheric distribution of nitrous oxide. *J. Geophys. Res.* 82, 5823-5832.

Pierotti, D., L. E. Rasmussen, and R. A. Rasmussen (1978). The Sahara as a possible sink for trace gases. *Geophys. Res. Lett.* 5, 1001-1004.

Pierotti, D., and R. A. Rasmussen (1980). Nitrous oxide measurements in the eastern tropical Pacific Ocean. *Tellus* 32, 56-72.

Pierson, D. H., R. S. Cambray, and G. S. Spicer (1966). Lead 212 and polonium 210 in the atmosphere. *Tellus* 18, 428-433.

Pierson, W. R., W. W. Brachaczek, T. J. Truex, J. W. Butler, and T. J. Korniski (1980). Ambient sulfate measurements on Allegheny Mountain and the question of atmospheric sulfate in the Northeastern United States. "Aerosols: Anthropogenic and Natural Sources and Transport" *In* (T. J. Kneip, and P. J. Lioy, eds.) *Ann. NY. Acad. Sci.* 338, 145-173.

Pittock, A. B. (1974). Ozone climatology, trends and the monitoring problem. *Proc. Int. Conf. Structure, Composition General Circulation Upper Lower Atmos. Possible Anthropogenic Perturbations, Melbourne, Australia*, pp. 455-461.

Pittock, A. B. (1977). Climatology of the vertical distribution of ozone over Aspendale (38°S, 145°E). *Q. J. R. Meteorol. Soc.* 103, 575-585.

Pitts, J. N., Jr., J. H. Sharp, and S. I. Chan (1964). Effects of wavelength and temperature on primary processes in the photolysis of nitrogen dioxide and a spectroscopic-photochemical determination of the dissociation energy. *J. Chem. Phys.* 40, 3655-3662.

Pitts, J. N., Jr., A. C. Lloyd, and J. L. Sprung (1975). Ecology, energy and economics. *Chem. Brit.* 11, 247-256.

Pitts, J. N., Jr., A. M. Winer, G. J. Doyle, and K. R. Darnall (1978). Correspondence. *Environ. Sci. Technol.* **12**, 100–102.

Pitts, J. N., H. W. Biermann, A. M. Winer, and E. C. Tuazon (1984a). Spectroscopic identification and measurement of gaseous nitrous acid in dilute auto exhaust. *Atmos. Environ.* **18**, 847–854.

Pitts, J. N., Jr., E. Sanhueza, R. Atkinson, W. P. L. Carter, A. M. Winer, G. W. Harris, and C. N. Plum (1984b). An investigation of the dark formation of nitrous acid in environmental chambers. *Int. J. Chem. Kinet.* **16**, 919–939.

Pitts, J. N., Jr., H. W. Biermann, R. Atkinson, and A. M. Winer (1984c). Atmospheric implications of simultaneous nighttime measurements of NO_3 radicals and HONO. *Geophys. Res. Lett.* **11**, 557–560.

Pitts, J. N., J. A. Sweetman, B. Zielinska, C. Winer, and R. Atkinson (1985). Determination of 2-nitrofluoranthene and 2-nitropyrene in ambient particulate organic matter: evidence for atmospheric reactions. *Atmos. Environ.* **19**, 1601–1608.

Plass, G. N. (1956). The carbon dioxide theory of climatic changes. *Tellus* **8**, 140–154.

Platt, U., D. Perner, and H. W. Pätz (1979). Simultaneous measurement of atmospheric CH_2O, O_3 and NO_2 by differential optical absorption. *J. Geophys. Res.* **84**, 6329–6335.

Platt, U., and D. Perner (1980). Direct measurement of atmospheric CH_2O, HNO_2, O_3, NO_2 and SO_2 by differential optical absorption. *J. Geophys. Res.* **85**, 7453–7458.

Platt, U., D. Perner, A. M. Winer, G. W. Harris, and J. N. Pitts, Jr. (1980a). Detection of NO_3 in the polluted troposphere by differential optical absorption. *Geophys. Res. Lett.* **7**, 89–92.

Platt, U., D. Perner, G. W. Harris, A. M. Winer, and J. N. Pitts (1980b). Observations of nitrous acid in an urban atmosphere by differential optical absorption. *Nature* **285**, 312–314.

Platt, U., D. Perner, J. Schröder, C. Kessler, and A. Toenissen (1981). The diurnal variation of NO_3. *J. Geophys. Res.* **86**, 11965–11970.

Poet, S. E., H. E. Moore, and E. A. Martell (1972). Lead 210, bismuth 210 and polonium 210 in the atmosphere: accurate ratio measurements and application to aerosol residence time determination. *J. Geophys. Res.* **77**, 6515–6527.

Pollack, J. B., and D. C. Black (1979). Implications of the gas compositional measurement of Pioneer Venus for the origin of planetary atmospheres. *Science* **205**, 56–59.

Pollack, J. B. and Y. L. Yung (1980). Origin and evolution of planetary atmospheres. *Annu. Rev. Earth Planet. Sci.* **8**, 425–487.

Pollard, R. T. (1977). "Hydrocarbons," *In* "Chemical Kinetics" (C. H. Bamford and C. F. H. Tipper, eds.), Gas-Phase Combustion, vol. 17, pp. 249–367. Elsevier, Amsterdam.

Pollock, W. L., L. E. Heidt, R. Lueb, and D. H. Ehhalt (1980). Measurement of stratospheric water vapor by cryogenic collection. *J. Geophys. Res.* **85**, 5555–5568.

Porter, L. K., F. G. Viets, Jr., and G. L. Hutchinson (1972). Air containing nitrogen-15 ammonia: Foliar absorption by corn seedlings. *Science* **175**, 759–761.

Postma, A. K. (1970). Effect of solubilities of gases on their scavenging by rain drops. *In* "Precipitation Scavenging 1970" (R. J. Engelmann and W. G. N. Slinn, coordinators), *AEC Symp. Ser.* **30**, 247–259. U.S. Atomic Energy Commission, Div. of Tech. Information, Oakridge, Tennessee.

Prager, M. J., E. R. Stephens, and W. E. Scott (1960). Aerosol formation from gaseous air pollutants. *Ind. Eng. Chem.* **52**, 521–524.

Prahm, L. P., U. Torp, and R. M. Stern (1976). Deposition and transformation rates of sulfur oxides during atmospheric transport over the Atlantic. *Tellus* **28**, 355–372.

Prather, M. J., M. B. McElroy, and S. C. Wofsy (1984). Reduction in ozone at high concentrations of stratospheric halogens. *Nature* **312**, 227–231.

Pratt, G. C., R. C. Hendrickson, B. I. Chevone, D. A. Christopherson, M. V. O'Brien, and S. V. Krupa (1983). Ozone and oxides of nitrogen in the rural upper-midwestern U.S.A. *Atmos. Environ.* **17**, 2013-2023.

Pressman, J., and P. Warneck (1970). The stratosphere as a chemical sink for carbon monoxide. *J. Atmos. Sci.* **27**, 155-163.

Preston, K. F., and R. F. Barr (1971). Primary processes in the photolysis of nitrous oxide. *J. Chem. Phys.* **54**, 3347-3348.

Pring, J. N. (1914). The occurrence of ozone in the upper troposphere. *Proc. R. Soc. A* **90**, 204-219.

Prinn, R. G., P. G. Simmonds, R. A. Rasmussen, R. D. Rosen, F. N. Alyea, C. A. Cardelino, A. J. Crawford, D. M. Cunnold, P. J. Fraser, and J. E. Lovelock (1983a). The atmospheric life time experiment. 1. Introduction, instrumentation and overview. *J. Geophys. Res.* **88**, 8353-8367.

Prinn, R. G., R. A. Rasmussen, P. G. Simmonds, F. N. Alyea, D. C. Cunnold, B. C. Lane, C. A. Cardelino, and A. J. Crawford (1983b). The atmospheric life time experiment. 5. Results for CH_3CCl_3 based on three years of data. *J. Geophys. Res.* **88**, 8415-8426.

Pruchniewicz, P. G., H. Tiefenau, P. Fabian, P. Wilbrandt, and W. Jessen (1974). The distribution of tropospheric ozone from world wide surface and aircraft observations. *Proc. Int. Conf. Structure, Composition General Circulation Upper Lower Atmos. Possible Anthropogenic Perturbations, Melbourne* (P. Goldsmith, A. D. Belmont, N. J. Derco, and E. J. Truhlar, eds.), Vol. 1, pp. 439-451.

Pruppacher, H. R., and J. D. Klett (1980). "Microphysics of Clouds and Precipitation." Reidel, Dordrecht, The Netherlands.

Pyle, J. A., and R. G. Derwent (1980). Possible ozone reductions and uv changes at the Earth's surface. *Nature* **286**, 373-375.

Pytkowicz, R. M. (1970). On the carbonate compensation depth in the Pacific Ocean. *Geochim. Cosmochim. Acta* **34**, 836-839.

Pytkowicz, R. M. (1975). Some trends in marine chemistry and geochemistry. *Earth Sci. Rev.* **11**, 1-46.

Radford, H. E. (1980). The fast reaction of CH_2OH with O_2. *Chem. Phys. Lett.* **71**, 195-197.

Radke, L. F., and P. V. Hobbs (1969). Measurement of cloud condensation nuclei, light scattering coefficient, sodium-containing particles and Aitken nuclei in the Olympic Mountains of Washington. *J. Atmos. Sci.* **26**, 281-288.

Radke, L. F., P. V. Hobbs, and J. E. Pinnons (1976). Observations of cloud condensation nuclei, sodium containing particles, ice nuclei and the light scattering coefficient near Barrow, Alaska. *J. Appl. Meteorol.* **15**, 982-995.

Radke, L. F., J. L. Stith, D. A. Hegg, and P. V. Hobbs (1978). Airborne studies of particles and gases from forest fires. *J. Air Pollut. Control Assoc.* **28**, 30-34.

Radke, L. F. (1981). Marine aerosol: simultaneous size distributions of the total aerosol and the sea-salt fraction from 0.1 to 10 μm diameter. *Proc. Int. Conf. Atmos. Aerosol, Condensation, Ice Nuclei, 9th, Galway, Ireland* (A. F. Roddy and T. C. O'Connor, eds.), pp. 487-491.

Rahn, K. A. (1975a). Chemical composition of the atmospheric aerosol. A compilation I. *Extern.* **4**, 286-313.

Rahn, K. A. (1975b). Chemical Composition of the atmospheric aerosol. A compilation II. *Extern.* **4**, 639-667.

Rahn, K. A., R. D. Borys, and R. A. Duce (1976). Tropospheric halogen gases: Inorganic and organic components. *Science* **192**, 549-550.

Rahn, K. A., R. D. Borys, and G. E. Shaw (1977). The Asian source of Arctic haze bands. *Nature* **268**, 713-715.

Rahn, K. A., and R. J. McCaffrey (1980). On the origin and transport of the winter Arctic aerosol. *In* "Aerosols: Anthropogenic and natural sources and transport" (T. J. Kneip and P. J. Lioy, eds.), *Ann. N.Y. Acad. Sci.* **338**, 486-503.

Ramanathan, V., and J. A. Coakley, Jr. (1978). Climate modeling through radioactive-convective models. *Rev. Geophys. Space Phys.* **16**, 465-689.

Randhawa, J. S. (1971). The vertical distribution of ozone near the equator. *J. Geophys. Res.* **76**, 8139-8142.

Raper, O. F., C. B. Farmer, R. A. Toth, and B. D. Robbins (1977). The vertical distribution of HCl in the stratosphere. *Geophys. Res. Lett.* **4**, 531-534.

Rasmussen, R. A., and F. W. Went (1965). Volatile organic material of plant origin in the atmosphere. *Proc. Natl. Acad. Sci. U.S.A.* **53**, 215-220.

Rasmussen, R. A. (1970). Isoprene: Identified as a forest-type emission to the atmosphere. *Environ. Sci. Technol.* **4**, 669-673.

Rasmussen, R. A. (1972). What do the hydrocarbons from trees contribute to air pollution. *J. Air. Pollut. Control Assoc.* **22**, 537-543.

Rasmussen, R. A., and C. A. Jones (1973). Emission of isoprene from leaf discs of Hammameis. *Phytochem.* **12**, 15-19.

Rasmussen, R. A., R. B. Chatfield, and M. W. Holdren (1977). Hydrocarbon species in rural Missouri air. Unpublished manuscript, Washington State University, Pullman, Washington.

Rasmussen, R. A., and D. Pierotti (1978). Inter-laboratory calibration of atmospheric nitrous oxide measurements. *Geophys. Res. Lett.* **5**, 353-355.

Rasmussen, R. A., L. E. Rasmussen, M. A. K. Khalil, and R. W. Dalluge (1980). Concentration distribution of methyl chloride in the atmosphere. *J. Geophys. Res.* **85**, 7350-7356.

Rasmussen, R. A., and M. A. K. Khalil (1980). Atmospheric halocarbons, measurements and analyses of selected trace gases. NATO *Adv. Study Inst.* "Atmospheric Ozone: Its Variation and Human Influences" (M. Nicolet and A. C. Aikin, eds.), pp. 209-231. **FAA-II-80-20**, U.S. Department of Transportation, Washington, D.C.

Rasmussen, R. A., and M. A. K. Khalil (1981a). Increase in the concentration of atmospheric methane. *Atmos. Environ.* **15**, 883-886.

Rasmussen, R. A., and M. A. K. Khalil (1981b). Inter-laboratory comparison of fluorocarbons, -11, -12, Methyl chloroform and nitrous oxide measurements. *Atmos. Environ.* **15**, 1559-1568.

Rasmussen, R. A., M. A. K. Khalil, and R. W. Dalluge (1981). Atmospheric trace gases in Antarctica. *Science* **211**, 285-287.

Rasmussen, R. A., and M. A. K. Khalil (1982). Latitudinal distribution of trace gases in and above the boundary layer. *Chemosphere* **11**, 227-235.

Rasmussen, R. A., M. A. K. Khalil, and S. D. Hoyt (1982a). The oceanic source of carbonyl sulfide. *Atmos. Environ.* **16**, 1591-1594.

Rasmussen, R. A., M. A. K. Khalil, R. Gunawardena, and S. D. Hoyt (1982b). Atmospheric Methyl Iodide (CH_3I). *J. Geophys. Res.* **87**, 3086-3090.

Rasmussen, R. A., M. A. K. Khalil, A. J. Crawford, and P. J. Fraser (1982c). Natural and anthropogenic trace gases in the southern hemisphere. *Geophys. Res. Lett.* **9**, 704-707.

Rasmussen, R. A., and M. A. K. Khalil (1983). Global production of methane by termites. *Nature* **301**, 700-702.

Rasmussen, R. A., and M. A. K. Khalil (1984). Atmospheric methane in the recent and ancient atmosphere: concentrations, trends and interhemispheric gradient. *J. Geophys. Res.* **89**, 11599-11605.

Rasool, S. I., and C. DeBergh (1970). The runaway greenhouse and the accumulation of CO_2 in the Venus atmosphere. *Nature* **226**, 1037-1039.

Ratner, M. J., and J. C. G. Walker (1972). Atmospheric ozone and the history of life. *J. Atmos. Sci.* **29**, 803-808.

Rebbert, R. E., and P. J. Ausloos (1975). Photodecomposition of $CFCl_3$ and CF_2Cl_2. *J. Photochem.* **4**, 419-434.

Rebbert, R. E., and P. J. Ausloos (1977). Gas-phase photodecomposition of carbon tetrachloride. *J. Photochem.* **6**, 265.

Rebbert, R. E., and P. Ausloos (1978). Decomposition of N_2O over particulate matter. *Geophys. Res. Lett.* **5**, 761-764.

Redfield, A. C., B. H. Ketchum, and F. A. Richards (1963). The influence of organisms in the composition of seawater. *In* "The Sea" (M. N. Hill, ed.), Vol. 2, pp. 26-77. Wiley (Interscience), New York.

Reed, R. J., and K. E. German (1965). A contribution to the problem of stratospheric diffusion by large-scale mixing. *Mon. Weather Rev.* **93**, 313-321.

Regener, E., and V. H. Regener (1934). Aufnahme des ultravioletten Sonnenspektrums in der Stratosphäre und vertikale Ozonverteilung. *Z. Phys.* **35**, 788-793.

Regener, V. H. (1957). Vertical flux of atmospheric ozone. *J. Geophys. Res.* **62**, 221-228.

Regener, V. H. (1964). Measurement of atmospheric ozone with the chemiluminescent method. *J. Geophys. Res.* **69**, 3795-3800.

Regener, V. H. (1974). Destruction of atmospheric ozone at the ocean surface. *Arch. Meteorol. Geophys. Bioklimatol. Ser. A* **23**, 131-135.

Reiners, W. A. (1973). Terrestrial detritus and the carbon cycle. *In* "Carbon and the Biosphere" (G. M. Woodwell and E. V. Pecan, eds.), *AEC Symp. Ser.* **30**, pp. 303-327. NTIS U.S. Dept. Commerce, Springfield, Virginia.

Reinsel, G. C. (1981). Analysis of total ozone data for the detection of recent trends and the effects of nuclear testing during the 1960's. *Geophys. Res. Lett.* **8**, 1227-1230.

Reinsel, G. K., G. C. Tiao, R. Lewis, and M. Bobkoski (1983). Analysis of upper stratospheric ozone profile data from the ground-based Umkehr method and the Nimbus-4 BUV satellite experiment. *J. Geophys. Res.* **88**, 5393-5409.

Reinsel, G. K., G. C. Tiao, J. J. Deluisi, C. L. Mateer, A. J. Miller, and J. E. Frederick (1984). Analysis of upper stratospheric Umkehr ozone profile data for trends and the effects of stratospheric aerosol. *J. Geophys. Res.* **89**, 4833-4840.

Reiss, H. (1950). The kinetics of phase transitions in binary systems. *J. Chem. Phys.* **18**, 840-848.

Reiter, E. R., and J. D. Mahlman (1965). Heavy radioactive fallout on the southern United States, November 1962. *J. Geophys. Res.* **70**, 4501-4520.

Reiter, E. R., M. E. Glasser, and J. D. Mahlman (1969). The role of the tropopause in stratospheric-tropospheric exchange processes. *Pure Appl. Geophys.* **12**, 183-221.

Reiter, R. (1970). On the causal relation between nitrogen-oxygen compounds in the troposphere and atmospheric electricity. *Tellus* **22**, 122-135.

Reiter, E. R. (1975). Stratospheric-tropospheric exchange processes. *Rev. Geophys. Space Phys.* **13**, 459-474.

Reiter, E. R., W. Carnuth, H. J. Kanter, K. Pötzl, R. Reiter, and R. Sladkovic (1975a). Measurements of stratospheric residence time. *Arch. Meteorol. Geophys. Bioklimatol. Ser. A* **24**, 41-51.

Reiter, R., R. Sladkovic, and K. Pötzl (1975b). Die wichtigsten chemischen Bestandteile des Aerosols über Mitteleuropa unter Reinluftbedingungen in 1800 m Seehöhe. *Meteorol. Rundsch.* **28**, 37-55.

Reiter, R., R. Sladkovic, and K. Pötzl (1976). Chemical components of aerosol particles in the lower troposphere above central Europe measured under pure-air conditions. *Atmos. Environ.* **10**, 841-853.

Reiter, E. R., H. J. Kanter, R. Reiter, and R. Sladkovic (1977). Lower tropospheric ozone of stratospheric origin. *Arch. Meteorol. Geophys. Bioklimatol. Ser. A* **26**, 179-186.

Reiter, R., R. Sladkovic, and K. Pötzl (1978). Chemische Komponente des Reinluftaerosols in Abhängigkeit von Luftmassenchrakter und meteorologischen Bedingungen. *Ber. Bunsen-Ges. Phys. Chem.* **82**, 1188-1193.

Revelle, R., and H. Suess (1957). Carbon dioxide exchange between atmosphere and oceans and the question of an increase of atmospheric CO_2 during the past decades. *Tellus* **9**, 18-27.

Rhode, H., and J. Grandell (1972). On the removal of aerosol particles from the atmosphere by precipitation scavenging. *Tellus* **24**, 442-454.

Richards, L. W., J. A. Anderson, D. L. Blumenthal, and J. A. McDonald (1983). Hydrogen peroxide and sulfur (IV) in Los Angeles cloudwater. *Atmos. Environ.* **17**, 911-914.

Richards, L. W. (1983). Comments on the oxidation of NO_2 to nitrate—day and night. *Atmos. Environ.* **17**, 397-402.

Ridley, W. P., L. J. Dizikesand, and J. M. Wood (1976). Biomethylation of toxic elements in the environment. *Science* **197**, 329-332.

Ridley, B. A., M. McFarland, J. T. Bruin, H. I. Schiff, and J. C. McConnell (1977). Sunrise measurements of stratospheric nitric oxide. *Can. J. Phys.* **55**, 212-221.

Ridley, B. A., and D. R. Hastie (1981). Stratospheric odd nitrogen: NO measurements of 51°N in summer. *J. Geophys. Res.* **86**, 3162-3166.

Ridley, B. A., and H. I. Schiff (1981). Stratospheric odd-nitrogen nitric oxide measurements at 31°N in autumn. *J. Geophys. Res.* **80**, 3167-3172.

Ringwood, A. E. (1979). "Origin of the Earth and Moon." Springer-Verlag, Berlin and New York.

Ritter, J., D. H. Stedman, and T. J. Kelly (1979). Ground level measurements of NO, NO_2 and O_3 in rural air. *In* "Nitrogeneous Air Pollutants—Chemical and Biological Implications" (D. Grosjean, ed.), pp. 325-343. Ann Arbor Science Publ., Ann Arbor, Michigan.

Roberts, J. M., F. C. Fehsenfeld, D. L. Albritton, and R. E. Sievers (1983). Measurement of monoterpene hydrocarbon at Niwot Ridge, Colorado. *J. Geophys. Res.* **88**, 10667-10678.

Roberts, J. M., F. C. Fehsenfeld, S. C. Liu, M. J. Bollinger, C. Hahn, D. L. Albritton, and R. E. Sievers (1984). Measurements of aromatic hydrocarbon ratios and NO_x concentrations in the rural troposphere: observation of air mass photochemical aging and NO_x removal. *Atmos. Environ.* **18**, 2421-2432.

Robbins, R. C., K. M. Borg, and E. Robinson (1968). Carbon monoxide in the atmosphere. *J. Air Pollut. Control Assoc.* **18**, 106-110.

Robbins, R. C., L. A. Cavanagh, L. J. Salas, and E. Robinson (1973). Analysis of ancient atmospheres. *J. Geophys. Res.* **78**, 5341-5344.

Robinson, E., and R. C. Robbins (1968a). Evaluation of CO data obtained on Eltanin cruises 27, 29 and 31. *Antarctic J. U.S.* **3**, 194-196.

Robinson, E., and R. C. Robbins (1968b). Sources, abundance and fate of gaseous atmospheric pollutants. Final Report, Project **PR-6755**, Stanford Research Institute, Palo Alto, California.

Robinson, E., and R. C. Robbins (1970a). Atmospheric background concentrations of carbon monoxide. *Ann. N.Y. Acad. Sci.* **174**, 89-95.

Robinson, E., and R. C. Robbins (1970b). Gaseous sulfur pollutants from urban and natural sources. *J. Air Pollut. Control Assoc.* **20**, 233-235.

Robinson, R. J., and R. H. Stokes (1970). "Electrolyte Solutions," 2nd ed. Butterworth, London.

Robinson, E., and R. C. Robbins (1972). Emissions, concentrations and fate of gaseous atmospheric pollutants. *In* "Air Pollution Control" (W. Strauss, ed.), Part II, pp. 1-93. Wiley (Interscience), New York.

Robinson, E., R. A. Rasmussen, H. H. Westberg, and M. W. Holdren (1973). Non-urban nonmethane low molecular weight hydrocarbon concentrations related to air mass identification. *J. Geophys. Res.* **78**, 5345-5351.

Robinson, E. R., R. A. Rasmussen, J. Krasnec, D. Pierotti, and M. Jakubovic (1977). Halocarbon measurements in the Alaskan troposphere and lower stratosphere. *Atmos. Environ.* **11**, 213-223.

Robinson, E. (1977). Review and analysis: *Int. Conf. Oxidants 1976.* "Analysis of Evidence and Viewpoints," Part V: The issue of Oxidant Transport, pp. 9-20. U.S. Environmental Protection Agency, Publ. No. **EPA-600/3-77-117.** Research Triangle Park, North Carolina.

Robinson, E. (1978). Hydrocarbons in the atmosphere. *Pure Appl. Geophys.* **116**, 372-384.

Robinson, E., and J. B. Homolya (1983). Natural and anthropogenic emission sources. "The Acidic Deposition Phenomenon and its Effects: Critical Assessment Review Papers" (A. P. Altshulter and R. A. Linthurst, eds.), Chap. A2. United States Environmental Protection Agency, **EPA-600/8-83-016A.** Office of Research and Development, Washington, D.C.

Robinson, E., D. Clark, and W. Seiler (1984). The latitudinal distribution of carbon monoxide across the Pacific from California to Antarctica. *J. Atmos. Chem.* **1**, 137-149.

Robson, R. L., and J. R. Postgate (1980). Oxygen and hydrogen in biological nitrogen fixation. *Annu. Rev. Microbiol.* **34**, 183-207.

Rodin, L. E., N. I. Bazilevich, and N. N. Rozov (1975). Productivity of the world's main ecosystems. *Proc. Seattle Symp., Nat. Acad. Sci., Washington, D.C.,* pp. 13-26.

Roedder, E. (1981). Are the 3800 Myr old Isua objects microfossils, limonite stained fluid inclusions, or neither? *Nature* **293**, 459-462.

Roessler, J. F., H. J. R. Stevenson, and J. S. Nader (1965). Size distribution of sulfate aerosols in the ambient air. *J. Air Pollut. Control Assoc.* **15**, 576-579.

Rohbock, E. (1982). Atmospheric removal of airborne metals by wet and dry deposition. *In* "Deposition of Atmospheric Pollutants" (H. W. Georgii and J. Pankrath, eds.), pp. 159-171. Reidel, Dordrecht, The Netherlands.

Ronov, A. B., and A. A. Yaroshevskiy (1969). Chemical composition of the earth's crust. *In* "The Earth's Crust and Upper Mantle", *Am. Geophys. Union Geophys. Monograph Ser.* **13**, 37-57.

Roscoe, H. K., J. R. Drummond, and R. F. Jarnot (1981). Infrared measurements of stratospheric composition, III. The daytime changes of NO and NO_2. *Proc. R. Soc., London* A, **375**, 507-528.

Rose, W. I., S. Bouis, R. E. Stoiber, M. Keller, and T. Bickford (1973). Studies of volcanic ash from two recent Central American eruptions. *Bull. Volcanol.* **37**, 338-364.

Rosen, J. M. (1969). Stratospheric dust and its relationship to the meteorite influx. *Space Sci. Rev.* **9**, 58-89.

Rosen, J. M. (1971). The boiling point of stratospheric aerosols. *J. Appl. Meteorol.* **10**, 1044-1046.

Rosen, J. M., and D. J. Hoffmann (1977). Balloonborne measurements of condensation nuclei. *J. Appl. Meteorol.* **16**, 56-62.

Rosen, J. M., and D. J. Hofmann (1978). Vertical profiles of condensation nuclei. *Amer. Meteorol. Soc.* **17**, 1737-1740.

Rosenfeld, W. D., and S. R. Silverman (1959). Carbon isotope fractionation in bacterial production of methane. *Science* **130**, 1658-1659.

Rosswall, T. (1976). The internal nitrogen cycle between microorganisms, vegetation and soil. *In* "Nitrogen, Phosphorus and Sulphur-Global Cycles" (B. H. Svensson and R. Söderlund, eds.), *SCOPE* **7**, *Ecol. Bull.* (*Stockholm*) **22**, 157-167.

Rosswall, T. (1981). The biogeochemical nitrogen cycle. *In* "Some Perspectives of the Major Biogeochemical Cycles" (G. E. Likens, ed.), *SCOPE* **17**, 25–49.

Rotty, R. M. (1981). Data for global CO_2 production from fossil fuels and cement. *In* "Carbon Cycle Modelling" (B. Bolin, ed.), *SCOPE* **16**, pp. 121–125.

Rotty, R. M. (1983). Distribution of and changes in industrial carbon dioxide production. *J. Geophys. Res.* **88**, 1301–1308.

Routhier, F., R. Dennet, P. D. Davis, A. Wartburg, P. Haagenson, and A. C. Delany (1980). Free tropospheric and boundary layer airborne measurements of ozone over the latitude range of 58°S to 70°N, *J. Geophys. Res.* **85**, 7307–7321.

Rowland, F. S., and M. J. Molina (1975). Chlorofluoromethanes in the environment. *Rev. Geophys. Space Phys.* **13**, 1–35.

Roy, C. R. (1979). Atmospheric nitrous oxide in the mid-latitudes of the southern hemisphere. *J. Geophys. Res.* **84**, 3711–3718.

Roy, C. R., I. E. Galbally, and B. A. Ridley (1980). Measurements of nitric oxides in the stratosphere of the southern hemisphere. *Q. J. R. Meteorol. Soc.* **106**, 887–894.

Rubey, W. W., (1951). Geological history of seawater, an attempt to state the problem. *Bull. Geol. Soc. Am.* **62**, 1111–1147.

Rudolph, J., D. H. Ehhalt, and G. Gravenhorst (1980). Recent measurements of light hydrocarbons in remote areas. *In* "Physico-Chemical Behaviour of Atmospheric Pollutants" (B. Versino and H. Ott, eds.), *Proc. Symp., 1st, Varese, Italy, 1979*, pp. 41–54. Commission the European Communities, Luxembourg.

Rudolph, J., and D. H. Ehhalt (1981). Measurements of C_2-C_5 hydrocarbon over the north Atlantic. *J. Geophys. Res.* **86**, 11959–11964.

Rüden, J., and R. Thofern (1976). Abscheidung von Schadstoffen und Mikroorganismen in Luftfiltern. *Staub Reinhalt. Luft* **36**, 33–36.

Rüden, H., E. Thofern, P. Fischer, and U. Mihm (1978). Airborne microorganisms: their occurrence, distribution and dependence on environmental factors—especially on organic compounds of air pollution. *Pure. Appl. Geophys.* **116**, 335–350.

Rust, F. (1957). Intramolecular oxidation. The autoxidation of some dimethyl alkanes. *J. Am. Chem. Soc.* **79**, 4000–4003.

Rutten, M. G. (1971). "The Origin of Life by Natural Causes." Elsevier, Amsterdam.

Ryaboshapko, A. G. (1983). The atmospheric sulphur cycle. *In* "The Global Biogeochemical Sulphur Cycle" (M. V. Ivanov and J. R. Freney, eds.), *SCOPE* **19**, 203–296.

Ryden, J. C., L. J. Lund, and D. D. Focht (1978). Direct in-field measurements of nitrous oxide flux from soils. *J. Am. Soc. Soil Sci.* **42**, 731–737.

Ryden, J. C. (1981). N_2O exchange between a grass land soil and the atmosphere. *Nature* **292**, 235–237.

Sadasivan, S. (1978). Trace elements in size separated aerosols over the sea. *Atmos. Environ.* **12**, 1677–1683.

Sakamaki, F., S. Hatakeyama, and H. Akimoto (1983). Formation of nitrous acid and nitric oxide in the heterogeneous dark reaction of nitrogen dioxide with water vapor in a smog chamber. *Int. J. Chem. Kinet.* **15**, 1013–1029.

Saltzman, B. E. (1954). Colorimetric microdetermination of nitrogen dioxide in the atmosphere. *Anal. Chem.* **26**, 1949–1955.

Saltzman, E. S., D. L. Savoie, R. G. Zika, and J. M. Prospero (1983). Methane sulfonic acid in the marine atmosphere. *J. Geophys. Res.* **88**, 10897–10902.

Samain, D., and P. C. Simon (1976). Solar flux determination in the spectral range 150–210 nm. *Sol. Phys.* **49**, 33–41.

Sanadze, G. A., and A. N. Kalandadze (1966). Light and temperature curves on the evolution of isoprene. *Sov. Plant Physiol.* **13**, 411–413.

Sandalls, F. J. and S. A. Penkett (1977). Measurements of carbonyl sulfide and carbon disulfide in the atmosphere. *Atmos. Environ.* **11**, 197-199.

Sandberg, D. V., S. G. Pickford, and E. F. Darley (1975). Emissions from slash burning and the influence of flame retardant chemicals. *J. Air Pollut. Control Assoc.* **25**, 278-281.

Sapper, K. (1927). Vulkankunde. Engelhorn Verlag, Stuttgart.

Savoie, D. L., and J. M. Prospero (1977). Aerosol concentration statistics for the northern tropical Atlantic. *J. Geophys. Res.* **82**, 5954-5964.

Savoie, D. L., and J. M. Prospero (1982). Particle size distribution of nitrate and sulfate in the marine atmosphere. *Geophys. Res. Lett.* **9**, 1207-1210.

Saxena, V. K., J. N. Burford, and J. L. Kassner (1970). Operation of a thermal diffusion chamber for measurements of cloud condensation nuclei. *J. Atmos. Sci.* **27**, 73-80.

Saxena, V. K., and A. H. Hendler (1983). In-cloud scavenging and resuspension of cloud active aerosols during winter storms over Lake Michigan. *In* "Precipitation Scavenging, Dry Deposition and Resuspension" (H. R. Pruppacher, R. G. Semonin, and W. G. N. Slinn, eds.), pp. 91-101. Elsevier, New York.

SCEP (1970). Man's impact on the global environment, Report of the study of critical environmental problems (SCEP). The MIT Press, Cambridge, Massachusetts.

Schidlowski, M. (1966). Beiträge zur Kenntnis der radioaktiven Bestandteile der Witwatersrand Konglomerate I-III. *N.Jb. Miner, Abh.* **107/106**, 183-202, 310-324, 55-71.

Schidlowski, M. (1970). Untersuchungen zur Metallogenese im südwestlichen Witwatersrand Becken (Oranje-Freistaat Goldfeld, Südafrika). *Beih. Geo. Jahrb.* **85**, 80-126.

Schidlowski, M. (1971). Probleme der atmosphärischen Evolution in Präkambrium. *Geol. Rundsch.* **60**, 1351-1384.

Schidlowski, M. (1978). A model for the evolution of photosynthetic oxygen. *Pure Appl. Geophys.* **116**, 234-238.

Schidlowski, M. (1982). Content and isotopic composition of reduced carbon in sediments. *In* "Mineral Deposits and the Evolution of the Biosphere" (H. D. Holland and M. Schidlowski, eds.), pp. 103-122. Dahlem Konferenzen, Springer-Verlag, Berlin and New York.

Schidlowski, M. (1983a). Biologically mediated isotope fractionations: biochemistry, geochemical significance and preservation in the earth's oldest sediments. *In* "Cosmochemistry and the Origin of Life" (C. Ponnamperuma, ed.), pp. 277-322. Reidel, Dordrecht, The Netherlands.

Schidlowski, M. (1983b). Evolution of photo-autotrophy and early atmospheric oxygen levels. *Precambrian Res.* **20**, 319-335.

Schidlowski, M., J. M. Hayes, and I. R. Kaplan (1983). Isotopic inferences of ancient biochemistries: carbon, sulfur and nitrogen. *In* "Earth's Earliest Biosphere: Its Origins and Evolution" (J. W. Schopf, ed.), pp. 149-186. Princeton Univ. Press, Princeton, New Jersey.

Schidlowski, M. (1984). Sedimentary organic matter: 3.8 billion years isotopic fingerprints of life. N. A. Bogdanov (ed.) *Proc. Int. Geol. Congr., 27th*, Vol. 19, Comparative Planetology, pp. 229-246. VNU Science Press, Utrecht, The Netherlands.

Schiff, H. I. (1972). Laboratory measurements of reactions related to ozone photochemistry. *Ann. Geophys.* **28**, 67-77.

Schiff, H. I., A. Pepper, and B. A. Ridley (1979). Tropospheric NO measurements up to 7 km. *J. Geophys. Res.* **84**, 7895-7897.

Schlegel, H. G. (1974). Production, modification, and consumption of atmospheric trace gases by microorganisms. *Tellus* **26**, 11-20.

Schlesinger, W. H. (1977). Carbon balance in terrestrial detritus. *Annu. Rev. Ecol. Syst.* **8**, 51-81.

Schmauss, A., and A. Wigand (1929). Die Atmosphäre als Kolloid. Vieweg & Sohn, Braunschweig.

Schmeltekopf, A. L., P. D. Goldan, W. R. Henderson, W. J. Harrop, T. L. Thompson, F. C. Fehsenfeld, H. I. Schiff, P. J. Crutzen, I. S. A. Isaksen, and E. E. Ferguson (1975). Measurements of stratospheric $CFCl_3$, CF_2Cl_2 and N_2O. *Geophys. Res. Lett.* **2**, 393-396.

Schmeltekopf, A. L., D. L. Albritton, P. J. Crutzen, P. D. Goldan, W. J. Harrop, W. R. Henderson, J. R. McAffee, M. McFarland, H. I. Schiff, T. L. Thompson, D. J. Hoffman, and N. T. Kjome (1977). Stratospheric nitrous oxide altitude profiles at various latitudes. *J. Atmos. Sci.* **34**, 729-736.

Schmidt, U. (1974). Molecular hydrogen in the atmosphere. *Tellus* **26**, 78-90; Erratum (1975). *Tellus* **27**, 1.

Schmidt, U. (1978). The latitudinal and vertical distribution of molecular hydrogen in the troposphere. *J. Geophys. Res.* **83**, 941-946.

Schmidt, U., G. Kulessa, and E. P. Röth (1980). The atmospheric H_2 cycle. *NATO Adv. Study Inst. Atmos. Ozone:* "Its Variations and Human Influences" (M. Nicolet and A. C. Aikin, eds.), pp. 307-322. **FAA-EE-80-20**. U.S. Department of Transportation, Washington, D.C.

Schmidt, U., J. Rudolph, F. J. Johnen, D. H. Ehhalt, A. Volz, E. P. Röth, R. Borchers, and P. Fabian (1981). The vertical distribution of CH_3Cl, $CFCl_3$ and CF_2Cl_2 in the midlatitude stratosphere. *Proc. Quadrenn. Int. Ozone Symp., Boulder, Colo., Natl. Center Atmos. Res.* pp. 816-823.

Schmidt, E. L. (1982). Nitrification in soil. *Agronomy* **22**, 253-288.

Schmidt, V., G. Y. Zhu, K. H. Becker, and E. F. Fink (1985). Study of OH reactions at high pressures by excimer laser photolysis-dye laser fluorescence. *Ber. Bunsen-Ges. Phys. Chem.* **89**, 321-322.

Schneider, S. H. (1975). On the carbon dioxide-climate confusion. *J. Atmos. Sci.* **32**, 2060-2066.

Schönbein, C. F. (1840). Beobachtungen über den bei der Elektrolysation des Wassers und dem Ausströmen der gewöhnlichen Elektrizität aus Spitzen eich entwickelnden Geruch. *Ann. Phys. Chem.* **50**, 616.

Schönbein, C. F. (1854). Über verschiedene Zustände des Sauerstoffs. *Liebigs Ann. Chem.* **89**, 257-300.

Schooley, A. H. (1969). Evaporation in the laboratory and at sea. *J. Marine Res.* **27**, 335-338.

Schopf, J. W. (1975). Precambrian paleobiology: problems and perspectives. *Annu. Rev. Earth Planet. Sci.* **3**, 213-249.

Schopf, J. W., and D. Z. Oehler (1976). How old are the Eucaryotes? *Science* **193**, 47-49.

Schubert, K. R., and H. J. Evans (1976). Hydrogen evolution: a major factor affecting the efficiency of nitrogen fixation in modulated symbionts. *Proc. Nat. Acad. Sci. U.S.* **73**, 1207-1211.

Schubert, B., U. Schmidt, and D. H. Ehhalt (1984). Sampling and analysis of acetaldehyde in tropospheric air. *In* "Physico-Chemical Behaviour of Atmospheric Pollutants" (B. Versino and G. Angeletti, eds.), *Proc. Eur. Symp., 3rd, Varese, Italy*, pp. 44-52. Reidel, Dordrecht, The Netherlands.

Schuck, E. A., H. W. Ford, and E. R. Stephens (1958). Air pollution effects of irradiated automobile exhaust as related to fuel composition, Rep. No. 26, Air Pollution Foundation, San Marino, Calif.

Schütz, K., C. Junge, R. Beck, and B. Albrecht (1970). Studies of atmospheric N_2O. *J. Geophys. Res.* **75**, 2230-2246.

Schütz, L., and R. Jaenicke (1974). Particle number and mass distributions above 10^{-4} cm radius in sand and aerosol of the Sahara Desert. *J. Appl. Meteorol.* **13**, 863-870.

Schütz, L. (1980). Long range transport of desert dust with special emphasis on the Sahara. *In* "Aerosols: Anthropogenic and Natural, Sources and Transport" (T. J. Kneip and P. J. Lioy, eds.), *Ann. N.Y. Acad. Sci.* **338**, 515-532.

Schütz, L., and K. A. Rahn (1982). Trace-element concentrations in erodible soils. *Atmos. Environ.* **16**, 171-176.

Schuetzle, D., A. L. Crittenden, and R. J. Charlson (1973). Application of computer controlled high resolution mass spectroscopy to the analysis of air pollutants. *J. Air Pollut. Control Assoc.* **23**, 704-709.

Schuetzle, D., D. Cronn, A. L. Crittenden, and R. J. Charlson (1975). Molecular composition of secondary aerosols and its possible origin. *Environ. Sci. Technol.* **9**, 838-845.

Schuetzle, D., and R. A. Rasmussen (1978). The molecular composition of secondary aerosol particles formed from terpenes. *J. Air Pollut. Control Assoc.* **28**, 236-240.

Schumacher, H. J. (1930). The mechanism of the photochemical decomposition of ozone. *J. Am. Chem. Soc.* **52**, 2377-2391.

Schumacher, H. J. (1932). Die Photokinetik des Ozons I: Der Zerfall im roten Licht. *Z. Phys. Chem. B* **17**, 405-416.

Schwartz, W. (1974). Chemical characterization of model aerosols. **EPA-650/3-76-085**. U.S. Environmental Protection Agency, Research Triangle Park, North Carolina.

Schwartz, R. M., and M. O. Dayhoff (1978). Origins of procaryotes, eucaryotes, mitochondria and chloroplasts. *Science* **199**, 395-403.

Schwartz, S. E. and J. E. Freiberg (1981). Mass-transport limitation to the rate of reactions of gases in liquid droplets: application to oxidation of SO_2 in aqueous solutions. *Atmos. Environ.* **15**, 1129-1144.

Schwartz, S. E., and W. H. White (1981). Solubility equilibria of the nitrogen oxides and oxiacids in dilute aqueous solution. *Adv. Environ. Sci. Eng.* **4**, 1-45.

Schwartz, S. E., and W. H. White (1983). Kinetics of reactive dissolution of nitrogen oxides with aqueous solution. *Adv. Environ. Sci. Technol.* **12**, 1-116.

Schwartz, S. E. (1986). Mass-transport considerations pertinent to aqueous phase reactions of gases in liquid water clouds. *In* "Multiphase Atmospheric Chemistry" (W. Jaeschke, ed.), pp. 415-471. NATO ASI Series Vol. G 6. Springer-Verlag, Berlin and New York.

Scott, W. E., E. R. Stephens, P. L. Hanst, and R. C. Doerr (1957). Further developments in the chemistry of the atmosphere. *Proc. Am. Petrol. Inst. (III)* **37**, 171-183.

Scott, W. D., and F. C. R. Cattell (1979). Vapor pressure of ammonium sulfates. *Atmos. Environ.* **13**, 307-317.

Scranton, M. I., W. R. Barger, and F. L. Herr (1980). Molecular hydrogen in the urban troposphere: measurements of seasonal variability. *J. Geophys. Res.* **85**, 5575-5580.

Sehmel, G. A., and S. L. Sutter (1974). Particle deposition rates on a water surface as a function of particle diameter and air velocity. *J. Rech. Atmos.* **8**, 911-918.

Sehmel, G. A., and W. H. Hodgson (1976). Predicted dry deposition velocities. *In* "Atmosphere-surface Exchange of Particulate and Gaseous Pollutants," *Proc. Symp. Richland, Washington, 1974, ERDA Symp. Ser.* **38**, 399-419. CONF 740921, NTIS, Springfield, Virginia.

Sehmel, G. A. (1980). Particle and gas dry deposition: a review. *Atmos. Environ.* **14**, 983-1011.

Seiler, W., and C. Junge (1969). Decrease of carbon monoxide mixing ratio above the polar tropopause. *Tellus* **21**, 447-449.

Seiler, W., and C. E. Junge (1970). Carbon monoxide in the atmosphere. *J. Geophys. Res.* **75**, 2217-2226.

Seiler, W., and P. Warneck (1972). Decrease of CO-mixing ratio of the tropopause. *J. Geophys. Res.* **77**, 3204-3214.

Seiler, W. (1974). The cycle of atmospheric CO. *Tellus* **26**, 116-135.

Seiler, W., and U. Schmidt (1974a). New aspects on CO and H_2 cycles in the atmosphere. *Proc. Int. Conf. Structure, Composition, General Circulation Upper Lower Atmos., Melbourne, Vol. 1*, pp. 192–222.

Seiler, W., and U. Schmidt (1974b). Dissolved nonconservative gases in seawater. *In* "The Sea" (E. P. Goldberg, ed.), Vol. 5, pp. 219–243. Wiley, New York.

Seiler, W., and J. Zankl (1975). Die Spurengase CO and H_2 über München. *Umschau* **75**, 735–736.

Seiler, W. (1978). The influence of the biosphere on the atmospheric CO and H_2 cycles. *In* "Environmental Biochemistry and Geomicrobiology" (W. E. Krumbein, ed.), pp. 773–810. Ann. Arbor Science Publ., Ann Arbor, Michigan.

Seiler, W., H. Giehl, and G. Bunse (1978a). Influence of plants on atmospheric carbon monoxide and dinitrogen oxide. *Pure Appl. Geophys.* **116**, 439–451.

Seiler, W., F. Müller, and H. Oeser (1978b). Vertical distribution of chlorofluorocarbons in the upper troposphere and lower stratosphere. *Pure Appl. Geophys.* **116**, 554–566.

Seiler, W., and P. Crutzen (1980). Estimates of the gross and natural flux of carbon between the biosphere and the atmosphere from biomass burning. *Clim. Change* **2**, 207–247.

Seiler, W., and J. Fishman (1981). The distribution of carbon monoxide and ozone in the free troposphere. *J. Geophys. Res.* **86**, 7275–7256.

Seiler, W., and R. Conrad (1981). Field measurements of natural and fertilizer-induced N_2O release rates from soils. *J. Air Pollut. Control Assoc.* **31**, 767–772.

Seiler, W., A. Holzapfel-Pschorn, R. Conrad, and D. Scharffe (1984a). Methane emissions from rice paddies. *J. Atmos. Chem.* **1**, 241–268.

Seiler, W., R. Conrad, and D. Scharffe (1984b). Field studies of methane emissions from termite nests into the atmosphere and measurements of methane uptake by tropical soils. *J. Atmos. Chem.* **1**, 171–186.

Seiler, W., H. Giehl, E. G. Brunke, and E. A. Halliday (1984c). The seasonality of CO abundance in the southern hemisphere. *Tellus* **36B**, 219–231.

Seiler, W. (1984). Contribution of biological processes to the global budget of CH_4 in the atmosphere. *In* "Current Perspectives in Microbial Ecology" (M. J. Klug and C. A. Reddy, eds.), pp. 468–477. Am. Soc. Meteorol., Washington, D.C.

Seiler, W., and R. Conrad (1987). Contribution of tropical ecosystems to the global budgets of trace gases especially CH_4, H_2, CO and N_2O. *In* "Geophysiology of Amazonia" (R. Dickinson, ed.), pp. 133–162. Wiley, New York.

Sellers, W. D. (1965). "Physical Climatology." Univ. of Chicago Press, Chicago.

Sexton, K., and H. Westberg (1984). Nonmethane hydrocarbon composition of urban and rural atmospheres. *Atmos. Environ.* **18**, 1125–1132.

Shardanand (1969). Absorption cross sections of O_2 and O_4 between 2000 and 2800 Å. *Phys. Rev.* **186**, 5–9.

Shardanand and A. D. P. Rao (1977). Collision-induced absorption of O_2 in the Herzberg continuum. *J. Quant. Spoctrosc. Radiat. Transfer* **17**, 433–439.

Shaw, R. W., Jr., R. K. Stevens, and J. Bowermaster (1982). Measurements of atmospheric nitrate and nitric acid: The denuder difference experiment. *Atmos. Environ.* **16**, 845–853.

Shaw, R. W., and R. J. Paur (1983). Measurements of sulfur in gases and particles during sixteen months in the Ohio River Valley. *Atmos. Environ.* **17**, 1431–1438.

Shaw, R. W., A. L. Crittenden, R. K. Stevens, D. R. Cronn, and V. S. Titov (1983). Ambient concentrations of hydrocarbons from conifers in atmospheric gases and aerosol particles measured in Soviet Georgia. *Environ. Sci. Technol.* **17**, 389–396.

Shepherd, E. S. (1925). The analysis of gases from volcanoes and from rocks. *J. Geol.* **33**, 289–370.

Shepherd, E. S. (1938). The gases in rocks and some related problems. *Am. J. Sci. 5th Ser.* **235a**, 311-351.

Shepherd, J. G. (1974). Measurements of the direct deposition of sulphur dioxide onto grass and water by the profil method. *Atmos. Environ.* **8**, 69-74.

Sheppard, J. C., H. Westberg, J. F. Hopper, K. Ganesan, and P. Zimmerman (1982). Inventory of global methane sources and their productions rates. *J. Geophys. Res.* **87**, 1305-1312.

Shneour, E. A. (1966). Oxidation of graphitic carbon in certain soils. *Science* **151**, 991-992.

Shugard, W. J., R. H. Heist, and J. J. Reiss (1974). Theory of water phase nucleation in binary mixtures of water and sulfuric acid. *J. Chem. Phys.* **61**, 5298-5307.

Sidebottom, H. W., C. C. Badcock, G. E. Jackson, J. G. Calvert, G. W. Reinhardt, and E. K. Damon (1972). Photooxidation of sulfur dioxide. *Environ. Sci. Technol.* **6**, 72-79.

Siegenthaler, U., and H. Oeschger (1978). Predicting future atmospheric carbon dioxide levels. *Science* **199**, 388-395.

Siegenthaler, U., M. Heimann, and H. Oeschger (1978). Model responses of the atmospheric CO_2 level and $^{13}C/^{12}C$ ratio to biogenic CO_2 input. *In* "Climate and Society" (J. Williams, ed.), pp. 79-84. Pergamon Press, New York.

Siegenthaler, U. and K. O. Münnich (1981). $^{13}C/^{12}C$ fractionation during CO_2 transfer from air to sea. *In* "Carbon Cycle Modelling" (B. Bolin, ed.), *SCOPE* **16**, 249-257.

Siegenthaler, U. (1983). Uptake of excess CO_2 by an outcrop-diffusion model of the ocean. *J. Geophys. Res.* **88**, 3599-3608.

Siever, R. (1968). Sedimentological consequences of a steady state ocean-atmosphere. *Sedimentology* **11**, 5-29.

Signer, P., and H. E. Suess (1963). Rare gases in the sun, in the atmosphere and in meteorites. *In* "Earth Science and Meteorites" (J. Geiss and E. D. Goldberg, eds.), pp. 241-272. North-Holland, Amsterdam.

Sigvaldason, G. E., and G. Elisson (1968). Collection and analysis of volcanic gases at Surtsey, Iceland. *Geochim. Cosmochim. Acta* **32**, 797-805.

Sillén, L. G. (1961). The physical chemistry of seawater. *In* "Oceanography" (M. Sears, ed.), *Am. Assoc. Adv. Sci. Publ.* **67**, 549-581.

Sillén, L. G. (1963). How has seawater got its present composition? *Sven. Kem. Tidskr.* **75**, 161-177.

Sillén, L. G. (1966). Regulation of O_2, N_2 and CO_2 in the atmosphere; thoughts of a laboratory chemist. *Tellus* **18**, 198-206.

Siman, G., and S. L. Janson (1976). Sulfur exchange between soil and atmosphere. *Swed. J. Agric. Res.* **6**, 37-45.

Simmonds, P. G., F. N. Ayea, C. A. Cardelino, A. J. Crawford, D. M. Cunnold, B. C. Lane, J. E. Lovelock, R. G. Prinn, and R. A. Rasmussen (1983). The atmospheric lifetime experiment 6. Results for carbon tetrachoride based on 3 years data. *J. Geophys. Res.* **88**, 8427-8441.

Simon, P. C. (1978). Irradiation solar flux measurements between 120 and 400 nm. Current position and future needs. *Planet. Space Sci.* **26**, 355-365.

Simoneit, B. R. T. (1977). Organic matter in eolian dust over the Atlantic ocean. *Mar. Chem.* **5**, 443-464.

Simoneit, B. R. T., R. Chester, and G. Eglington (1977). Biogenic lipids in particulates from the lower atmosphere over the eastern Atlantic. *Nature* **267**, 682-685.

Simoneit, B. R. T., and M. A. Mazurek (1981). Air pollution: The organic components. *CRC Crit. Rev. Environ. Control* **11**, 219-276.

Simoneit, B. R. T., and M. A. Mazurek (1982). Organic matter of the troposphere—II. Natural background of biogenic lipid matter in aerosols over the rural western United States. *Atmos. Environ.* **16**, 2139-2159.

Simoneit, B. R. T. (1984). Organic matter of the troposphere—III. Characterization and sources of petroleum and pyrogenic residues in aerosols over the western United States. *Atmos. Environ.* **18**, 51-57.

Simonsen, J., and D. H. R. Barton (1961). "The Terpenes," Vol. III: The Sesquiterpenes, Diterpenes and their Derivatives. Cambridge Univ. Press, London.

Sinanoglu, O. (1981). What size cluster is like a surface? *Chem. Phys. Lett.* **81**, 188-190.

Singh, H. B., D. P. Fowler, and T. O. Peyton (1976). Atmospheric carbon tetrachloride: another manmade pollutant. *Science* **192**, 1231-1234.

Singh, H. B. (1977a). Atmospheric halocarbons: evidence in favor of reduced average hydroxyl radical concentration in the troposphere. *Geophys. Res. Lett.* **3**, 101-104.

Singh, H. B. (1977b). Preliminary estimation of average tropospheric HO concentrations in the northern and southern hemispheres. *Geophys. Res. Lett.* **4**, 453-456.

Singh, H. B., L. J. Salas, and L. A. Cavanagh (1977a). Distribution, sources and sinks of atmospheric halogenated compounds. *J. Air. Pollut. Control Assoc.* **27**, 332-336.

Singh, H. B., L. Salas, H. Shigeishi, and A. Crawford (1977b). Urban-non-urban relationships of halocarbons, SF_6, N_2O and other atmospheric trace constituents. *Atmos. Environ.* **11**, 819-828.

Singh, H. B., F. L. Ludwig, and W. B. Johnson (1978). Tropospheric ozone: concentrations and variabilities in clean remote atmospheres. *Atmos. Environ.* **12**, 2185-2196.

Singh, H. B., L. J. Salas, and H. Shigeishi (1979a). The distribution of nitrous oxide in the global atmosphere and in the Pacific Ocean. *Tellus* **31**, 313-320.

Singh, H. B., L. J. Salas, H. Shigeishi, and E. Scribner (1979b). Atmospheric halocarbons, hydrocarbons, hexafluoride: Global distributions, sources and sinks. *Science* **203**, 899-903.

Singh, H. B., and P. L. Hanst (1981). Peroxyacetyl nitrate (PAN) in the unpolluted atmosphere: an important reservoir for nitrogen oxides. *Geophys. Res. Lett.* **8**, 941-944.

Singh, H. B., and L. J. Salas (1982). Measurement of selected light hydrocarbons over the Pacific Ocean: latitudinal and seasonal variations. *Geophys. Res. Lett.* **9**, 842-845.

Singh, H. B., and L. J. Salas (1983). Peroxyacetyl nitrate in the free troposphere. *Nature* **302**, 326-328.

Singh, H. B., L. J. Salas, and R. E. Stiles (1983a). Selected man-made halogenated chemicals in the air and oceanic environment. *J. Geophys. Res.* **88**, 3675-3683.

Singh, H. B., L. J. Salas, and R. E. Stiles (1983b). Methyl halides in and over the Eastern Pacific (40°N-32°S). *J. Geophys. Res.* **88**, 3684-3690.

Sjödin, A., and P. Grennfelt (1984). Regional background concentration of NO_2 in Sweden. *In* "Physico-chemical Behaviour of Atmospheric Pollutants" (B. Versino and G. Angeletti, eds.), *Proc. Eur. Symp., 3rd, Varese, Italy*, pp. 401-411. Reidel, Dordrecht, The Netherlands.

Skirrow, G. (1975). The dissolved gases—carbon dioxide. *In* "Chemical Oceanography", 2nd ed. (J. P. Riley and G. Skirrow, eds.), Vol. 2, Chap. 9, pp. 1-192. Academic Press, New York.

Skrabal, A., and R. Skrabal (1936). "Die Dynamik der Formaldehyd-Bisulfit Reaktion." *Monatsh. Chemi.* **69**, 11-41.

Slagle, I. R., R. E. Graham, and D. Gutman (1976). Direct identification of reactive routes and measurement of rate constants in the reactions of oxygen atoms with methanethiol, ethanethiol and methylsulfide. *Int. J. Chem. Kinet.* **8**, 451-458.

Slagle, I. R., F. Baiocchi, and D. Gutman (1978). Studies of the reaction of oxygen atoms with hydrogen sulfide, methanethiol, ethanethiol and methyl sulfide. *J. Phys. Chem.* **82**, 1333-1336.

Slater, D. H., and J. G. Calvert (1968). The photo-oxidation of 1,1'-azoisobutane. Reactions of the isobutyl free radical with oxygen. *Adv. Chem. Ser.* **76**, 58-68.

Slatt, B. J., D. F. S. Natush, J. M. Prospero, and D. L. Savoie (1978). Hydrogen sulfide in the atmosphere of the northern equatorial Atlantic Ocean and its relation to the global sulfur cycle. *Atmos. Environ.* **12**, 981-991.

Slemr, F., and W. Seiler (1984). Field measurements of NO and NO_2 emissions from fertilized and unfertilized soils. *J. Atmos. Chem.* **2**, 1-24.

Slinn, W. G. N., and J. M. Hales (1971). A reevaluation of the role of thermophoresis as a mechanism of in and below—cloud scavenging. *J. Atmos. Sci.* **28**, 1465-1471.

Slinn, W. G. N. (1974). Proposed terminology for precipitation scavenging. *In* "Precipitation Scavenging" (R. G. Semonin and R. W. Beadle, coordinators), *ERDA Symp. Ser.* pp. 813-818, **CONF-741003**, U.S. Dept. of Commerce, Springfield, Virginia.

Slinn, W. G. N. (1976). Some approximations for the wet and dry removal of particles and gases from the atmosphere. *J. Air Water Soil Pollut.* **7**, 513-543.

Slinn, W. G. N. (1977). Precipitation scavenging: some problems, approximate solutions and suggestions for future research. *In* "Precipitation Scavenging 1974" (R. G. Semonin and R. W. Beadle, eds.), *ERDA Symp. Ser.* **41**, pp. 1-60. Tech. Information Center Research and Development Energy Administration.

SMIC (Report of the Study of Man's impact on climate) (1971). "Inadvertent Climate Modification." MIT Press, Cambridge, Massachusetts.

Smith, R. A. (1872). "Air and Rain—The beginnings of a Chemical Climatology." Longmans, Green, London.

Smith, J. P., and P. Urone (1974). Static studies of sulfur dioxide reactions, effects of NO_2, C_2H_6 and H_2O. *Environ. Sci. Technol.* **8**, 742-746.

Smith, C. J., and P. M. Chalk (1979). Factors affecting the determination of nitric oxide and nitrogen dioxide evolution from soil. *Soil Sci.* **128**, 327-330.

Smithsonian Meteorological Tables (1951). 6th ed., Smithsonian Institution, Washington, D.C.

Smoluchowski, M. V. (1918). Versuch einer mathematischen Theorie der Koagulationskinetik kolloider Lösungen. *Z. Phys. Chem.* **92**, 129-168.

Söderlund, R., and B. H. Svensson (1976). The global nitrogen cycle. *In* "Nitrogen, Phosphorous and Sulfur Global Cycles" (B. H. Svensson and R. Söderlund, eds.), *SCOPE Rep.* 7, *Ecol. Bull.* (*Stockholm*) **22**, 23-73.

Sood, S. K., and M. R. Jackson (1970). Scavenging by snow and ice crystals. *In* "Precipitation Scavenging 1970" (R. J. Engelmann and W. G. N. Slinn, coordinators), pp. 121-134, **NTIS CONF 700601**. U.S. Atomic Energy Commission, Div. of Tech. Information.

Spedding, D. J. (1972). Sulphur dioxide absorption by seawater. *Atmos. Environ.* **6**. 583-586.

Spicer, C. W. (1980). The rate of NO_x reaction in transported urban air. *In* "Studies in Environmental Science" (M. M. Benarie, ed.), Vol. 8, pp. 181-186. Elsevier, Amsterdam, The Netherlands.

Spicer, C. W. (1982). Nitrogen oxide reactions in the urban plume of Boston. *Science* **215**, 1095-1092.

Spicer, C. W., J. E. Howes, Jr., T. A. Bishop, and L. H. Arnold (1982). Nitric acid measurement methods: an intercomparison. *Atmos. Environ.* **16**, 1476-1500.

Stallard, R. J., J. M. Edmond, and R. E. Newell (1975). Surface ozone in the south east Atlantic between Dakar and Walvis Bay. *Geophys. Res. Lett.* **2**, 289-292.

Stallard, R. F., and J. M. Edmond (1981). Geochemistry of the Amazon, 1. Precipitation chemistry and the marine contribution to the dissolved load at the time of peak discharge. *J. Geophys. Res.* **86**, 9844-9858.

Stanford, J. L. (1973). Possible sink for stratospheric water vapor at the winter Antarctic pole. *J. Atmos. Sci.* **30**, 1431-1436.

Starr, J. R., and B. J. Mason (1966). The capture of airborne particles by water drops and simulated snow crystals. *Q. J. R. Meteorol. Soc.* **92**, 490-499.

Stedman, D. H., W. Chameides, and J. O. Jackson (1975). Comparison of experimental and computed values for $j(NO_2)$. *Geophys. Res. Lett.* **2**, 22-25.

Stedman, D. H., and J. O. Jackson (1975). The photostationary state in photochemical smog. *Int. J. Chem. Kinet.* **7** (Symp. 1), 493-501.

Stedman, D. H., R. J. Cicerone, W. L. Chameides, and R. B. Harvey (1976). Absence of N_2O photolysis in the troposphere. *J. Geophys. Res.* **81**, 2003-2004.

Stedman, D. H., and M. J. McEwan (1983). Oxides of nitrogen at two sites in New Zealand. *Geophys. Res. Lett.* **10**, 168-171.

Stedman, D. H., and R. E. Shetter (1983). The global budget of atmospheric nitrogen species. "Trace Atmospheric Constituents, Properties, Transformations and Fates" (S. E. Schwartz, ed.), pp. 411-454. Wiley, New York.

Steed, J. M., A. J. Owens, C. Miller, D. L. Filkin, and J. P. Jesson (1982). Two dimensional modelling of potential ozone perturbation by chlorofluorocarbons. *Nature* **295**, 308-311.

Steiger, R. H., and E. Jäger (1977). Subcommission on geochronology: convention on the use of decay constants in Geo- and Cosmochronology. *Earth Planet. Sci. Lett.* **36**, 359-362.

Steinhardt, U. (1973). Input of chemical elements from the atmosphere. A tabular review of literature. *Göttinger Bodenkd. Ber.* **29**, 93-132.

Stelson, A. W., S. K. Friedlander, and J. H. Seinfeld (1979). A note on the equilibrium relationship between ammonia and nitric acid and particulate ammonium nitrate. *Atmos. Environ.* **13**, 369-371.

Stelson, A. W., and J. H. Seinfeld (1981). Chemical mass accounting of urban aerosol. *Environ. Sci. Technol.* **15**, 671-679.

Stelson, A. W., and J. H. Seinfeld (1982). Relative humidity and pH dependence of the vapor pressure of ammonium nitrate-nitric acid solutions at 25 degree Celsius. *Atmos. Environ.* **16**, 993-1000.

Stephens, E. R., P. L. Hanst, R. C. Doerr, and W. E. Scott (1956). Reactions of nitrogen oxides and organic compounds in air. *Ind. Eng. Chem.* **48**, 1498-1504.

Stephens, E. R., P. L. Hanst, R. C. Doerr, and W. E. Scott (1958). Auto exhaust: composition and photolysis products. *J. Air Pollut. Control Assoc.* **8**, 333-335.

Stephens, E. R., E. F. Darley, O. C. Taylor, and W. E. Scott (1961). Photochemical reaction products in air pollution. *Int. J. Air. Water Pollut.* **4**, 79-100.

Stephens, E. R. and F. R. Burleson (1967). Analysis of the atmosphere for light hydrocarbons. *J. Air Pollut. Control Assoc.* **17**, 147-153.

Stephens, E. R., and F. R. Burleson (1969). Distribution of light hydrocarbons in ambient air. *J. Air. Pollut. Control Assoc.* **19**, 929-936.

Stephens, E. R. (1971). Identification of odors from cattle feed lots. *Calif. Agric.* **25**, 10-11.

Stewart, N. G., R. N. Crooks, and E. M. R. Fisher (1956). The radiological dose to persons in the U.K. due to debris from nuclear test explosions prior to January 1956, **AERE HP/R-2017**. Atomic Energy Research Establishment, Harwell, England.

Stewart, R. W., S. Hameed, and J. P. Pinto (1977). Photochemistry of tropospheric ozone. *J. Geophys. Res.* **82**, 3134-3140.

Stith, J. C., P. V. Hobbs, and L. F. Radke (1978). Airborne particle and gas measurements in the emissions from six volcanoes. *J. Geophys. Res.* **83**, 4009-4017.

Stith, J. L., L. F. Radke, and P. V. Hobbs (1981). Particle emissions and the production of ozone and nitrogen oxides from the burning of forest slash. *Atmos. Environ.* **15**, 73-82.

Stix, E. (1969). Schwankungen des Pollen- und Sporengehaltes der Luft. *Umschau* **19**, 620-621.

St. John, D. S., S. P. Bailey, W. H. Fellner, J. M. Manor, and R. D. Snee (1981). Time series analysis for trends in total ozone measurements. *J. Geophys. Res.* **86**, 7299-7311.

Stockwell, W. R., and J. G. Calvert (1978). The near ultraviolet absorption spectrum of gaseous HONO and N_2O_3. *J. Photochem.* **8**, 193–203.

Stockwell, W. R., and J. G. Calvert (1983). The mechanism of NO_3 and HONO formation in the nighttime chemistry of the urban atmosphere. *J. Geophys. Res.* **88**, 6673–6682.

Stoiber, R. E., and A. Jepsen (1973). Sulfur dioxide contribution to the atmosphere by volcanoes. *Science* **182**, 577–578.

Stoiber, R. E., and G. B. Malone (1975). SO_2 emissions at the crater of Kilauea, at Mauna Ulu and Sulfur Banks, Hawaii. *EOS Trans. AGU* **56**, 461.

Stoiber, R. E., and G. Bratton (1978). Airborne correlation spectrometer measurements of SO_2 in eruption clouds from Guatemalian volcanoes. *EOS Trans. AGU* **59**, 1222.

Strobel, D. F. (1971). Odd nitrogen in the mesosphere. *J. Geophys. Res.* **76**, 8384–8393.

Strutt, R. J. (1918). Ultraviolet transparency of the lower atmosphere and its relative poverty in ozone. *Proc. R. Soc. A* **94**, 260–268.

Stuhl, F., and H. Niki (1972). Pulsed vacuum uv-photochemical study of reactions of OH with H_2, D_2 and CO using a resonance-fluorescence method. *J. Chem. Phys.* **57**, 3671–3677.

Stuiver, M. (1978). Atmospheric CO_2 increases related to carbon reservoir changes. *Science* **199**, 253–258.

Stuiver, M. (1981). C-14 distribution in the Atlantic Ocean. *J. Geophys. Res.* **85**, 2711–2718.

Stuiver, M. (1982). The history of the atmosphere as recorded by carbon isotopes. *In* "Atmospheric Chemistry" (E. D. Goldberg, ed.), pp. 159–179. *Dahlem Konferenzen, 1980*. Springer-Verlag, Berlin, New York.

Stull, D. R., and H. Prophet (1971). "JANAF Thermochemical Tables," 2nd ed., *Natl. Stand. Ref. Data Ser.*, Nat. Bur. of Stand. **37**, U.S. Government Printing Office, Washington, D.C.

Stumm, W., and P. A. Brauner (1975). Chemical speciation, Chapter 3. *In* "Chemical Oceanography," 2nd ed. (J. P. Riley and G. Skirrow, eds.), pp. 173–239. Academic Press, London.

Stumm, W., and J. J. Morgan (1981). "Aquatic chemistry." Wiley, New York.

Su, C. W., and E. D. Goldberg (1973). Chlorofluorocarbons in the atmosphere. *Nature* **245**, 27.

Suess, H. E. (1949). Die Häufigkeit der Edelgase auf der Erde und im Kosmos. *J. Geol. Res.* **57**, 600–607.

Suess, H. E. (1966). Some chemical aspects of the evolution of the terrestrial atmosphere. *Tellus* **18**, 207–211.

Sullivan, J. O., and P. Warneck (1967). Reaction of O^1D oxygen atoms, III. Ozone formation in the 1470 Å photolysis of O_2. *J. Chem. Phys.* **46**, 953–959.

Sundquist, E. T., L. N. Plummer, and T. M. L. Wigley (1979). Carbon dioxide in the ocean surface: The homogenous buffer factor. *Science* **204**, 1203–1205.

Sutton, O. G. (1953). "Micrometeorology." McGraw-Hill, New York.

Sutton, O. G. (1955). "Atmospheric Turbulence," 2nd ed. Wiley, New York.

Sverdrup, H. U., M. W. Johnson, and R. H. Fleming (1942). "The Oceans." Prentice-Hall, Englewood Cliffs, New Jersey.

Swinnerton, J. W., and V. J. Linnenboom (1967). Gaseous hydrocarbons in sea water: determination. *Science* **156**, 119–120.

Swinnerton, J. W., V. J. Linnenboom, and C. H. Cheek (1969). Distribution of methane and carbon monoxide between the atmosphere and natural waters. *Environ. Sci. Technol.* **3**, 836–838.

Swinnerton, J. W., V. J. Linnenboom, and R. A. Lamontagne (1970). Ocean: a natural source of carbon monoxide. *Science* **167**, 984–986.

Swinnerton, J. W., and R. A. Lamontagne (1974a). Carbon monoxide in the southern hemisphere. *Tellus* **26**, 136–142.

Swinnerton, J. W., and R. A. Lamontagne (1974b). Oceanic distribution of low-molecular weight hydrocarbons. *Environ. Sci. Technol.* **8**, 657–663.

Sze, N. D., M. K. W. Ko, M. Lifshitz, W. C. Wang, P. B. Ryan, R. E. Specht, M. B. McElroy, and S. C. Wofsy (1983). *Annu. Rep. Atmos. Chem., Radiat., and Dynam. Prog.*, Atmospheric and Environmental Research, Inc., Cambridge, Massachusetts.

Takahashi, T., W. S. Broecker, S. R. Werner, and A. E. Bainbridge (1980). Carbonate chemistry of the surface waters of the world oceans. *In* "Isotope Marine Chemistry" (E. D. Goldberg, Y. Hozibe, and K. Sarnhashi, eds.), pp. 291–326. Uchida Rokakuho, Tokyo.

Takahashi, T., W. S. Broecker, and A. E. Bainbridge (1981a). The alkalinity and total carbon dioxide concentration in the world oceans. *In* "Carbon Cycle Modelling" (B. Bolin, ed.), *SCOPE* **16**, 271–286.

Takahashi, T., W. S. Broecker, and A. E. Bainbridge (1981b). Supplement to the alkalinity and total carbon dioxide concentration in the world oceans. *In* "Carbon Cycle Modelling" (B. Bolin, ed.), *SCOPE* **16**, 159–199.

Takeuchi, H., M. Ando, and N. Kizawa (1977). Absorption of nitrogen oxides in aqueous sodium sulfite and bisulfite solutions. *Ind. Eng. Chem. Process Design. Devel.* **16**, 303–308.

Tanaka, Y., E. C. Y. Inn, and K. Watanabe (1953). Absorption coefficients of gases in the vacuum ultraviolet. Part. IV: Ozone. *J. Chem. Phys.* **21**, 1651–1653.

Tanaka, S., M. Durgi, and J. W. Winchester (1980). Short term effect of rain fall on elemental composition and size distribution of aerosols in north Florida. *Atmos. Environ.* **14**, 1421–1426.

Tang, I. N., and H. R. Munkelwitz (1977). Aerosol growth studies, III. Ammonium bisulfate aerosols in a moist atmosphere. *J. Aerosol Sci.* **8**, 321–330.

Tang, I. N., H. R. Munkelwitz, and J. G. Davis (1977). Aerosol growth studies, II. Preparation and growth measurements of monodisperse salt aerosols. *J. Aerosol Sci.* **8**, 149–159.

Tang, K. Y., P. W. Fairschild, and E. K. C. Lee (1979). Laser induced photodecomposition of formaldehyde ($\tilde{A}^1A_2$) from its single vibronic levels. Determination of quantum yields of H-atom by $HNO^*(\tilde{A}^1A'')$ chemiluminescence. *J. Phys. Chem.* **83**, 569–573.

Tang, I. N. (1980). On the equilibrium partial pressure of nitric acid and ammonia in the atmosphere. *Atmos. Environ.* **14**, 819–828.

Tanner, R., W. H. Marlow, and L. Newman (1979). Chemical composition correlations of size-fractionated sulfate in New York City. *Environ. Sci. Technol.* **13**, 75–78.

Tans, P. (1981). A computation of bomb C-14 data for use in global carbon model calculations. *In* "Carbon Cycle Modelling" (B. Bolin, ed.), *SCOPE* **16**, 131–157.

Taylor, O. C. (1969). Importance of peroxyacetyl nitrate (PAN) as a phytotoxic air pollutant. *J. Air Pollut. Control Assoc.* **19**, 347–351.

Taylor, P. S., and R. E. Stoiber (1973). Soluble material on ash from active Central American volcanoes. *Geol. Soc. Am. Bull.* **84**, 1031–1042.

Telegadas, K. (1972). Atmospheric radioactivity along the HASL ground-levels sampling network, 1968 to 1970 as an indicator of tropospheric and stratospheric sources. *J. Geophys. Res.* **77**, 1004–1011.

Temple, P. J., and O. C. Taylor (1983). World-wide ambient measurements of peroxyacetyl nitrate (PAN) and implications for plant injury. *Atmos. Environ.* **8**, 1583–1587.

Thompson, A. M. (1980). Wet and dry removal of tropospheric formaldehyde at a coastal site. *Tellus* **32**, 376–383.

Thorne, L., and G. P. Hanson (1972). Species differences in rates of vegetal ozone absorption. *Environ. Pollut.* **3**, 303–312.

Titoff, A. (1903). Beiträge zur Kenntnis der negativen Katalyse im homogenen System. *Z. Phys. Chem.* **45**, 641–683.

Tiefenau, H. K., and P. Fabian (1972). The specific ozone destruction at the ocean surface and its dependence on horizontal wind velocity from profile measurements. *Arch. Meteorol. Geophys. Bioklimatol. Ser. A* **21**, 399-412.

Tiefenau, H. K., P. G. Pruchniecwicz, and P. Fabian (1973). Meridional distribution of tropospheric ozone from measurements aboard commercial airliners. *Pure Appl. Geophys.* **106-108**, 1036-1040.

Tingey, D. T., M. Manning, L. C. Grothasu, and W. F. Burns (1979). The influence of light and temperature on isoprene emissions from live oak. *Physiol Plant* **47**, 117-118.

Tingey, D. T., M. Manning, L.G. Grothasu, and W. F. Burns (1980). The influence of light and temperature on monoterpene emission rates from slash pine. *Plant Physiol.* **65**, 797-801.

Tingey, D. T. (1981). The effect of environment factors on the emission of biogenic hydrocarbons from live oak and slash pine. *In* "Atmospheric Biogenic Hydrocarbons" (J. J. Bufalini and R. R. Arnts, eds.), Vol 1, Emissions, pp. 53-79. Ann Arbor Science, Ann Arbor, Michigan.

Tjepkema, J. D., R. J. Cartica, and H. F. Hemond (1981). Atmospheric concentration of ammonia in Massachusetts and deposition on vegetation. *Nature* **294**, 445-446.

Toba, Y. (1965). On the giant sea-salt particles in the atmosphere. *Tellus* **17**, 131-145.

Toba, Y., and M. Chaen (1973). Quantitative expression of the breaking of wind waves on the sea surface. *Rec. Oceanogr. Works Jpn.* **12** (1), 1-11.

Toon, G. C., C. B. Farmer, and R. H. Norton (1986). Detection of stratospheric N_2O_5 by infrared remote sounding. *Nature* **319**, 570-571.

Torres, A. L., P. J. Maroulis, A. B. Goldberg, and A. R. Bandy (1980). Atmospheric OCS measurements on project Gametag. *J. Geophys. Res.* **85**, 7357-7360.

Treinin, A., and E. Hayon (1970). Absorption spectra and reaction kinetics of NO_2, N_2O_3 amd N_2O_4 in aqueous solution. *J. Am. Chem. Soc.* **92**, 5821-5828.

Trenberth, K. E. (1981). Seasonal variations in global sea level pressure and the total mass of the atmosphere. *J. Geophys. Res.* **86**, 5238-5246.

Troe, J. (1974). Fall-off curves of unimolecular reactions. *Ber. Bunsen-Ges. Phys. Chem.* **78**, 478-488.

Troxler, R. F., and J. M. Dokos (1973). Formation of carbon monoxide and bile pigment in red and green algae. *Plant Physiol.* **51**, 72-75.

Tsunogai, S. (1971a). Oxidation rate of sulfite in water and its bearing on the origin of sulfate in meteoritic precipitation. *Geochem. J.* **5**, 175-185.

Tsunogai, S. (1971b). Ammonia in the oceanic atmosphere and the cycle of nitrogen compounds through the atmosphere and hydrosphere. *Geochem. J.* **5**, 57-67.

Tsuya, H. (1955). Geological and petrological studies of volcano Fuji. *Tokyo Daigaku Jishin Kenkyusho Iho* **33**, 341-382.

Tuazon, A. C., A. M. Winer, and J. W. Pitts, Jr. (1981). Trace pollutant concentrations in a multiday smog episode in the Californian south coast air basin by long path length Fourier transform infrared spectroscopy. *Environ. Sci. Technol.* **15**, 1232-1237.

Tuck, A. F. (1976). Production of nitrogen oxides by lightning discharges. *Q. J. R. Meteorol. Soc.* **102**, 749-755.

Tuesday, C. S. (1976). Sources of atmospheric hydrocarbons. *In* "Vapor-Phase Organic Pollutants." *Natl. Acad. Sci.*, Washington, D.C.

Turco, R. P., R. C. Whitten, O. B. Toon, J. B. Poleack, and P. Hamill (1980). OCS, stratospheric aerosols and climate. *Nature* **283**, 283-286.

Turco, R. P., R. J. Cicerone, E. C. Y. Inn, and L. A. Capone (1981a). Long wavelength carbonyl sulfide photodissociation. *J. Geophys. Res.* **86**, 5373-5377.

Turco, R. P., R. C. Whitten, O. B. Toon, E. C. Y. Inn, and P. Hamill (1981b). Stratospheric hydroxyl radical concentrations: new limitations suggested by observations of gaseous and particulate sulfur. *J. Geophys. Res.* **86**, 1129–1139.

Turekian, K. K. (1964). Degassing of argon and helium from the earth. *In* "The Origin and Evolution of Atmospheres and Oceans" (P. J. Brancazio and A. G. W. Cameron, eds.), pp. 74–82. Wiley, New York.

Turekian, K. K. (1971). Elements, geochemical distribution of. *In* "Encyclopedia of Science and Technology," 2nd ed., Vol. 4, pp. 627–630. McGraw-Hill, New York.

Turekian, K. K., Y. Nozaki, and L. K. Benninger (1977). Geochemistry of atmospheric radon and radon products. *Annu. Rev. Earth Planet. Sci.* **5**, 227–255.

Turman, B. N., and B. C. Edgar (1982). Global lightning distributions at dawn and dusk. *J. Geophys. Res.* **87**, 1191–1206.

Turner, N. C., P. E. Waggoner, and S. Rich (1974). Removal of ozone from the atmosphere by soil and vegetation. *Nature* **250**, 486–489.

Twenhofel, W. H. (1951). "Principles of Sedimentation." McGraw-Hill, New York.

Twomey, S. (1963). Measurements of natural cloud nuclei. *J. Rech. Atmos.* **1**, 101–104.

Twomey, S. (1971). The composition of cloud nuclei. *J. Atmos. Sci.* **28**, 377–381.

Twomey, S. (1977). "Atmospheric Aerosols." Elsevier, Amsterdam.

Tyson, B. J., J. C. Arvesen, and D. O'Hara (1978). Interhemispheric gradients of CF_2Cl_2, $CFCl_3$, CCl_4 and N_2O. *Geophys. Res. Lett.* **5**, 535–538.

Tyson, B. J., J. F. Vedder, J. C. Arvesen, and R. B. Brewer (1978). Stratospheric measurements of CF_2Cl_2 and N_2O. *Geophys. Res. Lett.* **5**, 369–372.

Umweltbundesamt (1980). Grossräumige Luftbelastung in der Bundesrepublik Deutschland, Bericht aus dem Messnetz des Unweltbundesamtes, Berlin, Germany.

Umweltbundesamt (1982). Grossräumige Luftbelastungin in der Bundesrepublik Deutschland, Bericht aus dem Messnetz des Umweltbundesamtes, Berlin, Germany.

United Nations (1978). "Yearbook of Industrial Statistics." 1976 edition.

Urey, H. C. (1952). "The Planets." Yale Univ. Press, New Haven, Connecticut.

U.S. Standard Atmosphere (1976). National Oceanic and Atmospheric Administration, Washington, D.C.

Uselman, W. M., and E. K. C. Lee (1976). A study of nitrogen dioxide (2^2B_2) photodecomposition to $O(^1D)$ and $NO(^2II)$ in its second predissociation region 2500–2139 Å. *J. Chem. Phys.* **65**, 1948–1955.

Van Cleemput, O., and L. Baert (1976). Theoretical considerations on nitrite self-decomposition reactions in soils. *Soil Sci. Soc. Am. J.* **40**, 322–323.

Van Cleemput, O., A. S. El-Sebaay, and L. Baert (1981). Production of gaseous hydrocarbons in soil. *In* "Physico-Chemical Behaviour of Atmospheric Pollutants" (B. Versino and H. Ott, eds.), *Proc. Eur. Symp., 2nd, Varese, Italy*, pp. 349–355. Reidel, Dordrecht, The Netherlands.

VanSickle, D. E., T. Mill, F. R. Mayo, H. Richardson, and C. W. Gould (1973). Intramolecular propagation in the oxidation of *n*-alkanes. Autoxidation of *n*-pentane and *n*-octane. *J. Org. Chem.* **38**, 4435–4440.

Van Vaeck, L., G. Broddin, and K. van Cauwenberghe (1979). Differences in particle size distributions of major organic pollutants in ambient aerosols in urban, rural and seashore areas. *Environ. Sci. Technol.* **13**, 1494–1502.

Várhelyi, G. (1978). On the vertical distribution of sulfur complunds in the lower troposphere. *Tellus* **30**, 542–545.

Várhelyi, G., and G. Gravenhorst (1983). Production rate of airborne sea-salt sulfur deduced from chemical analysis of marine aerosols and precipitation. *J. Geophys. Res.* **88**, 6737–6751.

Vassy, A. (1965). "Atmospheric Ozone," *Adv. Geophys.* **11**, 115-173.

Vassy, A. (1968). Concentration de l'air en ozone au niveau du sol a la station Port-Aux-Francais, Iles Kergulen (1961-1966), pub. 1968-5 Lab. de Phys. de l'Atmosphere, Paris.

Vassy, A. (1971). Concentration de l'air en ozone au niveau du sol. Pub. No. 1971-1 Lab. de Phys. de l'Atmosphere, Paris.

Vedder, J. F., B. J. Tyson, R. B. Brewer, C. A. Boitnott, and E. C. Y. Inn (1978). Lower stratospheric measurements of variation with latitude of CF_2Cl_2, $CFCl_3$, CCl_4 and N_2O profiles in the northern hemisphere. *Geophys. Res. Lett.* **5**, 33-36.

Vedder, J. F., C. E. Y. Inn, B. J. Tyron, C. A. Boitnott, and D. O'Hara (1981). Measurements of CF_2Cl_2, $CFCl_3$ and N_2O in the lower stratosphere between 2°S and 73°N latitude. *J. Geophys. Res.* **86**, 7363-7368.

Vidal-Madjar, A. (1975). Evolution of the solar Lyman-alpha flux during four consecutive years. *Solar Phys.* **40**, 69-86.

Vidal-Madjar, A. (1977). The solar spectrum of Lyman α. *In* "The Solar Output and Its Variation" (O. R. White, ed.), pp. 213-234. Colorado Assoc. Univ. Press, Boulder, Colorado.

Viemeister, P. E. (1960). Lightning and the origin of nitrates found in precipitation. *J. Meteorol.* **17**, 681-683.

Viggiano, A. A., and F. Arnold (1983). Stratospheric sulfuric acid vapor. *J. Geophys. Res.* **88**, 1457-1462.

Vilenskiy, V. D. (1970). The influence of natural radioactive atmospheric dust in determining the mean stay time of lead-210 in the troposphere. *Izv. Acad. Sci. USSR, Atmos. Oceanic Phys.* **6**, 307-310.

Vinogradov, A. P. (1959). "The Geochemistry of Rare and Dispersed Elements in Soils," 2nd ed. Consultants Bureau, Inc., New York.

Visser, S. A. (1961). Chemical composition of rainwater in Kampala, Uganda, and its relation to meteorological and topographic conditions. *J. Geophys. Res.* **66**, 3759-3765.

Visser, S. A. (1964). Origin of nitrates in tropical rainwater. *Nature* **201**, 35-36.

Volman, D. H. (1963). Photochemical gas phase reactions in the hydrogen–oxygen system. *Adv. Photochem.* **1**, 43-82.

Volmer, M., and A. Weber (1926). Keimbildung in übersättigten Gebilden. *Z. Phys. Chem.* **119**, 277-301.

Vohra, K. G., K. N. Vasuderan, and P. V. N. Nair (1970). Mechanisms of nucleus-forming reactions in the atmosphere. *J. Geophys. Res.* **75**, 2951-2960.

Volz, A., D. H. Ehhalt, and R. G. Derwent (1981a). Seasonal and latitudinal variation of ^{14}CO and the tropospheric concentration of OH radicals. *J. Geophys. Res.* **86**, 5163-5171.

Volz, A., U. Schmidt, J. Rudolph, D. H. Ehhalt, F. J. Johnson, and A. Khedim (1981b). Vertical profiles of trace gases at mid-latitudes, Kernforschungsanlage, Jülich, Fed. Repub. Germany, Rep. Nr. 1742.

Wade, T. C., and J. G. Quinn (1974). Transfer processes to the marine environment. *In* "Pollutant Transfer to the Marine Environment" (R. A. Duce, P. L. Parker, and G. S. Giam, eds.), p. 25. Univ. of Rhode Island, Kingston, Rhode Island.

Wade, T. C., and J. G. Quinn (1975). Hydrocarbons in the Sargasso Sea surface microlayer. *Mar. Pollut. Bull.* **6**, 54-57.

Wadleigh, C. H. (1968). Wastes in relation to agriculture and forestry, Miscellaneous Publ. No. 1065, Department of Agriculture, Washington, D.C.

Wadt, W. R., and W. A. Goddard, III (1975). Electronic structure of the Criegee intermediate. Ramifications for the mechanism of ozonolysis. *J. Am. Chem. Soc.* **97**, 3004-3021.

Wagener, K. (1978). Total anthropogenic CO_2 production during the period 1800-1935 from carbon-13 measurements in tree rings. *Radiat. Environ. Biophys.* **15**, 101-111.

Wagman, D. D., W. H. Evans, V. B. Parker, R. H. Schumm, I. Halow, S. M. Bailey, K. L. Churney, and R. L. Nutall (1982). The NBS tables of chemical thermodynamic properties. Selected values for inorganic and C_1 and C_2 organic substances in SI units. *J. Phys. Chem. Ref. Data* **2**, *Suppl.*, No. 2.

Waksman, S. A. (1938). "Humus, Origin Chemical Composition and Importance in Nature," 2nd Ed. Williams and Wilkins, Baltimore.

Walker, J. C. G. (1974). Stability of atmospheric oxygen. *Am. J. Sci.* **274**, 193-214.

Walker, J. C. G. (1977). "Evolution of the Atmosphere." Macmillan, New York.

Walker, J. F. (1975). "Formaldehyde," 3rd ed., Krieger, Huntington, New York.

Walter, H. (1973). Coagulation and size distribution of condensation aerosols. *J. Aerosol Sci.* **4**, 1-15.

Walter, M. R. ed. (1976). "Stromatolites," Elsevier, Amsterdam.

Walter, M. R., R. Buick, and J. S. R. Dunlop (1980). Stromatolites 3400-3500 Myr old from the North Pole area, Western Australia. *Nature* **284**, 443-445.

Walton, A., M. Ergin, and D. D. Markness (1970). Carbon-14 concentrations in the atmosphere and carbon dioxide exchange rates. *J. Geophys. Res.* **75**, 3089-3098.

Wang, C. C., and L. I. Davis (1974). Ground-state population distribution of OH determined with tunable laser. *Appl. Phys. Lett.* **25**, 34-35.

Wang, P. K., and H. R. Pruppacher (1977). An experimental determination of the efficiency with which aerosol particles are collected by water drops in subsaturated air. *J. Atmos. Sci.* **34**, 1664-1669.

Wang, P. K., S. N. Grover, and H. R. Pruppacher (1978). On the effect of electric charges on the scavenging of aerosol particles by clouds and small rain drops. *J. Atmos. Sci.* **35**, 1735-1743.

Wang, C. C., L. I. Davis, P. M. Selzer, and R. Munoz (1981). Improved airborne measurements of OH in the atmosphere using the technique of LIF. *J. Geophys. Res.* **86**, 1181-1186.

Warmbt, W. (1964). Luftchemische Untersuchungen des bodennahen Ozons 1952-1961, Methoden und Ergebnisse, Abhandlungen des Meteorologischen Dienstes der Deutschen Demokratischen Republik, Nr. 72 (Band X). Akademie Verlag, Berlin.

Warneck, P., F. F. Marmo, and J. O. Sullivan (1964). Ultraviolet absorption of SO_2: dissociation energies of SO_2 and SO. *J. Chem. Phys.* **40**, 1132-1136.

Warneck, P. (1972). Cosmic radiation as a source of odd nitrogen in the stratosphere. *J. Geophys. Res.* **77**, 6589-6591.

Warneck, P., C. Junge, and W. Seiler (1973). OH radical concentrations in the stratosphere. *Pure Appl. Geophys.* **106-108**, 1417-1430.

Warneck, P. (1974). On the role of OH and HO_2 radicals in the troposphere. *Tellus* **26**, 39-46.

Warneck, P. (1975). OH production rates in the troposphere. *Planet. Space Sci.* **23**, 1507-1518.

Warneck, P., W. Klippel, and G. K. Moortgart (1978). Formaldehyd in troposphärischer Reinluft. *Ber. Bunsen-Ges. Phys. Chem.* **82**, 1136-1142.

Warneck, P. (1985). The equilibrium distribution of atmospheric gases between the two phases of liquid water clouds. *In* "Multiphase Atmospheric Chemistry" (W. Jaeschke, ed.), pp. 473-499. NATO ASI Series, Vol. G6, Springer-Verlag, Berlin and New York.

Wasserburg, G. J., E. Mazor, and R. E. Zartman (1963). Isotopic and chemical composition of some terrestrial natural gases. *In* "Earth Science and Meteorites" (J. Geis and E. D. Goldberg, eds.), pp. 219-240.

Washida, N., Y. Mori, and I. Tanaka (1971). Quantum yield of ozone formation from photolysis of oxygen molecule at 1849 and 1931 Å. *J. Chem. Phys.* **54**, 1119-1122.

Watanabe, K., E. C. Y. Inn, and M. Zelikoff (1953). Absorption coefficients of oxygen in the vacuum ultraviolet. *J. Chem. Phys.* **21**, 1026-1030.

Waters, J. W., J. C. Hardy, R. F. Jarnot, and H. M. Pickett (1981). Chlorine monooxide radical ozone and hydrogenperoxide: stratospheric measurements by microwave limb sounding. *Science* **214**, 61-64.

Watson, E. R., and P. Lapins (1969). Losses of nitrogen from urine on soils from South-West Australia. *Aust. J. Exp. Agric. Anim. Husb.* **9**, 85-91.

Watson, A. J., J. E. Lovelock, and D. H. Stedman (1980). The problem of atmospheric methyl chloride. *In Proc. NATO Adv. Study Inst.*, "Atmospheric Ozone: Its Variation and Human Influences" (M. Nicolet and A. C. Aikin, eds.), Rep. No. **FAA-EE-80-20**. U.S. Department of Transportation, Federal Aviation Administration.

Wayne, L. G., and J. G. Romanofsky (1961). Rates of reaction of the oxides of nitrogen in the photooxidation of diluted automobile exhaust gases. *In* "Chemical Reactions in the lower and upper Atmosphere" (R. D. Cadle, ed.), pp. 71-86. Wiley (Interscience), New York.

Weast, R. C. ed. (1978). "CRC Handbook of Chemistry and Physics," 60th ed. CRC Press, Inc., Boca Raton, Florida.

Weber, E. (1969). Stand und Ziel der Grundlagenforschung bei der Nassentstaubung. *Staub* **29**, 272-277.

Wegener, A. (1911). "Thermodynamik der Atmosphäre." Barth, Leipzig.

Weickmann, H. (1957). Recent measurement of the vertical distribution of Aitken nuclei. *In* "Artificial Stimulation of Rain" (H. Weickmann and W. Smith, eds.), *Proc. Conf. Phys. Cloud Precip. Particles, 1st*, pp. 81-88. Pergamon, New York.

Weinstock, B. (1969). Carbon monoxide residence time in the atmosphere. *Science* **166**, 224-225.

Weinstock, B. (1971). *In* "Chemical Reactions in Urban Atmospheres" (C. S. Tuesday, ed.), pp. 54-55. Amer. Elsevier, New York.

Weinstock, E. M., M. J. Phillips, and J. G. Anderson (1981). *In situ* observations of ClO in the stratosphere: a review of recent results. *J. Geophys. Res.* **86**, 7273-7278.

Weiss, R. F., and H. Craig (1976). Production of atmospheric nitrous oxide by combustion. *Geophys. Res. Lett.* **3**, 751-753.

Weiss, R. F., and D. A. Price (1980). Nitrous oxide solubility in water and seawater. *Mar. Chem.* **8**, 347-359.

Weiss, R. F. (1981). The temporal and spatial distribution of tropospheric nitrous oxide. *J. Geophys. Res.* **86**, 7185-7195.

Welge, K. H. (1974). Photolysis of O_x, HO_x, CO_x and SO_x compounds. *Can. J. Chem.* **52**, 1424-1435.

Went, F. W. (1955). Air pollution. *Sci. Am.* **192**, 63-72.

Went, F. W. (1960a). Organic matter in the atmosphere and its possible relation to petroleum formation. *Proc. Nat. Acad. Sci. U.S.A.* **46**, 212-221.

Went, F. W. (1960b). Blue hazes in the atmosphere. *Nature* **187**, 641-643.

Went, F. W. (1964). The nature of Aitken nuclei in the atmosphere. *Proc. Nat. Acad. Sci. U.S.A.* **51**, 1259-1267.

Went, F. W. (1966). On the nature of Aitken condensation nuclei. *Tellus* **18**, 549-556.

Weschler, C. (1981). Identification of selected organics in the arctic aerosol. *Atmos. Environ.* **15**, 1365-1369.

Weschler, C. J., and T. E. Graedel (1982). Theoretical limitations on heterogeneous catalysis by transition metals in aqueous atmospheric aerosols. *In* "Heterogeneous Atmospheric Chemistry" (D. R. Schryer, ed.), *Geophys. Monogr.* **26**, 196-202. Amer. Geophys. Union, Washington, D.C.

Weseley, M. L., B. B. Hicks, W. P. Dannevick, S. Frissela, and R. B. Husar (1977). An eddy correlation measurement of particle deposition from the atmosphere. *Atmos. Environ.* **11**, 561-563.

Weseley, M. L., J. A. Eastman, D. H. Stedman, and E. D. Yalvac (1982). An eddy-correlation measurement of NO_2 flux to vegetation and comparison to O_3 flux. *Atmos. Environ.* **16**, 815-820.

Westberg, H. H. (1981). Biogenic hydrocarbon measurements. *In* "Atmospheric Biogenic Hydrocarbons" (J. J. Bufalini and R. R. Arnts, eds.), pp. 25-49. Vol. 2, "Ambient Concentrations and Atmospheric Chemistry." Ann Arbor Science, Ann Arbor, Michigan.

Wexler, H., L. Machta, D. H. Pack, and F. White (1956). Atomic Energy and Meteorology, *Proc. Int. Conf. Peaceful Uses Atomic Energy, 1st, Geneva, Switzerland, 1955* **13**, 333-344.

Wexler, H., W. Moreland, and W. Weyant (1960). A preliminary report on ozone observations in Little America, Antarctica. *Mon. Weather Rev.* **88**, 43-45.

Whelpdale, D. M., and L. A. Barrie (1982). Atmospheric monitoring network operations and results in Canada. *Water, Air, Soil Pollut.* **18**, 7-23.

Whitby, K. T. (1976). Electrical measurements of aerosols. *In* "Fine Particles: Aerosol Generation, Measurement, Sampling and Analysis" (B. Y. H. Liu, ed.). Academic Press, New York.

Whitby, R. A., and P. E. Coffey (1977). Measurement of terpenes and other organics in an Adirondack mountain pine forest. *J. Geophys. Res.* **82**, 5928-5934.

Whitby, K. T. (1978). The physical characteristics of sulfur aerosols. *Atmos. Environ.* **12**, 135-159.

Whitby, K. T., and G. M. Sverdrup (1980). California aerosols: their physical and chemical characteristics. *In* "The Characters and Origin of Smog Aerosols" (G. M. Hidy, P. K. Mueller, D. Grosjean, B. R. Appel and J. J. Wesolowski, eds.), *Adv. Environ. Sci. Technol.* **9**, 477-525.

White, D. E., J. D. Hem, and G. A. Waring (1963). Chemical composition of subsurface waters. *In* "Data of Geochemistry," 6th ed. (F. Chap, ed.), *U.S. Geol. Surv. Prof. Pap.* **440-F**.

Whitlaw-Gray, R., and H. S. Patterson (1932). Smoke: A Study of Aerial Disperse Systems." E. Arnold, London.

Whittaker, R. H., and G. E. Likens (1973). Carbon in the biota. *In* "Carbon and the Biosphere" (G. M. Woodwell and E. V. Pecan, eds.), *AEC Symp. Ser.* **30**, 281-302. NTIS US. Dept. Commerce, Springfield, Virginia.

Whittaker, R. H., and G. E. Likens (1975). The biosphere and man. *In* "Primary Productivity of the Biosphere" (H. Lieth and R. H. Whittaker, eds.), *Ecol. Stud.* **14**, 305-328. Springer-Verlag, New York.

Wiebe, H. A., A. Villa, T. H. Hellman, and J. Heicklen (1973). Photolysis of methyl nitrite in the presence of nitric oxide, nitrogen dioxide and oxygen. *J. Am. Chem. Soc.* **95**, 7-13.

Wilcox, R. W., G. D. Nastrom, and A. D. Belmont (1977). Periodic variations of total ozone and its vertical distribution. *J. Appl. Meteorol.* **16**, 290-298.

Wilhelm, E., R. Battino, and R. J. Wilcock (1977). Low pressure solubility of gases in liquid water. *Chem. Rev.* **77**, 219-262.

Wilkening, M. H. (1970). Radon 222 concentrations in the convective patterns of a mountain environment. *J. Geophys. Res.* **75**, 1733-1740.

Wilkins, E. T. (1954). Air pollution and the London fog of December 1952. *J. R. Sanit. Inst.* **74**, 1-21.

Wilkness, P., and D. J. Bressan (1972). Fractionation of the elements F, Cl, Na and K at the sea-air interface. *J. Geophys. Res.* **77**, 5307-5315.

Wilkness, P. E., R. A. Lamontagne, R. E. Larson, J. W. Swinnerton, C. R. Dickson, and T. Thompson (1973). Atmospheric trace gases in the Southern hemisphere. *Nature* **245**, 45–47.

Wilkness, P. E., J. W. Swinnerton, D. J. Bressan, R. A. Lamontagne, and R. E. Larson (1975). CO, CCl_4, Freon-11, CH_4 and Rn 222 concentrations at low altitude over the Arctic Ocean in January 1974. *J. Atmos. Sci.* **32**, 158–162.

Wilkness, P. E., R. A. Lamontagne, R. E. Larson, and J. W. Swinnerton (1978). Atmospheric trace gases and land and sea breezes at the Spiek River Coast of Papua, New Guinea. *J. Geophys. Res.* **83**, 3672–3674.

Wilks, S. S. (1959). Carbon monoxide in green plants. *Science* **129**, 964–966.

Willeke, K., K. T. Whitby, W. E. Clark, and V. A. Marple (1974). Size distribution of Denver aerosols—a comparison of two sites. *Atmos. Environ.* **8**, 609–633.

Williams, P. M., H. Oeschger, and P. Kinney (1969). Natural radiocarbon activity of the dissolved organic carbon in the North-East Pacific Ocean. *Nature* **224**, 256–258.

Williams, P. J. le B. (1975). Biological and chemical aspects of dissolved organic material in seawater. *In* "Chemical Oceanography" (J. P. Riley and G. Skirrow, eds.), pp. 301–366. 2nd ed. Academic Press, New York.

Williams, W. J., J. J. Kostus, A. Goldman, and D. G. Murcray (1976). Measurements of the stratospheric mixing ratio of HCl using an infrared absorption technique. *Geophys. Res. Lett.* **3**, 383–385.

Wilson, D. F., J. W. Swinnerton, and R. A. Lamontagne (1971). Production of carbon monoxide and gaseous hydrocarbons in seawater: Relation to dissolved organic carbon. *Science* **168**, 1577–1579.

Wilson, W. E., Jr., D. F. Miller, A. Levy, and R. K. Stone (1973). The effect of fuel composition on atmospheric aerosol due to auto exhaust. *J. Air. Pollut. Control Assoc.* **23**, 949–956.

Wilson, T. R. S. (1975). Salinity and major elements of sea water. *In* "Chemical Oceanography" (J. P. Riley and G. Skirrow, eds.), 2nd ed., Chap. 6, pp. 365–413. Academic Press, London.

Wilson, W. E., Jr. (1981). Sulfate formation in point source plumes: a review of recent field studies. *Atmos. Environ.* **15**, 2573–2581.

Winchester, J. W., and G. D. Nifong (1971). Water pollution in Lake Michigan by trace elements from pollution fall-out. *Water, Air, Soil Pollut.* **1**, 50–64.

Windom, H. L. (1969). Atmospheric dust records in permanent snow fields, implications to marine sedimentation. *Bull. Geol. Soc. Amer.* **80**, 761–782.

Wine, P. H., N. M. Kreutter, C. A. Gumb, and A. R. Ravishankara (1981). Kinetics of OH reactions with the atmospheric sulfur compounds H_2S, CH_3SH, CH_3SCH_3 and CH_3SSCH_3. *J. Phys. Chem.* **85**, 2660–2665.

Winer, A. M., K. R. Darnall, R. Atkinson, and J. N. Pitts, Jr. (1979). Smog chamber study of the correlation of hydroxyl radical rate constants with ozone formation. *Environ. Sci. Technol.* **13**, 822–826.

Winkler, P., and C. E. Junge (1972). The growth of atmospheric aerosol particles as a function of the relative humidity, Part I. Method and measurements at different locations. *J. Rech. Atmos.* **6**, 617–638. Memorial Henri Dessens.

Winkler, P. (1973). The growth of atmospheric particles as a function of relative humidity, Part II. Improved concept of mixed nuclei. *Aerosol Sci.* **4**, 373–387.

Winkler, P. (1974). Die relative Zusammensetzung des atmosphärischen Aerosols in Stoffgruppen. *Meteorol. Rundsch.* **27**, 129–136.

Wlotzka, F. (1972). Nitrogen. *In* "Handbook of Geochemistry" (K. H. Wedepohl, ed.), Chap. 7. Springer-Verlag, Berlin and New York.

WMO (1985). "Atmospheric Ozone 1985." World Meteorological Organization global ozone research and monitoring project, Rep. 16. World Meteorological Organization, Geneva.

Wofsy, S. C., J. C. McConnell, and M. B. McElroy (1972). Atmospheric CH_4, CO and CO_2. *J. Geophys. Res.* **77**, 4477-4493.

Wofsy, S. C., and M. B. McElroy (1973). On vertical mixing in the upper stratosphere and mesosphere. *J. Geophys. Res.* **78**, 2619-2624.

Wofsy, S. C., M. B. McElroy, and Y. L. Yung (1975). The chemistry of atmospheric bromine. *Geophys. Res. Lett.* **2**, 215-218.

Wofsy, S. C. (1978). Temporal and latitudinal variations of stratospheric trace gases: a critical comparison between between theory and experiment. *J. Geophys. Res.* **83**, 364-378.

Wofsy, S. C., and J. A. Logan (1982). Recent developments in stratospheric photochemistry, National Research Council. *In* "Causes and effects of Stratospheric Ozone Reduction: An Update," pp. 167-205. National Academy Press, Washington, D.C.

Wolff, G. T., P. J. Lioy, G. D. Wight, R. E. Meyers, and R. T. Cederwall (1977). An investigation of long-range transport of ozone across the midwestern and eastern United States. *Atmos. Environ.* **11**, 797-802.

Wolff, G. T., P. J. Lioy, R. E. Meyers, R. T. Cederwall, G. D. Wight, R. E. Pasceri, and R. S. Taylor (1977). Anatomy of the transport episodes in the Washington, D.C., to Boston, Mass., Corridor. *Environ. Sci. Technol.* **11**, 506-510.

Wood, J. M. (1974). Biological cycles for toxic elements in the environment. *Science* **183**, 1049-1052.

Woodcock, A. H. (1953). Salt nuclei in marine air as a function of altitude and midforce. *J. Meteorol.* **10**, 362-371.

Woodcock, A. H., D. C. Blanchard, and C. G. H. Rooth (1963). Salt-induced convection and clouds. *J. Atmos. Sci.* **20**, 159-169.

Woodcock, A. H. (1972). Small salt particles in oceanic air and bubble behavior in the sea. *J. Geophys. Res.* **77**, 5316-5321.

Wong, C. S. (1978). Atmospheric input of carbon dioxide from burning wood. *Science* **200**, 197-200.

Woodwell, G. M., R. H. Whittaker, W. A. Reiners, G. E. Likens, C. C. Delwiche, and D. B Botkin (1978). The biota and the world carbon budget. *Science* **199**, 141-146.

Wu, C. H., S. M. Japar, and H. Niki (1976). Relative reactivities of OH-hydrocarbon reactions from smog reaction studies. *J. Environ. Sci. Health, Part A*; *Environ. Sci. Eng. A* **11**, 191-200.

Wu, J. (1979). Sea spray in the atmospheric surface layer: Review and analysis of laboratory and oceanic results. *J. Geophys. Res.* **84**, 1683-1704.

Wuebbles, D. J., F. M. Luther, and J. E. Penner (1983). Effects of coupled anthropogenic perturbation on stratospheric ozone. *J. Geophys. Res.* **88**, 1444-1456.

Wulf, O. R., and L. S. Deming (1937). The distribution of atmospheric ozone in equilibrium with solar radiation and the rate of maintenance of the distribution. *Terr. Magn. Atmos. Electr.* **42**, 195-202.

Yokouchi, Y., M. Okaniwa, Y. Ambe, and K. Fuwa (1983). Seasonal variation of monoterpenes in the atmosphere of a pine forest. *Atmos. Environ.* **17**, 743-750.

Yoshinari, T. (1976). Nitrous oxides in the sea. *Mar. Chem.* **4**, 189-202.

Young, A. J., and A. W. Fairhall (1968). Radiocarbon from nuclear weapons tests. *J. Geophys. Res.* **73**, 1185-1200.

Zafiriou, O. C. (1974). Photochemistry of halogens in the marine atmosphere. *J. Geophys. Res.* **79**, 2730-2732.

Zafiriou, O. C. (1975). Reaction of methyl halides with seawater and marine aerosols. *J. Mar. Res.* **33**, 75-81.

Zafiriou, O. C., J. Alford, M. Herrera, E. T. Peltzer, and R. B. Gagosian (1980). Formaldehyde in remote marine air and rain: flux measurements and estimates. *Geophys. Res. Lett.* **7**, 341-344.

Zafiriou, O. C., and M. McFarland (1981). Nitric oxide from nitrite photolysis from the Central Equatorial Pacific. *J. Geophys. Res.* **86**, 3173-3182.

Zafonte, L., N. E. Hester, E. R. Stephens, and O. C. Taylor (1975). Background and vertical atmospheric measurements of fluorocarbon-11 and fluorocarbon-12 over Southern California. *Atmos. Environ.* **9**, 1007-1009.

Zander, R. (1975). Presence de HF dans la stratosphere superieure, *C. R. Acad. Sci., Paris B* **281**, 213-214.

Zander, R. (1981). Recent observations of HF and HCl in the upper stratosphere. *Geophys. Res. Lett.* **8**, 413-416.

Zartman, R. E., G. J. Wasserburg, and J. H. Reynolds (1961). Helium, argon and carbon in some natural gases. *J. Geophys. Res.* **66**, 277-306.

Zebel, G. (1966). Coagulation of aerosols. *In* "Aerosol Science" (C. N. Davis, ed.), Chap. 2, pp. 31-58. Academic Press, London.

Zeldovich, Ya. B., P. Ya Sadonikov, and D. A. Frank-Kamenetskii (1947). Oxidation of nitrogen in Combustion (M. Shelef, *Transl.*) *Acad. Sci. U.S.S.R., Inst. Chem. Phys. Moscow-Leningrad.*

Zellner, R. (1978). Recombination reactions in atmospheric chemistry. *Ber. Bunsen-Ges. Phys. Chem.* **82**, 1172-1179.

Zettlemoyer, A. C., ed. (1969). "Nucleation." Marcel Dekker, New York.

Zettwoog, P., and R. Haulet (1978). Experimental results on the SO_2 transfer in the Mediterranean obtained with remote sensing devices. *Atmos. Environ.* **12**, 795-796.

Zinder, S. H., and T. D. Brock (1978). Microbial transformations of sulfur in the environment. *In* "Sulphur in the Environment" (J. O. Nriagu, ed.), Part II, pp. 445-466. Wiley, New York.

Zimen, K. E., and F. K. Altenhein (1973). The future burden of industrial CO_2 on the atmosphere and the oceans. *Z. Naturforsch. A* **28**, 1747-1752.

Zimen, K. E., P. Offermann, and G. Hartmann (1977). Source functions of CO_2 and future CO_2 burden in the atmosphere. *Z. Naturforsch.* **32a**, 1544.

Zimin, A. G. (1964). Mechanisms of capture and precipitation of atmospheric contaminants by clouds and precipitation. *In* "Problems of Nuclear Meterology" (I. L. Karol and S. G. Malakhov, eds.), pp. 139-182. U.S. Atomic Energy Commission Rep. **AE-tr-6128**.

Zimmerman, P. R., R. B. Chatfield, J. Fishman, P. J. Crutzen, and P. L. Hanst (1978). Estimates on the production of CO and H_2 from the oxidation of hydrocarbon emissions from vegetation. *Geophys. Res. Lett.* **5**, 679-682.

Zimmerman, P. R. (1979a). Determination of emission rates of hydrocarbons from indigenous species of vegetation in the Tampa/St. Petersburg, Florida area, U.S. Environmental Protection Agency Rep. **90419-77-028**.

Zimmerman, P. R. (1979b). Testing of hydrocarbon emissions from vegetation, leaf litter and aquatic surfaces and development of a methodology for compiling biogenic emission inventories. U.S. Environmental Protection Agency, Research Triangle Park, North Carolina. **EPA-450/4-79-004**.

Zimmerman, P. R., J. P. Greenberg, S. O. Wandiga, and P. J. Crutzen (1982). Termite: a potentially large source of atmospheric methane, carbon dioxide and molecular hydrogen. *Science* **218**, 563-565.

Zoller, W. H., E. S. Gladney, and R. A. Duce (1974). Atmospheric concentration and sources of trace elements at the South Pole. *Science* **183**, 198-200.

Appendix: Supplementary Tables

Table A-1. *Temperature, Pressure, Density, and Number Density of Air as a Function of Altitude According to the U.S. Standard Atmosphere* (1976)[a]

z (km)	T (K)	p (mbar)	(kg/m^3)	n (cm^{-3})
0	288.15	1.01 (3)	1.23	2.55 (19)
2	275.15	7.95 (2)	1.00	2.09 (19)
4	262.17	6.17 (2)	8.19 0−1)	1.70 (19)
6	249.19	4.72 (2)	6.60 (−1)	1.37 (19)
8	236.22	3.57 (2)	5.26 (−1)	1.09 (19)
10	223.25	2.65 (2)	4.14 (−1)	8.60 (18)
12	216.65	1.94 (2)	3.12 (−1)	6.49 (18)
14	216.65	1.42 (2)	2.28 (−1)	4.74 (18)
16	216.65	1.04 (2)	1.67 (−1)	3.46 (18)
18	216.65	7.57 (1)	1.22 (−1)	2.53 (18)
20	216.65	5.53 (1)	8.89 (−2)	1.85 (18)
22	218.57	4.05 (1)	6.45 (−2)	1.34 (18)
24	220.56	2.97 (1)	4.69 (−2)	9.76 (17)
26	222.54	2.19 (1)	3.43 (−2)	7.12 (17)
28	224.53	1.62 (1)	2.51 (−2)	5.21 (17)
30	226.51	1.20 (1)	1.84 (−2)	3.83 (17)
32	228.49	8.89	1.36 (−2)	2.82 (17)
34	233.74	6.63	9.89 (−3)	2.06 (17)
36	239.28	4.99	7.26 (−3)	1.51 (17)
38	244.82	3.77	5.37 (−3)	1.12 (17)
40	250.35	2.87	4.00 (−3)	8.31 (16)
42	255.88	2.20	3.00 (−3)	6.23 (16)
44	261.4	1.70	2.26 (−3)	4.70 (16)
46	266.94	1.31	1.71 (−3)	3.57 (16)
48	270.65	1.03	1.32 (−3)	2.74 (16)
50	270.65	8.00 (−1)	1.03 (−3)	2.14 (16)
52	269.03	6.22 (−1)	8.10 (−4)	1.67 (16)
54	267.56	4.85 (−1)	6.39 (−4)	1.32 (16)
56	258.02	3.77 (−1)	5.04 (−4)	1.04 (16)
58	252.51	2.87 (−1)	3.96 (−4)	8.23 (15)
60	247.02	2.19 (−1)	3.10 (−4)	6.44 (15)
70	219.58	5.22 (−2)	8.28 (−5)	1.72 (15)
80	198.64	1.04 (−2)	1.85 (−5)	3.84 (14)
90	186.87	1.84 (−3)	3.42 (−6)	7.12 (13)

[a] Powers of 10 in parentheses.

Table A-2. *Saturation Vapor Pressure* (p_s) *over Water or Ice*[a]

Temperature (°C)	p_s (mbar)
Water	
−35	3.14 (−1)
−30	5.09 (−1)
−25	8.07 (−1)
−20	1.25
−15	1.91
−10	2.86
−5	4.21
0	6.11
5	8.72
10	1.23 (1)
15	1.70 (1)
20	2.34 (1)
25	3.17 (1)
30	4.24 (1)
35	5.62 (1)
40	7.38 (1)
Ice	
−90	9.67 (−5)
−85	2.35 (−4)
−80	5.47 (−4)
−75	1.22 (−3)
−70	2.62 (−3)
−65	5.41 (−3)
−60	1.08 (−2)
−55	2.09 (−2)
−50	3.93 (−2)
−45	7.20 (−2)
−40	1.28 (−1)
−35	2.23 (−1)
−30	3.80 (−1)
−25	6.32 (−1)
−20	1.03
−15	1.65
−10	2.6
−5	4.02
0	6.12

[a] From the Smithsonian Meteorological Tables (1951).

Table A-3. *Average Solar Photon Flux Outside Earth's Atmosphere According to Data of Samain and Simon* (1976) *and Neckel and Labs* (1981)

Wavelength interval (nm)	Solar flux (photons/ m^2 s nm)^a	Wavelength interval (nm)	Solar flux (photons/ m^2 s nm)^a
Ly121.567	3.0 (15)	173.9–175.4	5.9 (14)
116.3–117.0	9.8 (12)	175.4–177.0	7.1 (14)
117.0–117.6	4.7 (13)	177.0–178.6	8.4 (14)
117.6–118.3	1.0 (13)	178.6–180.2	1.2 (14)
118.3–119.0	9.3 (12)	180.2–181.0	1.2 (15)
119.8–120.5	2.0 (13)	181.0–182.0	2.0 (15)
120.5–121.2	9.0 (13)	182.0–183.0	1.9 (15)
121.2–122.0	2.9 (13)	183.0–184.0	2.0 (15)
122.0–122.7	1.8 (13)	184.0–185.0	1.7 (15)
122.7–123.5	1.1 (13)	185.0–186.0	1.9 (15)
123.5–124.2	1.7 (13)	186.0–187.0	2.4 (15)
124.2–125.0	7.4 (12)	187.0–188.0	2.7 (15)
125.0–125.8	8.4 (12)	188.0–189.0	2.8 (15)
125.8–126.6	2.3 (13)	189.0–190.0	2.9 (15)
126.6–127.4	5.7 (13)	190.0–191.0	2.9 (15)
127.4–128.2	1.0 (13)	191.0–192.0	3.3 (15)
128.2–129.0	7.5 (12)	192.0–193.0	3.5 (15)
129.0–129.9	7.9 (12)	193.0–194.0	2.5 (15)
129.9–130.7	8.6 (13)	194.0–195.0	4.5 (15)
130.7–131.6	2.0 (13)	195.0–196.0	4.3 (15)
131.6–132.4	1.0 (13)	196.0–197.0	4.9 (15)
132.4–133.3	3.7 (13)	197.0–198.0	4.9 (15)
133.3–134.2	5.8 (13)	198.0–199.0	4.9 (15)
134.2–135.1	9.5 (12)	199.0–200.0	5.5 (15)
135.1–136.0	2.3 (13)	200.0–202.0	7.0 (15)
136.0–137.0	1.4 (13)	202.0–204.1	8.0 (15)
137.0–137.9	1.6 (13)	204.1–206.2	9.9 (15)
137.9–138.9	1.4 (13)	206.2–208.3	1.2 (16)
138.9–140.8	1.6 (13)	208.3–210.5	1.9 (16)
140.8–142.8	3.8 (13)	210.5–212.8	3.1 (16)
142.8–144.9	3.2 (13)	212.8–215.0	3.5 (16)
144.9–147.0	6.5 (13)	215.0–217.4	3.5 (16)
147.0–149.2	5.0 (13)	217.4–219.8	4.6 (16)
149.2–151.5	5.3 (13)	219.8–222.2	5.0 (16)
151.5–153.8	7.7 (13)	222.2–224.7	6.6 (16)
153.8–156.2	1.2 (14)	224.7–227.3	4.9 (16)
156.2–158.7	9.6 (13)	227.3–229.9	5.0 (16)
158.7–161.3	1.1 (14)	229.9–232.6	5.6 (16)
161.3–163.9	1.4 (14)	232.6–235.3	4.9 (16)
163.9–166.7	2.2 (14)	235.3–238.1	5.4 (16)
166.7–169.5	2.8 (14)	238.1–241.0	4.6 (16)
169.5–172.4	4.6 (14)	241.0–243.9	7.0 (16)
172.4–173.9	4.9 (14)	243.9–246.9	6.1 (16)

Table A-3 (*continued*)

Wavelength interval (nm)	Solar flux (photons/ m^2 s nm)[a]	Wavelength interval (nm)	Solar flux (photons/ m^2 s nm)[a]
246.9–250.0	6.1 (16)	274.0–277.8	2.7 (17)
250.0–253.2	5.7 (16)	277.8–281.7	2.0 (17)
253.2–256.4	7.0 (16)	281.7–285.7	3.7 (17)
256.4–259.7	1.4 (17)	285.7–289.9	5.2 (17)
259.7–263.2	1.2 (17)	289.9–294.1	8.2 (17)
263.2–266.7	3.2 (17)	294.1–298.5	7.7 (17)
266.7–270.3	3.3 (17)	298.5–303.0	7.2 (17)
270.3–274.0	2.9 (17)	303.0–307.7	9.4 (17)

Wavelength (nm)	Solar flux (photons/ m^2 s nm)[a]	Wavelength (nm)	Solar flux (photons/ m^2 s nm)[a]
310	9.9 (17)	460	4.8 (18)
315	1.2 (18)	465	4.8 (18)
320	1.2 (18)	470	4.8 (18)
325	1.4 (18)	475	4.9 (18)
330	1.7 (18)	480	5.0 (18)
335	1.6 (18)	485	4.6 (18)
340	1.8 (18)	490	4.8 (18)
345	1.7 (18)	495	5.0 (18)
350	1.7 (18)	500	4.8 (18)
355	1.8 (18)	505	4.9 (18)
360	1.6 (18)	510	5.0 (18)
365	2.1 (18)	515	4.6 (18)
370	2.2 (18)	520	4.8 (18)
375	1.9 (18)	525	4.8 (18)
380	2.2 (18)	530	5.1 (18)
385	1.8 (18)	535	5.1 (18)
390	2.4 (18)	540	5.0 (18)
395	1.9 (18)	545	5.1 (18)
400	3.4 (18)	550	5.1 (18)
405	3.4 (18)	555	5.1 (18)
410	3.7 (18)	560	5.0 (18)
415	3.9 (18)	565	5.1 (18)
420	3.9 (18)	570	5.2 (18)
425	3.6 (18)	575	5.3 (18)
430	3.3 (18)	580	5.3 (18)
435	4.0 (18)	585	5.4 (18)
440	4.0 (18)	590	5.2 (18)
445	4.4 (18)	595	5.4 (18)
450	4.7 (18)	600	5.3 (18)
455	4.6 (18)	605	5.4 (18)

Table A-3 (*continued*)

Wavelength interval (nm)	Solar flux (photons/ m^2 s nm)[a]	Wavelength interval (nm)	Solar flux (photons/ m^2 s nm)[a]
610	5.3 (18)	730	4.9 (18)
615	5.2 (18)	735	4.9 (18)
620	5.2 (18)	740	4.8 (18)
625	5.2 (18)	745	4.8 (18)
630	5.2 (18)	750	4.8 (18)
635	5.2 (18)	755	4.8 (18)
640	5.3 (18)	760	4.8 (18)
645	5.2 (18)	765	4.7 (18)
650	5.1 (18)	770	4.6 (18)
655	5.0 (18)	775	4.6 (18)
660	5.1 (18)	780	4.7 (18)
665	5.2 (18)	785	4.7 (18)
670	5.2 (18)	790	4.6 (18)
675	5.2 (18)	795	4.6 (18)
680	5.1 (18)	800	4.5 (18)
685	5.2 (18)	805	4.5 (18)
690	5.0 (18)	810	4.4 (18)
695	5.2 (18)	815	4.4 (18)
700	5.2 (18)	820	4.4 (18)
705	5.0 (18)	825	4.4 (18)
710	5.0 (18)	830	4.3 (18)
715	5.0 (18)	835	4.3 (18)
720	4.9 (18)	840	4.2 (18)
725	5.0 (18)		

Table A-4. Rate Coefficients at 298 K and Arrhenius Parameters for Gas–Phase Atmospheric Reactions[a]

Number	Reaction	k_{298} (cm^3/molecule s)	A (cm^3/molecule s)	E_a/R_g (K)	Reference[b]
		Reactions of O(^{3}P) atoms			
1	$O + O_2 + M \rightarrow O_3 + M$	See Table A-5	1.8 (-11)	1,163	b
2	$O + O_3 \rightarrow O_2 + O_2$	8.4 (-15)	2.2 (-11)	-117	a
3	$O + OH \rightarrow O_2 + H$	3.3 (-11)	3.0 (-11)	-200	b
4	$O + HO_2 \rightarrow OH + O_2$	5.9 (-11)	1.6 (-11)	2,230	b
5	$O + H_2 \rightarrow OH + H$	3.5 (-18)	1.4 (-12)	2,000	b
6	$O + H_2O_2 \rightarrow OH + HO_2$	1.7 (-15)	1.3 (-10)	38,000	b
7	$O + N_2 \rightarrow NO + N$	5.4 (-66)			d
8	$O + NO + M \rightarrow NO_2 + M$	See Table A-5	9.3 (-12)	±0	b
9	$O + NO_2 \rightarrow NO + O_2$	9.3 (-12)	1.0 (-11)	±0	b
10	$O + NO_3 \rightarrow O_2 + NO_2$	1.0 (-11)	—		b
11	$O + HNO_3 \rightarrow OH + NO_3$	<3.0 (-17)	7.0 (-11)	3,370	b
12	$O + HO_2NO_2 \rightarrow products$	8.6 (-16)	2.1 (-11)	2,200	b
13	$O + COS \rightarrow SO + CO$	1.3 (-14)			d
14	$O + SO_2 + M \rightarrow SO_3 + M$	See Table A-5	4.7 (-11)	50	b
15	$O + ClO \rightarrow Cl + O_2$	4.0 (-11)	1.0 (-11)	3,340	b
16	$O + HCl \rightarrow OH + Cl$	1.4 (-16)	1.0 (-11)	2,200	b
17	$O + HOCl \rightarrow OH + ClO$	6.0 (-15)	3.0 (-12)	808	b
18	$O + ClONO_2 \rightarrow products$	2.0 (-13)	3.0 (-11)	±0	b
19	$O + BrO \rightarrow Br + O_2$	3.0 (-11)			d
20	$O + CO + M \rightarrow CO_2 + M$	See Table A-5	3.5 (-11)	4579	c
21	$O + CH_4 \rightarrow OH + CH_3$	1.7 (-17)	4.2 (-11)	3,221	c
22	$O + C_2H_6 \rightarrow OH + C_2H_5$	9.1 (-16)	—		e
23	$O + C_3H_8 \rightarrow OH + C_3H_7$	1.1 (-14)	2.5 (-11)	2,100	e
24	$O + n\text{-}C_4H_{10} \rightarrow OH + n\text{-}C_4H_9$	2.2 (-14)	5.5 (-12)	554	c
25	$O + C_2H_4 \rightarrow products$	8.1 (-13)	4.2 (-12)	38	c
26	$O + C_3H_6 \rightarrow products$	3.7 (-12)			c

(continued)

Table A-4 (*continued*)

Number	Reaction	k_{298} (cm³/molecule s)	A (cm³/molecule s)	E_a/R_g (K)	Reference[b]
27	O + cis-butene → products	1.7 (−11)	9.8 (−15)	−151	c
28	O + C₂H₂ → products	1.4 (−13)	2.9 (−11)	1,600	b
29	O + benzene → products	2.4 (−14)	3.3 (−11)	2,013	c
30	O + toluene → products	7.5 (−14)	1.4 (−11)	1,560	c
31	O + HCHO → OH + HCO	1.6 (−13)	3.0 (−11)	1,550	b
	Reactions of O(¹D) atoms				
32	O* + N₂ → O + N₂	2.6 (−11)	1.8 (−11)	−107	b
33	O* + O₂ → O + O₂	4.0 (−11)	3.2 (−11)	−67	b
34	O* + O₃ → O₂ + O₂	1.2 (−10)	1.2 (−10)	±0	b
	→ O₂ + O + O	1.2 (−10)	1.2 (−10)	±0	b
35	O* + H₂O → OH + OH	2.2 (−10)	2.2 (−10)	±0	b
36	O* + H₂ → OH + H	1.0 (−10)	1.0 (−10)	±0	b
37	O* + N₂O → N₂ + O₂	4.9 (−11)	4.9 (−11)	±0	b
	→ NO + NO	6.7 (−11)	6.7 (−11)	±0	b
38	O* + CO₂ → O + CO₂	1.1 (−10)	7.4 (−11)	−117	b
39	O* + CH₄ → OH + CH₃	1.4 (−10)	1.4 (−10)	±0	b
	→ H₂ + HCHO	1.4 (−11)	1.4 (−11)	±0	b
40	O* + CCl₄ → products	3.3 (−10)	3.3 (−10)	±0	b
41	O* + CCl₃F → products	2.3 (−10)	2.3 (−10)	±0	b
42	O* + CCl₂F₂ → products	1.4 (−10)	1.4 (−10)	±0	b
43	O* + CF₄ → products	1.8 (−13)	1.8 (−13)	±0	b
44	O* + HCl → products	1.5 (−10)	1.8 (−10)	±0	b
45	O* + HF → OH + F	1.4 (−10)	1.4 (−10)	±0	b
	Reactions involving H atoms and HO₂ radicals				
46	H + O₂ + M → HO₂ + M	See Table A-5			b
47	H + O₃ → OH + O₂	2.9 (−11)	1.4 (−10)	470	b

No.	Reaction				
48	$H + HO_2 \rightarrow OH + OH$ $\rightarrow H_2O + O$ $\rightarrow H_2 + O_2$	7.4 (−11)	7.4 (−11)	±0	b
49	$OH + OH \rightarrow H_2O + O$	1.9 (−12)	4.2 (−12)	242	b
50	$OH + OH + M \rightarrow H_2O_2 + M$	See Table A-5			b
51	$OH + HO_2 \rightarrow H_2O + O_2$	7.0 (−11)	1.7 (−11)	−416	b
52	$HO_2 + HO_2 \rightarrow H_2O_2 + O_2$	1.7 (−12)	2.3 (−13)	−590	b
53	$OH + O_3 \rightarrow HO_2 + O_2$	6.8 (−14)	1.6 (−12)	940	b
54	$HO_2 + O_3 \rightarrow OH + 2O_2$	2.0 (−15)	1.4 (−14)	580	b
55	$HO_2 + NO \rightarrow OH + NO_2$	8.3 (−12)	3.7 (−12)	−240	b
	$HO_2 + NO_2 \rightarrow HO_2NO_2$	See Table A-5			b
	Reactions involving N atoms, NO_x, and NH_2 radicals				
56	$N + O_2 \rightarrow NO + O$	8.9 (−17)	4.4 (−12)	3,220	b
57	$N + O_3 \rightarrow NO + O_2$	1.0 (−15)	—	—	b
58	$N + NO \rightarrow N_2 + O$	3.4 (−11)	3.4 (−11)	±0	b
59	$N + NO_2 \rightarrow N_2O + O$	3.0 (−12)	—	—	b
60	$NO + O_3 \rightarrow NO_2 + O_2$	1.8 (−14)	1.8 (−12)	1,370	b
61	$NO + NO_3 \rightarrow 2NO_2$	3.0 (−11)	1.3 (−11)	−250	b
62	$O_3 + NO_2 \rightarrow NO_3 + O_2$	3.2 (−17)	1.2 (−13)	2,450	b
63	$NO_3 + NO_2 \rightarrow N_2O_5$	See Table A-5	See Table A-5		g
64	$NO + NO_2 + H_2O \rightarrow 2HNO_2$	See Table A-5	See Table A-5		g
65	$2NO_2 + H_2O \rightarrow HNO_3 + HNO_2$	See Table A-5	See Table A-5		g
66	$N_2O_5 + H_2O \rightarrow 2HNO_3$	<2.0 (−21)			d
67	$2NO + O_2 \rightarrow 2NO_2$	See Table A-5	—	—	b
68	$NH_2 + O_2 \rightarrow products$	<3.0 (−18)	—	—	b
69	$NH_2 + O_3 \rightarrow products$	2.1 (−13)	4.8 (−12)	930	b
70	$NH_2 + NO \rightarrow products$	1.7 (−11)	3.8 (−12)	−450	b
71	$NH_2 + NO_2 \rightarrow products$	1.9 (−11)	2.1 (−12)	−650	b
	Methane oxidation reactions				
72	$OH + CH_4 \rightarrow H_2O + CH_3$	7.7 (−15)	2.4 (−12)	1,710	b
73	$CH_3 + O_2 + M \rightarrow CH_3O_2 + M$	See Table A-5			b
74	$CH_3O_2 + NO \rightarrow NO_2 + CH_3O$	7.6 (−12)	4.2 (−12)	−180	b

(continued)

725

Table A-4 (*continued*)

Number	Reaction	k_{298} (cm³/molecule s)	A (cm³/molecule s)	E_a/R_g (K)	Reference[b]
75	$CH_3O_2+CH_3O_2 \rightarrow HCHO+CH_3OH+O_2$	2.2 (−13)	1.6 (−13)	−220	a, b
	$\rightarrow 2CH_3O+O_2$	1.5 (−13)			
	$\xrightarrow{M} CH_3OOCH_3+O_2$	<3 (−14)			
76	$CH_3O_2+NO_2 \rightarrow CH_3O_2NO_2$	See Table A-5			b
77	$CH_3O_2+HO_2 \rightarrow CH_3OOH+O_2$	6.0 (−12)	7.7 (−14)	−1,300	b
78	$CH_3O+O_2 \rightarrow HCHO+HO_2$	1.5 (−15)	8.4 (−14)	1,200	b
79	$CH_3O+NO_2 \rightarrow CH_3ONO_2$	1.2 (−11)	1.2 (−11)	±0	b
80	$OH+HCHO \rightarrow HCO+H_2O$	1.0 (−11)	1.0 (−11)	±0	b
81	$HCO+O_2 \rightarrow CO+HO_2$	5.5 (−12)	3.5 (−12)	−140	b
	Halogen atom and radical reactions				
82	$Cl+O_3 \rightarrow ClO+O_2$	1.2 (−11)	2.8 (−11)	257	b
83	$Cl+O_2+M \rightarrow ClO_2+M$	See Table A-5			b
84	$Cl+H_2 \rightarrow HCl+H$	1.6 (−14)	3.7 (−11)	2,300	b
85	$Cl+CH_4 \rightarrow HCl+CH_3$	1.0 (−13)	9.6 (−12)	1,350	b
86	$Cl+C_2H_6 \rightarrow HCl+C_2H_5$	5.7 (−11)	7.7 (−11)	90	b
87	$Cl+C_3H_8 \rightarrow HCl+C_3H_7$	1.6 (−10)	1.4 (−10)	−40	b
88	$Cl+C_2H_2 \rightarrow$ products	1 (−12)	—	—	b
89	$Cl+CH_3Cl \rightarrow HCl+CH_2Cl$	4.9 (−13)	3.4 (−11)	1,260	b
90	$Cl+HCHO \rightarrow HCl+HCO$	7.3 (−11)	8.2 (−11)	34	b
91	$Cl+HOCl \rightarrow$ products	1.9 (−12)	3.0 (−12)	130	b
92	$Cl+H_2O_2 \rightarrow HCl+HO_2$	4.1 (−13)	1.1 (−11)	980	b
93	$Cl+HNO_3 \rightarrow$ products	1.7 (−14)	—	—	b
94	$Cl+HO_2 \rightarrow HCl+O_2$	3.2 (−11)	1.8 (−11)	−170	b
	$\rightarrow OH+ClO$	9.1 (−12)	4.1 (−11)	450	b
95	$Cl+ClO_2 \rightarrow Cl_2+O_2$	1.4 (−10)	1.4 (−10)	±0	b
	$\rightarrow ClO+ClO$	8.0 (−12)	8.0 (−12)	±0	b
96	$Cl+ClONO_2 \rightarrow$ products	1.2 (−11)	6.8 (−12)	−160	b

726

(continued)

No.	Reaction				
97	ClO+NO → NO₂+Cl	1.7 (−11)	6.2 (−12)	−294	b
98	ClO+NO₂ $\xrightarrow{M}$ ClONO₂	See Table A-5			b
99	ClO+HO₂ → HOCl+O₂	5.0 (−12)	4.6 (−13)	−710	b
100	ClO+O₃ → products	<1.0 (−18)	—	—	b
101	Br+O₃ → BrO+O₂	1.1 (−12)	1.4 (−11)	755	b
102	Br+H₂O₂ → HBr+HO₂	2.0 (−15)	—	—	b
103	Br+HCHO → HBr+HCO	1.1 (−12)	1.7 (−11)	800	b
104	BrO+NO → NO₂+Br	2.1 (−11)	8.7 (−12)	−265	b
105	BrO+NO₂ $\xrightarrow{M}$ BrONO₂	See Table A-5			b
106	BrO+HO₂ → products	5.0 (−12)	—	—	b
107	F+H₂ → HF+H	2.7 (−11)	1.6 (−10)	525	b
108	F+CH₄ → HF+CH₃	8.0 (−11)	3.0 (−10)	400	b
109	F+H₂O → HF+OH	1.1 (−11)	4.2 (−11)	400	b
110	F+O₃ → FO+O₂	1.3 (−11)	2.8 (−11)	226	b
111	FO+NO → F+NO₂	2.6 (−11)	2.6 (−11)	±0	b
	Reactions of OH radicals				
112	OH+H₂ → H₂O+H	6.7 (−15)	6.1 (−12)	2,030	b
113	OH+H₂O₂ → H₂O+HO₂	1.7 (−12)	3.1 (−12)	187	b
52	OH+O₃ → HO₂+O₂	6.8 (−14)	1.6 (−12)	940	b
114	OH+NO $\xrightarrow{M}$ HNO₂	See Table A-5			b
115	OH+NO₂ $\xrightarrow{M}$ HNO₃	See Table A-5			b
116	OH+HNO₂ → H₂O+NO₂	6.6 (−12)	—	—	c
117	OH+HNO₃ → H₂O+NO₃	1.3 (−13)	1.5 (−14)	−650	b
118	OH+HO₂NO₂ → products	4.6 (−12)	1.3 (−12)	−380	b
119	OH+CH₃CO₃NO₂ → products	<2 (−13)	—	—	a
120	OH+NH₃ → H₂O+NH₂	1.6 (−13)	3.3 (−12)	900	a
121	OH+HCl → H₂O+Cl	8.0 (−13)	2.6 (−12)	350	b
122	OH+HOCl → H₂O+ClO	1.8 (−12)	3.0 (−12)	150	b
123	OH+ClONO₂ → products	3.9 (−13)	1.2 (−12)	333	b
124	OH+HBr → H₂O+Br	1.1 (−11)	1.1 (−11)	±0	b

Table A-4 (*continued*)

Number	Reaction	k_{298} (cm³/molecule s)	A (cm³/molecule s)	E_a/R_g (K)	Reference[b]
125	OH+H₂S → H₂O+HS	4.7 (−12)	5.9 (−12)	65	b
126	OH+COS → products	1.0 (−15)	3.9 (−13)	1,780	b
127	OH+CS₂ → products	See discussion in Section 10.2.3			
128	OH+CH₃SH → products	3.4 (−11)	1.2 (−11)	−338	q
129	OH+CH₃SCH₃ → products	4.2 (−12)	6.8 (−12)	−138	q
130	OH+CH₃SSCH₃ → products	2.1 (−10)	5.9 (−11)	−380	q
131	OH+SO₂ $\xrightarrow{M}$ SO₃	See Table A-5			b
132	OH+CO → CO₂+H	2.8 (−13)	$k = (1.5 \times 10^{-13})(1 + 0.6p)$		b
72	OH+CH₄ → H₂O+CH₃	7.7 (−15)	2.4 (−12)	1,710	b
133	OH+C₂H₆ → H₂O+C₂H₅	2.8 (−13)	1.1 (−11)	1,090	b
134	OH+C₃H₈ → H₂O+C₃H₇	1.1 (−12)	1.6 (−11)	800	b
135	OH+n-C₄H₁₀ → H₂O+n-C₄H₉	2.7 (−12)	1.8 (−11)	554	f
136	OH+i-C₄H₁₀ → H₂O+i-C₄H₉	2.5 (−12)	8.7 (−12)	387	f
137	OH+n-C₅H₁₂ → H₂O+n-C₅H₁₁	3.7 (−12)	—	—	f
138	OH+i-C₅H₁₂ → H₂O+i-C₅H₁₁	3.8 (−12)	—	—	f
139	OH+neo-C₅H₁₂ → H₂O+neo-C₅H₁₁	8.3 (−13)	1.4 (−11)	844	f
140	OH+n-C₈H₁₈ → H₂O+n-C₈H₁₇	8.4 (−12)	3.0 (−11)	364	f
141	OH+C₂H₄ → products	7.8 (−12)			f
142	OH+C₃H₆ → products	2.5 (−11)	4.1 (−12)	−546	f
143	OH+1-butene → products	3.5 (−11)	7.6 (−12)	−468	f
144	OH+isobutene → products	5.1 (−11)	9.2 (−12)	−503	f
145	OH+cis-butene → products	5.4 (−11)	1.0 (−11)	−488	f
146	OH+trans-butene → products	7.0 (−11)	1.1 (−11)	−549	f
147	OH+1-pentene → products	2.9 (−11)	—	—	f
148	OH+2-pentene → products	9.0 (−11)	—	—	f
149	OH+1,3-butadiene → products	6.8 (−11)	1.4 (−11)	−468	f

728

No.	Reaction				Ref.
150	OH + isoprene → products	9.3 (−11)	2.4 (−11)	−409	h
151	OH + C$_2$H$_2$ → products	8.5 (−13)	2.0 (−12)	252	f
152	OH + benzene → products	1.2 (−12)		—	f
153	OH + toluene → products	6.4 (−12)		—	f
154	OH + o-xylene → products	1.4 (−11)		—	f
155	OH + m-xylene → products	2.4 (−11)		—	f
156	OH + p-xylene → products	1.5 (−11)		—	f
157	OH + ethylbenzene → products	7.9 (−12)		—	f
158	OH + α-pinene → products	6.0 (−11)	1.4 (−11)	−446	h
159	OH + β-pinene → products	7.8 (−11)	2.4 (−11)	−358	h
160	OH + d-limonene → products	1.4 (−10)		—	h
161	OH + myrcene → products	2.3 (−10)		—	h
162	OH + Δ3-carene → products	8.7 (−11)		—	h
80	OH + HCHO → H$_2$O + HCO	1.0 (−11)	1.0 (−11)	±0	b
163	OH + CH$_3$CHO → H$_2$O + CH$_3$CO	1.6 (−11)	6.9 (−12)	−257	f
164	OH + C$_6$H$_5$CHO → products	1.3 (−11)		—	f
165	OH + CH$_3$COCH$_3$ → products	5.0 (−13)		—	s
166	OH + CH$_3$COC$_2$H$_5$ → products	3.4 (−12)		—	f
167	OH + CH$_3$OH → CH$_3$O + H$_2$O	1.0 (−12)		—	f
168	OH + C$_2$H$_5$OH → products	2.5 (−12)		—	f
169	OH + n-C$_4$H$_9$OH → products	7.6 (−12)		—	f
170	OH + CH$_3$Cl → H$_2$O + CH$_2$Cl	4.3 (−14)	1.8 (−12)	1,112	b
171	OH + CH$_3$Br → H$_2$O + CH$_2$Br	3.8 (−14)	6.1 (−13)	825	b
172	OH + CH$_2$Cl$_2$ → H$_2$O + CHCl$_2$	1.4 (−13)	4.5 (−12)	1,032	b
173	OH + CHCl$_3$ → H$_2$O + CCl$_3$	1.0 (−13)	3.3 (−12)	1,034	b
174	OH + CCl$_4$ → products	1.0 (−15)		—	f
175	OH + CH$_2$FCl → H$_2$O + CHFCl	4.4 (−14)	2.0 (−12)	1,134	b
176	OH + CHFCl$_2$ → H$_2$O + CFCl$_2$	3.0 (−14)	8.9 (−13)	1,013	b
177	OH + CHF$_2$Cl → H$_2$O + CF$_2$Cl	4.6 (−15)	7.8 (−13)	1,530	b
178	OH + CFCl$_3$ → products	<5.0 (−18)		—	b
179	OH + CF$_2$Cl$_2$ → products	<6.5 (−18)		—	b
180	OH + CH$_3$CH$_2$Cl → products	3.9 (−13)		—	b
181	OH + CH$_2$ClCH$_2$Cl → products	2.2 (−13)		—	b

(continued)

Table A-4 (*continued*)

Number	Reaction	k_{298} (cm³/molecule s)	A (cm³/molecule s)	E_a/R_g (K)	Reference[b]
182	$OH + CH_3CCl_3 \rightarrow CH_2CCl_3 + H_2O$	1.2 (−14)	5.4 (−12)	1,820	b
183	$OH + CHCl{=}CCl_2 \rightarrow$ products	2.2 (−12)	5.0 (−13)	−445	b
184	$OH + CCl_2{=}CCl_2 \rightarrow$ products	1.7 (−13)	9.4 (−12)	1,200	b
	Ozone reactions				
	(see also reactions 2, 34, 47, 52, 53, 57, 60, 62, 69, 82, 100, 110)				
185	$O_3 + CH_4 \rightarrow$ products	<7 (−24)			c
186	$O_3 + CO \rightarrow CO_2 + O_2$	<4 (−25)			i
187	$O_3 + SO_2 \rightarrow SO_3 + O_2$	<2 (−22)			c
188	$O_3 + C_2H_4 \rightarrow$ products	1.7 (−18)	9.0 (−15)	2,566	c
189	$O_3 + C_3H_6 \rightarrow$ products	1.1 (−17)	6.1 (−15)	1,912	c
190	$O_3 + 1\text{-butene} \rightarrow$ products	1.1 (−17)	2.9 (−15)	1,711	c
191	$O_3 + \text{isobutene} \rightarrow$ products	1.2 (−17)	3.2 (−15)	1,661	c
192	$O_3 + trans\text{-butene} \rightarrow$ products	2.0 (−16)	6.0 (−15)	2,100	c
193	$O_3 + cis\text{-butene} \rightarrow$ products	1.2 (−16)	3.1 (−15)	1,900	c
194	$O_3 + 1\text{-pentene} \rightarrow$ products	1.1 (−17)			j
195	$O_3 + 2\text{-pentene} \rightarrow$ products	2.1 (−17)			j
196	$O_3 + 1,3\text{-butadiene} \rightarrow$ products	8.4 (−18)			j
197	$O_3 + \text{isoprene} \rightarrow$ products	1.2 (−17)	1.6 (−14)	8,193	t
198	$O_3 + C_2H_2 \rightarrow$ products	3.8 (−20)			k
199	$O_3 + \text{benzene} \rightarrow$ products	7 (−23)			k
200	$O_3 + \text{toluene} \rightarrow$ products	1.5 (−22)			k
201	$O_3 + o\text{-xylene} \rightarrow$ products	7 (−22)			k
202	$O_3 + p\text{-xylene} \rightarrow$ products	4 (−22)			k
203	$O_3 + m\text{-xylene} \rightarrow$ products	6 (−22)			k
204	$O_3 + \text{ethylbenzene} \rightarrow$ products	1.2 (−20)			k
205	$O_3 + \alpha\text{-pinene} \rightarrow$ products	8.4 (−17)	9.4 (−16)	731	t
206	$O_3 + \beta\text{-pinene} \rightarrow$ products	2.1 (−17)			t
207	$O_3 + d\text{-limonene} \rightarrow$ products	6.4 (−16)			j

730

No.	Reaction				Note
208	$O_3 + myrcene \rightarrow products$	$1.2(-15)$	—		j
209	$O_3 + \Delta3$-carene $\rightarrow$ products	$9.0(-17)$	—		j
210	$O_3 + HCHO \rightarrow products$	$2.1(-24)$	—		l

NO_3 reactions (see also reactions 10, 61, 63)

No.	Reaction				Note
211	$NO_3 + CO \rightarrow CO_2 + NO_2$	$<4(-16)$	—		m
212	$NO_3 + SO_2 \rightarrow SO_3 + NO_2$	$<7(-21)$	—		n
213	$NO_3 + HCHO \rightarrow HNO_3 + HCO$	$3.2(-16)$	—		o
214	$NO_3 + CH_3CHO \rightarrow HNO_3 + CH_3CO$	$1.3(-15)$	—		o
215	$NO_3 + C_2H_4 \rightarrow products$	$6.1(-17)$	—		o
216	$NO_3 + C_3H_6 \rightarrow products$	$4.2(-15)$	—		o
217	$NO_3 + 1$-butene $\rightarrow$ products	$5.4(-15)$	—		o
218	$NO_3 + isobutene \rightarrow products$	$1.7(-13)$	—		o
219	$NO_3 + cis$-butene $\rightarrow$ products	$1.8(-13)$	—		o
220	$NO_3 + trans$-butene $\rightarrow$ products	$2.1(-13)$	—		o
221	$NO_3 + 1,3$-butadiene $\rightarrow$ products	$5.3(-14)$	—		o
222	$NO_3 + isoprene \rightarrow products$	$3.2(-13)$	—		o
223	$NO_3 + benzene \rightarrow products$	$2.3(-17)$	—		o
224	$NO_3 + toluene \rightarrow products$	$1.8(-17)$	—		o
225	$NO_3 + \alpha$-pinene $\rightarrow$ products	$6.1(-12)$	—		o
226	$NO_3 + \beta$-pinene $\rightarrow$ products	$2.5(-12)$	—		o
227	$NO_3 + d$-limonene $\rightarrow$ products	$1.4(-11)$	—		o
228	$NO_3 + myrcene \rightarrow products$	$1.1(-11)$	—		o
229	$NO_3 + \Delta3$-carene $\rightarrow$ products	$1.1(-11)$	—		o
230	$NO_3 + CH_3SCH_3 \rightarrow products$	$5.4(-13)$	—		o

Miscellaneous reactions

No.	Reaction				Note
231	$CH_3 + O_3 \rightarrow CH_3O + O_2$	$2.6(-12)$	$5.4(-12)$	220	a
232	$CH_2OH + O_2 \rightarrow HCHO + HO_2$	$2(-12)$	—	—	a
233	$C_2H_5 + O_2 \rightarrow C_2H_5O_2$	$5(-12)$	(1 atm pressure)	—	
234	$C_2H_5O_2 + NO \rightarrow C_2H_5O + NO_2$	$7(-12)$	(see Table A-5)	—	a
76	$CH_3O + NO_2 \xrightarrow{M} CH_3ONO_2$	$4(-12)$	—	—	a
235	$\rightarrow HCHO + HNO_2$	$3.0(-14)$	—	—	a

(continued)

Table A-4 (*continued*)

Number	Reaction	k_{298} (cm³/molecule s)	A (cm³/molecule s)	E_a/R_g (K)	Reference[b]
236	$CH_3(CO)O_2 + NO \rightarrow CH_3 + CO_2 + NO_2$	1.4 (−11)	—	—	a
237	$CH_3(CO)O_2 + NO_2 \xrightarrow{M} CH_3(CO)O_2NO_2$	6 (−12)	(1 atm pressure)		a
238	$n\text{-}C_3H_7 + O_2 \xrightarrow{M} n\text{-}C_3H_7O_2$	6 (−12)	(1 atm pressure)		a
239	$i\text{-}C_3H_7 + O_2 \xrightarrow{M} i\text{-}C_3H_7O_2$	1.5 (−12)	(1 atm pressure)		
240	$n\text{-}CH_3H_7O_2 + NO \rightarrow n\text{-}C_3H_7O + NO_2$	7.6 (−12)	—	—	a
241	$CH_3O + SO_2 \xrightarrow{M} CH_3OSO_2$	5.5 (−13)	(1 atm pressure)		n
242	$CH_3O_2 + SO_2 \xrightarrow{M} CH_3O_2SO_2$	1.4 (−14)	(1 atm pressure)		n
	$\rightarrow CH_3O + SO_3$	<1 (−18)			n
243	$CH_3(CO)O_2 + SO_2 \rightarrow CH_3 + CO_2 + SO_3$	<7 (−19)	—		n
244	$S + O_2 \rightarrow SO + O$	2.3 (−12)	2.3 (−12)	±0	b
245	$SO + O_2 \rightarrow SO_2 + O$	8.4 (−17)	2.3 (−13)	2,370	b
246	$SH + O_2 \rightarrow$ products	<4 (−17)	—	—	p
247	$SH + O_3 \rightarrow HSO + O_2$	1 (−13)	—		b
248	$HOSO_2 + O_2 \rightarrow HO_2 + SO_3$	1 (−13)	—		r
249	$SO_3 + H_2O \rightarrow H_2SO_4$	9 (−13)	—		a
250	$N_2O_5 + SO_2 \rightarrow N_2O_4 + SO_3$	<4 (−23)	—		n

[a] The Arrhenius equation is $k_{bim} = A \exp(-E_a/R_g T)$. Powers of 10 are shown in parentheses.

[b] References: (a) Baulch et al. (1980, 1982, 1984); (b) DeMore et al. (1985); (c) L. C. Anderson (1976); (d) Hampson (1980); (e) Herron and Huie (1969) and Atkinson et al. (1977); (f) Atkinson et al. (1979); (g) Kaiser and Wu (1977), McKinnon et al. (1979); (h) Kleindienst et al. (1982), Pitts et al. (1977); (i) Arin and Warneck (1972); (j) Japar et al. (1974), Grimsrud et al. (1975); (k) Pate et al. (1976), Roberts et al. (1984); (l) Braslavski and Heicklen (1976); (m) Burrows et al. (1985); (n) Calvert and Stockwell (1984); (o) Atkinson et al. (1984a, b, c); (p) Black (1984); (q) Wine et al. (1981); (r) Schmidt et al. (1985); (s) Cox et al. (1980); (t) Atkinson et al. (1982b).

Table A-5. *Rate-Coefficient Parameters for Gas-Phase Association Reactions, with Air Providing the Third-Body Molecule* M^a

Number	Reaction	k_0^{300}	n	k_∞^{300}	m	k_{atm}^{300}
1	$O + O_2 + M$	6.0 (−34)	2.3	—	—	1.5 (−14)
8	$O + NO + M$	9.0 (−32)	1.5	3.0 (−11)	0	1.6 (−12)
14	$O + SO_2 + M$	7.1 (−34)	−4.1	—	—	1.7 (−14)
20	$O + CO + M$	3.3 (−36)	−6.3	—	—	8.1 (−17)
46	$H + O_2 + M$	5.5 (−32)	1.6	7.5 (−11)	0	1.2 (−12)
49	$OH + OH + M$	6.9 (−31)	0.8	1.0 (−11)	1.0	9.9 (−12)
55	$HO_2 + NO_2 + M$	2.0 (−31)	2.7	4.2 (−12)	2.0	1.4 (−12)
63	$NO_3 + NO_2 + M$	2.2 (−30)	4.3	1.5 (−12)	0.5	1.2 (−12)
64	$NO + NO_2 + H_2O$	4.4 (−40)	—	—	—	—
65	$NO_2 + NO_2 + H_2O$	8.0 (−38)	—	—	—	—
67	$NO + NO + O_2$	2.0 (−38)	1.3	—	—	1.0 (−19)
73	$CH_3 + O_2 + M$	4.5 (−31)	2.0	1.8 (−12)	1.7	4.2 (−12)
76	$CH_3O_2 + NO_2 + M$	1.5 (−30)	4.0	6.5 (−12)	2.0	4.0 (−12)
83	$Cl + O_2 + M$	2.0 (−33)	1.4	—	—	4.9 (−14)
98	$ClO + NO_2 + M$	1.8 (−31)	3.4	1.5 (−11)	1.9	2.3 (−12)
105	$BrO + NO_2 + M$	5.0 (−31)	2.0	1.0 (−11)	1.0	3.3 (−12)
114	$OH + NO + M$	7.0 (−31)	2.6	1.5 (−11)	0.5	4.8 (−12)
115	$OH + NO_2 + M$	2.6 (−30)	3.2	2.4 (−11)	1.3	1.1 (−11)
131	$OH + SO_2 + M$	3.0 (−31)	3.3	1.5 (−12)	0	3.0 (−12)

^{a} The second-order compound rate coefficient at ambient temperature T (K) and number density $n(M)$ may be calculated from $k = (0.6)^y k_0(T)n(M)/[1 + k_0(T)n(M)/k_\infty(T)]$ where $y = \{1 + \log_{10}[k_0(T)n(M)/k_\infty(T)]^2\}^{-1}$. The low-pressure rate coefficient is $k_0(T) = k_0^{300}(T/300)^{-n}$ in units of cm^6/molecule^2 s; the high-pressure limiting rate coefficient is $k_\infty(T) = k_\infty^{300}(T/300)^{-m}$ in units of cm^3/molecule s. Given are k_0^{300}, n, k_∞^{300}, and m as far as known, and the rate coefficient k_{atm}^{300} at 300 K for a pressure of 1 atm = 1013 mbar. Powers-of 10 are shown in parentheses.

INDEX

A

Acetaldehyde
  PAN production, pathway, 470
  photochemical behavior in atmosphere, 81
  photodecomposition, 87
Acetone
  PAN production, pathway, 470
  photochemical behavior in atmosphere, 81
  photodecomposition, 87
Acetylene, mixing ratios
  over continents in rural sites, 237, 239–241
  over Pacific ocean, meridional distribution, 246–248
Aerosol particles
  chemical constitution
    inorganic fraction, *see also specific elements and ions*
      alkali and alkaline earth elements from rocks and seawater, 341–348
      in marine aerosol, comparison with seawater, 338–341
      mass concentration at several locations, 332–338

NH_4^+/SO_4^{2-} molar ratios for continental aerosols, 336–337
      sulfate, 333–338
      trace elements, 347–350
    main fractions in continental air, 330–332
    organic fraction
      petroleum-derived n-alkanes, 359
      solvent extractions and chromatography, 350–351
      terrestrial material, contribution to marine aerosol, 353–354
      in urban, rural, and maritime locations, 351, 352, 354–358
      from vascular plant waxes, 354–355, 359–360
  definition, continental, maritime, and tropospheric types, 278–279
  global production rate from various sources, 326–330
  physicochemical behavior
    coagulation
      effect on size distribution, 290–293
      mathematical description, 287–290

735

International Geophysics Series

EDITED BY

J. VAN MIEGHEM
(1959–1976)

ANTON L. HALES
(1972–1979)

WILLIAM L. DONN
Lamont-Doherty Geological Observatory
Columbia University
Palisades, New York
(1980–1986)

Current Editors

RENATA DMOWSKA
Division of Applied Science
Harvard University

JAMES R. HOLTON
Department of Atmospheric Sciences
University of Washington
Seattle, Washington